T0234569

Lecture Notes in Computer Science　　　8588

Commenced Publication in 1973
Founding and Former Series Editors:
Gerhard Goos, Juris Hartmanis, and Jan van Leeuwen

De-Shuang Huang   Vitoantonio Bevilacqua
Prashan Premaratne (Eds.)

Intelligent Computing Theory

10th International Conference, ICIC 2014
Taiyuan, China, August 3-6, 2014
Proceedings

 Springer

Volume Editors

De-Shuang Huang
Tongji University
Machine Learning and Systems Biology Laboratory
School of Electronics and Information Engineering
4800 Caoan Road, Shanghai 201804, China
E-mail: dshuang@tongji.edu.cn

Vitoantonio Bevilacqua
Politecnic of Bari
Electrical and Electronics Department
Via Orabona, 4, 70125 Bari, Italy
E-mail: vitoantonio.bevilacqua@gmail.com

Prashan Premaratne
University of Wollongong
School of Electrical, Computer and Telecommunications Engineering
North Wollongong, NSW 2522, Australia
E-mail: prashan@uow.edu.au

ISSN 0302-9743   e-ISSN 1611-3349
ISBN 978-3-319-09332-1   e-ISBN 978-3-319-09333-8
DOI 10.1007/978-3-319-09333-8
Springer Cham Heidelberg New York Dordrecht London

Library of Congress Control Number: 2014943570

LNCS Sublibrary: SL 3 – Information Systems and Application, incl. Internet/Web
and HCI

Typesetting: Camera-ready by author, data conversion by Scientific Publishing Services, Chennai, India

Printed on acid-free paper

Springer is part of Springer Science+Business Media (www.springer.com)

Preface

The International Conference on Intelligent Computing (ICIC) was started to provide an annual forum dedicated to the emerging and challenging topics in artificial intelligence, machine learning, pattern recognition, bioinformatics, and computational biology. It aims to bring together researchers and practitioners from both academia and industry to share ideas, problems, and solutions related to the multifaceted aspects of intelligent computing.

ICIC 2014, held in Taiyuan, China, during August 3–6, 2014, constituted the 10th International Conference on Intelligent Computing. It built upon the success of ICIC 2013, ICIC 2012, ICIC 2011, ICIC 2010, ICIC 2009, ICIC 2008, ICIC 2007, ICIC 2006, and ICIC 2005 that were held in Nanning, Huangshan, Zhengzhou, Changsha, China, Ulsan, Korea, Shanghai, Qingdao, Kunming, and Hefei, China, respectively.

This year, the conference concentrated mainly on the theories and methodologies as well as the emerging applications of intelligent computing. Its aim was to unify the picture of contemporary intelligent computing techniques as an integral concept that highlights the trends in advanced computational intelligence and bridges theoretical research with applications. Therefore, the theme for this conference was "**Advanced Intelligent Computing Technology and Applications**". Papers focused on this theme were solicited, addressing theories, methodologies, and applications in science and technology.

ICIC 2014 received 667 submissions from 21 countries and regions. All papers went through a rigorous peer-review procedure and each paper received at least three review reports. Based on the review reports, the Program Committee finally selected 235 high-quality papers for presentation at ICIC 2013, included in three volumes of proceedings published by Springer: one volume of *Lecture Notes in Computer Science* (LNCS), one volume of *Lecture Notes in Artificial Intelligence* (LNAI), and one volume of *Lecture Notes in Bioinformatics* (LNBI).

This volume of *Lecture Notes in Computer Science* (LNCS) includes 92 papers.

The organizers of ICIC 2014, including Tongji University and North University of China, Taiyuan Normal University, Taiyuan University of Science and Technology, made an enormous effort to ensure the success of the conference. We hereby would like to thank the members of the Program Committee and the referees for their collective effort in reviewing and soliciting the papers. We would like to thank Alfred Hofmann, executive editor from Springer, for his frank and helpful advice and guidance throughout and for his continuous support in publishing the proceedings. In particular, we would like to thank all the authors for contributing their papers. Without the high-quality submissions from the authors, the success of the conference would not have been possible. Finally, we are especially grateful to the IEEE Computational Intelligence Society, the International Neural Network Society, and the National Science Foundation of China for their sponsorship.

May 2014

De-Shuang Huang
Vitoantonio Bevilacqua
Prashan Premaratne

ICIC 2014 Organization

General Co-chairs

De-Shuang Huang, China
Vincenzo Piuri, Italy
Yan Han, China

Jiye Liang, China
Jianchao Zeng, China

Program Committee Co-chairs

Kang Li, UK
Juan Carlos Figueroa, Colombia

Organizing Committee Co-chairs

Kang-Hyun Jo, Korea
Valeriya Gribova, Russia

Bing Wang, China
Xing-Ming Zhao, China

Award Committee Chair

Vitoantonio Bevilacqua, Italy

Publication Chair

Phalguni Gupta, India

Workshop/Special Session Co-chairs

Jiang Qian, USA
Zhongming Zhao, USA

Special Issue Chair

M. Michael Gromiha, India

Tutorial Chair

Laurent Heutte, France

International Liaison

Prashan Premaratne, Australia

Publicity Co-chairs

Kyungsook Han, Korea   Abir Hussain, UK
Ling Wang, China   Zhi-Gang Zong, China

Exhibition Chair

Chun-Hou Zheng, China

Program Committee Members

Khalid Aamir, Pakistan   Minrui Fei, China
Andrea F. Abate, USA   Juan Carlos Figueroa-García,
Sabri Arik, Korea   Colombia
Vasily Aristarkhov, Australia   shan Gao, China
Costin Badica, Japan   Liang Gao, China
Waqas Bangyal, Pakistan   Dun-wei Gong, India
Vitoantonio Bevilacqua, Italy   Valeriya Gribova, China
Shuhui Bi, China   Michael Gromiha, China
Jair Cervantes, Mexico   Xingsheng Gu, China
Yuehui Chen, China   Kayhan Gulez, USA
Qingfeng Chen, China   Ping Guo, China
Wen-Sheng Chen, China   Phalguni Gupta, India
Xiyuan Chen, China   Kyungsook Han, Korea
Guanling Chen, USA   Fei Han, China
Yoonsuck Choe, USA   Laurent Heutte, France
Ho-Jin Choi, Korea, Republic of   Wei-Chiang Hong, Taiwan
Michal Choras, Colombia   Yuexian Hou, China
Angelo Ciaramella, China   Jinglu Hu, China
Youping Deng, Japan   Tingwen Huang, Qatar
Primiano Di Nauta, Italy   Peter Hung, Taiwan
Salvatore Distefano, USA   Abir Hussain, UK
Ji-Xiang Du, China   Saiful Islam, India
Jianbo Fan, China   Li Jia, China

Zhenran Jiang, China
Kang-Hyun Jo, Korea
Dah-Jing Jwo, Korea
Seeja K.R, India
Vandana Dixit Kaushik, India
Gul Muhammad Khan, Pakistan
Sungshin Kim, Korea
Donald Kraft, USA
Yoshinori Kuno, Japan
Takashi Kuremoto, Japan
Jaerock Kwon, USA
Vincent Lee, Australia
Shihua Zhang, China
Guo-Zheng Li, China
Xiaodi Li, China
Bo Li, China
Kang Li, UK
Peihua Li, China
Jingjing Li, USA
Yuhua Li, UK
Honghuang Lin, USA
Meiqin Liu, USA
Ju Liu, China
Xiwei Liu, China
Shuo Liu, China
Yunxia Liu, China
Chu Kiong Loo, Mexico
Zhao Lu, USA
Ke Lu, China
Yingqin Luo, USA
Jinwen Ma, USA
Xiandong Meng, China
Filippo Menolascina, Italy
Ivan Vladimir Meza-Ruiz, Australia
Tarik Veli Mumcu Mumcu, Turkey
Roman Neruda, Turkey
Ben Niu, China
Seiichi Ozawa, Korea
Paul Pang, China
Francesco Pappalardo, USA
Surya Prakash, India
Prashan Premaratne, Australia
Daowen Qiu, China
Angel Sappa, USA
Li Shang, China

Dinggang Shen, USA
Fanhuai Shi, China
Shitong Wang, China
Wilbert Sibanda, USA
Jiatao Song, China
Stefano Squartini, Italy
Badrinath Srinivas, USA
Zhan-Li Sun, China
Evi Syukur, USA
Joaquín Torres-Sospedra, Spain
Rua-Huan Tsaih, USA
Antonio Uva, USA
Jun Wan, USA
Yong Wang, China
Ling Wang, China
Jim Jing-Yan Wang, USA
Xuesong Wang, China
Bing Wang, China
Ze Wang, USA
Junwen Wang, HK
Hong Wei, UK
Wei Wei, Norway
Yan Wu, China
QingXiang Wu, China
Junfeng Xia, China
Shunren Xia, China
Bingji Xu, China
Gongsheng xu, China
Yu Xue, China
Xin Yin, USA
Xiao-Hua Yu, USA
Zhigang Zeng, China
Shihua Zhang, China
Jun Zhang, China
Xing-Ming Zhao, China
Hongyong Zhao, China
Xiaoguang Zhao, China
Zhongming Zhao, USA
Bojin Zheng, China
Chunhou Zheng, China
Fengfeng Zhou, China
Yongquan Zhou, China
Hanning Zhou, China
Li Zhuo, China
Xiufen Zou, China

Reviewers

Jakub Šmíd	Giuseppe Carbone
Pankaj Acharya	Raffaele Carli
Erum Afzal	Jair Cervantes
Parul Agarwal	Aravindan Chandrabose
Tanvir Ahmad	Yuchou Chang
Musheer Ahmad	Deisy Chelliah
Syed Ahmed	Gang Chen
Sabooh Ajaz	Songcan Chen
Haya Alaskar	Jianhung Chen
Felix Albu	David Chen
Dhiya Al-Jumeily	Hongkai Chen
Israel Alvarez Villalobos	Xin Chen
Muhammad Amjad	Fanshu Chen
Ning An	Fuqiang Chen
Mary Thangakani Anthony	Bo Chen
Masood Ahmad Arbab	Xin Chen
Soniya be	Liang Chen
Sunghan Bae	Wei Chen
Lukas Bajer	Jinan Chen
Waqas Bangyal	Yu Chen
Gang Bao	Junxia Cheng
Donato Barone	Zhang Cheng
Silvio Barra	Feixiong Cheng
Alex Becheru	Cong Cheng
Ye Bei	Han Cheng
Mauri Benedito Bordonau	Chi-Tai Cheng
Simon Bernard	Chengwang Xie
Vitoantonio Bevilacqua	Seongpyo Cheon
Ying Bi	Ferdinando Chiacchio
Ayse Humeyra Bilge	Cheng-Hsiung Chiang
Honghua Bin	Wei Hong Chin
Jun Bo	Simran Choudhary
Nora Boumella	Angelo Ciaramella
Fabio Bruno	Azis Ciayadi
Antonio Bucchiarone	Rudy Ciayadi
Danilo Caceres	Danilo Comminiello
Yiqiao Cai	Carlos Cubaque
Qiao Cai	Yan Cui
Guorong Cai	Bob Cui
Francesco Camastra	Cuco Curistiana
Mario Cannataro	Yakang Dai
Kecai Cao	Dario d'Ambruoso
Yi Cao	Yang Dan

Farhan Dawood
Francesca De Crescenzio
Kaushik Deb
Saverio Debernardis
Dario Jose Delgado-Quintero
Sara Dellantonio
Jing Deng
Weilin Deng
M.C. Deng
Suping Deng
Zhaohong Deng
Somnath Dey
Yunqiang Di
Hector Diez Rodriguez
Rong Ding
Liya Ding
Sheng Ding
Sheng Ding
Shihong Ding
Dayong Ding
Xiang Ding
Salvatore Distefano
Chelsea Dobbins
Xueshi Dong
Vladislavs Dovgalecs
Vlad Dovgalecs
Gaixin Du
Dajun Du
Kaifang Du
Haibin Duan
Durak-Ata Lutfiye
Malay Dutta
Tolga Ensari
Nicola Epicoco
Marco Falagario
Shaojing Fan
Ming Fan
Fenghua Fan
Shaojing Fan
Yaping Fang
Chen Fei
Liangbing Feng
Shiguang Feng
Guojin Feng
Alessio Ferone

Francesco Ferrise
Juan Carlos Figueroa
Michele Fiorentino
Carlos Franco
Gibran Fuentes Pineda
Hironobu Fujiyoshi
Kazuhiro Fukui
Wai-Keung Fung
Chun Che Fung
Chiara Galdi
Jian Gao
Yushu Gao
Liang Gao
Yang Gao
Garcia-Lamont Farid
Garcia-Marti Irene
Michele Gattullo
Jing Ge
Na Geng
Shaho Ghanei
Rozaida Ghazali
Rosalba Giugno
Fengshou Gu
Tower Gu
Jing Gu
Smile Gu
Guangyue Du
Weili Guo
Yumeng Guo
Fei Guo
Tiantai Guo
Yinan Guo
Yanhui Guo
Chenglin Guo
Lilin Guo
Sandesh Gupta
Puneet Gupta
Puneet Gupta
Shi-Yuan Han
Fei Han
Meng Han
Yu-Yan Han
Zhimin Han
Xin Hao
Manabu Hashimoto

Tao He
Selena He
Xing He
Feng He
Van-Dung Hoang
Tian Hongjun
Lei Hou
Jingyu Hou
Ke Hu
Changjun Hu
Zhaoyang Hu
Jin Huang
Qiang Huang
Lei Huang
Shin-Ying Huang
Ke Huang
Huali Huang
Jida Huang
Xixia Huang
Fuxin Huang
Darma Putra i Ketut Gede
Haci Ilhan
Sorin Ilie
Saiful Islam
Saeed Jafarzadeh
Alex James
Chuleerat Jaruskulchai
James Jayaputera
Umarani Jayaraman
Mun-Ho Jeong
Zhiwei Ji
Shouling Ji
Yafei Jia
Hongjun Jia
Xiao Jian
Min Jiang
Changan Jiang
Tongyang Jiang
Yizhang Jiang
He Jiang
Yunsheng Jiang
Shujuan Jiang
Ying Jiang
Yizhang Jiang
Changan Jiang

Xu Jie
Jiening Xia
Taeseok Jin
Jingsong Shi
Mingyuan Jiu
Kanghyun Jo
Jayasudha John Suseela
Ren Jun
Fang Jun
Li Jun
Zhang Junming
Yugandhar K.
Tomáš Křen
Yang Kai
Hee-Jun Kang
Qi Kang
Dong-Joong Kang
Shugang Kang
Bilal Karaaslan
Rohit Katiyar
Ondrej Kazik
Mohd Ayyub Khan
Muhammed Khan
Sang-Wook Kim
Hong-Hyun Kim
One-Cue Kim
Duangmalai Klongdee
Kunikazu Kobayashi
Yoshinori Kobayashi
Takashi Komuro
Toshiaki Kondo
Deguang Kong
Kitti Koonsanit
Rafal Kozik
Kuang Li
Junbiao Kuang
Baeguen Kwon
Hebert Lacey
Chunlu Lai
Chien-Yuan Lai
David Lamb
Wei Lan
Chaowang Lan
Qixun Lan
Yee Wei Law

Tien Dung Le
My-Ha Le
Yongduck Lee
Jooyoung Lee
Seokju Lee
Shao-Lun Lee
Xinyu Lei
Gang Li
Yan Li
Liangliang Li
Xiaoguang Li
Zheng Li
Huan Li
Deng Li
Ping Li
Qingfeng Li
Fuhai Li
Hui Li
Kai Li
Longzhen Li
Xingfeng Li
Jingfei Li
Jianxing Li
Keling Li
Juan Li
Jianqing Li
Yunqi Li
Bing Nan Li
Lvzhou Li
Qin Li
Xiaoguang Li
Xinwu Liang
Jing Liang
Li-Hua Zhang
Jongil Lim
Changlong Lin
Yong Lin
Jian Lin
Genie Lin
Ying Liu
Chenbin Liu
James Liu
Liangxu Liu
Yuhang Liu
Liang Liu

Rong Liu
Liang Liu
Yufeng Liu
Qing Liu
Zhe Liu
Zexian Liu
Li Liu
Shiyong Liu
Sen Liu
Qi Liu
Jin-Xing Liu
Xiaoming Liu
Ying Liu
Xiaoming Liu
Bo Liu
Yunxia Liu
Alfredo Liverani
Anthony Lo
Lopez-Chau Asdrúbal
Siow Yong Low
Zhen Lu
Xingjia Lu
Junfeng Luo
Juan Luo
Ricai Luo
Youxi Luo
Yanqing Ma
Wencai Ma
Lan Ma
Chuang Ma
Xiaoxiao Ma
Sakashi Maeda
Guoqin Mai
Mario Manzo
Antonio Maratea
Erik Marchi
Carlos Román Mariaca Gaspar
Naoki Masuyama
Gu Meilin
Geethan Mendiz
Qingfang Meng
Filippo Menolascina
Muharrem Mercimek
Giovanni Merlino
Hyeon-Gyu Min

Martin Renqiang Min
Minglei Tong
Saleh Mirheidari
Akio Miyazaki
Yuanbin Mo
Quanyi Mo
Andrei Mocanu
Raffaele Montella
Montoliu Raul
Tsuyoshi Morimoto
Mohamed Mousa Alzawi
Lijun Mu
Inamullah Muhammad
Izharuddin Muhammed
Tarik Veli Mumcu
Francesca Nardone
Fabio Narducci
Rodrigo Nava
Patricio Nebot
Ken Nguyen
Changhai Nie
Li Nie
Aditya Nigam
Evgeni Nurminski
Kok-Leong Ong
Kazunori Onoguchi
Zeynep Orman
Selin Ozcira
Cuiping Pan
Binbin Pan
Quan-Ke Pan
Dazhao Pan
Francesco Pappalardo
Jekang Park
Anoosha Paruchuri
Vibha Patel
Samir Patel
Lizhi Peng
Yiming Peng
Jialin Peng
Klara Peskova
Caroline Petitjean
Martin Pilat
Surya Prakash
Philip Pretorius

Ali Qamar
Xiangbo Qi
Shijun Qian
Pengjiang Qian
Bin Qian
Ying Qiu
Jian-Ding Qiu
Junfeng Qu
Junjun Qu
Muhammad Rahman
Sakthivel Ramasamy
Tao Ran
Martin Randles
Caleb Rascon
Muhammad Rashid
Haider Raza
David Reid
Fengli Ren
Stefano Ricciardi
Angelo Riccio
Alejo Roberto
Abdus Samad
Ruya Samli
Hongyan Sang
Michele Scarpiniti
Dongwook Seo
Shi Sha
Elena Shalfeeva
Li Shang
Linlin Shen
Yehu Shen
Haojie Shen
Jin Biao Shen
Ajitha Shenoy
Jiuh-Biing Sheu
Xiutao Shi
Jibin Shi
Fanhuai Shi
Yonghong Shi
Yinghuan Shi
Atsushi Shimada
Nobutaka Shimada
Ji Sun Shin
Ye Shuang
Raghuraj Singh

Dushyant Kumar Singh
Haozhen Situ
Martin Slapak
Sergey Smagin
Yongli Song
Yinglei Song
Meiyue Song
Bin Song
Rui Song
Guanghua Song
Gang Song
Sotanto Sotanto
Sreenivas Sremath Tirumala
Antonino Staiano
Jinya Su
Hung-Chi Su
Marco Suma
Xiaoyan Sun
Jiankun Sun
Sheng Sun
Yu Sun
Celi Sun
Yonghui Sun
Zengguo Sun
Jie Sun
Aboozar Taherkhani
Shinya Takahashi
Jinying Tan
Shijun Tang
Xiwei Tang
Buzhou Tang
Ming Tang
Jianliang Tang
Hissam Tawfik
Zhu Teng
Sin Teo
Girma Tewolde
Xiange Tian
Yun Tian
Tian Tian
Hao Tian
Gao Tianshun
Kamlesh Tiwari
Amod Tiwari
Kamlesh Tiwari

Andrysiak Tomasz
Torres-Sospedra Joaquín
Sergi Trilles
Yao-Hong Tsai
Naoyuki Tsuruta
Fahad Ullah
Pier Paolo Valentini
Andrey Vavilin
Giuseppe Vettigli
Petra Vidnerová
Villatoro-Tello Esaú
Ji Wan
Li Wan
Lin Wan
Quan Wang
Zixiang Wang
Yichen Wang
Xiangjun Wang
Yunji Wang
Hong Wang
Jinhe Wang
Xuan Wang
Xiaojuan Wang
Suyu Wang
Zhiyong Wang
Xiangyu Wang
Mingyi Wang
Yan Wang
Zongyue Wang
Huisen Wang
Yongcui Wang
Xiaoming Wang
Zi Wang
Jun Wang
Aihui Wang
Yi Wang
Ling Wang
Zhaoxi Wang
Shulin Wang
Yunfei Wang
Yongbo Wang
Zhengxiang Wang
Sheng-Yao Wang
Jingchuan Wang
Qixin Wang

Yong Wang

Fang-Fang Wang

Tian Wang

Zhenzhong Wang

Panwen Wang

Lei Wang

Qilong Wang

Dong Wang

Ping Wang

Huiwei Wang

Yiqi Wang

Zhixuan Wei

Zhihua Wei

Li Wei

Shengjun Wen

Shiping Wen

Di Wu

Yonghui Wu

Chao Wu

Xiaomei Wu

Weili Wu

Hongrun Wu

Weimin Wu

Guolong Wu

Siyu Xia

Qing Xia

Sen Xia

Jin Xiao

Min Xiao

Qin Xiao

Yongfei Xiao

Zhao Xiaoguang

Zhuge Xiaozhong

Minzhu Xie

Zhenping Xie

Jian Xie

Ting Xie

Chao Xing

Wei Xiong

Hao Xiong

Yi Xiong

Xiaoyin Xu

Dawen Xu

Jing Xu

Yuan Xu

Jin Xu

Xin Xu

Meng Xu

Li Xu

Feng Xu

Zhenyu Xuan

Hari Yalamanchili

Atsushi Yamashita

Mingyuan Yan

Yan Yan

Zhigang Yan

Bin Yan

Zhile Yang

Dan Yang

Yang Yang

Wankou Yang

Wenqiang Yang

Yuting Yang

Chia-Luen Yang

Chyuan-Huei Yang

Deng Yanni

Jin Yao

Xiangjuan Yao

Tao Ye

Xu Ye

Fengqi Yi

Wenchao Yi

Kai Yin

James j.q. Yu

Helen Yu

Jun Yu

Fang Yu

Xu Yuan

Lin Yuan

Jinghua Yuan

Lin Yuling

Faheem Zafari

Xue-Qiang Zeng

Haimao Zhan

Yong-Wei Zhang

Jing Zhang

Shuyi Zhang

Kevin Zhang

Ming Zhang

Zhenmin Zhang

Xuebing Zhang
Minlu Zhang
Jianping Zhang
Xiaoping Zhang
Yong Zhang
Qiang Zhang
Guohui Zhang
Chunhui Zhang
Yifeng Zhang
Wenxi Zhang
Xiujun Zhang
Long Zhang
Jian Zhang
Boyu Zhang
Hailei Zhang
Hongyun Zhang
Jianhua Zhang
Chunjiang Zhang
Peng Zhang
Jianhai Zhang
Hongbo Zhang
Lin Zhang
Xiaoqiang Zhang
Haiying Zhang
Jing Zhang
Wei Zhang
Qian Zhang
Hongli Zhang
Guohui Zhang
Liping Zhang
Hongbo Zhang
Sen Zhao
Yaou Zhao
Jane Zhao
Liang Zhao

Yunlong Zhao
Miaomiao Zhao
Xinhua Zhao
Xu Zhao
Guodong Zhao
Liang Zhao
Feng Zhao
Juan Zhao
Junfei Zhao
Changbo Zhao
Yue Zhao
Min Zheg
Dong Zhen
Guang Zheng
Xiaolong Zheng
Huanyu Zheng
Shenggen Zheng
Shan Zhong
Qi Zhou
Mian Zhou
Yinzhi Zhou
Jiayin Zhou
Songsheng Zhou
Bo Zhou
Qiang Zhou
Lei Zhu
Lin Zhu
Yongxu Zhu
Nanli Zhu
Xiaolei Zhu
Majid Ziaratban
Xiangfu Zou
Brock Zou
Tanish
Qiqi Duan

Table of Contents

Evolutionary Computation and Learning

Swarm Intelligence and Optimization

Machine Learning

Social and Natural Computing

Neural Networks

Biometrics Recognition

Image Processing

Information Security

Virtual Reality and Human-Computer Interaction

Knowledge Discovery and Data Mining

Signal Processing

Pattern Recognition

Biometric System and Security for Intelligent Computing

Self-tuning Performance of Database Systems with Neural Network

Conghuan Zheng, Zuohua Ding, and Jueliang Hu

Lab of Intelligent Computing and Software Engineering
Zhejiang Sci-Tech University
Hangzhou, Zhejiang, China, 310018
997351601@qq.com, zouhuading@hotmail.com, hujlhz@163.com

Abstract. Performance self tuning in database systems is a challenge work since it is hard to identify tuning parameters and make a balance to choose proper configuration values for them. In this paper, we propose a neural network based algorithm for performance self-tuning. We first extract Automatic Workload Repository report automatically, and then identify key system performance parameters and performance indicators. We then use the collected data to construct a Neural Network model. Finally, we develop a self-tuning algorithm to tune these parameters. Experimental results for oracle database system in TPC-C workload environment show that the proposed method can dynamically improve the performance.

Keywords: Performance Tuning, Oracle Database, Neural Network.

1   Introduction

Database management systems (DBMS) have a wide range of applications such as in online transaction systems, decision support systems, data mining applications and e-commerce applications. It is critical to maintain a good performance of the system when it is running. In general, there are hundreds of configuration parameters that can affect the performance. It is extremely difficult for database administrators (DBA) to manually tune these parameters for a good performance. Hence, many performance self-tuning methods have been proposed. For example, a self-tuning configurable cache that performs transparent runtime cache tuning for reduced energy is introduced in [7]. A tuning algorithm based on new script estimated tuning parameters for enhance performance is presented in [15]. Oh *et al.* [10] use the correlation coefficient to quantify the incremental or decremental relationship between the size of parameters and the performance indicators, thus can automatically select parameters that affect the system performance. Debnath *et al.* [6] rank various tuning parameters based on statistical analysis. Agarwal *et al.* [2] use Materialized views and Indexes, Pruning table and column sets to automatically tune the DBMS to deliver enhanced performance. Zhang *et al.* [17] take different tuning strategies for different types of applications that can achieve better tuning effect. Dageville *et al.* [5] and Belknap et al. [4] describe that the Oracle provides a way to perform automatically performance

D.-S. Huang et al. (Eds.): ICIC 2014, LNCS 8588, pp. 1–12, 2014.
© Springer International Publishing Switzerland 2014

diagnosis and tuning, and solving some performance problems without any DBA intervention. Alhadi *et al.* [1] presents an improvement process of query tuning which can reduce queries execution time. In order to enhance query response time, they modified query scripts to fit the tuning process.

In this paper, we present a neural network based method for self-tuning performance of DBMS in commercial application system, which can be classified into online transaction processing (OLTP) and decision support system (DSS) [9]. Here is the sketch of our method. We first extract AWR report automatically, and then analyze AWR report and collect data of key performance parameters, then use this information to train neural network so that to find the relationship between performance indicators and performance parameters, then we develop a self-tuning algorithm to tune parameters. Finally, we update oracle script file to alter those parameters for enhance performance. The benefits we can get from this method are that 1) it can tune multi-parameters simultaneously to get better performance and 2) the tuning process is automatic.

The rest of this paper is organized as follows: Section 2 gives performance tuning architecture. Section 3 describes how to construct a neural network model. Section 4 develops a performance tuning algorithm. Section 5 is the evaluation of experiments. Section 6 discusses related work. Section 7 concludes the paper.

2   Performance Tuning Architecture

The tuning architecture contains the following components as shown in Fig. 1: DBMS, AWR, Neural Network, Tuning Rule, Tuner and Script.

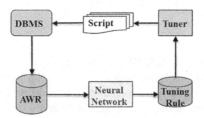

Fig. 1. Script based Tuning Architecture

2.1   Database Management System

The DBMS was used here is oracle database system. The memory structure of Oracle database system were composed by system globe area (SGA) and private globe area (PGA) as shown in Fig. 2, SGA mainly contains shared pool, data buffer cache, redo log buffer, large pool, java pool and streams pool. On the other hand, PGA is a memory region that contains data and control information for a server process, it mainly contains Private SQL area, Cursors and SQL areas.

We use the TPC-C benchmarks [14] to construct a standard workload environment. The TPC-C benchmark represents the workloads of OLTP environments. Under the TPC-C workloads, resource usages, which are represented by three performance indicators, are investigated in response to the changes of three performance parameters (db_cache_size, shared_pool_size and pga_aggregate_target). We utilize three performance indicators to measure the change of above three parameters:

— **The Data Buffer Hit Ratio:** A measure of the effectiveness of the Oracle data block buffer.
— **The disk writes with checkpoints:** It measures the amount of disk writes that have checkpointing and would have occurred.
— **The redo log size:** Total amount of redo logs generated in bytes.

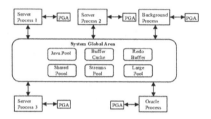

Fig. 2. The Memory Structure of Oracle

We will study how to tune performance by reconfiguring the performance parameters. We will consider two cases: (1) only one control parameter change at one time while the others are unchanged. For example, the value of shared pool size increases from 30MB to 480MB by 30MB intervals, while the other two parameters remain unchanged; and (2) all performance parameters change.

2.2 Automatic Workload Repository

The structure of AWR report mainly consists of six fields, these six fields are the core of the whole report and record the key performance parameters of the database system as follows:

- **Instance information** provides the instance name, number, snapshot ids, total time the report was taken for and the database time during this elapsed time.
- **Cache Sizes** shows the size of each SGA region after AMM (Automatic Memory Management) has changed them. This information can be compared to the original init.ora parameters at the end of the AWR report.
- **Load Profile** shows important rates expressed in units of per second and transactions per second. This has to be compared to baseline report to understand the expected load on the machine and the delta during bad times.
- **Instance Efficiency Percentages** talks about how close are the vital ratios like buffer cache hit, library cache hit, parses etc. These can be taken as indicators, but should not be a cause of worry if they are low.
- **Shared Pool Statistics** summarizes changes to the shared pool during the snapshot period.

- **Top 5 Timed Events** shows the most serious of the five waiting events based on the proportion of waiting time.

2.3   Neural Network

An artificial neural network consists of simple processing units, which has a natural propensity for storing experiential knowledge and making it available for use. The building block inside an ANN is called a perceptron or neuron was illustrated in Fig.3. We identify three basic elements of the neural model [8]:A set of synapses, An adder and An activation function.

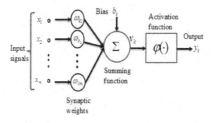

Fig. 3. Nonlinear Model of a Neuron

As is known to all, neural networks are learning mechanisms that can approximate any non-linear continuous function based on the given data [11].We can describe the neural network in a mathematical form, the output of a perceptron can be represented as

$$u_k = \sum_{j=1}^{m} \omega_{kj} x_j , \qquad\qquad y_k = \varphi(u_k + b_k) \qquad (1)$$

where $x_1, x_2, ..., x_m$ are the input signals; $\omega_{k1}, \omega_{k2}, ..., \omega_{km}$ are the synaptic weights of neuron k, respectively; u_k is the linear combiner output due to the input signals; b_k is the bias; $\varphi(\cdot)$ is the activation function; and y_k is the output signal of the neuron. By far the most common form of an activation function is a sigmoid function, which is defined as follows:

$$\varphi(x) = \frac{1}{1+e^{ax}} \qquad (2)$$

where a is the slope parameter of the sigmoid function. Most of the currently used ANNs are layered feed-forward networks, also called multilayer perceptron.

3   Constructing the Neural Network Model

There are two steps to construct neural network model: collecting sample data and training neural network.

3.1   Collecting Sample Data

In order to extract the AWR reports, we use a command script which is awrrpt.sql. The following will show how to export the AWR report: Firstly, we enter the $DBHOME 1/RDBMS/admin directory, to use PL/SQL Developer to run @E:/app/li/product/11.2.0/dbhome_1/RDBMS/ADMIN/awrrpt.sql script, and then there are two kinds export format to choose which are TXT and HTML format, finally according to clew to accomplish task step by step. In this paper, the workload data is collected 3375 times at intervals of ten minutes during 9:00 to 20:00 per day about four weeks in TPC-C workload environment.

Firstly, we extract all the data with respect to key performance parameters and performance indicators from the AWR reports, and then construct a sample training data set for neural network. The entire sample training data sets consist of three performance parameters and three performance indicators and it includes a total of 3375 records. A part of sample training data set was shown in Table 1.

Table 1. A Part of Sample Training Data Set

Db_cache _size(MB)	Shared_pool _size(MB)	Pga_aggregate _target(MB)	Data buffer hit ratio (%)	Library cache hit ratio (%)	In memory sort ratio (%)
30	30	30	76.02	84.51	92.24
60	90	30	78.64	89.42	91.32
60	30	60	77.89	84.87	93.55
90	60	120	82.67	88.34	96.63
90	90	120	82.21	90.17	95.54
120	120	120	85.43	91.62	95.98
150	30	60	87.19	86.36	92.76
150	90	120	86.34	91.00	96.31
180	150	60	89.75	93.48	95.25
210	150	120	91.32	94.01	97.72
90	210	150	83.89	95.33	98.77
150	240	60	88.48	96.09	94.04
240	120	150	94.56	92.22	97.65

3.2   Training Neural Network

A sample training data set of size 2000 were used to train the neural network, while the rest were used to test the trained neural network. In our experiments, the neural network is a back propagation neural network as shown in Fig. 4. It has 3 input neurons which are data buffer hit ratio and library cache hit ratio and in memory sort ratio, respectively; also has 3 output neurons which are db_cache_size, shared_pool _size and pga_aggregate_target while has 7 nodes in the hidden layer, respectively. The activation function used in hidden layer is sigmoid function and in output layer is linear function.

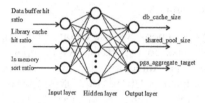

Fig. 4. Back-Propagation Neural Network Model

The training process was carried on MATLAB 7.0 with an epoch value of 100, learning rate of 0.2. Where training error performance curve was shown in Fig. 5.

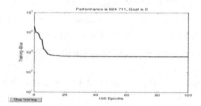

Fig. 5. Training Error Performance Curve

According to the training of the neural network, we can obtain the weight as follows:

$$
X = \begin{pmatrix}
0.0382 & 0.0626 & 0.0092 \\
-27.6409 & 44.0273 & -12.8464 \\
-13.1777 & -14.9651 & 22.0582 \\
-23.7565 & 78.5184 & -71.3239 \\
-27.3925 & 55.5363 & -29.6557 \\
55.9076 & -61.5203 & 2.3222 \\
209.8902 & -126.2826 & -57.6884
\end{pmatrix}
\qquad
Y = \begin{pmatrix}
2917.9 & 2456.0 & 514.3 \\
-151.6 & -80.9 & 43.1 \\
-144.6 & -52.8 & 17.0 \\
36.0 & 118.7 & 83.8 \\
-44.6 & 63.5 & -46.7 \\
67.5 & -1.8 & 29.3 \\
43.0 & -21.4 & -100.1
\end{pmatrix}^{T}
$$

$X = (x_{ih})$ and $Y = (y_{ho})$ refer to the weight matrices of the neural network, where x_{ih} is the synaptic weight from the input neuron i to the hidden neuron h, and y_{ho} is the synaptic weight from the hidden neuron h to the output neuron o. Thereby, we can adjust the output value to control the input values.

4   Performance Tuning Algorithm

We know that one performance indicator of the database can be affected by a number of parameters, although the degree is different. The data buffer hit ratio and in memory sort ratio and library cache hit ratio are performance indicators that can be utilized to tune parameters for enhance performance. These indicators reveal whether the size of parameters is appropriate or not. By using trained neural network model, we obtained a non-linear function relationship between performance indicators and parameters. In this section, we give a tuning algorithm as shown in Table 2.

Table 2. Performance Tuning Algorithm

Algorithm: Performance Tuning.

Input: // Performance Indicators values

    float p_1 (=data buffer hit ratio), p_2 (=library cache hit ratio), p_3 (=in memory sort ratio).

Output: // Tuned Parameters values

    V_1 (=db_cache_size), V_2 (=shared_pool_size), V_3 (=pga_aggregate_target).

Begin

    float $p_1^* = 95\%$, $p_2^* = 95\%$, $p_3^* = 98\%$; // p_i^* is a better indicator value for p_i

    for i = 1 to 3

      $u_i = f_1(p_i, p_i^*)$;

      if $5\% \le u_i < 25\%$

        for $j = 1$ to 3

          $change\_size1 = f_2(K_p, \omega_{ij}, QOS)$; $V_j = V_j + change\_size1$;

        end for

      else if $25\% \le u_i < 75\%$

        for j = 1 to 3

          $change\_size2 = f_3(K_p, \omega_{ij})$; $V_j = V_j + change\_size2$;

        end for

      else if $75\% \le u_i < 100\%$

        for $j = 1$ to 3

          $change\_size3 = f_4(K_p)$; $V_j = V_j + change\_size3$;

        end for

      else if $u_i < 5\%$

        continue;

      end if

    end for

End

The function $f_1(p_i, p_i^*)$, $f_2(K_p, \omega_{ij}, QOS)$, $f_3(K_p, \omega_{ij})$ and $f_4(K_p)$ can be represented as

$$f_1(p_i, p_i^*) = \frac{p_i^* - p_i}{p_i^*} \times 100\% \tag{3}$$

and

$$f_2(K_p, \omega_{ij}, QOS) = K_p \cdot (1 - u_i) \cdot \omega_{ij} \cdot (1 - \frac{QOS - QOS_0}{QOS}) \tag{4}$$

and

$$f_3(K_p, \omega_{ij}) = K_p \cdot (1 - u_i) \cdot \omega_{ij} \qquad f_4(K_p) = K_p \tag{5}$$

where K_p is the adjustment unit value of parameters which can be set manually, while ω_{ij} is correlation value of a parameter j for the performance indicator i. If the relationship between performance indicator and parameter is positive correlation, it set 1, otherwise, it set -1. The QOS of database system is defined as

$$QOS = r_1 \cdot (1 - \frac{R_0}{R_{max}}) + r_2 \cdot (1 - \frac{p_0}{p_{max}}) + r_3 \cdot (1 - \frac{X_{min}}{X_0}) +$$
$$r_4 \cdot (1 - \frac{CPU_{min}}{CPU_0}) + r_5 \cdot \sum_{i=1}^{n} a_i \cdot Rate_i \tag{6}$$

where $r_i, r_i \geq 0$ and $\sum_{i=1}^{5} r_i = 1$, present the weight indicating the relative importance of each performance indicator, while R_0 is the average response time and R_{max} is the longest one tolerated , p_0 is the average refusing ratio of request and p_{max} is the maximum one tolerated, X_0 is the average throughput and X_{min} is the minimum one tolerated, CPU_0 is the average ratio of CPU utilization and CPU_{min} is the minimum one tolerated, $Rate_i$ are the various cache hit ratios and a_i are the correlation value and $\sum_{i=1}^{n} a_i = 1$, respectively. And QOS_0 is the average quality of service of database system. According to the expert experience, p_i^* are commonly set as 95%, 95%,98% for performance indicators data buffer hit ratio, library hit ratio and in memory sort hit ratio, respectively.

5   Experiment Evaluation

In order to demonstrate the effectiveness and feasibility of our self-tuning algorithm, we conduct experiments. The experiments are carried on Oracle Database 11g release 11.2.0.1.0. The main tool for database operations is PL/SQL Developer with the version is 8.0.4.1514. This tool runs in PC with Operating System Microsoft Windows 8 Professional Version (2012 Microsoft corporation), Processor Intel(R) Pentium(R) 4 CPU 3.00GHz and Memory 2.75GB of RAM.

5.1   Experiment 1 - One Parameter Tuning

In this experiment, we only consider tuning one performance parameter by fixed other two parameters to analyze the effectiveness of our self-tuning algorithm. Fig. 6 show the data buffer hit ratio in the TPC-C workloads in which the indicator were not affected by tuning parameters likes shared_pool_size and pga_aggregate_target. Fig. 7 show the disk writes with checkpoints in which the indicator was affected by tuning parameters db_cache_size, while were not affected by other two parameters. As shown in Fig. 8, we can see that the redo log size seem to decline gradually with tuning shared_pool_size after change number was 3, while there is little influence by other two parameters. In a word, the above results by using our self-tuning algorithm for one parameter tuning coincide with resource usage analysis of the work [10].

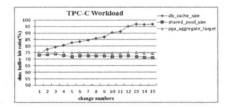

Fig. 6. Data buffer hit ratio for tuning different parameters

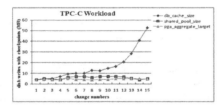

Fig. 7. Disk writes with checkpoints for tuning different parameters

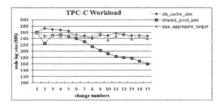

Fig. 8. Redo log size for tuning different parameters

5.2   Experiment 2 - Multi-parameters Tuning

In this experiment, we simulate an OLTP type of application based on the TPC-C workload environment, resulting in pressure on the oracle database. In order to testify the effectiveness of our self-tuning algorithm, we monitored the quality of the response time and throughput of oracle database, and then compare our tuning algorithm with the tuning algorithm proposed by Rodd [12]. A1 represents our tuning algorithm and A2 represents the algorithm by Rodd. Fig. 9 shows the effect of the number of users on the response time. When the number of users is lower than 60, there is no difference for the response time between A1 and with A2, and they are all better than the one without tuning algorithm. If the number of users is between 60 and 140, the response time of A2 tuning algorithm is better than A1 tuning algorithm. But as the number of users increases beyond 140, the response time of A2 tuning algorithm starts rises, while A1 tuning algorithm starts decline rapidly. However, when the number of users is beyond 220, the response time rises gradually with A1 tuning algorithm, but it is always lower than that with A2 tuning algorithm.

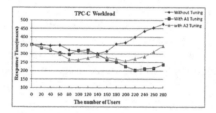

Fig. 9. The Response Time for different users

As Fig. 10 illustrates below, the transaction per second of database system with A1 tuning algorithm is higher than the one with A2 tuning algorithm as the number of transactions executed increases. When the number of different transaction executed is lower than 11000, the transaction per second with tuning A1 algorithm is almost less than the one with A2 tuning algorithm. But, when the number of different transaction executed is beyond 13000, the transaction per second with tuning A1 algorithm increases faster than the one with A2 tuning algorithm. The value of throughput with A1 tuning algorithm is very large as the number of transaction executed is increasing until up to 25000. After that, the throughput declines gradually with tuning A1 algorithm, but it still larger than the one with A2 tuning. In summary, our tuning algorithm is better than A2 tuning algorithm, and can enhance throughput and reduce response time significantly. Hence, the performance of oracle database has obtained great improvements by using our performance tuning algorithm.

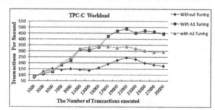

Fig. 10. The Throughput for different Transactions

6   Related Work

Yoo *et al.* [16] proposes a non-linear model based on an artificial neural network to explore the relationship between workload configurations and performance characteristics, they applied such model to a 3-tier web service with response time restrictions and providing a substantial assistance to the performance tuning efforts. But the limitation of this model is hard to perform a quantitative analysis for a complete understanding of individual contribution of a particular feature to the output. Bigus [3] presents results of applying neural networks and techniques from control systems theory to computer system performance tuning. A neural network model-based adaptive system tuning framework were used in a simulated multi programmed computer system with a time-varying workload comprising four job classes, and results shown that back-propagation and radial basis function neural network controllers can be trained on-line to tune memory allocations for enhance performance. Rodd *et al.* [13] proposes a novel technique that combines learning ability of the artificial neural network and the ability of the fuzzy system to deal with imprecise inputs are employed to estimate the extent of tuning required. Furthermore, the estimated values are moderated based on knowledge base built using experimental findings. The experimental results show significant performance improvement as compared to built in self tuning feature of the DBMS.

Compared with all of the above methods, our work emphasized on how to construct neural network model and how to tune multi performance parameters for enhance performance. We proposed a self-tuning algorithm for tuning multi parameters based on neural network, and compare the effectiveness of our self-tuning algorithm with other self-tuning algorithm. Experimental results shown that our self-tuning algorithm can effectively enhance throughput and decline response time.

7   Conclusions and Future Work

This paper describes how to extract AWR report and gather key system performance parameters and indicators. A Neural Network model was built to estimate the performance characteristic. A self-tuning algorithm for how to tune multi-parameters was proposed to enhance system performance. We also compared the proposed tuning algorithm with other tuning algorithm regarding the response time and throughput in the TPC-C workload environment. Experimental results shown that our self-tuning algorithm can effectively enhance throughput and decline response time.

In this paper, we only consider to tune parameters for enhance performance in the TPC-C workload environment. As a future work, we will study the relationship between performance indicators and parameters in the TPC-W workload environment, and also investigate the impacts on the performance of database with more parameters.

Acknowledgements. This work is supported by NSFC under Grants No. 61210004 and 61170015, and Philosophy and Social Science Research Project of the Ministry of Education under Grant No. 10JZD0006.

References

1. Alhadi, N., Ahmad, K.: Query tuning in oracle database. Journal of Computer Science 8(11), 1889–1896 (2012)
2. Agrawal, S., Chaudhuri, S., Narasayya, V.R.: Automated Selection of Materialized Views and Indexes in SQL Databases. In: 2000 International Conference on Very Large Data Base, pp. 496–505 (2000)
3. Bigus, J.P.: Applying neural networks to computer system performance tuning. In: 1994 IEEE International Conference on Neural Networks, vol. 4, pp. 2442–2447 (1994)
4. Belknap, P., Dageville, B., Dias, K., et al.: Self-tuning for SQL performance in Oracle database 11g. In: 2009 IEEE International Conference on Data Engineering, pp. 1694–1700 (2009)
5. Dageville, B., Dias, K.: Oracle's Self-Tuning Architecture and Solutions. IEEE Data Engineer. Bulletin 29, 24–31 (2006)
6. Debnath, B.K., Lilja, D.J., Mokbel, M.F.: SARD: A Statistical Approach for Ranking Database Tuning Parameters. In: 2008 IEEE International Conference on Data Engineering Workshop, pp. 11–18 (2008)
7. Gordon-Ross, A., Vahid, F.: A self-tuning configurable cache. In: ACM Proceedings of the 44th Annual Design Automation Conference, pp. 234–237 (2007)
8. Haykin, S.S.: Neural networks and learning machines, 3rd edn. Pearson Education, Upper Saddle River (2009)
9. Panayotakis, K.: Decision Support Systems, OLTP vs DSS systems (2006), http://ezinearticles.com/?expert=Kostis-Panayotakis
10. Oh, J.S., Lee, S.H.: Resource selection for autonomic database tuning. In: 2005 IEEE International Conference on Data Engineering Workshops, p. 1218 (2005)
11. Perlovsky, L.I.: Neural Networks and Intellect: using model-based concepts. Oxford University Press, New York (2001)
12. Rodd, S.F., Kulkrani, U.P., Yardi, A.R.: Neural Network based Database Tuning Architecture. International Journal of Recent Trends in Engineering (2009)
13. Rodd, S.F., Kulkrani, U.P., Yardi, A.R.: Adaptive neuro-fuzzy technique for performance tuning of database management systems. Evolving Systems 4(2), 133–143 (2013)
14. TPC Benchmark C Specification, C, Revision 5.0 (2001), http://www.tpc.org/tpcc/defalut.asp
15. Verma, A.: Enhanced Performance of Database by Automated Self-Tuned Systems. International Journal of Computer Science (2011), 2231–5268
16. Yoo, R.M., Lee, H., Chow, K., et al.: Constructing a non-linear model with neural networks for workload characterization. In: 2006 IEEE International Symposium on Workload Characterization, pp. 150–159 (2006)
17. Zhang, G., Chen, M., Liu, L.: A Model for Application-Oriented Database Performance Tuning. In: 2012 6th IEEE International Conference on New Trends in Information Science and Service Science and Data Mining, pp. 389–394 (2012)

GA-EAM Based Hybrid Algorithm

Ashish Tripathi, Divya Kumar, Krishn Kumar Mishra, and Arun Kumar Misra

Computer Science and Engineering Department
Motilal Nehru National Institute of Technology Allahabad
Allahabad, Uttar Pradesh, India
ashish.mnnit44@gmail.com,
{divyak,kkm,akm}@mnnit.ac.in

Abstract. The methods of searching optimal solutions are distinct in different evolutionary algorithms. Some of them do search by exploiting whereas others do by exploring the whole search space. For example Genetic Algorithm (GA) is good in exploitation whereas the Environmental Adaption Method (EAM) performs well in exploring the whole search space. Individually these algorithms have some limitations. In this paper a new hybrid algorithm has been proposed, which is created by combining the techniques of GA and EAM. The proposed algorithm attempts to remove the limitations of both GA and EAM and it is compared with some state-of-the-art algorithms like Particle Swarm Optimization-Time Variant Acceleration Coefficient (PSO-TVAC), Self-Adaptive Differential Evolution (SADE) and EAM on six benchmark functions with experimental results. It is found that the proposed hybrid algorithm gives better results than the existing algorithms.

Keywords: evolutionary algorithm, environmental adaption method, genetic algorithm, phenotypic plasticity.

1 Introduction

GA is a well-known evolutionary algorithm that has been used in a variety of optimization problems. It is a stochastic search and optimization technique, in which only the fittest solutions will survive and reproduce better offspring [1, 2]. But the problem with the GA is that it may stagnate in local optimal solutions [3]. EAM is also an evolutionary algorithm [4, 5, 6]. It is based on the principle of adaptive learning [7], which states that only those candidate solutions will survive in the consecutive generations which are capable to adapt the new environmental changes. Problem with EAM is that its operators do search for optimal solution by exploring the whole search space and it does not exploit the search space near the best solutions that are obtained in the previous generations. So its performance may degrade in uni-modal problems. To overcome the limitation of GA and EAM, a hybrid algorithm GEHA is created. Performance of this algorithm has been checked over the well-known state-of-the-art algorithms like PSO-TVAC [8], SADE [9] and EAM [10] on five benchmark functions. Result analysis shows that the proposed algorithm does not

D.-S. Huang et al. (Eds.): ICIC 2014, LNCS 8588, pp. 13–20, 2014.

have the problem of stagnation and perform exploration and exploitation of all good regions in the search space.

The rest of the paper is organised in three sections; section 2 discusses the background details, section 3 explains the proposed approach and result analysis and finally the paper is concluded in section 4.

2   Background Details

2.1   Genetic Algorithm

GA is based on Charles Darwin's theory of evolution and survival of the fittest. It is a search algorithm that works mainly using a process of natural selection, crossover and mutation. GA was first discovered by Holland in 1960s and further enhanced by Goldberg [2, 3]. GA needs a genetic representation of the solution space (usually represented as an array of binary bits) and a fitness function to calculate the fitness of every individual in the search space. It begins with a sample set of potential solutions which then evolves toward a set of more optimal solutions. Better solutions within the sample set are participated in mating and also transmit their good traits form one generation to another generation whereas the poor solutions tend to die out. The total size of the solution space remains constant. For every new solution added in the solution space, an old and unfit solution is removed. GA has some supporting elements, which are equally important and participated in the working of GA, such as encoding [12, 13, 15], fitness function [11], selection [14, 16], crossover [16], mutation [17] and elitism [17].

2.2   Environmental Adaption Method (EAM)

The working of EAM is based on three operators namely adaption, alteration and selection [5]. Adaption operator is used to update the phenotypic structure (PS) of every solution of the current generation. During adaption an individual Pin adapts itself and generates its offspring such as P'in = (α*(Decoded value in decimal of binary coding of P'in)|F(Pin)/Favg| + β)%2l. This adaption is directed by two random parameters α, β. The ratio of the current fitness of the individual to the average fitness of the population for the current generation is F(Pin)/Favg and it depends on the length l of the bits used in the encoded individual. Alteration operator may result due to environmental noise. Selection operator is used to select the best individuals. This process will be continued until either it gives the expected solution or the maximum iteration has been encountered.

3   Proposed Approach

Like GA and EAM, the proposed algorithm is also a population based algorithm and the initial population is generated randomly. The selection of the solution is based on the environmental fitness. The solution whose fitness is higher than environmental fitness will be moved into GA pool otherwise it will move into EAM pool. Only good solutions go through crossover and mutation and the bad ones try to adapt themselves according to their surrounding environmental conditions. Here good solutions are those solutions which have good Phenotypic Structure (PS) i.e. greater than Favg. Here Favg is the average fitness value of current population and represents the current environmental fitness. PS is the value of traits of an organism that is affected by the changes in environmental conditions such as weather, temperature, pollution etc. In GA, before applying crossover and mutation, elitism is applied on the solutions and a copy of these solutions is stored in a buffer, while in EAM adaption operator is used on the solutions which are not as good as the solutions used for GA. Adaption is the first operator in EAM which is used to update the PS of each individual solution of the current generation by evolving its current fitness towards environmental fitness and it helps these solutions to be selected for GA in the successive generations. After applying crossover and adaption operators on the solutions of the current generation, mutation and alteration are performed respectively. The termination criterion is checked, and if it is satisfied, the algorithm will terminate, else it will continue with the next iteration. Finally this algorithm gives the final set of optimal solutions (F*). The working model of proposed algorithm is given in fig 1.

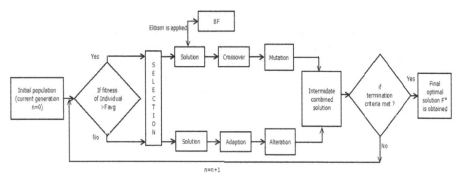

Fig. 1. Working model of proposed algorithm

3.1   Proposed Algorithm

Input Variables	MAXGEN – Maximum number of generation. POP_SIZE – Total population size.	
Output Variable	F* - Final optimal set of solutions.	
Other Variables		
n – Number of generations.	$POPn$ – population for nth generation.	
$SOLGn$ – Available solution for GA.	SOLEn – Available solution for EAM.	
$Favg$ – Average of the current fitness of solutions.	CFT – Current fitness of solution.	
EFT – Environmental fitness of a solution.	CP – Crossover probability.	
MP – Mutation probability.	IO_C – Intermediate offspring after applying crossover.	

IO_M: Intermediate offspring after applying Mutation.	IO_AD - Intermediate offspring after applying adaption.
IO_AL - Intermediate offspring after applying alteration.	IO_GA – Population obtained after applying crossover and mutation.
IO_EAM - Population obtained after applying adaption and alteration.	PGE_n – The final population obtained after each iteration.
Favg = (sum of current fitness of all individuals) / total no individuals.	

Steps:
1. Start
2. Set n=0
3. Generate POP_0, taking random numbers form POP_SIZE.
4. Apply selection using F_{avg}, If CFT > F_{avg} Then Case1. go for crossover else Case2. go for adaption
5. In case 1 Apply elitism before using crossover and make a copy of good solution in a buffer named BF.
6. Apply crossover as per the crossover probability CP sexually on $SOLG_n$ and get the IO_C_n.
7. After completing crossover perform mutation with mutation probability MP asexually on IO_C_n and get IO_M_n.
8. After crossover and mutation IO_GA_n is obtained.
9. In case 2, the adaption operator is applied to each member of the population $SOLE_n$ asexually, which improves the CFT up to EFT and forms IO_AD_n.
10. After adaption, alteration takes place on IO_AD_n to form IO_AL_n.
11. After adaption and alteration IO_EAM_n is obtained.
12. After every iteration of the GA and EAM in parallel, IO_GA_n and IO_EAM_n are obtained.
13. PGE_n = {IO_GA_n U IO_EAM_n}
14. 14. n = n + 1
15. If n == MAXGEN
16. Set F*= PGE_{MAXGEN}
     Then go to 16 else go to 4
17. Return F*
18. End

3.2   Result Analysis

Both uni-modal and multi-modal functions [18, 19] have been taken in this paper to check the performance of the proposed algorithm with PSO-TVAC, SADE and EAM. For each benchmark function population size (P = 40 and 100), generation (G = 1000) and dimensions (D = 7, 10, 20, 30) have been taken for experimental analysis. For each experiment, statistical data like mean, minimum and standard deviation has been calculated. But only minimum value has been shown in the figures. Distinct colours are used in the chart and each colour represents performance of different algorithm. Below, we can see the performance of algorithms with proper explanation. For Sphere function, the minimum value obtained by GEHA is comparable to PSO-TVAC, SADE and EAM. Graphs also confirm this analysis.

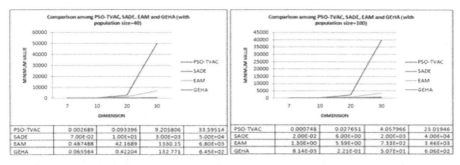

PSO-TVAC	0.002689	0.093396	9.205806	33.59514	PSO-TVAC	0.000748	0.027651	4.057966	23.01946
SADE	7.00E-02	1.00E+01	3.00E+03	5.00E+04	SADE	2.00E-02	6.00E+00	2.00E+03	4.00E+04
EAM	0.487488	42.1689	1330.15	6.80E+03	EAM	1.30E+00	5.59E+00	7.33E+02	3.46E+03
GEHA	0.065564	0.42204	132.771	6.45E+02	GEHA	8.14E-03	2.21E-01	5.07E+01	6.06E+02

Fig. 2. Comparison of PSO-TVAC, SADE, EAM and GEHA on Sphere for minimum value on different dimensions

For Shifted Schwefel function, at dimension 20, GEHA gives good result than EAM while at dimension 10 and 30, minimum value is obtained by EAM is better than GEHA. GEHA gives better results than PSO-TVAC in all dimensions (7, 10, 20, and 30) for population size 40. For population size 100, GEHA gives better results than PSO-TVAC and SADE in all dimensions. The minimum value received by GEHA and EAM almost same but for the dimensions 20 and 30, GEHA has gained a slight improvement over EAM.

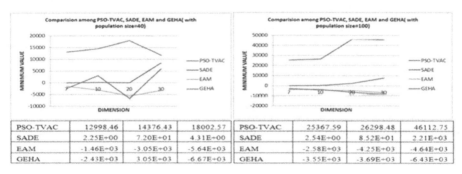

PSO-TVAC	12998.46	14376.43	18002.57	PSO-TVAC	25367.59	26298.48	46112.75
SADE	2.25E+00	7.20E+01	4.31E+00	SADE	2.54E+00	8.52E+01	2.21E+03
EAM	-1.46E+03	-3.05E+03	-5.64E+03	EAM	-2.58E+03	-4.25E+03	-4.64E+03
GEHA	-2.43E+03	3.05E+03	-6.67E+03	GEHA	-3.55E+03	-3.69E+03	-6.43E+03

Fig. 3. Comparison of PSO-TVAC, SADE, EAM and GEHA on Shifted Schwefel for minimum value on different dimensions

For Rotated Bent Cigar function, minimum value generated by both SADE and GEHA are almost same dimension 7 and population size 40. SADE dominates GEHA for the dimension sizes 10 and 20 but suddenly GEHA improve its performance and shows its strength over the dimension 30 where it achieves good result than SADE. For the population size 100, the value accomplished by GEHA is very impressive and better than rest of the algorithms up to the dimension 20 but at 30 SADE dominates GEHA.

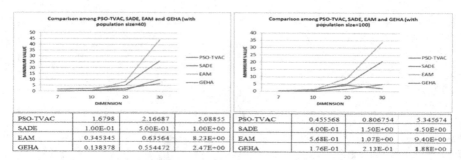

PSO-TVAC	1.6798	2.16687	5.08855	PSO-TVAC	0.455568	0.806754	5.345674
SADE	1.00E-01	5.00E-01	1.00E+00	SADE	4.00E-01	1.50E+00	4.50E+00
EAM	0.345345	0.63564	8.23E+00	EAM	5.68E-01	1.07E+00	9.40E+00
GEHA	0.138378	0.554472	2.47E+00	GEHA	1.76E-01	2.13E-01	1.88E+00

Fig. 4. Comparison of PSO-TVAC, SADE, EAM and GEHA on Rotated Bent Cigar for minimum value on different dimensions

It is found that the minimum value obtained by GEHA is very less as compared to other algorithms for all dimensions with population size 40 and 100 for Shifted Rastrigin function. GEHA has given good results for the population size 40 as compared to the population size 100.

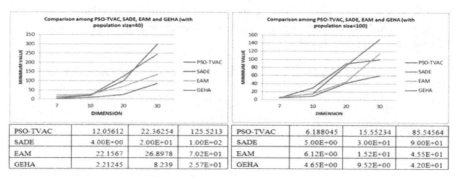

PSO-TVAC	12.05612	22.36254	125.5213	PSO-TVAC	6.188045	15.55234	85.54564
SADE	4.00E+00	2.00E+01	1.00E+02	SADE	5.00E+00	3.00E+01	9.00E+01
EAM	22.1567	26.8978	7.02E+01	EAM	6.12E+00	1.52E+01	4.55E+01
GEHA	2.21245	8.239	2.57E+01	GEHA	4.65E+00	9.52E+00	4.20E+01

Fig. 5. Comparison of PSO-TVAC, SADE, EAM and GEHA on Shifted Rastrigin for minimum value on different dimensions

For Shifted Rosenbrock function the minimum value obtained by SADE, EAM and GEHA are looking similar up to the dimension number 20. For dimension 30 the major changes can be seen in the figures.

The proposed algorithm, GEHA gives minimum value with population size 40 for all dimensions starting from 7, 10, 20 and 30 respectively for **Shifted Rotated Rastrigin function**. While GEHA dominated by SADE with population size 100 for the dimension 7 and 10. But in the next dimensions 20 and 30, GEHA result has been improved over SADE, EAM and PSO-TVAC. Overall results gained by GEHA are very positive.

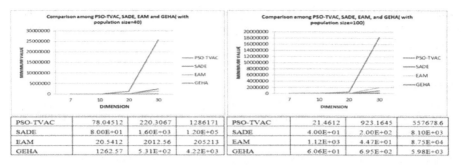

PSO-TVAC	78.04512	220.3067	1286171		PSO-TVAC	21.4612	923.1645	557678.6
SADE	8.00E+01	1.60E+03	1.20E+05		SADE	4.00E+01	2.00E+02	8.10E+03
EAM	20.5412	2012.56	205213		EAM	1.12E+03	4.47E+01	8.75E+04
GEHA	1262.57	5.31E+02	4.22E+03		GEHA	6.06E+01	6.95E+02	5.98E+03

Fig. 6. Comparison of PSO-TVAC, SADE, EAM and GEHA on Shifted Rosenbrock for minimum value on different dimensions

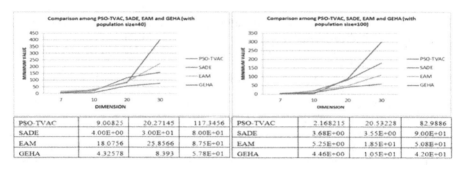

PSO-TVAC	9.00825	20.27145	117.3456		PSO-TVAC	2.168215	20.53228	82.9886
SADE	4.00E+00	3.00E+01	8.00E+01		SADE	3.68E+00	3.55E+00	9.00E+01
EAM	18.0756	25.8566	8.75E+01		EAM	5.25E+00	1.85E+01	5.08E+01
GEHA	4.32578	8.393	5.78E+01		GEHA	4.46E+00	1.05E+01	4.20E+01

Fig. 7. Comparison of PSO-TVAC, SADE, EAM and GEHA on Shifted Rotated Rastrigin for minimum value on different dimensions

To make comparison of the significance of the algorithms Wilcoxon rank sum test method [20] has been applied which is shown in table 1.

Table 1. Wilcoxon's rank sum test

PERFORMANCE	SADE	PSO-TVAC	*EAM*
-	12	14	10
+	7	6	8
≈	6	5	7

The mean error, standard deviation and minimum values are calculated for each algorithm for 30 runs. Wilcoxon's rank sum test is performed between GEHA, SADE, PSO-TVAC and EAM at 0.05 significance level. The operator "-"denotes the performance of an algorithm worse than, "+", better than, and "≈", similar to the performance of the proposed GEHA.

4   Conclusion

The results obtained by the experiment show that, the proposed algorithm GEHA performs well as compared to the performance of PSO-TVAC, SADE and EAM for different dimensions and two different population sizes. This algorithm performs well for both uni-modal and multi-modal problems. This algorithm can be used to solve complex real life problems.

References

1. Whitley, D.: A genetic algorithm tutorial. Statistics and Computing 4(2), 65–85 (1994)
2. Goldberg, D.: Genetic Algorithms in Search, Optimization and Machine Learning, pp. 1–25. Addison-Wesley, Reading (1989)
3. Gen, M., Cheng, R.: A survey of penalty techniques in genetic algorithms. In: Proceedings of IEEE International Conference on Evolutionary Computation, pp. 804–809. IEEE (1996)
4. Baldwin, J.M.: A new factor in evolution (Continued). The American Naturalist 30(355), 536–553 (1896)
5. Holland, J.H.: Adaptation in natural and artificial systems: An introductory analysis with applications to biology, control, and artificial intelligence. U. Michigan Press (1975)
6. Price, K.V.: An introduction to differential evolution. In: New Ideas in Optimization, pp. 79–108. McGraw-Hill Ltd., UK (1999)
7. Wund, M.A.: Assessing the impacts of phenotypic plasticity on evolution. Integrative and Comparative Biology 52(1), 5–15 (2012)
8. Chaturvedi, K.T., Pandit, M., Srivastava, L.: Particle swarm optimization with time varying acceleration coefficients for non-convex economic power dispatch. International Journal of Electrical Power and Energy Systems 31, 249–257 (2009)
9. Omran, M.G.H., Salman, A., Engelbrecht, A.P.: Self-adaptive Differential Evolution. In: Hao, Y., Liu, J., Wang, Y.-P., Cheung, Y.-M., Yin, H., Jiao, L., Ma, J., Jiao, Y.-C. (eds.) CIS 2005. LNCS (LNAI), vol. 3801, pp. 192–199. Springer, Heidelberg (2005)
10. Mishra, K.K., Tiwari, S., Misra, A.K.: A bio inspired algorithm for solving optimization problems. In: 2011 2nd International Conference on Computer and Communication Technology (ICCCT), pp. 653–659. IEEE (2011)
11. http://www.obitko.com/tutorials/geneticalgorithms/
12. http://nc25.troja.mff.cuni.cz/~soustruznik/GA.html#encoding
13. Back, T.: Evolutionary Algorithms in Theory and Practice. Oxford University Press, New York (1996)
14. Back, T., Schwefel, H.P.: An Overview of Evolutionary Algorithms for pa-rameter optimization. Evolutionary Computation 1(1), 1–23 (1993)
15. Jin, Y.: A comprehensive survey of fitness approximation in evolutionary computation. Soft Computing 9, 3–12 (2005)
16. Fogel, D.B.: Evolutionary Computation. IEEE Press (1995)
17. Yao, X.: Evolutionary computation: A gentle introduction. In: Sarkar, R., Mohammadian, M., Yao, X. (eds.) Evolutionary Optimization, ch. 2, pp. 27–53. Kulwer Academic Publishers (2002)
18. Suganthan, P.N., et al.: Problem definitions and evaluation criteria for the CEC 2005 special session on real-parameter optimization. KanGAL Report 2005005 (2005)
19. Liang, J.J., et al.: Problem definitions and evaluation criteria for the CEC 2013 special session on real-parameter optimization. Computational Intelligence Laboratory, Zhengzhou University, Zhengzhou, China and Nanyang Technological University, Singapore. Technical Report 201212 (2013)
20. Ng, H.K.T., Balakrishnan, N.: Wilcoxon-Type Rank-Sum Precedence Tests. Australian & New Zealand Journal of Statistics 46(4), 631–648 (2004)

Hybrid ITO Algorithm for Solving Numerical Optimization Problem

Yunfei Yi[1,2], Xiaodong Lin[2], Kang Sheng[1], Lin Jiang[1], Wenyong Dong[1], and Yongle Cai[1]

[1] Computer School, Wuhan University, Wuhan 430079
[2] College of Computer and Information Engineering, Hechi University, Yizhou, 546300

Abstract. A hybrid algorithm that integrates ITO algorithm with PSO algorithm is proposed for solving numerical optimization problem in this paper. We design a strategy that exert a synchronized wave process to move and wave operator so as to improve the ability of finding global optimization of the proposed algorithm. Specially, we employ the mechanism of PSO algorithm, thus mapping the wave coefficient to c_1 and the move coefficient to c_2 where c_1 and c_2 are two parameters in PSO algorithm, while the inertia weight decreases linearly with temperature synchronously, finally we use the individual optimization to exert an additional wave process to the population, which lead the particles to jump out of the local optimal solution and move in the direction of global optimal solution. Experiments have been performed on several benchmark test functions, which usually are intractable to achieve good solutions , then comparison is conducted to show the performance difference to standard PSO and ITO. The experimental results demonstrate that the proposed algorithm is feasible, possesses fast convergence speed, strong robustness and stability.

Keywords: ITO algorithm, particle swarm optimization(PSO) algorithm, numerical optimization problem, move operator, wave operator.

1 Introduction

Particle Swarm Optimization[1] was first proposed by Kennedy et al in the IEEE International Conference on neural network in 1995. It is a global optimization algorithm based on the global swarm intelligence which is inspired from animal behavior and social psychology. In fact, it is a simulation research on the group behavior of birds in nature. As for standard PSO, each problem to be optimized in the search space is regarded as a particle, and a large number of particles come to a population, in which a new generation get updated by tracking two extreme values, and in the process of updating, each individual take full advantage of the characteristics of the population and itself to constantly learn and adjust to move towards the global optimal solution and eventually get satisfactory solutions. While PSO algorithm has been widely applied and achieved desirable results in many areas such as in function optimization, fuzzy system control, combinatorial optimization, intrusion detection and

D.-S. Huang et al. (Eds.): ICIC 2014, LNCS 8588, pp. 21–31, 2014.

decision scheduling and so on, sometimes it tend to be trapped in local optimal solution which leads to the low convergence precision and drawback of not converging to the global optimal solution.

Professor Dong Wenyong and Professor Li Yuanxiang proposed the ITO algorithm[2-10] (Ito Algorithm or Evolutionary Algorithm inspired by Ito stochastic process, abbreviated to ITO), which is a new class of evolutionary algorithm. It analyzes the movement of particles from the microscopic point of view, and mimic the kinetic studies of particles' interactive colliding in the particle system for designing algorithm and solving problems. ITO algorithm reflects the "group search" characteristics of bionic evolutionary algorithm, because the solution consists of particles and all of the particles constitute a particle system. The performance of this algorithm is heavily dependent on the settings of two parameters: move operator and wave operator, effective design of these two operators is crucial for achieving the global optimization solution. The move rate and wave rate are designed according to the law of macromolecules' movement, which is proposed by Einstein and Langevin, and the thermal law of particles' motion as well. As a result, it owns some typical characteristics of simulated annealing. Currently, ITO algorithm performs well and is able to achieve better solution in function optimization[3,4], time series modeling[5] and combinatorial optimization[6-10] etc.

In this paper, we first make an analysis of PSO algorithm and ITO, then a hybrid approach which combines these two algorithms is proposed. In the hybrid algorithm, the advantages of the two algorithms are taken and a helpful strategy is employed to solve function optimization problems, that is we simultaneously exert an additional wave process to move operator and wave operator. The experimental results are presented by implementing the proposed algorithm for three typical complex benchmark functions, which show desirable results are achieved.

2   Hybrid ITO Algorithm for Function Optimization Problems

In PSO[11-13], the first operation is to initialize the population, then try to find the optimal solution according to equation (1) and (2) by successive iteration. In the running process of every generation, particles get updated to move in the direction of global optimal solution by tracking two extreme values as follows.

$$v_{id}^{t+1} = wv_{id}^t + c_1 r_1 (p_{id} - x_{id}^t) + c_2 r_2 (p_{gd} - x_{id}^t) \qquad (1)$$

$$x_{id}^{t+1} = x_{id}^t + v_{id}^{t+1} \qquad (2)$$

In the above formulae, $i = 1,2,...N$, N is number of particles in the population; t=1,2,...M, M represents the maximum iteration count of the population; v_{id}^t denotes the d-dimensional component of the ith particle's velocity vector in the tth iteration and x_{id}^t tells position vector, p_{id} represents the d-dimensional component of the ith particle's individual extreme value(the particle's historical optimal value); p_{gd} refers to the d-dimensional component of the current global optimal value; r1 and r2 are random numbers that generated subject to U(0,1) distribution; c_1 and c_2 are learning factors, generally $c_1 = c_2 = 2$; w is the inertia factor.

The deficiency of learning factors c_1 and c_2 in formula (1) is lack of dynamic adaptation, which is due to that the factors are usually set based on empirical knowledge. In this paper, however, we will set the two operators of ITO algorithm respectively as the learning factors c_1 and c_2 for simultaneous optimization, which can leads to adaptive ability on setting the parameters so as to compensate the defect on setting learning factors. Additionally, it further enable the populations to move in the direction of global optimal solution and wave towards the optimal solution of current generation. When a particle moves to a new location, it takes the individual extreme value as its current move destination, then an additional move is implemented to the population to make it move in accordance with the individual extreme value. It is worth mentioning that as we can learn from the essence of Brownian motion: particles are doing random movement and colliding randomly with aerosols constantly, that means particles are never going to be stopped in the process of movement. The particle will not be in static state after finding a feasible solution, but doing constant fluctuations in their neighborhood areas, which is in line with the laws of Brownian motion. As a consequence, the individual extreme in this paper is designed to be the wave operator in order to improve the overall performance of the particle and make the population move towards the global optimal solution.

The first two problems to be solved are about how the Ito stochastic process be mapped to the optimization algorithms: one is how to design the macro trend term of Ito process namely the move process, making the algorithm move toward the optimal solution at the macro level; another is how to design the wave process to make each particle able to search in the local space so as to improve search efficiency. The move operator and wave operator are closely associated with the particle's radius and temperature. It will have a positive impact on the performance if there exists effective design on move and wave operator. In the next part an investigation is conducted on several key designs of the algorithm, including definition of the particle radius, cooling schedule, move operator and wave operator. Finally PSO is executed by utilizing the approach of exerting additional fluctuation to move and wave processes simultaneously.

3   The Design of Particle Radius

Each particle in the search space owns a radius that its size determines particle's moving ability. According to Einstein and Langevin's equation, we can see that the larger the particle's radius is the weaker of its moving ability gets, similarly, the smaller the more intense. Therefore, effective design of the radius plays a key role in the movement or further to the extent that enhance the capability of moving toward global optimal solution. A series of functions can be utilized to design the particle radius, such as: linear transformation function, exponential function, sorting, etc. In this paper, we employ the approach based on sorting for designing particle radius. The procedure is as follows:

First, sort all the particles of current population in accordance with the objective function value in ascending order, then update the size of radius according to the following formula.

$$r_i = r_{min} + \frac{(r_{max} - r_{min})}{N} * i \ (i \in [0, N])$$ (3)

The generated particle radius based on sorting is averagely distributed between $[r_{min}, r_{max}]$, that is, the radius is in even distribution so as to be able to find better solutions in the search space. Besides, more consistency is shown in line with the objective function value when in even more complex circumstance. To simplify the computational complexity of the algorithm we typically assume $r_{min} = 0$, $r_{max} = 1$. Then formula (3) can be converted to the formula below:

$$r_i = \frac{1.0}{N} * i \ (i \in [0, N])$$ (4)

4   The Design of Cooling Schedule

According to the displacement equation of ideal gas giant molecular systems which was proposed by Einstein and Langevin, we can see that athletic ability of particle is affected desperately by temperature. More specifically, particle does strenuous exercise under high temperature and limited movement under low temperature. Obviously, effectively designed temperature can exert great impact on the algorithm's performance of finding optimization solution; In this paper, an approach of simulated annealing algorithm is utilized, we use cooling schedule as the control parameter of temperature change. Cooling schedule is an important factor that affects the experimental performance heavily, which should be selected reasonably in order to apply the algorithm. The following formula shows the setting of the annealing[14] function:

$$T^{t+1} = \rho T^t$$ (5)

where $0 \leq \rho \leq 1$, and usually ρ is set 0.99.

5   The Move Operator

The move operator is mainly utilized to control the particle's exploration ability, that is, the particle's motion in the macro view is moving in the direction of the attractor. There are two aspects need to be considered when designing this operator: one is the move intensity factor namely the move coefficient u ; another is the process of moving, which leads particle to move in neighborhood in accordance with the move intensity. u is effected by its own radius r of particle individual, the current ambient temperature T and other factors. Dong and other researchers have proposed an approach of separated variables, which goes as follows:

$$f(r,T) = u_{min} + f_1(r) * f_2(T) * (u_{max} - u_{min}) \tag{6}$$

Where u_{max} and u_{min} denote the particle's maximum and minimum move intensity respectively. To facilitate the design of the algorithm, we assume $u_{max} = 1$, $u_{min} = 0$ in this paper. Function $f_1(r)$ is the move intensity determined by the radius of particle, a feasible solution is as follows:

$$f_1(r) = \frac{e^{-\lambda r} - e^{-\lambda r \max}}{e^{-\lambda \min} - e^{-\lambda r \max}} \tag{7}$$

Function $f_2(T)$ is the activity intensity affected by ambient temperature, a feasible approach is imitating the Metropolis criterion in simulated annealing algorithm that structured as follows:

$$f_2(T) = \exp^{-1/T} \tag{8}$$

Where T represents the current ambient temperature. After the move intensity being available, particle moves in line with the intensity, making the particle move towards the attractor. As mentioned before that $r_{max} = 1, r_{min} = 0, u_{max} = 1, u_{min} = 0$, the formula goes as follows:

$$f(r,T) = \frac{e^{-\lambda r} - e^{-\lambda}}{1 - e^{-\lambda}} * \exp^{-1/T} \tag{9}$$

6   The Wave Operator

Wave operator is used to control particle's movement in the microscopic view and lead it to produce local disturbance in neighborhood area so as to raise their exploitation capability. There are also two issues need to be considered like move operator: wave intensity, namely the coefficient of wave γ; another is the wave process, which is about how to make effective fluctuation in neighborhood according to the wave intensity. To address the problem of poor search capability when particle waves rapidly toward the attractor on the macro level, an approach of wave operator is firstly designed as follows:

$$f_r = u * f_2(T) \tag{10}$$

7   Updating Particle's Status

Based on the respective advantages and disadvantages of PSO and ITO, we in this paper make a compromise choice. More specifically, we implement the move operator and wave operator as the factor c_1 and c_2 in PSO algorithm simultaneously.

Additionally, the inertia weight and temperature were designed to decrease synchronously and linearly; Once the population gets updated, an additional wave process which utilizes the individual extreme value is used as current wave operator. In this way, the search process can be very likely to jump out of the local optimal solution and lead the population to move in the direction of global optimal solution. As a result, the accuracy and convergence speed could be improved greatly. The updating formulas are as follows:

$$v_{id}^{t+1} = wv_{id}^{t} + ur_1(p_{bd} - x_{id}^{t}) + \gamma r_2(p_{gd} - x_{id}^{t}) + \gamma r_3(p_{id} - x_{id}) \quad (11)$$

$$x_{id}^{t+1} = x_{id}^{t} + v_{id}^{t+1} \quad (12)$$

Where p_{bd} represents the best individual in current population, the other parameters are consistent with former settings.

8   Execution Steps

Step 1. Initialize the related parameters, including initial radius, move rate and wave rate, etc., read the constraints of test functions from the file "input.txt", randomly initialize the population according to these constraints, select one arbitrary location in solution space to be the initial position of the particle and set its initial velocity.

Step 2. Calculate the fitness value of all particles then update the global optimal value and the worst value in current population and individual extreme value according to the objective function (test function),.

Step 3. Update ambient temperature and particle's inertia weight in accordance with the simulated annealing table namely the formula (5); Meanwhile, utilize the sort-based approach which represented by formula (4) to update the particle radius, use the formula (9) and (10) to update move rate and wave rate; Finally update the velocity and location of the population according to equation (11) and (12).

Step 4. Determine that if the stop conditions are satisfied, which is measured according to that if 100 consecutive generations are not updated or the maximum evolution limit is satisfied. If not, go to Step 2; Otherwise, the running process comes to halt, and the global optimal solution at the time is the optimal solution of the test functions.

9   Experimental Results and Analysis

In order to verify the feasibility of the algorithm, we carefully employ three typical intricate benchmark test functions for testing the proposed algorithm. The specific characteristics and global optimal solution of these functions refer to reference [15]. By comparing the standard PSO algorithm with standard ITO algorithm, the experimental statistical results are shown in Table 1, the results of the 30-dimensional vector corresponding to the optimal solution of the three test functions are presented in Table 2. The comparisons among the proposed algorithm, PSO and ITO in convergence speed, global optimal solution, average solution and mean square deviation are shown in Figure 1~4, the convergence rates when searching for global optimal solution among these three benchmark functions are shown in Figure 5~7.

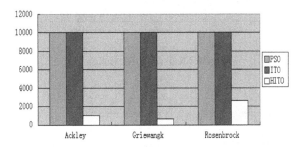

Fig. 1. Convergence rate

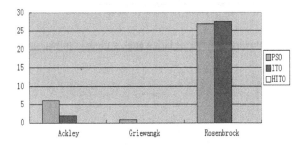

Fig. 2. Global optimal solution

The simulation environment: Intel (R) Core (TM) i3 M 370@2.40GHz, 2GB RAM , W7 system. Experiment simulation software: Eclipse Helios Service Release. The related parameters are set as follows: population size N = 200, the maximum size of evolution generation M = 100000, particle dimension Dim = 30, initial temperature of annealing table T = 800, annealing table length TLength = 200, annealing rate ρ = 0.99, parameters relate to move λ = 1.5, initial move rate μ = 2, initial wave rate = 1.5, inertia weight w = 0.9, initial radius R = 1.

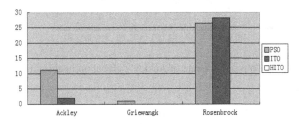

Fig. 3. Average solution

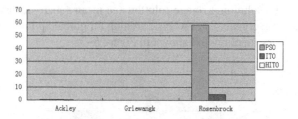

Fig. 4. Mean square deviation

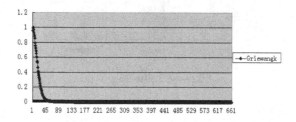

Fig. 5. Convergence rate of Griewangk's global optimal solution

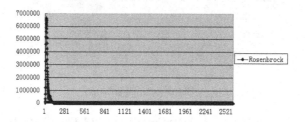

Fig. 6. Convergence rate of Rosenbrock's global optimal solution

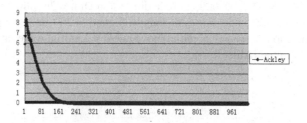

Fig. 7. Convergence rate of Ackley's global optimal solution

Table 1. Comparison between the PSO, ITO and HITO

FUN	AL	IT	OS	AS	ASE
	PSO	10000	6.1157	11.1402	0.2605
Ackley	ITO	9052	1.8997	1.9270	0.1516
	HITO	1029	7.9936 E-15	1.1104 E-11	1.5726 E-10
	PSO	10000	0.9215	1.0957	0.0048
Griewangk	ITO	10000	7.0331 E-8	1.3838 E-6	4.9555 E-6

Table 2. The 30-dimensional vector to three functions

(coordinate) Ackley	Rosenbrock	Griewangk
position(1) = -2.9906E-16	5.3970E-8	0.999998522128918
position(2) = -2.0902E-15	-1.3612E-7	0.9999988423045173
position(3) = -2.0713E-15	-7.0862E-8	0.99999972606998379
position(4) = -1.2972E-16	-2.3712E-7	0.9999920041627416
position(5) = -1.0644E-15	2.0698E-7	1.000003395559262
position(6) = -1.5012E-15	-9.1553E-7	1.0000164524313018
position(7) = 5.1287E-15	6.4422E-8	1.000005569895165
position(8) = 6.7888E-16	-4.9498E-7	0.9999982335655769
position(9) = 1.3031E-15	-3.5155E-7	0.9999910738425085
position(10) = 3.5624E-15	-1.3084E-6	0.999991557060363
position(11) = 3.3112E-15	-1.0016E-7	0.9999886813291057
position(12) = -6.1656E-16	6.3586E-7	0.999990769576629
position(13) = 2.6051E-15	-1.2189E-6	0.9999919101717967
position(14) = -1.7757E-15	-4.4544E-7	0.9999810431283936

With the experimental data in Table 1 and the three diagrams in Figures 1·-4, it could be easily concluded that the hybrid algorithm outperforms single PSO algorithm or single ITO algorithm in the number of iterations, the optimal solution, average solution and mean square deviation when we integrate the PSO with ITO (HITO)and exert the additional synchronic wave operation after previous move and wave process. What's more, the accuracy of solution is much superior to PSO and ITO. As a result, the proposed algorithm can effectively overcome the shortcomings of local optimum and approach the global optimal solution. From the three diagrams of the test functions in Fig 5~7 the statement can be supported that the proposed algorithm owns faster convergence speed. Generally speaking, the approach we proposed in this paper possesses fast convergence, robustness and strong stability in solving function optimization.

10   Conclusion

To make ITO algorithm more suitable for the law of Brownian motion, we in this paper have made a few improvements on the traditional IT0 algorithm. For instance, the move and wave processes are implemented simultaneously, when a feasible solution is found there goes an additional wave process in vary degrees. Additionally, We also employ PSO algorithm for mapping the move operator and wave operator to the learning factors c_1 and c_2 of PSO algorithm. Besides, we utilize the mechanism which make inertia weight decrease linearly and simultaneously with the ambient temperature. When a particle moves to a new position, the individual extreme value is acted as the wave object at that time, then one more wave process is executed in order to make the algorithm jump out of the local optimal solution and move in the direction of global optimal solution. Obviously, the settings of parameters in this paper could be further adjusted, thus enhancing the efficiency and accuracy of the proposed algorithm. Something need to be further studied is on setting the parameters, since most of the time this essential part is based on empirical assumption and analysis of large amounts of data, we haven't yet find enough scientific rules for the settings. Therefore, our future work will concentrate on searching for fit and reasonable parameters.

Acknowledgement. This work was supported by the National Natural Science Foundation of China(No.60873114, 61170305), the National Undergraduate Training Programs for Innovation and Entrepreneurship(No. 201310605017, 201310605018), and the Education Scientific Research Foundation of Guangxi Province (No.200103YB136).

References

1. Kennedy, J., Eberhart, R.C.: Particle Swarm Optimization. In: Proceedings of IEEE International Conference on Neural Networks, vol. 4(2), pp. 1942–1948 (1995)
2. Dong, W.Y., Zhang, W.S., Yu, R.G.: Convergence and Runtime Analysis of ITO Algorithm for One Class of Combinatorial Optimization. Chinese Journal of Computers 34(4), 636–646 (2011) (in Chinese)
3. Dong, W., Lei, M., Yu, R.: A New Evolutionary Algorithms for Global Numerical Optimization Based on Ito Process. In: Cai, Z., Tong, H., Kang, Z., Liu, Y. (eds.) ISICA 2010. CCIS, vol. 107, pp. 57–67. Springer, Heidelberg (2010)
4. Dong, W.Y., et al.: BBOB-benchmarking: A new evolutionary algorithms inspired by ITO process for noiseless function tested. Journal of Computational Information Systems, 2195–2203 (2011)
5. Dong, W.Y.: Time Series Modeling Based on ITO Algorithm. In: Proc. Int. Conf. on Natural Computation, pp. 398–402 (2007)
6. Dong, W., Yu, R., Lei, M.: Merging the Ranking and Selection into ITO Algorithm for Simulation Optimization. In: Cai, Z., Tong, H., Kang, Z., Liu, Y. (eds.) ISICA 2010. CCIS, vol. 107, pp. 87–96. Springer, Heidelberg (2010)
7. Dong, W.Y.: Simulation Optimization Based on the Hypothesis Testing and ITO Process. In: Third International Conference on Natural Computation, pp. 1210–1221 (2007)
8. Yi, Y.F., Cai, Y.L., Dong, W.Y., et al.: Improved ITO Algorithm for Solving the CVRP. Computer Science 40(5), 213–216 (2013) (in Chinese)
9. Dong, W., Zhang, D., Weicheng, Z., Leng, J.: The Simulation Optimization Algorithm Based on the Ito Process. In: Huang, D.-S., Heutte, L., Loog, M. (eds.) ICIC 2007. CCIS, vol. 2, pp. 563–573. Springer, Heidelberg (2007)

10. Dong, W., Zhang, D., Li, Y.: The Multi-objective ITO Algorithms. In: Kang, L., Liu, Y., Zeng, S. (eds.) ISICA 2007. LNCS, vol. 4683, pp. 53–61. Springer, Heidelberg (2007)
11. Hung, H., Qin, H., Hao, Z.F.: Example-based Learning Particle Swarm Optimization for Continuous Optimization Sciences. Information Sciences 182(1), 125–138 (2012)
12. Eberhart, R.C., Kennedy, J.: A new optimizer using particle swarm theory. In: The 6th International Symposium on Micro Machine and Human Science, Nagoya, Japan (1995)
13. Ling, S.H., Iu, H.H.C., Chan, K.Y., et al.: Hybrid particle swarm optimization with wavelet mutation and its industrial applications. IEEE Transactions on Systems, Man, and Cybernetics, Part B: Cybernetics 38(3), 743–764 (2008)
14. Corana, A., Marchesi, M., Martini, C., et al.: Minimizing multimode functions of continuous variables with the Simulated Annealing Algorithm. ACM Transactions on Mathematical Software 13(3), 262–280 (1987)
15. He, Y.C., Zhang, C.J., Wang, P.H., et al.: Comparison between Particle Swarm Optimization and Guo Tao algorithm on function optimization problems. Computer Engineering and Applications 43(11), 100–103 (2007) (in Chinese)

Performance Analysis of a (1+1) Surrogate-Assisted Evolutionary Algorithm

Yu Chen[1,2] and Xiufen Zou[2]

[1] School of Science, Wuhan University of Technology, 430070, Wuhan, China
ychen@whut.edu.cn, chymath@gmail.com
[2] School of Mathematics and Statistics, Wuhan University, 430072, Wuhan, China

Abstract. Although surrogate-assisted evolutionary algorithms have been widely utilized in the science and engineering fields, efficient management of surrogates is still a challenging task because of the complicated action mechanisms of surrogates. Trying to investigate how surrogates influence the convergence of evolutionary algorithms, this paper performs an analysis of a (1+1) surrogate-assisted evolutionary algorithm with Gaussian mutations, where a linear surrogate model is employed to approximately evaluate the candidate solutions. Numerical results demonstrate that the linear surrogate model greatly improves performance of evolutionary algorithms on a unimodal problem when the standard deviation of Gaussian mutation σ is small. However, when σ is set big, the improving functions of linear surrogate models are not significant. However, for a multi-modal problem, the positive function of the linear surrogate model is not significant, because the employed Gaussian mutation cannot keep a balance between exploration and exploitation.

1 Introduction

Nowadays surrogate models have been widely utilized to improve the performances of evolutionary algorithms (EAs). By approximately evaluating individuals via surrogate models, surrogate-assisted evolutionary algorithms (SAEAs) can reduce computation time for expensive functions, handle black-box problem, and reduce function evaluations (FEs) for robust optimal solutions [5,6,10,7]. However, SAEAs must employ surrogates together with the real fitness functions, because approximation errors of surrogate models can sometimes mislead the evaluation processes of SAEAs to "false optimal solutions"[1]. Thus, how to appropriately manage surrogate models is critical to efficient SAEAs.

Jin [6] divided techniques of managing surrogates into three categories, namely, the individual-based, generation-based and population-based methods. Due to the surrogates' complicated mechanism of action, most of existing literatures are devoted to incorporate various manage strategies in SAEAs, and then, test the proposed SAEAs by numerical results on selected problems [9,1,2,4,8,12,11]. However, few

---

[1] Here the so-called ``false optimal solutions" could be local optimal solutions or non-optimal solutions of the investigated optimization problems.

D.-S. Huang et al. (Eds.): ICIC 2014, LNCS 8588, pp. 32–40, 2014.

studies are reported to investigate how surrogate models influence the convergence processes of SAEAs. Chen *et al.* [3] performed a theoretical analysis on a (1+1) surrogate-assisted evolutionary algorithm ((1+1)SAEA) by case studies. The theoretical investigations on a 1-D unimodal problem and a multimodal problem show that the pre-selection strategy sometimes accelerates the convergence of the (1+1)SAEA, but sometimes also misleads its evolving process for both unimodal and multi-modal test problems.

However, how do SAEAs function on high-dimensional problems? As an initial work, a numerical analysis is performed in this paper. The remainder of this paper is organized as follows. Section 2 presents a (1+1)SAEA for investigations. Then, the performance analysis is carried out for a unimodal problem and a multimodal problem in Section 3 and Section 4, respectively. Finally, we draw the conclusion in Section 5.

2   The (1+1) Surrogate-assisted Evolutionary Algorithm

For a minimization problem (MP)

$$\min \quad f(\mathbf{x}), \quad \mathbf{x} = (x_1, x_2, \cdots, x_n) \in S_x \subset R^n, \tag{1}$$

the (1+1) SAEA is illustrated by Algorithm 1, which consists of a one-individual population $\mathbf{x}$ and an archive $\mathbf{X}_A = (\mathbf{x}_1, \mathbf{x}_2, \cdots \mathbf{x}_\mu)$ of size μ. At the beginning, $\mathbf{x}$ is randomly initialized, and the archive $\mathbf{X}_A$ is constructed by

$$\mathbf{x}_i = \mathbf{x} + N(0, \sigma^2) \cdot \mathbf{x}, \quad i = 1, 2, \cdots, \mu$$

where $N(0, \sigma^2)$ is a Gaussian random number. Then, the linear surrogate model

$$f_A(\mathbf{x}) = c_0 + c_1 x_1 + c_2 x_2 + \cdots + c_n x_n, \quad \forall \mathbf{x} = (x_1, x_2, \cdots, x_n) \tag{2}$$

is constructed via $\mathbf{X}_A$, where $c_0, c_1, c_2, \cdots, c_n$ are n+1 parameters of the linear model. After a candidate solution $\mathbf{x}'$ is generated by $\mathbf{x}$, $\mathbf{x}'$ is pre-evaluated by $y' = f_A(\mathbf{x}')$. Then, there are two cases to be distinguished.

1. If $y' < f(\mathbf{x})$, perform the precise evaluation $y' = f(\mathbf{x}')$ every *GenPer* generation;
2. otherwise, perform the precise evaluation $y' = f(\mathbf{x}')$ with probability p.

Whether $\mathbf{x}'$ is precisely evaluated or not, replace $\mathbf{x}$ with $\mathbf{x}'$ when $y' < f(\mathbf{x})$. Once the candidate $\mathbf{x}'$ is precisely evaluated better than $\mathbf{x}_w$, the worst solution in the archive $\mathbf{X}_A$, replace $\mathbf{x}_w$ with $\mathbf{x}'$. In this way, the archive $\mathbf{X}_A$ is always precisely evaluated, and no approximation error will be introduced when constructing the surrogate model. After the archive $\mathbf{X}_A$ is updated, sort it by the members' fitness values, and denote the worst member by $\mathbf{x}_w$. Repeat the aforementioned process until the stopping criterion is satisfied. Then, the ultimate $\mathbf{x}$ is the obtained approximate optimal solution of MP (1).

Algorithm 1: The (1+1) surrogate-assisted evolutionary algorithm
  Parameters: σ, $GenPer$, p;

Begin
  Step 1: Set t=1;
  Step 2: Randomly initialize the individual $\mathbf{x}$, and initialize the archive $\mathbf{X}_A = (\mathbf{x}_1,\mathbf{x}_2,\cdots,\mathbf{x}_\mu)$ by

$$\mathbf{x}_i = \mathbf{x} + N(0,\sigma^2)\cdot\mathbf{x}, \quad i = 1,2,\cdots,\mu.$$

Sort $\mathbf{X}_A$ via fitness values and denote the worst one to be $\mathbf{x}_w$.
Repeat
  Step 3: Construct the surrogate model $f_A(\mathbf{x})$ by $\mathbf{X}_A$; Generate a new candidate solution $\mathbf{x}'$ and approximately evaluate it via $y'=f_A(\mathbf{x}')$;
  Step 4: If $y'\geq f(\mathbf{x})$, go to Step 5; otherwise, evaluate $\mathbf{x}'$ by $y'=f(\mathbf{x}')$ when $t\,(\mathrm{mod}\,GenPer)\equiv 0$; If $y'< f(\mathbf{x}_w)$, replace $\mathbf{x}_w$ with $\mathbf{x}'$. Sort the archive $\mathbf{X}_A$ via fitness values and denote the worst one to be $\mathbf{x}_w$.
  Step 5: If $rand < p$, evaluate the candidate solution $\mathbf{x}'$ by $y'=f(\mathbf{x}')$; If $y'< f(\mathbf{x}_w)$, replace $\mathbf{x}_w$ with $\mathbf{x}'$. Sort the archive $\mathbf{X}_A$ via fitness values and denote the worst one to be $\mathbf{x}_w$;
  Step 6: If $y'< f(\mathbf{x})$, replace $\mathbf{x}$ with $\mathbf{x}'$. Set t=t+1;
  until the stopping criterion is satisfied
end

In the (1+1)SAEA illustrated by Algorithm 1, the recombination strategy for candidate solution generation is optional, that is, various recombination strategies can be employed to generate the candidate solutions. As an initial work, we employ the adaptive Gaussian mutation to generate candidate solutions, that is,

$$\mathbf{x}'= \mathbf{x}+N(0,\sigma^2)*\mathbf{x},$$

where $N(0,\sigma^2)$ is a Gaussian random number.

3   Performance Analysis of (1+1)SAEA for a Unimodal Problem

In this paper, we first investigate the performance of (1+1)SAEA on the minimization problem of the positive definite quadratic function

$$\min \quad f(x) = \mathbf{x}\mathbf{A}\mathbf{x}^T \tag{3}$$

where $\mathbf{A}$ is a positive definite matrix, $\mathbf{x} = (x_1, x_2, \cdots, x_n)$. When the objective function is approximated by the linear surrogate model (2), we take n+1 samples $(\mathbf{x}^{(1)}, f(\mathbf{x}^{(1)})), (\mathbf{x}^{(2)}, f(\mathbf{x}^{(2)})), \cdots, (\mathbf{x}^{(n+1)}, f(\mathbf{x}^{(n+1)}))$ to   confirm   the   parameters $\mathbf{c} = (c_0, c_1, \cdots, c_n)^T$ via

$$\begin{pmatrix} f(\mathbf{x}^{(1)}) \\ f(\mathbf{x}^{(2)}) \\ \vdots \\ f(\mathbf{x}^{(n+1)}) \end{pmatrix} = \begin{pmatrix} 1 & \mathbf{x}^{(1)} \\ 1 & \mathbf{x}^{(2)} \\ \vdots & \vdots \\ 1 & \mathbf{x}^{(n+1)} \end{pmatrix} \begin{pmatrix} c_0 \\ c_1 \\ \vdots \\ c_n \end{pmatrix}.$$

Then,

$$\mathbf{c} = \begin{pmatrix} c_0 \\ c_1 \\ \vdots \\ c_n \end{pmatrix} = \begin{pmatrix} 1 & \mathbf{x}^{(1)} \\ 1 & \mathbf{x}^{(2)} \\ \vdots & \vdots \\ 1 & \mathbf{x}^{(n+1)} \end{pmatrix}^{-1} \begin{pmatrix} f(\mathbf{x}^{(1)}) \\ f(\mathbf{x}^{(2)}) \\ \vdots \\ f(\mathbf{x}^{(n+1)}) \end{pmatrix} \tag{4}$$

when $\begin{pmatrix} 1 & \mathbf{x}^{(1)} \\ 1 & \mathbf{x}^{(2)} \\ \vdots & \vdots \\ 1 & \mathbf{x}^{(n+1)} \end{pmatrix}$ is nonsingular. Thus, the approximation error is

$$Err_f(\mathbf{x}') = (f(\mathbf{x}') - f_A(\mathbf{x}'))^2 = (\mathbf{x}'\mathbf{A}\mathbf{x}'^T - (1 \quad \mathbf{x}')\mathbf{c})^2,$$

and its partial derivative is

$$\frac{\partial}{\partial \mathbf{x}'} Err_f(\mathbf{x}')$$

$$= 2(\mathbf{x}'\mathbf{A}\mathbf{x}'^T - (\mathbf{x}' \quad 1) \cdot \mathbf{c}) \cdot \frac{\partial}{\partial \mathbf{x}'}(\mathbf{x}'\mathbf{A}\mathbf{x}'^T - (\mathbf{x}' \quad 1) \cdot \mathbf{c})$$

$$= 2(\mathbf{x}'\mathbf{A}\mathbf{x}'^T - (\mathbf{x}' \quad 1) \cdot \mathbf{c}) \cdot (2\mathbf{A}\mathbf{x}'^T - \mathbf{c})$$

In case that the sample points are selected, the zero value of approximation error can be reached when $\mathbf{x}'\mathbf{A}\mathbf{x}'^T = (1 \quad \mathbf{x}')\mathbf{c}$, which holds for the sample points $(\mathbf{x}^{(1)}, f(\mathbf{x}^{(1)})), (\mathbf{x}^{(2)}, f(\mathbf{x}^{(2)})), \cdots, (\mathbf{x}^{(n+1)}, f(\mathbf{x}^{(n+1)}))$. Meanwhile, when $\mathbf{x}'$ satisfies

$$2\mathbf{A}\mathbf{x}'^T - \mathbf{c} = 0, \tag{5}$$

the approximation error is maximized because the Hessian matrix $\mathbf{A}$ is positive definite. Thus, the approximation error of the surrogate model is small when the candidate solutions are generated near the present individual $\mathbf{x}$. For this case, the surrogate model is helpful to reduce the number of FEs.

For numerical validation, we perform numerical experiments on the minimization problem of sphere function

$$\min \quad f(\mathbf{x}) = \sum_{i=1}^{n} x_i^2 \qquad (6)$$

where $\mathbf{x} = (x_1, x_2, \cdots, x_n) \in [-10,10]^{10}, n = 10$. Taking $f(\mathbf{x}) < 1.0E - 10$ as the stopping criterion, we set perform numerical experiments by setting $p = 0.1, 0.2, \cdots, 1$. Numerical results show that whatever the values of σ and *GenPer* are, the best performance of (1+1)SAEA is always obtained when $p = 0.1$. Then, the changing curves of expected FEs for 1000 independent runs of the (1+1)SAEA are illustrated in Fig. 1, where σ is set to be 0.1, 0.5, 1.0, 1.5, 2.0, 2.5, and *GenPer* varies from 1 to 21 with step size 2. For the case $\sigma = 0.1$, the expected values of FEs are transferred to their natural logarithms plus 70.

Numerical results indicate that if surrogates evaluate $\mathbf{x}'$ worse than the present individual $\mathbf{x}$, it should be precisely re-evaluated with a small probability. The reason is that Gaussian mutation focusing on local exploitation generates candidate solutions adjacent to the present individual $\mathbf{x}$, and to some extent the approximate evaluation can function well for unimodal problems. When σ is set to be 0.1 and 0.5, the expected values of FEs increase with the parameter *GenPer* . Because Gaussian mutation focuses on local exploitation when σ is small, errors of approximate evaluations are relatively small, and consequently, the pre-evaluation scheme of surrogates can filter bad candidate solutions. Therefore, the expected values of FEs can be greatly reduced, for reason that the approximate evaluation can function nearly as well as the precise function evaluation. However, when *GenPer* increases, the surrogate model cannot be timely updated, which postpones the improvement of approximation errors. As a consequence, the pre-evaluation scheme of surrogates cannot function well, and more unnecessary precise FEs is needed to obtain a satisfactory solution.

However, when the mutation strength σ is endowed with relatively big values 1.0, 1.5, 2.0 and 2.5, the approximation errors of surrogates go up. Thus, if *GenPer* is small, most of the frequent evaluations are not necessary, which leads to the increase of required FEs for the given stopping criterion. Meanwhile, increase of *GenPer* also postpones update of surrogate models, which frustrates the improvement of approximation errors. As a comprehensive effect, the numbers of required FEs first decrease with rise of *GenPer* , and then, slowly increase after their minimum values have been obtained.

To illustrate the positive effect of the linear surrogate model, we compare the best results (minimum expected FEs) of (1+1)SAEA with those of the corresponding (1+1) evolutionary algorithm ((1+1)EA) in Tab. 1. It is shown that with the increase of σ , the improvement rate (IR) of FEs decreases, which validates that the theoretical results obtained in [3] can also be extended to the high-dimensional unimodal problems, that is, the promising effect of linear surrogate model will decrease with mutation strength σ when local Gaussian mutation is employed. Meanwhile, both (1+1)SAEA and (1+1)EA reach their minimum values of expected FEs when $\sigma = 0.1$, because both algorithms can keep a balance between exploration and exploitation for this case.

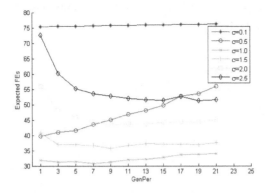

Fig. 1. Simulation results of (1+1)SAEA with $p = 0.1$ for minimization of Sphere function

Table 1. Comparison of expected FEs between (1+1)SAEA and (1+1)EA for the minimization problem of Sphere function

σ	0.1	0.5	1.0	1.5	2.0	2.5
(1+1)SAEA	337.0780	45.9670	32.2980	36.2650	44.0710	53.0960
(1+1)EA	227.2350	39.8050	30.7725	35.7597	43.5515	51.2996
IR	0.674	0.866	0.953	0.986	0.988	0.966

4   Performance Analysis of (1+1)SAEA for Multi-modal Problem

In this section, we investigate the promising function of linear surrogate model (2) for a multimodal problem, the minimization problem of the Rastrigin's function

$$g(\mathbf{x}) = \sum_{i=1}^{n} (x_i^2 - 10\cos(2\pi x_i) + 10) \tag{7}$$

where $\mathbf{x} = (x_1, x_2, \cdots, x_n) \in [-10,10]^n$. Similarly, the linear surrogate model

$$g_A(\mathbf{x}) = p_0 + p_1 x_1 + p_2 x_2 + \cdots + p_n x_n, \quad \forall \mathbf{x} = (x_1, x_2, \cdots, x_n) \tag{8}$$

can be confirmed by n+1 samples $(\mathbf{x}^{(1)}, g(\mathbf{x}^{(1)})), \ldots, (\mathbf{x}^{(n+1)}, g(\mathbf{x}^{(n+1)}))$, and

$$\mathbf{p} = \begin{pmatrix} p_0 \\ p_1 \\ \vdots \\ p_n \end{pmatrix} = \begin{pmatrix} 1 & \mathbf{x}^{(1)} \\ 1 & \mathbf{x}^{(2)} \\ \vdots & \vdots \\ 1 & \mathbf{x}^{(n+1)} \end{pmatrix}^{-1} \begin{pmatrix} g(\mathbf{x}^{(1)}) \\ g(\mathbf{x}^{(2)}) \\ \vdots \\ g(\mathbf{x}^{(n+1)}) \end{pmatrix}$$

when $\begin{pmatrix} 1 & \mathbf{x}^{(1)} \\ 1 & \mathbf{x}^{(2)} \\ \vdots & \vdots \\ 1 & \mathbf{x}^{(n+1)} \end{pmatrix}$ is nonsingular.

For a candidate solution $\mathbf{x}' = (x'_1, x'_2, \cdots, x'_n)$, the approximation error introduced by the surrogate $g_A(\mathbf{x}')$ is

$$Err_g(\mathbf{x}') = (g(\mathbf{x}') - g_A(\mathbf{x}'))^2$$

$$= \left(\sum_{i=1}^{n} (x'^2_i - 10\cos(2\pi x_i) + 10) - (\mathbf{x}' \quad 1)\mathbf{p} \right)^2,$$

and its partial derivative is

$$\frac{\partial}{\partial \mathbf{x}'} Err_g(\mathbf{x}')$$

$$= 2 \left(\sum_{i=1}^{n} (x'^2_i - 10\cos(2\pi x_i) + 10) - (\mathbf{x}' \quad 1)\mathbf{p} \right) \cdot \begin{pmatrix} 2x'_1 + 20\pi \sin 2\pi x'_1 - p_1 \\ 2x'_2 + 20\pi \sin 2\pi x'_2 - p_2 \\ \vdots \\ 2x'_n + 20\pi \sin 2\pi x'_n - p_n \end{pmatrix}$$

Thus, the stationary points of the error include:

1. points $\mathbf{x}' = (x'_1, x'_2, \cdots, x'_n)$ satisfying

$$\sum_{i=1}^{n} (x'^2_i - 10\cos(2\pi x_i) + 10) - (\mathbf{x}' \quad 1)\mathbf{p} = 0$$

where the surrogate model intersects with the Rastringin's function, and the error is zero;

3. points $\mathbf{x}' = (x'_1, x'_2, \cdots, x'_n)$ satisfying

$$2x'_i + 20\pi \sin 2\pi x'_i - p_i = 0, \quad \forall i = 1, 2, \cdots, n$$

where the approximation error could reach its local maximum values.

For numerical investigations, we take a 10-D Rastrigin's function as the case study, where the stopping criterion is $g(\mathbf{x}) < 1.0e - 10$. After 1000 independent runs of the (1+1)SAEA for each set of parameters settings, the numerical results are illustrated by Fig. 2 and Tab. 2, where the best performances of (1+1)SAEA are also obtained when $p = 0.1$.

Fig. 2 illustrates the changing trend of expected FEs with increase of *GenPer*. When $\sigma = 0.5$, the average FEs increase quickly with increase of *GenPer*, which implies that *GenPer* significantly influences the efficiency of the (1+1)SAEA. However, for other cases, influences of *GenPer* on average FEs are not remarkable. The reason could be that when $\sigma = 0.5$, the mutation of individual mainly focuses on local exploitation, and then the changing trends of expected FEs for Rastrigin's function is similar to that for Sphere function. However, when σ is relatively big, its exploration ability increases, but, approximation errors of the linear surrogate models get out of control. Consequently, the value of *GenPer*, which mainly determines the frequency of surrogate update, does not influence the performance of (1+1)SAEA significantly.

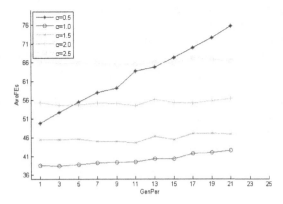

Fig. 2. Simulation results of (1+1)SAEA with $p = 0.1$ for minimization of Rastrigin's function

Meanwhile, results in Tab. 2 indicate that when σ is endowed with a big value, the best performances are obtained when $GenPer = 11$. For this case, the precise evaluation should not be frequently performed, because the linear surrogate model can ignore the influence of the multi-modal landscape of the Rastrigin's function. But, similar to the unimodal case, when σ increases, the reduced promising function of surrogate leads to the decrease of IR, which show that if the candidate solutions are generated by local mutations, linear surrogates cannot greatly improve performances of (1+1)SAEA on the multimodal problem, even if we set the mutation radius σ great to strengthen its global exploration ability. It is because when the global exploration ability is obtained by setting σ great, the positive function of linear surrogate model is increased, but, local exploitation ability is reduced at the same time.

Table 2. Comparisons of expected FEs between (1+1)SAEA and (1+1)EA for the minimization problem of Rastrigin's function

$\sigma(p, GenPer)$	0.5(0.1,1)	1.0(0.1,3)	1.5(0.1,11)	2.0(0.1,11)	2.5(0.1,11)
(1+1)EA	62.5970	39.6080	45.2300	56.1200	66.2170
(1+1)SAEA	49.8010	38.3750	44.6150	54.4720	65.4950
IR	0.796	0.969	0.986	0.971	0.989

5   Conclusion and Future Work

To investigate the promising function of surrogates in EAs, this paper presents an analysis on a (1+1)SAEA via a unimodal problem and a multi-modal problem. Numerical results show that when a linear surrogate model is employed in the (1+1)SAEA, it functions well when local mutations are employed, because approximation errors of surrogates for candidate individuals are relatively small. However, local mutations focus on local exploitation instead of global exploration, even if the mutation strength is set great. Consequently, for multi-modal problems the (1+1)SAEA cannot perform well, because strong exploitation ability and positive functions of the linear surrogate model cannot be obtained at the same time. Thus, it could be appropriate to design a more flexible adaptive mutation strategy, or to introduce crossover to propose an efficient surrogate-assisted evolutionary algorithm.

Acknowledgment. This work was partially supported by the Natural Science Foundation of China (No. 61303028) as well as the Fundamental Research Funds for the Central Universities (WUT: 2013-Ia-001).

References

1. Branke, J., Schmidt, C.: Fast Convergence by Means of Fitness Estimation. Soft Comp. 9, 13–20 (2005)
2. Buche, D., Schraudolph, N.N., Koumoutsakos, P.: Accelerating Evolutionary Algorithms with Gaussian Process Fitness Function Models. IEEE Trans. on Sys., Man, and Cyb., Part C: App. and Rev. 35, 183–194 (2005)
3. Chen, Y., Xie, W., Zou, X.: How Can Surrogates Influence the Convergence of Evolutionary Algorithms? Swar. & Evol. Comp. 12, 18–23 (2013)
4. Emmerich, M., Giannakoglou, K.C., Naujoks, B.: Single- and Multiobjective Evolutionary Optimization Assisted by Gaussian Random Field Metamodels. IEEE Trans. on Evol. Comp. 10(4), 421–439 (2006)
5. Jin, Y., Sendhoff, B.: Fitness Approximation in Evolutionary Computation - A Survey. In: 2002 Genetic and Evolutionary Computation Conference, pp. 1105–1112. ACM, San Francisco (2002)
6. Jin, Y.: A Comprehensive Survey of Fitness Approximation in Evolutionary Computation. Soft Comp. 9, 3–12 (2005)
7. Jin, Y.: Surrogate-assisted Evolutionary Computation: Recent Advances and Future Challenges. Swar. & Evol. Comp. 1, 61–70 (2011)
8. Knowles, J.: PAREGO: A Hybrid Algorithm with On-line Landscape Approximation for Expensive Multiobjective Optimization Problems. IEEE Trans. on Evol. Comp. 10, 50–66 (2006)
9. Ong, Y.S., Nair, P.B., Keane, A.J.: Evolutionary Optimization of Computationally Expensive Problems via Surrogate Modeling. AIAA Journal 41, 687–696 (2003)
10. Tenne, Y., Goh, C.K. (eds.): Computational Intelligence in Expensive Optimization Problems. Springer (2009)
11. Zhang, Q., Liu, W., Tsang, E., Virginas, B.: Expensive Multiobjective Optimization by MOEA/D with Gaussian Process Model. IEEE Trans. on Evol. Comp. 14, 456–474 (2010)
12. Zhou, Z., Ong, Y.S., Lim, M.H., Lee, B.S.: Memetic Algorithm Using Multisurrogates for Computationally Expensive Optimization Problems. Soft Comp. 11, 957–971 (2007)

Parameters Selection for Support Vector Machine Based on Particle Swarm Optimization

Jun Li[1,2,3] and Bo Li[1,2]

[1] College of Computer Science and Technology,Wuhan University of Science and Technology,
WUST, Wuhan, 430065,China
[2] Hubei Province Key Laboratory of Intelligent Information Processing
and Real-Time Industrial System, WUST, Wuhan,430065, China
[3] State Key Laboratory of Software Engineering (Wuhan University), Wuhan, 430072, China
lijun@wust.edu.cn

Abstract. In this paper, an SVM classification system based on particle swarm optimization (PSO) is proposed to improve the generalization performance of the SVM classifier. Authors have optimized the SVM classifier design by searching for the best value of the parameters that tune its discriminant function. The experiments are conducted on the basis of benchmark dataset. Fourteen obtained results clearly confirm the superiority of the PSO-SVM approach.

Keywords: high-dimentional classfication, support vector machine(SVM), particle swarm optimization(PSO).

1   Introduction

Classification problems have been extensively studied. SVM which is an emerging data classification technique proposed by Vapnik [1], and has been widely adopted in various fields of high-dimensional classification problems in recent years.An important question is how to optimize the parameters of the support vector machine (SVM) to improve its classification accuracy.

In related research, the grid search proposed in the literature [2] is an alternative search method that is simple and easy to use, but its search ability is poor. Genetic algorithm proposed in [3] and ant colony optimization algorithm proposed in [4-5] are for optimization selection of SVM parameters respectively, although these intelligent methods reduce the reliance on initial value selection, yet the algorithm principles and the thought is more complex, and it is easy to fall into part optimum.

The particle swarm optimization (PSO), originally developed by Kennedy and Eberhart [6], is a method for optimizing hard numerical functions on metaphor of social behaviors of flocks of birds and schools of fish. In PSO algorithm, each individual will be seen as a particle with no weight and volume in n-dimensional search space, flying in the search space at a certain speed. The speed of the flight dynamically adjusts from flying experience of individuals and population. PSO algorithm is initialized to a group of random particles (random solutions), and then

D.-S. Huang et al. (Eds.): ICIC 2014, LNCS 8588, pp. 41–47, 2014.
© Springer International Publishing Switzerland 2014

find the optimal solution through evolution. In each evolution, the particles update themselves by tracking two extreme values.The first extremum is the optimal solution experienced by the particles themselves, which is known as individual extremum. Another extremum is the best solution currently experienced by the entire population, which is called a global optimum. The algorithm using the sharing mechanism of information in the birds', is the movement of the population in the problem solving space processing evolution from disorderly to orderly to obtain the optimal solution.

The structure of the paper is as the following. Part 2 introduces the basic principles of the support vector machine (SVM). Part 3 introduces the particle swarm optimization algorithm. Part 4 elaborates the proposed SVM parameter selection algorithm based on particle swarm optimization. Part 5 is the experimental results. Finally Part 6 is the conclusion.

2   Svm

The basic idea of support vector machine (SVM) is through a nonlinear mapping to map the input space into a high-dimensional even infinite dimensional feature space. In the feature space seek for the optimal separating hyperplane, so that it can separate two classes of data points correctly as much as possible, while making twoclasses of data point furthest distance from classification surface[7].

Provided sample set as $\{(x_1,y_1),\ (x_2,\ y_2),...,(x_l,y_l)\} \in (x \times y)^l$, among them, $x_i \in x \subset R^n$ is the input vector, $y_i = \{-1\ ,\ 1\}$ is the corresponding output values for x_i, i=1,2, ...,l.

When the sample points approximately linear classification, the problem boils down to the following optimization problem:

$$\min \frac{1}{2}\| w \|^2 + C\sum_{i=1}^{l} \zeta_i$$
$$\text{s.t.}\quad y_i((w \cdot x_i)+b)+\zeta_i \geq 1(\zeta_i \geq 0, i=1,2,...,l) \tag{1}$$

Among this, $\| w \|^2$ is called structural risk which is on behalf of the complexity of the model and   make the function smoother, thereby can improve the generalization ability; $C\sum_{i}^{l}\zeta$ is known as the empirical risk which on behalf of the error of the model, and C is known as punishment parameter.

When the sample points are shown to exhibit non-linear relationship, the original sample set will be mapped to the high dimensional feature space through a nonlinear mapping $\varphi(x)$, then make the linear classification in the high-dimensional feature space. The inner product operation on the high dimensional feature space can be defined as kernel function:

$$K(x_i,\ x_j)= \varphi(x_i)\ \varphi(x_j) \tag{2}$$

only need to use the function operate the variables in the original low dimension.

Using the duality theory in the optimization theory can get the final decision function:

$$f(x) = \text{sgn}(\sum_{i=1}^{l} y_i a_i^* K(x_i, x) + b^*) \tag{3}$$

Commonly used kernel function includes the linear, polynomial kernel, Sigmoid and Gaussian RBF kernel (RBF). In this paper, the widely used Gaussian RBF kernel is adopted, that is

$$K(x, y) = e^{-\frac{\|x-y\|^2}{\sigma^2}} \quad \text{or} \quad K(x, y) = e^{-\delta\|x-y\|^2} \tag{4}$$

The major parameters that affect the performance of support vector machine are punished parameter C and RBF kernel parameter δ.

3   Particle Swarm Optimization

Particle swarm optimization comprises a set of search techniques, inspired by the behavior of natural swarms, for solving optimization problems [8]. A swarm consists of individuals, which are called particles, which change their positions over time. Each particle represents a potential solution to the problem. In a PSO system, particles fly around in a multidimensional searching space. During its flight, each particle adjusts its position according to its own experience and the experience of its neighboring particles making use of the best position encountered by itself and its neighbors. The effect is that particles move towards the better solution areas, while still having the ability to search a wide area around the better solution areas. The performance of each particle is measured according to a pre-defined fitness function, which is related to the problem being solved. The PSO has been found to be robust and fast in solving nonlinear, non-differentiable and multi-modal problems[9]. The mathematical description and executive steps of the PSO are as follows. Let the i th particle in a D-dimensional space be represented as $x_i=(x_{i1},\ldots x_{id},\ldots x_{iD})$.The best previous position of the ith particle is recorded and represented as $p_i=(p_{i1},\ldots p_{id},\ldots p_{iD})$, which gives the best fitness value and is also called pbest. The index of the best pbest among all the particles is represented by the symbol g. The location Pg is also called gbest. The velocity for the ith particle is represented as $v_i=(v_{i1},\ldots v_{id},\ldots v_{iD})$. The concept of the particle swarm optimization consists of changing the velocity and location of each particle towards its pbest and gbest locations at each time step:

$$v_{id} = \bullet wv_{id} + c_1 r_1(p_{id} - x_{id}) + c_2 r_2(p_{gd} - x_{id}) \tag{5}$$

$$x_{id} = x_{id} + v_{id} \tag{6}$$

where w is the inertia coefficient which is a constant in the interval [0, 1] and can be adjusted in the direction of linear decrease, c_1 and c_2 are learning rates which are nonnegative constants, r_1 and r_2 are generated randomly in the interval [0, 1]. The termination criterion for iterations is determined according to whether the maximum generation or a designated value of the fitness is reached.

4   Conclusions

This research proposes a particle-swarm-optimization-based approach, denoted as PSO-SVM model, to obtain the optimal parameter settings for SVM kernel parameters which result in a better classification accuracy rate.

Figure 1 shows the architecture of the developed PSO-based parameter selection approach for SVM.

The outline of the proposed algorithm lists as follows:

Step1. Initialize population size NP; initialize population X randomly; initialize velocity and random position of each particle.

Step2. Train individuals in the population X with svm to get each individual's cross-validation accuracy. Then find the best cross-validation accuracy temp_BestAccuracy and its corresponding temp_c and temp_g. If BestAccuracy<temp_BestAccuracy, then BestAccuracy = temp_BestAccuracy, Bestc = temp_c, Bestg = temp_g.

Step3. Determine whether the current accuracy and number of iterations reaches the end of the condition, if reached, go to step 7), otherwise the next step into the next generation iterative process.

Step4. For every particle, compare the cross validation accuracy of the current position to the cross validation accuracy of the best position(Bestpi) it has ever experienced. If the former is better, the current position can be regarded as the best position.

Step5. For every particle, compare the cross validation accuracy of the current position to the cross validation accuracy of the best position(Bestpg) it has experienced in the global space. If the former is better, the current position can be regarded as the best position.

Step6. Update the position and velocity of the particles in the population. Then return to step 3).

Step7. Get the best cross-validation accuracy temp_BestAccuracy in the population X and its corresponding temp_c and temp_g, If BestAccuracy<temp_BestAccuracy, then BestAccuracy = temp_BestAccuracy, Bestc = temp_c, Bestg = temp_g. Then go to step 3).

Step8. Training data set to obtain learning model, use the model to predict the test data set to get the prediction accuracy of the test data set.

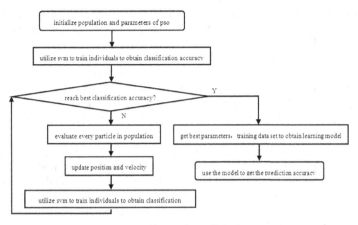

Fig. 1. The architecture of the PSO-SVM model

5   Experiment Results

The platform adopted to develop the PSO-SVM approach is Dell desktop with the following features: Intel Pentium Cores duo CPU E7200 2.53 GHz main frequency, 2GMB RAM, a Windows XP sp3 operating system. Our implementation is carried out on the Matlab2010a development environment by extending the LIBSVM which is originally designed by Chang and Lin [10].

In order to measure the performance of the developed PSO-SVM approach, the following datasets in UC Irvine Machine Learning Repository[11] are used: balance, blood, glass,   diabetes, haberman, kr_v_kp, liver, liverdisorder,   parkinsons, soybean, splice, tic_tac_toc, waveform2, wdbc, wine, zoo.   Table 1 presents the properties of these datasets.

Table 1. Datasets from the UCI repository

Dataset	Number of instances	Number of Features	Number of classes
balance	625	4	3
blood	748	4	2
diabetes	768	8	2
glass	214	9	6
haberman	306	3	2
kr_v_kp	3196	36	2
liver	327	6	2
liverdisorder	345	6	2
parkinsons	195	22	2
soybean	47	35	4
splice	1000	60	2
tic_tac_toc	958	9	2
waveform2	5000	21	3
wdbc	569	30	2
wine	178	12	3
zoo	101	16	7

To guarantee valid results for making predictions regarding new data, the dataset is further randomly partitioned into training sets and independent test sets via a k-fold cross validation. Each of the k subsets acts as an independent holdout test set for the model trained with the remaining k-1 subsets. The advantages of cross validation are that all of the test sets were independent and the reliability of the results could be improved. This study used k=10. Each of the 10 subset is used as test data sets in turn, so the program runs 10 times. The final classification accuracy is expressed in the form "mean ± standard deviation". The smaller the variance, the volatility of the data is smaller.

The searching range of parameters C and δ of SVM is between 10^7 and 10^7. Parameter v of cross validation is 5. In the particle swarm optimization algorithm, the parameters are initially set as follows. Iterations GEN is 200; population size NP is 40. Acceleration constant c1 is 1.5, c2 is 1.7. C1 is step length which is to adjust the particle to flying towards its own from the best location direction. C2 is the step length adjusting the particle to flying towards its own from the direction of the best location of the global space.

Table 2. Experimental Results

Dataset	SVM(%)	PSO-SVM(%)
balance	98.71±2.12	94.52±3.50
blood	75.41±7.23	75.68±4.08*
diabetes	76.32±3.34	75.53±3.93
glass	45.24±9.05	45.72±13.51*
haberman	68.00±8.63	79.33±9.03*
kr_v_kp	99.40±0.54	99.47±0.33*
liver	66.88±11.04	69.07±5.79*
liverdisorder	64.12±8.53	65.59±7.97*
parkinsons	82.63±11.91	82.63±11.91
soybean	100.00±0.00	97.50±7.91
splice	88.10±1.52	88.60±3.10*
tic_tac_toc	100.00±0.00	95.47±4.00
waveform2	85.62±2.54	85.86±2.26*
wdbc	87.68±8.89	87.68±10.56
wine	87.06±14.36	92.24±10.74*
zoo	93.00±8.23	93.00±6.75

To verify the excellence of PSO-SVM for parameters optimization, authors use the approach for comparing the performance of default parameter SVM classifier algorithm. The experimental result is shown in Table 2. It can be distinctly seen that in most of the datasets, the mean test accuracy of the pso-svm algorithm is better than the original svm.

Among total 16 testing data sets, 9 test results of PSO-SVM are better (* is put at the end of the test results), 3 test results of PSO-SVM are nearly the same with the original svm(the mean test accuracy is the same, only the standard deviation is slightly different), the other 4 test results are poorer .

The obtained results clearly confirm the superiority of the PSO-SVM approach compared to default parameters SVM classifier.

6   Conclusions

From the obtained experimental results, we can strongly recommend the use of the SVM approach for classifying high-dimensional data on account of their superior generalization capability. This study presents a particle-swarm-optimization based approach, capable of searching for the optimal parameter values for SVM. Comparison of the obtained results with default parameters SVM classifier demonstrates that the developed PSO-SVM approach has better classification accuracy than others tested.

Acknowledgement. This work was supported by National Natural Science Foundation of China under Grant NO. 61273303 and by the Open Project Program of the National Laboratory of Pattern Recognition under Grant NO. 201104212.

References

1. Vapnik, V.N.: The nature of statistical learning theory. Statistics for engineering and information science. Springer, New York (2000)
2. Hsu, C.W., Chang, C.C., Lin, C.J.: A practical guide to support vector classification. Technical report, University of National Taiwan, Department of Computer Science and Information Engineering, pp. 1–12 (July 2003)
3. Chunhong, Z., Licheng, J.: Automatic parameters selection for SVM based on GA. In: 5th IEEE World Congress on Intelligent Control and Automation, pp. 1869–1872 (2004)
4. Chun-bo, L., Xian-fang, W., Feng, P.: Parameters selection and stimulation of support vector machines based on ant colony optimization algorithm. Journal of Central South University: Science and Technology 39(6), 1309–1313 (2008)
5. Zhang, X.L., Chen, X.F., He, Z.J.: An ACO-based algorithm for parameter optimization of support vector machines. Expert Systems with Applications 37(9), 6618–6628 (2010)
6. Kennedy, J., Eberhart, R.: Particle swarm optimization. In: IEEE International Conference on Neural Networks, pp. 1942–1948. IEEE Neural Networks Society, Perth (1995)
7. Scholkopf, B., Smola, A.J.: Learning with kernels: support vector machines, regularization, optimization, and beyond. MIT Press (2001)
8. Correa, E., Freitas, A., Johnson, C.: Particle swarm for attribute selection in Bayesian classification: an application to protein function prediction. Journal of Artificial Evolution and Applications, 1–12 (2008)
9. Clerc, M.: Particle swarm optimization. John Wiley & Sons (2010)
10. Chih-Chung, C., Chih-Jen, L.: LIBSVM: A library for support vector machines (2001), Software available at http://www.csie.ntu.edu.tw/~cjlin/libsvm
11. Hettich, S., Blake, C.L., Merz, C.J.: UCI repository of machine learning databases, Department of Information and Computer Science, University of California, Irvine, CA, http://www.ics.uci.edu/~mlearn/MLRepository.html

Adaptive Hybrid Artificial Fish School Algorithm for Solving the Real Roots of Polynomials

Pan Xue

Guangxi University for Nationalities, Nanning 530006, China
2897136684@qq.com

Abstract. An adaptive hybrid artificial fish swarm algorithm (AHAFSA) is proposed, which uses the Hooke-Jeeves pattern search method as a local search operator embedded in artificial fish swarm algorithm (AFSA) to speed up the local search. Then the paper Uses AHAFSA to solve the real roots of polynomials, the numerical experiment results show that the AHAFSA not only can effectively locate the global optimum, but also have a rather high convergence speed. It is a promising approach for solving the real roots of polynomials.

Keywords: artificial fish school algorithm, Hooke-Jeeves method, hybrid algorithm, adaptive, the real roots of polynomials.

1   Introduction

In a large number of scientific and engineering computing, many questions have to be involved in solving real roots of polynomials, so the study of algorithms for solving real roots of polynomials has important theoretical and applied significance. Currently, the traditional dichotomy polynomial rooting algorithms include Secant method, iterative method, Newton method, etc., but these algorithms have some limitations [1-2]. For example, Newton's method is a straight curl solving method, its convergence rate is fast, but the results depend on solving the initial value selection. If the initial value is not good, it will not get satisfactory solution, or even does not converge. Therefore, so far, more than five times of roots of a polynomial problem still do not have a good solution. So the study of solving the roots of a polynomial algorithm has practical significance.

Artificial fish swarm algorithm (AFSA) [3-4] which was proposed by scholar Xiao-Lei Li in 2002s is a novel method to search global optimum. Because AFSA has a good ability of global convergence and is not sensitivity to the initial guess of the solution, it has been used successfully in many fields [5-6]. However, along with the far-going applications of AFSA, the shortcoming of the algorithm is found. Particularly, the main shortage is that the algorithm's local convergence rate is low although it has a good ability of global convergence. In order to improve the shortage of AFSA, this paper proposes a new improved algorithm which is called adaptive hybrid artificial fish swarm algorithm (AHAFSA). Firstly, adaptively reduce the

D.-S. Huang et al. (Eds.): ICIC 2014, LNCS 8588, pp. 48–54, 2014.

perception scope of artificial fish in the search process, it can improve the convergence speed; Secondly, the Hooke-Jeeves algorithm[7] which has a strong local search ability is used as an operator of AFSA to improve the local convergence speed of AFSA. These are the basic idea of the new algorithm. Finally, AHAFSA is used for solving the real roots of polynomials, the numerical experiment results show that the AHAFSA not only can effectively locate the global optimum, but also have a rather high convergence speed. It is a promising approach for solving the real roots of polynomials.

2   Problem Presentation

n - times polynomial is expressed as:

$$f(x) = a_0 x^n + a_1 x^{n-1} + \cdots + a_{n-1} x + a_n \tag{1}$$

Let $x_i (i = 1, \cdots, n)$ as $f(x)$'s root, then exists a real number $R > 0$ meeting this formula $| x_i | < R$, $(i = 1, \cdots, n)$, generally, we let $R = 1 + \max\limits_{1 \le i \le n}\{| \frac{a_k}{a_0} |\}$.

Let $y = \dfrac{x}{R}$, the formula (1) is conversed as :

$$f(x) = g(y) = b_0 y^n + b_1 y^{n-1} + \cdots + b_{n-1} y + b_n \tag{2}$$

Obviously, the root of $g(y)$ meets $| y_i | < 1$, and the roots of $f(x)$ is $x_i = R y_i (i = 1, \cdots, n)$, The aim of the conversion is to determine the root of the scope, and let the problem easy to solve. In order to use AHAFSA to get all real roots of $g(y)$, specific practices are as follows: Firstly, use AHAFSA to get a root of $g(y)$ which is marked as y_1, then Let $g(y)$ reduce the time, which is marked $g'(y) = \dfrac{g(y)}{y - y_1}$, and $g'(y)$ is $n-1$ times polynomial, and then calculates the root of $g'(y)$ which is marked as y_2 in the same way [8]. Repeat this process, we can obtain all the real roots of $g(y)$ in turn by AHAFSA.

3   Adaptive Hybrid Artificial Fish School Algorithm

3.1   Artificial Fish Swarm Algorithm

Artificial fish swarm algorithm (AFSA) is a random search algorithm based on simulating fish swarm behaviors which contain food seeking behavior, swarming behavior and following behavior. It constructs the behaviors of every artificial fish firstly, and then makes the global optimum come true finally based on fish individuals' local searching behaviors. Firstly, we introduce some definitions about AFSA [3]-[4].

Assuming $\{X_1, X_2 \cdots, X_N\}$ is a set of artificial fish and $Y = F(X_i)$ represents the fitness function at position X_i which is an arbitrary individual fish in the set. Let $d_{ij} = \| X_i - X_j \|$ stands for the distance between artificial fish X_i and artificial fish X_j; *Visual* represents the field of vision of the artificial fish; *Step* represents the maximum step of the artificial fish moving; δ represents the density coefficient and try-number stands for the maximum attempt times of the artificial fish moving in its food seeking behavior. The typical behavior of artificial fish is expressed as followings:

1)Behavior of Food Searching
Assume X_i is the artificial fish state at present, and X_j is an arbitrary individual fish within X_i's field of view, then we comparing with their fitness function value. If Y_j is better than Y_i, then X_i will move one step forward towards the direction of X_j within its field of view. Otherwise, X_i will randomly move one step within its field of view when it has tried try-number times to find a better position within its field of view but failed. The behavior of food searching can be expressed in Formula 3.

$$\begin{cases} X_{inext} = X_i + Random(Step)\dfrac{X_j - X_i}{\|X_j - X_i\|} & , \; \text{if} \;\; Y_j > Y_i \\[2mm] X_{inext} = X_i + Random(Step) & \text{el se} \end{cases} \tag{3}$$

2) Behavior of Swarming
Assume X_i is the artificial fish state at present, nf is the number of its fellow within its field of view, and $X_c = \sum X_j / nf$. The behavior of swarming can be expressed in Formula 4.

$$\begin{cases} X_{inext} = X_i + Random(step)\dfrac{(X_c - X_i)}{\|X_c - X_i\|} & , \; \text{if} \;\; Y_c / nf > \delta Y_i \\[2mm] AF - prey() & else \end{cases} \tag{4}$$

3)Behavior of Following
In the process of the artificial fish swimming, when one fish discovers more food, the other one will move towards the direction. Assume X_i is the artificial fish state at present, X_{max} is the artificial fish which fitness function value is the maximum within its field of view and Y_{max} is the fitness function of X_{max}. The behavior of following can be expressed in Formula 5.

$$\begin{cases} X_{inext} = X_i + Random(step)\dfrac{(X_{max} - X_i)}{\|X_{max} - X_i\|}, & \text{if } \dfrac{Y_{max}}{nf} > \delta Y_i \\ AF - prey() & \text{el se} \end{cases} \tag{5}$$

According to the character of the problem, the artificial fish will select an appropriate behavior to carry out. For example, every artificial fish simulates behaviors of swarming and following, then selects the better one to carry out actually.

3.2 Solving the Real Roots of Polynomials Based on AHAFSA

1) Adaptive Hybrid Artificial Fish School Algorithm
Large numbers of studies have shown that AFSA has good global search ability, but it also has some shortcomings. For example, the field view of artificial fish remains unchanged in the search process. With the artificial fish approaching the individual merits, if *visual* is too large will reduce the convergence rate, if *visual* is too small, the algorithm in the search process is very easy to fall into local optimum. Therefore, in order to improve the convergence rate of AFSA and avoid local optimum, we propose an adaptive artificial fish school algorithm (AAFSA).

The main idea is: Initialization to the algorithm in a relatively large $visual$, which is conducive to the initial global search, as the algorithm progresses, the adaptive decrease, which will help improve the convergence speed. We use the adaptive reduction formula is:

$$visual_{k+1} = (\left| ybest - y \right| \big/ ybest) \cdot visual_k \tag{6}$$

ybest is the theory of optimal value, y is the average value of fitness function each iteration and k is the number of iterations. Clearly, as the algorithm progresses, the average fitness function value will gradually approach the optimal theory value, we propose a formula which is called adaptive reduction formula.

2) The step of Solving Real Roots of Polynomials by AHAFSA
Numerical experiments show that the adaptive artificial fish school algorithm (AAFSA) has good global search ability and convergence speed has been greatly improved, but the ability of local search has yet to be improved. Therefore, in order to improve the local search ability of AAFSA, Therefore, in order to improve the local search ability of AAFSA, we introduced an algorithm called Hooke-Jeeves method [7] which has strong local search ability as AAFSA a local search operator. It can enhance the local search ability of AAFSA, so we propose a hybrid algorithm called adaptive hybrid artificial fish school algorithm (AHAFSA). In this paper, we use AHAFSA to solve all the real roots of polynomials, the specific steps of the algorithm is as follows:

- Step 1: Initialize some variables, N- number of fishes; *Visual* - the field of vision of the fish; *Step* - the maximum step of the fish moving; δ - the density coefficient; p -the probability of local search.
- Step 2: Compare the food concentration of every fish, record down the best one which are signed by $\max Y$, $\max X$.
- Step 3: Behavior choices: Simulation of artificial fish are all behavior of swarming and behavior of following, by comparing the objective function to choose the best acts to perform.
- Step 4: Update bulletin board: After the first iteration of each artificial fish, more a function of their own value and the value of bulletin boards, bulletin boards such as the value of batter than replaced, otherwise the value of the same bulletin board.
- Step 5: Execute the Hooke-Jeeves method: All artificial fish are generated random number $r \in (0,1)$, if $r < p$, Use the artificial fish value as the initial of the solution to execute Hooke-Jeeves method. Record down the best solution *besty* which Hooke-Jeeves method search when the algorithm meets the end condition of Hooke-Jeeves method. Then exit operation of the Hooke-Jeeves method.
- Step 6: Execute the adaptive operation: Adaptively updated the field of vision of the fish *Visual* by the formula.
- Step 7: Termination of conditional: Repeat step 3 and step 6, until the bulletin board to reach the optimal solution within a satisfactory error bound up.
- Step 8: Algorithm terminated: Output the optimal solution (that is, the artificial fish bulletin board state and its function value).

4 Numerical Experiments and Discussion

In order to evaluate the effectiveness and efficiency of AHAFSA in solving the real roots of polynomials, numerical experiments have been carried out on a number of problems. Then the result of AHAFSA is compared with the result of [10-11]. Parameters of AHAFSA are defined as following:

N=50, *Visual* =1.5, *Step* =0.3, δ =2, try-number=6, P_1 =0.8, p =0.1, the maximum number iterations of Hooke-Jeeves K_t =60.

P 1: $f_1(z) = z^4 - 5z^2 + 4$

The true roots of this polynomial are -2, 2, 1, -1. In this paper, in order to compare the different optimization approaches for solving real roots of polynomial, 10 computations are simulated by AHAFSA. The results are compared with the literature [10] shown in table 1.

Table 1. Comparisons among the results of [10] and AHAFSA

Algorithm	Z_1	Z_2	Z_3	Z_4	Iterations	time(s)
RCRM	-2.00000011	2.00000098	0.99999873	-0.99999960	373266	7.64
RMM	-2.00000085	2.00000169	0.99999871	-0.99999955	1791239	3.32
RRMM	-2.00000103	2.00000186	0.99999871	-0.99999954	2826496	2.96
AHAFSA	-2.00000000	2.00000000	1.00000000	-0.99999999	27	1.04

We can see from table 1 that AHAFSA can find the more accurate roots of polynomial in less time and less number of iterations compared with the literature [10].

P 2: $f_2(z) = z^5 - 4z^4 + 4.43z^3 - 0.164z^2 - 1.464z + 0.144$

The essence of true roots of this polynomial are 0.1, 1.2, 1.2, -0.5, 2. In order to compare the different optimization approaches for solving real roots of polynomial, 10 computations are simulated by AHAFSA. The results are compared with the literature [10] shown in table 2.

Table 2. Comparisons among the results of [10] and AHAFSA

Algorithm	Z_1	Z_2	Z_3	Z_4	Z_5	Iterations	time(s)
RCRM	0.0999973	1.2005921	1.1995214	-0.5000003	2.0000004	12392	121.24
Newton	0.09999988	1.2003105	1.1996134	-0.5000002	2.0000011	123247	156.35
RRMM	0.1000001	1.2002169	1.1998809	-0.5000001	2.0000021	71526	7.21
AHAFSA	0.10000000	1.2000000	1.2000000	-0.5000000	1.9999999	38	6.871

From table 2, the results of the compared shown that AHAFSA gets more accurate roots of polynomial in less time and less number of iterations.

P 3: $f_3(z) = z^8 - 10z^6 + 33z^4 - 40z^2 + 16$

The true roots of this polynomial are -2, 1, -2, 1, -1, 2, -1, 2. 10 computations are simulated by AHAFSA, the results are compared with the literature [11] shown in table 3.

Table 3. Comparisons among the results of [11] and AHAFSA

algorithm	Approximate real roots	Average number of iterations
Recursive fast neural roots of control	-2.00069220676009 1.00000693410678 -1.99933983350431 0.9999938058267339 -0.999938058267339 2.00057729143192 -1.00006988438118 1.99940899375285	The first block:8621 The second block:2805 The third block:930 2-order direct method:564
Roots of direct Neural control	-2.00008812923599 1.00006032497794 -1.99989207805059 0.999965089833384 -0.99998993865839 1.99999042628593 -1.00003612027881 1.99999042628593	29735
AHAFSA	-2.00000925230651 0.99999999198987 -2.00000925230651 0.99999999560006 -0.99999040568555 2.00000000388276 -1.00000961442870 2.00000000446462	106

From table 3, we know that AHAFSA is a good method for solving the real roots of polynomial. It not only has a high convergence rate, but also gets more accurate solution.

5   Conclusions

In this paper, we propose a hybrid algorithm-AHAFSA which contains AFSA and Hooke-Jeeves method. AHAFSA fully uses the global convergence of AFSA and the strong local convergence ability of Hooke-Jeeves method. The numerical experiment which uses AHAFSA to solving real roots of polynomials results show that the AHAFSA can find the more accurate roots of polynomial in less time and less number of iterations compared with the literatures. Because of the good performance of AHAFSA, solving roots of polynomials in the complex domain by AHAFSA will be the next step.

Acknowledgement. This work is supported by the Foundation of Guangxi Department of education research project (No.201010LX088), and the Opening Foundation of Guangxi Key Laboratory of Hybrid Computation and IC Design Analysis.

References

1. Modem Applied Mathematics Handbook Editorial Board, Handbook of modem applied mathematics and numerical analysis of calculated volume. Qinghua University Press (2009)
2. Cao, D.Q., Zhang, M.: Work out the roots of polynomial based on the evolution strategy. Guangxi Sciences 14(2), 98–102 (2007)
3. Tu, X.Y.: Artificial Animals for Computer Animation: Biomechanics, Locomotion, Perception, and Behavior, Dissertation for doctor degree (1999)
4. Li, X.L., Shao, Z.J., Qian, J.X.: An optimizing method based on autonomous animate: fish swarm algorithm. System Engineering Theory and Practice 22(11), 32–38 (2002)
5. Ma, J.W., Zhang, G.L., Xie, H.: Optimization of feed-forward neural networks based on artificial fish-swarm algorithm. Journal of Computer Applications 24(10), 21–23 (2004)
6. Huang, G.Q., Lu, Q.Q.: An approach to air leakage points identification in branches of ventilation system based on fish-swarm algorithm. Journal of System Simulation 19(12), 2677–2682 (2007)
7. Tang, H.W.: Practical optimization methods. Dalian University of Technology (1994)
8. Yang, L., Xia, B.S.: Selections symbolic computation. Qinghua University Press (2003)
9. Yu, Y., Yin, Z.F.: Multiuser detector based on adaptive artificial fish school algorithm. Journal of Electronics & Information Technology 29(1), 121–124 (2007)
10. Huang, D.-S.: A Constructive Approach for Finding Arbitrary Roots of Polynomials by Neural Networks. IEEE Transactions on Neural Networks 15(2) (2004)
11. Huang, D.S., Chi, Z.R.: Sub-neural network-based fast recursive method for solving the root of any polynomial order. Science in China (series E) 33(12), 1115–1124 (2003)

Optimization of Moving Objects Trajectory Using Particle Filter

Yangweon Lee

Honam University, 417, Eodeung-daero, Gwangsan-gu, Gwangju, Korea
ywlee@honam.ac.kr

Abstract. This paper suggested the guidance algorithm that enable the unmanned flying objects (UFOs) to track trajectories. It increase the amount of information provided by the measurements and improve overall estimation observability. We assume that small UFOs equipped with camera and navigation sensors are using for improvement of target tracking and an accurate target location estimate. The UFO trajectory optimization is performed for stationary targets and dynamic targets. We considered the Particle Filter for estimation algorithm. The suggested algorithm shows flying object trajectories that increase filter convergence and overall estimation accuracy, illustrating the importance of information-based trajectory design for target localization using small flying objects.

Keywords: particle filter, object tracking.

1 Introduction

The main purposes involved with target localization include: maintaining the target in the field of view, and developing a vehicle trajectory such that the target location estimation error is minimized [1, 5, 6]. When the target is moving, it is difficult for an operator to recognize what the best vehicle path is. Therefore, autonomous flight path generation and trajectory design are essential to successful target localization [2].

In figure 1, is shown the overall system configuration of optimal trajectory estimation of unmanned flying object system. The camera and navigation sensors are mounted under the flying objects. Most modern small unmanned vehicles are designed to be in expensive and therefore include low quality sensors and poor navigation filtering algorithms, leading to achieve a high level of precision in the target location estimate since the errors in the vehicle state propagate into the targeting algorithms. The resulting target estimation error is a combination of errors in the vehicle position, vehicle orientation, and sensor noise.

The purpose of this paper is to find methods to reduce this target location estimation error given noisy and inaccurate vehicle navigation. The vehicle navigation system is provided and certain amounts of vehicle state errors are unavoidable, but the trajectory of the unmanned flying object can be controlled to reduce the uncertainty or entropy about the target estimation and mitigate the effect of the noisy vehicle state estimates.

D.-S. Huang et al. (Eds.): ICIC 2014, LNCS 8588, pp. 55–60, 2014.
© Springer International Publishing Switzerland 2014

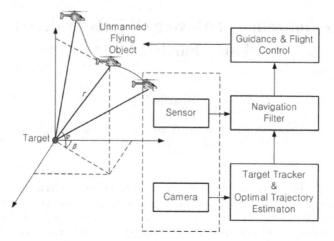

Fig. 1. System configuration of target localization with vision

2 Modeling of Particle Filter for Trajectory Estimation

The particle filter approach to track trajectory of flying objects, also known as the condensation algorithm [3,4] and Monte Carlo localization [7,8,9,10], uses a large number of particles to explore the state space. Each particle represents a hypothesized target location in state space. Initially the particles are uniformly randomly distributed across the state space, and each subsequent frame the algorithm cycles through the steps illustrated in figure 2:

1. Initially to initialize, draw N samples from the proposal distribution $p(x_0)$. Update the importance weights up to a normalized constant and equal.

$$x_0^i \sim p(x_0), \quad i = 1{:}N \tag{1}$$

$$m_0^i = \frac{1}{N}, \quad i = 1{:}N \tag{2}$$

2. Propagate the particle set according to a process model of target dynamics.

$$x_k^i = f(x_{k-1}^i) + w_{k-1}^i, \quad i = 1{:}N \tag{3}$$

Here $f(x_{k-1}^i)$ is the dynamics model and w_{k-1}^i is the noise.

3. Update probability density function (PDF) based on the received measurement z_k and the likelihood function of receiving that measurement given the current particle value.

$$m_k^i = m_{k-1}^i \, p(z_k | x_k^i), \quad i = 1{:}N \tag{4}$$

4. Normalize the weights.

$$m_k^i = \frac{m_k^i}{\sum_{i=1}^{N} m_k^i} \tag{2}$$

5. Compute the effective sample size. The bounds on the effective sample size are given by $1 \leq N_{eff} \leq N$.

$$N_{eff} = \frac{1}{\sum_{i=1}^{N}(m_k^i)^2} \tag{3}$$

6. Compare N_{eff} to the resampling threshold. Resample if necessary.

$$N_{eff} < N_{thr} \rightarrow \text{Resample and GO TO STEP B}$$

$$N_{eff} \geq N_{thr} \rightarrow \text{GO TO STEP B} \tag{4}$$

Here, N_{eff} is used to determine if the particles are unevenly distributed. Also if a few particles have large weighting and the other have small weighting values, resulting in N_{eff} will be close to 1. In contrast, if all the particles have equal weighting values, N_{eff} will be equal to N .

At any time step, the value of the estimate is as follows:

$$\bar{x}_k = \sum_{i=1}^{N}(x_k^i - \bar{x}_k)^2 m_k^i \tag{5}$$

The covariance can be computes using

$$P_k = \sum_{i=1}^{N}(x_k^i - \bar{x}_k)^2 m_k^i \tag{6}$$

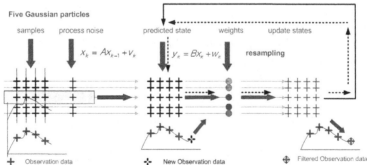

Fig. 2. Particle Filter Calculation Process

One consideration to keep in mind while using particle filters (PF) is the insufficiency of the Cramer-Rao Lower Bound (CRLB) in assessing the complete performance of the PF algorithm. For the particle filtering algorithm the mean and covariance of the estimation can be computed for comparison with CRLB, which for Gaussian processes is enough. However, using the particle filtering algorithm, the posterior densities obtained for the nonlinear filtering problem are usually non-Gaussian, requiring order moments to characterize the density.

3   Flying Object Trajectory Optimization

The particle filtering algorithm, mentioned in Section 2, is implemented. The target dynamics model is assumed to be linear but the measurement model is nonlinear, giving the following system dynamics,

$$x_{k+1} = \Phi_{k+1,k} \, x_k + w_k \tag{7}$$

$$z_k = h(x_k) + v_k \tag{8}$$

Where $\Phi_{k+1,k}$ is the state transition matrix of the system from time k to k+1 and w_k and v_k are the process and measurement noises, which are uncorrelated, Gaussian and white with zero mean and covariance Q_k and $\mathcal{R}_k$ respectively.

The number of particles is set to $N = 200$, which is usually low for a particle filter. The sample particle are initialized from a Gaussian distribution $p(x_0) = \mathcal{N}(\hat{x}_0, \, P_0)$ with mean and covariance given by

$$\hat{x}_0 = \begin{bmatrix} 5 \\ 5 \\ 5 \end{bmatrix}, \quad P_0 = \begin{bmatrix} 10 & 0 & 0 \\ 0 & 20 & 0 \\ 0 & 0 & 10 \end{bmatrix} \tag{9}$$

The proposed algorithm is simulated under the following scenarios: a stationary target, a target exhibiting random walk behavior, a constant velocity target and a target following a semi-circular trajectory.

For the stationary target case, the particle propagation model is assumed to be constant. The noise model is given by $w_k \sim \mathcal{N}(0, \sigma_w)$ and the tuning parameters are set to $\sigma_w = 0.1, N_{thr} = 20$. The measurement likelihood function is based on two bearing measurements and its equation is given by

$$p(z_k|x_k^i) = \frac{1}{\sqrt{2\pi\sigma^2}} \exp\{-\frac{1}{\sigma^2}[z_k - h(x_k^i)]^T [z_k - h(x_k^i)]\} \tag{10}$$

Where $z_k = [\theta \quad \varphi]_k^T$ and the measurement model is

$$h(x_k^i) = \begin{bmatrix} \tan^{-1}(\frac{p_{k_x} - x_{k_x}^i}{p_{k_y} - x_{k_y}^i}) \\ \tan^{-1}(\frac{p_{k_z} - x_{k_z}^i}{\sqrt{(p_{k_x} - x_{k_x}^i)^2 + (p_{k_y} - x_{k_y}^i)^2}}) \end{bmatrix} = \begin{bmatrix} \tan^{-1}(\frac{r_{k_x}}{r_{k_y}}) \\ \tan^{-1}(\frac{r_{k_z}}{\sqrt{r_{k_x}^2 + r_{k_y}^2}}) \end{bmatrix} \tag{11}$$

The PF results for the stationary target case are presented in Figure 3. The final estimation errors is 0.25m, but the variance is still high.

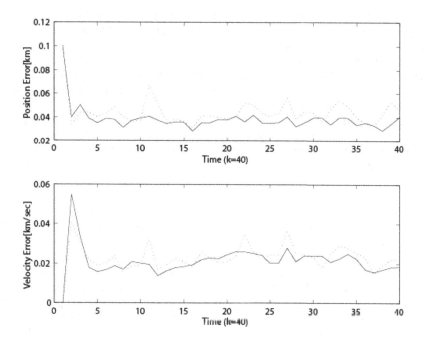

Fig. 3. Localization of a stationary target using a particle filter

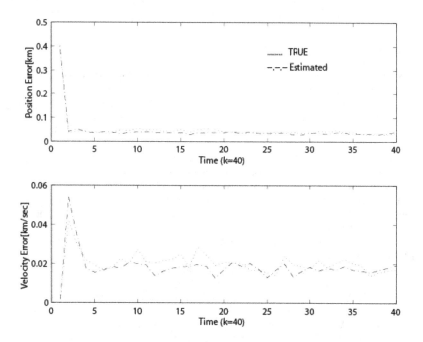

Fig. 4. Localization of a turning target using a particle filter

The next case considered is for a slow moving target starting at the origin and following a semi-circular trajectory. The particle propagation model for this case is also assumed constant and the tuning parameters are set to $\sigma_w = 1.2, N_{thr} = 10$.

The result for the moving target are presented in Figure 4. The tracking performance is noisier and the estimation error is about 3m with a variance 20.5. Decreasing σ_w would make the PF results less noisy, but for this moving target case, lower values of σ_w would decrease the stability of the filter.

Overall, the particle filtering algorithm performs as desired yielding good tracking performance for the stationary target case and for the cases of slow moving targets with unknown target dynamics, even though tracking in the fast moving target case is not achievable.

4   Conclusions

This paper presents a target trajectory estimation algorithm for unmanned flying object in the 3-D environment.

We recognized that the particle filter algorithm is more stable than the EKF and is more robust than the existing baysian estimation filters. It is also able to handle the nonlinear, non-Gaussian distribution and if there are given enough particles it supply the more accurate estimation of the target motion. On the other hand, the particle filter is more sensitive to its number of particles, requiring more design effort to properly select these particles, and, in many cases, it is not obvious what the best selection of these should be.

We recognized that the result of estimation accuracy of target trajectories shown in this paper can be directly applied to small UFO with stabilized camera systems.

References

1. Lee, Y.: Implementation of An Interactive Interview System Using Hand Gesture Recognition. Neurocomputiong 116, 272–279 (2013)
2. Aidala, V., Hammel, S.: Utilization of Modified Polar Coordinates for Bearings Only Tracking. IEEE Transactions on Automatic Control AES-15(1), 29–39 (1979)
3. Black, M.J., Jepson, A.D.: A Probabilistic Framework for Matching Temporal Trajectories: CONDENSATION-Based Recognition of Gestures and Expressions. In: Burkhardt, H.-J., Neumann, B. (eds.) ECCV 1998. LNCS, vol. 1406, pp. 909–924. Springer, Heidelberg (1998)
4. Huang, D.: A Constructive Approach for Finding Arbitrary Roots of Polynomials By Neural Networks. IEEE Trans. Neural Networks 15(2), 477–491 (2004)
5. Lee, Y.-W.: Adaptive Data Association for Multi-target Tracking Using Relaxation. In: Huang, D.-S., Zhang, X.-P., Huang, G.-B. (eds.) ICIC 2005. LNCS, vol. 3644, pp. 552–561. Springer, Heidelberg (2005)
6. Lee, Y., Seo, J., Lee, J.: A study on the TWS tracking filter for Multi-Target Tracking. J. KIEE 41(4), 411–421 (2004)
7. Jia, W., Huang, D., Zhang, D.: Palm Print Verification Based on Robust Line Orientation Code. Pattern Recognition 41(5), 1521–1530 (2008)
8. Huang, D., Jia, W., Zhang, D.: Palm Print Verification Based on Principal Lines. Pattern Recognition 41(4), 1316–1328 (2008)
9. Isard, M., Blake, A.: A Mixed-State Condensation Tracker with Automatic Model Switching. In: Proceedings of the Sixth International Conference on Computer Vision, pp. 107–112 (1998)
10. Huang, D., Ip, H., Chi, Z.: A Neural Root Finder of Polynomials Based on Root Moments. Neural Comput. 16(8), 1721–1762 (2004)

Handing the Classification Methodology ORCLASS by Tool OrclassWeb

Plácido Rogerio Pinheiro, Thais Cristina Sampaio Machado, and Isabelle Tamanini

University of Fortaleza (UNIFOR), Graduate Program in Applied Informatics,
Av. Washington Soares, 1321 - Bl J Sl 30 - 60.811-905, Brazil
placido@unifor.br,
{thais.sampaio,isabelle.tamanini}@gmail.com

Abstract. The Decision Making is present in every activity of the human world, as in simple day-by-day problems, as in complex situations inside of an organization. In the human world, emotions and reasons become hard to separate; therefore, Decision Support Systems (DSS) were created to help decision makers to take complex decisions. The paper presents the development of a new tool, called OrclassWeb, which reproduces the procedure to apply the Verbal Decision Analysis (VDA) Methodology ORCLASS that supports the process of the mentioned DSS method.

Keywords: Decision Support Systems, ORCLASS, OrclassWeb, Verbal Decision Analysis.

1   Introduction

Presented in every human activity, the decision may be defined because of a process of choice, given an identified problem or when the decision maker faces an opportunity of creation, optimization or improvement in an environment. Aiming to solve complex problems, qualitatively and difficultly to be resolved by human process capacity a decision maker to rank or classify multi-attribute alternatives, according to [7]. The methods that constitute the Verbal Decision Analysis framework are: ZAPROS-III, ZAPROS-LM, PACOM and ORCLASS, as well as their characteristics and applications [9, 10, 13 and 22]. Objecting to solve complex problems and provide better utilization of ORCLASS, the new tool OrclassWeb was developed to reproduce the mentioned methodology automatically [14 and 15]. According to [5], the choice of the method should be the result of an evaluation of the chosen parameters, the type and accuracy of the data, the way of thinking of the decision maker, and his/her knowledge of the problem. It is noteworthy that the direct consequence of the choice between various methods is that the results can be uneven and even contradictory.

In the issue in question, the application of method ORCLASS came to acceptance by the decision maker, which meant that the issues that were being presented to the decision maker made sense to him, and he had confidence in answering them. Still in the view of the authors, this stressed that the methodology for multicriteria decision has several methods that can be applied in various problems.

D.-S. Huang et al. (Eds.): ICIC 2014, LNCS 8588, pp. 61–72, 2014.

2   ORCLASS Method - Overview and Structure

The ORCLASS method (Ordinal Classification) is focused in classifying multi-criteria alternatives, which are the options for solving the problem and can be characterized according to the criteria provided. The method aims at categorizing the alternatives into a small number of decision classes or groups, which are preordered according to the decision maker's preferences [8].

In fact, many different methods for solving multicriteria classification problems are widely known [[16]. ORCLASS methodology aims at classifying the alternatives in a given set: the decision maker needs these alternatives to be categorized into a small number of decision classes or groups, usually two. According to [1], presents a flowchart with steps to apply the method ORCLASS, that visualized in Fig. 1. The application of the method can be divided in three stages: Problem Formulation, Structuring of the Classification Rule and Analysis of the Information Obtained.

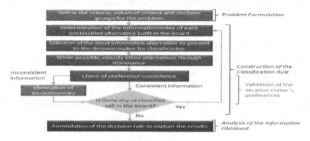

Fig. 1. Procedure to apply the ORCLASS method

2.1   Formal Statement of the Problem

The methodology follows the same problem formulation proposed in (8):

1. $K = 1, 2,..., N$, representing a set of N criteria;

2. n_q represents the number of possible values on the scale of q-th criterion, $(q \in K)$; For the ill-structured problems, as in this case, usually $n_q \leq 4$;

3. $X_q = \{x_{iq}\}$ represents a set of values to the q-th criterion, which is this criterion scale; $|X_q| = n_q (q \in K)$;

The values of the scale are ranked from best to worst, and this order does not depend on the values of other scales;

4. $Y = X_1 * X_2 * ... * X_n$

represents a set of vectors yi (every possible alternative: hypothetical alternatives + real alternatives) such that: $y_i = (y_{i1}; y_{i2}; ...; y_{iQ})$, and $y_i \in Y$, $y_{iq} \in X_q$ and $Q = |Y|$, such

that $|Y| = \prod_{q=1}^{Q} n_q$

5. A = {a$_i$}∈Y , i=1,2,...,t. such that the set of t vectors represents the description of the real alternatives.

6. L: number of ordered decision classes.

Required: The multicriteria alternatives classification based on the decision maker's preferences (judgment) to build a reflection.

2.2 Method's Application

The construction of the classification rule will be carried out based on the decision maker's preferences. In order to present the method's application in a comprehensive way, we use the same concepts presented on [6], on which a classification task is presented as a set of boards.

Each cell is composed by a combination of values for each criterion defined to the problem, which represents a possible alternative to the problem. Let us consider a problem containing the criteria A, B and C, the criteria values: A = {A1, A2, A3}, B = {B1, B2, B3}, C = {C1, C2, C3}, and two decision classes (I and II), so that, according to the decision maker, the first class is preferable to the second one. For the given problem, we can say that: An alternative composed by the best characteristics ([A1, B1, C1]) will always belong to Class I and an alternative composed by the worst characteristics ([A3, B3, C3]) will always belong to Class II.

As defined in [8] if a possible alternative composed by the criterion values [A1, B1, C3] is presented to the decision maker for judgment. For example, and supposing that the decision maker classifies the alternative on the first group, then, we can infer that a better alternative as [A1, B1, C2] certainly belongs to the first group, since the last one is naturally more preferable than the previous. This response would fill two cells of the classification board. However, if the decision maker judges that the alternative from the example [A1, B1, C3] is classified into the second group, and then the worst alternatives will be classified as group 2. This way, the alternatives [A1, B2, C3], [A1, B3, C3], [A2, B1, C3], [A2, B2, C3], [A2, B3, C3], [A3, B1, C3], [A3, B2, C3] will belong to the second decision group, since they are naturally less preferable than the previous.

2.3 Tools Based on Verbal Decision Analysis

There are a few quantities of tools for automating Verbal Decision Analysis (VDA) methodologies, which are described in detail in [3 and 4] illustrates an implementation, in Visual C++ 6.0, a different version of ZAPROS-III, however the implementation does not apply Formal Index of Quality (FIQ) for alternatives comparing process. UniComBOS [1] is a tool for VDA, but does not intend to reproduce a VDA Framework methodology. The tool implements a new procedure for comparison and choice of multi-criteria alternatives. For ZAPROS-III, [20] implements a tool focused on treatment of problems incomparability, generating the new approach methodology ZAPROS III-*i*.

2.4   A Tool for the ORCLASS Method - The OrclassWeb Tool

In order to facilitate the decision making process using ORCLASS and perform it consistently, observing its rules and aiming at making it accessible, a tool developed in platform Java Web that applies the methodology is presented. The tool was made in a web environment in Platform Java 1.6, using JSF 2 and runs in server Tomcat 6. OrclassWeb tool was proposed to automate the comparison process of alternatives and to provide the decision maker a concrete result for the problem, according to ORCLASS definition. OrclassWeb was developed divided into four stages: Criteria and criteria values Definition, Alternatives Definition, Construction of the Classification Rule and Presentation of Results Obtained.

The manual application of the system ORCLASS is usually made with a maximum of three criteria and three criteria values for each one, because the complexity of the application increases exponentially with the growth of these numbers. This way, the main advantage of OrclassWeb is that the tool processes the complexity of the application, which means that the user can apply ORCLASS with a greater number of criterion and criteria values. OrclassWeb was developed adapting the rules to identify the most informative cell, after applying the rules defined by [11 and 12]. Other applications using the SCRUM OrclassWeb applied are found in [11, 14 and 17], with a purpose to verify robusteness of the tool. After the identification of the most informative index according to the rules, the tool verifies among all the others alternatives, which one presents the highest values for both indexes; being this most informative cell of the board. The adaptation was necessary to increase the method's comparison capacity, without giving away the adherence to the ORCLASS method. For research purposes, OrclassWeb is a tool developed for web, the System can be accessed through the website: **http://www2.unifor.br/OrclassWeb**

2.5   OrclassWeb: Interfaces and Availability

First of all, at the beginning of the application it is necessary to define the criteria and criteria values associated that the alternatives will be compared against, presented in the problem. Fig. 2 presents the tool's interface for defining criteria and criteria values. This is the stage named "Problem Formulation" from the procedure. The second stage of OrclassWeb is the definition of the alternatives. This is the step which the decision maker states the alternatives and characterize them according to the criteria values. Fig. 3 presents the tool's interface for defining alternatives. This is the phase named "Structuring of The Classification Rule" from the procedure.

The next step of OrclassWeb is the elicitation of preference stage. The interface for elicitation of preferences presents questions that can be easily answered by the decision maker in order to obtain the scale of preferences. According to the decision maker's responses, OrclassWeb initiates the division of alternatives into the classes/groups. Fig. 4 presents the tool's interface for capturing preferences. The current step also belongs to "Structuring of The Classification Rule" phase from the procedure. As the last step for applying the method, at the end of the comparison process OrclassWeb presents an interface composed by the groups and the related alternatives preferred for each one in order to allow a complete and more detailed analysis of the results obtained. Fig. 6 presents the tool's interface for obtaining the analysis results. This interface provides the visualization of "Analysis of Information Obtained" stage from procedure.

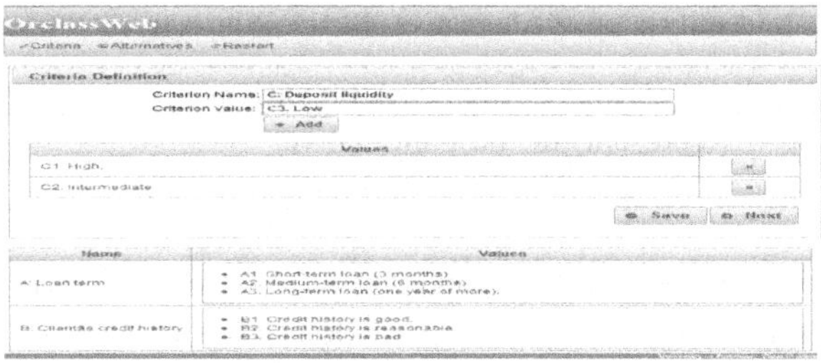

Fig. 2. Criteria Definition Interface: Who is worth the loan?

3 Tests of Reliability

Functional tests were elaborated and executed aiming to prove that OrclassWeb is a reliable and effective tool. The tests were done by applying ORCLASS system manually and were also applied through the tool. The results are compared in order to identify that OrclassWeb attends to all the requirements imposed by the methodology.

The current reliability test was elaborated according to specific criteria and criteria values and determined alternatives. The problem stated for the tests is the loan applications evaluation defined in [11]. The situation consists in a group of businesses submitting a loan application to a commercial bank. In what cases the loan should be rejected or accepted? The bank is the decision maker and provides a policy for granting loans. The decision maker is the person to whom the decision will be taken, and the elicitation of preferences in OrclassWeb will provide the same preferences from the bank. Table 1 presents the list of criteria and criteria values, which will be base to apply the methodology. The criteria values are described from the naturally most preferable to the least preferable one.

Table 1. Criteria and associated values: Who is worth the loan?

Criteria	Criteria Values
A: Loan term	A1. Short-term loan (3 months).
	A2. Medium-term loan (6 months).
	A3. Long-term loan (one year of more).
B: Client's credit history	B1. Credit history is good.
	B2. Credit history is reasonable.
	B3. Credit history is bad.
C: Deposit liquidity	C1. High.
	C2. Intermediate.
	C3. Low.

The alternatives for the application will be a hypothetical situation, since the problem definition in [11] does not describe the alternatives. Table 2 presents the list of alternatives for this case.

Table 2. Identification of Alternatives Board: Who is worth the loan?

ID	Alternatives
Alt 1	Represents the best alternative
Alt 2	Represents the worst alternative

Criteria and criteria values are required by the tool according to its preference degrees, at "Criteria Definition Interface" from OrclassWeb, visualized in Table 3.

Table 3. Criteria and associated values: Selecting practices of Framework SCRUM

Criteria	Criteria Values
A: Difficult degree for implemention	A1. Low: Its implementation does not require experience with the framework SCRUM. A2. Medium: Its implementation requires a little experience with the framework SCRUM or can be learned on the job. A3. High: Its implementation requires experience (maturity) about framework SCRUM.
B: Time consumption	B1. Gain: The consumption of time in the project for executing the activity is less than the process defined. B2. Not changed: There is no extra time in project for executing the activity than the process defined. B3. Lose: There is extra time in project for executing the activity comparing to the process defined.
C: Cost for the project	C1. Gain: The new activities are able to provide to the project an economy of cost. C2. Not changed: The new activities do not change the cost of the project. C3. High cost: The new activities are able to increase the project new costs.

Alternatives are required by the tool according to its characterization in criteria values, at "Alternatives Definition Interface" from OrclassWeb, visualized in Fig. 3

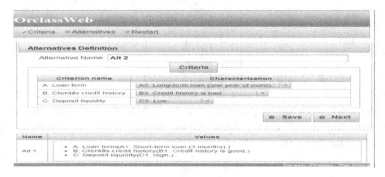

Fig. 3. Alternatives Definition Interface: Who is worth the loan?

OrclassWeb tool in parallel to the application made manually, through "Preferences Elicitation Interface", as presented in Fig. 4, conducts the process of Elicitation of preferences.

Fig. 4. Preferences Elicitation Interface: Who is worth the loan?

The tool conducts the entire process of Elicitation of preferences, and manually, the process can be applied using a table, that will be filled according to the decision maker preferences. Manually, the application of ORCLASS is made calculating the most informative index, which will provide the next question to be posed to the decision maker. Afterwards, the most informative cell from the board is exactly the next question presented to the user of OrclassWeb. The conclusion of elicitation of preferences is reached in both tools: OrclassWeb and manually. Comparing the result obtained in both situations, it is the same and can be visualized in the followed figure 5.

Fig. 5. Result OrclassWeb - reliability tests

Considering that agile methodologies, in focus Framework SCRUM [2, 17, 18, and 19] are each time more common in Development Software Companies, noticing that the mentioned companies cannot always apply every approach of the framework, the research apply ORCLASS System aiming to select the approaches to be applied in the company, considering the elicitation of preferences of a decision maker.

The application consists in the classification of SCRUM approaches in groups or classes. The first group was chosen to support the practices of framework SCRUM which should be selected after the application of ORCLASS, to be implanted by the organizations, and the second class will support the set of approaches that should not be implanted by the organization which desire to implement part of SCRUM. Table 3 presents the list of criteria and criteria values, which will be base to apply the methodology, which are required by the tool according to its preference degrees, at "Criteria Definition Interface" from OrclassWeb, visualized in Fig. 6.

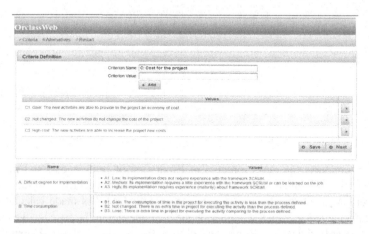

Fig. 6. Criteria Definition Interface: Selecting practices of Framework SCRUM

The alternatives for the application will be SCRUM approaches and they are listed in Table 4.

Table 4. Identification of Alternatives: Selecting practices of Framework SCRUM

ID	Alternatives
Prac1	Sprints (or iterations) with 1 to 4 weeks
Prac2	A product backlog and a sprint backlog creation and prioritization
Prac3	Planning meeting – part 1
Prac4	Planning meeting – part 2
Prac5	Daily Meeting
Prac6	Burn down chart and visible activities board
Prac7	Sprint Review
Prac8	Sprint Retrospective
Prac9	Release Planning

Alternatives are required by the tool according to its characterization in criteria values, at "Alternatives Definition Interface" from OrclassWeb, visualized in Fig. 7.

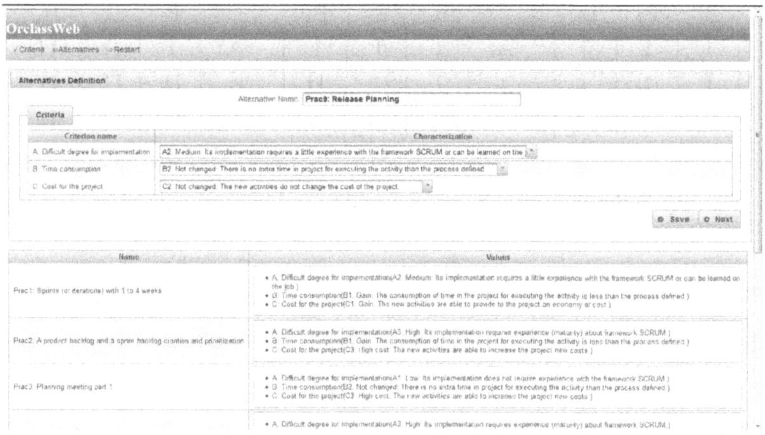

Fig. 7. Alternatives Definition Interface: Selecting practices of Framework SCRUM

The process of Elicitation of preferences is conducted by OrclassWeb, tool in parallel to the application made manually, through "Preferences Elicitation Interface", as presented in Fig. 8.

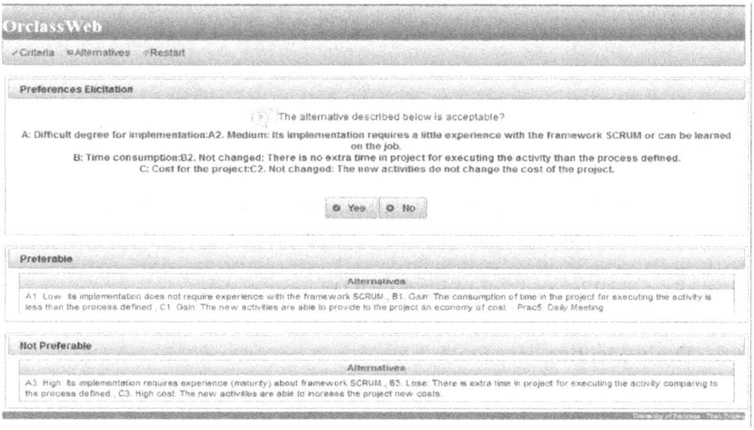

Fig. 8. Preferences Elicitation Interface: Selecting practices of Framework SCRUM

The tool conducts the entire process of Elicitation of preferences, and manually, the process can be applied using a table, that will be filled according to the decision maker preferences. Manually, the application of ORCLASS is made calculating the most informative index, which will provide the next question to be posed to the decision maker. Afterwards, the most informative cell from the board is exactly the next question presented to the user of OrclassWeb. The elicitation of preferences is reached in both tools: OrclassWeb and manually. The result can be visualized in the followed figures [9 and 10]

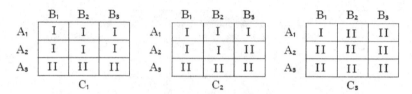

	B₁	B₂	B₃
A₁	I	I	I
A₂	I	I	I
A₃	II	II	II

C₁

	B₁	B₂	B₃
A₁	I	I	I
A₂	I	I	II
A₃	II	II	II

C₂

	B₁	B₂	B₃
A₁	I	II	II
A₂	II	II	II
A₃	II	II	II

C₃

Fig. 9. Result matrix: Selecting practices of Framework SCRUM

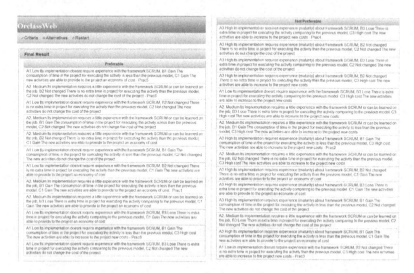

Fig. 10. Result OrclassWeb: Selecting practices of Framework SCRUM

Comparing the result obtained in both situations, the first and the second group are composed by the SCRUM approaches as described below:

- Class 1 is composed by the following SCRUM approaches:
  — Prac1 - Sprints (or iterations) with 1 to 4 weeks
  — Prac5 - Daily Meeting
  — Prac6 - Burn down chart and visible activities board
  — Prac7 - Sprint Review
  — Prac8 - Sprint Retrospective
  — Prac9 - Release Planning

- Class 2 is composed by SCRUM approaches
  — Prac2 - A product backlog and a sprint backlog creation and prioritization
  — Prac3 - Planning meeting – part 1
  — Prac4 - Planning meeting – part 2

Conclusions and Future Works

The decision may be defined as a result of a process of choice, given an identified problem or when the decision maker faces an opportunity of creation, optimization or improvement in an environment. Many decisions involve several factors that can be measured or not and influence on it. It means that the decision is taken according to the decision maker preferences. The decision maker is capable of manifesting his preferences and interests sufficient to solve simple problems. If the decision maker needs to solve complex problems covering many alternatives and many information that cannot be measured and easily compared, there are some methodologies to be applied in order to support the decision making process. The main contribution described in this work was OrclassWeb Tool. The software was developed and reliability tests were provided aiming to proof its adherence to Classification Methodology ORCLASS from VDA framework. It was available for anyone through a website for research. The tests of reliability, the developed tool OrclassWeb is proved as truly, reliable and strong enough to support the methodology of Verbal Decision Analysis and perform it consistently. First, as future works, another tool can be developed in order to reproduce others methodologies not provided automatically yet. Furthermore, more research can be done applying the OrclassWeb tool with more than three criteria and criteria values in known published problems. There is also the intention to refine the software in order to provide more treatment of incorrect insertions, storage of results, visual presentation of results, and much more.

Acknowledgements. The authors are thankful to the National Counsel of Technological and Scientific for the Developments (CNPq) for the support received for this project.

References

1. Ashimhmin, I., Furems, E.: Unicombos - Intelligent Decision Support System for Multi-Criteria Comparison and Choice. Journal of Multi-Criteria Decision Analysis 13(2/3), 147–157 (2006)
2. Beck, K., et al.: Manifesto for Agile Software Development (2001), http://Agilemanifesto.org
3. Dimitriadi, G.G., Larichev, O.I.: Decision Support System and the ZAPROS-III Method for Ranking the Multi-Attribute Alternatives With Verbal Quality Estimates. European Journal of Operational Research (December 2002)
4. Dimitriadi, G.G., Larichev, O.: Decision Support System and the ZAPROS III Method for Ranking the Multiattribute Alternatives With Verbal Quality Estimates. Automation And Remote Control 66(8), 1322–1335 (2005)
5. Gomes, L.F.A.M., Moshkovich, H., Torres, A.: Marketing Decisions in Small Businesses: How Verbal Decision Analysis Can Help. International Journal Management and Decision Making 11(1), 19–36 (2010)
6. Larichev, O.I.: Ranking Multicriteria Alternatives: The Method ZAPROS III. European Journal of Operational Research 131(3), 550–558 (2001)

7. Larichev, O.I.: Method ZAPROS for Multicriteria Alternatives Ranking And The Problem of Incomparability. Informatica 12, 89–100 (2001)
8. Larichev, O.I., Moshkovich, H.M.: Verbal Decision Analysis for Unstructured Problems. Kluwer Academic Publishers, The Netherlands (1997)
9. Machado, T.C.S., Menezes, A.C., Tamanini, I., Pinheiro, P.R.: A Hybrid Model in the Selection of Prototypes for Educational Tools: An Applicability in Verbal Decision Analysis. In: IEEE Symposium Series on Computational Intelligence in Multicriteria Decision-Making (2011)
10. Machado, T.C.S., Menezes, A.C., Lívia, F.R.P., Tamanini, I., Pinheiro, P.R.: Applying Verbal Decision Analysis in Selecting Prototypes for Educational Tools. In: IEEE International Conference on Intelligent Computing And Intelligent Systems, pp. 531–535 (2011)
11. Machado, T.C.S.: Towards Aided by Multicriteria Support Methods and Software Development: A Hybrid Model of Verbal Decision Analysis for Selecting Approaches of Project Management, Master Thesis, Graduate Program in Applied Informatics, University of Fortaleza (2012)
12. Machado, T.C.S., Pinheiro, P.R., Albuquerque, A.B., de Lima, M.M.L.: Applying Verbal Decision Analysis in Selecting Specific Practices of CMMI. In: Li, T., Nguyen, H.S., Wang, G., Grzymala-Busse, J., Janicki, R., Hassanien, A.E., Yu, H. (eds.) RSKT 2012. LNCS, vol. 7414, pp. 215–221. Springer, Heidelberg (2012)
13. Machado, T.C.S., Pinheiro, P.R., Tamanini, I.: Towards the Selection of Prototypes for Educational Tools: A Hybrid Model in The Verbal Decision Analysis. International Journal of Information Science, Scientific & Academic Publishing 2(3), 23–32 (2012)
14. Machado, T.C.S., Pinheiro, P.R., Tamanini, I.: Project Management Aided by Verbal Decision Analysis Approaches: A Case Study for the Selection of The Best SCRUM Practices. International Transactions in Operational Research (2014)
15. Pinheiro, P.R., Machado, T.C.S., Tamanini, I.: Verbal Decision Analysis Applied on the Choice of Educational Tools Prototypes: A Study Case Aiming At Making Computer Engineering Education Broadly Accessible. International Journal of Engineering Education 30(3), 585–595 (2014)
16. Machado, T.C.S., Pinheiro, P.R., de Lima Marcelo, M.L., Landim, H.F.: Towards a Verbal Decision Analysis on the Selecting Practices of Framework SCRUM. In: Liu, B., Chai, C. (eds.) ICICA 2011. LNCS, vol. 7030, pp. 585–594. Springer, Heidelberg (2011)
17. Machado, T.C.S., Pinheiro, P.R., Tamanini, I.: Dealing the Selection of Project Management Through Hybrid Model of Verbal Decision Analysis. Procedia Computer Science 17, 332–339 (2013)
18. Schwaber, K.: Agile Project Management With Scrum. Microsoft (2004)
19. Schwaber, K.: SCRUM Guide. SCRUM Alliance (2009)
20. Tamanini, I., Pinheiro, P.R.: Challenging The Incomparability Problem: An Approach Methodology Based on ZAPROS. In: Le Thi, H.A., Bouvry, P., Dinh, T.P. (eds.) MCO 2008. CCIS, vol. 14, pp. 338–347. Springer, Heidelberg (2008)
21. Tamanini, I., De Castro, A.K.A., Pinheiro, P.R., Pinheiro, M.C.D.: Verbal Decision Analysis Applied on the Optimization of Alzheimer's Disease Diagnosis: A Study Case Based on Neuroimaging. Advances in Experimental Medicine and Biology 696, 555–564 (2011)
22. Tamanini, I., Pinheiro, P.R.: Reducing Incomparability in Multiciteria Decision Analysis: An Extension of the ZAPROS Methods. Pesquisa Operacional 31, 251–270 (2011)

Solving TSP Problems with Hybrid Estimation of Distribution Algorithms

Xiaoxia Zhang and Yunyong Ma

College of Software Engineering, University of Science and Technology Liaoning, China
aszhangxx@163.com

Abstract. In this paper, a hybrid Estimation of Distribution Algorithms is proposed to solve traveling salesman problem, and a greedy algorithm is used to improve the quality of the initial population. It sets up a Bayes probabilistic model of the TSP. The roulette method is adopted to generate the new population. In order to prevent falling into local optimum, the mutation and limit were proposed to enhance the exploitation ability. At the same time, three new neighborhood search strategies and the second element optimization method are presented to enhance the ability of the local search. The simulation results and comparisons based on benchmarks validate the efficiency of the proposed algorithm.

Keywords: Estimation of distribution algorithm, Traveling salesman problem,Neighborhood search.

1   Introduction

Traveling Salesman Problem (TSP) is a classic route problem. The complexity of the problem grows exponentially with the expansion of TSP problem's scale, so TSP problem is a typical NP-Hard problem. In recent years, many intelligence computations have been used to solve the problem, such as ant colony algorithm [1], genetic algorithm [2], particle swarm optimization [3], artificial bees colony algorithm [4], estimation of distribution algorithms [5], etc.

Estimation of Distribution Algorithms (EDA) is developed from the basis of Genetic Algorithm (GA). As a relatively new paradigm, EDA is widely used in NP-Hard problems' research .According to its complexity and sampling method, it can be roughly divided into dependency-free, bivariate dependencies, and multivariate dependencies [6].EDA is evolved from the genetic algorithm. The algorithm is different from genetic algorithm's complex operators such as the crossover and mutation. In the EDA, a probability model of the most promising area is built by statistical information based on the searching experience, and then the probability model is used for sampling to generate the new individuals.   In such an iterative way, the optimal solution is obtained by evolving eventually. At present Estimation of Distribution Algorithm has been widely applied to industries, such as feature selection, engineering optimization [7], machine learning [8], flow-shop [9], multidimensional knapsack problem (MKP) [10], etc.

D.-S. Huang et al. (Eds.): ICIC 2014, LNCS 8588, pp. 73–81, 2014.

In the algorithms, greedy algorithm is used to obtain a relatively optimal solution, which enhances the quality of the initial solutions. In the literature 11,two elements optimization algorithm (2 - optimization, 2 - OPT) is introduced into the traveling salesman problem which enhances the capability of the search apparently .At the same time, three new neighborhood search strategy are proposed in this paper to improve the local search capability with little damage to the building blocks. In order to improve the global search ability of the algorithm, the negative feedback is introduced in this algorithm to control the change of the probability. The mutation operator is built into the proposed strategy to maintain population diversity and improve the ability of global exploration.

The advantage of this algorithm is put forward some new neighborhood search strategies which can improve the local search speed and balance the exploration and the exploitation abilities very well. The second element optimization method is also presented to enhance the quality of the local optimal solution. Negative feedback strategy and mutation operations are introduced to guarantee the global search ability of the algorithm, which can prevent the algorithm trapping in local optimal solution. And simulation results show that the algorithm has higher search capabilities, and the iterative algebra is significantly reduced.

2   Estimation of Distribution Algorithms

The notion of Estimation of Distribution Algorithms burnout was advocated in 1996, around 2000, the algorithm was developed rapidly. In recent years, EDA has been discussed as an important project in the world's famous academic conferences in the field of evolutionary computation, such as ACM SIGEVO, IEEE CEC [12]. EDA abandons the cross in the genetic algorithm (GA) such as restructuring operation. EDA builds the probability model for the next generation. The general framework of the EDA is illustrated in Fig. 1.

The critical step of the under procedure is to estimate the probability distribution [13]. The probability model used to describe the distribution of the solution space in each generation. The updating process reflects the evolutionary trend of the population. The probability distribution and updating process are both the most critical step in the Estimation of Distribution Algorithms. With respecting to several types of problem, a proper probability model and a suitable updating mechanism should be well developed to estimate the underlying probability distribution. However, more attention was paid to global exploration while the searching ability of Estimation of Distribution algorithm is relatively limited in the EDA. So, an effective EDA should balance the exploration and the exploitation abilities.

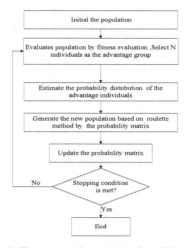

Fig. 1. The general framework of the EDA

2.1   Initial Population

For case of description, this article uses decimal-strings encode. Every individual of the population is a solution of the TSP. Suppose that there are L cities, let π_i be a natural number, Π ($\pi_1, \pi_2, ..., \pi_L$) be a city permutation. To ensure the diversity of initial population, we generate n city permutations as initial population randomly.

2.2   Advantage Group

According to the information about two cities' distance information, calculate the fitness of individuals in the population. Advantage group is the best of p individuals by the fitness.

2.3   Probability Model

Establish a frequency matrix k, let $k_{i,j}$ be the frequency in the advantage group from city i to city j.

$$k_{i,j} = \frac{1}{L} \sum_{k=1}^{L} \delta(x_k)$$

(1)

The value of $\delta(x_k)$ is decided by whether city i to city j is appeared in the route, if appeared $\delta(x_k)$ is 1, otherwise 0. When the probability model is constructed, Literature 14 uses the machine learning Heb rule to optimization the probability model:

$$P_{i,j} = \frac{1-\partial}{L} + \partial * k_{i,j} \qquad (2)$$

Because the initial group is generated randomly, the probability model of the initial value set to be $1 / L$, $\partial \in (0, 1)$, and ∂ is the learning efficiency. The bigger ∂ is , the faster learning rate is, and the greater the parent's influence on children and vice versa.

2.4   Generate the New Population

Bayes' theorem [15] is an important theorem in statistical inference which belongs to the conditional probability problem. According to a probability model as a prior probability is constructed, we can estimate the left posterior distribution of the occurrence of the event. Bayesian has widely applied to solve many combinatory problems now, such as shop scheduling problem [16] etc. In this paper, it is used to solve the TSP problem.

According to the matrix of probability model, taking the roulette wheel to obtain new species, using Bayesian probability model for the next point of probability distribution:

$$p(B_i \mid A) = \frac{p(A \mid B_i)p(B_i)}{\sum_{j=1}^{j=L} p(A \mid B_j)p(B_j)} \qquad (3)$$

$1 \le i \le L$, A represents the set of reached cities, B_i means the next to reached city is B_i. Roulette method is used to produce the next point until all cities are chosen, according to the probability distribution. The basic process can be expressed in the type [17]:

$$p(\pi_1, \pi_2, ..., \pi_L) = p(\pi_1) * p(\pi_2 \mid \pi_1) *, ..., p(\pi_L \mid \pi_1, \pi_2, ..., \pi_{L-1}) \qquad (4)$$

Algorithms produce some routes as the new initial population based on the same way. In order to keep the search speed, increase the algorithm's global search ability, reduce the loss of the quality of individual, paper will retain the best individual in this generation, which Adds directly to the next generation of initial population and takes part in the evolution of next generation.

Determine the convergence of the new population. If the population is convergent, algorithm is end, otherwise proceeds to next step.

2.5   Update the Probability Model

When the populations evolve to $t + 1$ generation, the probability model is as following:

$$p^{t+1}{}_{i,j} = (1-\partial)p^{t}{}_{i,j} + \partial * k_{i,j} \qquad (5)$$

p^t shows the population when the first generation t matrix, similarly p^{t+1} says the population in the first generation $t + 1$. Population will be optimized completely by updating the probability model constantly and iterative optimization.

3   Local Search Strategy

3.1   Greedy Solution

In order to improve the quality of the initial population, in this paper, the greedy algorithm is applied to the generation of initial population. Generate a point randomly as a starting point and then choose the nearest city as the second, and so on until you walk all the cities. Add the routes generated by the greedy algorithm to the initial population.

3.2   Neighborhood Search

In order to improve the local search ability of the algorithm, the paper puts forward three kinds of neighborhood search mode:

1. Because the new individual is produced by the conditional probability and every two connected cities' distance is not restricted with direction, therefore if we exchange the order of the adjacent two cities, there is small effect on groups .This article chooses one position and exchanges its next adjacent city randomly. such as p (1,4,5,3,2,6), the selected location is 3, the exchanged individuals are p (1,4,3,5,2,6).
2. Select a city randomly and places it between any two cities, such as p (1,4,5,3,2,6), the selected location is 3, then it will be inserted into randomly the rest route (1,2,4,5,6).
3. Regard two adjacent cities as a module. Randomly select two cities, respectively for strategy 2 operation .Such as p (1,4,5,3,2,6), the selected location 3, the city 5 and 3 are randomly inserted into the rest route(1,4,2,6).

All the three strategies take effect when the population evolution is to a certain degree and the best individual is not optimized in a certain generation. Frist and second strategies are used for all the population. Third strategy is used for the optimal individual. If the individual fitness is increased after three strategies operations, reserved the new individual, abandoned the original individual, otherwise unchanged.

In order to improve the local search speed, the algorithm also introduces 2-OPT algorithm. The general framework of 2-OPT is illustrated in Fig. 2.

Fig. 2. 2-OPT operation

2 - OPT neighborhood search method is a kind of common neighborhood search heuristic method, which disconnected two cities' cables, and cross linking cities. If the fitness of the individual is greater than the original individual, retain new individual and abandon the original individual, otherwise unchanged.

4   Global Search Strategy

4.1   Mutation Operation

As the evolution of the population, the algorithm is likely to be trapped in local optimal solution, and can't get the global optimal solution. The essay introduces mutation operation to ensure the population diversity and to improve the global search ability of the algorithm. The basic idea of Mutation is to select a percentage of the individuals in the initial solution randomly and exchange the positions of the two cities randomly.

In the initial stages of the evolution, population diversity is very good, and almost doesn't need a mutation, so the mutation only in when the best individual is not optimized in a certain generation to take effect.

4.2   Negative Feedback Operation

With the evolution of the population, polarization phenomenon occurs in probability matrix. It is easy to lose the information which lead to reduce of the global search ability.

Compare the probability value with the learning rate, if two adjacent cities' probability is closes to ∂ /L, reduces the learning rate, and by the same token, when one of the two adjacency cities probability is very small, increases the learning rate.

In order to keep the population diversity, this article also limits the scope of probability, and the probability controls within a certain range, so the probability is controlled under a certain scope.

5   Experiments

In order to verify the effectiveness of the algorithm, we code the procedure on Visual C++ 2012 on Lenovo G480 processor and use the well-known data sets TSPLIB for testing.

As a result of the algorithm running time limit, this paper set up an algorithm obtained that when the best solution provided by TSPLIB or run to maximum the algebra operation stop. All of the following tests' largest number of iterations is set to 5000 generations. ∂ =0.15. The Minimum probability is 0.0001*∂ /L. The size of initial population is 100. Advantage of population's size is 30. The largest probability is 0.9999*∂ /L, if the optimal value is not evolution for 30generations, local search strategy and variation operation will be take effect.

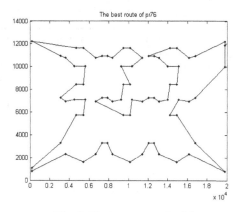

Fig. 3. The best route of pr76

HEDA find the best route of   pr76, which was shown in fig3.

Make gr24 as the test example, and the algorithm (HEDA) runs 20 times, the results was shown in table 1. In order to verify the effectiveness of the algorithm, the algorithm is compared with other algorithms in this paper, which include EHBSA algorithm, OX algorithm and eER algorithm from the literature 14; the test data of algorithm (MMIC) and improved algorithm (IMMIC) from the literature 17. The simulation results of the gr48 were shown in table 2. This algorithm is same to gr24 problem in Settings. Table 3 shows the results of the algorithm to solve the some TSP problems, and every case was run 20 times.

In the three tables, Scale means the scale of initial population, Times means the times of simulation run. Best means the best solution provided by TSPLIB. We ran the program 20 times, took the average and the best, Op is the best, Average is the average. Algebra means the number of iterations while Op is equal to Best. The pr136 case' best route is not found

Table 1. gr24 test results

Algorithm	Scale	Average	Times
EHBSA/WO	60	1281	10
OX	60	1345	10
eER	60	1299	10
MMIC	2000	1284	20
IMMIC	300	1273.3	20
HEDA	100	1272	20

Table 2. gr48 test results

Algorithm	Scale	Average	Times
EHBSA/WO	60	5212	10
OX	60	5527	10
eER	60	5653	10
MMIC	10000	5234.8	20
IMMIC	1000	5065.4	20
HEDA	100	5051	20

Table 3. The TSP test results

The case	Best	Op	Average	Algebra
Burm14	3323	3323	3324.4	100
Gr24	1272	1272	1272	190
Gr48	5046	5046	5051	466
Pr76	108159	108159	108880	1000
Pr136	96772	9700	9720	—

Through the test data, we can get the following conclusion:

(i) Table 1 and table 2 show that the algorithm effective is very good in solving gr24 gr48 problem. The Average is better than other algorithms, which suggest that HEDA has the higher searching ability.

(ii) The algebra to obtain best route can be seen from table 3. We can see that algorithm has a high search speed, less number of iterations. This suggests that the neighborhood strategy can be very good to speed up the search speed and improve the quality of evolution.

(iii) According to the three table above, we can see that most of the testing example' best solution can be found and can maintain the stability. This suggests that the algorithm has high global search ability.

Conclusions

In this paper, an effective estimation of distribution algorithm based on Bayes probability model is proposed for solving the TSP problem. For local exploitation, three effective Variable Neighborhood Search (VNS) were incorporated into EDA. For global exploration, we design mutation operation and negative feedback operation to improve the diversity of population. The simulation results and comparisons with some existing algorithms demonstrated the effectiveness of the HEDA. The future work is to design EDA based algorithm for the massive TSP problem.

References

1. Xu, C., Chang, H.-Y., Xu, J.: Novel Ant Colony Optimization Algorithm with Estimation of Distribution. Computer Science 32 (2010)
2. Liang, Q.J., Shu, J., Fan, X.: TSP Modeling Method Based on Genetic Algorithm. Computer Engineering 37 (2011), Wang, D., Wu, X.-B., Mao, X.-C.: Improved Hybrid Particle Swarm Optimization Algorithm for Solving TSP. Computer Engineering 34 (2008)
3. Hu, Z., Zhao, M.: Simulation on Traveling Salesman Problem (TSP) Based on Artificial Bees Colony Algorithm. Transactions of Beijing Institute of Technology 29 (2009)
4. Huang, B.-Z., Xiao, J.: Solving traveling salesman problem with improved MIMIC algorithm. Computer Engineering and Design 31 (2010)
5. Zhou, S.-D., Sun, Z.-Q.: A Survey on Estimation of Distribution Algorithms. Acta Automatica Sinica (2007)
6. Simionescu, P.A., Beale, D.G., Dozier, G.V.: Teeth-number synthesis of a multispeed planetary transmission using an estimation of distribution algorithm. Journal of Mechanical Design 128 (2006)
7. Joaquin, R., Roberto, S.: Improving the discovery component of classifier systems by the application of estimation of distribution algorithms. In: Proceedings of Students Sessions, ACAI 1999, Chania, Greece (1999)
8. Wang, L., Wang, S., Xu, Y., et al.: A bi-population based estimation of distribution algorithm for the flexible job-shop scheduling problem. Computers & Industrial Engineering 62 (2012)
9. Wang, L., Wang, S.-Y., Xu, Y.: An effective hybrid EDA-based algorithm for solving multidimensional knapsack problem. Expert Systems with Applications 39 (2012)
10. Wang, X., Li, Y.-X.: A Solution to Traveling Salesman Problem by Using Local Evolutionary Algorithm. Computer Engineering 32 (2006)
11. He, X. J., Zeng, J.-C.: Solving TSP Problems with Estimation of Distribution Algorithm Based on Superiority Pattern Junction. PTA&AT 24 (2011)
12. Sheng, J., Xie, S.-Q.: Probability and mathematical statistics. Higher Education Press, BeiJing (2008)
13. He, X.-J., Zeng, J.-C.: Solving flexible job-shop scheduling problems with Bayesian statistical inference-based estimation of distribution algorithm. Systems Engineering Theory & Practice 32 (2012)
14. Hauschild, M., Pelikan, M.: An introduction and survey of estimation of distribution algorithms. Swarm and Evolutionary Computation 1 (2011)
15. Hao, C.-W., Gao, H.-M.: Modified Decimal MIMIC Algorithm for TSP. Computer Science 39 (2012)
16. Xiao-Juan, H., Jian-Chao, Z.: Solving flexible job-shop scheduling problems with Bayesian statistical inference-based estimation of distribution algorithm. Systems Engineering Theory & Practice 32, 380–388 (2012)
17. Hauschild, M., Pelikan, M.: An introduction and survey of estimation of dis-tribution algorithms. Swarm and Evolutionary Computation 1, 111–128 (2011)
18. Cheng-Wei, H., Hui-Min, G.: Modified Decimal MIMIC Algorithm for TSP. Computer Science 39, 233–236 (2012)

A Path Planning Algorithm Based on Parallel Particle Swarm Optimization

Weitao Dang[1], Kai Xu[1], Quanjun Yin[1], and Qixin Zhang[2]

[1] College of Information System and Management
National Univ. of Defense Technology
Changsha, 410073, China
dangwt08@163.com
[2] Nanchang Military Academy
Nanchang, 330103, China

Abstract. A novel path planning algorithm based on Parallel Particle Swarm Optimization (PSO) is proposed in this paper to solve the real-time path planning problem in dynamic multi-agent environment. This paper first describes the advantages of PSO algorithm in real time search problems, i.e. path finding problems. Then considering the development trend of multiprocessors, the parallel PSO (PPSO) was proposed to speed up the search process. Due to the above mentioned advantages, we in this paper adopt the PPSO to distribute particles onto different processors. By exchanging data upon the shared memory, these processors collaborate to work out optimal paths in complicated environment. The resulting simulation experiments show that when compared with traditional PSO, using PPSO could considerably reduce the searching time of path finding in multi-agent environment.

Keywords: Path Planning, Particle Swarm Optimization, Parallelism, Multiprocessor, Multi-agent environment.

1 Introduction

Path planning has become an unelectable issue in multiple domains, including computer games, robotics and simulation. The algorithm of path planning greatly limits the scale and development of training simulation and computer games which have relatively high real-time requirements. Researches from different directions have been done to enhance the efficiency of path planning.

A*[1] is one of the most widely used algorithms for path planning. It is a kind of best-first heuristic search method, which guarantees optimal path return when the heuristic is admissible. A* could always find the optimal path by expanding lots of nodes. However, it is time consuming and unpractical in the real-time applications.

Finding an optimal path is normally time consuming, thus people work on finding a sub-optimal path to decrease the computation time to an acceptable level. Many optimal algorithms which based on Evolutionary Computation (i.e., Genetic Algorithms (GA) and Particle Swarm Optimization (PSO)) are proposed in path

D.-S. Huang et al. (Eds.): ICIC 2014, LNCS 8588, pp. 82–90, 2014.

planning. These algorithms can fleetly find a feasible path via stochastic searches, but still needs exponential time for finding an optimal one.

With the development of multi-core and relevant techniques, more and more attentions have been paid to the idea of parallel computing, which can extremely decrease the computing time across multiprocessors working collaboratively. Parallel computing have been quickly developed and widely applied. The former fundamental path planning algorithm associated with parallelism is based on A*, such as Parallel Bidirectional Search (PBS)[2], Parallel Retracting A* (PRA*)[3], Parallel Local A* (PLA*)[4] and so on.

Although these works could improve planning efficiency, there is too much synchronization overhead in the parallelism algorithms based on A*, which leads to the fact that efficiency couldn't improve linearly with the increase of processors. The overhead couldn't be eliminated because A* has to maintain an Open and a Close list and access the Open and Close List continually during the whole search process. Therefore, the synchronization is virulent for the improvement of the efficiency across parallelism.

PSO[5] is an evolutionary computation technique developed by Dr. Eberhart and Dr. Kennedy in 1995, inspired by social behaviors of bird flocking or fish schooling. PSO optimizes an objective function by conducting population-based search. The population consists of potential solutions, called particles, similar to birds in a flock. The particles are randomly initialized and then freely fly across the multi-dimensional search space. While flying, every particle updates its velocity and position based on its own best experience and that of the entire population.

Compared to A*, there is no data exchange between the particles of PSO, the only shared data is the best position of the entire particles. There will be little synchronization overhead to maintain the position by all the particles in the parallelism of PSO[6-8]. Because the computation of every particle is the same, we can distribute the particles among the cores averagely to obtain load balance. The characteristics of PSO make it suit for parallelism, and the speedup will increase along with the number of processors which is hardly for other algorithm like A*.

The organization of this paper is as follows. Firstly, we give a brief introduction about the PSO and how to apply it to path planning. Then we discussed the parallelism of the algorithm. Next, the implementation of the parallelism of PSO is presented and compared with the serial PSO to analyze about the optimization, speedup and the scalability of the parallel algorithm.

2   PSO and Path Planning

2.1   PSO

Particle Swarm Optimization is based on the social behavior that a population of individuals adapts to its environment by returning to promising regions that were previously discovered. This adaptation to the environment is a stochastic process that depends on both the memory of each individual as well as the knowledge gained by the population as a whole[9].

Each particle in the PSO represents a solution which has a function to estimate its fitness value, and the solution with the less fitness value is closer to the optimal solution. Each particle is initialized with a random position as well as the velocity. The algorithm then repeatedly updates the particles' position.

The position of each particle at iteration k+1 is calculated from Equation (1):

$$x_{k+1}^i = x_k^i + v_{k+1}^i \Delta t \tag{1}$$

where x_{k+1}^i is the position of particle i at iteration k+1, v_{k+1}^i is the corresponding velocity, and Δt is the time step of an iteration.

The velocity of the particle is calculated from Equation (2):

$$v_{k+1}^i = \omega v_k^i + c_1 r_1 \frac{(p^i - x_t^i)}{\Delta t} + c_2 r_2 \frac{(p_k^g - x_t^i)}{\Delta t} \tag{2}$$

Here, p^i is the best experience of particle i, where p_k^g is the best position amongst all the particles at iteration k. r_1 and r_2 are random values between 0 and 1. c_1 and c_2 are constant values, representing the degree of how v_{k+1}^i affected by p^i and p_k^g respectively. ω is the inertia coefficient of the particle, with a larger value facilitating a more global search and a smaller value facilitating a more local search.

2.2   Path Planning Using PSO

Path planning is a search problem in a map from start location to the goal location. Path planning using PSO[10] means searching a path without crossing the obstacles. Based on this, the path should be as far as possible optimal, namely to be as shorter or less cost.

Fig. 1 shows the path planning environment in which the black polygons represent the obstacles, and the point S is the start location while point G is the target location. Fig. 2 explains how to search a path using PSO.

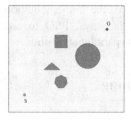

Fig. 1. The environment for path planning

To the path planning problem, one m-dimensions particle which consists of m points represents a path from the start point S to the goal point G. The fitness of the particle can be evaluated by the length of the path.

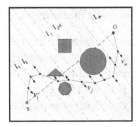

Fig. 2. Path Planning using PSO

As Fig. 2 shows, the segment $\overrightarrow{SG}$ is divided into m+1 parts. The search space of these dimensions is constrained by the m vertical lines (i.e. l_1 to l_m) respectively. Then connect S to the first point and G to the last point, we can obtain the path from S to G(the solid line in Fig. 2). The particle i for the iteration k can be represented as:

$$x_k^i = (x_k^i(1), x_k^i(2), \ldots, x_k^i(m)) \tag{3}$$

$x_k^i(1)$ to $x_k^i(m)$ in Equation (3) is the point on the vertical lines l_1 to l_m respectively.

The velocity of the particle i for the iteration k can be represented as:

$$v_k^i = (k_1 * \overline{e_1}, k_2 * \overline{e_2}, \ldots, k_m * \overline{e_m}) \tag{4}$$

$\overline{e_1}$ to $\overline{e_m}$ is the unit vector of the lines l_1 to l_m which indicate the direction of the velocity and k_1 to k_m is the coefficient to indicate the value. This makes every dimension of particles can only move on the vertical lines.

Based on above mentioned preconditions, the search process can be described as follow:

First, the process generates positions and velocities of the particles randomly. Then it iterates the updating process using Equation (1) and Equation (2). At every iteration, the best experience of every particle and that of all particles (p^i and p_k^g in Equation (2)) are also updated bases on the fitness value. The fitness of a particle is calculated by the evaluation function as Equation (5) :

$$f(x_k^i) = L(x_k^i) + k * L'(x_k^i) \tag{5}$$

The evaluation function consists of two parts: the length of path which the particle represented and the additional castigatory value. $L(x_k^i)$ in Equation (5) is the length of path which the particle represented, and $L'(x_k^i)$ is the total length of the path cross the obstacles, k is the castigatory coefficient.

When the terminate condition has been satisfied, the iteration process will be finished and the best experience of all particles generates the final selected path.

2.3   Parallelism of the PSO for Path Planning

The Parallelism of the PSO mainly distributes the particles to multi-cores so as to improve the efficiency of the algorithm. The particles in multiprocessors search simultaneously and they maintain a global best path together. For each iteration step, all of the particles compare their path with the global one. If one of the local paths is better, the global one will be updated. The detail steps of the algorithm are as follows:

- Step 1: coordinate transformation

For the convenience of computation, we need a coordinate transformation to transform the segment of S to G into X axis. Origin of the new coordinate is start point. Fig. 3 shows the transformed coordinate of Fig. 2.

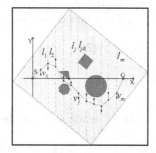

Fig. 3. The transformed coordinate

- Step 2: calculate the search space

The range of value in each dimension is calculated. The segment $\overline{SG}$ is divided into m+1 equal parts. Then the vertical lines (i.e. l_1 to l_m) which pass the dividing points constraints the range of possible values for these dimensions.

- Step 3: initialization

Initialize position and velocity for each particle. Generate the position and velocity for all dimensions of each particle. Then calculate the fitness value of the particles. Choose the particle with the least fitness as global best and regard the initial position of each particle as their own local best, then start iterate.

- Step 4: parallel search

Distribute particles to multiprocessors averagely and search parallel. Each iterate process contains several parts: calculate the velocity using the Equation (2), update the positions using the Equation (1) and evaluate the fitness and select the local best and global best at last.

- Step 5: terminate planning process

Detect if terminating conditions are satisfied. Had been satisfied, transform the global best in new coordinate to the original coordinate and the result is the path which has been found.

The process of the algorithm could be described with the following pseudo code:

```
Procedure Parallel_Path_Palnning(Map, startPnt, endPnt)
    PSOInit()
    {
      CoordinateConversion( startPnt, endPnt )
      CalculateYRange()
      InitParticlePos()
      InitParticleVel()
      InitPBest()
      InitGBest()
    }
    ParallelPSO( CoreNum )
    {
      for i = 1 to CoreNum
        BeginThread(i)
      end
      For every Thread
        while( isRunning )
          CalculateVel()
          UpdatePos()
          AssessParticleFitness()
          UpdatePBest()
          UpdateGBest()
        end
    }
```

3   Implementation and Experimental Results

This section presents the results of the parallel PSO path planning algorithm described in section 2 and compared it with the serial algorithm for performance analysis. The test system used for the experiments is a 12-core Intel E5620(2.40 GHz with 12 MB Cache) processor, 16 GB RAM and Windows Server 32-bit OS.

We implement the algorithm with Visual C++, regard the whole window frame as the map, one pixel in the window represents one meter in reality. Polygons in the window simulate obstacles. We build a map with fixed width and height, set the start location and target location and add different number of obstacles to the map so as to simulate the environment with different complexities.

The parameters of functions for experiments are set as follow: the maximum number of iterations was 300, the value of c_1 and c_2 in Equation (2) are both set to be 2, ω in Equation (2) decrease from 0.9 to 0.4 with the function:

$$\omega = \omega_{max} - (\omega_{max} - \omega_{min}) * i / n$$

in which ω_{max} is 0.9 and ω_{min} is 0.4, n is the max number of iterations while i is the current number of iterations. The number of dimensions is set to be 20 and the number of particles is set to be 40.

We will compare the parallel and serial algorithm in several aspects to evaluate performance of the parallel algorithm based on the parameters above.

3.1   Length of Path

Firstly, we compare the average length of the resulting paths obtained by our algorithm and the serial one respectively. As shown in Fig. 4, the number of the obstacles located randomly in the map ranged from 3 to 30. For each scenario both the two algorithms have been executed for 20 times, taking the average length as the result. X axis is the number of obstacles and Y axis is length of the path found by the two algorithms respectively. The blue line represents the length of parallel algorithm while the red is the serial one.

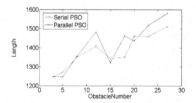

Fig. 4. Average length of the two algorithms

From the line chart, we could see that the average lengths of the resulting paths are almost the same. As a whole, the paths found by parallel are a little longer than which found by serial algorithm. But the longer is less than 5%.

3.2   Speedup

We research on the efficiency by comparing the speedup of parallel algorithm. Fig. 5 shows the searching time of two algorithms in each scenario. The red line represents the time spent for searching using serial algorithm while the blue one represents the time using parallel algorithm implementing on two threads. X axis is the number of obstacles and Y axis is the time spent for searching.

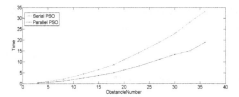

Fig. 5. The speedup of parallel algorithm with two threads

To have an intuitionistic understand of the speedup, Table 1 shows the speedup of the parallel algorithm implement with two threads. Speedup represents the time ratio of time consuming of serial algorithm and that of parallel algorithm.

Table 1. The speedup of the parallel algorithm

Obstacle	5	8	12	15	20	23	27	33	36
Time ratio	1.70	1.73	1.69	1.74	1.74	1.76	1.70	1.80	1.75

3.3   Scalability

We implement the parallel with different number of threads to study the scalability. The experiments are implemented repeatedly in different scenarios. Fig. 6 is the speedup of parallel algorithm with different number of threads. X axis is the number of cores the algorithm implements in, Y axis is the speedup compared with the serial algorithm. The red line represents the speedup of parallel algorithm. The blue dashed is an ideal linear speedup.

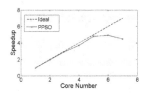

Fig. 6. The speedup of parallel algorithm with different threads

Fig. 6 shows that the parallel algorithm has a near ideal speedup when the number of cores is two or three. While with the increase of the number of cores, the speedup decreases obviously. So much as the speedup of 7 cores is worse than that of 6 cores. Synchronization overhead that all threads have to access the shared memory makes the decrease of speedup with the increase of core number.

3.4   Summary

Three aspects of comparison are presented in this section. The length of path found by parallel PSO algorithm is a little longer than that of the serial one. And the parallel algorithm can obtain a near linear speedup when the number of cores is less than 5. Because of the synchronization overhead, the speedup will be worse with the increase of the number of cores.

4   Conclusions and the Future Work

In this paper, we present the parallelism of PSO used in path planning. General experiments are presented to analyze the parallel PSO algorithm in several aspects, showing that the parallel PSO could obtain a near linear speedup and the final path found by parallel algorithm is only a little longer than that of the serial one. The parallel algorithm has a good scalability while the number of cores should not be large.

We expect to using parallelism in the whole path planning process. Firstly, parallel construct the environment, we plan to partition the map average and generate the waypoint in each area. Secondly, search the key node of path among waypoints. Finally, a refined planning between the key nodes is executed parallel.

References

1. Rich, E.: Artificial intelligence. McGraw-Hill (1983)
2. Brand, S.: Efficient obstacle avoidance using autonomously generated navigation meshes (Doctoral dissertation, MSc Thesis, Delft University of Technology, The Netherlands) (2009)
3. Evett, M., Hendler, J., Mahanti, A., Nau, D.: PRA*: Massively parallel heuristic search. Journal of Parallel and Distributed Computing 25(2), 133–143 (1995)
4. Dutt, S., Mahapatra, N.R.: Parallel A* algorithms and their performance on hypercube multiprocessors. In: Proceedings of Seventh International Parallel Processing Symposium, pp. 797–803. IEEE (1993)
5. Eberhart, R.C., Kennedy, J.: A new optimizer using particle swarm theory. In: Proceedings of the Sixth International Symposium on Micro Machine and Human Science, vol. 1, pp. 39–43 (1995)
6. Venter, G., Sobieszczanski-Sobieski, J.: Parallel particle swarm optimization algorithm accelerated by asynchronous evaluations. Journal of Aerospace Computing, Information, and Communication 3(3), 123–137 (2006)
7. Schutte, J.F., Reinbolt, J.A., Fregly, B.J., Haftka, R.T., George, A.D.: Parallel global optimization with the particle swarm algorithm. International Journal for Numerical Methods in Engineering 61(13), 2296–2315 (2004)
8. Chu, S.C., Roddick, J.F., Pan, J.S.: Parallel particle swarm optimization algorithm with communication strategies. Submitted to IEEE Transactions on Evolutionary Computation (2003)
9. Kennedy, J., Spears, W.M.: Matching algorithms to problems: an experimental test of the particle swarm and some genetic algorithms on the multimodal problem generator. In: Proceedings of the IEEE International Conference on Evolutionary Computation, Anchorage, AK, USA, pp. 78–83 (1998)
10. Raja, P., Pugazhenthi, S.: Path planning for mobile robots in dynamic environments using particle swarm optimization. In: International Conference on Advances in Recent Technologies in Communication and Computing, ARTCom 2009, pp. 401–405. IEEE (2009)

An Optimal Probabilistic Transformation of Belief Functions Based on Artificial Bee Colony Algorithm

Yafei Song, Xiaodan Wang, Lei Lei, and Aijun Xue

School of Air and Missile Defense, Air Force Engineering University,
Xi'an 710051, P.R. China
yafei_song@163.com

Abstract. One of the key issues in the applications of Dempster-Shafer evidence theory is how to make a decision based on Basic Probability Assignment (BPA). Besides the widely known pignistic transformation of belief functions, another conventional method is Plausibility Transformation, which transforms BPA to probability distribution by normalizing plausibility function of every singleton proposition. However, these two methods are stuck with the problem of impreciseness. To overcome this drawback, a new transformation method based on Artificial Bee Colony Algorithm is proposed. The obtained probability has the maximum correlation coefficient with the original BPA. Numerical examples are used to illustrate the efficiency of the proposed method.

Keywords: Belief function, Dempster-Shafer theory, Probabilistic transformation, Correlation Coefficient, Artificial Bee Colony Algorithm.

1   Introduction

Since it was firstly presented by Dempster [1], and was later extended and refined by Shafer [2], the Dempster-Shafer theory of evidence, or the D-S theory for short, has generated considerable interest. Its application has extended to many areas such as expert systems [3], diagnosis and reasoning [4,5], pattern classification [6-8], information fusion [9], knowledge reduction [10], global positioning system [11], regression analysis [12], and data mining [13]. However, for its counterintuitive results in some cases, especially when there is high conflict among bodies of evidence, various alternatives to D-S theory have been presented [14-21].

Belief functions are useful to combine different sources of evidence. However, they are not directly suited to derive decisions. A decision is better justified by probabilities for each elementary proposition out of the discriminant frame. For this reason, it is necessary to transform the resulting evidence representation by a general belief function to manifestation of probability. Such a probability should be consistent with the original belief function. In fact, we can consider it as a belief function of a special type, called Bayesian belief function. We call such a transformation as a probabilistic transformation.

D.-S. Huang et al. (Eds.): ICIC 2014, LNCS 8588, pp. 91–100, 2014.
© Springer International Publishing Switzerland 2014

Pignistic transformation, as a special case of probabilistic transformation, has been widely used. In the last years, several papers on alternative probabilistic transformations have been published [22-24], and a new justification of pignistic transformation has appeared [25, 26]. In this paper, a new probabilistic transformation will be proposed, by the maximizing the correlation coefficient between two belief functions, i.e. Bayesian belief function and the original belief function.

The remaining of this paper is organized as follows. Section 2 recalls Dempster-Shafer theory and a series of probabilistic transformations from various sources. Definition of the correlation coefficient between two belief functions is proposed in Section 3. A new definition of the general probabilistic transformation is presented in Section 4. Section 5 is dedicated to validate the performance of proposed method. In the last section, this paper is concluded by a discussion about which transformation should be applied in applications.

2   D-S Theory and Probabilistic Transformations

Dempster-Shafer theory of evidence was modeled based on a finite set of mutually exclusive elements, called the frame of discernment denoted by Ω. The power set of Ω, denoted by 2^{Ω}, contains all possible unions of the sets in Ω including Ω itself. Singleton sets in a frame of discernment Ω will be called atomic sets because they do not contain nonempty subsets. It is assumed that only one atomic set can be true at any one time. If a set is assumed to be true, then all supersets are considered true as well.

An observer who believes that one or several sets in the power set of Ω might be true can assign belief masses to these sets. Belief mass on an atomic set $A \in 2^{\Omega}$ is interpreted as the belief that the set in question is true. Belief mass on a non-atomic set $A \in 2^{\Omega}$ is interpreted as the belief that one of the atomic sets it contains is true, but that the observer is uncertain about which of them is true. Definitions and terminologies such as frame of discernment and basic probability assignment (BPA) can be found in references [1] and [2].

Beliefs are useful to combine different sources of evidence as described above, but they are not directly suited to derive decisions. A decision is better justified by probabilities for each elementary proposition out of the discernment frame. For this reason, the probabilistic transformation has attracted much attention in past decades.

The pignistic transformation distributes $m(A)$ equally among all elements of A. It was named and justified by Smets for Transferable Belief Model (TBM).

Definition 1 [25,26]. The pignistic transformation maps a belief structure m to so called pignistic probability function. The pignistic transformation of a belief structure m on $\Omega = \{A_1, A_2, \ldots, A_n\}$ is given by

$$BetP(A) = \sum_{B \subseteq \Omega} \frac{|A \cap B|}{|B|} \frac{m(B)}{1 - m(\varnothing)}, \forall A \subseteq \Omega \tag{1}$$

where $|A|$ is the cardinality of set A.

Definition 2. Suppose m is a belief structure on Ω. Let Pl_m and Pl_P_m, respectively, denote the plausibility function and probability function corresponding to m. The plausibility transformation method can be expressed as:

$$Pl\_P_m(A) = \frac{Pl(A)}{\sum_{B\in\Omega} Pl(B)} = \frac{\sum_{A\in X\subseteq\Omega} m(X)}{\sum_{B\in\Omega}\sum_{B\in X\subseteq\Omega} m(X)} \tag{2}$$

This transformation is called "the pignistic probability proportional to normalized plausibility" (Pr*NPl*) by Sudano [24].

3   Correlation Coefficient between BPAs

Jousselme has defined a distance of evidence in [27] as follows:

$$d_{BPA}(m_1, m_2) = \sqrt{\frac{1}{2}(m_1 - m_2)^T \underline{D}(m_1 - m_2)} \tag{3}$$

where m_1 and m_2 are the BPAs , and Jaccard matrix D is a $2^{|\Omega|} \times 2^{|\Omega|}$ matrix, whose elements are:

$$D(A, B) = \frac{|A \cap B|}{|A \cup B|} \tag{4}$$

where $A, B \subseteq \Omega$.

To described the similarity between belief functions, we defined correlation coefficient as

$$cor(m_1, m_2) = \frac{\langle m_1, m_2\rangle}{\|m_1\| \cdot \|m_2\|} \tag{5}$$

In order to take into account the similarity among the subsets of Ω, we take matrix D to modify the BPA:

$$\begin{cases} m_1' = m_1 D \\ m_2' = m_2 D \end{cases} \tag{6}$$

And $cor\ (m_1, m_2)$ is redefined as

$$cor(m_1, m_2) = \frac{\langle m_1', m_2'\rangle}{\|m_1'\| \cdot \|m_2'\|} \tag{7}$$

Apparently, $cor\ (m_1, m_2)$ satisfies the following properties:

- $0 \le cor\ (m_1, m_2) \le 1$;
- $cor\ (m_1, m_2) = cor\ (m_2, m_1)$;

- $cor\ (m_1, m_2)=1 \Leftrightarrow m_1=m_2;$
- $cor\ (m_1, m_2) = 0 \Leftrightarrow (\cup A_i) \cap (\cup B_j)=\varnothing,\ A_i,\ B_j$ are focal elements for $m_1,\ m_2$ respectively.

Next example will show its rationality to describe the conflict between two BPAs.

Example 1. We suppose the discriminant frame $\Omega=\{A, B, C\}$, four BPAs are defined as follows :

$$m_1 : m_1(A) = 0.99, m_1(B) = 0.01, m_1(C) = 0$$

$$m_2 : m_2(A) = 0.99,\ m_2(B) = 0, m_2(C) = 0.01$$

$$m_3 : m_3(A) = 0.8,\ m_3(B) = 0.1, m_3(C) = 0.1$$

$$m_4 : m_4(A) = 0.6,\ m_4(B) = 0.2, m_4(C) = 0.2$$

The correlation coefficient between m_1 and m_2, m_1 and m_3, m_1 and m_4, m_3 and m_4 are:

$$cor(m_1, m_2) = 0.999,\ \ cor(m_1, m_3) = 0.986$$

$$cor(m_1, m_4) = 0.908,\ \ cor(m_3, m_4) = 0.965$$

Intuitively, m_1 and m_2 is closer than m_1 and m_3, m_1 and m_4. The similarity between m_3 and m_4 should be higher than that between m_1 and m_4. We can get the conclusion that the correlation coefficients can reflect the similarity between evidences well. Such conclusion prompts us to take correlation coefficient as a quantitative measure of conflict between two belief functions.

4   An Optimal Probability Transformation

4.1   Proposed Transformation

For a BPA m, we can get many probability distributions $P(x)$ by different methods, whenever:

$$\begin{cases} Bel(x) \le P(x) \le Pl(x) \\ \sum_{x \in \Omega} P(x) = 1 \end{cases} \tag{8}$$

It was described in [28] that we need to select an optimized probability distribution from all the consistent alternatives according to certain criterion and use the corresponding probability distribution as the results of probability transformation. For the probability is get from the BPA, it must respect the original information in BPA. So, if we take probability as a special BPA, the distance between them should be the minimized.

The conflict degree k was taken as optimal object to choose the optimized probability transformation [28]. It is well known that not all k indicate a conflict that can be satisfactorily interpreted qualitatively. So we need find new parameter to optimize. Xu Peida proposed an optimized transformation based information entropy

[29], but the relation between entropy and similarity of evidence is not so clear.

Here we choose $cor(m_1, m_2)$ to be the distance between two evidences. The original BPA is denoted as m_1, and the probability after transformation is m_2. Then, the problem of probability transformation of belief can be regarded as an optimization problem as following:

$$\max\{cor(m_1, m_2)\}$$
$$s.t\begin{cases} Bel(x) \le P(x) \le Pl(x) \\ \sum_{x \in \Omega} P(x) = 1 \end{cases} \tag{9}$$

This is a typical nonlinear programming problem. There are many optimization algorithms for solving such problem, such as Genetic Algorithm (GA), Differential Evolution Algorithm (DE), Particle Swarm Optimization (PSO) [27], Ant colony optimization (ACO) [30], artificial bee colony (ABC) algorithm [31-33], etc. In our simulation, the artificial bee colony algorithm is used to solve the optimization problem.

Artificial Bee Colony (ABC) algorithm was proposed for optimization based on the intelligent foraging behavior of honey bee swarm's behaviors [31-33]. Therefore, ABC is more successful and most robust on multimodal functions included within the set with respect to DE, PSO, and GA. The ABC algorithm provides a solution in the organized form by dividing the bee objects into different tasks such as employed, onlooker, and scout bees. These three bees/tasks determine the objects of problems by sharing information to other's bees. Fig. 1shows the flow chart of the ABC algorithm.

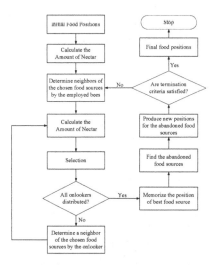

Fig. 1. Flowchart of the Artificial Bee Colony algorithm

4.2   Probability Transformation Based on ABC

In this section, we apply ABC algorithm to the optimization of correlation coefficient for probability transformation. In ABC algorithm, the position of a food source represents a possible solution to the optimization problem and the nectar amount of a food source corresponds to the quality (fitness) of the associated solution. The number of employed bees is equal to the number of solutions since each employed bee is associated with one and only one solution. To apply ABC algorithm, the considered optimization problem is first converted to the problem of finding the best parameter vector which maximizes a fitness function. The artificial bees randomly discover a population of initial solution vectors and then iteratively improve them by employing the strategies: moving towards better solutions by means of a neighborhood search mechanism while abandoning poor solutions.

Suppose the population size of the artificial bee colony is SN, the population size of employed bees and onlooker bees are L_e and L_o (we assume $L_e = L_o$). The searching dimension equals to the number of elements in discriminant frame N, where $N = |\Omega|$. And the searching range is $[X_{min}, X_{max}]$, where $X_{min} = [Bel(x_1), Bel(x_2), ..., Bel(x_N)]$ and $X_{max} = [Pl(x_1), Pl(x_2), ..., Pl(x_N)]$ are the belief function and plausibility function of the evidence, respectively. Each food source X_i^N is an N-dimensional solution vector to the optimization problem which is to maximize the correlation coefficient.

The code of the ABC algorithm proposed to optimize the probability transformation is given below:

```
input original BPA m₁;
Calculate the Jaccard matrix D;
Get the modified BPA m´₁= m₁D;
Set m₂ =[P(x₁), P(x₂,),…,P(xₙ)]ᵀ=X;
Initialize the population of solutions xᵢⱼ, i=1,2,…,SN,
j=1,2,…,N, set modification rate (MR) as a control
parameter;
Evaluate the population;
cycle = 1;
Repeat;
Produce a new solution vᵢ for each employed bee by using
Eq. (15) and evaluate it;
Apply a greedy selection process between vᵢ and xᵢ;
Calculate the probability values pᵢ for the solutions
using correlation coefficient as fitness;
For each onlooker bee, produce a new solution vᵢ by Eq.
(16) in the neighborhood of the solution selected
depending on pᵢ and evaluate it;
Apply a greedy selection process between vᵢ and xᵢ;
If Scout Production Period (SPP ) is completed, determine
the abandoned solutions by using "limit" parameter for
the scout, if it exists, replace it with a new randomly
produced solution by Eq.(17);
```

```
Memorize the best solution achieved so far;
cycle = cycle+1;
until cycle = MCN;
end.
```

5   Simulation Results

To calculate the performance of the proposed method, two numerical examples are provided to illustrate the proposed method, comparing with the pignistic probability transformation and plausibility transformation.

Example 2   Suppose frame of discriminant $\Omega=\{A, B, C\}$, the corresponding BPA is as follows:

$m(a) = 0.2, m(b) = 0.1, m(c) = 0.1, m(a,b) = 0.3, m(b,c) = 0.3$

We can get the belief function for each singleton is as:
$Bel(a) = 0.2, Bel(b) = 0.1, Bel(c) = 0.1$

And the plausibility distribution is
$Pl(a) = 0.5, Pl(b) = 0.7, Pl(c) = 0.4$

The obtained probability interval is:
$p(a) \in [0.2, 0.5]$, $p(b) \in [0.1, 0.7]$, $p(c) \in [0.1, 0.4]$

The probability distribution obtained by different methods is shown in Table 1. For the original BPA, intuitively, we can know that proposition $\{b\}$ should has the largest probability value and the proposition $\{c\}$ has the least. Comparing the results of three methods, the proposed method gets the correct and clearer result for decision making and show the focus property.

Table 1. Probability distribution for Example 1

Method	Probability distribution
Pignistic Transformation	$P(a)$=0.35, $P(b)$=0.4, $P(c)$=0.25
Plausibility Transformation	$P(a)$=0.31, $P(b)$=0.44, $P(c)$=0.25
Our proposed Transformation	$P(a)$=0.32, $P(b)$=0.45, $P(c)$=0.23

Example 3. In a target identification system, we need to recognize a group of targets. Assume that there are four targets, the information about them is too limited to recognize them. We must identify them by fusing results of multi sensors. Denote the four targets as A, B, C, and D. Then the frame of discriminant is: $\Omega=\{A, B, C, D\}$.The identification result is given in the form of belief structure as:

$m(A) = 0.2, m(B) = 0.1, m(D) = 0.2, m(AC) = 0.2, m(ABC) = 0.3$

The belief function and plausibility function for each singleton are, respectively, as:

$Bel(A) = 0.2, Bel(B) = 0.1, Bel(C) = 0, Bel(D) = 0.2$
$Pl(A) = 0.7, Pl(B) = 0.4, Pl(C) = 0.5, Pl(D) = 0.2$

The obtained probability distributions calculated by different methods are shown in Table 2. We can get $p(D)=0.2$ effortlessly from $Bel(D) \leq p(D) \leq Pl(D)$. The pignistic transformation and our proposed transformation can get $p(D)=0.2$, coinciding with intuitive analysis. Comparing the results of two methods, the result of pignistic transformation is not able to get a clear decision. Using plausibility transformation, we can get the probability order $p(A)> p(C)> p(B)> p(D)$. While the probability order from our proposed transformation is $p(A)> p(C)> p(B)> p(D)$, where $\{A\}$ get the largest probability. So the decision maker has enough confidence to classify the target to A. Moreover, the order of $p(A)> p(C)> p(D)> p(B)$ coincides with the intuition of human beings.

Table 2. Probability distribution for Example 2

Method	Probability distribution
Pignistic Transformation	$P(A)=0.4, P(B)=0.2, P(C)=0.2, P(D)=0.2$
Plausibility Transformation	$P(A)=0.39, P(B)=0.22, P(C)=0.28, P(D)=0.11$
Proposed Transformation	$P(A)=0.43, P(B)=0.14, P(C)=0.23, P(D)=0.2$

6   Conclusions

Decision making in D-S framework is an important step in the applications of evidence theory. After analyzing two usually used probability transformations, called pignistic probability transformation and plausibility transformation, we propose a novel probability transformation method using ABC algorithm to optimize the correlation coefficient. Experimental results show the efficiency of the proposed method. However, decision making is crucial significant in practical applications, especially in military application. So decision makers must be even cautious in practice.   In most cases, the priori knowledge is too limited to make a proper decision. We should make uncertain or fuzzy decision for further analysis by additional techniques, rather than making crisp decision which may cause severe damage. We must note the fact that none of the probabilistic transformations is appropriate for all situations. Therefore, the choose of probabilistic transformation is still an open problem need further research.

Acknowledgments. This work is supported by National Science Foundation of China under grants 61273275 and 60975026.

References

1. Dempster, A.P.: Upper and Lower Probabilities Induced by a Multivalued Mapping. The Annals of Mathematical Statistics 38(2), 325–339 (1967)
2. Shafer, G.: A Mathematical Theory of Evidence. Princeton U. P., Princeton (1976)

3. Beynon, M., Cosker, D., Marshall, D.: An expert system for multi-criteria decision making using Dempster-Shafer theory. Expert Systems with Applications 20, 357–367 (2001)
4. Benferhat, S., Saffiotti, A., Smets, P.: Belief functions and default reasoning. Artificial Intelligence 122, 1–69 (2000)
5. Jones, R.W., Lowe, A., Harrison, M.J.: A framework for intelligent medical diagnosis using the theory of evidence. Knowledge-Based Systems 15, 77–84 (2002)
6. Liu, Z.G., Pan, Q., Dezert, J.: Evidential classifier for imprecise data based on belief functions. Knowledge-Based Systems 52, 246–257 (2013)
7. Reformat, M., Yager, R.R.: Building ensemble classifiers using belief functions and OWA operators. Soft Comput. 12, 543–558 (2008)
8. Laha, A., Pal, N.R., Das, J.: Land cover classification using fuzzy rules and aggregation of contextual information through evidence theory. IEEE Transactions on Geoscience and Remote Sensing 44(6), 1633–1641 (2006)
9. Telmoudi, A., Chakhar, S.: Data fusion application from evidential databases as a support for decision making. Information and Software Technology 46(8), 547–555 (2004)
10. Wu, W.Z., Zhang, M., Li, H.Z.: Knowledge reduction in random information systems via Dempster-Shafer theory of evidence. Information Sciences 174, 143–164 (2005)
11. Aggarwal, P., Bhatt, D., Devabhaktuni, V.: Dempster Shafer neural network algorithm for land vehicle navigation application. Information Sciences 253, 26–33 (2013)
12. Su, Z.G., Wang, Y.F., Wang, P.H.: Parametric regression analysis of imprecise and uncertain data in the fuzzy belief function framework. International Journal of Approximate Reasoning 54, 1217–1242 (2013)
13. Denoeux, T.: Maximum likelihood estimation from uncertain data in the belief function framework. IEEE Transactions on Knowledge and Data Engineering 25(1), 119–130 (2013)
14. Yang, Y., Han, D.Q., Han, C.Z.: Discounted combination of unreliable evidence using degree of disagreement. International Journal of Approximate Reasoning 54, 1197–1216 (2013)
15. Jousselme, A.-L., Maupin, P.: Distances in evidence theory: Comprehensive survey and generalizations. International Journal of Approximate Reasoning 53, 118–145 (2012)
16. Lefevre, E., Colot, O., Vannoorenberghe, P.: Belief functions combination and conflict management. Information Fusion 3(2), 149–162 (2002)
17. Florea, M.C., Jousselme, A.-L., Bosse, E.: Robust combination rules for evidence theory. Information Fusion 10(2), 183–197 (2009)
18. Schubert, J.: Conflict management in Dempster-Shafer theory using the degree of falsity. International Journal of Approximate Reasoning 52(3), 449–460 (2011)
19. Baroni, P., Vicig, P.: Transformations from Imprecise to Precise Probabilities. In: Nielsen, T.D., Zhang, N.L. (eds.) ECSQARU 2003. LNCS (LNAI), vol. 2711, pp. 37–49. Springer, Heidelberg (2003)
20. Cobb, B.R., Shenoy, P.P.: A Comparison of Methods for Transforming Belief Function Models to Probability Models. In: Nielsen, T.D., Zhang, N.L. (eds.) ECSQARU 2003. LNCS (LNAI), vol. 2711, pp. 255–266. Springer, Heidelberg (2003)
21. Daniel, M.: Transformations of Belief Functions to Probabilities. In: Vejnarova, J. (ed.) Proceedings of 6th Workshop on Uncertainty Processing (WUPES 2003), pp. 77–90. V SE Oeconomica Publishers (2003)
22. Daniel, M.: Consistency of Probabilistic Transformations of Belief Functions. In: Proceedings of the Tenth International Conference IPMU, pp. 1135–1142 (2004)

23. Sudano, J.J.: Pignistic Probability transforms for Mixes of Low and High Probability Events. In: Proc. of the 4th Int. Conf. on Information Fusion (Fusion 2001), Montreal, Canada, TUB3, pp. 23–27 (2001)

24. Sudano, J.J.: Equivalence Between Belief Theories and Naive Bayesian Fusion for Systems with Independent Evidential Data: Part II, the Example. In: Proc. of the 6th Int. Conf. on Information Fusion (Fusion 2003), Cairns, Australia, pp. 1357–1364 (2003)

25. Smets, P.: Decision Making in a Context where Uncertainty is Represented by Belief Functions. In: Srivastava, R.P., Mock, T.J. (eds.) Belief Functions in Business Decision, pp. 17–61. Physica-Verlag, Heidelberg (2002)

26. Smets, P.: Decision Making in the TBM: the Necessity of the Pignistic Transformation. International Journal of Approximative Reasoning 38, 133–147 (2005)

27. Jousselme, A.-L., Grenier, D., Bosse, E.: A new distance between two bodies of evidence. Information Fusion 2, 91–101 (2001)

28. Xu, P.D., Wu, J.Y., Li, Y.: A method for translating belief function models to probability models. Journal of Information and Computational Science 8, 1817–1823 (2011)

29. Xu, P.D., Han, D.Q., Deng, Y.: An optimal transformation of basic probability assignment to probability. Acta Electronica Sinica 39, 121–125 (2011)

30. Dorigo, M., Maniezzo, V., Colorni, A.: Ant system: Optimization by a colony of cooperating agents. IEEE Transactions on Systems, Man and Cybernetics 26, 29–41 (1996)

31. Karaboga, D.: An idea based on honey bee swarm for numerical optimization. Erciyes University, Engineering Faculty, Computer Engineering Department (2005)

32. Karaboga, D., Basturk, B.: A powerful and efficient algorithm for numerical function optimization: Artificial bee colony (ABC) algorithm. Journal of Global Optimization 39, 459–471 (2007)

33. Karaboga, D., Basturk, B.: A comparative study of artificial bee colony algorithm. Applied Mathematics and Computation 214, 108–132 (2009)

Cost-Sensitive Bayesian Network Classifiers and Their Applications in Rock Burst Prediction

Ganggang Kong, Yun Xia, and Chen Qiu

Department of Computer Science, China University of Geosciences
Wuhan, Hubei, China 430074
kongganggang16@gmail.com

Abstract. Bayesian learning provides a simple but efficient method for classification by combining the sample information with the prior knowledge and the dependencies with probability estimates. However, Bayesian network classifiers that minimize the number of misclassification errors ignore different misclassification costs. For example, in rock burst prediction, the cost of misclassifying a rock which happens to burst as a rock which doesn't burst is much higher than the opposite type of error. This paper studies the cost-sensitive learning and then applies it to different Bayesian Network classifiers, and the resulted algorithms are called cost-sensitive Bayesian Network classifiers. The experimental results on 36 UCI datasets validate their effectiveness in terms of the total misclassification costs. Finally, we apply the cost-sensitive Bayesian Network classifiers to some real-world rock burst prediction examples and achieve good results.

Keywords: cost-sensitive learning, Bayesian network classifiers, misclassification cost, rock burst prediction.

1 Introduction

Bayesian network becomes one of the most popular classification methods due to its the unique uncertainty knowledge expression form, the abundant ability of probability representation and the incremental learning for comprehensive prior knowledge.

Learning a Bayesian classifier is a process of constructing a special Bayesian network from a given set of training instances with class labels. A test instance x is represented by an attribute vector $<a_1, a_2, \cdots, a_n>$, and a_i is the value of the ith attribute A_i. The probability that instance x belongs to class c is calculated as equation (1), using Bayes theorem.

$$P(c \mid a_1, a_2, \cdots, a_n) = \frac{P(a_1, a_2, \cdots, a_n \mid c)P(c)}{P(a_1, a_2, \cdots, a_n)},$$

(1)

where $P(a_1, a_2, \cdots, a_n/c)$ is the probability of instance x $<a_1, a_2, \cdots, a_n>$ given the class c. However, the calculation of it is an NP-hard problem [1] and only correct on large data sets.

D.-S. Huang et al. (Eds.): ICIC 2014, LNCS 8588, pp. 101–112, 2014.

Faced with the NP-hard problem, Naive Bayes classifier (NB) has been a simple but effective method although its unrealistic assumption that all of the attributes are independent given the class. Thus, the probability $P(a_1, a_2, \cdots, a_n/c)$ estimated by Naive Bayes is:

$$P(a_1, a_2, \cdots, a_n \mid c) = \prod_{j=1}^{n} P(a_j \mid c), \qquad (2)$$

where $P(a_j/c)$ denotes the probability of jth attribute value is a_j given the class c.

NB simplifies the class probability estimation by the attribute independence assumption. In order to weaken the effect of its weakness assumption and finally get the target of the minimum number of misclassification errors, researchers are devoted to improving the performance of Naive Bayes via the methods of structure extension, attribute selection, attribute weighted, instance weighted and instance selection, also called local learning [2].

Traditional NB and its improved algorithms are all based on the minimum number of misclassification errors but ignore that the differences between different misclassification errors can be quite large. Actually, misclassification costs on different classes are different in many real-world domains. For example, the cost that classifying an useful email as a spam email is significantly greater than the opposite type of error. In addition, it's meaningless and unrealistic to simply aim at the classifiers with the minimum number of misclassification errors in class-imbalanced data. For example, in medical diagnosis prediction where ninety-nine percent of patients under test are healthy, classifier simply classifies all patient as health and its accuracy will be 99%, in other word its misclassification rate is only 1%. Obviously this classifier model isn't suitable and cant't effectively diagnose patient in medical diagnosis. Thus, compared with the classifiers based on the minimum number of misclassification errors, cost-sensitive learning based on the minimum total misclassification costs is more reasonable.

There are various kinds of cost in classification, such as misclassification cost between class variables, cost for getting data and so on. For example, there will be some economic cost and other cost for getting the comprehensive data in rock burst prediction. Chai et al. (2004) [3] focus on the cost of getting missing data and study the method of minimum misclassification cost by a strategy named test-cost. Elkan C (2001) [4] points out that the correct classification has some benefits. For example, approving a good customer will bring benefits while rejecting a bad customer will avoid some losses in bank credit. Ling et al. (2008) [5] research the theory of misclassification cost, and review current methods of cost-sensitive learning which provides us with a great reference.

Rock burst has received many researchers' increased attention due to its huge catastrophes in under-ground coal mines and hydroelectric tunnels. Traditional rock burst predictions are all depended on the mechanical methods, and very difficult to get accuracy classification because of poor understanding of the mechanism. Recent years, a new method named intelligent rock mechanics [6] has been proposed and provides an effective direction for rock burst prediction.

In this paper, we pay our attention to the misclassification costs between class variables and restrict our discussion to the cost-sensitive learning algorithms of the binary classification without loss of generality. The rest of this paper is organized as follows. In section 2, we review the theory of cost-sensitive learning. In section 3, we incorporate the learning method based on minimizing misclassification costs into Bayesian network classifiers, such as Naive Bayes (NB), tree augmented Bayesian network (TAN) [7], SuperParent (SP) [8], averaged one-dependence estimators (AODE) [9], weighted averaged one-dependence estimators (WAODE) [10] and hidden Naive Bayes (HNB) [11]. In section 4, we describe the experiment setup and analyze the results. In section 5, we apply the coat-sensitive Bayesian network classifiers to the real-world rock burst prediction. In section 6, we draw conclusions.

2   Cost-Sensitive Learning Methods

Misclassification cost-sensitivity denotes that the loss of classifying an instance which has actual class A as class B is different from the opposite type error. Therefore, we should consider the misclassification costs between class variables while constructing classifiers for prediction.

In cost-sensitive learning, the cost matrix $C[k,k]$ denotes the misclassification costs among k class variables of test instances, and $C(i,j)$ is the cost of misclassifying an instance belonging to class j as class i. It's correct when i is equal to j, otherwise it's uncorrect. According to the cost matrix, we classify an instance as the class that has the minimum excepted cost which called the minimum excepted cost principle. The classifier uses equation (3) to calculate the excepted cost $L(i/x)$ of classifying instance x as class i.

$$L(i \mid x) = \sum_{i} P(j \mid x)C(i \mid j), \qquad (3)$$

where $P(j/x)$ is the probability of classifying instance x as class j and $C(i,j)$ is the cost of misclassifying an instance belonging to class j as class i.

A classifier would predict an instance based on the principle of minimum excepted cost in cost-sensitive learning. For example, given a test instance x, $L(i/x)$ represents the cost loss of classifying instance x as class i and a cost-sensitive classifier will classify instance x as a class which has minimum value of $L(i/x)$.

Without loss of generality, our cost-sensitive learning is discussed in binary classification (positive and negative) and we often denote the minority class as positive class and the majority class as negative (we use 0 for positive and 1 for negative). Table 1 shows significant difference between different class variables in binary classification, and $C(0,0)$ is the cost of classifying positive class as positive class, $C(0,1)$ is the cost of classifying negative class as positive class, $C(1,0)$ is the cost of classifying positive class as negative class and $C(1,1)$ is the cost of classifying negative class as negative class.

Table 1. Cost matrix for binary classification

	Actual Positive	Actual Negative
Predict Positive	C(0,0)	C(0,1)
Predict Negative	C(1,0)	C(1,1)

From the cost matrix, We usually set correct classification $C(0,0)$ and $C(1,1)$ to 0 when we don't consider the benefits of correct classification. According to the minimum excepted cost principle described by equation (3), a classifier classifies instance x as positive class only if equation (4) is met. Otherwise, the classifier classifies it as negative class.

$$P(1 \mid x)C(0,1) \leq P(0 \mid x)C(1,0) \tag{4}$$

In cost-sensitive classification, the confusion matrix about binary classification is shown as Table 2. In Table 2, FP denotes the number of false positive, TP denotes the number of true positive, FN denotes the number of false negative, and TN denotes the number of true negative.

Table 2. A confusion matrix for a binary classification problem.

	Actual Positive	Actual Negative
Predict Positive	TP	FP
Predict Negative	FN	TN

Obviously, the criteria such as accuracy or error rate are not suitable in cost-sensitive learning and the criterion for evaluating cost-sensitive classifiers is defined by equation (5), which named as the total misclassification cost [12].

$$Costs = FP \times C(0,1) + FN \times C(1,0), \tag{5}$$

where $C(0,1)$ and $C(1,0)$ is given by Table 1. The cost matrix is currently defined by users or experts [13-15].

3   Cost-Sensitive Bayesian Network Classifiers

The cost-sensitive learning algorithms can be broadly divided into two main categories. one is the direct cost-sensitive learning based on the models such as decision tree, neural networks and SVM etc, the other is the cost-sensitive meta-learning based on the methods such as sampling [16], instance weighted and threshold adjusting etc.

In this paper, we incorporate cost-sensitive learning into Bayesian network classifiers (BNC) to learn cost-sensitive Bayesian network classifiers. First, we construct Bayesian network classifiers on training data and get class membership probabilities of test instances. Then, we classify the test instances based on the minimum excepted cost principle. Now, let us give the detailed description of our algorithm as follows.

Algorithm: A Cost-Sensitive Bayesian Network Classifier
Input: a training data D and a cost matrix $C[k,k]$
    1.construct a Bayesian network classfier (BNC) using D
    2. For each test instance x
       a. estimate its class membership probability $P(c_i/x)$ using the constructed BNC
       b. get the class c_k which has the minimum value of L according to   Equation 4
Output: return c_k as the class label of x

Please note that, the Bayesian network classifier above can be anyone of Bayesian network classifiers. In this paper, we research the classifiers including NB, TAN [7], SP [8], AODE [9], WAODE [10], and HNB [11] and denote the resulted classifiers CSNB, CSSP, CSTAN, CSAODE, CSWAODE and CSHNB respectively. The experimental results in section 4 validate their effectiveness in terms of the total misclassification costs.

4   Experiments and Results

The purpose of this section is to validate the effectiveness of our cost-sensitive Bayesian network classifiers. Therefore, we conduct our experiments to compare our cost-sensitive Bayesian network classifiers with the original Bayesian network classifiers in terms of the total misclassification costs. We ran our experiments on 36 widely used UCI datasets [17] published on the main website of WEKA platform [18], which represent a wide range of domains and data characteristics. In our experiment, missing values are replaced with the modes and means from the available data. Numeric attribute values are discretized using the unsupervised ten-bin discretization implemented in WEKA platform. Besides, we transform each multi-class dataset into binary one by taking the two largest classes and manually delete three useless attributes: the attribute "Hospital Number" in the data set "colic.ORIG", the attribute "instance name" in the data set "splice", and the attribute "animal" in the data set "zoo".

In our experiments, the total misclassification costs each classifier on each dataset are obtained via 10 runs of 10-fold cross-validation. Runs with the various algorithms are carried out on the same training sets and evaluated on the same test sets. In particular, the cross-validation folds are the same for all the experiments on each data set. Finally, we conducted a corrected paired two-tailed t-test with 95% confidence level to compare each pair of cost-sensitive Bayesian network classifier and the original Bayesian network classifier.

At first, we design a group of experiments to compare NB and CSNB in terms of the total misclassification costs. In this group of experiments, we set the cost ratio $C(1,0)/C(0,1)$ to 10, and $C(0,1)$ is a unit cost of 1. Table 3 shows the detailed compared results. Besides, the averaged total misclassification costs and the $W/T/L$ values are summarized at the bottom of the table. Each entry $W/T/L$ in the table means that CSNB wins on W data sets, ties on T data sets, and loses on L data sets, compared to NB.

Table 3. Comparison of the total misclassification costs: NB versus CSNB. v and *
respectively denote statistically significant improvement or degradation over NB.

Dataset	NB	CSNB	
anneal	7.25±8.38	1.74±1.70	
anneal.ORIG	35.85±15.26	14.80±6.48	v
audiology	0.01±0.10	0.19±0.42	
autos	15.92±10.09	11.73±9.19	
balance-scale	2.52±5.12	20.93±2.14	*
breast-cancer	50.67±14.13	29.12±12.62	v
breast-w	5.40±6.67	3.58±4.11	
colic	37.03±13.99	31.19±12.35	
colic.ORIG	55.03±17.81	21.48±9.85	v
credit-a	62.55±20.00	33.46±13.06	v
credit-g	158.26±25.86	67.86±16.60	v
diabetes	104.90±26.61	41.89±12.09	v
glass	13.05±8.70	8.94±5.17	
heart-c	29.68±13.45	16.78±11.59	v
heart-h	30.64±12.84	17.25±9.99	v
heart-statlog	26.43±12.50	15.88±8.98	v
hepatitis	10.75±8.49	7.21±6.09	
hypothyroid	152.17±17.73	144.89±22.01	
ionosphere	21.48±12.50	20.29±11.54	
iris	0.00±0.00	0.00±0.00	
kr-vs-kp	232.25±44.04	90.22±8.59	v
labor	1.54±3.85	1.56±3.15	
letter	9.56±6.47	10.54±3.70	
lymph	13.94±9.68	10.52±9.20	
mushroom	336.20±61.27	143.30±39.66	v
primary-tumor	13.55±9.00	6.11±4.49	v
segment	0.00±0.00	0.00±0.00	
sick	49.86±19.13	50.95±16.42	
sonar	28.06±15.22	21.55±13.40	v
soybean	1.19±2.21	1.88±1.10	
splice	35.37±18.49	20.84±10.36	v
vehicle	35.06±16.05	30.07±13.74	
vote	16.68±10.84	11.77±8.83	
vowel	12.26±10.46	3.80±1.79	v
waveform-5000	46.29±8.26	52.11±6.57	*
zoo	0.00±0.00	0.00±0.00	
Average	45.87	26.79	
W/T/L	-	15/19/2	

From the experimental results above,in terms of total misclassification costs,
CSNB significantly outperforms NB on 15 datasets and only loses on 2 datasets.
Besides, the averaged total misclassification costs of CSNB on 36 datasets of CSNB
is only 26.79, which is much lower than that of NB (45.87).

Table 4. Comparison of the total misclassification costs: different Bayesian network classifiers (NB, TAN, SP, AODE, WAODE, and HNB) under various cost ratios (2, 5, 10, 15, and 20)

a. Comparisons for NB versus CSNB			
Cost-Ratio	NB	CSNB	W/T/L
2	12.44	11.96	4/29/3
5	24.98	19.22	13/20/3
10	45.87	26.79	15/19/2
15	66.77	32.64	22/13/1
20	87.66	36.65	22/13/1

b. Comparisons for TAN versus CSTAN			
Cost-Ratio	TAN	CSTAN	W/T/L
2	10.26	9.77	4/32/0
5	21.67	16.24	13/23/0
10	40.68	23.13	21/15/0
15	59.96	28.00	23/13/0
20	78.71	32.05	24/12/0

c. Comparisons for SP versus CSSP			
Cost-Ratio	SP	CSSP	W/T/L
2	8.98	8.98	1/33/2
5	18.07	15.31	10/24/2
10	33.22	21.89	15/20/1
15	48.37	26.34	19/16/1
20	63.52	29.73	20/15/1

d. Comparisons for AODE versus CSAODE			
Cost-Ratio	AODE	CSAODE	W/T/L
2	9.39	8.91	4/31/1
5	19.62	14.84	12/23/1
10	36.67	20.76	16/19/1
15	53.72	24.42	21/14/1
20	70.76	27.05	22/13/1

e. Comparisons for WAODE versus CSWAODE			
Cost-Ratio	WAODE	CSWAODE	W/T/L
2	9.06	8.63	3/32/1
5	19.22	14.44	10/25/1
10	36.61	20.00	14/21/1
15	53.10	23.87	20/16/0
20	70.04	26.50	21/15/0

f. Comparisons for HNB versus CSHNB			
Cost-Ratio	HNB	CSHNB	W/T/L
2	9.20	8.73	2/33/1
5	19.06	14.17	13/22/1
10	35.49	19.91	20/16/0
15	51.92	23.75	22/14/0
20	68.35	26.70	24/12/0

Secondly, we designed a series of experiments to investigate the effectiveness of different Bayesian network classifiers (i.e. NB, TAN, SP, AODE, WAODE, and HNB) under various cost ratios (i.e. 2, 5, 10, 15, and 20). The detailed compared results on 36 UCI datasets are shown in Table 4 above. From the experimental results, we can conclude that: the cost-sensitive learning method based on the minimum excepted cost principle can significantly reduce the total misclassification costs of Bayesian network classifiers.

1. Under the situation of different cost ratios, in terms of the average total misclassification costs, cost-sensitive Bayesian network classifiers are much lower than Bayesian network classifiers (except that the cost ratio is 2), and almost each number of wins is much larger than that of loses in two-tailed t-test (except that the cost ratio is 2). In terms of the performance that reducing total misclassification costs, the overall performance of each cost-sensitive Bayesian network classifier outperforms the original Bayesian network classifier.
2. With the increase of the cost ratio, each cost-sensitive Bayesian network classifier is obviously on a growing ability to reduce the total misclassification costs, compared to the original Bayesian network classifier. For example, compared to NB, CSNB wins on 4 datasets and loses on 3 datasets and the difference on average total misclassification costs between them is just 0.48 when the cost ratio is 2, but CSNB wins on 22 datasets and loses on 1 datasets and the difference on average total misclassification costs between them is surprisedly 51.01 when the cost ratio is 20.
3. With the increase of the cost ratio, the speed of growth on each cost-sensitive Bayesian network classifier is far slower than the original Bayesian network classifier. For example, in terms of the total misclassification costs, HNB has increased 59.15 from 9.20 to 68.35 but CSHNB has only increased 17.97 from 8.73 to 26.70 when the cost ratio increases from 2 to 20.

5   Cost-Sensitive Bayesian Networks for Rock Burst Prediction

Current studies on rock burst prediction described as section 1 are all in the situation of cost-insensitive learning. However, in real-word application, rock burst prediction is a cost-sensitive problem in terms of two aspects. One is the cost of collecting rock burst data, and another is the misclassification cost of misclassifying between rock burst and no rock burst. In this paper, we focus our attention on the misclassification cost and proposed the cost-sensitive Bayesian network to predict rock burst problems.

Table 5. The detail information of three rock burst datasets

Dataset	Instances	Attributes	Classes
VCR	104	9	2
CCR	60	5	2
TCR	79	9	2

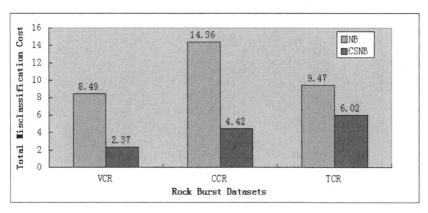

Fig. 1. The compared results of NB versus CSNB on three rock burst datasets

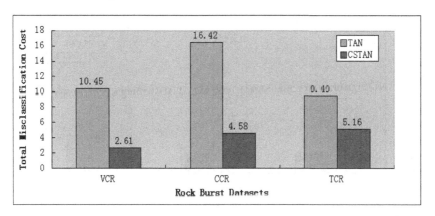

Fig. 2. The compared results of TAN versus CSTAN on three rock burst datasets

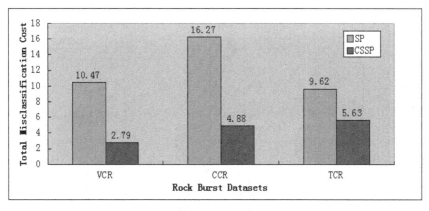

Fig. 3. The compared results of SP versus CSSP on three rock burst datasets

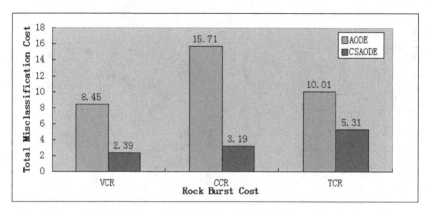

Fig. 4. The compared results of AODE versus CSAODE on three rock burst datasets

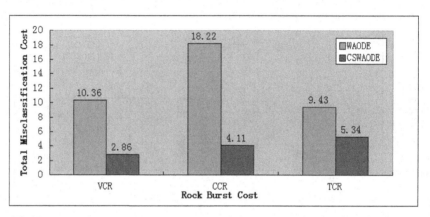

Fig. 5. The compared results of WAODE and CSWAODE on three rock burst datasets

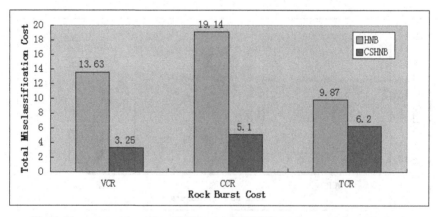

Fig. 6. The compared results of HNB versus CSHNB on three rock burst datasets

Our experimental rock burst datasets are all from [19] and we call them VCR, CCR and TCR datasets. The detailed description about them is shown in Table 5. We replace the missing values of the rock burst datasets with the averaged values, and assume that rock burst is positive class. The method of our experiment is the same as section 4 and the cost ratio is set to 10.

Figures 1-6 present the compared results between Bayesian network classifiers (i.e. NB, TAN, SP, AODE, WAODE, and HNB) and cost-sensitive Bayesian network classifiers (i.e. CSNB, CSTAN, CSSP, CSAODE, CSWAODE, and CSHNB) on three rock burst datasets. The compared results show that, in terms of the total misclassification costs, cost-sensitive Bayesian network classifiers are all much lower than those of original Bayesian network classifiers on three rock burst datasets. From all the classifiers we compared, CSAODE has the best performance in terms of the value of misclassification costs which is 10.89, sum of three rock burst datasets.

6   Conclusions

In this paper, we first summarize the methods of cost-sensitive learning and incorporate it into Bayesian network classifiers to learn cost-sensitive Bayesian network classifiers. The experimental results on 36 widely used UCI datasets show that cost-sensitive Bayesian network classifiers can significantly reduce the total misclassification costs of the original versions. Finally, we apply the cost-sensitive Bayesian network classifiers to three cost-sensitive rock burst prediction problems and achieve good results. Based on the experimental results above, we can draw a conclusion: cost-sensitive Bayesian network classifiers provide a new effective method for rock burst prediction, which has great values for the prediction and prevention of rock burst.

Acknowledgement. The authors would like to thank Prof. Liangxiao Jiang at China University of Geosciences for his kindly help and valuable suggestions. This work was supported by the National college students' innovation entrepreneurial training plan of China University of Geosciences (WuHan) (201310491057).

Reference

1. Chickering, D.M.: Learning Bayesian Networks Is NP-Complete. In: Learning from Data: Artificial Intelligence and Statistics, pp. 121–130 (1996)
2. Jiang, L., Cai, Z., Zhang, H., Wang, D.: Not So Greedy: Randomly Selected Naive Bayes. Expert Systems With Applications 39(12), 11022–11028 (2012)
3. Chai, X., Deng, L., Yang, Q., et al.: Test-Cost Sensitive Naive Bayes Classification. In: Proceedings of the 4th IEEE International Conference on Data Mining, pp. 51–58 (2004)
4. Elkan, C.: The Foundations of Cost-Sensitive Learning. In: Proceedings of 17th International Joint Conference on Artificial Intelligence, vol. 17(1), pp. 973–978. Morgan Kaufmann (2001)
5. Ling, C.X., Sheng, V.S.: Cost-Sensitive Learning and the Class Imbalance Problem. In: Encyclopedia of Machine Learning, pp. 231–235 (2008)

6. Feng, X.T., et al.: A New Direction-Intelligent Rock Mechanics and Rock Engineering. International Journal of Rock Mechanics and Mining Sciences 34(1), 135–141 (1997)

7. Friedman, N., Geiger, D., Goldszmid, T.M.: Bayesian Network Classifiers. Machine Learning 29(2-3), 131–163 (1997)

8. Keogh, E., Pazzani, M.: Learning Augmented Bayesian Classifiers: A Comparison of Distribution-Based and Classification-Based Approaches. In: Proceedings of the Seventh International Workshop on Artificial Intelligence and Statistics, pp. 225–230 (1999)

9. Webb, G.I., Boughton, J.R., Wang, Z.: Not So Naive Bayes: Aggregating One-Dependence Estimators. Machine Learning 58(1), 5–24 (2005)

10. Jiang, L., Zhang, H., Cai, Z., Wang, D.: Weighted Average of One-Dependence Estimators. Journal of Experimental & Theoretical Artificial Intelligence 24(2), 219–230 (2012)

11. Jiang, L., Zhang, H., Cai, Z.: A Novel Bayes Model: Hidden Naive Bayes. IEEE Transactions on Knowledge and Data Engineering 21(10), 1361–1371 (2012)

12. Nigam, K., Mccallum, A.K., Thrun, S., et al.: Text Classification from Labeled and Unlabeled Documents Using EM. Machine Learning 39(2-3), 103–134 (2000)

13. Margineantu, D.D.: Active Cost-Sensitive Learning. In: Proceedings of the 19th International Joint Conference on Artificial Intelligence, vol. 5, pp. 1622–1623 (2005)

14. Liu, A., Jun, G., Ghosh, J.: Spatially Cost-Sensitive Active Learning. In: Proceedings of the 9th SIAM International Conference on Data Mining, pp. 814–825 (2009)

15. Wang, T., Qin, Z., Zhang, S., et al.: Cost-Sensitive Classification With Inadequate Labeled Data. Information Systems 37(5), 508–516 (2012)

16. Jiang, L., Li, C., Cai, Z., Zhang, H.: Sampled Bayesian Network Classifiers for Class-Imbalance and Cost-Sensitive Learning. In: Proceedings of the 25th IEEE International Conference on Tools With Artificial Intelligence (ICTAI 2013), Herndon, Virginia, USA, pp. 512–517 (2013)

17. Frank, A., Asuncion, A.: UCI Machine Learning Repository. University of California. School of Information and Computer Science, Irvine, CA, 213 (2010), http://Archive.Ics.Uci.Edu/Ml

18. Witten, I.H., Frank, E., Hall, M.A.: Data Mining: Practical Machine Learning Tools and Techniques, 3rd edn. Morgan Kaufmann (January 2011) ISBN 978-0-12-374856-0

19. Feng, X.T., Wang, L.N.: Rock-Burst Prediction Based on Neural Networks. Transactions of Nonferrous Metals Society of China 4(1), 7–14 (1994)

Nonparametric Discriminant Multi-manifold Learning

Bo Li[1,2,3] , Jun Li[1,2,4], and Xiao-Ping Zhang[3]

[1] School of Computer Science and Technology, Wuhan University of Science
and Technology, 430081, Wuhan, China
[2] Hubei Province Key Laboratory of Intelligent Information Processing
and Real-time Industrial System, Wuhan China
[3] Department of Electrical and Computer Engineering, Ryerson University,
Toronto, Ontario, M5B 2K3, Canada
[4] State Key Lab. of Software Engineering, Wuhan, China
liberol@126.com

Abstract. In this paper, a nonparametric discirminant multi-manifold learning (NDML) method is presented for dimensionality reduction. Based on the assumption that the data with same label locate on the same manifold and those belonging to varied classes are resided on the corresponding manifolds, the traditional classification problem can be deduced to multi-manifold identification in low dimensional space. So this paper presents a discriminant learning algorithm to distinguish different manifolds, where a novel nonparametric manifold-to-manifold distance is defined. Moreover, an optimization function is modeled to explore a subspace with maximum manifold-to-manifold distances and minimum locality preserving. Experiments on AR face data and YaleB face data validate that NDML is of better performance than some other dimensionality reduction methods, such as Unsupervised Discriminant Projection (UDP), Constrained Maximum Variance Mapping (CMVM) and Linear Discriminant Analysis (LDA).

Keywords: dimensionality reduction, multi-manifold, feature extraction.

1 Introduction

During the last decade, as one kind of efficient dimensionality reduction methods, manifold learning has attracting more and more attentions. By applying k Nearest Neighborhoods (KNN) criterion, any point and its k nearest neighbors are viewed as located on a super-plane, where all the computational rules in linear space can be handled. So the locality can be constructed by linear tricks. When mapping all the localities to a global framework, although the local geometry is linear, the corresponding global structure still shows its nonlinearity. Among all the manifold learning based methods, Isometric Mapping (ISOMAP) [1], Locally Linear Embedding (LLE) [2-3], Laplacian Eigenmap (LE) [4], Local Tangent Space Alignment (LTSA) [5] and Maximum Variance Unfolding (MVU) [6] are representative manifold learning algorithms. It was shown by many examples that these methods have yielded impressive results on artificial and real-world data sets.

D.-S. Huang et al. (Eds.): ICIC 2014, LNCS 8588, pp. 113–119, 2014.
© Springer International Publishing Switzerland 2014

Generally speaking, manifold learning based methods have two characteristics. One is that the locality or the local structure can be preserved, and then the global structure can be kept unchanged; the other is that manifold learning based approaches can explore the essential dimensions of manifold distributed data embedded in high dimensional space. Due to the characteristics mentioned above, manifold learning based methods are enough to visualize data because it is the goal to map the original data set into a (2-D or 3-D) space that can preserve the intrinsic structure information as much as possible, especially for those data resided on a manifold. However, when confronting to the task of classification, manifold learning based methods will show their limitations on classification. One reason is that they can yield an embedding directly based on the training data set. Nevertheless, because of the implicitness of nonlinear projection, when applied to a new sample, the original manifold learning methods cannot easily obtain the sample's image in the embedding space by utilizing the low-dimensional embedding results of the training data set. This limits the applications of the original manifold learning based algorithms to pattern recognition problems greatly. In order to avoid this out-of-sample problem [7], linearization, kernelization, tensorization and some other techniques [8-10] were proposed, which are validated to be efficient to find the images of test data on the basis of the mapping results of the training samples.

Secondly, the original manifold learning based methods were constructed under circumstance that the data distributed on a manifold have the same class label. That is to say, it is unproblematic for existing manifold learning algorithms as they try to model a simple manifold. For example, in ref. [1-2], the left-right pose, up-down and lighting direction are three essential dimensions by performing the classical manifold learning methods, but it must be noted that all the images are taken from one person, i.e. this images are distributed on a manifold. However, if there are many manifolds, it is still a problem to map the different manifolds data more separately. For instance, if face images belonging to many persons do exist in a high dimensional space, then different persons' face images should lie on corresponding manifold. So it would be necessary to distinguish individual images from different manifolds. In order to achieve an optimal recognition result, the recovered embeddings corresponding to different manifolds should be as separable as possible in the final embedding space. This poses a problem that might be called "classification-oriented multi-manifolds learning" [10]. The problem cannot be solved by current manifold learning algorithms, including some supervised versions [11-18] because they are all based on the characterization of "locality."

In this paper, a new dimensionality reduction method, named nonparametric discirminant multi-manifold learning (NDML), is proposed to overcome the problems mentioned above. Not only the locality is preserved for manifold learning, but also a novel nonparametric manifold-to-manifold distance is defined according to both the label information and the local information of manifold, which are introduced for the task of manifolds identification. Moreover, the linearization trick is also adopted to avoid the out-of-sample problem.

2   Method

After the presence of manifold learning methods, many supervised extensions have been boomed for dimensionality reduction. Some supervised versions adjusted neighborhood weights by taking the class information into account when constructing the KNN graph, other combined Linear Discriminant Analysis (LDA) [20] for discriminant dimensionality reduction. However, most of these supervised versions pay more attentions to local graph construction instead of distances between manifolds. Until to now, to our knowledge, very few researches have been reported on the distance or dissimilarity for manifolds. Yang proposed manifolds dissimilarity to be non-locality of manifolds without considering the label information [10]. Later, Li presented a supervised manifold distance defined by the distances between any two points with different labels [19]. Wang put forward a manifold-manifold distance modeled by the distance between subspace-to-subspace [21].However, any definition mentioned above either focuses on locality without label information or concentrates on global distance. So in the proposed algorithm, we will offer a novel nonparametric manifold distance, where both local and supervised information are all taken into account.

2.1   Manifolds Distance

Based on the assumption that the data with same label locate on the same manifold and those belonging to varied classes are resided on the corresponding manifold, how to measure the manifold distance becomes more and more important for data classification. In this Subsection, a novel definition is stated below.

$$D = \sum_i \left(X_i^{c(i)} - \overline{X}_{ij}^{c(ij)} \right) \left(X_i^{c(i)} - \overline{X}_{ij}^{c(ij)} \right)^T \tag{1}$$

where D denotes manifold variance matrix, $c(i)$ and $c(ij)$ are the class label of any sample point X_i and its k nearest neighbors $X_{ij}^{c(ij)}$, respectively. $\overline{X}_{ij}^{c(ij)}$ is the mean value of the k nearest neighbors with different class labels to X_i, i.e.

$$\overline{X}_{ij}^{c(ij)} = \frac{1}{k} \sum_j^k X_{ij}^{c(ij)} \tag{2}$$

where $c(i) \neq c(ij)$.

2.2   Manifold Locality

Manifold learning aims to project the high dimensional data into a low dimensional space by locality preserving. In LLE, the locality of any point can be expressed by its k nearest neighbors and their least reconstruction error weights, which can be obtained from Eq (3).

$$\varepsilon\left(W_{int\,ra}\right) = \min\left\|X_i - \sum_{j=1}^{k} W_{ij} X_j\right\|^2 \tag{3}$$

Suffered the sum-to-one constraint for the reconstruction weights, Eq (4) can also be rewritten as follows.

$$\varepsilon(X) = \arg\min \sum_i \left\|X_i - \sum_{j=1}^{k} W_{ij} X_j\right\|^2 = \arg\min \sum_{ij} X_i^T (\delta_{ij} - W_{ij})^T (\delta_{ij} - W_{ij}) X_j \tag{4}$$
$$= X(I-W)(I-W)^T X^T$$

2.3   Linear Dimensionality Reduction

In the proposed NDML algorithm, we aim to find a low dimensional subspace where manifolds will locate far and manifold locality can be preserved. So the following objective function can be constructed for dimensionality reduction.

$$S(Y) = \max \frac{\sum_i \left(Y_i^{c(i)} - \overline{Y}_{ij}^{c(ij)}\right)\left(Y_i^{c(i)} - \overline{Y}_{ij}^{c(ij)}\right)^T}{Y(I-W)(I-W)^T Y^T} \tag{5}$$

In order to overcome out-of-sample problem, a linear transformation $Y_i = A^T X_i$ is introduced to Eq(5).

$$S(A) = \max \frac{\sum_i \left(A^T X_i^{c(i)} - A^T \overline{X}_{ij}^{c(ij)}\right)\left(A^T X_i^{c(i)} - A^T \overline{X}_{ij}^{c(ij)}\right)^T}{A^T X(I-W)(I-W)^T X^T A} = \max \frac{A^T DA}{A^T X(I-W)(I-W)^T X^T A} \tag{6}$$

For Eq(6), Lagrange multiplier method can be adopted to find the linear transformation A as follows.

$$DA = \lambda X(I-W)(I-W)^T X^T A \tag{7}$$

Thus the linear transformation matrix A is composed of eigenvectors associated to the top d eigenvalue of generalized equation mentioned above.

3   Experiments

In this Section, AR face data set, ORL face data a are applied to evaluate the performance of the proposed NDML algorithm, which is compared with those of UDP, CMVM and LDA.

3.1   Experiments on AR Face Data

The AR face [22] contains over 4,000 color face images of 126 people (70 men and 56 women), including frontal views of faces with different facial expressions, lighting conditions, and occlusions. The pictures of 120 individuals were taken in two sessions (separated by two weeks) and each section contains 13 color images. In this experiment, 14 grayscale face images (each session containing 7) of these 120 individuals are selected. The face portion of each image is manually cropped and then normalized to be size of 40×50 pixels, which is displayed in Fig 2. and occlusions. The pictures of 120 individuals (65 men and 55 women) were taken in two sessions (separated by two weeks) and each section contains 13 color images. In this experiment, 14 grayscale face images (each session containing 7) of these 120 individuals are selected and involved. The face portion of each image is manually cropped and then normalized to be size of 40×50 pixels.

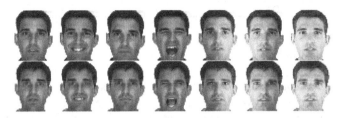

Fig. 1. The sample images for one person in AR database

Table 1. Performances comparison on AR database

Methods	Performances	Dimensions
NDML	0.9833	174
CMVM	0.9722	110
UDP	0.9714	126
LDA	0.9569	152

Shown in Table.1 is the maximum recognition rate comparison on AR subset data set by performing NDML, CMVM, UDP and LDA respectively. From the AR data set, we randomly select 8 samples per class as training set and the remaining 6 samples as test. Moreover, the nearest neighbor classifier is used to identify the features extracted by NDML, CMVM, UDP and LDA. From the experimental results, it can be found that the proposed NDML outperforms the other three methods.

3.2   Experiments on ORL Face Data

The ORL database contains 400 images of 40 individuals, including variation in facial expression, pose and other details like glasses/non-glasses. Fig.2 illustrates a sample subject of the ORL database along with all ten views, which are cropped to 64-by-64 size.

Shown in Table.2 is the maximum performance comparison on ORL data set by performing NDML, CMVM, UDP and LDA respectively. The experiments validates that the proposed NDML superior to the others.

Fig. 2. One sample images from ORL database

Table 2. Performances comparision on high-grade glioma data

Methods	Performances	Dimensions
NDML	0.9917	32
CMVM	0.9833	32
UDP	0.9817	16
LDA	0.9667	36

4   Conclusions

In this paper, a nonparametric discriminant multi-manifold learning method is proposed for dimensionality reduction. In the proposed method, a manifold distance is locally or nonparametric defined which can be expressed to distance between any point and the mean of its K nearest neighbors with varied class labels. Moreover, an objective function is constructed to explore the low dimensional subspace with maximum manifold distance and minimum locality, which is suit for multiply manifolds identification. The result is validated either from the theoretical analysis or from experiments on benchmark data set.

Acknowledgments. This work was partly supported by the grants of National Natural Science Foundation of China (61273303, 61273225 &61373109), China Postdoctoral Science Foundation (20100470613 & 201104173), Natural Science Foundation of Hubei Province (2010CDB03302), Research Foundation of Education Bureau of Hubei Province (Q20121115) and Open Project Program of the National Laboratory of Pattern Recognition (201104212).

References

1. Tenenbaum, J.B., de Silva, V., Langford, J.C.: A Global Geometric Framework for Nonlinear Dimensionality Reduction. Science 290, 2319–2323 (2000)
2. Roweis, S.T., Saul, L.K.: Nonlinear Dimensionality Reduction by Locally Linear Embedding. Science 290, 2323–2326 (2000)
3. Saul, L.K., Roweis, S.T.: Think Globally, Fit Locally: Unsupervised Learning of Low Dimensional Manifolds. J. Mach. Learning Res. 4, 119–155 (2003)
4. Belkin, M., Niyogi, P.: Laplacian Eigenmaps for Dimensionality Reduction and Data Representation. Neural Comput. 15(6), 1373–1396 (2003)

5. Zhang, Z., Zha, H.: Principal Manifolds and Nonlinear Dimensionality Reduction by Local Tangent Space Alignment. SIAM Journal of Scientific Computing 26(1), 313–338 (2004)
6. Weinberger, K.Q., Saul, L.K.: Unsupervised Learning of Image Manifolds by Semi-definite Programming. International Journal of Computer Vision 70(1), 77–90 (2006)
7. Bengio, Y., Paiement, J., Vincent, P.: Out-of-Sample Extensions for LLE, Isomap, MDS, Eigenmaps, and Spectral Clustering. Technical Report 123, Univ. de Montreal (2003)
8. Brand, M.: Charting a Manifold. In: Proc. 15th Conf. Neural Information Processing Systems (2002)
9. Yan, S., Xu, D., Zhang, B., Zhang, H.J.: Graph Embedding: A General Framework for Dimensionality Reduction. IEEE Trans. Pattern Analysis and Machine Intelligence 29(1), 40–51 (2007)
10. Yang, J., Zhang, D., Yang, J.Y., Niu, B.: Globally Maximizing, Locally Minimizing: Unsupervised Discriminant Projection with Application to Face and Palm Biometrics. IEEE Trans. Pattern Analysis and Machine Intelligence 29(4), 650–664 (2007)
11. Zhao, Q., Zhang, D., Lu, H.: Supervised LLE in ICA Space for Facial Expression Recognition. In: ICNNB, pp. 1970–1975 (2005)
12. Han, P.Y., Beng, A.T.J., Kiong, W.E.: Neighborhood Discriminant Locally Linear Embedding in Face Recognition. In: CGIV 2008, pp. 223–228 (2008)
13. Zhang, Z., Zhao, L.: Probability-based Locally Linear Embedding for Classification. In: FSKD, pp. 243–247 (2007)
14. Zhao, L., Zhang, Z.: Supervised Locally Llinear Embedding with Probability-based Distance for Classification. Computation Mathematics Application 7(6), 919–926 (2009)
15. Zhang, J., Shen, H., Zhou, Z.-H.: Unified Locally Linear Embedding and Linear Discriminant Analysis Algorithm (ULLELDA) for Face Recognition. In: Li, S.Z., Lai, J.-H., Tan, T., Feng, G.-C., Wang, Y. (eds.) SINOBIOMETRICS 2004. LNCS, vol. 3338, pp. 296–304. Springer, Heidelberg (2004)
16. Zhang, J., He, L., Zhou, Z.-H.: Ensemble-Based Discriminant Manifold Learning for Face Recognition. In: Jiao, L., Wang, L., Gao, X.-B., Liu, J., Wu, F. (eds.) ICNC 2006. LNCS, vol. 4221, pp. 29–38. Springer, Heidelberg (2006)
17. Ridder, D., de Loog, M., Reinders, M.J.T.: Local Fisher embedding. In: the 17th International Conference on Pattern Recognition, vol. 2, pp. 295–298 (2004)
18. Li, B., Zheng, C., Huang, D.S.: Locally linear Discriminant Embedding: An Efficient Method for Face Recognition. Pattern Recognition 41(12), 3813–3821 (2008)
19. Li, B., Huang, D.S., Wang, C., Liu, K.H.: Feature Extraction Using Constrained Maximum Variance Mapping. Pattern Recognition 41(11), 3287–3294 (2008)
20. Kim, T.K., Stenger, B., Kittler, J., Cipolla, R.: Incremental Linear Discriminant Analysis Using Sufficient Spanning Sets and Its Applications. International Journal of Computer Vision 91(2), 216–232 (2011)
21. Wang, R., Shan, S., Chen, X., Gao, W.: Manifold-manifold Distance with Application to Face Recognition based on Image Set. In: CVPR 2008, pp. 2940–2947 (2008)
22. The AR Face Database,
   http://rvl1.ecn.purdue.edu/aleix/~aleix_face_DB.html

Orthogonal Neighborhood Preservation Projection Based Method for Solving CNOP

Bin Mu, Shicheng Wen, Shijin Yuan, and Hongyu Li

School of Software Engineering, Tongji Univeristy
{binmu,2013_wenshicheng,yuanshijin,hyli}@tongji.edu.cn

Abstract. Conditional nonlinear optimal perturbation has been widely used in predictability and sensitivity studies of nonlinear numerical models. The main solution for CNOP is the adjoint-based method. However, many modern numerical models have no adjoint models which thus lead to a limitation of CNOP applications. To alleviate the limitation, we propose an ensemble projection method based on the orthogonal neighborhood preservation projection. To demonstrate the validity, we apply our method to CNOP of the Zebiak-Cane model and make a comparison with the adjoint-based method. Experimental results show that the proposed method can obtain similar results with the adjoint-based method.

Keywords: ensemble projection, ONPP, CNOP, ZC model.

1 Introduction

Predictability and sensitivity of atmospheric or oceanic motions play a critical role in both theoretical research and numerical weather and climate prediction. One possible approach to these problems is to determine the evolution of initial perturbations which usually represents the initial uncertainties. To achieve this purpose, Lorenz proposed the concepts of linear singular vector (LSV) and linear singular value (LSVA) [1]. Such a linear approach has been widely adopted to compute the fastest-growing perturbations [2-3]. However, the theories of LSV and LSVA are linear and cannot satisfy the requirements of describing nonlinear evolution of finite-amplitude initial perturbation.

To study the nonlinear mechanism of amplification of initial perturbations, Mu [4] proposed the concept of conditional nonlinear optimal perturbation (CNOP) for the initial perturbation with the largest nonlinear evolution. Subsequently, CNOP is adopted to study the predictability of El Nino-Southern Oscillation (ENSO) and the effect of nonlinearity on error growth for ENSO [5]; and conduct ensemble forecast [6].

The popular solution of CNOP is based on adjoint models. However, many modern numerical models have no corresponding adjoint models. Furthermore, developing a new one for CNOP is a quite huge work. The widespread use of CNOP, thus, is limited. To alleviate the limitation, Wang and Tan used an ensemble technique for

D.-S. Huang et al. (Eds.): ICIC 2014, LNCS 8588, pp. 120–126, 2014.

CNOP [7], and Chen and Duan further proposed an ensemble projection method based on Singular Value Decomposition (SVD) [8] to obtain CNOP. The both methods require a base dataset generated by principal component analysis (PCA) [9]. It is known that PCA can just get the linear features rather than nonlinear features which the complicated atmospheric and oceanic data have. Therefore, it is necessary to explore a method of extracting nonlinear feature for CNOP.

In this paper, we propose utilizing orthogonal neighborhood preserving projections (ONPP) [10] for solving CNOP. Specifically, in the training process, a set of features are extracted through ONPP. And then the optimization algorithm is performed in the space expanded by the features to search CNOP. The experimental results demonstrate that the ensemble projection method based on ONPP is an approximate solution to the adjoint-based method.

The rest of the paper is organized as follows: Section 2 introduces some related knowledge. In section 3, we come up the ensemble projection based on ONPP. Section 4 reports the experimental results. This paper ends with the conclusion and future work in Section 5.

2   CNOP and ZC Model

2.1   CNOP

CNOP is the perturbation $x_{0\delta}^{*}$ which makes the target function $J(x_{0})$ achieve the maximum with a condition of $\|x_{0}\| \leq \delta$, i.e.

$$J(x_{0\delta}) - \max_{\|x_{0}\|_{A} \leq \delta} \|J_{NT}\|^{2},$$
$$J_{NT} = M_{t_{0} \to t}(\overline{x}_{0} + x_{0}) - M_{t_{0} \to t}(\overline{x}_{0}),$$
(1)

where $M_{t_{0} \to t}$ means the propagator of a nonlinear model from the initial time t_{0} to the prediction time t, J_{NT} the nonlinear evolution of the initial perturbation, x_{0} the initial perturbation, and δ is the size of uncertainty. Here, both $\|\cdot\|_{A}$ and $\|\cdot\|$ have the same meaning, the L^{2} norm. For the convenience of optimization, Eq. (1) can also be converted to a minimum problem as follows:

$$J(x_{0\delta}) = \min_{\|x_{0}\| \leq \delta}(-\|J_{NT}\|^{2})$$
(2)

Therefore, the gradient of J can further obtained as:

$$\nabla J = -2 * H^{T} * J_{NT}$$
(3)

where H is the tangent matrix of the nonlinear model $M_{t_0 \to t}$, and its transpose H^T is so-called Jacobian matrix of J_{NT}. As mentioned previously, many numerical models without the adjoint models is difficult in computing CNOP.

2.2   ZC Model

The ZC model [11] is a mesoscale air-sea coupled model for the Tropical Pacific, which is famous for the good prediction of ENSO in 1986. It manly consists of three sub-models: the atmospheric model, ocean model and coupled model. The whole region of the ZC coupled model is shown in Fig.1.

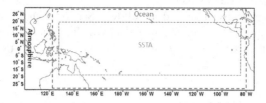

Fig. 1. The region of the ZC coupled model

3   ONPP Based Ensemble Projection Method

It is concluded in previous studies [12] that a driven dissipative system can get into a steady state that consists of attractors with low dimension after a long evolution. Any state, that is, can be projected onto the space of attractors. Based on this conclusion, we make an assumption that CNOP can also be projected on the space. Therefore, we require computing a group of orthogonal basis in the space of the attractors first. Due to nonlinear character of model data, we adopt the ONPP method which is a neighborhood preserving projection rather than PCA. Then, an optimization algorithm is chosen to obtain CNOP.

3.1   Neighborhood Preserving Feature Extraction

Suppose $x_1, x_2 \ldots x_n, x_i \in R^l$ $(i = 1, 2 \ldots n)$ are a training dataset from a long-run evolution of the models, and compose a matrix $X \in R^{l \times n}$. Before proceeding, we should perform a preprocessing including dimensionless and centering. The dimensionless process aims to get rid of the influence of dimension to describe real physical laws better.

$$x_i = x_i / a \, (i = 1, \ldots n),$$ (4)

where a is a positive coefficient. And the centering makes the data average equal to zero:

$$x_i = x_i - \overline{x} \ (i = 1, \ldots n), \ \overline{x} = \frac{1}{n} \sum x_i \tag{5}$$

Due to nonlinear characters of model data, we adopt the ONPP algorithm to extract features. It begins by building an affinity matrix by computing optimal weights, which will show the relationship of a given point with its neighbors in some locally optimal. The basic assumption is that each data sample, along with its k-NNs approximately lies on a locally linear manifold. Thus, each sample x_i is reconstructed by a linear combination of its k-NNs. The reconstruction errors are measured by minimizing the objective function:

$$\varepsilon_D = \sum_i \left\| x_i - \sum_j w_{ij} x_j \right\|_2^2 . \tag{6}$$

The weights w_{ij} denote the linear coefficients for restricting the sample x_i from its neighbors $\{x_j\}$. It is noted that the weights should be constrained:

- $w_{ij} = 0$, if x_j is not in the k-NNs of x_i.
- $\sum_j w_{ij} = 1$, that is, x_i is approximated by a convex combination of its neibors.

Define the k-by-k local Gram matrix G_i for x_i as $G_i[j,l] = (x_i - x_{N(j)}^{(i)})^T (x_i - x_{N(l)}^{(i)})$, that is $G_i = (x_i \vec{1}^T - X^{(i)})^T (x_i \vec{1}^T - X^{(i)})$, where the k-by-1 column vector $\vec{1}$ consists of 1 and the columns of the D-by-k matrix $X^{(i)}$ contain x_i's k nearest neighbors. Setting the k-by-1 column vector $w_i = (w_1^{(i)}, w_2^{(i)}, \cdots w_k^{(i)})^T$, Eq. (6) can be rewritten as

$$\varepsilon_D = \sum_{i=1}^{n} w_i^T G_i w_i \tag{7}$$

Given the constraint on the weights, we solve the least squares problem and obtain the closed-form solution

$$w_i = \frac{G_i^{-1} \vec{1}}{\vec{1}^T G_i^{-1} \vec{1}} \tag{8}$$

Assume that each data point $x_i \in R^l$ is mapped to a lower dimensional point $y_i \in R^d$, $d \ll l$. In the mapping stage, the representations of those intrinsic weights are preserved to maintain the nonlinear manifold geometry in the low dimensional space. With the same weights, the cost function can be defined in the embedding space as

$$\varepsilon_d = \sum_{i=1}^{n} \left\| y_i - \sum_{j}^{k} w_{ij} y_j \right\|_2^2 . \tag{9}$$

This optimization problem is formulated under the following constraints in order to the problem well posed:

- $\sum_i y_i = 0$, that is, the mapped coordinates are centered at the origin.

- $\frac{1}{n} \sum_i y_i y_i^T = I$, that is the embedding vectors have unit covariance.

Therefore, the reconstruction error is minimized by solving an eigenvalue problem. Take the smallest d eigenvectors as the desired feature basis after discarding the bottom eigenvector.

3.2   Optimization Algorithm

After getting the feature space, the next step is to project the initial perturbation x_0 onto it. The problem (2) can be converted to the following optimization problem with a constraint,

$$J(\omega_{0\delta}) = \min_{\|C \cdot \omega\| \leq \delta} - \left\| M_{t_0 \to t}(\overline{x}_0 + P \cdot \omega) - M_{t_0 \to t}(\overline{x}_0) \right\| \tag{10}$$

We utilize the spectral gradient projected method (SPG) for the optimization problem. The target function can be calculated through the model propagator directly. Also, the gradient can be calculated according to the definition.

4   Experiments

To show the feasibility of the proposed method, CNOP of the ZC model has been studied as a case in this paper. In our experiments, the Spectral Projecting Gradient Version 2 (SPG2) algorithm [13] is chosen as the optimization algorithm. The integration time span is 9 months, and the training set is generated by the integration of the ZC model over 200 years. To demonstrate the validity, the proposed method is compared with the adjoint-method [14] (for short, the XD method).

4.1   Comparison

The more eigenvectors are taken, the better the experimental results are. But more eigenvectors correspondingly lead to consuming more running time. Hence, we make a compromise between the computing efficiency and effects, taking 30 eigenvectors. The CNOP of the XD method and our method are shown in Fig. 3(a) and 3(b), respectively. The three rows in each subfigure denote the sea surface anomaly

(SSTA), thermocline anomaly (THA) and SSTA evolution during the time span of 9 months. Comparing Fig. 3(a) and 3(b), it is easily found that they have similar patterns of the SSTA and SSTA evolution but with different magnitudes. Our method has smaller magnitudes than the XD method. The most important reason is that only 30 base vectors are taken. In addition, the THA generated by our method is quite different from that in the XD method. The latter is smooth while the former, containing more details, is coarse. This may be caused by removing parts of eigenvectors corresponding to noises in our method. On the whole, our method can get similar results with the XD method.

4.2   Efficiency Analysis

In our method, the computing time consists of two parts: calculating feature basis by the ONPP algorithm and computing CNOP in the low dimensional space. Removing the marginal part, each data sample, consisting of SSTA and THA, has a dimension of 1080, and there are 2400 samples. As for this training data, the calculation of the feature basis by the ONPP algorithm requires about 450.5s. However, this process is off-line. That is, the feature basis requires being computed just once in many times of numerical simulations. Therefore, this part of consuming time can be ignored.

As for computing CNOP, the XD method is faster than our method, since the proposed method has to calculate the nonlinear models more times than the XD method. However, our method is free of the adjoint models. We do not need to implement an adjoint model which is really a huge work, and can apply our method to solving CNOP of other models without adjoint models easily.

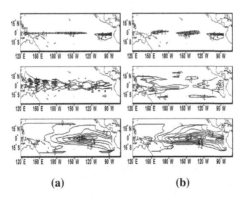

(a)   (b)

Fig. 2. CNOP results of the XD method and our method. The three rows in each subfigure are the SSTA, THA and SSTA evolution during the time span of 9 months. The initial month is May. (a) The XD method; (b) our method.

5   Conclusions and Future Work

In this paper, we propose an ONPP based ensemble projection method for solving the CNOP. The ONPP algorithm is a neighborhood preserving projection method, so it can explore some nonlinear features of data. Experimental results show that our method can get similar results with the XD method. Furthermore, our method is free of adjoint models. That is, our method can be used for CNOP of other models without adjoint models, which is the topic of our future work.

Acknowledgements. This work is supported by the Knowledge Innovation Program of the Chinese Academy of Sciences under Grant KZCX2-EW-201.

References

1. Lorenz, E.N.: A study of the predictability of a 28-variable atmosphere model. Tellus 17, 321–333 (1965)
2. Farrell, B.F.: Optimal excitation of baroclinic waves. J. Atmos. Sci. 46, 1193–1206 (1989)
3. Farrell, B.F., Moore, A.M.: An adjoint method for obtaining the most rapidly growing perturbation to oceanic flows. J. Phys. Oceanogr. 22, 338–349 (1992)
4. Mu, M., Duan, W.S.: A new approach to studying ENSO predictability: Conditional nonlinear optimal perturbation. Chin. Sci. Bull. 48, 1045–1047 (2003)
5. Duan, W.S., Mu, M., Wang, B.: Conditional nonlinear optimal perturbation as the optimal precursors for El Nino Southern oscillation events. J. Geophys. Res. 109, 1–12 (2004)
6. Mu, M., Jiang, Z.N.: A new approach to the generation of initial perturbations for ensemble prediction: Conditional nonlinear optimal perturbation. Chin. Sci. Bull. 53(13), 2062–2068 (2008)
7. Wang, B., Tan, X.W.: Conditional Nonlinear Optimal Perturbations: Adjoint-Free Calculation Method and Preliminary Test. American Meteorological Society 138, 1043–1049 (2010)
8. Chen, L., Duan, W.S.: A SVD-based Ensemble Projection Algorithm for Calculating conditional nonlinear optimal perturbation. Chinese Journal of Atmospheric Sciences (submmited, 2014) (in Chinese)
9. Wold, S., Esbensen, K., Geladi, P.: Principal component analysis. Chemometrics Intell. Lab. Syst. 2, 37–52 (1987)
10. Kokiopoulou, E., Saad, Y.: Orthogonal neighborhood preserving projections: A projection-based dimensionality reduction technique. IEEE Transactions on Pattern Analysis and Machine Intelligence 29(12), 2143–2156 (2007)
11. Zebiak, S.E., Cane, M.A.: A model El Nino-Southern Osillation. Mon. Wea. Rev. 115, 2262–2278 (1987)
12. Osborne, A.R., Pastorello, A.: Simultaneous occurence of low-dimensional chaos and colored random noise in nonlinear physical systems. Phys. Lett. A 181(2), 159–171 (1993)
13. Birgin, E.G., Martinez, J.M., Raydan, M.: Nonmonotone Spectral Projected Gradient Methods on Convex Sets. Society for Industrial and Applied Mathematics Journal on Optimization 10, 1196–1211 (2000)
14. Xu, H., Duan, W.S., Wang, J.C.: The Tangent Linear Model and Adjoint of a Coupled Ocean- Atmosphere Model and Its Application to the Predictability of ENSO. In: Procs: IEEE Int'l. Conf. on Geoscience and Remote Sensing Symposium, pp. 640–643 (2006)

A Safety Requirement Elicitation Technique of Safety-Critical System Based on Scenario

Junwei Du, Jiqiang Wang, and Xiaogang Feng

College of Information Science and Technology, Qingdao University of Science and Technology, Qingdao, China, 266061
djwqd@163.com

Abstract. Safety requirement of software is a fundamental part of the system safety requirements to safety critical system. The majority of the existing safety analysis techniques try to analyze the potential safety problems from system level; therefore, it is difficult to pinpoint hidden factors to software behavior and state level. In this paper, a safety requirement elicitation technique combined with scenario is proposed to refine the system-level safety analysis into software behaviors in specific scenarios, which would affirm those software behaviors and states that affect the system safety and, furthermore, formulate safety requirement model which could be directly applied to safety requirements validation and verification. The feasibility of the method presented in this paper was practically demonstrated by metro traffic control system's safety analysis.

Keywords: safety critical system, safety requirement elicitation, scenario analysis, failure analysis.

1 Introduction

Safety, an important property of a system, emphasizing on the application of a system will not bring an accident or loss [1]. Safety Critical System, abbreviated as SCS refers to the computer software and hardware system in safety field, which are directly responsible for safety and have higher safety requirements, therefore, the strategy of safety analysis in the system must be carefully studied [2]. When failure or fault occurs in this kind of system would incur loss and even catastrophic consequences to human life, property and the environment. SCS is directly related to the safety of human life and property. Since the increasingly widespread software has been integrated into the railway, nuclear power, aerospace and other safety-critical systems, how to evaluate and assure safety of software becomes the key problem [3]. All possible hazards will be identified by safety analysis method, potential factors will be found out and protective measures will be taken, and all the details will be recorded in the system safety requirements specification. Software safety requirements will further identify factors associated with software behavior and state from system safety requirements.

D.-S. Huang et al. (Eds.): ICIC 2014, LNCS 8588, pp. 127–136, 2014.

How to get the impact on system safety due to software behaviors and form accurate software safety requirements have already been the basis for protecting the system's safety. The available safety analysis algorithms, including FTA, FME (C) A, HAZOP, EA [4], try to analyze the potential safety problems from system level, therefore, it is difficult for them to locate hidden factors to software behavior and state level, thus, the analysis results could not be directly applied to safety requirements validation and verification [5]. This paper analyzes some of the shortcomings of the existing safety analysis techniques in software safety analysis, the proposed method combining with concrete scenario could refine the safety requirements analysis into software behaviors and states by system modeling language and provide property model for the validation and verification of safety requirements designing model[6,7], furthermore, lay solid foundation for further realization of formal verification method into system safety verification.

2   Safety Requirement Combined with Scenario

2.1   System Safety and Software Safety

Software safety is not only a software problem itself, but rather a system one. Obviously, the software itself could not cause the accident [5]. When the software used to control potentially hazardous systems, it then becomes the safety-related component.

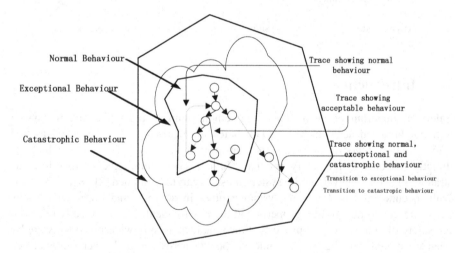

Fig. 1. Conversion of safety critical system behavior

Figure 1 depicts the behaviors of safety-critical system converting relationship, as is shown, both normal behavior and exceptional behavior would not cause harm, but the exceptional behavior is often the way to catastrophic behaviors[2]. Therefore, only the execution within the normal range and the implementation of exceptional behavior which can turn into normal behavior are considered as acceptable, on the contrary, those who lead to catastrophic behaviors are seen as inacceptable. Traditional safety analysis focuses on the relationship of system faults ,failures and accidents which could be observed, and most of those observed system faults stay in the integrated software's system or subsystem level[3], therefore, it is impossible to analyze software's logical function failures, not to mention what kind of behaviors or states trigger the system failure. Then, the safety analysis results based on the above theorems, obviously, could not be directly applied to requirement verification in software designing.

2.2   Access to Safety Requirement

Hazard could be identified and possible accident could be predicted through risk analysis. Potential hazardous factors of system-level, subsystem-level, component-level and hazard tolerant rate could be analyzed according to the system components, this method of risk and hazard analysis is called safety analysis technique. All possible harms to the system are treated as research object, it aims to answer why they exist and find out the corresponding control methods. System safety requirement is to record risk and hazards analysis. Safety analysis techniques often use top-down or bottom-up searching method. Top-down strategy is used to search a hazard which could be decomposed into sub-events, while bottom-up method is applied to analyze possible harms and what kind of components or combination of events incur the harms. These two techniques are suitable to apply to the structured system model, that is, high-level abstraction of the system can refine into the components level. In order to perform a comprehensive hazards and risks analysis, it is necessary to combine multiple safety analysis techniques of different searching strategy, it is able to discover new hazard and specify the possible reasons of the hazard.

Safety requirement obtained by the safety analysis describes all the possible factors may lead to hazards and their protection measures. Safety analysis adopting Fault Tree Analysis[8], abbreviated as FTA, starts from the faults and failures of system, subsystem and components, find out possible factors which may lead to system accident, pinpoint what kind of events or events combination could incur system failure or fault. The results of such an analysis actually constitute safety constraints for system requirements analysis and system designing. However, these results do not directly work as property model of the designing model, further refinement combined with formal language is still necessary. Meanwhile, in addition to the constraint which can not incur hazardous accident, there are safety requirements for each component or subsystem assigned security-related functions necessary to achieve the requirements that need to combine the results of the formal language of the safety analysis to be extended and fine technology.

3   Safety Requirement Elicitation Framework with Scenairo

Safety requirements come from the risks and hazards analysis, and base on a variety of strategies of safety analysis [6]. A typical example is the combination of HAZOP and FTA [9], on the one hand, it could identify potential hazards, on the other hand, it is also able to find out the possible reasons of each hazard. However, these fault patterns, which are frequently performed as software failure, environmental factors and human errors, cannot be refined into system behavior and state level, and therefore the results cannot be directly applied to safety verification of system-level or software designing level, that means, it is necessary to analyze hazards and potential problems combined with system's behaviors and states and find out the exceptional behaviors in concrete scenario. In the study of system behavior or state level analysis, state chart, Petri Nets are often used to show the states transformation [10]. Since the reachability of bounded Petri nets and finite state machines are isomorphic, state chart, in essence, is a kind of finite automata, using state chart to express its component's external states and states of the combination to achieve refined fault mode which could analyze system's harmful behaviors in specified scenario[11].

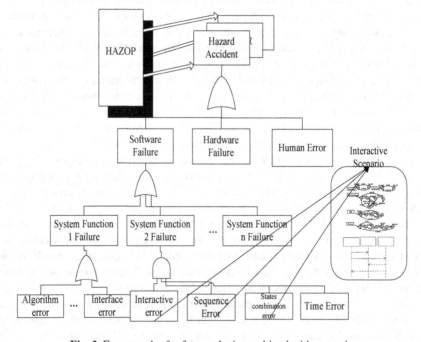

Fig. 2. Framework of safety analysis combined with scenario

The basic events obtained from the fault tree are software failure behaviors, but the failure behaviors of the software are very complicated. In different scenarios, failures are not the same. If fault tree is the only method to analyze hazardous states or behaviors, it will be difficult to express accurately or completely formalize, software safety requirement obtained by the above way will be impossible to use formal verification techniques for automatic verification at the designing stage. Scenario-based analysis method is effective conversation and transformation technical system in requirements analysis phase and system designing phase, and the scenario model can be automatically converted to state behavior of designing model, which provides a possible way for fault mode to software behavior and state. Based on this, a safety analysis techniques combined with scenario is proposed in this paper to research fault tree refinement problem from 2 levels: failure state and failure behavior. Specific implementation framework is shown in Figure 2.

3.1   Fault Tree Refinement Technology with Scenario

In FTA, there are many factors due to software itself which lead to a lot of accidents, the traditional failure, fault and error does not distinguish, but be seen as event. Events of fault tree are divided into basic event sand middle event, Where basic event A is defined as the state of system, subsystem or component. When a state occurs, A = True, on the opposite, A = False, While middle event is defined as a failure, fault or error mode caused by the underlying events. Gate is a logical operator, it could be used to express how bottom events organize to make top event occur.

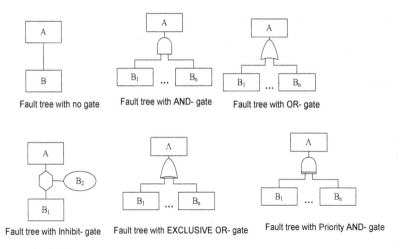

Fig. 3. Basic structures of fault tree

It mainly includes: no gate, AND, OR, Inhibit, Exclusive-OR and Priority-AND[12], their semantics are as following:

1) no gate: $A = B$

2) AND-gate: $A = B_1 \wedge B_2 \wedge \ldots \wedge B_n$

3) OR-gate: $A = B_1 \vee B_2 \vee \ldots \vee B_n$

4) Inhibit-gate $A = B_1 \wedge B_2$

5) Exclusive OR-gate: $A = (B_1 \wedge \neg(B_2 \vee \ldots \vee B_n))$

$$\vee$$

$$\ldots$$

$$\vee (B_n \wedge \neg(B_1 \vee \ldots \vee B_{n-1}))$$

6) Priority AND-gate: $A = B_1 \wedge \Diamond (B2 \wedge \Diamond (B2 \ldots \wedge \Diamond Bn) \ldots)$

Fault tree analysis is usually executed by analyzing the fault tree cut sets [13]. There are a variety of algorithms to compute minimal cut sets, the traditional methods are based on the top-down, bottom-up or BDD (Binary Decision Diagrams)to generate fault tree disjunctive normal form. In the state-oriented safety analysis [14], fault tree is used to obtain the incorrect states set of components system, that is, the temporal relationship of states set formed by minimal cut set is the safety constraint specification.

Definition 1 Fault tree safety requirement condition: fault tree safety requirement constraints use temporal logics to describe the system safety requirements specification analyzed by fault tree. The constraint condition of top events for the fault tree is R cannot occur. that is$\square \neg$ R, logical expression of all bottom events connected by the logical operators $\wedge$ and $\vee$.

Definition 2 fault tree safety requirement constraints disjunctive normal form: if safety requirement constraints formula such as$\square\neg$(B1 $\vee$ B2 $\vee$...$\vee$ B_n),where $B_i = l_{i1} \wedge$ li2 $\wedge$...$\wedge$ l_{im} ($1 \leq i \leq n$, $m \geq 1$)and ($l_{i1}, l_{i2}, ..., l_{im}$) is a fault tree minimal cut sets, then it is fault tree safety requirement constraints disjunctive normal form

Definition 3 safety requirement fault set: If system safety requirement constraints disjunctive normal form is: $\square\neg$($B_1 \vee B_2 \vee ... \vee B_n$), where $B_i = l_{i1} \wedge$ li2 $\wedge$...$\wedge$ l_{im} ($1 \leq i \leq n$, $m \geq 1$), then system safety requirement fault set is $\hat{E} = \{(l_{i1}, l_{i2}, ..., l_{im}) | l_{i1} \wedge$ li2 $\wedge$ ...$\wedge$ $l_{im}$is the basic conjunctive item of fault tree safety requirement constraints disjunctive normal form $\}$, that is, the set of all the fault tree minimal cut sets.

Definition 4 safety requirements fault set based on the state: the system fault set $\hat{E} = \{(l_{i1}, l_{i2}... l_{im}) | l_{i1} \wedge$ li2 $\wedge$ ...$\wedge l_{im}$is the basic conjunctive item of fault tree safety requirement constraints disjunctive normal form$\}$, lik $\wedge$... $\wedge$ $l_{ij}$is system-level basic events combination ($1 <= k <= j <= m$), is defined as refined temporal relationship of software states from the events combination. safety requirements fault set based on the state $\hat{E}_s = \{T(s_{i1}, s_{i2}, ..., s_{in}) | T$ is temporal formula of states set$\}$

Considering with the definition given above, the followings are the safety requirements construction algorithm combined with specific scenario:

Step 1:Initialize the system, construct system safety requirement specification fault tree model by requirement analysis, and compute safety requirement constraints disjunctive normal form and system safety requirement fault set $\hat{E}$.

Step 2: Combined with concrete scenario model, analyze the behavior and state of multi-system interaction, draw a combination of state and timing behavior set S in system execution space S.

Step 3: Get a x from set $\hat{E}$, analyze the evoluation model of every assembly or temporal combination of events, if there exsits $X' \subseteq X$, set $T(S_{kx'})$ is the refinement behaviors of X', then put $T(S_{kx'})$ into S_x, Continue analyze the rest compose events.

Step 4: If every event of event assembly can refine the specified behaviors or states, then compute the refinement fault states of every minmal cut set. If events or events assembly can not translate the software behaviors of states , they must be the reflection of external environment or human factors, then $T(S_{kx'})$ may be the abnormal behaviors which is need to check the external scenaios.

4   Case Study

The following example describes how to combine the fault tree analysis with component state chart to produce a design-oriented safety constraint. Figure 4 shows the fault tree model at the station where train collision would occur.

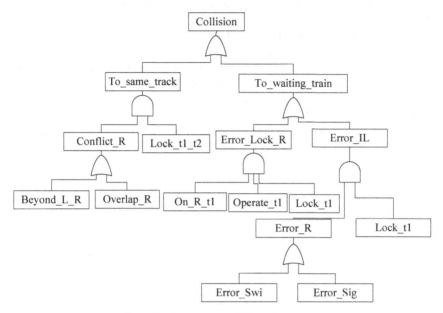

Fig. 4. Fault tree of possible train collision

The specific meaning explained in Table 1 below.

Table 1. Event meaning of fault tree in Fig.4

Node	Meaning
Collision	Train collision accident
To_same_track	Train goes to the conflict track section
To_waiting_train	Train goes to a waiting train
Lock_t1_t2	Lock the route t1,t2 at the same time
Conflict_R	Conflict routes
Error_Lock_R	Lock route R failed
Error_IL	Incorrect interlocking
Beyond_L_R	Over limit conflict of 2 routes
Overlap_R	2 routes has overlap
On-R_t1	t1 route has a train
Operate_t1	Create route t1
Lock_t1	t1was created successfully
Error_R	Route was locked, signal device state error
Error_Swi	Point position error
Error_Sig	Signal error

According to the fault tree formal semantic interpretation given above, fault tree temporal logic formulas are obtained:

Collision= ((Beyond_L_R ∨ Overlap_R) ∧ Lock_t1_t2) ∨ ((On_R_t1 ∧ Operate_t1 ∧ Lock_t1) ∨ ((Error_Swi ∨ Error_Sig) ∧ Lock_t1))

Because the safety invariant expression is the negative formal of the result of fault tree analysis, therefore, it is the safety constraint conditions of safety requirements.

Safe_Collsion=□¬(Collision)=

□¬ (((Beyond_L_R ∨ Overlap_R) ∧ Lock_t1_t2) ∨ ((On_R_t1 ∧ Operate_t1 ∧ Lock_t1) ∨ ((Error_Swi ∨ Error_Sig) ∧ Lock_t1)))

Fault tree safety requirement constraints disjunctive normal form can be obtained after equivalent transformation of logical formula

□¬((Beyond_L_R ∧ Lock_t1_t2) ∨ (Overlap_R ∧ Lock_t1_t2) ∨ (On_R_t1 ∧ Operate_t1 ∧ Lock_t1) ∨ ((Error_Swi ∧ Lock_t1)∨(Error_Sig ∧ Lock_t1))

safety requirement fault set of the fault tree is formed:∑={(Beyond_L_R, Lock_t1_t2),   (Overlap_R,Lock_t1_t2),   (On_R_t1,Operate_t1,Lock_t1), (Error_Swi,Lock_t1), (Error_Sig,Lock_t1)}

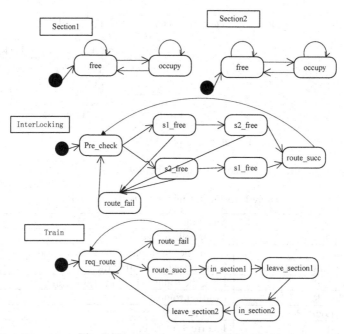

Fig. 5. The state chart of operating track

The above safety requirement fault set describes the possible collision of two trains hazard mode, but the event in the minimal cut set could not be refined into designing level states, that is, it is just an abstract state combined mode. The pattern obtains different state combination under different scenarios. In order to achieve automatic verification in designing level of safety requirement model, hazard state space should be researched in different scenario and fault tree analysis should be expressed in state chart to depict components' functions. Figure.5 depicts a specific scenario: Every train in-route contain two sections. These transition graphs below show how do these section components, interlocking checking component, and train requesting component work, when an in-route is requested to set up by train.

According to the minimal cut set formed by the fault tree event sets (On_R_t1, Operate_t1, Lock_t1), combining state chart, when a train is On_R_t1, and meanwhile executingOperate_t1 would lead to collision. Build another train into the path may lead to accidents. Analyzing the state chart in Figure 5 ,any occupied track section would fail the request to lock road, therefore, the event set can be refined :

(Train.req_route ∧ ◇ ((Section1.occupy ∧ Section2.free ∧ Interlocking.succ) ∨ (Section1.free ∧ Section2.occupy ∧ Interlocking.succ) ∨ (Section1.occupy ∧ Section2.occupy ∧ Interlocking.succ))

The analysis of fault tree events are refined by refining the state chart, the refined fault tree was shown in Figure 6.

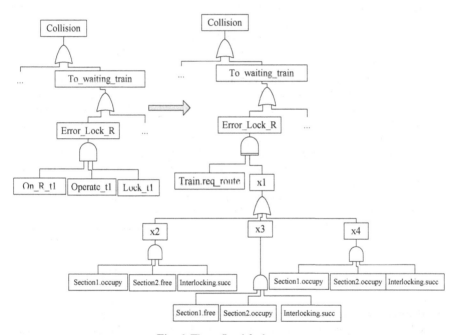

Fig. 6. The refined fault tree

5   Conclusion

This paper studied the traditional safety analysis methods used in software safety requirement, and proposed a safety requirement elicitation technique combined with scenario, which would make the safety analysis methods express accurately hazardous states or states combination of software. The proposed method compensate for the disadvantages on software interaction behaviors, lay solid foundation for further realization of formal verification method into system safety verification.

Acknowledgement. The research work was supported by National Natural Science Foundation of China under Grant No. 61273180, and Natural Science Foundation of Shandong Provincial under Grant No. ZR2011FQ005, ZR2012FL17, No.2011BSB01194.

References

1. CENELEC, EN50126: Railway Applications – The Specification and Demonstration of Reliability, Availability, Maintainability and Safety(RAMS). The European Standard (1999)
2. CENELEC, EN50128:Railway Applications- Software for Railway Control and Protection Systems. The European Standard (2001)
3. CENELEC, EN50129: Railway applications- Safety Related Electronic Systems for Signaling. The European Standard (2003)
4. Hansen, K.M., Ravn, A.P.: From Safety Analysis to Software Requirement. IEEE Transactions on Software Engineering 24(7), 573–584 (1998)
5. Abdullah, B., Liu, S.Y.: Hazard Analysis for Safety-Critical Systems Using SOFL. In: 2013 IEEE Symposium on Computational Intelligence for Engineering Solutions (CIES), pp. 133–140 (2013)
6. Du, J.W., Xu, Z.W., Mei, M.: Verification of Scenario-Based Safety Requirement Specification on Components Composition. In: IEEE Proceedings International Conference on Computer Science and Software Engineering, (2), pp. 686–689 (2008)
7. Du, J.W., Liu, G.Z.: Verification Framework of Scenario-Based Safety Requirement. In: Software Engineering, WCSE 2009, (4), pp. 154–158 (2009)
8. Helmer, G., Wong, J., Slagell, M., Honavar, V., Miller, L., Lutz, R.: A Software Fault Tree Approach to Requirements Analysis of An Intrusion Detection System. Requirements Engineering 7(4), 207–220 (2002)
9. Cha, S., Yoo, J.: A Safety-Focused Verification Using Software Fault Trees. Future Generation Computer Systems 28(8), 1272–1282 (2012)
10. Angela, A., David, H.: Sequential Failure Analysis Using Counters of Petri Net Models. IEEE Transactions on Systems, Man, and Cybernetics-Part A: System and Humans 33(1), 1–11 (2003)
11. Du, J.W., Xu, Z.W., Jiang, F.: Research on Verification of Behavior Requirement Patterns Based on Action Sequences. Journal of China Institute of Communications 32(1), 94–105 (2011)
12. Tiwari, S., Rathore, S.S., Gupta, S.: Analysis of Use Case Requirements Using SFTA and SFMEA Techniques. In: 2012 17th International Conference on Engineering of Complex Computer Systems (ICECCS), pp. 29–38. IEEE (2012)
13. Rao, S.: A Foundation for System Safety Using Predicate Logic. In: 3rd Annual IEEE Systems Conference 2009, pp. 42–47 (2009)
14. El, A.O., Xu, D., Wong, W.E.: Integrating Safety Analysis with Functional Modeling. IEEE Transactions on Systems, Man and Cybernetics, Part A: Systems and Humans 41(4), 610–624 (2011)

Modeling Information Diffusion in Social Networks Using Latent Topic Information

Devesh Varshney[1], Sandeep Kumar[1], and Vineet Gupta[2]

[1] Dept. of Computer Science, IIT Roorkee, India
[2] Adobe Research India Labs, Bangalore, India
{devu.var,vineetgupta10}@gmail.com, sgargfec@iitr.ac.in

Abstract. In present scenario, social networking and microblogging sites have become dynamic and widely used media for communication. Here people share information on various topics through which they express their likes and interests. Analyzing the process of information diffusion on these platforms not only helps in understanding the underlying social dynamics, but is also important for various applications like marketing and advertising. In this paper, we explore a novel problem in social network analysis which is to identify the active edges in the diffusion of a message in the social network. We cast this task as a binary classification problem of detecting whether a link in the social network participates in the propagation of a given message. We propose a learning-based framework which uses user interests and content similarity modeled using latent topic information, along with the features related to the social network. We evaluate our model on data obtained from a well-known social network platform - Twitter. The experiments show a significant improvement over the existing methods.

Keywords: Social Network Analysis, Information Diffusion, Diffusion Network, Topic Modeling, Social Media Analysis.

1 Introduction

In the recent past, social media has significantly contributed to the increase in online presence of users. As a result, social networking sites like Facebook[1] and Twitter[2] have shown a tremendous potential to make content go viral in no time. Today, social networks are hosts to a number of viral phenomena: breaking news propagation, information dissemination during emergency, marketing campaigns, etc. This potential has caught the eyes of researchers as well as marketers, prompting them to focus on the word-of-mouth marketing strategy using these social platforms. Word-of-mouth marketing is significantly different from mass marketing and direct marketing since it is based on trust among individuals, hence, it takes into consideration the value that a customer is capable of creating in his online social

---

[1] www.facebook.com
[2] www.twitter.com

D.-S. Huang et al. (Eds.): ICIC 2014, LNCS 8588, pp. 137–148, 2014.

circle. Studies show that people trust more on the information obtained from their close circles than the information obtained from advertising channels [1].

Studying and modeling the information propagation through social networks is helpful in making effective use of social platforms. Especially for marketers who wish to promote their products and services through viral marketing campaigns and for government authorities to quickly spread information during emergencies.

The information diffusion through online social networks has been studied by various researchers in the past. These works can be classified into two major focus areas - the problem of influence maximization using diffusion models like influence cascade model and linear threshold model, originally proposed by Kempe et. al. [2], and the problem of identifying active links in social networks through which diffusion can occur. Most of these works use network dynamics for creating solution models; however, these dynamics alone cannot accurately capture user interests and other relevant features. It has been shown that information dissemination processes are homophily-driven [3] and message propagation occurs more frequently between users having common interests [4]. Recently, Romero et. al. [5] showed that the textual content plays a significant role in information propagation through social networks.

In this paper, we present a novel approach of detecting active links in social networks for propagation of a message. We use textual content of messages and diffusion history of the network in addition to the network and user characteristics. Our method exploits latent information extracted from the textual content for various features like user similarity, content similarity and user interests as described in section 3.2. Given a social network and its diffusion history, our task is to find the edges in the network which are active in the propagation of a given message.

The rest of this paper is organized as follows. We review the related works on information diffusion and influence maximization in Section 2. In Section 3, we explain our data collection process, followed by extraction of various features. In Section 4, we present the experiments performed to evaluate our method and their results along with comparing them with existing approaches. Finally, Section 5 concludes the paper and provides the scope of future work.

2   Related Work

The process of information diffusion through social networks has been studied by a number of researchers in the past. Some of the recent works aim at utilizing diffusion history to predict the information propagation in social platforms. Galuba et. al. [6] studied the patterns in information propagation through social networks and discovered statistical regularities like power-laws in user activity, the exponential fall of cascade depth and log-normal distribution of the propagation delays. They also presented a propagation model to predict which users will tweet which URLs using a historical set of URL mentions by the users. However, their method fails to learn certain significant parameters effectively for 45% of positive instances. Petrovic et. al. [7] presented an approach to predict whether a tweet will be retweeted or not. They trained one model on entire dataset and 24 models on data of each 1-hour window of the day. Their best performing model was based only on social features including number of followers and friends, status count, list count, favorite count for the user and has a low F1 score of 39.6%.

Fei et. al. [4] used a multi-task learning approach to predict a user's response (like or comment) to a post of his friend making use of features representing content similarity and user interests. Lin et. al. [8] proposed a probabilistic model named TIDE (text-based information diffusion and evolution) to track how a topic evolves with time in social communities and what are the diffusion paths for that topic. Their model extracts features from text of posts and captures implicit features using Gaussian random field. But this model ignores the features related to social connections.

Zhu et. al. [9] studied the retweeting behavior of users and presented a logistic regression model to predict the retweeting probability of the incoming tweet by the target user. They used features related to the network like the number of friends and followers, status count of the tweet author and number of mutual friends, mutual followers, mutual mentions, mutual retweets to capture relationship among users. The model also takes into account the timing of tweet. They modeled content influence using URLs, mentions and hash-tags in the tweet. They captured topic similarity between the incoming tweet and the tweets of the user as the cosine similarity of their term frequency vectors. However, term frequency vectors are very sparse due to diverse vocabulary of interacting users.

The work of Kuo et. al. [10] is close to ours, they presented a method to predict the links active in the diffusion of a topic. Their best performing model uses topic similarity, indegree and number of different topics propagated through a link as the features. The difference in their and our work is that we consider the tweet content for our prediction instead of defining the diffusion specific to a topic. Thus, different content belonging to same topic category may have different diffusion networks in our method.

Another research problem in the study of information diffusion is of influence maximization. With the motivation of applications in marketing, Domingos and Richardson [11, 12] posed the problem of influence maximization in social networks for the first time. They proposed a probabilistic model using Markov random fields. Kempe, Kleinberg, and Tardos [2] modeled the same as discrete optimization problem. They proved that influence maximization is a NP-hard problem for two basic diffusion models namely - independent cascade and linear threshold, and propose a greedy approximation algorithm. The main idea behind these models is to start with a seed set of active nodes which try to activate their neighbors and then these neighbors activate their neighbors and so on.

Several studies aim at improving the efficiency and speed of these greedy algorithms [13-16]. Specifically, Chen et. al. [15, 16] proposed another heuristic method which restricts the influence propagation computation only to local influence regions of nodes and presented the MIA (maximum influence arborescence) model. Their model is more than six orders of magnitude faster than previous greedy algorithms. Chen et. al. [17] extended the influence cascade model to take into account the emergence and propagation of negative opinions. They modeled negative opinions based on product defects using the quality factor.

Most of the above mentioned works on influence maximization build the information diffusion model based on network features and do not make use of historical diffusion data.

On the other hand, our work focuses on utilizing the content of the messages diffused through the network along with the diffusion history data and network dynamics for detecting the links active in the propagation of the given message.

3   Proposed Approach

In this section, we describe the details of our approach to tackle the problem of identifying active links for diffusion in a social network. This includes the description of our data collection process, which is followed by our approach for feature extraction.

3.1   Data Collection

As there are no data corpus available publicly comprising of tweet content, retweet information and user network information for studying information diffusion in the network, we created our own. We collected over one million publicly available tweets and retweets from Twitter. This was done using REST API[3] provided by Twitter. Since, the API provides only a small fraction of tweets randomly; we collected the retweets of all the tweets separately. These tweets belong to various topics like shopping, politics, online messaging, and festive season as specified by the search keywords used (as mentioned in Table 1). We also collected the user information like name, followers count, friends count, status count, location, etc.

The resulting dataset contained over 80 thousand users (nodes) and over 203 thousand retweets (links). The topic-wise distribution of tweets is shown in Table 2. Here, the number of retweets indicates the total number of tweets collected which are retweets. Thus, if a tweet gets retweeted multiple times, it will be counted that many times.

For our analysis, we use the retweet graph created using collected nodes, tweets and retweets. In order to construct the training corpus, for each topic, the retweet links were taken to be positive instances and an equal number of negative instances were sampled by randomly selecting absence of retweet links of that topic between the nodes in the graph.

3.2   Feature Extraction

Here, we describe our approach of selecting features for designing an efficient detection of active edges for propagation of a message in the social network. Each edge in the network is represented as triplet having the source user, the target user and the message diffused from the source to the target user as its elements. Thus, we

---

[3] https://dev.twitter.com/docs/api/1.1

consider the edges to be directed. To build the classification model for each topic category, we model an edge using features corresponding to the source user, the target user and the diffused message.

Table 1. Keywords related to topics used for data collection

Topic Category	Keywords
Shopping	shopping, sales, deals, buy, gosf[4]
Politics	politics, democracy, elections, delhipoll, aap[5], bjp, congress
Social Media	socialmedia, facebook, twitter, whatsapp, smm, linkedin
Festive Season	holidays, chrismas, xmas, newyear, gifts, presents
Cricket	cricket, indiavssa, asiacup, indvsnz, ipl

Table 2. Number of tweets and retweets in the collected dataset

Topic Category	No. Of Tweets	No. Of Retweets
Shopping	227203	39087
Politics	240710	59612
Social Media	219099	40328
Festive Season	181404	35649
Cricket	174814	28173

3.2.1   Delta-TFIDF (d-TFIDF)

An important property of a tweet is its textual content. Most of the works on studying information diffusion done in the past do not consider this for extracting features. Here, we propose that the content of the tweet plays a significant role in determining the likelihood of a link being active. Thus, the textual content of tweets is modeled using bag-of-words approach and Delta-TFIDF features are extracted from it. Delta-TFIDF is a technique to efficiently weigh and score words for classification tasks. It has been shown to be more effective than TFIDF in binary classification of data using unigrams, bigrams and trigrams [18]. For any term w in tweet t, the Delta-TFIDF score $V(w,t)$ is computed as follows:

$$V(w,t) = n(w,t) * log_2 \left(\frac{Pw}{Nw}\right)$$
(1)

Here $n(w,t)$ is the frequency count of term w in tweet t. Pw and Nw are the number of tweets that contain word w belonging to the topic under consideration and the other topics respectively.

3.2.2   Latent Topic Information (LTI)

Topic modeling identifies the distribution of latent topics in the text, which is useful in modeling the interest distribution in conversations. Though Twitter has a hash-tag based mechanism to identify the topics of a tweet but it is not possible to keep track of

---

[4] Gosf is an abbreviation for Great Online Shopping Festival.
[5] Aap, bjp and congress are three popular political parties in India.

huge number of hash-tags. Further, due to the large variations in hash-tags, the distribution will be sparse. Thus, we use Latent Dirichlet Allocation (LDA) [19] which is widely used for topic modeling for extracting the hidden topics from the textual data. LDA is a Bayesian probabilistic model of text documents. It assumes a collection of k topics and each topic defines a multinomial distribution over the vocabulary. For this task, we use fast implementation of online LDA [20] provided by the *gensim*[6] module. To make the topic model generic, topic modeling is performed using all the tweets in the corpus. The topic model thus obtained is applied on the content of the message and the resulting topic vector is used as a feature-set.

3.2.3   Latent User Preference (LUP)

Most of the users post tweets and retweets on variety of topics of their interest [21, 22]. In order to model the interest of a user on these topics, we collected past tweets and retweets of the user and compute the topic distribution of his tweets using the LDA topic model obtained in the section 3.2.2. The resulting topic vector gives us latent user interests and preferences. This computation is done for both of the users involved in the link formation, thus we obtain two real valued vectors giving us a preference based feature-set.

3.2.4   Network Features (NF)

Features described above make use of the content of tweets, but to study the information diffusion, it is necessary to include features providing information about network dynamics.

Connections
It has been shown that the features related to social network like number of friends and followers, number of user mentions are good indicators of 'retweetability' [23] . For this purpose, we include number of followers, number of friends for both users of the directed link.

Activity and Response
To capture the activity of the source user, we include his total number of tweets, the number of his tweets belonging to topic under consideration. To take the response of other users in the network towards the activity of source user into account, we include the number of retweets of source user's tweets in topic under consideration and the total number of retweets of his tweets.

To consider the responsive activity of the target user towards other users, we include total number of retweets done by him, number of retweets by him that belong to the topic under consideration.

Interaction
To capture the interaction between source and target user, we include the number of times target user has reweeted the tweets of source user.

---

[6] http://radimrehurek.com/gensim/

Thus, we obtain several features representing the social network dynamics related to the users involved in the link.

3.2.5   Similarity between Users (SU)

It has been shown that content propagation across links occurs more frequently between the users having common interests [4] and thus, the content dissemination process is driven by the interest based homophily in the social network [3]. For computing this feature, we use the latent user preferences, as modeled in section 3.2.3, to represent the interest distribution of a user. To find the interest similarity between the two users, we calculate the cosine distance between the interest distributions of the users. The similarity between users is defined as:

$$S(u, v) = \frac{Tu.Tv}{\|Tu\|.\|Tv\|} \tag{2}$$

Here, Tu, Tv represent the interest distribution as obtained using topic model for user u and user v respectively. Thus, we obtain a real valued feature representing the interest similarity between the users involved in the link formation.

3.2.6   Content User Similarity (CUS)

It has also been demonstrated that the probability of diffusion through a node increases with increase in the similarity between the textual content of the message under consideration and the user interests [5]. For computing this feature, we use the latent user preferences, as modeled in section 3.2.3, to represent the interest distribution of a user and latent topic information, as modeled in section 3.2.2, to represent the topic distribution of the message content. To find the similarity between the message content and the user interest, we calculate the cosine distance between the topic distribution of the message content and the interest distribution of the user. The content user similarity is defined as:

$$S(m, u) = \frac{Tm.Tu}{\|Tm\|.\|Tu\|} \tag{3}$$

Here, Tm, Tu represent the topic distribution of message content and the interest distribution of the user respectively.

This computation is done for both of the users involved in the link formation, thus we obtain two real valued features representing the similarity of message content with the interests of the users.

4   Experiments and Results

4.1   Data Preprocessing

After collecting the data as described in section 3.1, several pre-processing steps are performed for data cleansing. The users who are disconnected from the social networks are removed from the data. The textual content obtained is also processed by stemming the words and removing stop-words and symbols like '@' and 'RT'. This processed data is then used for model formulation and experiments.

4.2   Experiments

For our experiments, we use various classifiers namely, support vector machine, artificial neural network and random forest. We perform experiments using different sets of features and evaluate the incremental performance improvement they provide on each topic category. These feature combinations are:

1. Content Features = d-TFIDF + LTI + LUP
2. Content + Network Features = d-TFIDF + LTI + LUP + NF
3. ALL Features = Content + Network + Similarity Features = d-TFIDF + LTI + LUP + NF + SU + CUS

Here, we report the results which show significant improvement in performance, i.e. namely the results of SVM with linear kernel provided by LIBSVM[7] toolkit [24] and ANN with 50 hidden units provided by PyBrain toolkit [25].

4.3   Model Evaluation

To better evaluate the performance of the classifiers, we use F1-score and the area under the curve (AUC) of a receiver operating characteristic (ROC) curve as the measures of accuracy [26]. The F1-score is harmonic mean of precision and recall and is a standard metric of performance evaluation in the information retrieval field. The ROC curve has "True positive rate (sensitivity)" along y-axis and "False positive rate" along x-axis. Larger values of F1-score and area under the ROC curve indicate better classifier performance. We used a 10-fold cross validation process for performance evaluation for each topic category data to remove any bias of over-fitting.

4.4   Baseline

We compare the performance of our classifier with baseline models. Firstly, we assign the normalized total number of diffusions as the probability of propagation for each link and exploit the independent cascade (IC) model [2]. This baseline is independent of any topic category. Secondly, we use the model proposed by Kuo et. al. [10] with their best performing feature set of topic similarity (TS), indegree (ID), and number of distinct topic propagated through the link (NDT). We apply this model for each topic category.

---

[7] www.csie.ntu.edu.tw/~cjlin/libsvm

4.5   Results

The performance of independent cascade (IC) based baseline is shown in Table 3. Table 4 to 7 report the results of baseline method proposed by Kuo et. al. [10], and performance of our classifiers using different sets of features on each topic dataset. These results show that our models outperform the exiting methods by significant margins. Also, our models perform consistently across all topic categories as observable from the result tables. The variation in classifier performance across different topic categories can be attributed to the variation in topic modeling by LDA for these different topic categories.

Figure 1 shows the comparison of our model with baselines for the 'Shopping' category data. Figure 2 shows the performance of our SVM classifiers using different combinations of features for 'Festive Season' topic category data to present the incremental change in performance with feature selection. Experiments on other topic category datasets also produce similar ROC curves.

Table 3. Results for Independent Cascade (IC) based Baseline

Features	F1-score	AUC
Normalized Diffusion	37.26%	53.84%

Table 4. Results for Shopping category data

Features	F1-score	AUC
Kuo et. al. (TS+ID+NDT) Baseline	52.17%	74.49%
SVM (d-TFIDF+LTI+LUP)	67.86%	86.10%
SVM (All features)	77.40%	93.46%
ANN (d-TFIDF+LTI+LUP)	78.01%	92.94%
ANN (All features)	83.09%	95.59%

Table 5. Results for Politics category data

Features	F1-score	AUC
Kuo et. al. (TS+ID+NDT) Baseline	53.92%	76.80%
SVM (d-TFIDF+LTI+LUP)	68.05%	87.64%
SVM (All features)	78.96%	94.16%
ANN (d-TFIDF+LTI+LUP)	79.83%	93.63%
ANN (All features)	84.71%	95.48%

Table 6. Results for Social Media category data

Features	F1-score	AUC
Kuo et. al. (TS+ID+NDT) Baseline	48.76%	72.04%
SVM (d-TFIDF+LTI+LUP)	55.67%	83.01%
SVM (All features)	69.91%	90.74%
ANN (d-TFIDF+LTI+LUP)	69.32%	89.13%
ANN (All features)	74.62%	93.57%

Table 7. Results for Cricket category data

Features	F1-score	AUC
Kuo et. al. (TS+ID+NDT) Baseline	47.34%	72.42%
SVM (d-TFIDF+LTI+LUP)	49.47%	83.46%
SVM (All features)	67.04%	90.55%
ANN (d-TFIDF+LTI+LUP)	68.38%	89.17%
ANN (All features)	76.57%	93.98%

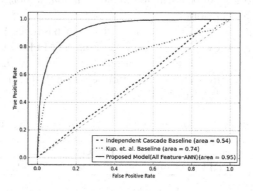

Fig. 1. The receiver operating characteristic (ROC) curve comparing proposed classifier with existing baselines on 'Shopping' category data

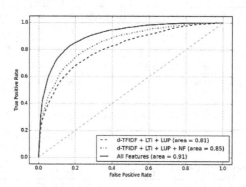

Fig. 2. The receiver operating characteristic (ROC) curve of proposed SVM classifier on 'Festive Season' category data comparing different feature combinations

5   Conclusion

This paper studies a novel problem of identifying the links active in the diffusion of a message through the social network. We carry out an analysis of the social network structure and exploit the latent information from the content of messages and present a feature extraction method that captures the diffusion activity and user interests effectively. We then train a classifier using these features to classify various links into active or inactive for the message propagation. We believe that our work can provide important insights to applications using information diffusion process in social networks.

Our future research would focus on understanding and using more specific characteristics of social media like *friend network* to make better utilization of available information. We would also make efforts in designing a method for calculating the diffusion probability of the links instead of just classifying them as active or inactive links.

References

1. Nail, J.: The consumer advertising backlash. Forrester Research and Intelliseek Market Research Report (2004)
2. Kempe, D., Kleinberg, J., Tardos, E.: Maximizing the spread of influence through a social network. In: Proceedings of the Ninth ACM SIGKDD International Conference on Knowledge Discovery and Data Mining. ACM (2003)
3. Golder, S.A., Wilkinson, D.M., Huberman, B.A.: Rhythms of social interaction: Messaging within a massive online network. In: Communities and Technologies 2007. Springer (2007)
4. Fei, H., Jiang, R., Yang, Y., Luo, B., Huan, J.: Content based social behavior prediction: a multi-task learning approach. In: Proceedings of the 20th ACM International Conference on Information and Knowledge Management. ACM (2011)
5. Romero, et al.: Differences in the mechanics of information diffusion across topics: idioms, political hashtags, and complex contagion on twitter
6. Petrovic, S., Osborne, M., Lavrenko, V.: Rt to win! predicting message propagation in twitter. In: ICWSM (2011)
7. Lin, C., Mei, Q., Jiang, Y., Han, J., Qi, S.: Inferring the diffusion and evolution of topics in social communities. Social Network Mining and Analysis 3(d4), d5 (2011)
8. Zhu, J., Xiong, F., Piao, D., Liu, Y., Zhang, Y.: Statistically modeling the effectiveness of disaster information in social media. In: 2011 IEEE Global Humanitarian Technology Conference (GHTC). IEEE (2011)
9. Kuo, T.-T., Hung, S.-C., Lin, W.-S., Peng, N., Lin, S.-D., Lin, W.-F.: Exploiting latent information to predict diffusions of novel topics on social networks. In: Proceedings of the 50th Annual Meeting of the Association for Computational Linguistics: Short Papers-Volume 2. Association for Computational Linguistics (2012)
10. Domingos, P., Richardson, M.: Mining the network value of customers. In: Proceedings of the Seventh ACM SIGKDD International Conference on Knowledge Discovery and Data Mining. ACM (2001)
11. Richardson, M., Domingos, P.: Mining knowledge-sharing sites for viral marketing. In: Proceedings of the eighth ACM SIGKDD International Conference on Knowledge Discovery and Data mining. ACM (2002)
12. Kimura, M., Saito, K.: Tractable models for information diffusion in social networks. In: Fürnkranz, J., Scheffer, T., Spiliopoulou, M. (eds.) PKDD 2006. LNCS (LNAI), vol. 4213, pp. 259–271. Springer, Heidelberg (2006)
13. Chen, W., Wang, C., Wang, Y.: Scalable influence maximization for prevalent viral marketing in large-scale social networks. In: Proceedings of the 16th ACM SIGKDD International Conference on Knowledge Discovery and Data Mining, pp. 1029–1038. ACM (2010)

14. Leskovec, et al.: Costeffective outbreak detection in networks
15. Chen, et al.: Efficient influence maximization in social networks
16. Blei, D.M., Ng, A.Y., Jordan, M.I.: Latent dirichlet allocation. The Journal of Machine Learning Research 3, 993–1022 (2003)
17. Hoffman, M.D., Blei, D.M., Bach, F.R.: Online learning for latent dirichlet allocation. NIPS 2(3), 5 (2010)
18. Banerjee, N., Chakraborty, D., Dasgupta, K., Mittal, S., Joshi, A., Nagar, S., Rai, A., Madan, S.: User interests in social media sites: an exploration with micro- blogs. In: Proceedings of the 18th ACM Conference on Information and Knowledge Management, pp. 1823–1826. ACM (2009)
19. Java, A., Song, X., Finin, T., Tseng, B.: Why we twitter: understanding microblogging usage and communities. In: Proceedings of the 9th WebKDD and 1st SNA-KDD 2007 Workshop on Web Mining and Social Network Analysis, pp. 56–65. ACM (2007)
20. Suh, B., Hong, L., Pirolli, P., Chi, E.H.: Want to be retweeted? large scale analytics on factors impacting retweet in twitter network. In: 2010 IEEE Second International Conference on Social Computing (socialcom), pp. 177–184. IEEE (2010)
21. Chang, C.-C., Lin, C.-J.: Libsvm: a library for support vector machines. ACM Transactions on Intelligent Systems and Technology (TIST) 2(3), 27 (2011)
22. Schaul, T., Bayer, J., Wierstra, D., Sun, Y., Felder, M., Sehnke, F., Rückstie, T., Schmidhuber, J.: PyBrain. Journal of Machine Learning Research 11, 743–746 (2010)
23. Davis, J., Goadrich, M.: The relationship between precision-recall and roc curves. In: Proceedings of the 23rd International Conference on Machine Learning, pp. 233–240. ACM (2006)

PRDiscount: A Heuristic Scheme of Initial Seeds Selection for Diffusion Maximization in Social Networks

Yufeng Wang[1], Bo Zhang[1], Athanasios V. Vasilakos[2], and Jianhua Ma[3]

[1] Nanjing University of Posts and Telecommunications, China
wfwang@njupt.edu.cn
[2] University of Western Macedonia, Greece
vasilako@ath.forthnet.gr
[3] Hosei University, Japan
jianhua@hosei.ac.jp

Abstract. This paper focused on seeking a new heuristic algorithm for the influence maximization problem in complex social network, that is, how to effectively select a subset of individuals to trigger a large cascade of further adoptions of new behavior under certain influence cascade models. In literature, degree and other centrality-based heuristics are commonly used to estimate the influential power of nodes in social networks. One of major issues with degree-based heuristics is that they are derived from the uniform IC (independent cascade) model, where propagation probabilities on all edges are the same, which is rarely the case in reality. Based on the general weighted cascade model (WC), this paper proposes Pagerank-like heuristic scheme, PRDiscount, in which discounting the influence power is explicitly adopted to alleviate the "overlapping effect" occurred in behavior diffusion. Then, we use both the artificially constructed social network graphs (with the features of power-law degree distribution and small-world characteristics) and the real-data traces of social networks to verify the performance of our proposal. Simulations illustrate that PRDiscount can advantage over the existing degree-based discount algorithm, DegreeDiscount, and achieve the comparable performance with greedy algorithm.

Keywords: Influence maximization, Pagerank, Overlapping effect, Social networks.

1 Introduction

Generally, there is a wide range of situations (epidemiology, computer virus, marketing, political science, and agriculture etc.), in which users coordinate their decisions and form conventions through being influenced by behaviors of their friends or neighbors, e.g., either to adopt a new behavior or product or not. Consider the following hypothetical scenario as a motivating example. A small company develops a new product and wants to market it on real social network through the effect of word of mouth (WOM). Due to the limited budget for marketing, it can only select a

D.-S. Huang et al. (Eds.): ICIC 2014, LNCS 8588, pp. 149–161, 2014.

small number of initial users in the network to use it (by giving them gifts or payments). The company wishes that these initial users would start influencing their friends on the social network to use it, and their friends would influence their friends' friends and so on, and thus through so-called contagion process, a large population in the social network would finally adopt the product. The problem is whom to be selected as the initial users so that they eventually influence the largest number of people in the network.

The above problem, called influence maximization, was firstly stated by Domingos and Richardson [1]: if we can try to convince a subset of individuals to adopt a new product and the goal is to trigger a large cascade of further adoptions, which set of individuals should we target in order to achieve a maximized influence? It was shown that, finding the influential set of nodes is NP-hard problem. Only for submodular function of diffusion model, the greedy algorithm can approximately work, with bounded threshold [2]. Kempe et al. have proved that a simple greedy algorithm (choosing the nodes with maximal marginal gain) can approximate the optimal solution by a (1-1/e), i.e., within 63% of optimal. However, the simple greedy-based approach has a heavy computation load. Specifically, the Greedy-based algorithm calculates the influence power precisely by enumeration. The more rounds the enumeration takes, the more accurate the result will get. However, when the network size increases, the computational time will increase dramatically, which prevents the greedy algorithm to become a feasible solution for the influence maximization problem in the real world. The research community has greatly studied the algorithmic aspects of maximizing influence in social networks, basically from two directions to solve the problem of efficiency: improving the greedy algorithm to further reduce its running time; the second is to propose new heuristics.

Ref. [3] presented an optimized greed algorithm, which is referred as the "Cost-Effective Lazy Forward (CELF)" scheme. The CELF optimization uses the submodular property of the influence maximization objective to reduce the number of evaluations on the influence spread of nodes. Their experimental results demonstrate that, in comparison with simple greedy-based approach, CELF optimization could achieve as much as 700 times speedup in selecting seeds.

Considering the fundamental step of the greedy algorithm is to pick a node in each iteration from the remaining nodes, which attempts to make the maximum marginal contribution to the process of spread of information, Ref. [4] proposed a SPIN (ShaPley value based Influential Nodes) algorithms for computing the marginal contributions using the concept of Shapley (a well known solution concept in cooperative game theory). Specifically, the Shapley value of a coalitional game provides the marginal contributions of the individual players to the overall value that can be achieved by the grand coalition of all the players.

Even with those improved greedy algorithms, their running time is still large and may not be suitable for large social network graphs. A possible alternative is to use heuristics. In sociology literature, degree and other centrality-based heuristics are commonly used to estimate the influence of nodes in social networks. Degree is frequently used for selecting seeds in influence maximization (so-called "degree centrality"): simply, the nodes with highest degree in remaining nodes are chosen.

Surprisingly, if we choose all target users with high centrality, the resultant scheme only outperforms random selection for small target sets, due to overlapping effect [5]. The main reason lies in that, the selection approach simply based on degree centrality does not take into account the neighborhood overlapping (so-called overlapping effect): A given group of connected nodes may have a high degree, but if there adjacent nodes are overlapped then the information may not propagate through the rest of the network.

A degree discount heuristic algorithm called DegreeDiscount was proposed in [6], which, based on the traditional degree-based scheme, discounts the degree of each node by removing the neighbors that are already in the activation seed set. They also note that the performance of this heuristic algorithm is comparable to that of the greedy algorithm. The major issue is that the above degree discount heuristics are derived from the uniform IC model where propagation probabilities on all edges are the same, which is rarely the case in reality.

Instead of focusing our effort in further improving the running time of the greedy algorithm, we argue that fine-tuned heuristics may provide truly scalable solutions to the influence maximization problem with satisfying influence spread and blazingly fast running time. Specifically, in this paper, we proposed Pagerank-like heuristic scheme of initial seeds selection, PRDiscount. PRDiscount has the following features. First, similar as Pagerank for ranking Web pages, when measuring one's influential power, PRDiscount takes into account its neighbors' influential power (in iterative way); Second, discounting individual's influence power is adopted to alleviate the "overlapping effect" occurred in behavior diffusion. Then, we use both the artificially constructed social network graphs (with the features of power-law degree distribution and small-world characteristics) and the real-data traces of social networks to verify the performance of our proposal (in terms of the eventually influenced users with the number of initial seeds).

The paper is organized as follows. Section 2 briefly introduces related heuristic schemes of initial seeds selection. In section 3, inspired from Pagerank algorithm, we proposed new heuristic scheme, PRDiscount under the general WC model, which considers the effect of whole network when inferring individual's influential power, and alleviates the "overlapping effect". And for the purpose of comparison, we re-design the simple DegreeDiscount to make it suitable for WC model. In section 4, we present detailed experimentation, and comprehensive comparisons among PRDiscount, greedy-based and DegreeDiscount. Finally, we briefly conclude this paper.

2   Related Work

Assuming there is a new behavior in the social network. We focus on its spreading among the population. Each user, classified as either an "active (behavior 1)" or a "potential (behavior 0)" consumer, is represented by a node in the social network structure, in which users are communicated one with another through certain channels determined by the social systems. More precisely, each user interacts only with her fixed group of neighbors, i.e., direct connections. These interactions represent personal and professional contact. A potential consumer is an agent that does not have the product but is susceptible of obtaining it if exposed to someone who does. An active consumer is a user that has already adopted the product and so can influence her neighbors in favor of obtaining it.

To describe the social system formally, consider a directed and weighted network having N nodes. The weight of the link from node i to j is denoted by $w(i,j)$ and assumed to be nonnegative. $w(i,j)>0$ represents the strength with which node i governs node j (that is, node j is influenced by a neighbor node i according to a weight $w(i,j)$). $w(i,j)$ and $w(j,i)$ are generally different from each other. There are two popular operational diffusion models in the literature that capture the underlying dynamics of the diffusion process: the general threshold model and the cascade model. It was proven in [2] that, for some conditions, these two are in fact equivalent. Especially, in our paper, we consider the so-called weighted cascade model (WC). In detail, let d_v be the indegree of v in social graph G, and let uv be a directed (or undirected) edge in G; in WC model, if u is activated in round t, then with probability $1/d_v$, v is activated by u in round $(t+1)$.

As described above, due to the huge computational overhead in greedy-based schemes, we seek to propose new heuristic method that could quickly infer users' influential power, and meanwhile achieve comparable performance with greedy-based schemes. In fact, Ref. [6] proposed a degree discount heuristic algorithm called DegreeDiscount, which nearly matched the performance of the greedy algorithms for the Independent Cascade (IC) model, and improved upon the pure degree heuristic in other cascade models. The general idea is given as follows. Let v be a neighbor vertex u. if u has been selected as a seed, then when considering selecting v as new seed based on its degree, the edge vu towards its degree is not counted, i.e., discounting v's degree by one due to the presence of u in the seed set, and the same discount will be done on v's degree of every neighbor of v that is already in the seed set. Simulations show that the performance of this heuristic algorithm is comparable to that of the greedy algorithm while its running time is much less than that of the greedy algorithm. The major issue is that the degree discount heuristics are derived from the uniform IC model where propagation probabilities on all edges are the same, which is rarely the case in reality.

Our paper deeply investigates how to select the initial seeds based on the more general weighted cascade model then the uniform independent cascade model.

3   PRDiscount: Pagerank-inspired Initial Seeds Selection Scheme

Intuitively, one's influence is not only embodied by the number of friends but also reflected by the kind of friends he has. In other words, the neighbors' influence should also be considered when measuring one's influential power.

The above intuition is very similar to the idea used in Pagerank algorithm [7], which was originally developed for ranking Web pages. Basically, the Pagerank of a page is large when the page receives many links from important pages that do not have too many outgoing links. Traditionally, a Web structure is represented as a directed graph $G=(V;E)$, in which, V is the set of pages, and the edges E represent direct links between pages. The adjacent matrix W of the Web graph is: $w(i,j)$ equals 1 if there is a hyperlink from page i to page j, or 0, otherwise. Specifically, edge $(i, j) \in E$ pointing from peer i and peer j, implicitly conveys the positive vote of page i on page j.

In our paper, the row normalized adjacent matrix, denoted as RW, is defined as follows: the sum of each row in adjacent matrix W equals 1. That is:

$$rw(i, j) = w(i, j) / \sum_j w(i, j) \text{ if } \sum_j w(i, j) > 0 \tag{1}$$

Basically, in Pagerank algorithm, a specific page's ranking value depends on how many pages points the very page and what are ranking values of those pages. Specifically, the following power iteration equation can be used to calculate the Pagerank values of pages:

$$PR^{(k+1)} = \alpha \cdot RW^T \times PR^k + (1 - \alpha) \cdot p \tag{2}$$

The constant parameter α always equals 0.85. p is the random jump N-vector, where the destination of the random jump is chosen from all the nodes with equal probability $1/N$.

The above iteration formulation can intuitively be interpreted as follows: on the Web graph, Pagerank algorithm follows a random walk where at each step, with probability α, the algorithm follows a randomly chosen link from the current page, and with probability $(1-\alpha)$, jumps to a randomly chosen page. The introduction of random jump vector is to solve the known problem that traps a random walker inside a local neighborhood when the graphs are disconnected or loosely connected.

However, Pagerank algorithm is originally designed for web graphs. Therefore, receiving links increases Pagerank value of each page, which is opposite to the contention of the influence diffusion. To relate the Pagerank to the influence, we artificially reverse all the links in the original social network. Then, based on the reversed network graph, PageRank algorithm can be directly to calculate each user's influential power.

In detail, the column normalized adjacent matrix of social network graph, denoted as CW, can be defined as follows:

$$cw(i, j) = w(i, j) / \sum_i w(i, j) \text{ if } \sum_i w(i, j) > 0 \tag{3}$$

Then the following iteration equation can be used to calculate the influential power of each user (denoted as IP):

$$IP^{(k+1)} = \alpha \cdot \left(CW^T\right)^T \times IP^k + (1 - \alpha) \cdot p = \alpha \cdot CW \times IP^{(k)} + (1 - \alpha) \cdot p \tag{4}$$

Similar as Pagerank, the constant parameter α always equals 0.85. p is the random jump N-vector. It was pointed out by [5], that, for degree or centrality based initial seeds selection schemes, if all initial seeds with high degree or centrality are selected in descending order (independent of node overlapping), the resultant scheme only outperforms random selection for small target sets, due to overlapping effect. By overlapping effect it means that a given group of connected nodes may have high degree (or centrality), but if there adjacent nodes are overlapped then behavior may not propagate through the rest of the network.

A simple degree discount heuristic algorithm, called DegreeDiscount, was proposed in [6], which based on the traditional degree-based scheme, discounts the degree of each node by removing the neighbors that are already in the activation seed set. The major issue with DegreeDiscount heuristic is that it is derived from the uniform IC (independent cascade) model, where propagation probabilities on all edges are the same, which is rarely the case in reality.

Inspired by the work in [6], using WC model, our proposal PRDiscount heuristic, explicitly discounts the influence power of users (who points to the users that already have been select into the seeds set), to alleviate the "overlapping effect". Note that, for the purpose of comparison, we also re-design the simple DegreeDiscount to make it suitable for WC model.

The following pseudocode offers the algorithmic process of PRDiscount heuristic.

PRDiscount(*numberofseeds*, *IP*)

Input parameters:

numberofseeds: the number of seeds that should be chosen;

IP: vector of users' influential power calculated by Pagerank on reverse social network graph

Output parameter:

candidateseeds[]: set of finally chosen initial seeds

FOR *i*=1 TO *numberofseeds* DO
    //find out the user with maximal influential power, and name it *chosenseed*
    chosenseed=MAX(*IP*);
    //put the user *chosenseed* into the seeds set *candidateseeds*[]
    candidateseeds[*i*]=*chosenseed*;
    //find out all users that point to the seed *chosenseed*, put those users into the vector
    //*neighborsofchosenseed*, and correspondingly reduce those users' influential power
    FOR *j* = 1 TO LENGTH(*neighborsofchosenseed*) DO
        neighbor=*neighborsofchosenseed*(*j*);
        $IP(neighbor) =$

$$IP(neighbor) - IP(neighbor) * w(neighbor, maxpeer) \Big/ \sum_k w(neighbor, k)$$

    ENDFOR
ENDFOR

4   Simulation Setting and Results

We adopt general WC to model the diffusion of behavior, and evaluate the performance of the proposed algorithm PRDiscount experimentally. For our experiments, we consider both synthetic data sets and real-world data sets. We first describe the data sets used in our paper, and then demonstrate the performance of the proposed algorithm in comparison with greedy algorithm and DegreeDiscount algorithm. We use the following convention while plotting the performance curves for various algorithms: the X-axis represents the number of initial seeds that can be chosen, and Y-axis represents the number of active nodes, that is, the number of influenced nodes at the end of the diffusion process.

4.1   Simulation Settings

1) Synthetic Data sets: There are several popular models to generate synthetic data sets with specific structural characteristics. Usually, social networks are characterized by the following two key features: scale-free degree distributions (i.e., the degree distribution complies with the power-law rule) and "Small World" effect used in [8] to denote these systems that have a high average clustering coefficient, such as regular lattices, yet a small characteristic path length, such as random graphs. We consider the following two types of synthetic data sets.

Scale-free (SF) Networks: Preferential attachment is a model proposed by Barabasi and Albert [9] for generating random graphs with heavy-tailed degree distribution (roughly comply with scale-free degree distribution). Consider a graph with N vertices. The N vertices of the graph are added one at a time, and for each of them, a fixed number of edges connecting to previously created vertices with probability proportional to their degree are added. The above preferential attachment (B-A model) does generate scale-free degree distribution, but does not exhibit any clustering. In our experiments, we use the following two processes to generate scale-free networks.

① Growth: Starting with t_0=50 nodes, at each round we add 10 new nodes, each with 10 edges. The total number of peers is 2000;

② Preferential Attachment: The probability that a new edge attaches to any of the peer with degree d is $dp_d / \sum_d dp_d$, where p_d is the fraction of nodes in the social network with degree d. Here, we explicitly distinguished three cases.

- The newly formed edge is uni-directional (directed), and points to the node that has already existed in the network. Obviously, in this case, the in-degree distribution complies with power-law rule, while the out-degree is almost identical for all nodes (equals 10). We called in-degree SF network;
- The newly formed edge is uni-directional, and points to the node that newly joins the network. Obviously, in this case, the out-degree distribution complies with power-law rule, while the in-degree is almost identical for all nodes (equals 10). We called out-degree SF network;

● The newly formed edge is two-directional (undirected). In this case, each node's out-degree and in-degree is same, and complies with power-law rule scale-free network. We called symmetric SF network.

A search-based model of small-world (SW) network formation was proposed by [10] that presents a dynamic model of network formation where nodes find other nodes with whom to form links in two ways: some are found uniformly at random, while others are found by searching locally through the current structure of the network (e.g., meeting friends of friends). This combination of meeting processes results in a spectrum of features exhibited by large social networks, including the presence of more high and low degree nodes than when links are formed independently at random, having low distances between nodes in the network, and having high clustering of links on a local level. In our experiments, for each newly added peer, the number of random neighbors is set as 2, and the other 8 neighbors are obtained through search. All links are undirected and unweighted.

2) Real World Social Network Datasets: We now present details about two real world datasets that we have experimented with. Note that, those dataset are provided at the website   (http://www-personal.umich.edu/~mejn/netdata/)   for   educational   and scientific purposes.

● Political blogs: A directed network of hyperlinks between weblogs on US politics, recorded in 2005 by Adamic and Glance [11].
● Netscience Data Set: This is a co-authorship network of scientists working on network theory, compiled by Newman in May 2006 [12].

We explicitly make the following two notes: First, due to the well-known fact the running time of greedy-like algorithms is extremely larger than heuristic schemes of initial seeds selection, in our simulations, we did not compare the greedy-based algorithm with heuristic schemes including DegreeDiscount and our proposal PRDiscount; Second, considering the inherent stochastic property of behavior diffusion in social networks, the following results are the average values of 100 times runs.

4.2   Simulation Results

Fig.1 illustrated the number of finally influenced nodes with the change of the number of chosen initial seeds in in-degree scale-free network (generated by B-A model), under three schemes, PRDiscount, DegreeDiscount, and Greedy algorithm. We can infer that, firstly, PRDiscount can achieve almost same performance as greedy algorithm; in in-degree scale-free network, the performance of DegreeDiscount is extremely bad. The reason lies in that, basically, DegreeDiscount simple uses node's out-degree as measurement of its influential power, however, in in-degree scale-free network, nodes' out-degrees are identical (equals 10 in our experiments). Thus, in this scenario, DegreeDiscount is analogous to the method of randomly selecting initial seeds, which, in turn, leads to the bad performance of DegreeDiscount.

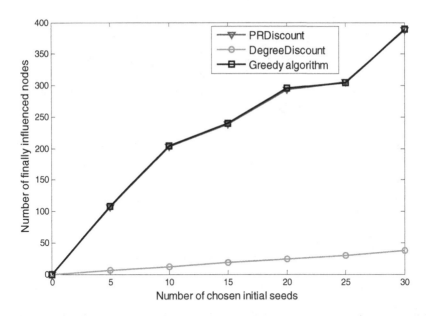

Fig. 1. # of finally influenced nodes vs. # of chosen initial seeds in in-degree SF network

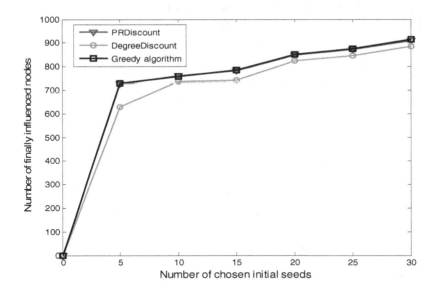

Fig. 2. # of finally influenced nodes vs. # of chosen initial seeds in out-degree SF network

Fig.2 illustrated the number of finally influenced nodes with the change of the number of chosen initial seeds in out-degree scale-free network (generated by B-A model), under three schemes, PRDiscount, DegreeDiscount, and Greedy algorithm. Similarly, we can draw the conclusion that PRDiscount performs always better than DegreeDiscount, and achieves same performance as Greedy algorithm.

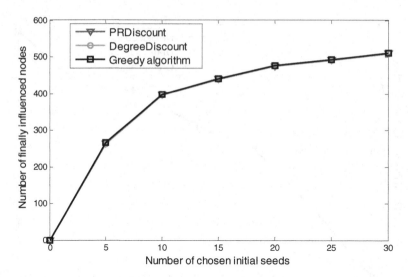

Fig. 3. # of finally influenced nodes vs. # of chosen initial seeds in symmetrical SF network

Fig.3 illustrated the number of finally influenced nodes with the change of the number of chosen initial seeds in out-degree scale-free network (generated by B-A model), under three schemes, PRDiscount, DegreeDiscount, and Greedy algorithm. Interestingly, those three schemes achieve almost same performance, which, in our opinion, is resulted by the undirected and unweighted network topology generated by the symmetrical B-A model. In detail, according to the WC model used in our paper, the column normalized adjacent matrix of social network graph in undirected and unweighted network graph is symmetrical. In this case, those three schemes choose same initial seeds, and have extremely similar performance.

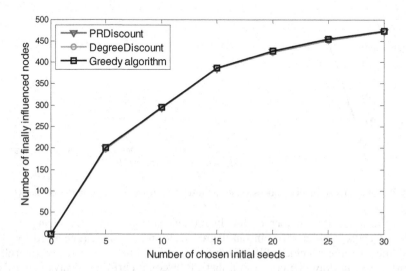

Fig. 4. # of finally influenced nodes vs. # of chosen initial seeds in search-based SW network

Fig.4 illustrated the number of finally influenced nodes with the change of the number of chosen initial seeds in search-based small-world network, under three schemes, PRDiscount, DegreeDiscount, and Greedy algorithm. Similar as the reason for the phenomenon shown in Fig. 3 (the symmetrical, undirected and unweighted network graph), those three schemes achieve almost same performance.

Then, we use real data traces to illustrate the performance of PRDiscount.

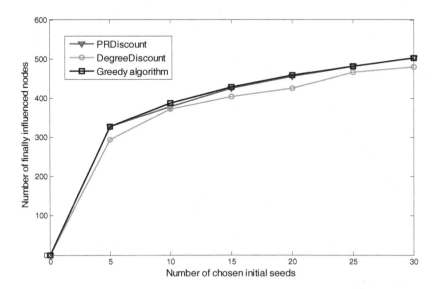

Fig. 5. # of finally influenced nodes vs. # of chosen initial seeds in Political blogs

Fig.5 and Fig. 6 respectively illustrated the number of finally influenced nodes with the change of the number of chosen initial seeds in Political blogs dataset and Netscience Data Set, under three schemes, PRDiscount, DegreeDiscount, and Greedy algorithm. From those two figures, we can observe clearly that PRDiscount always performs better than DegreeDiscount, and achieves the comparable performance as Greedy algorithm.

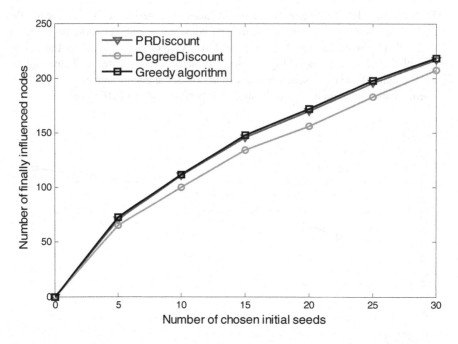

Fig. 6. # of finally influenced nodes vs. # of chosen initial seeds in Netscience

5   Conclusion

Our paper investigates the phenomenon of influence propagation in social networks, specifically, focuses on the problem of influence maximization: if a subset of individuals can be convinced to adopt a new product and the goal is to trigger a large cascade of further adoptions, which set of individuals should we target in order to achieve a maximized influence? Inspired by Pagerank algorithm (for ranking Web page), we proposed heuristic scheme of initial seeds selection, PRDiscount, which takes into account its neighbors' influential power (when measuring one's influential power), and explicitly discounts individual's influence power to alleviate the "overlapping effect" occurred in behavior diffusion. Simulations using both artificially generated and real-trace network data illustrate that, our proposed scheme PRDiscount, can achieve larger diffusion range than DegreeDiscount heuristic scheme, and achieve almost same performance as the brute-force greedy algorithm.

Acknowledgments. This research is partially support by the NSFC Grant 61171092 and JiangSu 973 project BK2011027.

References

1. Domingos, M., Richardson, M.: Mining the Network Value of Customers. In: Proceedings of the 7th ACM SIGKDD International Conference on Knowledge Discovery and Data Mining (KDD) (2001)
2. Kempe, D., Kleinberg, J.M., Tardos, E.: Maximizing the Spread of Influence Through a Social Network. In: Proceedings of the 9th ACM SIGKDD Conference on Knowledge Discovery and Data Mining (2003)
3. Leskovec, J., Krause, A., Guestrin, C., Faloutsos, C., VanBriesen, J., Glance, N.S.: Cost-effective Outbreak Detection in Networks. In: Proceedings of the 13th ACM SIGKDD Conference on Knowledge Discovery and Data Mining (2007)
4. Narayanam, R., Narahari, Y.: A Shapley Value based Approach to Discover Influential Nodes in Social Networks. IEEE TASE 8(1), 130–147 (2011)
5. Han, B., Srinivasan, A.: Your Friends Have More Friends Than You Do: Identifying Influential Mobile Users Through Random Walks. In: Proceedings of the 13th ACM International Symposium on Mobile Ad Hoc Networking and Computing (MOBIHOC) (2012)
6. Chen, W., Wang, Y.: Yang. S.: Efficient Influence Maximization in Social Networks. In: Proceedings of KDD (2009)
7. Page, L., Brin, S., Motwani, R., Winograd, T.: The PageRank Citation Ranking: Bringing Order To the Web, Technical report, Stanford Digital Library Technologies Project (1998)
8. Watts, D.J., Strogatz, S.H.: Collective Dynamics of 'Small-World' Networks. Nature 393(6684) (1998)
9. Barabasi, A.L., Albert, R.: Emergence of Scaling in Random Networks. Science 286, 509–512 (1999)
10. Jackson, M.O., Rogers, B.W.: Meeting Strangers and Friends of Friends: How Random are Socially Generated Networks? American Economic Review 97(3) (2007)
11. Adamic, L.A., Glance, N.: The Political Blogosphere and the 2004 US Election. In: Proceedings of WWW Workshop on the Weblogging Ecosystem (2005)
12. Newman, M.E.J.: Finding Community Structure in Networks Using the Eigenvectors of Matrices. Physical Review E74(036104) (2006)
13. Latora, V., Marchiori, M.: Economic Small-World Behavior in Weighted Networks. Eur. Phys. J. B 32(2), 249–263 (2003)

Quantifying the Evolutions of Social Interactions

Zhenyu Wu[1], Yu Liu[1], Deyi Li[1], and Yan Zhuang[2]

[1] State Key Laboratory of Software Development Environment, Beihang University, China
zhenyu.wu@cse.buaa.edu.cn, liuyu8014@163.com,
deyi_li@nlsde.buaa.edu.cn
[2] Beijing Emerald Education & Technology Company, China
zhuangyan@feicuiedu.com

Abstract. Understanding social interactions and their evolutions has important implications for exploring the collective intelligence embedded in social networks. However, the dynamic patterns in social interactions are not well investigated, and the time when the interactions take place is ignored in the existing studies. In this paper, a graph-based model incorporating a decay function is proposed to study the evolutions of social interactions quantitatively. In the experiments, the proposed model is applied to Digg dataset. The results show that the node degree of the social interaction graph follows the power law distribution, and the users' interactions have locality property. Furthermore, the results demonstrate that the evolutions of social interactions are useful for tracking the trends of topics.

Keywords: Social computing, Social interaction, Collective intelligence.

1 Introduction

Online social networks have become a popular way for users to connect, express themselves, and share content. Thus, social network studies have attracted increasing interests in recent decades. In fact, real social networks are dynamic and evolving due to the joining of new users, the leaving of old users, and the formation of new connections [1]. In social networks, the connections between users are formed because they participate in the same topic, such as a set of scholars coauthoring a paper and a person making phone calls with his friends. Thus, social interaction is a natural way to study the dynamics and evolutions of the social network. Moreover, the temporal evolutions of social interaction networks could be used to reveal the structure of the system [2-3]. In fact, a network is the most common way to capture information on social interactions [4] because multiple connections can be formed in the networks depending on the number of participants in a topic.

However, to date, few studies have been able to examine the detailed interactions over time, and few have studied the problem of modeling user interactions [5]. There are two aspects that are not well considered in existing studies: Firstly, most of the existing studies have not considered the evolutions of social interactions, especially from the angle of a specific topic. Secondly, the time when the interactions have

D.-S. Huang et al. (Eds.): ICIC 2014, LNCS 8588, pp. 162–172, 2014.
© Springer International Publishing Switzerland 2014

occurred is ignored. Although the strength of each connection is characterized by the frequency of interactions, the interactions cannot be differentiated in time scale.

To be able to investigate the evolutions of social interactions in the context of a topic, the interaction data is divided into the successive time slices firstly. Then, since the time when the first message is posted, the interaction graph is constructed by collecting all messages and interaction data about this topic in each time slice. With the increasing of the number of the posted messages and interactions, the evolutions of social interactions could be studied. To differentiate the interactions in time scale, the social interaction graph does not include removal of nodes or edges. Instead, a decay function is used to measure the influence of the interactions.

To summarize, the key contributions of this work are as follows. (1) To the best of our knowledge, a social interaction graph with decay function is first proposed to model the evolutions of social interactions quantitatively. (2) The dynamic social interaction patterns including degree distribution of social interaction graph and interaction locality of social interaction are investigated. (3) The process during which a topic becomes popular is clarified using the proposed method.

The rest of this paper is organized as follows. Section 2 introduces the related work about this paper. Section 3 defines the problems. Section 4 presents the social interaction graph with a decay function. Section 5 describes the dataset and presents experiment results. Finally, Section 6 discusses some potential future work and concludes this paper.

2   Related Work

This work is closely related to the evolution of social network. There are two broad classes of models for generating social networks: static and dynamic. Static models try to reproduce a single static network snapshot [6]. Dynamic models can provide insights on how nodes arrive and create links, such as preferential attachment [7], nearest neighbor [8]. Because of the dynamic characteristics of social networks, the dynamic models have been studied numerously. Sala et al. evaluate such models using both network metrics and application benchmarks and show that the nearest neighbor model outperforms others [9]. The dynamic model proposed by Leskovec et al. mimics the nearest neighbor model in a dynamic setting [10].

The study of evolution is one of the important aspects of dynamic network. Recent research has pointed out temporal evolution at two levels: the connectivity-level at which nodes enter or leave the system, and at the interaction-level capturing processes whose timing and duration takes place at a shorter scale, typically reflecting interaction between the units forming the network [3-4]. The connectivity-level models focus on capturing network growth. Baojun et al. propose an event-driven framework for creating network growth models, and they notice that in evolving networks, both the behavior of the whole network and the behavior of nodes evolve over time [1]. In addition, a variety of factors, including the attachment factor (the degree of nodes) and the locality factor (distance between nodes) are used to discover the growth schemes of social networks [11]. The interaction-level models focus on the interactions among individuals. The rich set of interactions between individuals in the social networks lead to the complex community structure capturing highly connected circles of friends, families, or professional cliques in a social network [12]. Renato et al. study the interaction network evolutions focusing on topological aspects [13].

Our work is different from these works in many aspects. Firstly, we explicitly present the social interactions with temporal decay function which is ignored by the existing studies. Secondly, we study the dynamic social interaction patterns and the evolutions of the social interactions quantitatively.

Another related work is about topic detection and tracking [14]. However, these existing studies depend on the mining of text. In addition to the content of the topic, our work focuses on the social interactions of users and investigates its effects to the process during which a topic becomes popular.

3   Problem Definitions

The structural effects imply that the presence of links is highly dependent on the presence of other links. Similarly, the historical interactions might influence the upcoming interactions in social interaction graph. For example, the interactions which have occurred a long time ago still have influence on the popular status of the topic. That is, the influence of social interaction has temporal property. We argue that the interactions would be much more effective if the time it occurs is much closer to the current time. In contrast, an interaction which is much farther from now has much smaller influence.

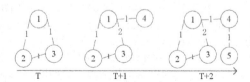

Fig. 1. Traditional social interaction graph

In this paper, the challenges of studying the evolutions of social interactions could be summarized as follows. (1) How to model the social interactions in the case of one topic. We argue that just considering the interaction frequency cannot describe the current state of the topic well. For example, in Fig. 1, social interactions are evolving from time slice T to time slice T+2. However, the interaction frequency between node 4 and node 5 is totally different from that between node 1 and node 2. The former case can reflect the status of the topic more accurately. Thus, the temporal information of social interaction is necessary. (2) The definition and evolution of local interactions. Are there any local effect exiting? For example, transitivity called by sociologists: if u has an interaction with v, and v has an interaction with w, then u would interact with w. (3) The roles of the historic interaction data. The historic interaction data might influence other users' upcoming interactions. Moreover, the historic data should influence the process during which the topic becomes popular.

Some brief definitions are as follows.

DEFINITION 1: Social interaction graph is an undirected graph. The nodes stands for users and the weighted edges stand for interaction frequencies.

DEFINITION 2: Evolution of social interaction, the changes of the interaction graph $G(t)$ $(t{\geq}0)$ in a sequence of discrete time slices. In each time slice, new nodes stand for the new comers, and new edges stand for the interactions at time t. Weights of edges should be updated by a decay function.

DEFINITION 3: Evolution of a topic, given a topic ξ, the social interaction graph evolves with the increases of the new posted messages and the new interactions.

DEFINITION 4: Given a topic ξ, seed user is the one who posts a new message about this topic.

4   Our Method

In order to describe the social interactions, each user is represented by a node in the graph, and there is an edge between two nodes if a social interaction occurs. Edges are commonly weighted by frequencies of interactions.

In time slice T, if there is one social interaction between user u and user v for the first time, an edge weighted 1 is added between these two nodes. Although the interactions in previous time slices have influence on the process during which the topic become popular, they should be differentiated from the interactions in this time slice. Thus, the earlier interactions should be updated by a decay function. Assuming that, this value is W_{ij} in time slice T. In time slice $T+1$, this value should be updated by the following rules:

$$W_{ij}(T+1)=\begin{cases}W_{ij}\times f(\Delta t) & if\ \ no\ interaction\\ W_{ij}\times f(\Delta t)+1 & if\ \ exist\ interaction\end{cases} \qquad (1)$$

Where, f is the decay function which is defined as follows:

$$f(\Delta t)=\frac{1}{\Delta t+\omega} \qquad (2)$$

Where, Δt is the time interval between time $T+1$ and the time of the last interaction. And, ω is the parameter to avoid the value of Δt is 0. The evolutions of social interactions could be described in Fig. 2.

If the number of nodes in the social interaction graph is N, the trend of the topic is quantified by GS:

$$GS=\sum_{n\in N}strength(n) \qquad (3)$$

Where, *strength* is a function to measure the interactions of nodes. Assuming that node i has M neighbors, *strength* is defined as follows:

$$\text{strength(i)} = \sum_{j \in M} W(ij) \tag{4}$$

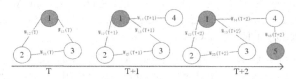

Fig. 2. The social interaction graph with temporal information

In summary, our method includes the following steps: (1) Given the topic ξ, get all the post messages and interaction data in time slice T; (2) Construct or update the interaction graph according to the interactions in this time slice; (3) Get the active users and investigate the dynamic interaction patterns; (4) Calculate the strength of the social interactions and identify the relationship between the topic status and the interaction strength; (5) Handle the data in the next time slice $T+1$ repeatedly.

5   Experiments

5.1   Dataset

Digg is a popular social news aggregator, and it allows users to submit links to and rate news stories by voting on, or digging them. Digg is appealing for investigating the evolutions of social interactions for the following reasons. Firstly, the latest news could be found in Digg, and there is collective behavior in Digg. Users could comment or reply to the topics which they are interested in. Secondly, the news dug by users is categorized into the topic or sub-topic. Thus, it is convenient to investigate the social interactions in a specific topic.

Munmun et al. use the Digg dataset to predict the social synchrony [15]. Our work is based on the same dataset. The statistics of the dataset are listed in Table 1. Every story belongs to a specific topic. There are 51 topics totally.

Table 1. Dataset Statistic

Users	Stories	Comments	Replies
41902	1488	104082	88940

5.2   Evaluations

Previous works have shown that a 24-hour window may be regarded as an inclusive bound of the popularity of a piece of information over time on different online social networks [16]. Thus, 24-hour is used as the time interval to investigate the evolutions of social interactions.

We focus on two topics "US Election" and "Comedy" respectively. Especially, the former topic is used to investigate the global property of social interaction graph. The rationale behind choosing these two topics is to analyze the social interactions in different background. The topic "US Election" is discussed during a specific period of time while the topic "Comedy" is discussed consistently.

Given a topic, all the interaction data are collected by the descending order. The user data includes the users posting the messages, users commenting the messages and users replying to the comments. Then, the time when the first message is posted is regarded as the start time. In this way, we could divide all the data into the successive time slices. In each time slice, we could get an interaction record according to the timestamps. For the users who posting new messages, a new node is added in the social interaction graph. For the users who comment a message, if this user is a new user, a new node is added. Then, we find the owner of the message, and add an edge with weight between them to present the interaction frequency. Moreover, if there is already an edge between these two users, we just update the frequency of the edge. Similarly, for the users who reply to other users, the social interaction graph is updated. Thus, a social interaction graph in one time slice could be constructed. And, the evolutions of social interactions in the successive time slices could be investigated.

We study the social interactions and their evolutions in the following three aspects: global property, local property and topic property.

Global Property
In order to observe how the growth rates of social interactions change as the network sizes increasing over time, some informative metrics of network structure, such as degree distribution, are used in this paper. As these properties are the reflections of the whole social interaction graph, they are called the global property.

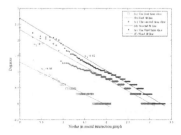

Fig. 3. Degree distributions of social interaction graph

As edges of one node stand for the number of users they interact with, the degree distributions reflect the trends of interactions. The social interaction graphs in three successive time slices (called the first time slice, the second time slice and the third time slice respectively) are selected to investigate the degree distributions. The results are shown in Fig. 3 where the horizontal axis stands for the nodes in social interaction graph, and the vertical axis stands for the degrees of the nodes. In Fig. 3, curves (a), (c), (e) show the degree distributions of the first time slice, the second time slice and the third time slice respectively. Moreover, the lines (b), (d), (f) show the linear fit of the above three curves. According to the experiment results, the degree distributions follow the power law distributions and the exponents of power law are 0.68, 0.77 and 0.82 respectively. The increase of the exponents demonstrates that the increase of the social interactions about this topic.

Are there any relationships between the new interactions and nodes degrees? That is, what's kind of users tending to launch an interaction? To answer this question, the following rule is used:

$$F_d(k) = M_k / N_k \tag{5}$$

Where, N_k denotes the number of node with degree equal to k in the last time slice, and M_k is the number of nodes among them that form new interactions in the current time slice.

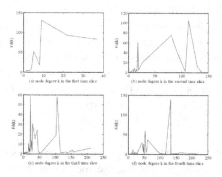

Fig. 4. Interactions and node degrees in successive time slices

In Fig. 4, the horizontal axis stands for the node degree in the last time slice, and the vertical axis stands for the value of $F_d(k)$. Subfigures (a), (b), (c) and (d) show values of $F_d(k)$ for the first time slice, the second time slice, the third time slice respectively. It is obvious that the maximum node degrees increase with time. In Fig. 4(a), the interactions are at the beginning, thus, the maximum node degree is just 39. And the maximum value of $F_d(k)$ is about 140 when the node degree is 9. This shows that nodes with degree 9 launch the maximum interactions in the first time slice. The common characteristic in the following time slices is that the value of $F_d(k)$ reaches the maximum when the node degree is in the range of 100 and 150. However, the nodes with much larger degrees launch fewer interactions. This shows that the nodes with higher degree not always interact with others more frequently.

In each time slices, there will be plenty of users joining in the interactions. How about the interactions of these new comers? Will they just interact with those users with higher node degree since they have just joined? To answer this question, the following rule is used:

$$F_n(k) = L_k / N_k \tag{6}$$

Where, N_k denotes the number of node with degree equal to k in the last time slice, and L_k denotes the number of interaction launched by the new comers with the nodes of degree k in the current time slice.

The results are shown in Fig. 5 where the horizontal axis stands for the node degrees in a time slice, and the vertical axis stands for the value of $F_n(k)$. Subfigures (a), (b), (c), (d) show the results of the first, the second, the third and the fourth time slices respectively. The distributions of $F_n(k)$ is similar to that of $F_d(k)$ in these four time slices. New comers do not always select those users with highest degree to interact with. In contrast, the nodes with degree between 100 and 150 are selected frequently. Moreover, the new comers show the same behaviors to whole set of users. This demonstrates that the continuous arrivals of new users are very important to the evolutions of the social interactions.

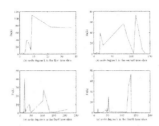

Fig. 5. Interactions and node degrees for new comers in successive time slices

Local Property

Local knowledge is one of the important features of social and complex networks where nodes have only limited influence on the network topology. The local property means users prefer to interact with others who are their neighbors. To investigate the local property of social interactions, the number of hops of the interactions is studied.

Here, the hop means the number of connections between two users in social interaction graph. If there are new interactions in time slices $T+1$, are the new interactions formed among the neighbors of a node? Moreover, the hops between which two users tend to form new interactions should also be clarified.

We first consider the following notion of social interaction locality: for each new interaction (u, w) in social interaction graph, we measure the number of hops it spans (the length of the shortest path between nodes u and w) immediately before the edge is created in social interaction graph in time slice $T+1$.

In time slice T, every social interaction is investigated. We will clarify the social interaction locality using the relationship between interaction hops and the number of the interactions in this hop. Formally, the rule could be defined as:

$$F_r(r) = g\left(M_r, N_r\right) \tag{7}$$

Where, N_r denotes the number of node pairs at distance r, and M_r is the number of interactions that form among the node pairs in the current time slice. Function g is used to present the relationship between N_r and M_r.

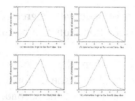

Fig. 6. The locality property of social interactions

The experiment results are shown in Fig. 6 where the horizontal axis stands for the interaction hops, and the vertical axis stands for the number of the interactions. Subfigures (a), (b), (c) and (d) show the results of the first, the second, the third and the fourth time slice respectively.

We could observe that the numbers of interactions taking placing within 1 hop is below 100 in these successive time slices. Moreover, the number of interactions is not increased with the hops. It reaches the maximum when the value of hops is equal 3. Then, it begins to decrease. This means that most interactions occur locally between the nodes that are about 3 hops.

As we know, there is the transitivity characteristic in social science. That is, if user u is a friend of v, and v is a friend of w, then u will be the friend of w with higher probability. In the case of social interactions, although the number of new interactions occur between two users with two hops is larger, maximum value occurs when the hop equals 3 indeed. Moreover, this value is not changed in the four successive time slices.

This observation is useful to understand the social interactions. The users we have interacted with (1 hop users) might not be selected as our future target of interactions. So does the users with 2 hops. One reason for this might be we have been familiar with those users that are too closer to us. And, users with larger hops, such as 5 hops, might be relatively far away from us.

Topic Property

Some real-world events would experience the process of occurrence, prosperity and depression. For example, the "US Election" is rarely discussed in the next or previous years. In contrast, it is discussed most during the period of the election. Other events, like "Comedy", might be discussed every day.

Thus, these two different kinds of topics are chosen to investigate the social interactions. For every topic, the interaction strengths of all nodes are calculated to present the interaction degree in every time slices. To compare with our method, the traditional method which just considers the interaction frequency is used.

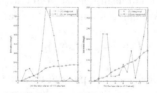

Fig. 7. Social interaction strengths in successive time slices

The results of social interaction strength are shown in Fig. 7 where the horizontal axis stands for the time slices of one specific topic, and the vertical axis stands for the social interaction strength in that time slices respectively. And, curve 1 and curve 2 show the results of our method and the traditional method respectively.

Fig. 7(a) shows the result of "US Election". In the traditional method, the interaction strength is the accumulation of social interaction frequency. Thus, it keeps increasing in the successive time slices. Moreover, we could observe that topic is rarely discussed in the last several time slices as the curve becomes relatively flat. However, we cannot clarify the development of the event. Our method shows a different result. As shown by curve 2 in Fig. 7(a), the interaction strength increase in the second and third time slice while decrease in the next two time slice. But it reaches the maximum value in the seven time slice. There is a burst phase which is similar to the real-world event. In the last several time slices, the interaction strength decays. This demonstrates that the influence of social interactions has the temporal property.

The results of Fig. 7(b) are different. The interaction strength increases in the successive time slices. This could not describe the "Comedy" topic well as this topic is consistently discussed. Our method could be expressed by curve 1 in Fig. 7(b). The interaction strength in every time slice shows that this topic a continuous one, although the values are fluctuant.

In addition, we observe that there are new seed users in every time slices. However, the social interactions are not determined by them. This is because few users interact with them when the topic is not popular any more.

6   Conclusions

The social network is dynamic as lots of users keep expressing their ideas and communicating with others. For an interesting topic, there will be a large amount of discussions since some users post related messages. Moreover, users may comment the messages, and they also reply to other users. Lots of previous works have studied the dynamics of social network. Compared with these studies, clarifying the evolution of social interactions could be much more helpful for understanding users' interests as well as their behaviors.

Social interaction graph is a useful tool to study the interactions among users. However, the temporal information is ignored by traditional social interaction graph, especially in the case of one topic. We argue that the interactions have consistent influence on the process during which the topic becomes popular. Thus, we consider the temporal influence of social interaction using a decay function when we investigate the interactions in Digg dataset.

In this paper, we conclude that the rules of social interactions could be summarized as follows. (1) The distribution of social interaction graph follows the power law distribution. And the exponents increase with time. (2) Users are not always interact with others who own larger degree. In our analysis, the nodes with degree between 100 and 150 are interacted frequently. (3) Social interaction has local effect. Two users with 3 hops tend to form a new interaction easily.

The experiment results of describing two different kinds of topics show that our method outperforms the traditional social interaction graph method.

Acknowledgement. This work is supported by the National Natural Science Foundation of China under Grant Nos.61035004, 61273213. State Key Laboratory of Software Development Environment No.SKLSDE-2011ZX-08.

References

1. Qiu, B., Ivanova, K., Yen, J., Liu, P.: Behavior evolution and event-driven growth dynamics in social networks. In: SocialCom, pp. 217–224 (2010)
2. Perra, N., Gongalves, B., Pastor-Satorras, R., Vespignani, A.: Activity driven modeling of time varying networks. Nature 2, 469 (2012)
3. Holme, P., Saramaki, J.: Temporal networks. Physics Reports 519(3), 97–125 (2012)
4. Newman, M.E.J.: The structure and function of complex networks. SIAM Review 45, 167–256 (2003)
5. Yang, Z., Xue, J., Zhao, X., Wang, X., Zhao, B.Y., Dai, Y.: Unfolding dynamics in a social network: co-evolution of link formation and user interaction. In: WWW (2013)
6. Mahadevan, P., Krioukov, D., Fall, K., Vahdat, A.: Systematic topology analysis and generation using degree correlations. In: SIGCOMM (2006)
7. Barabasi, A.: Emergence of scaling in random networks. Science 286(5439), 509–512 (1999)
8. Vazquez, A.: Growing network with local rules: Preferential attachment, clustering hierarchy, and degree correlations. Physical Review E (2003)
9. Sala, A., Cao, L., Wilson, C., Zablit, R., Zheng, H., Zhao, B.Y.: Measurement-calibrated graph models for social network experiments. In: WWW (2010)
10. Leskovec, J., Backstrom, L., Kumar, R., Tomkins, A.: Microscopic evolution of social networks. In: KDD, pp. 462–470 (2008)
11. Jin, E., Girvan, M., Newman, M.E.J.: Structure of growing social networks. Physical Review E 64, 046132 (2001)
12. Palla, G., Barabasi, A.L., Vicsek, T.: Quantifying social group evolution. Nature 446, 664–667 (2007)
13. Fabbri, R., Junior, V., Fabbri, R., Antunes, D., Pisani, M., Costa, L.: On the evolution of interaction networks: primitive typology of vertex and prominence of measures. arXiv:1310.7769 (2013)
14. Wang, X., Zhai, C., Hu, X., Sproat, R.: Mining correlated bursty topic patterns from correlated text stream. In: KDD (2007)
15. Choudhury, M.D., Sundaram, H., John, A., Seligmann, D.D.: Social synchrony: predicting mimicry of user actions in online social media. In: SocialCom (2009)
16. Banos, R.A., Borge-Holthoefer, J., Moreno, Y.: The role of hidden influentials in the diffusion of online information cascades. EPJ Data Science (2013)

Computing the Kirchhoff Index of Some xyz-Transformations of Regular Molecular Graphs

Yujun Yang

School of Mathematics, Shandong University, Jinan, Shandong 250100 P.R. China
School of Mathematics and Information Science, Yantai University,
Yantai, Shandong 264005 P.R. China
yangyj@yahoo.com

Abstract. Let G be a connected molecular graph. The resistance distance between any two vertices of G is defined as the effective resistance between the two corresponding nodes in the electrical network constructed from G by replacing each edge of G with a unit resistor. The Kirchhoff index of G is defined as the sum of resistance distances between all pairs of vertices. Gao et al. (2012) and You et al. (2013) gave formulae for the Kirchhoff index of two types of $xyz-$transformations, namely, the subdivision graph and the total graph, of regular graphs. In this paper, we compute the Kirchhoff index of some other $xyz-$transformations of regular (molecular) graphs, with explicit formulae for the Kirchhoff index of these transformation graphs being given in terms of parameters of the original graph.

Keywords: regular graph, resistance distance, $xyz-$transformation, Kirchhoff index, Laplacian matrix.

1 Introduction

Let G be a connected molecular graph with vertex set $V(G) = \{v_1, v_2, \ldots, v_n\}$ and edge set $E(G)$. The *resistance distance* [1] between any two vertices v_i and v_j, denoted by $\Omega_G(v_i, v_j)$, is defined to be the effective resistance between the two corresponding nodes in the electrical network constructed from G by replacing each edge of G with a unit resistor. The *Kirchhoff index* [1] of G, denoted by $Kf(G)$, is defined as the sum of resistance distances between all pairs of vertices of G, i.e.

$$Kf(G) = \sum_{\{v_i, v_j\} \subseteq V(G)} \Omega_G(v_i, v_j). \tag{1}$$

The Kirchhoff index is an important resistance distance-based graph invariant, which is also known as the *total effective resistance* [11] or the *effective graph resistance* [8]. In theoretical chemistry, a graph invariant applicable in chemistry to model physico-chemical properties of chemical compounds is called a *topological index* (or *molecular structure descriptor*). Since American physico-chemist Harold

D.-S. Huang et al. (Eds.): ICIC 2014, LNCS 8588, pp. 173–183, 2014.
© Springer International Publishing Switzerland 2014

Wiener [22] proposed the first topological index-the *Wiener index* (i.e., the sum of distances between all pairs of vertices) in 1947, a variety types of such indices were designed, capturing different aspects of molecular structure. As an alternative of the famous Wiener index, the Kirchhoff index also serves as an important molecular structure-descriptor, which turns out to be very useful in the study of Quantitative structure-activity relationship (QSAR) and Quantitative structure-property relationship (QSPR). Consequently, the Kirchhoff index is widely studied in both mathematical and chemical literatures. For more information, the reader is referred to [1-5, 7, 9, 10, 14-17, 19-21, 23-30], and references therein.

We proceed to introduce some notions, concepts, and preliminary results. We follow the notation of [6]. For a graph $G=(V(G),E(G))$, the *complement* G^c of G is the graph with vertex set $V(G^c)=V(G)$ and $v_iv_j \in E(G^c)$ if and only if $v_iv_j \notin E(G)$. The *line graph* G^l of G is the graph with vertex set $E(G)$ and two vertices are adjacent in G^l if and only if the corresponding edges in G are adjacent. Let G^0 be the empty graph with $V(G^0)=V(G)$, G^1 the complete graph with $V(G^1)=V(G)$, $G^+=G$, and $G^-=G^c$. Let $B(G)$ (resp., $B^c(G)$) be the graph with vertex set $V(G) \cup E(G)$ such that (v_i,e) is an edge in $B(G)$ (resp., in $B^c(G)$) if and only if $v_i \in V(G), e \in E(G)$, and vertex v_i is incident (resp., not incident) to edge e in G.

Then the definition of the $xyz-$ transformation G^{xyz} of G is given in the following:

Definition 1. *[6] Given a graph $G=(V(G),E(G))$ and three variables $x,y,z \in \{0,1,+,-\}$, the $xyz-$ transformation G^{xyz} of G is the graph with $V(G^{xyz})=V(G) \cup E(G)$ and $E(G^{xyz})=E(G^x) \cup (E(G^l)^y) \cup E(W)$, where $W=B(G)$ if $z=+$, $W=B^c(G)$ if $z=-$, W is the graph with $V(W)=V(G) \cup E(G)$ and with no edges if $z=0$, and W is the complete bipartite graph with parts $V(G)$ and $E(G)$ if $z=1$.*

The *degree* $d(v_i,G)$ of a vertex $v_i \in V(G)$ is the number of vertices adjacent to v_i. G is called $r-$ regular if every vertex of G has degree r. The *adjacency matrix* $A(G)=(a_{ij})$ is an $n \times n$ matrix with $a_{ij}=1$ if v_i and v_j are adjacent and otherwise $a_{ij}=0$. Let $D(G)=\text{diag}(d(v_1,G),d(v_2,G),...,d(v_n,G))$ be the diagonal matrix of degrees of vertices. Then the Laplacian matrix of G is defined as $L(G)=D(G)-A(G)$. The *Laplacian polynomial* of G is the characteristic polynomial $L(G,\lambda)=\det(\lambda I-L(G))$, where I is the identity matrix. Let $\lambda_0,\lambda_1,...,\lambda_{n-1}$ be the eigenvalues of $L(G)$, called the *Laplacian eigenvalues* of G. Suppose that λ_0 is the smallest Laplacian eigenvalue, then $\lambda_0=0$ and $\lambda_k>0$ for each $k > 0$ [18]. As given in the following elegant result, the Kirchhoff index could be computed in terms of positive Laplaician eigenvalues of G.

Theorem 1. *[12,31] For any connected* $n-vertex$ *graph* G, $n \geq 2$,

$$Kf(G) = n \sum_{k=1}^{n-1} \frac{1}{\lambda_k}. \tag{2}$$

Suppose that $L(G,\lambda) = \lambda^n + a_1\lambda^{n-1} + \cdots + a_{n-2}\lambda^2 + a_{n-1}\lambda$. Then according to relations between roots and coefficients of $L(G,\lambda)$, it directly leads to the following result.

Theorem 2. *[26] For any connected* $n-vertex$ *graph* G, $n \geq 2$,

$$\frac{Kf(G)}{n} = -\frac{a_{n-2}}{a_{n-1}}. \tag{3}$$

The most important problem in the study of the Kirchhoff index is the computation of this graph invariant, which is attracting more and more attention. To reduce the computational time and complexity, much effort has been made. One the one hand, analytical formulae for the Kirchhoff index have been given for some special classes of graphs, especially those graphs with some sort of symmetry, such as complete graphs, cycles, circulant graphs, linear hexagonal chains, folded hypercubes and related complex networks, etc (see, e.g. [16,17,26,29]). On the other hand, for some large composite graphs obtained from certain binary operations on small graphs, formulae for the Kirchhoff index are given in terms of parameters of the parent graphs (see, e.g. [24]). Indeed, it is also worth considering the same sort of problems for graphs derived from a single graph by unitary operations, such as the line graph, the subdivision graph, the triangulation graph, the total graph, and $xyz-$ transformations, etc. To this end, in [10], Gao et al. obtained formulae for the Kirchhoff index of G^{0++} and G^{00+} (the subdivision graph) of a regular graph G. Then the present author [24] generalized their result by giving a formula for the Kirchhoff index of G^{00+} of a general graph G. In [21], Wang et al. gave formula for the triangulation graph and of a regular graph G. In [28], You et al obtained formula for the Kirchhoff index of G^{+++} (the total graph) of a regular graph G. Along this line, in this paper, we compute the Kirchhoff index of some new unitary operations obtained from a regular (molecular) graph G, namely, G^{0++}, G^{+0+}, G^{0+1}, G^{0+-}, and G^{-+}, with explicit formulae being given for these $xyz-$ transformations in terms of parameters of the original graph G. The main technique used in this study, as commonly used in pervious papers, is making use of the Laplacian spectrum of these $xyz-$ transformation graphs and the elegant relation between Laplacian spectrum and the Kirchhoff index in Theorem 1.

2   Main Results

In this section, we compute the Kirchhoff index of G^{0++}, G^{+0+}, G^{0+1}, G^{0+-}, and G^{--+} of a regular graph G. Throughout this section, G is assumed to be r − regular with n vertices, $m = nr/2$ edges, and Laplacian eigenvalues $0, \lambda_1, \ldots, \lambda_{n-1}$.

2.1   Kirchhoff Index of G^{0++}

In [6], the Laplacian polynomial of G^{0++} is characterized, which is expressed in terms of the Laplacian polynomial of G as given in the following result.

Lemma 1. *[6] Let G be an r − regular graph ($r \geq 2$) with n vertices and m edges. Then*

$$L(\lambda, G^{0++}) = (\lambda - r - 2)^n (\lambda - 2r - 2)^{m-n} L\left(\frac{\lambda^2 - (r+2)\lambda}{\lambda - r - 1}, G\right).$$

By Lemma 1, a formula for the Kirchhoff index of G^{0++} could be obtained.

Theorem 3. *Let G be an r − regular graph ($r \geq 2$) with n vertices and m edges. Then*

$$Kf(G^{0++}) = \frac{(r+2)^2}{2} Kf(G) + \frac{n(r+2)[n(r+2)-4]}{8(r+1)} + \frac{n}{2}. \tag{4}$$

Proof. First recall that the Laplacian polynomial of G is
$$L(G, \lambda) = \lambda^n + a_1\lambda^{n-1} + \cdots + a_{n-2}\lambda^2 + a_{n-1}\lambda.$$
By Lemma 1, we get

$$L(\lambda, G^{0++}) = (\lambda - r - 2)^n (\lambda - 2r - 2)^{m-n} L\left(\frac{\lambda^2 - (r+2)\lambda}{\lambda - r - 1}, G\right)$$

$$= (\lambda - r - 1)^n (\lambda - 2r - 2)^{m-n} [\lambda^n \left(\frac{\lambda - r - 2}{\lambda - r - 1}\right)^n + a_1\lambda^{n-1}\left(\frac{\lambda - r - 2}{\lambda - r - 1}\right)^{n-1}$$

$$+ \cdots + a_{n-2}\lambda^2 \left(\frac{\lambda - r - 2}{\lambda - r - 1}\right)^2 + a_{n-1}\lambda\frac{\lambda - r - 2}{\lambda - r - 1}]$$

$$= (\lambda - 2r - 2)^{m-n} [\lambda^n(\lambda - r - 2)^n + a_1\lambda^{n-1}(\lambda - r - 2)^{n-1}(\lambda - r - 1) + \cdots$$

$$+ a_{n-2}\lambda^2 (\lambda - r - 2)^2 (\lambda - r - 1)^{n-2} + a_{n-1}\lambda(\lambda - r - 2)(\lambda - r - 1)^{n-1}].$$

To obtain the Kirchhoff index of G^{0++}, we need to compute the coefficients of λ and λ^2 in $L(\lambda, G^{0++})$, so that the Kirchhoff index of G^{0++} can be derived according to Theorem 2. Denote the coefficients of λ and λ^2 in $L(\lambda, G^{0++})$ by $H_1(\lambda)$ and $H_2(\lambda)$, respectively. Then according to the above expression for $L(\lambda, G^{0++})$, we have

$$H_1(\lambda) = (-2r-2)^{m-n}(-r-2)(-r-1)^{n-1}a_{n-1},$$

and

$$H_2(\lambda) = (-2r-2)^{m-n}[a_{n-2}(-r-2)^2(-r-1)^{n-2} + a_{n-1}(-r-1)^{n-1}$$

$$+a_{n-1}(n-1)(-r-2)(-r-1)^{n-2}] + (m-n)(-2r-2)^{m-n-1}(-r-2)(-r-1)^{n-1}a_{n-1}.$$

Hence by Theorem 2, we have

$$\frac{Kf(G^{0++})}{m+n} = -\frac{H_2(\lambda)}{H_1(\lambda)} = \frac{r+2}{r+1}\left(-\frac{a_{n-2}}{a_{n-1}}\right) + \frac{1}{r+2} + \frac{m+n-2}{2r+2}.$$

Consequently,

$$Kf(G^{0++}) = \frac{(r+2)(m+n)}{r+1}\left(-\frac{a_{n-2}}{a_{n-1}}\right) + \frac{m+n}{r+2} + \frac{(m+n-2)(m+n)}{2r+2}$$

$$= \frac{r+2}{r+1}\frac{m+n}{n}Kf(G) + \frac{m+n}{r+2} + \frac{(m+n-2)(m+n)}{2r+2}$$

$$= \frac{(r+2)^2}{2}Kf(G) + \frac{n(r+2)[n(r+2)-4]}{8(r+1)} + \frac{n}{2},$$

as desired.   ∎

2.2   Kirchhoff Index of G^{+0+}

The relation between the Laplacian polynomial of G^{+0+} and that of G is established as given in the following result.

Lemma 2. *[6] Let G be an $r-$ regular graph ($r \geq 2$) with n vertices and m edges. Then*

$$L(\lambda, G^{+0+}) = (\lambda-3)^n(\lambda-2)^{m-n}L\left(\frac{\lambda^2-(r+2)\lambda}{\lambda-3}, G\right).$$

Thus, we have the following.

Theorem 4. *Let G be an $r-$ regular graph ($r \geq 2$) with n vertices and m edges. Then*

$$Kf(G^{+0+}) = \frac{(r+2)^2}{6}Kf(G) + \frac{(r+2)(n^2-n)}{6} + \frac{(r-4)n^2}{8} + \frac{n}{2}. \tag{5}$$

Proof. By Lemma 2, we know that

$$L(\lambda, G^{+0+}) = (\lambda-3)^n(\lambda-2)^{m-n} L\left(\frac{\lambda^2-(r+2)\lambda}{\lambda-3}, G\right)$$

$$= \frac{(\lambda-3)^n}{(\lambda-2)^{n-m}}\left[\lambda^n\left(\frac{\lambda-r-2}{\lambda-3}\right)^n + \cdots + a_{n-2}\lambda^2\left(\frac{\lambda-r-2}{\lambda-3}\right)^2 + a_{n-1}\lambda\left(\frac{\lambda-r-2}{\lambda-3}\right)\right]$$

$$= (\lambda-2)^{m-n}[\lambda^n(\lambda-r-2)^n + \cdots + a_{n-2}\lambda^2(\lambda-r-2)^2(\lambda-3)^{n-2}$$

$$+ a_{n-1}\lambda(\lambda-r-2)(\lambda-3)^{n-1}].$$

Then from the above expression, it follows that the coefficient of λ in $L(\lambda, G^{+0+})$ is

$$H_1(\lambda) = (-2)^{m-n}(-r-2)(-3)^{n-1}a_{n-1},$$

and the coefficient of λ^2 is

$$H_2(\lambda) = (-2)^{m-n}[a_{n-2}(r+2)^2(-3)^{n-2}] + a_{n-1}(-3)^{n-1} + a_{n-1}(n-1)(-r-2)(-3)^{n-2}$$

$$+ (m-n)(-2)^{m-n-1}(-r-2)(-3)^{n-1}a_{n-1}.$$

Hence we have

$$\frac{Kf(G^{+0+})}{m+n} = -\frac{H_2(\lambda)}{H_1(\lambda)} = \frac{r+2}{3}\left(-\frac{a_{n-2}}{a_{n-1}}\right) + \frac{1}{r+2} + \frac{m-n}{2} + \frac{n-1}{3}.$$

By Theorem 2, it follows immediately that

$$Kf(G^{+0+}) = \frac{r+2}{3}\frac{m+n}{n}Kf(G) + \frac{m+n}{r+2} + \frac{(m-n)(m+n)}{2} + \frac{(n-1)(m+n)}{3}$$

$$= \frac{(r+2)^2}{6}Kf(G) + \frac{n(n-1)(r+2)}{6} + \frac{n^2(r^2-4)}{8} + \frac{n}{2}.$$

This completes the proof.   ∎

2.3   Kirchhoff Index of G^{0+1}

The Laplacian polynomial of G^{0+1} relates the Laplaican polynomial of G in the following way.

Lemma 3. *[6] Let G be an $r-$regular graph ($r \geq 2$) with n vertices and m edges. Then*

$$L(\lambda, G^{0+1}) = \lambda(\lambda-m-n)(\lambda-m)^{n-1}(\lambda-n)^{-1}(\lambda-2r-n)^{m-n}L(\lambda-n, G).\,.$$

As a consequence, we have the following formula for $Kf(G^{0+1})$.

Theorem 5. *Let G be an $r-$regular graph ($r \geq 2$) with n vertices and m edges. Then*

$$Kf(G^{0+1}) = \frac{(n-1)(r+2)}{r} + \frac{n^2(r^2-4)}{8r+4n} + \sum_{i=1}^{n-1} \frac{n(2+r)}{2(n+\lambda_i)} + 1. \tag{6}$$

Proof. Note first that

$$L(\lambda - n, G) = (\lambda - n)\prod_{i=1}^{n-1}(\lambda - n - \lambda_i).$$

Then by Lemma 3, it is easily seen that the Laplacian polynomial of G^{0+1} is

$$L(\lambda, G^{0+1}) = \lambda(\lambda - m - n)(\lambda - m)^{n-1}(\lambda - n)^{-1}(\lambda - 2r - n)^{m-n}L(\lambda - n, G)$$

$$= \lambda(\lambda - m - n)(\lambda - m)^{n-1}(\lambda - n)^{-1}(\lambda - 2r - n)^{m-n}(\lambda - n)\prod_{i=1}^{n-1}(\lambda - n - \lambda_i)$$

$$= \lambda(\lambda - m - n)(\lambda - m)^{n-1}(\lambda - 2r - n)^{m-n}\prod_{i=1}^{n-1}(\lambda - n - \lambda_i).$$

Hence the positive Laplacian eigenvalues of G^{0+1} are

$$m+n, \underbrace{m, \cdots, m}_{n-1}, \underbrace{2r+n, \ldots, 2r+n}_{m-n}, \lambda_1 + n, \lambda_2 + n, \ldots, \lambda_{n-1} + n.$$

Then by Theorem 1, we have

$$Kf(G^{0+1}) = (m+n)\left(\frac{1}{m+n} + \frac{n-1}{m} + \frac{m-n}{2r+n} + \sum_{i=1}^{n-1}\frac{1}{n+\lambda_i}\right)$$

$$= 1 + \frac{(n-1)(m+n)}{m} + \frac{(m-n)(m+n)}{2r+n} + \sum_{i=1}^{n-1}\frac{m+n}{n+\lambda_i}$$

$$= \frac{(n-1)(r+2)}{r} + \frac{n^2(r^2-4)}{8r+4n} + \sum_{i=1}^{n-1}\frac{n(2+r)}{2(n+\lambda_i)} + 1.$$

The proof is complete.   ■

2.4   Kirchhoff Index of G^{0+-}

To compute $Kf(G^{0+-})$, we need the Laplacian polynomial of G^{0+-} as given in the following result.

Lemma 4. *[6] Let G be an $r-regular$ graph ($r \geq 2$) with n vertices and m edges. Then*

$$L(\lambda, G^{0+-}) = \lambda(\lambda - m - n + r + 2)(\lambda - n - 2r + 2)^{m-n} \times$$

$$\prod_{i=1}^{n-1}[(\lambda - m + r)(\lambda - n + 2 - \lambda_i) - 2r + \lambda_i].$$

Now we are ready to give a formula for the Kirchhoff index of G^{0+-}.

Theorem 6. *Let G be an $r-$regular graph ($r \geq 2$) with n vertices and m edges. Then*

$$Kf(G^{0+-}) = \frac{n}{n-2} + \frac{n^2(r^2-4)}{4(n+2r-2)} + \frac{r+2}{2}\sum_{i=1}^{n-1}\frac{(n-2)(r+2)+2\lambda_i}{[(n-2)r+2]\lambda_i+nr(n-2)}. \tag{7}$$

Proof. For $1 \leq i \leq n-1$, let

$$P_i(x) = (\lambda - m + r)(\lambda - n + 2 - \lambda_i) - 2r + \lambda_i.$$

Then simple calculation yields

$$P_i(x) = \lambda^2 - (m+n-r-2+\lambda_i)\lambda + (m-r+1)\lambda_i + 2mn - 4m.$$

Suppose the two roots of $P_i(x)$ are μ_i^1 and μ_i^2 . Then by relations between coefficients and roots of $P_i(x)$, it follows that

$$\mu_i^1 + \mu_i^2 = m+n-r-2+\lambda_i,$$

$$\mu_i^1\mu_i^2 = (m-r+1)\lambda_i + 2mn - 4m,$$

which gives that

$$\frac{1}{\mu_i^1} + \frac{1}{\mu_i^2} = \frac{\mu_i^1 + \mu_i^2}{\mu_i^1\mu_i^2} = \frac{m+n-r-2+\lambda_i}{(m-r+1)\lambda_i + 2mn - 4m}.$$

Note, by Lemma 4, that the positive Laplacian eigenvalues of G^{0+-} are

$$m+n-r-2, \underbrace{n+2r-2, \ldots, n+2r-2}_{m-n}, \mu_1^1, \mu_1^2, \mu_2^1, \mu_2^2, \ldots, \mu_{n-1}^1, \mu_{n-1}^2.$$

Hence by Theorem 1, we get

$$Kf(G^{0+-}) = (m+n)\left[\frac{1}{m+n-r-2} + \frac{m-n}{n+2r-2} + \sum_{i=1}^{n-1}\left(\frac{1}{\mu_i^1} + \frac{1}{\mu_i^1}\right)\right]$$

$$= \frac{m+n}{m+n-r-2} + \frac{(m+n)(m-n)}{n+2r-2} + \sum_{i=1}^{n-1}\frac{(m+n)(m+n-r-2+\lambda_i)}{\lambda_i(m-r+1)+2mn-4m}$$

$$= \frac{n}{n-2} + \frac{n^2(r^2-4)}{4(n+2r-2)} + \frac{r+2}{2}\sum_{i=1}^{n-1}\frac{(n-2)(r+2)+2\lambda_i}{[(n-2)r+2]\lambda_i+nr(n-2)}.$$

Thus we have done.   ∎

2.5   Kirchhoff Index of G^{--+}

As before, we first introduce the formula for the Laplacian polynomial of G^{--+}, which plays an essential rule in the computation of the Kirchhoff index of G^{--+}.

Lemma 5. *[6] Let G be an $r-$ regular graph ($r \geq 2$) with n vertices and m edges. Then*

$$L(\lambda, G^{--+}) = \lambda(\lambda - r - 2)(\lambda - m + 2r - 2)^{m-n} \prod_{i=1}^{n-1} [(\lambda - n - r + \lambda_i)(\lambda - m - 2 + \lambda_i) - 2r + \lambda_i].$$

Making use of Lemma 5, we have the following result.

Theorem 7. *Let G be an $r-$ regular graph ($r \geq 2$) with n vertices and m edges. Then*

$$Kf(G^{--+}) = \frac{n^2(r^2-4)}{2nr+8-8r} + \frac{nr+2n}{2(r+2)} + \sum_{i=1}^{n-1} \frac{n(r+2)(nr+2n+2r+4-4\lambda_i)}{2(n+r-\lambda_i)(nr+4-2\lambda_i)-8r+4\lambda_i}. \tag{8}$$

Proof. For $1 \leq i \leq n-1$, let

$$Q_i(x) = (\lambda - n - r + \lambda_i)(\lambda - m - 2 + \lambda_i) - 2r + \lambda_i,$$

and let v_i^1 and v_i^2 be the two roots of $Q_i(x)$. Then according to relations between coefficients and roots of $Q_i(x)$, we have

$$v_i^1 + v_i^2 = m + n + r + 2 - 2\lambda_i,$$

$$v_i^1 v_i^2 = (n + r - \lambda_i)(m + 2 - \lambda_i) - 2r + \lambda_i.$$

Hence,

$$\frac{1}{v_i^1} + \frac{1}{v_i^2} = \frac{v_i^1 + v_i^2}{v_i^1 v_i^2} = \frac{m+n+r+2-2\lambda_i}{(n+r-\lambda_i)(m+2-\lambda_i)-2r+\lambda_i}.$$

Clearly, by Lemma 5, the positive Laplacian eigenvalues of G^{--+} are

$$r+2, \underbrace{m+2-2r, \ldots, m+2-2r}_{m-n}, v_1^1, v_1^2, v_2^1, v_2^2, \ldots, v_{n-1}^1, v_{n-1}^2.$$

Then by Theorem 1, we have

$$Kf(G^{-+}) = (m+n)\left[\frac{m-n}{m+2-2r} + \frac{1}{r+2} + \sum_{i=1}^{n-1}\left(\frac{1}{v_i^1} + \frac{1}{v_i^1}\right)\right]$$

$$= \frac{(m+n)(m-n)}{m+2-2r} + \frac{m+n}{r+2} + \sum_{i=1}^{n-1}\frac{(m+n)(m+n+r+2-2\lambda_i)}{(n+r-\lambda_i)(m+2-\lambda_i)-2r+\lambda_i},$$

$$= \frac{n^2(r^2-4)}{2nr+8-8r} + \frac{nr+2n}{2(r+2)} + \sum_{i=1}^{n-1}\frac{n(r+2)(nr+2n+2r+4-4\lambda_i)}{2(n+r-\lambda_i)(nr+4-2\lambda_i)-8r+4\lambda_i}.$$

This completes the proof. ∎

3   Concluding Remarks

In the present paper, simple formulae for the Kirchhoff index of some types of xyz − transformations of a regular graph G are derived in terms of parameters of G.

However, for a general graph G, such type of formula is only known for G^{00+} (the subdivision graph). Thus it remains challenging to establish formulae for the Kirchhoff index of other types of xyz − transformations of a general graph G in terms of parameters of G.

Acknowledgement. The support of the National Science Foundation of China (through Grant No. 11201404), China Postdoctoral Science Foundation (through Grant Nos. 2012M521318 and 2013T60662), and Special Funds for Postdoctoral Innovative Projects of Shandong Province (through Grant No. 201203056), is greatly acknowledged.

References

1. Bianchi, M., Cornaro, A., Palacios, J.L., Torriero, A.: Bounds for the Kirchhoff index via majorization techniques. J. Math. Chem. 51, 569–587 (2013)
2. Das, K.C.: On the Kirchhoff index of graphs. Z. Naturforsch. 68a, 531–538 (2013)
3. Das, K.C., Güngör, A.D., Çevik, A.S.: On Kirchhoff index and resistance-distance energy of a graph. MATCH Commun. Math. Comput. Chem. 67, 541–556 (2012)
4. Das, K.C., Xu, K., Gutman, I.: Comparison between Kirchhoff index and the Laplacian-energy-like invariant. Linear Algebra Appl. 436, 3661–3671 (2012)
5. Deng, Q., Chen, H.: On the Kirchhoff index of the complement of a bipartite graph. Linear Algebra Appl. 439, 167–173 (2013)
6. Deng, A., Kelmans, A., Meng, J.: Laplacian spectra of regular graph transformations. Discrete Appl. Math. 161, 118–133 (2013)
7. Deng, Q., Chen, H.: On extremal bipartite unicyclic graphs. Linear Algebra Appl. 444, 89–99 (2014)
8. Ellens, W., Spieksma, F.M., Van Mieghem, P., Jamakovic, A., Kooij, R.E.: Effective graph resistance. Linear Algebra Appl. 435, 2491–2506 (2011)
9. Feng, L., Yu, G., Xu, K., Jiang, Z.: A note on the Kirchhoff index of bicyclic graphs. Ars Combin. (in press)

10. Gao, X., Luo, Y., Liu, W.: Kirchhoff index in line, subdivision and total graphs of a regular graph. Discrete Appl. Math. 160, 560–565 (2012)
11. Ghosh, A., Boyd, S., Saberi, A.: Minimizing effective resistance of a graph. SIAM Rev. 50, 37–66 (2008)
12. Gutman, I., Mohar, B.: The Quasi-Wiener and the Kirchhoff indices coincide. J. Chem. Inf. Comput. Sci. 36, 982–985 (1996)
13. Klein, D.J., Randić, M.: Resistance distance. J. Math. Chem. 12, 81–95 (1993)
14. Kuang, X., Yan, W.: The Kirchhoff indices of some graphs. J. Jimei Univ. 17, 65–70 (2012)
15. Li, R.: Lower bounds for the Kirchhoff index. MATCH Commun. Math. Comput. Chem. 70, 163–174 (2013)
16. Liu, J., Cao, J., Pan, X., Elaiw, A.: The Kirchhoff index of hypercubes and related complex networks. Discrete Dyn. Nat. Soc. 2013, Article ID 543189, 7 pages (2013)
17. Liu, J., Pan, X., Wang, Y., Cao, J.: The Kirchhoff Index of Folded Hypercubes and Some Variant Networks. Math. Probl. Eng. 2014, Article ID 380874, 9 pages (2014)
18. Merris, R.: Laplacian matrix of graphs: A survey. Linear Algebra Appl. 197–198, 143–176 (1994)
19. Nikseresht, A., Sepasdar, Z.: On the Kirchhoff and the Wiener Indices of Graphs and Block Decomposition. Electron. J. Combin. 21, #P1.25 (2014)
20. Shirdareh-Haghighi, M.H., Sepasdar, Z., Nikseresht A.: On the Kirchhoff index of graphs and some graph operations. Proc. Indian Acad. Sci. (in press)
21. Wang, W., Yang, D., Luo, Y.: The Laplacian polynomial and Kirchhoff index of graphs derived from regular graphs. Discrete Appl. Math. 161, 3063–3071 (2013)
22. Wiener, H.: Structural determination of paraffin boiling points. J. Amer. Chem. Soc. 69, 17–20 (1947)
23. Yang, Y.: Bounds for the Kirchhoff index of bipartite graphs. J. Appl. Math. 2012, Article ID 195242, 9 pages (2012)
24. Yang, Y.: The Kirchhoff index of subdivisions of graphs. Discrete Appl. Math. 171, 153–157 (2014)
25. Yang, Y., Jiang, X.: Unicyclic graphs with extremal Kirchhoff index. MATCH Commun. Math. Comput. Chem. 60(1), 107–120 (2008)
26. Yang, Y., Zhang, H.: Kirchhoff index of linear hexagonal chains. Int. J. Quantum Chem. 108, 503–512 (2008)
27. Yang, Y., Zhang, H., Klein, D.J.: New Nardhaus-Gaddum-type results for the Kirchhoff index. J. Math. Chem. 49(8), 1587–1598 (2011)
28. You, Z., You, L., Hong, W.: Comment on "Kirchhoff index in line, subdivision and total graphs of a regular graph". Discrete Appl. Math. 161, 3100–3103 (2013)
29. Zhang, H., Yang, Y.: Resistance distance and Kirchhoff index in circulant graphs. Int. J. Quantum Chem. 107(2), 330–339 (2007)
30. Zhang, H., Yang, Y., Li, C.: Kirchhoff index of composite graphs. Discrete Appl. Math. 107, 2918–2927 (2009)
31. Zhu, H.-Y., Klein, D.J., Lukovits, I.: Extensions of the Wiener number. J. Chem. Inf. Comput. Sci. 36, 420–428 (1996)

Parameters Selection for Genetic Algorithms and Ant Colony Algorithms by Uniform Design

Wenyong Dong and Xueshi Dong

Computer School, Wuhan University, Wuhan Hubei 430072, China
xueshi.wh@gmail.com

Abstract. This paper presents a method based on Uniform Design (UD) to choose appropriate parameters of Genetic Algorithm (GA) and Ant Colony Algorithm (ACA) for solving Traveling Salesmen Problem (TSP). Although meta-heuristic algorithms, such as GA and ACA, are ideal optimizer for engineering practice, one of their drawbacks focuses on performance dependency on the parameters, which differ from one problem to another. It is interesting and significant to properly choose algorithms and their parameters to adapt different problems. Thus this paper utilizes UD to choose the limited representative parameter combinations and check the effectiveness with the traditional exploratory method. Experiments show the performance of GA and ACA with different parameter configuration and problems changes drastically.

Keywords: Uniform Design, Genetic Algorithm, Ant Colony Algorithm, TSP.

1 Introduction

Uniform design, known as the uniform design experiment method or space filling design, is an experimental design method. It is the experimental design method which tests points of the test range in uniform spread. Professor Fang Kaitai and mathematician Wang Yuan put it forward together in 1978. The uniform design is one of the methods of statistical experimental design, the method and many other experimental design methods can complement each other, such as orthogonal design, optimum design, design and Bias rotation design. The essence of experimental design method is selecting representative points in all the test range. Fang Kaitai and Wang Yuan completed the "uniform design theory, methods and Applications", first establish the uniform design theory and method, reveal the internal relations of uniform design and classical factor design, optimal design, modern saturation design and assembly design, and prove that the uniform design has better robustness than the traditional experimental design, the work involves number theory, function theory, experimental design, stochastic optimization and computational complexity, and opens a new research areas.

Genetic algorithm and ant colony algorithm are proposed as a bionic algorithm, inspired by biological evolution, the difference is that genetic algorithm is simulation of the evolution of genes, and ACA simulates the behavior of the group composed by simple individuals. The two algorithms are applied to solve combination optimization problems, and have achieved some achievement. The traveling salesman problem, or known as TSP,

D.-S. Huang et al. (Eds.): ICIC 2014, LNCS 8588, pp. 184–195, 2014.

is a famous problem in mathematics; it is one of the most important research fields in the field or in geographic information science. Assuming there is a travel businessman to visit n city, must choose the path to go, the restriction is to visit each city only once, and finally go back to the original departure city. The target of selection path is to get the distance to the minimum value among all paths, or called shortest path.

2   Solve TSP with Genetic Algorithm

Genetic Algorithm, or named GA, is a computational model for the simulation of Darwin's genetic selection and natural elimination of the biological evolution process. It stems from the idea of genetics and the natural law of survival of the fit; it is the search algorithm of random probability with "survival examination" and iterative process. Genetic algorithm takes each individual of a group as the object, and use randomization techniques to guide efficient search. Among them, selection, crossover and mutation operation of genetic algorithm compose of genetic operators; there are parameter coding, setting of initial population, fitness function design, and design of genetic operation and setting of control parameter, these 5 elements constitute the core content of the genetic algorithm.

The design steps of genetic algorithm:

(1) Determine the data coding scheme, randomly generated a set of initial individuals consists of initial population.

(2) Give adaptive function of evaluating individual, and assess each individual match value or fitness. Individual evaluation is the calculation of the path for each individual's length, and the length as the individual fitness function, then set the fitness function for the $f(x)$:

$$f(x) = \sum_{i=1}^{n-1} d(i, i+1) + d(1, n) \tag{1}$$

Among them, $d(i, j)$ is the distance between the city i and city j. The fitness individual is smaller, the path to the individual is shorter, and the individual is better.

(3) Judge whether the algorithm convergence criterion is satisfied, if meet the criterion, output the search results, otherwise perform the following steps.

(4) Perform a copy operation based on the adaptive size in a certain way.

(5) According to crossover probability, implement the crossover operations.

(6)According to mutation probability p, do crossover operations by using inversion mutation. Namely, randomly select two points in chromosomes and reverse sub-string between them.   (7) Back to step 3.

3   Solve TSP with Ant Colony Algorithm

Ant colony algorithm, referred as ACA, first proposed by Italian scholar Dorigo.M in 1996. Ant colony algorithm simulates food search process of artificial ants (Through the exchange of information between the individual and mutual cooperation, it is to find the shortest distance from the nest to a food source) to solve the complex optimization problems. There is left pheromone on the traversed path when ants are foraging, while pheromone is volatile over time. The more ants walk on a path, the more pheromone is left; the other hand, high concentrations of pheromone on the path will attract more ants. Taking TSP for example, the solution of the n city TSP of ant colony system model is followed:

$$
p_{ij}^{k}(k) = \begin{cases} \dfrac{\tau_{ij}^{\alpha}\eta_{ij}^{\beta}(t)}{\sum_{s}\tau_{is}^{\alpha}\eta_{is}^{\beta}(t)} & s, j \notin tabu_{k} \\ 0 & otherwise \end{cases} \tag{2}
$$

m is the number of ant, $\tau_{ij}(t)$ and $p_{ij}^{k}(t)$ respectively represent the amount of information of the road between city i and city j as well as the probability of the transfer of ant k from city i to city j in the time t. After n times moving, each ant completes one cycle, the shortest one is recorded in closed paths walked by all ants, α 、 β respectively, accumulated information of the kth ant in the process of motion and the different roles of inspiration factor when the ants select the path; $\eta_{ij}(t)$ represents the degree of expectation of moving from the city i to city j determined by some heuristic algorithm; tabuk (k = 1,2, ..., m)record the cites traversed by kth ant, The list is dynamically adjusted at any time. In addition, the amount of information of the path between cities can be updated by the style:

$$
\tau_{ij}(t+n) = (1-\rho)(t) + \Delta\tau_{ij}, \Delta\tau_{ij} = \sum_{k=1}^{m}\Delta\tau_{ij}^{k} \tag{3}
$$

The parameter $1-\rho$ represents the change of information, $\Delta\tau_{ij}^{k}$ as the increment of the amount of information of kth ant in this cycle of the path ij, $\Delta\tau_{ij}$ indicates the total increment on the amount of information of the path ij of this loop.

Dorigo.M had given three different models known as Ant Cycle System, Ant Quantity System, Ant Density System; their difference is the different ways of expression $\Delta\tau_{ij}^{k}$. In the Ant Cycle System model:

$$
\Delta\tau_{ij}^{k} = \begin{cases} \dfrac{Q}{L_{K}} & if \quad the \quad k_{th} \quad ant \quad uses \quad edge(i, j) \quad in \quad this \quad cycle \\ 0 & otherwise \end{cases} \tag{4}
$$

This experiment of the paper adopts Ant Cycle System model.

4   Experiment Simulation and Analysis

In order to compare the validity and practicability of the two basic algorithms in solving the TSP problem, the system simulation interface of GA and ant ACA for solving TSP problems is in the following based on C++. The codes of the experiment are provided by Peter Kohout from Australia, the website is www.codeproject.com. Figure 1 is the interface of GA and ACA for solving TSP.

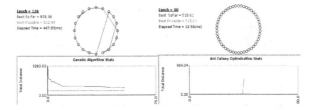

Fig. 1. The interface of GA and ACA for solving TSP

Figure 1 is divided into two parts by the middle of the horizontal line, the upper part is a running simulation of genetic algorithm and ant colony algorithm for solving TSP problems; the lower part is the changes of current solution approaching the optimal solution in the solving process, when the dynamic changes lines coincide with below one, the system obtains the optimal solution; if they do not coincide in a long period of time, means that it is not the optimal solution or is trapped into local optimal solution.

In this Figure, the part above the middle line, there are some important terms, "Best So Far" is current solution of the system; "Best Possible" shows the best solution of algorithm; "Elapsed Time" stands for the running time of the algorithm in the simulating system. In the part of below this middle line, if the above line overlaps with the line under itself, it means that the simulation gets the best solution; if the upper line doesn't coincide with the line under itself, it shows that it isn't the best solution or it falls into the locally optimal solution.

Parameters of genetic algorithm are the population size N, crossover probability Pc and mutation probability Pm. Different combinations of the three parameters affect the performance of algorithm, because the number of combinations is large scale, the experiment has the certain difficulty. There is an idea which is from the combination of parameters selected a representative combinatorial experiment, this is the uniform design method. Using uniform design to set genetic algorithm parameter is followed:

(1) Determine each value range of the three parameters;

(2) Remove the n level of each parameter; find out the uniform design table Un (nn-1), according to the recommended table, use the s column of n-1 column, and obtain Un (NS);

(3) According to the uniform table, make test plan, test many times for each group, and statistic the data to find out the optimal combination, table 1 is uniform scheme of GA[9].

The table below is GA uniform design; the number of experiments is a total of 11, parameters settings are as shown in table 3.

Table 1. Uniform table of GA

Group	1	2	3	4	5	6	7	8	9
1	1	2	3	4	5	6	7	8	9
2	2	4	6	8	10	1	3	5	7
3	3	6	9	1	4	7	10	2	5
4	4	8	1	5	9	2	6	10	3
5	5	10	4	9	3	8	2	7	1
6	6	1	7	2	8	3	9	4	10
7	7	3	10	6	2	9	5	1	8
8	8	5	2	10	7	4	1	9	6
9	9	7	5	3	1	10	8	6	4
10	10	9	8	7	6	5	4	3	2
11	11	11	11	11	11	11	11	11	11

Table 2. The usage table of $U_{11}(11^{10})$

Factors			Colum			
2	1	5				
3	1	5	7			
4	1	2	5	7		
5	1	2	3	5	7	
6	1	2	3	5	7	10

Table 3. The uniform design of GA

Group	N	Pc	Pm
1	1(10)	5(0.41)	7(0.16)
2	2(20)	10(0.86)	3(0.12)
3	3(30)	4(0.32)	10(0.19)
4	4(40)	9(0.77)	6(0.15)
5	5(50)	3(0.23)	2(0.11)
6	6(60)	8(0.68)	9(0.18)
7	7(70)	2(0.14)	5(0.14)
8	8(80)	7(0.59)	1(0.1)
9	9(90)	1(0.05)	8(0.17)
10	10(100)	6(0.5)	4(0.13)
11	11(110)	11(0.95)	11(0.2)

Choose the groups from 1th to 11th group of listed data in Table 3, the experiment runs 15 times in the simulation system, the experimental results obtained are as shown in table 4.

Table 4. GA experimental results based on uniform design (20 cities)

Group	Best time	Worst time	Average time	Local solutions
1	603.39	1365.70	1085.45	9
2	1021.40	1929.78	1398.24	8
3	566.92	1239.65	905.96	8
4	488.64	1285.9	825.84	9
5	702.30	2341.97	1302.13	7
6	561.32	1276.42	922.89	5
7	716.82	1288.93	1016.05	10
8	881.09	2544.13	1591.31	9
9	728.01	1027.19	868.06	10
10	481.67	1742.55	957.84	7
11	582.69	942.37	775.24	6

Table 4 just show us one case with city number as 20, in order to make it convincing, we add the example of 30 cities, the experiment results are in the table 5.

Table 5. GA experimental results based on uniform design (30 cities)

Group	Best Time	Worst Time	Average Time	Local Solutions
1	2430.11	3845.21	3063.49	12
2	3912.54	5070.45	4503.58	11
3	2228.46	5001.78	3388.06	12
4	2581.25	2734.39	2657.82	13
5	3105.52	5761.45	4820.06	13
6	2886.70	3407.54	3147.12	13
7	∞	∞	∞	15
8	5540.33	6863.40	6201.65	13
9	5408.58	5408.58	5408.58	14
10	3167.40	4516.52	3926.10	11
11	1460.85	2469.13	1907.90	11

According to the many experiments in the paper, if the number of city is more 50, it is very easy to fall into local solutions. From table 4 and table 5 we can see, under the experimental conditions, the genetic algorithm for solving the TSP problem is that the characteristic of algorithm is easy to fall into local optimal solution, the numbers of local solutions is all more than 10, the worst time and optimal time of best running solution fluctuates greatly, and the relation between the population quantity and time is not obvious. In this paper, through the verification, the population is 1.5 to 5 times than the number of city, it use less time by using genetic algorithm to solve the TSP; from the chart we can see, the population is during the range, the algorithm can get better results. According to the experimental data, the probability of crossover genetic algorithm is generally larger, when the crossover probability of the algorithm is during 0.7 ~ 1, the algorithm can obtain better results; when the mutation probability is from 0.15 to 0.2, the experiment result of the system is better; the population is identified as city number 1.5 ~ 5 times, the algorithm can obtain better results.

According to the analysis of experimental parameters settings above, we choose appropriate parameters for solving TSP of different city, the simulation experiment parameters setting of genetic algorithm are as followed: crossover rate Pc as 0.95, mutation rate Pm as 0.20, population N as 110, the results of running 15 times of each problem is shown in table 6.

Table 6. Time change of GA for solving TSP (time unit: ms)

City number	Best time	Worst time	Average time	Local solutions
5	0.49	0.52	0.51	0
10	49.92	149.42	93.95	0
15	197.85	616.54	387.73	1
20	592.84	981.66	795.61	11
25	650.95	2005.67	1510.75	11
30	1460.85	2469.13	1907.90	11
35	2359.87	3401.17	2769.33	12
40	3876.09	5061.23	4468.66	13
45	4837.59	4837.59	4837.59	14
50	∞	∞	∞	15
75	∞	∞	∞	15
100	∞	∞	∞	15
150	∞	∞	∞	15
250	∞	∞	∞	15

Table 6 represents the worst time, average time and the number of local solutions under running 15 times different TSP problems to get optimal solution, "∞" means that the experiment falls into local optimal solution with 15 times in the table, which cannot get the optimal solution.

From table 6 we can see when the city number is small, the time spent on solving TSP is very little, with the number becoming larger, the time increases quickly, while the number is more than 50, GA of solving the problem is very easy to fall the local solutions.

From the beginning of the table 7 below, it is the part of ant colony algorithm based on uniform design for solving TSP. In order to validate the uniform design, the exploratory experiment is also used in simulation experiment in order to verify the practicability and validity of the uniform design. Through the test, we can draw conclusion that utilizing uniform design to choose the proper parameters of ACA is possible and valid in this paper.

The experimental simulation system test the combination of the different parameters of the experiment by using exploratory experiment method[11], to verify the effectiveness of the application of uniform experimental design and the application feasibility in the algorithm, α as 2, β as 4 and ρ as 0.67 are as fixed parameters to solve the TSP problem of 50 cities, three of parameters are changed, the ranges are that alpha is from 0 to 5, beta is during 0 to 15, P is among 0.1 to 0.9, the number of 50 cities is fixed, the running result is shown in the table 7.

Table 7. The experimental results of exploratory experiment

Parameter type	Parameter setting	Best time	Worst time	Average time	Local solutions
	0	11.14	24.02	14.23	0
α	1	11.45	24.62	13.67	0
	2	11.67	13.38	12.18	0
	3	11.60	24.76	16.19	0
	4	11.59	24.05	13.18	0
	5	9.4	24.96	13.75	0
	0	∞	∞	∞	15
β	1	∞	∞	∞	15
	2	24.70	26.57	25.60	0
	5	10.69	13.09	11.79	0
	10	11.02	12.33	11.68	0
	15	11.47	12.52	11.91	0
	0.1	11.03	24.95	12.76	0
ρ	0.3	11.48	23.48	13.02	0
	0.5	11.44	23.90	12.97	0
	0.7	8.84	12.82	11.67	0
	0.9	10.88	13.02	12.15	0

According to table 7 we can see when α is 2 and other parameters are fixed, and the running result is best one in the table; when beta is 5, it can obtain better results; when the value p is 0.7, the simulation experimental results can get best result. Uniform design scheme of ACA is shown in the table 8.

According the fifteen groups in table 8, the running results of the experiment are followed in table 9.

Table 8. Uniform design of ACA

Group	m	ρ	α	β
1	1(8)	8(0.666)	6(2.3)	12(4.1)
2	2(9)	2(0.438)	12(4.1)	6(2.3)
3	3(10)	12(0.818)	3(1.4)	2(1.1)
4	4(11)	15(0.932)	9(3.2)	9(3.2)
5	5(12)	4(0.514)	8(2.9)	15(5.0)
6	6(13)	9(0.704)	15(5.0)	4(1.7)
7	7(14)	6(0.590)	1(0.8)	7(2.6)
8	8(15)	13(0.856)	13(4.4)	14(4.7)
9	9(16)	1(0.400)	4(1.7)	10(3.5)
10	10(17)	7(0.628)	10(3.5)	1(0.8)
11	11(18)	14(0.894)	5(2.0)	5(2.0)
12	12(19)	5(0.552)	14(4.7)	11(3.8)
13	13(20)	10(0.742)	2(1.1)	13(4.4)
14	14(21)	3(0.476)	7(2.6)	3(1.4)
15	15(22)	11(0.780)	11(3.8)	8(2.9)

Table 9. The experimental results of ACA based on uniform design (50 cities)

Group	Best time	Worst time	Average time	Local solutions
1	11.11	15.32	13.88	0
2	21.92	47.87	33.41	3
3	13.81	101.56	67.52	12
4	18.50	39.99	27.64	0
5	18.39	21.88	20.05	0
6	39.94	46.88	44.84	0
7	17.33	49.26	38.76	0
8	22.33	27.93	24.41	0
9	18.43	28.01	25.53	0
10	∞	∞	∞	15
11	57.51	63.51	60.00	0
12	28.66	32.58	30.57	0
13	30.06	33.39	32.01	0
14	96.16	135.49	108.96	11
15	34.86	72.90	54.32	0

When the city number is 100, the experimental results are in the table below.

Table 10. The experimental results of ACA based on uniform design (100 cities)

Group	Best time	Worst time	Average time	Local solutions
1	40.98	99.34	54.57	0
2	104.81	162.76	121.52	7
3	∞	∞	∞	15
4	62.18	134.68	121.71	0
5	57.91	73.80	69.75	0
6	∞	∞	∞	15
7	∞	∞	∞	15
8	83.06	173.42	103.98	0
9	92.02	189.49	159.99	0
10	∞	∞	∞	15
11	174.31	190.47	185.16	0
12	113.12	225.80	201.81	0
13	110.14	226.34	121.11	0
14	∞	∞	∞	15
15	176.28	226.86	219.05	0

From table 10 we can see that when the group is 15, 1 and 14, ACA for solving TSP can get better results, it is just only one example with city number as 50, in order to choose proper parameters to get better results, we add more cases such as the 100 cities and 150 cities.

When the city number is 100, the experimental results are in the table 10; and city number as 150 is shown in the table 11.

Table 11. The experimental results of ACA based on uniform design (150 cities)

Group	Best time	Worst time	Average time	Local solutions
1	93.70	214.47	155.28	0
2	∞	∞	∞	15
3	∞	∞	∞	15
4	243.77	290.48	272.04	0
5	137.16	160.02	152.94	0
6	∞	∞	∞	15
7	∞	∞	∞	15
8	181.36	370.39	228.05	0
9	286.53	374.95	346.63	0
10	∞	∞	∞	15
11	334.78	539.01	400.40	0
12	181.27	414.78	370.27	0
13	183.01	261.43	244.59	0
14	∞	∞	∞	15
15	391.44	474.50	461.62	0

According to the mathematical model of ant colony algorithm, parameters which influence ant colony system are heuristic factor, expectation factor and the pheromone, the large number of ant does not mean better result, the number of ants in this experiment is during 8 to 12, the experiment results are better from the data, we can see from the table when the value is 8 it can get the best results. The combination configuration of inspiration factorα, expected heuristic factor β, information retention$\rho$are key important, if the configuration is improper, it will cause to the slow solving speed, and lead to bad solution. Judging the 15 group experiments in table 9, table 10 and table 11, the experimental results in the first group is the better, heuristic factorα, expected heuristic factor β and information retain degree ρ are as 2.3, 4.1 and 0.666, in order to make the setting convenient, the final parameters are changed to be 2, 4 and 0.67. From the experiment we can know appropriate configuration is 2, 4 and 0.67, the experiment can get better result, the higher pheromone Q is, the more fast speed of convergence of ACA is, but the convergence speed of the algorithm can influence the global search. Therefore, according to actual situation, the Q is during 50 to 150 [12]. According uniform design, we can obtain the better setting of parameters of ACA for solving TSP as following: α is as 2, β as 4, ρ as 0.67 and m as 8, the simulation system runs fifteen times, and experimental results are in figure 2.

Table 9, table 10, table 11 and figure 2 show the results of the different TSP problems with ACA, including the optimal time, the worst time, and the average time of each problem by running 15 times. The three parameters α, β, ρ of ant colony algorithm have great influence on the performance of the algorithm. α means the degree of attention of information amount in each node, the value of α is more, the possibility of ants choosing the previous path is bigger, but the big α makes search prematurely into local solution; β represents the attention degree of heuristic information, the value is bigger, the possibility of ants choosing city near themselves is larger; ρ is the pheromone retention rate, if the value obtained is not appropriate, the results will be poor.

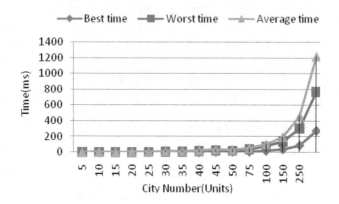

Fig. 2. The time of ACA for solving different problems

In this paper, according to the mathematical model and the corresponding experimental results, we can know ACA and GA are to accumulate and converge to the optimal path through information pheromone, inadequate initial pheromone can lead to the slow speed; Genetic algorithm has rapid global search capability, but the feedback information of the system is not used, often lead to redundant iteration inaction, the solution efficiency is low. According to table 6 and figure 2, when the city number is very large, GA for solving TSP problem is very easy to fall into local solutions and the time used by ACA for solving TSP problem is increasing very quickly; for the TSP, ACA is effective and robust than GA.

From the experiment we can conclude that it is useful and effective the two algorithms solve TSP by using uniform design; under the test condition and parameters setting, if the values ofαβρare 2, 4, 0.67 in the experiment, the population is from 8 to 12, ACA can get better result; The crossover between 0.7 and 1 and mutation in 0.15 and 0.2, the number is 1.5 to 5 times than the number of city, generation algorithm for solving this problem is better.

5   Conclusions

Parameters directly affect the performance of algorithm, parameters setting problem is an important work of simulation experiment. Uniform design used in the experiment is to choose limited representative parameter combinations of the algorithms, which can reduce the amount of the number of trials and improve the experimental efficiency. The parameters choose method based on uniform design is simple to be used, and it has practical application value and significance. In this paper, we utilize uniform design to choose parameter combinations and prove the method is effective, based on the data analysis of experiment draw some useful conclusions, which has some reference significance for relevant work.

Acknowledgements. We are gratefully acknowledge the financial support from the National Natural Science Foundation of China under Grant No. 61170305.

References

1. Rudolph, C.: Convergence properties of canonical Genetic Algorithms. IEEE Trans. on Neural Networks 5(1), 96–101 (1994)
2. Thomas, S., Holger, H.H.: MAX-MIN ant system. Fulture Generation Computer Systems 16(8), 889–914 (2000)
3. Cao, L., Luo, B., Qing, M.: A Genetic Algorithm for Finding Shortest Paths. Journal of Hefei University of Technology (Natural Science) 19(3), 112–116 (1996)
4. Wang, Y., Xie, J.: An Adaptive Ant Colony Algorithm and Simulation. Journal of System Simulation 2, 39–47 (2002)
5. Xu, J., Cao, X., Wang, X.: Polymorphic Ant Colony Algorithm. Journal of University of Science and Technology of China 35(1), 59–65 (2005)
6. Ding, J., Chen, G.: On the Combination of Genetic Algorithm and Ant Algorithm. Journal of Computer Research and Development 40(9), 1531–1536 (2003)
7. Huang, G., Cao, X., Wang, X.: An ANT Colony Optimization Algorithm Based on Pheromone Diffusion. ACTA Electronica Sinica 32(5), 866–868 (2004)
8. Hu, C., Zhu, Y., Gao, S.: A Hybrid Algorithm Based on Artificial Immune Algorithm and Ant Colony Algorithm for Solving Traveling Salesman Problem. Computer Engineering and Applications 40(34), 60–63 (2004)
9. Fang, K.: Uniform Design and Uniform Table, pp. 30–45. Science Press, Beijing (1994)
10. Fang, K.: Uniform Design-the application of number theoretic method in experiment design. Acta Mathematicae Applicatae Sinica (3), 363–371 (1980)
11. Ye, Z., Zheng, Z.: Configuration of Parameters α, β, ρ in Ant Colony Algorithm. Geomatics and Information Science of Wuhan University 27(7), 597–600 (2004)
12. Haibin, D.: Ant colony algorithm theory and its applications, pp. 112–116. Science Press, Beijing (2005)
13. He, D., Wang, F., Zhang, C.: Establishment of Parameters of Genetic Algorithm Based on Uniform Design. Journal of Northeastern University(Natural Science) 24(5), 419–411 (2003)
14. Huang, Y., Liang, C., Zhang, X.: Parameter Establishment of an Ant System Based on Uniform Design. Control and Decision 21(1), 93–96 (2006)
15. He, D., Wang, F., Jia, M.: Uniform Design of Initial Population and Operational Parameters of Genetic Algorithm. Journal of Northeastern University(Natural Science) 26(9), 828–831 (2005)

Result-Controllable Dendritic Cell Algorithm

Song Yuan[1] and He Zhang[1,2]

[1] College of Computer Science and Technology,
Wuhan University of Science and Technology, Wuhan, China
[2] Hubei Province Key Laboratory of Intelligent Information Processing
and Real-time Industrial System, Wuhan, China
yuansong_2002@163.com

Abstract. To realize that the false positive rate and false negative rate can be adjusted and improve the detection accuracy of the classical Dendritic Cell Algorithm which contains various uncertain elements, the concept of Tendency Factor and a Result-Controllable Dendritic Cell Algorithm are proposed by analyzing the signal processing function, weight matrixes and the other random parameters involved. The new algorithm has the higher detection accuracy and better robustness, in which the Tendency Factor can be obtained according to different contexts in order to control the detection results. Simulation experiments are performed using different parameters and multiple data sets and the Tendency Factor and the Result-Controllable Dendritic Cell Algorithm are proved to be reasonable and effective.

Keywords: dendritic cell algorithm, anomaly detection, tendency factor, artificial immune.

1 Introduction

Being the latest research result of the danger theory [1] in artificial immunology, Dendritic Cell Algorithm (DCA) [2] has been the focus of a lot of researchers and successfully applied to a range of problems, such as Nmap portscan detection [3], computer virus detection [4], stealthy malware detection [5] and web server aging detection [6], etc.

DCA is a highly random algorithm which is inspired by the function of Dendritic Cell (DC) in natural immune system, there are many influencing factors in the biological mechanisms of DC, including danger signal fusion and internal signal generation that are very complex, and there are still many unknown factors so far [7], so DCA is difficult to be analyzed and controlled. At present the researches of DCA mainly focus on improving the accuracy, but how to coordinate the relationship between false positive rate and false negative rate has not been presented. In order to make DCA have better robustness to adapt to more applications, qualitative and quantitative analysis is given on some uncertain factors in the algorithm, the classical DCA is improved, Tendency Factor (TF) and the Result-Controllable Dendritic Cell Algorithm (RCDCA) are put forward, and the simulation experiments are performed. The experimental results show that, compared with the classical DCA, the RCDCA

D.-S. Huang et al. (Eds.): ICIC 2014, LNCS 8588, pp. 196–205, 2014.
© Springer International Publishing Switzerland 2014

can not only improve the detection accuracy through reducing the influence of uncertain parameter assignments, but also control the detection accuracy through adjusting the TF in a certain range, when a lower false negative rate is required, the false positive rate will be compromised, and vice versa.

2   Biological Mechanism of DC

DC, which will have many dendritic protrusions when it matures, is the most important part of the biological immune system. DC is the bond of natural immune and adaptive immune, it can effectively collect and process antigens, process the environment signals at the same time, and release the specific cytokines to affect the differentiation of immune T cells.

DC exists in one of three different states of maturity, i.e. immature, semi-mature or mature states. The environment signals are categorized either as pathogen associated molecular patterns (PAMP), danger signals (DS) which are released as results of unplanned cell death, safe signals (SS) which are released as results of healthy tissue cell function, or inflammatory cytokines (IC). Immature DC (iDC) has no effect on T cells, it may turn to mature DC (mDC) or semi-mature DC (smDC) according to the environment signal type and concentration, the conversion process is shown in Fig.1.

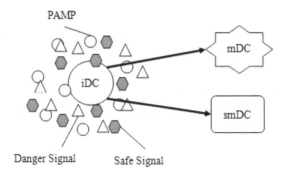

Fig. 1. Conversion process of DC

3   Classical DCA

In the classical DCA, DC is abstracted as a data structure similar to a signal processor, which records the sampled antigen quantity and type, the cumulative concentration of co-stimulatory molecules (CSM), the cumulative concentration of mature cytokine (Mature), the cumulative concentration of semi-mature cytokines (semi_Mature), etc. The maximum antigen quantity each DC can sample is determined by the CSM. Mature and semi_Mature influence the environment and the transition state of the DC. Antigens in the environment and their characteristic value will become input signals after pretreatment, iDC processes input signals to get output signals, i.e. concentrations of CSM, Mature and semi_Mature. When the cumulative CSM reaches the migration threshold, iDC will transform into mDC or smDC according to the concentrations of Mature and semi_Mature.

In [8] the UCI Wisconsin Breast Cancer standard data set is used to validate the classical DCA and it is a well understood two-class data set. The UCI consists of 700 items with 240 items of them are labeled as class 1 (normal) and the other 460 items are labeled as class 2 (anomalous). Although the UCI data is a static data set, it can be used for the DCA, if the facility to read in data in an event driven manner is provided. In the data set, each data item has ten attributes, among them five attributes i.e. Clump Thickness (CT), Cell Size (CS), Cell Shape (CH), Bare Nuclei (BN), Normal Nucleoli (NN) are used to calculate three input signals i.e. PAMP, SS and DS.

Select CT to calculate PAMP and SS, after pretreatment the middle of the CT (midCT) is 4, then the calculation process of PAMP and SS is as follows:

```
if(CT>midCT)
{
    SS=|CT-midCT|;
    PAMP=0;
}
else
{
    PAMP=|CT-midCT|;
    SS=0;
}
```

Select CS, CH, BN, NN four attributes to calculate DS, after pretreatment to the normal items in the data set, the averages of the four attributes are obtained respectively, the four attribute values of each item subtract the corresponding averages respectively to get the absolute differences, which will be added together and then divided by four to get DS.

For example, the DS of the first item in the data set is calculated as follows:
ID CS CH BN NN
1   8   8   4   7
CS CH BN NN = 8   8   4   7
average = 6.59 6.56 7.62 5.88
absolute difference = 1.41 1.44 3.62 1.12
DS = (1.41+1.44+3.62+1.12)/4=1.898

Immunologists research the DCs in biological immune, and draw the conclusion that a high level of safe signals will increase the output signal value for the semi_Mature, and reduce the cumulative value of the Mature output signal, which will be incremented by the receipt of either PAMPs or DS. The internal signal process is very complex, in order to simplify the calculation, the effect of IC is ignored. Three input signals of each item will be calculated to get three corresponding output signals through Equation (1) [2], the coefficients are determined by the weight matrix A.

$$O_{ip} = \sum_{j=0}^{2} S_{ij} W_{jp}$$

(1)

i- the sequence numbers of the sampled antigens in the data set;

j=0,1,2 - three input signals of PAMP, DS, SS;
p=0,1,2 - three output signals of CSM, semi_Mature, Mature;
O_{ip} - the p output signal concentration of the i antigen;
S_{ij} - the j input signal concentration of the i antigen;
W_{jp} - the transforming weight from S_{ij} to O_{ip}.

Table 1. Weight matrix A

	PAMP	Danger	Safe
CSM	2	1	2
semi_Mature	0	0	2
Mature	2	1	-2

More details of the classical DCA refer to [2].

In the classical DCA, whether the iDC transforms into smDC or mDC is determined by the semi_Mature output signal concentration and Mature output signal concentration, i.e. if Mature > semi_Mature, the iDC turns to mDC, otherwise, the iDC turns to smDC. DCA can be used in various fields, the application environments are different and the uncertain factors are also different. Therefore, the method of the classical DCA to determine whether the iDC will transform to smDC or mDC has a certain limitation. If a new way can be found to assess the safety of the environment according to different applications, it can not only effectively improve the accuracy of the algorithm, but also can coordinate the false positives and false negatives according to the actual needs.

4   TF and RCDCA

The classical DCA has a high detection accuracy, but can not control the false positive rate and false negative rate flexibly to meet actual needs. In order to realize the flexible control of the detection rate of DCA, firstly, we need to figure out the functions of various probabilistic ingredients through qualitative analysis, such as Data sets, the signal processing equation, etc. which can affect the detection accuracy; secondly, we need to figure out the influences of random-based parameters through quantitative analysis, such as weight matrix, the assessment criterion of DC context, etc. which can affect the detection accuracy with different values.

4.1   Clustering Analysis on Antigens

Calculate the input signals, put the input signals and weight matrix into the signal processing function to obtain 3 columns of output signals according to the 240 normal items in the standard data set and then figure out the average for each column of out put signals. We name these three averages as the Safe Antigen Clustering Signal (SACS). Conduct the same operations on the 460 anomalous items to obtain the Danger Antigen Clustering Signal (DACS). The data of SACS and DACS are shown in Table 2.

Table 2. Antigen clustering signals

Ave	CSM	semi_Mature	Mature
Safe	9.53	6.69	-3.85
Danger	8.6	0.59	7.42

For each row in Table 2, subtract semi_Mature from Mature to obtain a corresponding Danger Level (DL).

DL of SACS (DLS): DLS= -3.85-6.69 = -10.54

DL of DACS (DLD): DLD = 7.42-0.59 = 6.83

Based on the biological mechanism of DC, DL affects the safety of context and the transition state of DC directly. Based on the signal processing function, we can find that the safety of context and the probability of the transition from DC into mDC are positively correlated with the size of DL.

4.2   TF

- Definition 1: In DCA, after working out three averages of three kinds of output signals for the two classes of data in the bipartition training samples set, the result of merging the same type output signals in the above two group averages can be called Standard Merging Data (SMD).
- Definition 2: In DCA, the result of subtracting semi_Mature from Mature in SMD can be called Tendency Factor of SMD (TF_SMD).

Deal with the data in Table 2 to obtain SMD K (Table 3) and the TF of K is TF_SMD=-3.71. We can draw a conclusion that DLD > TF_SMD and DLS < TF_SMD by comparing TF_SMD with DLD and DLS in paragraph 4.1. Apparently, TF_SMD can distinguish DACS and SACS easily. In other words, DACS can be seen as the typical representative of danger antigens output signals, SACS can be seen as the typical representative of safe antigens output signals and TF_SMD can be seen as the most fair boundary between SACS and DACS. The signals in the context of a DC are determined by all the antigens that sampled by the DC, and the amount of antigens a DC can sample is determined by the value of CSM. According to the CSM of K, we can figure out that there are about n=CSM/18.13 pairs of antigens collected by a DC when the DC performs state transition, i.e., the n pairs of antigens will affect the context. Therefore, TF_SMD can be utilized as the standard to evaluate the safety of the context, and the evaluation result can be used to classify the antigens. The specific method is removing the influence of TF_SMD from the accumulation of DL values of all the antigens collected by a DC, and then classifying the antigens, namely compare the accumulation of DL and n*TF_SMD, if DL is greater than n*TF_SMD, the characteristic of the signals in the context tends to Danger, i.e., the antigens collected by the DC are probably danger antigens, otherwise the characteristic of the signals in the context tends to Safe, i.e., the antigens collected by the DC are probably safe antigens. If we want to coordinate the relationship between false negative rate and false positive rate, what we need to do is controlling λ——the multiple of TF_SMD.

Table 3. SMD K

ave	CSM	semi_Mature	Mature
K	18.13	7.28	3.57
TF_SMD	Mature-semi_Mature = 3.57-7.28 = -3.71		

4.3   Procedure of RCDCA

input: antigen and signals feature vectors
output: antigen plus context values vectors

```
1    input λ;
2    input UCI Wisconsin Breast cancer data set
3    calculate TF_SMD;
4    initialize an antigens pool
5    select 20 antigens from antigens vector;
6    initialize a result pool;
7    for antigens vector do
8       initialize a DC;
9       for CSM out put signal < migration threshold do
10         randomly select an antigen from the antigens
pool;
11         store antigen;
12         get signals;
13         update cumulative output signals;
14      end
15      for antigens collected by DC do
16         if DC is mature then
17            update cumulative mature values;
18         end
19         if current antigen has been valued 10 times then
20            calculate MCAV of current antigen;
21            current antigen context is assigned;
22            current antigen removed from the antigens pool;
23            new antigen added to the antigens pool;
24         end
25      end
26   end
27   collate the 10 context per antigen ID;
28   generate MCAV per antigen type;
```

Line 1st: set the size of λ;

Line 2nd - line 3rd: carry out the data preprocessing and work out the TF_SMD;

Line 4th - line 6th: initialization part of the RCDCA. Initialize an antigens pool which has 20 spaces and a result pool to save the evaluation results of antigens;

Line 7th - line 26th: the key part of the RCDCA. In line 16th, a DC processes all the signals of antigens collected by the DC, compares the difference between Mature and semi_Mature with $n*\lambda*TF\_SMD$ to evaluate the safety of the context and determine the target state of iDC to transform into;

Line 27th: collate all the antigens evaluation results;

Line 28th: calculate the MCAV of each type of antigens and then classify antigens;

5   Simulation and Result Analysis

We conduct experiments on context-shifted-1-time dataset (CS1, 240 normal items -460 anomalous items), context-shifted-2-times dataset (CS2, 230 anomalous items -240 normal items-230 anomalous items), context-shifted-4-times dataset (CS4, 11 5 anomalous items-120 normal items-115 anomalous items-120 normal items-230 anomalous items). Fig.2 is the final result of the experiments conducted on conte xt-shifted-1-time dataset, the abscissa is λ; the ordinate is false negative rate (dott ed line) and false positive rate (solid line). We conduct 20 times experiments and calculate the average of the result for each λ. Fig.3 and Fig.4 are the final resul ts of the experiments conducted on context-shifted-2-times dataset and context-shi fted-4-times dataset respectively.

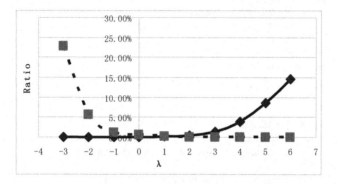

Fig. 2. Weight matrix A, CS1

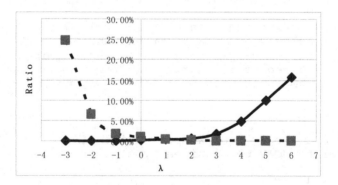

Fig. 3. Weight matrix A, CS2

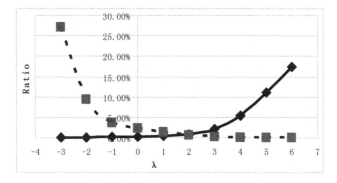

Fig. 4. Weight matrix A, CS4

To verify the robustness of RCDCA, we conduct the same experiments by using weight matrix B [9]. After the preprocessing of data, we can work out that TF_SMD = -11.

Table 4. Weight matrix B

	PAMP	Danger	Safe
CSM	2	1	3
semi_Mature	0	0	3
Mature	2	1	-3

The experimental results of using weight matrix B are shown in the 3 following tables:

Table 5. Weight matrix B, CS1

λ	-1	-0.5	0	0.5
false positive rate	0.00%	0.00%	0.06%	0.09%
false negative rate	31.24%	6.54%	1.06%	0.54%
Accuracy	68.76%	93.46%	98.88%	99.37%
λ	1	1.5	2	2.5
false positive rate	0.37%	0.60%	2.00%	5.64%
false negative rate	0.16%	0.08%	0.05%	0.00%
Accuracy	99.47%	99.32%	97.95%	94.36%

Table 6. Weight matrix B,CS2

λ	-1	-0.5	0	0.5
false positive rate	0.00%	0.00%	0.06%	0.11%
false negative rate	33.93%	8.47%	1.85%	0.98%
Accuracy	66.07%	91.53%	98.09%	98.91%
λ	1	1.5	2	2.5
false positive rate	0.36%	0.92%	2.51%	6.79%
false negative rate	0.47%	0.16%	0.05%	0.00%
Accuracy	99.17%	98.92%	97.44%	93.21%

Table 7. Weight matrix B,CS4

λ	-1	-0.5	0	0.5
false positive rate	0.00%	0.01%	0.06%	0.14%
false negative rate	35.46%	11.07%	4.09%	2.28%
Accuracy	64.54%	88.92%	95.85%	97.58%
λ	1	1.5	2	2.5
false positive rate	0.43%	0.97%	2.89%	7.59%
false negative rate	1.13%	0.62%	0.24%	0.00%
Accuracy	98.44%	98.41%	96.87%	92.41%

Through comparing Table 1 and Table 4, we can find that the difference between weight matrix A and weight matrix B is the values in the 3rd column. According to the formulas in the algorithm and the assessment standard, the 3 attribute values in the 3rd column respectively affect the quantity of antigens that a DC can collect, and the two context signals of DC: the Mature and semi_Mature output signals. By adjusting the weight matrix, the DS has a higher effect on the Mature output signals and the SS has a higher effect on the semi_Mature output signals, however, the TF_SMD is changed at the same time. As a result, the assessment standard of whether a DC is mature should be adjusted through TF_SMD. In the classical DCA, the assessment method that comparing the Mature output signals and the semi_Mature output signals corresponds to set TF_SMD=0 permanently, and raises the false negative rate or the false positive rate in different application environments to destroy the balance between the false negative rate and the false positive rate as a result.

Data in Fig.2, Fig.3, Fig.4 and data in Table 5, Table 6, Table 7 have the same tendency. The curves of false negative rate and false positive rate in Fig.2, Fig.3, Fig.4 are concave and symmetric with respect to the line $\lambda=1$. These curves illustrate that TF_SMD is capable of being the evaluation criteria to value the safety of an antigen, when $\lambda=1$, the false negative rate and the false positive rate are closest to each other and the accuracy is up to 99.5%. In Table 5, Table 6 and Table 7, we can also see that false negative rate and the false positive rate are closest to each other and the accuracy is up to 99.47%. As DCA is a highly random algorithm which is originated from biology and contains various probabilistic ingredients, it is impossible to find the accurate TF_SMD value to make false negative rate equals to false positive rate all the time, but we can see that the λ of the accurate TF_SMD is around 1. As evidenced by experiments data, we can control the accuracy by adjusting λ to meet the needs of different application environments.

The step size of λ in the experiments that uses weight matrix A is 1 and that uses weight matrix B is 0.5. The reason is that the absolute values of TF_SMD in different weight matrixes are varied. We can fine-tune the specific step size according to the actual application environment to fulfill our needs.

The above data prove that the accuracy of RCDCA when the λ is around 1 is higher than the classical DCA [8]. More importantly, RCDCA can calculate the TF_SMD of current environment and has the ability to coordinate the false negative rate and the false positive rate by adjusting λ. The above two points verify that the concept of Tendency Factor is reasonable and efficient.

6   Conclusions

Aiming at some problems in the classical DCA, the functions of several probabilistic elements are clarified, the concept of TF which affects the evaluation result of antigens is presented, RCDCA which is based on the TF and can coordinate the relationship between false negative rate and false positive rate according to actual requirements is put forward. Simulation results demonstrate RCDCA is easy to implement and has great utility. Certainly, the TF and the RCDCA are necessary to be given further research. The specific research work includes combining the TF with DDCA [10], and comparing RCDCA with the other technology to analyze the merits and shortcomings of RCDCA and improve it.

Acknowledgements. This work was supported in part by the National Natural Science Foundation of China (61273225).

References

1. Matzinger, P.: Friendly and Dangerous Signals: Is the Tissue in Control? Nature Immunology 8(1), 11–13 (2007)
2. Greensmith, J.: The Dendritic Cell Algorithm. PhD thesis, School of Computer Science, University Of Nottingham (2007)
3. Fang, X.J., Song, D.J.: Dendritic Cells Algorithm and Its Application to Nmap Portscan Detection. China Communications 9(3), 145–152 (2012)
4. Ou, C.M.: Multiagent-based Computer Virus Detection Systems: Abstraction from Dendritic Cell Algorithm with Danger Theory. Telecommunication Systems 52(2), 681–691 (2013)
5. Fu, J., Yang, H.: Introducing Adjuvants to Dendritic Cell Algorithm for Stealthy Malware Detection. In: The 5th International Symposium on Computational Intelligence and Design, Hangzhou, China, pp. 18–22 (2012)
6. Yang, H., Yi, S.J., Liang, Y.W., Fu, J., Tan, C.Y.: Dendritic Cell Algorithm for Web Server Aging Detection. In: 2012 International Conference on Automatic Control and Artificial Intelligence, Xiamen, China, pp. 760–763 (2012)
7. Ni, J.C., Li, Z.S., Sun, J.R., Zhou, L.P.: Research on Differentiation Model and Application of Dendritic Cells in Artificial Immune System. Acta Electronica Sinica 36(11), 2210–2215 (2008)
8. Greensmith, J., Aickelin, U., Cayzer, S.: Introducing Dendritic Cells as a Novel Immune-Inspired Algorithm for Anomaly Detection. In: Jacob, C., Pilat, M.L., Bentley, P.J., Timmis, J.I. (eds.) ICARIS 2005. LNCS, vol. 3627, pp. 153–167. Springer, Heidelberg (2005)
9. Gu, F., Greensmith, J., Aickelin, U.: Further Exploration of the Dendritic Cell Algorithm: Antigen Multiplier and Time Windows. In: Bentley, P.J., Lee, D., Jung, S. (eds.) ICARIS 2008. LNCS, vol. 5132, pp. 142–153. Springer, Heidelberg (2008)
10. Greensmith, J., Aickelin, U.: The Deterministic Dendritic Cell Algorithm. In: Bentley, P.J., Lee, D., Jung, S. (eds.) ICARIS 2008. LNCS, vol. 5132, pp. 291–302. Springer, Heidelberg (2008)

Feature Analysis of Uterine Electrohystography Signal Using Dynamic Self-organised Multilayer Network Inspired by the Immune Algorithm

Haya Alaskar[1], Abir Jaafar Hussain[1], Fergus Hussain Paul[1], Dhiya Al-Jumeily[1], Hissam Tawfik[2], and Hani Hamdan[3]

[1] Liverpool John Moores University, Byroom Street, Liverpool, L3 3AF, UK
{h.alaskar,a.hussain,d.aljumeily,P.fergus}@2011.ljmu.ac.uk
[2] Liverpool Hope University, UK
tawfikh@hope.ac.uk
[3] Supélec, Department of Signal Processing & Electronic Systems,
Plateau de Moulon, 3 rue Joliot-Curie, 91192 Gif sur yvette Cedex, France
Hani.Hamdan@supelec.fr

Abstract. Premature birth is a significant worldwide problem. There is little understanding why premature births occur or the factors that contribute to its onset. However, it is generally agreed that early detection will help to mitigate the effects preterm birth has on the child and in some cases stop its onset. Research in mathematical modelling and information technology is beginning to produce some interesting results and is a line of enquiry that is likely to prove useful in the early prediction of premature births. This paper proposes a new approach which is based on a neural network architecture called Dynamic Self-organised Multilayer Network Inspired by the Immune Algorithm to analyse uterine electrohystography signals. The signals are pre-processed and features are extracted using the neural network and evaluated using the Mean Squared Error, Mean absolute error, and Normalized Mean Squared Error to rank their ability to discriminate between term and preterm records.

Keywords: Uterine EMG Signals, Artificial Immune Systems, Time Series Data, Features extraction.

1 Introduction

Feature extraction is known as the transformation of signals to features that characterize the properties of the signal [1]. Feature extraction is an essential task to improve the performance of classifiers. A number of techniques have been used to represent and extract features from biological signals in order to improve the classification process. An example of feature extraction approaches include autoregressive coefficients (AR), Wavelet transform, and Fourier transformation, which divides the signals into sinusoidal components of different frequencies[2], [3]. However, these methods suffer from high noise sensitivity. Fele et al. [4] used Power Spectral Densities (PSD)

D.-S. Huang et al. (Eds.): ICIC 2014, LNCS 8588, pp. 206–212, 2014.
© Springer International Publishing Switzerland 2014

to generate features from uterine signals. PSD represents only the assessed power across a range of frequencies [5]. Moreover, these current methods have suffered from a limited ability to obtain both spatial and temporal information from biological signals [5] as   they ignore the essential non-linear dynamic behaviour of signals [6]. The success of classification depends on utilizing signal representations that capture the crucial information needed to identify the right classes. Biological and biomedical signals exhibit non-linear dynamic properties with complex behaviours [1]. These types of signals are considered chaotic [7]. Therefore, non-linear dynamical measures can be used to compute clinically useful parameters and salient information about the signals themselves. During the last few decades, the use of non-linear analysis techniques to determine the properties of biological signals has increased significantly.

The Dynamic Self-organised Multilayer network Inspired by the Immune Algorithm (DSMIA) is used in this paper for feature extractions. This network was designed to deal with stationary and nonstationary behaviour in financial time series. It has proved its ability to model dynamical behavior of financial time series [9]. In this paper, this network is used to extract the uterine contraction articulation pattern for a given patient. DSMIA has been used to model EHG signals in order to extract some useful features to detect preterm subject. The features are evaluated on their ability to distinguish between different classes of Uterine EHG signals using the linear discriminant analysis classifiers (LDA) and nonlinear classifier (MLP). The accuracy of the classification is evaluated using the sensitivity and specificity measures as the performance measures.

The remainder of this paper is organized as follows. Section 2 describes Uterine EHG signals. The methods used for forecasting uterine EHG signals are described in section 5 before our DSMIA is described in 3-4. The paper is concluded in Section 6.

2   Term-Preterm ElectroHysteroGram (TPEHG) Data

The data used in this research was recorded at the Department of Obstetrics and Gynaecology, Medical Centre, Ljubljana between 1997 and 2006 and provided as an open dataset called Term-Preterm ElectroHysteroGram (TPEHG) [10]. The TPEHG database contains 300 records of patients that were collected at the 22nd or 32nd week of gestation. The records were 30 minutes long, had a sampling frequency (fs) of 20Hz, and had a 16-bit resolution over a range of ±2.5 millivolts. The signals were filtered using a 0.3-4Hz 4-pole digital Butterworth filter. Signals have been filtered twice in both backwards and forwards directions in order to overcome the phase-shifting that can occur when utilizing this filter.

Different features have been used by several authors to classify the uterine EHG signals into preterm using time domain features [12], frequency domain features [13], [14] and wavelet coefficients [15], [16]. However, it is possible to find an exploratory method so that it encompasses wider range of the information from signals. In this paper, we have taken into consideration and examined dynamic changes during the contraction process. New features are used to identify term and preterm records using the mean absolute error (MAE), normalized mean squared error (NMSE), mean squared error (MSE) and the power of the predicted signals (MSY). These measurements are extracted from the DSMIA forecasting signals and used to classify preterm and term records.

3   Dynamic Self-organised Multilayer Network Inspired by the Immune Algorithm (DSMIA)

The Dynamic Self-organised Multilayer network Inspired by the Immune Algorithm (DSMIA) is used to capture some of the complex patterns found in EHG. The structure of the DSMIA network is shown in Fig. 1. The DSMIA network has three or more layers: the input, the self-organising hidden layer, and the output layers with feedback connections from the output layer to the input layer. The input layer holds copies of the current inputs as well as the previous output produced by the network. This provides the network with memory. As such, the previous behaviour of the network is used as an input affecting current behaviour. Similar to the Jordan recurrent network [19], the output of the network is fed back to the input through the context units.

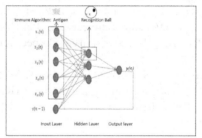

Fig. 1. The structure of the proposed DSMIA network

Suppose that N is the number of external inputs $x(n)$ to the network, and yk(n-1) is the output of the network from the previous time step $(n-1)$ and O refers to the number of outputs. In the proposed *DSMIA*, the overall input to the network will be the component of $x(n)$ and $y_k(n-1)$ and the number of inputs to the network is $N+O$ defined as U where

$$U(n) = \begin{cases} x_i(n) & i = 1,....N \\ y_i(n-1) & i = 1,....,O \end{cases} \tag{1}$$

The output of the hidden layer is computed as

$$v_{hj}(n) = \alpha \sqrt{\sum_{i=1}^{N}(w_{hj} - x_{hj}(n))^2} \tag{2}$$

$$z_{hj}(n) = \beta \sqrt{\sum_{k=1}^{O}(wz_{hjk} - y_k(n-1))^2} \tag{3}$$

$$D_{hj}(n) = v_{hj}(n) + z_{hj}(n) \tag{4}$$

$$x_{hj}(n) = f_{ht}(D_{hj}(n)) \tag{5}$$

$$\hat{y}_k = f_{ot}\left(\sum_{j=1}^{N_H} w_{ojk}\, x_{Hj} + B_{ok}\right) \tag{6}$$

Where f_{ht}, f_{ot} are nonlinear activation functions, N is the number of external inputs, O is the number of output units. w_{ojk} is the weight corresponds to the external input while wz_{hjk} is the weight corresponding to the previous output, and n is the current time step, while α,β are selected parameters with 0< α and   0< β.

The first layer of the DSMIA is a self-organised hidden layer trained similar to the recursive self-organized map RecSOM [20]. In this case, the training rule for updating the weights is based on the same technique for updating the weights of the self-organized network inspired by the immune algorithm (SONIA) network [17]. The change in this network is that the weights of the context nodes wz_{hjk} are also updated in the same way as the weights of the external inputs w_{hj}. This is done by first finding D, which is the distance between the input units and the centroid of the jth hidden units:

$$D_{hj}(n) = \alpha\sqrt{\sum_{i=1}^{N}(w_{hji} - x_{hj}(n))^2} + \beta\sqrt{\sum_{k=1}^{O}(wz_{hjk} - y_k(n-1))^2} \tag{7}$$

From $D_{hj}(n)$, the position of the closest match will be determined as :

$$c(n) = argmin(D_{hj}(n)) \tag{8}$$

If the shortest distance D_c is less than the stimulation level value, $s_l \in (0, 1)$, then the weight from the external input vector and the context vector are updated as follows:

$$W_{hji}(n+1) = W_{hji}(n) + \gamma D_c(n) \tag{9}$$

$$Wz_{hji}(n+1) = Wz_{hjk}(n) + \gamma D_c(n) \tag{10}$$

Where Wz_{hji} is the weight of the previous output and W_{hji} is the weight for the external inputs, and γ is the learning rate which is updated during the epochs.

4   Modelling EHG by DSMIA

The experiments have been applied on 76 EHG signals with 38 preterm and 38 term values. The first experiment used the DSMIA to model EHG signals before filtering. While the second experiment modelled the EHG signals with the DSMIA after using a band-pass filter configured between 0.3 Hz and 4 Hz as been used by [4].

We evaluate the forecasting performance in the two experiments by splitting the time series signals into three data sets, forming the training, validation, and finally out-of-sample sets, of 25%, 25%, and 50% of the total data, respectively. The DSMIA was trained for 30 times with 500 epochs to obtain an averaged result.

The performance of network is evaluated using the Mean Squared Error (MSE), and Signal-to-Noise Ratio (SNR). Table 1 shows the average results using 30 simulations for the MSE, and signal-noise-ratio (SNR) using 76 Uterine EHG signals. The result shows that DSMIA are successful able to model the nonlinear relation in the EHG signals.

Table 1. Averge Results for the prediction of EHG signals

	SNR	MSE
DSMIA (Before filtering)	19.1211	3.8528e-04
DSMIA (after Filtering 0.3Hz to 4Hz)	14.7493	0.0000256

The features are obtained by computing the power of the prediction signals, the mean error signals, normalized mean squared error and mean absolute error.

4.1   Classification

The aforementioned forecasting error measures have been computed from the DSMIA network. These features of the 76 signals can be thought of as a feature vector. Since the data used in this experiment consist of 76 records only, the linear classifier has been used to classify the data into preterm and term classes. It has a low computational complexity and is less affected by over-fitting [21]. In the final step of the experiments, the error features are passed to the linear discriminate analysis (LDA) and MLP classifier to classify data as preterm or term. The results for the features used in our work were compared to the original features provided in [4]   which were Root Mean Square (RMS), Median Frequency (MF), and Peak Frequency (PF) and Sample Entropy (SpEn).

5   Discussion

In these experiments, we have demonstrated that DSMIAs are able to forecast single step ahead in time for the Uterine EHG signals, achieving a mean squared error (MSE) of 0.00038. The DSMIA model is able to capture temporal behaviour of the signals.

Furthermore, the features that have been extracted are the Mean Squared Error (MSE), Normalised Mean Squared Error (NMSE), the mean squared of the predicted signals (MSY), and Mean Absolute Error (MAE). They have been used to distinguish between term and preterm subjects in EHG signals. The extracted feature values from 76 EHG signals are passed to a LDA and MLP classifiers. The classification results of the proposed methods show that using forecasting errors as features could clearly distinguish between preterm and term. The difference between   [4] and our work is that we only selected 38 preterm and 38 term patient records to avoid bias in our classification and feature extraction results.

The performance of each features is evaluated using Sensitivity (SV), Specificity (SP), Negative, Positive predicted values and the area under the curve (AUC) using 30 simulations. The accuracies achieved by the proposed method before and after filtering, are 46.26% and 50%, respectively. These are improved results in comparison to the results achieved when using the original features with only 39% accuracy. The averaged classification accuracies obtained from LDA for the proposed features and the original features are presented in Table 2. The results show that RMS has the highest classification accuracy, followed by MAE and MSE.

Table 2. Averaged result for classification of each features

Features	SV	SP	negative	positive	MSE	accuracy
MSE	0.6579	0.2632	0.4348	0.4717	0.2506	46.0526
MAE	0.3158	0.6316	0.4800	0.4615	0.2505	47.3684
NMSE	0.4474	0.3947	0.4167	0.4250	0.2601	42.1053
MYS	0.7368	0.1316	0.3333	0.4590	0.2532	43.4211
RMS	0.55263	0.4210	0.484	0.4883	0.25334	48.68
MF	0.1578	0.605	0.418	0.2857	0.2950	38.15
PF	0.0526	0.815	0.4626	0.2222	0.2507	43.42
SpEn	0.5263	0.3947	0.4545	0.46518	0.26479	46.05

6   Conclusion

This paper presents a novel feature extraction procedure for extracting features from the Uterine EHG signals. In this experiment, we propose and evaluate a feature extraction method based on the nonlinear predictability of the EHG signals to detect term and preterm records. This study applies a nonlinear dynamic method, which is the DSMIA in forecasting the signal for one step ahead. The experiments have demonstrated that DSMIA is able to model the Uterine EHG signals, achieving a low mean squared error (MSE).

The features that have been extracted from the modeling signal are four forecasting error functions. These four functions have been used successfully to distinguish between term and preterm records.

References

1. Meng, Q., Zhou, W., Chen, Y., Zhou, J.: Feature analysis of epileptic EEG using nonlinear prediction method. In: Conf. Proc. IEEE Eng. Med. Biol. Soc., vol. 2010, pp. 3998–4001 (January 2010)
2. Sunde, C.: Noise Diagnostics of Stationary and Non-Stationary Reactor Processes. Chalmers University of Technology, Sweden (2007)
3. Rojas, R.: Neural Networks Introduction, A Systematic, 1st edn., p. 502. Springer (1996)

4. Fele-Zorz, G., Kavsek, G., Novak-Antolic, Z., Jager, F.: A comparison of various linear and non-linear signal processing techniques to separate uterine EMG records of term and pre-term delivery groups. Med. Biol. Eng. Comput. 46(9), 911–922 (2008)
5. Forney, E.M., Anderson, C.W.: Classification of EEG during imagined mental tasks by forecasting with Elman Recurrent Neural Networks. In: 2011 Int. Jt. Conf. Neural Networks, pp. 2749–2755 (July 2011)
6. Übeyli, E.D., Guler, I.: Statistics over Lyapunov Exponents for Feature Extraction: Electroencephalographic Changes Detection Case. World Acad. Sci. Eng. Technol., 625–628 (2007)
7. Guler, N., Ubeyli, E., Guler, I.: Recurrent neural networks employing Lyapunov exponents for EEG signals classification. Expert Syst. Appl. 29(3), 506–514 (2005)
8. Szkoła, J., Pancerz, K., Warchoł, J.: Recurrent neural networks in computer-based clinical decision support for laryngopathies: An experimental study. Comput. Intell. Neurosci. 2011, 289398 (2011)
9. Al-Jumeily, D., Hussain, A., Alaskar, H.: Recurrent Neural Networks Inspired by Artificial Immune Algorithm for Time Series Prediction. In: Proceedings of International Joint Conference on Neural Networks, Dallas, pp. 2047–2054 (2013)
10. Physionet.org, The Term-Preterm EHG Database (TPEHG DB) (2010), http://www.physionet.org/pn6/tpehgdb/
11. Baghamoradi, S., Naji, M., Aryadoost, H.: Evaluation of cepstral analysis of EHG signals to prediction of preterm labor. In: 18th Iranian Conference on Biomedical Engineering, pp. 1–3 (December 2011)
12. Diab, A., Hassan, M., Karlsson, B., Marque, C.: Effect of decimation on the classification rate of nonlinear analysis methods applied to uterine EMG signals. In: IRBM, pp. 12–14 (2013)
13. Bazregar, S., Mahdinejad, K.: Preterm Birth Detection Using EMG Signal Processing. Life Sci. J. 10(3), 25–30 (2013)
14. Moslem, B., Karlsson, B., Diab, M.O., Khalil, M., Marque, C.: Classification performance of the frequency-related parameters derived from uterine EMG signals. In: Conf. Proc. IEEE Eng. Med. Biol. Soc., vol. 2011, pp. 3371–3374 (January 2011)
15. Hassan, M., Terrien, J., Alexandersson, A., Marque, C., Karlsson, B.: Nonlinearity of EHG signals used to distinguish active labor from normal pregnancy contractions. In: 32nd Annual International Conference of the IEEE EMBS, pp. 2387–2390 (2010)
16. Diab, M.O., Moslem, B., Khalil, M., Marque, C.: Classification of uterine EMG signals by using Normalized Wavelet Packet Energy. In: 2012 16th IEEE Mediterr. Electrotech. Conf., pp. 335–338 (March 2012)
17. Widyanto, M.R., Nobuhara, H., Kawamoto, K., Hirota, K., Kusumoputro, B.: Improving recognition and generalization capability of back-propagation NN using self-organized network inspired by immune algorithm. Appl. Soft Comput. 6, 72–84 (2005)
18. Shen, X., Gao, X.Z., Bie, R.: Artificial Immune Networks:Model and Application. Int. J. Comput. Intell. Syst. 1, 168–176 (2008)
19. Jordan, M.I.: Attractor dynamics and parallelism in a connectionist sequential machine. In: Artificial Neural Networks, pp. 112–127. IEEE Press, Piscataway (1990)
20. Voegtlin, T.: Recursive self-organizing maps. Neural Netw. 15(8-9), 979–991 (2002)
21. Balli, T., Palaniappan, R.: Classification of biological signals using linear and nonlinear features. Physiol. Meas. 31(7), 903–920 (2010)

An Improved Ensemble of Extreme Learning Machine Based on Attractive and Repulsive Particle Swarm Optimization

Dan Yang[*] and Fei Han

School of Computer Science and Communication Engineering, Jiangsu University, Zhenjiang, Jiangsu, China
Darlene913ydan@gmail.com

Abstract. Although ensemble neural networks perform more effectively than individual one in many cases, it is easy to fall into the situation of comparatively low overall convergence accuracy while the individual members obtain high classification accuracies. To improve the convergence performance of the ensemble system, a modified ensemble of extreme learning machine (ELM) based on attractive and repulsive particle swarm optimization (ARPSO) is proposed in this paper. In the proposed method, ARPSO is applied to select the ELMs to build the ensemble system by optimizing the ensemble weights of the candidate ELMs. Experiment results on two data verify that the proposed method could obtain better convergence performance than some classical ELMs and ensemble ELMs.

Keywords: Ensemble neural networks, extreme learning machine, attractive and repulsive particle swarm optimization.

1 Introduction

Early in 1990, the method of neural networks ensembles was inaugurated by Hansen and Salamon [1]. Their research indicated that an ensemble of neural network synthesized by a plurality consensus scheme could acquire better performance than a single network. Since then, a lot of researchers have done great deal of fundamental studies on the ensemble of neural networks, and the ensembles of neural networks are still a quite active topic in the field of machine learning.

Traditional learning algorithms for feedforward neural networks (FNNs) such as backpropagation (BP) algorithm have some defects including long training time, being trapped into local minimum, large network redundancy and some instability. Recently, extreme learning machine (ELM), a new learning scheme for single hidden layer FNNs (SLFNs), which randomly chooses the input weights and hidden biases and analytically determines the output weights of SLFNs was proposed by Huang et al [2,3]. Different from the traditional gradient-based learning algorithms, ELM not only has fast learning speed but also achieves better generalization performance. Meanwhile non-differentiable activation functions and straightforward solution are features and

---

[*] Corresponding author.

D.-S. Huang et al. (Eds.): ICIC 2014, LNCS 8588, pp. 213–220, 2014.

advantages of the ELM. Although ELM has many merits, the issues such as predictive instability and over-fitting still remain to be solved [4] and batch learning which is required in the process of training with consuming more works and time need to be improved [5].

As a global search algorithm, particle swarm optimization (PSO) has many advantages such as easy to implement, less parameters and fast convergence rate. These features of PSO make it widely to be used in many fields such as process control, dynamic optimization, adaptive control and power system optimization. However, traditional PSO still has the problem of premature convergence and easily fall into local minima [6]. An improved PSO called attractive and repulsive particle swarm optimization (ARPSO) [7] was proposed to keep the diversity of swarm effectively in the search process, so it obtained better convergence accuracy than traditional PSO.

It is evident that the key to improve the performance of ensembles is to choose the right sub-classifiers, and there is no deterministic approach to achieve this. How to find accurate and efficient ensemble ELMs with simple methods are the jobs of this study. For the good search ability of ARPSO, we use ARPSO to select the optimal ELM subsets to form the ensemble system in this study. Thus, an improved ensemble of extreme learning machine based on ARPSO (E-ARPSOELM) is proposed in this paper. In the proposed algorithm, the optimal ELM subset is determined by optimizing the ensemble weights of the candidate ELMs. Finally, the experiment results show that E-ARPSOELM achieves better performance than original ELM, E-ELM, OS-ELM and E-OSELM.

2   Preliminaries

2.1   Extreme Learning Machine

Different from traditional implementations where all the parameters of the feedforward networks need to be tuned, ELM claims that the hidden node learning parameters are randomly allocated and the output weights are determined analytically [2,3]. The learning speed of ELM is faster with better generalization performance than that of some gradient-based learning algorithms.

For N arbitrary distinct samples (x_i, t_i), where $x_i = [x_{i1}, x_{i2}, \cdots, x_{in}] \in R^n$ and $t_i = [t_{i1}, t_{i2}, \cdots, t_{im}]^T \in R^m$, the standard SLFN with $\tilde{N}$ hidden neurons and an activation function g(x) is mathematically modeled as

$$\sum_{i=1}^{\tilde{N}} \beta_i g(w_i \cdot x_j + b_i) = o_j, j = 1, \cdots, N, \tag{1}$$

where $w_i = [w_{i1}, w_{i2}, \cdots, w_{in}]^T$ is the weight vector connecting the i-th hidden neuron and the input neurons, $\beta_i = [\beta_{i1}, \beta_{i2}, \cdots, \beta_{im}]^T$ is the weight vector connecting the i-th hidden neuron and the output neurons and b_i is the bias of the i-th hidden neuron.

The above N equations can be written compactly as

$$H\beta = T \qquad (2)$$

where

$$H\left(w_1, w_2, \cdots, w_{\tilde{N}}, b_1, b_2, \cdots, b_{\tilde{N}}, x_1, x_2, \cdots, x_N\right)$$

$$= \begin{bmatrix} g(w_1 \cdot x_1 + b_1) & \cdots & g\left(w_{\tilde{N}} \cdot x_1 + b_{\tilde{N}}\right) \\ \vdots & \cdots & \vdots \\ g(w_1 \cdot x_N + b_1) & \cdots & g\left(w_{\tilde{N}} \cdot x_N + b_{\tilde{N}}\right) \end{bmatrix}_{N \times \tilde{N}} \quad \beta = \begin{bmatrix} \beta_1^T \\ \cdot \\ \cdot \\ \cdot \\ \beta_{\tilde{N}}^T \end{bmatrix}_{\tilde{N} \times m} \quad \text{and} \quad T = \begin{bmatrix} t_1^T \\ \cdot \\ \cdot \\ \cdot \\ t_N^T \end{bmatrix}_{N \times m}$$

As in [8], H is called the hidden layer output matrix; the i-th column of H is the i-th hidden neuron output vector with respect to inputs $x_1, x_2, \cdots, x_N$.

To find a solution for (2), Huang et al. [2] proposed a means of randomly assigning values for parameters w_i and b_i. To train a SLFN is simply equivalent to find a least-squares solution $\hat{\beta}$ of the linear system $H\beta = T$ as follows:

$$\hat{\beta} = H^+ T \qquad (3)$$

where H^+ is the Moore–Penrose generalized inverse of the hidden layer output matrix H.

2.2   Attractive and Repulsive Particle Swarm Optimization

Although the global search ability of the PSO algorithm is considerable, it is easy to lose the diversity of the swarm [9]. Low diversity may lead the particles in the swarm to be trapped into the local minima which is called "premature convergence". To avoid the problem of premature convergence, Riget [7] proposed the attractive and repulsive particle swarm optimizer (ARPSO) which guarantees the diversity of the swarm. Suppose that the position of the i-th particle is $X_i = (x_{i1}, x_{i2}, ..., x_{iD})$, $i = 1, 2, ..., N$ and the velocity of the i-th particle can be expressed as $V_i = (v_{i1}, v_{i2}, ..., v_{iD})$; $P_i = (p_{i1}, p_{i2}, ..., p_{iD})$ is denoted as the best historical position of the i-th particle which is called pbest; $P_g = (p_{g1}; p_{g2}; ...; p_{gD})$ as the best solution of the swarm is called gbest [9]. In ARPSO, the velocity and position of the particles are updated as follows:

$$V_i(t+1) = W * V_i(t) + dir * (c_1 * r_1 * (P_i - X_i(t)) + c_2 * r_2 * (P_g - X_i(t))) \qquad (4)$$

$$X_i(t+1) = X_i(t) + V_i(t+1) \qquad (5)$$

where

$$dir = \begin{cases} -1, & dir > 0, diversity < d_{low} \\ 1, & dir < 0, diversity > d_{high} \end{cases} \tag{6}$$

c_1, c_2 are the acceleration constants which usually value between 0 and 2; r_1, r_2 are the generated uniformly random numbers in the range of $(0,1)$; w is the inertia weight.

In ARPSO, a function was proposed to calculate the diversity of the populations as:

$$diversity(S) = \frac{1}{|S| \cdot |L|} \cdot \sum_{i=1}^{|S|} \sqrt{\sum_{j=1}^{D} (p_{ij} - \overline{p_j})^2} \tag{7}$$

where S is the swarm, $|S|$ is the number of all the particles, $|L|$ is the length of the radius in the search space, D is the dimension of the problem, p_{ij} is the j-th value of the i-th particle and $\overline{p_j}$ is the average j-th value over of all particles.

In the attraction phase ($dir = 1$), the swarm is attractive, and consequently the diversity decreases. When diversity drops below the lower bound, d_{low}, the swarm switches to the repulsion phase ($dir = -1$). When the diversity reaches the upper bound d_{high}, the swarm switches back to the attraction phase. ARPSO alternates between phases of exploiting and exploring – attraction and repulsion – low diversity and high diversity and thus improve its search ability [10].

3   The Proposed Algorithm

In this study, the ELMs for ensemble are determined adaptively by ARPSO. The detail steps of the proposed ensemble algorithm are summarized as follows:

Step 1: Bagging method is used to randomly assign different training sets with the same size [11]. With a training set, a corresponding ELM is randomly generated to train the training data.

Step 2: Initialize the position of X_i ($i = 1,2,...,N$) and its velocity V_i in the range of $(0,1)$; X_i represents the coefficient vector of D ELMs. Thus x_{ij} needs to be rounded as 0 or 1 which indicates the j-th ELM is selected or not for ensemble.

Step 3: Optimize the population by ARPSO. The ensemble accuracy represents the fitness value. The fitness function of ARPSO can be computed by the following equation:

$$fitness(X_i) = 1 - (\sum_{k=1}^{n} \|T_{ik} - TT_{ik}\|_2^2)/n \tag{8}$$

where n is the number of samples in the validation, T_{ik} is the actual output class of the ensemble and TT_{ik} is the desired output class.

(a) Set X_i representing the current ensemble classifiers as pbest, compute each fitness value of n classifiers by utilizing equation (11) and find the global optimum position gbest which builds the best ensemble classifier owning the highest accuracy.

(b) Update v_{ij} and x_{ij} according to Eqs. (4-7). At the same time, x_{ij} needs rounding operation and new ensemble classifiers are generated.

(c) Calculate new fitness value and compare it to previous results. If the new fitness value of the current position is better than before update the current position as pbest. Simultaneously, if the best fitness value of the current position exceeds the value of gbest, update the corresponding position as gbest.

(d) Go to Step 4 if the end condition of the algorithm is met, or else return sub-step (a).

Step 4: Get the ultimate ensemble of ELM established by the gbest and it is used for testing samples by decision rules including the majority voting, weighting voting or average voting methods. The weight w_j of j-th ELM in the weighting method is computed by the following equation:

$$w_j = Ac_j \bigg/ \sum_{j=1}^{D} Ac_j \tag{9}$$

where Ac_j denote the validation accuracy of the j-th ELM.

4   Experiment Results

In this section, the performance of the proposed E-ARPSOELM is compared with the original ELM, OSELM [5], and other ensemble learning algorithms including E-ELM [12] and EOS-ELM [13]. The experiments are conducted on two datasets including Diabetes and Satellite Image. All experiments are carried out in MATLAB 2012b in an INTER 2 Duo Core processor with 2.93 GHz and 2 GB RAM.

Table 1. The description of the two data

Datasets	Training sets	Testing sets	Categories	Attributes
Diabetes	576	192	2	8
Satellite Image	4435	2000	6	36

The two datasets used in the experiment are taken from the UCI Machine Learning Repository. The description of the two datasets is given in Table 1. The training sets are further divided into the training sets and validation ones where the 70% and 30% of the whole training sets are the final training and validation sets respectively.

In the experiments, the ELMs for ensemble on the same data have the same hidden nodes. The numbers of the nodes on the Diabetes and Satellite Image are set as 25 and 400 respectively. The inertia weight W value starts from 0.9 and ends with 0.1. According to [9], the diversity controlling parameters d_{low} and d_{high} are taken as $5*10^{-6}$ and 0.25 respectively; the velocity of a particle V_i is set in the range of (0, 1) and X_{max} is set as 2; c1 and c2 are equal to 2. In E-ARPSOELM, the maximum number of iterations is set at 100. The experiment results are the average of 20 independent runs for each algorithm.

Tables 2-4 show the classification accuracies among E-ARPSOELM, ELM, E-ELM, OS-ELM and E-EOSELM with different decision rules which are applied to build the ensemble classifier. In E-ARPSOELM the number of candidate ELMs is 30, and the swarm size is 50 (N=50, D=30). In E-ELM and EOS-ELM, the size of the ensemble will be set to 10 component models (D=10).

Table 2. Classification accuracy of five algorithms on two data with the majority voting

Datasets	Algorithm	Training Accuracy	Testing Accuracy	Standard Deviation
Diabetes	ELM	0.7850	0.8146	0.0121
	E-ELM	0.7886	0.8276	0.0109
	OS-ELM	0.7836	0.8182	0.0126
	EOS-ELM	0.7906	0.8276	0.0088
	E-ARPSOELM	0.8223	0.8359	0.0077
Satellite Image	ELM	0.9209	0.8913	0.0028
	E-ELM	0.9221	0.8919	0.0032
	OS-ELM	0.9211	0.8914	0.0033
	EOS-ELM	0.9215	0.8923	0.0026
	E-ARPSOELM	0.9316	0.8976	0.0023

From Table 2, it can be found that the testing accuracy of E-ARPSOELM on two datasets is the best one among that of all algorithms. Moreover, the standard deviation of the proposed algorithms is less than that of other algorithms on the two data. It is readily demonstrated that the generalization performance of the proposed ensemble algorithm is better and the system is more stable than original ELM, OS-ELM, E-ELM and E-OSELM. From Table 3 and Table 4, similar conclusion is easily drawn with the weight voting and average voting methods.

Table 3. Classification accuracy of three algorithms on two data with the weight voting

Datasets	Algorithm	Training Accuracy	Testing Accuracy	Standard Deviation
Diabetes	E-ELM	0.7905	0.8247	0.0085
	EOS-ELM	0.7878	0.8214	0.0129
	E-ARPSOELM	0.8205	0.8271	0.0081
Satellite Image	E-ELM	0.9254	0.8963	0.0023
	EOS-ELM	0.9258	0.8969	0.0026
	E-ARPSOELM	0.9304	0.8972	0.0019

Table 4. Classification accuracy of three algorithms on two data with the average voting

Datasets	Algorithm	Training Accuracy	Testing Accuracy	Standard Deviation
Diabetes	E-ELM	0.7820	0.8090	0.0051
	EOS-ELM	0.7820	0.8063	0.0113
	E-ARPSOELM	0.8022	0.8118	0.0051
Satellite Image	E-ELM	0.9177	0.8881	6.3753e-04
	EOS-ELM	0.9163	0.8883	0.0034
	E-ARPSOELM	0.9215	0.8887	0.0013

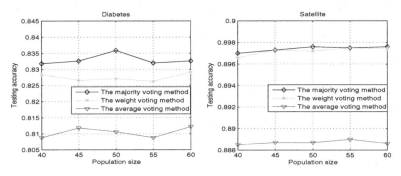

Fig. 1. Testing accuracies on two data with different number of particles

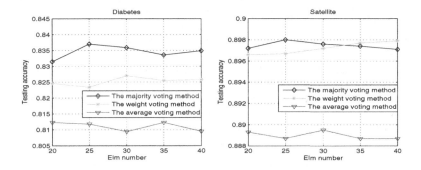

Fig. 2. Testing accuracies on two data with different number of ELMs

Fig 1 depicts the testing accuracy curve with the number of particles on two datasets. The three curves sign different performance of E-ARPSOELM with three decision rules. The number of particles N changes between 40 and 60. Meanwhile, the number of ELMs D is 30. From Fig 1 the testing accuracy calculated by the majority voting method is better than other two ways on Diabetes data. On the Satellite Image data, the testing accuracy of the weight voting method is about the same as the majority method which outperforms the average method.

Fig 2 depicts the testing accuracy obtained by the proposed algorithm with the different number of ELMs which is set between 20 and 40. The number of particles is fixed as 50. Similarly, from Fig 2 the majority voting method is the best of the three decision rules on Diabetes and the performance of the average voting method is inferior to other two methods on two data.

5   Conclusions

In this study, ARPSO was applied to select ELMs for ensemble to improve the generalization ability of the ensemble system. The number of ELMs for ensemble is determined adaptively in the proposed algorithm. The experiment results on two data verified that E-ARPSOELM outperforms the original ELM, OS-ELM algorithm and other ensemble learning algorithms. Future work will include how to improve the diversity of the ensemble system and apply the ARPSO-based ensemble ELM on more complex data.

Acknowledgements. This work was supported by the National Natural Science Foundation of China (Nos. 61271385, 60702056) and the Initial Foundation of Science Research of Jiangsu University (No.07JDG033).

References

1. Hansen, L.K., Salamon, P.: Neural network ensembles. IEEE Transactions on Pattern Analysis and Machine Intelligence 12(10), 993–1001 (1990)
2. Huang, G.B., Zhu, Q.Y., Siew, C.K.: Extreme learning machine: a new learning scheme of feedforward neural networks. In: IEEE International Joint Conference on Neural Networks, vol. 2, pp. 985–990. IEEE Press (2004)
3. Huang, G.B., Zhu, Q.Y., Siew, C.K.: Extreme learning machine: theory and applications. Neurocomputing 70(1), 489–501 (2006)
4. Zhai, J., Xu, H., Wang, X.: Dynamic ensemble extreme learning machine based on sample entropy. Soft Computing 16(9), 1493–1502 (2012)
5. Liang, N.Y., Huang, G.B., Saratchandran, P., Sundararajan, N.: A fast and accurate online sequential learning algorithm for feedforward networks. IEEE Transactions on Neural Networks 17(6), 1411–1423 (2006)
6. Kennedy, J., Eberhart, R.: Particle swarm optimization. In: IEEE International of First Conference on Neural Networks (1995)
7. Riget, J., Vesterstrøm, J.S.: A diversity-guided particle swarm optimizer-the ARPSO. Dept. Comput. Tech. Rep. (2002)
8. Huang, G.B.: Learning capability and storage capacity of two-hidden-layer feedforward networks. IEEE Transactions on Neural Networks 14(2), 274–281 (2003)
9. Han, F., Zhu, J.S.: Improved Particle Swarm Optimization Combined with Backpropagation for Feedforward Neural Networks. International Journal of Intelligent Systems 28(3), 271–288 (2013)
10. Zeng, J.C., Jie, J., Cui, Z.H.: Particle Swarm Optimization. Science Press (2004)
11. Liang, Y.Y.: Review of ensemble learning [EB/OL] (2006)
12. Liu, N., Wang, H.: Ensemble based extreme learning machine. IEEE Signal Processing Letters 17(8), 754–757 (2010)
13. Lan, Y., Soh, Y.C., Huang, G.B.: Ensemble of online sequential extreme learning machine. Neurocomputing 72(13), 3391–3395 (2009)

Application of Random Forest, Rough Set Theory, Decision Tree and Neural Network to Detect Financial Statement Fraud – Taking Corporate Governance into Consideration

Fu Hsiang Chen, Der-Jang Chi*, and Jia-Yi Zhu

Department of Accounting, Chinese Culture University
No. 55, Hwa-Kang Road, Taipei, 11114, Taiwan
derjang@yahoo.com.tw

Abstract. Given that corporate financial statement fraud cases have been increasing in recent years, establishment of a model to effectively detect financial statement fraud has turned out to be one of the crucial issues. This study has integrated RF, RST, decision tree and back propagation network (BNP) to construct a corporate financial statement fraud detection model, so as to help auditor detect the signs of financial statement fraud before a fraud occurs in an enterprise. The study has established a new financial statement fraud detection model, for which the empirical result shows that, while analyzing the corporate financial statement fraud indexes considered and measured by RF, corporate governance indexes should also be included in financial statement fraud detection other than other financial indexes. In addition, it is also found that the corporate financial statement fraud detection model constructed by integrating RF and RST could also effectively enhance classification accuracy.

Keywords: Random Forest(RF), Rough Set Theory(RST); Decision Tree, Neural Networks, Financial Statement Fraud, Corporate Governance.

1   Introduction

As a result of increasing corporate fraud cases in recent years [1, 2, 3], stock prices are sagging and trading volumes are shrinking, which has affected the market capital raising function. Many studies suggest that, when a company is in a poor operating condition, financial statement fraud is more likely to occur [4, 5, 6]. Other than affecting the market capital raising function, any financial statement fraud may further lead to investors' loss of confidence in the financial statements audited and certified by the public certified accountant (CPA), which increases the degree of information asymmetry between investors and administrators. Hence, how to effectively detect the possibility of financial statement fraud, take required measures in advance and reduce the chance of financial statement fraud has turned out to be an issue of importance. According to the study conducted by Basely [7], non-financial information, e.g. the corporate governance mechanism, is associated with financial

---

* Corresponding author.

D.-S. Huang et al. (Eds.): ICIC 2014, LNCS 8588, pp. 221–234, 2014.

statement fraud, i.e. financial statement fraud is less likely to occur to a company with a sound financial condition and corporate governance mechanism. As such, when evaluating financial statement fraud, non-financial information shall also be included.

Most studies in relation to financial statement fraud have used the logistic regression method to explore empirical verification, but in so doing, the hypothesis of some conventional analysis shall be sustained, for instance: the data have to conform to normal distribution, but most of financial ratio data do not meet the terms of normal distribution [8].

As suggested by Taiwan's Statement on Auditing Standards No. 43, auditing personnel may adopt computer assistive techniques, e.g. data mining, in their auditing execution. Detection of financial statement fraud is an issue of classification. Out of numerous diagnosis classification methods, the RST is popular among many academics, for instance, Yeh, et al. [9] classified data with the RST. Furthermore, when using the RST for analysis, different from general conventional statistics which require specific basic assumption [10], it is not restrained by any hypotheses. In order to raise the accuracy of classification results, the study first screened the variables, followed by putting the selected variables into RST for classification. RF is also a variable screening application method frequently discussed in recent years. It mainly constructs collective tree-like classifiers to put variables into each tree-form classifier, so as to acquire the optimal result. By using the oob error to rank important variables [11], RF can also reserve important variables so as to increase classification accuracy and cut down data processing time. When using data mining for classification, if too many variables are input, it may result in influence and interference. Thus, in order to reduce the interference caused by too many variables, the study screened variables with RF, followed by using the RST for classification, so as to identify the signs before fraud occurs to an enterprise, and take required prevention measures.

2   Literature Review

2.1   Financial Statement Fraud

Definition of financial statement fraud: financial statement fraud can be defined and described as fraud or error resulting from unfaithful representation of financial statements. The difference between fraud and error lies in whether the motivation of unfaithful representation of financial statements is deliberate.

Table 1 shows the summary of financial statement fraud related definitions:

Table 1. Summary of financial statement fraud related definitions

	Definition of major content
Statement on Auditing Standards (SAS) No. 99	Unfaithful representation of financial statements is one of the fraud types. It refers to deliberate misstatement, amount omission or disclosure, or preparation of the financial statements causing others to misunderstand.
Taiwan's Statement on Auditing Standards No. 43	According to No. 43 of the Statement on Auditing Standards, the unfaithful representation of financial statements may result from fraud or error. The difference between fraud and error lies in whether the motivation of the unfaithful representation of financial statements is deliberate.

2.2   Summary of Financial Statement Fraud Related Studies

(1)   Study methods related to financial statement fraud

In terms of study methodology, logistic regression is the method most frequently applied [5, 6, 12, 13, 14]. However, this statistic method often requires satisfying specific basic assumption in statistics (e.g. data distribution shall meet the normal distribution) in order to be applicable. Hence, the data mining method not requiring any statistic assumption for data combination has come to rise. Currently, some academics have adopted the data mining method to detect financial statement fraud and acquired pretty good classification accuracy [14]. In recent years, many studies have suggested that data mining is a good classification method, which may solve the problem encountered by conventional statistic methods, such as logistic regression analysis, for which the data require meeting the normal distribution. As such, the study tried to use the data mining method to detect financial statement fraud.

(2)   Variable selection and results

For variable selection, some academics adopt either financial variables or non-financial variables, while others use both financial variables and non-financial variables. It is found that the classification accuracy after adding non-financial variables is higher than that without adding non-financial variables [15, 16]. As such, in addition to financial variables, the study also added non-financial variables, in an attempt to give more well-rounded detection of financial statement fraud.

2.3   Corporate Governance

The "OECD Corporate Governance Rules" released by Organization of Economic Cooperation and Development (OECD) in 1999 defined corporate governance as a corporate management and control system, in which the distribution of responsibilities and authority of the related parties covering the board of directors, managers, shareholders and other interested parties have been regulated. Basely [7] and Dechow et al. [17] put forth that supervision is stronger when the ratio of independent directors or external directors is higher. It shows that the constitution of a board of directors is related to supervision power, in which when corruption occurs, the function of the board of directors must also have problems. Thus, giving better understanding about constitution of a board of directors may help corruption prevention.

3   Study Methodology

The purpose of the study is to integrate RF and RST so as to construct a detection model to detect corporate financial statement fraud. The procedure was first to screen the financial and non-financial variables related to financial statement fraud with RF, followed by putting the selected variables in RST for analysis, so as to acquire classification results.

3.1   Random Forest

Random forest [18] is a tree-like classifier developed from the bootstrap aggregation (bagging) method and includes several decision trees. Its output category is determined by the mode of respective trees' output categories. It may assess the importance of a variable when deciding the category. When selecting bootstrap samples, out of bag samples (i.e. the samples not being selected) can be used to assess the error as the reference for variable selection. Hence, RF can also be used for screening variables [19, 20].

3.2   Rough Set Theory

The RST is a classification-like method, which is used to identify uncertain hidden knowledge. When using the rough set theory to process fuzzy and imprecise information, it does not require pretesting the data set or having extra information. The steps to use RST to investigate the rule knowledge existing between the input variables and the bad position are as follows: data collection, processing, discretization, screening, generating rules, screening rules, classification, calculation of classification accuracy, and establishment of the diagnosis system. The greatest advantage of the rough set theory is that it could effectively pick up information in a fuzzy and uncertain environment. It is often used for prediction of company's failure, bankruptcy and financial crisis, etc. [10, 21].

4   Empirical Study

4.1   Study Structure

In order to practically investigate the influence of corporate governance on financial statement fraud and verify the validity of detection of financial statement fraud with the random forest and rough set mentioned in this paper, this study analyzed financial statement fraud with the samples of companies having financial statement fraud in the period between 1998 and 2008. This was the period when financial turmoil in the US led to a global financial crisis where most corporate financial performance was significantly reduced. To avoid this particular economic condition affecting the information as to normal financial status, data was selected for the sample only up to 2008 for empirical study. And an attempt to help auditor effectively detect financial statement fraud in their audit of financial statements. In so doing, auditor will have less risk in being accused for unfaithful audit. Fig. 1 below shows the study's structure:

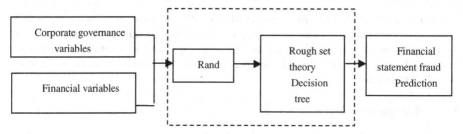

Fig. 1. Study structure

4.2   Data Sources

The study samples were collected from companies having financial statement fraud in the period between 1998 and 2008. As listed by the Securities and Futures Investors Protection Center, they were claimed to have "unfaithful financial statements" and were subject to "Article 20 of the Securities and Exchange Act". In addition, those companies were investigated by Financial Supervisory Commission for suspicion of false trading and unfaithful financial statements. A total of 47 companies had financial statement fraud. Moreover, the study adopted the recommendations given by Fanning and Cogger [13] and Chen, et al [15] for the sample matching principle, with which the companies falling in the same industry as the financial statement fraud companies and having the total asset amount close to the financial statement fraud companies in the year before the year that financial statement fraud occurred were deemed the matching samples. In the end, 94 samples in total were selected.

4.3   Variable Screening

The study variables were selected according to the financial statement fraud related measure indexes adopted by local and foreign academics. As shown in Table 2, 21 financial variables and 11 corporate governance variables were included.

The study integrated RF and RST to detect financial statement fraud. When processing the prefix data, the study used RF in software R to screen variables, followed by using C5.0 of SPSS clementine and PC Neuron 4.0 as well as Rough Set Exploration System 2.2 RSES for data analysis. In order to show the correlation degree between corporate governance and financial statement fraud, in addition to the financial variables, the study also added corporate governance variables in its overall experimental design, so as to identify how corporate governance could influence financial statement fraud.

4.4   Random Forest Analysis

First of all, the study used RF's variable screening function as the rule and method for reducing variables, so as to acquire more precise input variables. As a total, 21 financial variables and 11 corporate governance variables were selected in accordance with financial statement fraud measure indexes chosen by this paper. After calculation of the selected variables with RF, it was found that below: 13 financial variables and 6 corporate governance variables from the aforesaid 21 financial variables and 11 corporate governance variables for prediction of financial statement fraud. The selected variables are: F02; F03; F05; F06; F09; F10; F11; F13; F17; F18; F19; F20; F21; C01; C02; C04; C05; C10.

Table 2. Summary of measure variables' names and sources

	No.	Variable name	Reference
Financial variables	F01	Ratio of accounts receivable to total asset amount	Persons [5]
	F02	Ratio of cost of goods sold to net sales income	Kaminski et al.[12]
	F03	Debt ratio	Persons [5]
	F04	Ratio of fixed assets to total asset amount	Kaminski et al.[12]
	F05	Ratio of after-tax net profit to net sales income	Kaminski et al. [12]
	F06	Ratio of interest expense to total debt amount	Kaminski et al. [12]
	F07	Ratio of inventory to net sales income	Fanning and Cogger [13]
	F08	Ratio of net sales income to total asset amount	Persons [5]
	F09	Ratio of operating cash flow to net sales income	Persons [5]
	F10	Current ratio	Kaminski et al. [12]
	F11	Net debt value ratio	Kaminski et al. [12]
	F12	Ratio of operating profit to net sales income	Kaminski et al. [12]
	F13	EPS	Persons [5]
	F14	Acid-test ratio	Kirkose et al. [14]
	F15	Cash flow ratio	Persons [5]
	F16	Inventory turnover ratio	Kathleen et al. [12]
	F17	Operating income growth rate	Kirkose et al. [14]
	F18	Pre-tax income rate	Sphathis et al.[1]
	F19	Gross profit margin	Kaminski et al. [12]
	F20	Operating expense rate	Kaminski et al. [12]
	F21	Interest coverage ratio	Kaminski et al. [12]
Corporate governance variables	C01	Size of the board of directors	Chen et al. [15]
	C02	Internal director percentage	Lin et al. [16]
	C03	Institutional director percentage	Chen et al. [15]
	C04	General manager's duality	Chen et al. [15]
	C05	Director's shareholding ratio	Lin et al. [16]
	C06	Director's real shareholding ratio	Lin et al. [16]
	C07	Major shareholder's shareholding ratio	Chen et al. [15]
	C08	Manager's shareholding ratio	Lin et al. [16]
	C09	Size of the board of supervisors	Lin et al. [16]
	C10	Supervisor's shareholding ratio	Chen et al. [15]
	C11	Institutional supervisor percentage	Lin et al. [16]

The above selected variables hold important messages for prediction of financial statement fraud. The impact incurred from financial variables and corporate governance variables on prediction of financial statement fraud is hereby respectively explained as follows:

(1) Financial variables:

Compared with normal companies, fraud companies' sales grow rapidly with a higher return on investment, and the variables of return on assets, debt ratio and the ratio of operating cash to sales income show a significant correlation with financial statement fraud. A small current ratio represents a company's lack of liquidity or insufficiency of solvency. In other words, it does not have the fraud impetus because of capital deficiency. A small interest coverage ratio represents that a company does not have sufficient working capital to pay its interest. A small ratio of cash flow to net sales income represents that a company could only get a small cash flow from its sales income. The most direct way for a company to commit financial statement fraud is to falsely increase its sales income, so as to overestimate its earnings and raise its EPS.

(2) Corporate governance variables:

The number of the directors in a board shows the size of the board of directors. When the board of directors has a larger size, its organization will become more complicated and involved interests will be wider, and therefore the board of directors will be less likely to be manipulated by the management. The members of a board of directors may affect a company's fraud tendency. When there is a higher percentage of independent directors and a higher shareholding ratio from independent institutional investors, the company will be less likely to have financial statement fraud [15]. On the other hand, when the president is concurrently the general manager, the company will be more likely to have financial statement fraud. When internal directors also take posts in the company, they will have more information that could be used to discipline the management's behavior. However, it is also possible for them to collude with the management and make decisions unfavorable to shareholders .

4.5   Rough Set Analysis

By putting the selected financial variables and corporate governance variables into the rough set, the study constructed the financial statement fraud prediction model, in which 80% of the data fell in the training group and the remaining 20% went to the test group, to find out the classification accuracy by comparing model 1: financial variables and corporate governance variables with model 2: financial variables.

First of all, the study divided the SPSS discretization data into four zones. The division was made by deducting the minimum value from the maximum value, followed by dividing the resulting value by 4 (max − min/4). With the genetic algorithm (GA) of RSES, the study could get the decision rules. The decision rules are also called classification knowledge, which are the rules induced from the training data, and can be used to describe the samples and their classification. At the same time, by applying the classification knowledge to the new acquired samples, e.g. when a company's data are the same as a kind of classification knowledge, the company's possibility for occurrence of financial statement fraud could be figured out according to the classification knowledge in question. Table 3 below shows part of the classification knowledge picked up from the RSES analysis:

Table 3. Classification knowledge of the training data

Rule No. (1-750)	Match	Decision rules
1	9	(F17=3)&(F21=4)=>(Fraud={0[9]})
2	8	(F02=1)&(C02=1)=>(Fraud={0[8]})
3	8	(F02=1)&(F03=1)=>(Fraud={0[8]})
4	8	(C02=2)&(F19=3)=>(Fraud={0[8]})
5	8	(C01=1)&(C02=1)=>(Fraud={0[8]})
6	8	(F03=1)&(C01=1)=>(Fraud={0[8]})
7	8	(F03=2)&(F19=3)=>(Fraud={0[8]})
8	8	(C02=1)&(C04=1)&(F17=3)=>(Fraud={0[8]})
9	8	(F03=1)&(F05=4)&(F17=3)=>(Fraud={0[8]})
10	8	(F11=1)&(C04=1)&(F17=3)=>(Fraud={0[8]})

For the 1st column of Table 3: rule no. (1-750) elaborates on the 750 decision rules acquired after analyzing the training data with RSES. In order to give a specific explanation, only decision rules 1 – 10 were chosen for description. For the 2nd column of Table 3: match, it refers to the same decision rules of the data of several companies as the ones listed in Table 3, e.g. for rule 1, nine companies' attribute is same as the decision rule. For the final column of Table 3: decision rules, it represents when the ranking combination of the attribute of a company's data is the same as the decision rule, the company's decision attribute could be inferred, e.g. decision rule no. 1 explains that, out of test data 1, a total of nine companies match the rule of F17 (operating income growth rate) = 3, F21 (interest coverage ratio) = 4 and fraud = 0 (dummy variable, with fraud = 1, without fraud = 0). The prediction results acquired after the RSES analysis are explained as per Table 4 and Table 5 below:

Table 4. Model 1 – Training group's prediction results

Model 1 – training group's prediction results (financial variables + corporate governance variables)					
Company category	Number of tested company samples	Test results (number of tested companies/ percentage)		Accuracy	Ensemble accuracy
		Fraud	Normal		
Fraud	36	36 (100%)	0 (0%)	100%	100%
Normal	39	0 (0%)	39 (100%)	100%	

As shown in Table 4, a total of 36 fraud companies were accurately predicted by RSES as fraud companies and there was no fraud company being mistakenly predicted as the no-fraud (normal) company. The prediction accuracy is 100%. On the other hand, a total of 39 normal companies were accurately predicted as normal companies and there was no normal companies being mistakenly predicted as the fraud company, representing 100% prediction accuracy.

Table 5. Model 1 – Test group's prediction results

Model 1 – test group's prediction results (financial variables + corporate governance variables)				
Company category	Number of tested company samples	Test results (number of tested companies/ percentage)		Accuracy	Ensemble accuracy
		Fraud	Normal		
Fraud	11	8 (72.7%)	3 (27.3%)	72.7%	78.9%
Normal	8	1 (12.5%)	7 (87.5%)	87.5%	

As shown in Table 5, a total of 8 fraud companies were accurately predicted by RSES as fraud companies, and three fraud companies were mistakenly predicted as normal companies. The prediction accuracy is 72.7%. On the other hand, a total of 8 normal companies were accurately predicted as normal companies and one normal company was mistakenly predicted as a fraud company, representing a prediction accuracy of 87.5%.

Table 6. Model 2 – Training group's prediction results

Model 2 – training group's prediction results (financial variables)				
Company category	Number of tested company samples	Test results (number of tested companies/ percentage)		Accuracy	Ensemble accuracy
		Fraud	Normal		
Fraud	40	40 (100%)	0 (0%)	100%	100%
Normal	35	0 (0%)	35 (100%)	100%	

As shown in Table 6, a total of 40 fraud companies were accurately predicted by RSES as fraud companies and there was no fraud company being mistakenly predicted as the no-fraud company. The prediction accuracy is 100%. A total of 35 no-fraud companies were accurately predicted as no-fraud companies and there was no no-fraud company being mistakenly predicted as the fraud company, representing 100% prediction accuracy.

Table 7. Model 2 – Test group's prediction results

Model 2 – test group's prediction results (financial variables)				
Company category	Number of tested company samples	Test results (number of tested companies/ percentage)		Accuracy	Ensemble accuracy
		Fraud	Normal		
Fraud	7	6 (85.7%)	1 (14.3%)	85.7%	73.7%
Normal	12	4 (33.3%)	8 (66.7%)	66.7%	

As shown in Table 7, a total of 6 fraud companies were accurately predicted by RSES as fraud companies, and 1 fraud company was mistakenly predicted as a no-fraud company. The prediction accuracy is 85.7%. A total of 12 no-fraud companies were accurately predicted as no-fraud companies, and 4 no-fraud companies were mistakenly predicted as fraud companies, representing a prediction accuracy of 66.7%.

4.6   Decision Tree C5.0

Decision tree is one of the methods to establish classification models. The study used C.50 as its study methodology. The decision tree C.50 is mainly divided into two parts. The first part is the classification standard with which a decision tree can be completely constructed by calculating the acquired gain ratio, whereas the second part is the pruning standard for which the pruning calculation is based on the error rate and the decision tree is considered to be required to be pruned, so as to raise the classification accuracy. The study used C5.0 of SPSS climentine to execute the decision tree. The selected financial variables and corporate governance variables were put into the decision tree to construct the financial statement fraud prediction model, in which 80% of the data fell in the training group and the remaining 20% went to the test group, to find out the classification accuracy by comparing model 3: financial variable and corporate governance variables with model 4: financial variables.

Table 8 shows model 3's prediction results, in which the prediction accuracy for the training group is 82.67% while it is 78.95% for the test group, and the prediction error rate for the training group is 17.33% while it is 21.05% for the test group. The prediction results are shown in Table 8.

Table 8. Decision tree C5.0 model 3's prediction results

Model 3's prediction results (financial variables + corporate governance variables)				
	Training group		Test group	
Accuracy	62	82.67%	15	78.95%
Error rate	13	17.33%	4	21.05%
Total	75		19	

Table 9 shows model 4's prediction results, in which the prediction accuracy for the training group is 81.33% while it is 73.68% for the test group, and the prediction error rate for the training group is 18.67% while it is 26.32% for the test group. The prediction results are shown in Table 9.

Table 9. Decision tree C5.0 model 4's prediction results

Model 4's prediction results (financial variables)				
	Training group		Test group	
Accuracy	61	81.33%	14	73.68%
Error rate	14	18.67%	5	26.32%
Total	75		19	

4.7   Neural Networks

Neural networks refer to the information processing system simulating bio-neural networks. The study's prediction model was also constructed by using back propagation network (BPN), in which the calculation was made by transmitting the data from the input layer to the hidden layer, and, after calculation and conversion, the results were sent to the out layer. Generally speaking, the number of input layer's neurons is the number of the variables to be input. The number of hidden layers can be one layer or multiple layers. The output result of the neurons in the out layer is the network's final output value. The study used PC Neuron 4.0 to execute neural networks, and the selected financial variables and corporate governance variables were put into the neural network to build a financial statement fraud prediction model, in which 80% of the data fell in the training group, and the remaining 20% went to the test group.

Table 10. Prediction results of the combinations of different parameters under the BPN model

Node of the hidden layer	Learning rate	RMSE training group	RMSE test group
36	0.02	0.20783	0.33617
	0.025	0.19571	0.34562
	0.03	0.20447	0.33204
37	0.02	0.20075	0.34475
	0.025	0.20169	0.34861
	0.03	0.19186	0.34665
38	0.02	0.20162	0.35515
	0.025	0.19129	0.32915
	0.03	0.19237	0.34813
39	0.02	0.20410	0.33726
	0.025	0.19392	0.35383
	0.03	0.20581	0.33122
40	**0.025**	0.21182	0.32767
	0.025	**0.19961**	**0.32723**
	0.03	0.20305	0.33756

Table 10 below shows the comparison of the BNP diagnosis models combining different neurons and learning rates. When the network structure is {19-40-1}, the input layer includes 19 neurons, the hidden layer covers 40 neurons and the out layer has 1 neuron, whereas there is the minimum test data RMSE value when the learning rate is 0.025. Moreover, Table 11 shows the summary of the results of the corporate financial statement fraud diagnosis processed by the neural network structure {19-40-1}.

Table 11. Results of BPN diagnosis analysis

	Model 5 (financial variables + corporate governance variables)		Model 6 (financial variables)	
Sample type	Training group	Test group	Training group	Test group
Accuracy	79.271%	67.093%	75.424%	65.397%
Error rate	20.729%	32.907%	24.576%	34.603%

Table 11 shows the prediction results of model 5, in which the accurate prediction on the training group is 79.271% while the test group is 67.093%. The error prediction on the training group is 20.729% and the test group is 32.907%. For model 6, it is found that the accurate prediction on the training group is 75.424% while the test group is 65.397%. The error prediction on the training group is 24.576% while the test group is 34.603%.

4.8   Comparison of Analysis Results

After comparing the classification results of RST, C5.0 and BPN, it is found that the classification accuracy of rough sets in the aspect of financial statement fraud is higher than that of C5.0 and BPN, whereas, after adding corporate governance variables, the accurate prediction for the training group could effectively rise. Also, corporate governance variables are effective in prediction of financial statement fraud. Table 12 below shows the classification results.

Table 12. Classification results from comparison of RST, C5.0 and BPN

	Model 1 (financial variables and corporate governance variables)			Model 2 (corporate governance variables)		
method	RST	C5.0	BPN	RST	C5.0	BPN
Training group	100%	82.67%	79.271%	100%	81.33%	75.424%
Test group	78.9%	78.95%	67.093%	73.7%	73.68%	65.397%

To verify the validity of the model integrating RF and RST as put forth by the study, the results of the prediction made by simply using the RST model were also analyzed. The prediction results are shown in Table 13.

Table 13. Results of the classification without using RF

	RST	C5.0	BPN
Training group s	100%	77.33%	72%
Test group	73.7%	68.42%	63.16%

It is found that the classification accuracy without using RF for variable importance ranking and screening is less than that using RF for variable importance ranking and screening. Thus, the diagnosis model integrating RF and RST as put forth by the study is effective.

5   Conclusion and Recommendations

With increasing financial fraud cases in recent years, the risk faced by auditing personnel in auditing failure has also been increasing accordingly. Hence, establishing an effective corporate financial statement fraud detection model has turned out to be one of the crucial issues in practice and in the academia. By considering the fact that most of local and foreign financial statement fraud detection studies have adopted the regression analysis, together with the suggestion given by the Statement of Auditing Standards No. 43 that auditing personnel may adopt the data mining method to detect financial statement fraud, the study adopted decision tree C5.0, BPN and RST in the data mining method as the tools to detect financial statement fraud. By integrating RF, RST, decision tree and BPN, the study constructed the models, with which the major purpose is to use RF's variable screening function to screen variables, followed by putting the selected variables into RST, C5.0 and BPN and comparing the prediction accuracy of the three methods. When constructing the financial statement fraud diagnosis model, other than financial variables, the study also added corporate governance variables so as to find out the financial statement fraud prediction accuracy by comparing financial variables plus corporate governance variables with merely financial variables.

The study results suggest that, for the causes of corporate financial statement fraud found through screening of financial statement fraud's major indexes with the RF analysis, other than conventional financial construct indexes, corporate financial statement fraud is also likely to be affected by corporate governance construct indexes. As for the model integrating RF and RST as put forth by the study, the diagnosis indicates that its effect is better than that of simply using RST, i.e. the model constructed by integrating RF, RST, decision tree and BPN could indeed have a better effect. In terms of financial statement fraud detection accuracy, the integrated model also has better effect compared with the model individually using RST only. Moreover, after adding corporate governance variables, the ensemble classification accuracy of financial statement fraud becomes higher than that of simply using financial variables. It illustrates that when detecting financial statement fraud, auditing personnel shall take corporate governance variables into account.

References

1. Sphathis, C., Doumpos, M., Zopounidis, C.: Detecting Falsified Financial Statement: A Comparative Sstudy Using Multivariate Statistical Techniques. The European Accounting Review 11(3), 509–535 (2002)
2. Rezaee, Z.: Causes, Consequences, and Deterrence of Financial Statement Fraud. Critical Perspectives on Accounting 16(3), 277–298 (2005)
3. Ugrin, J., Kovar, S., Pearson, J.: An Examination of the Relative Deterrent Effects of Legislated Sanctions on Attitudes about Financial Statement Fraud: A policy Capturing Approach. Journal of Accounting, Ethics, and Public Policy 14(3), 693–734 (2013)
4. Kinney, W.R., McDaniel, L.S.: Characteristics of Firms Correcting Previously Reported Quarterly Earnings. Journal of Accounting and Economics 11, 71–93 (1989)
5. Persons, O.S.: Using Financial Statement Data to Identify Factors Associated with Fraudulent Financial Reporting. Journal of Applied Business Research 11(3), 38–46 (2011)
6. Ugrin, J.C., Odom, M.D., Ott, R.L.: Examining the Effects of Motive and Potential Detection on the Anticipation of Consequences for Financial Statement Fraud. Journal of Forensic & Investigative Accounting 6(1), 151–180 (2014)

7. Basely, M.: An Empirical Analysis of the Relation between the Board of Director Composition and Financial Statement Fraud. The Accounting Review 71(4), 443–466 (1996)
8. Min, J.H., Lee, Y.C.: Bankruptcy Prediction Using the Support Vector Machine with Optimal Choice of Kernel Function Parameters. Expert System with Applications 28(4), 603–614 (2005)
9. Yeh, C.C., Chi, D.J., Lin, Y.R.: Going-concern Prediction Using Hybrid Random Forests and Rough Set Approach. Information Sciences 254, 98–110 (2014)
10. Dimitras, A.I., Slowinski, R., Susmaga, R., Zopounidis, C.: Business Failure Predicition Using Rough Sets. European Journal of Operational Research 114(2), 263–280 (1999)
11. Speed, T.: Statistical Analysis of Gene Expression Microarray Data. Chapman & Hall/CRC Press, New York (2003)
12. Kaminski, A., Kaminski, T., Wetzel, S., Guan, L.: Can Financial Ratios Detect Fraudulent Financial Reporting? Managerial Auditing Journal 19(1), 15–28 (2004)
13. Fanning, K., Cogger, K.: Neural Network Detection of Management Fraud Using Published Financial Data. International Journal of Intelligent Systems in Accounting, Finance and Management 7(1), 21–24 (1998)
14. Kirkos, S., Spathis, C., Manolopoulos, Y.: Data Mining Techniques for the Detection of Fraudulent Financial Statements. Expert Systems with Application 32(4), 995–1003 (2007)
15. Chen, G., Firth, M., Gao, D.N., Rui, O.M.: Ownership Structure, Corporate Governance, and Fraud: Evidence from China. Journal of Corporate Financial 12(3), 424–448 (2006)
16. Lin, Y.C., Lin, C.S., Chen, H.J., Zhuang, J.H.: The Application of Neuro-Fuzzy Expert System in Litigation Prediction from the View of Corporate Governance. The International Journal of Accounting Studies 44, 95–126 (2007)
17. Dechow, P.M., Sloan, R.G., Sweeney, A.P.: Cause and Consequences of Earnings Manipulation: A Analysis of Firms Subject to Enforcement Actions by the SEC. Contemporary Accounting Research 13, 1–36 (1996)
18. Breiman, L.: Random Forests. Machine Learning 45, 5–32 (2001)
19. Harb, R., Yan, X., Radwan, E., Su, X.: Exploring Precrash Maneuvers Using Classification Trees and Random Forests. Accident Analysis and Prevention 41, 98–107 (2009)
20. Robin, G., Jean-Michel, P., Christine, T.M.: Variable Selection Using Random Forests. Pattern Recognition Letters (2010)
21. Thomas, E.M.: Developing a Bankruptcy Prediction Model via Rough Sets Theory. International Journal of Intelligent Systems in Accounting, Finance and Management 9(3), 159–173 (2000)

SAE-RNN Deep Learning for RGB-D
Based Object Recognition

Jing Bai and Yan Wu[*]

College of Electronics & Information Engineering, Tongji University, 201804, Shanghai, China
yanwu@tongji.edu.cn

Abstract. RGB-D image is a multimodal data. Previous works have proved that using color and depth images together can dramatically increase the RGB-D based object recognition accuracy, but most of them either simply take all modalities as input, ignoring information about specific modalities, or train a first layer representation for each modality separately and concatenate them ignoring correlated modality information. In this paper, we use a variant of the sparse auto-encoder (SAE) which can specify how mode-sparse or mode-dense the features should be. A new deep learning network combining the variant SAE with the recursive neural networks (RNNs) was proposed. Through it, we got very discriminating features and obtained state of the art performance on a standard RGB-D object dataset.

Keywords: RGB-D Based Object Recognition, Sparse Auto-Encoder, Recursive Neural Networks, Feature Extracting, Deep Learning.

1   Introduction

Previous works [1-6] have proved that combining RGB features with depth features for the object recognition can dramatically increase the object recognition accuracy. A naive way (called model 1 later) to apply feature learning algorithm to multimodal data is to simply combine all modalities as a concatenated vector and use the vector as input of feature extracting network, ignoring information about specific modalities, as shown in the left of Fig 1. This approach may lead to overfit or fail to learn associations between modalities with very different underlying statistics.

Instead of concatenating multimodal input as a vector, Ngiam et al. [8] proposed feature representation for each modality separately (called model 2 later), as seen in the middle of Fig. 1. They learned lower-level features separately from each modality, while learning higher-layer features from pure multimodal. However, for RGB-D data, separating first layer features may get bad performance.

To solve these problems, Lenz et al. [9] proposed a variant of SAE (called model 3 later), by incorporating a structured penalty term, which can specify how mode-sparse or mode-dense the features should be, as shown in the right of Fig. 1. In this paper, we introduced a new deep learning network by combining the variant SAE with recursive neural networks (RNNs) base on their work for RGB-D based object recognition task,

---

[*] Corresponding author.

D.-S. Huang et al. (Eds.): ICIC 2014, LNCS 8588, pp. 235–240, 2014.

with which, we get state of the art performance on a standard RGB-D object dataset while being more accurate and timesaving during training and testing than comparable works. In summer, there are two contribution of this paper:

First: we proposed a new deep learning model.

Second: we extracted features from RGB and depth modality simultaneous.

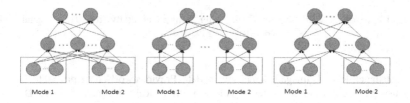

Fig. 1. Three models for multimodal deep learning

2   Related Work

In this paper, we focus on the problem of RGB-D based object recognition. In this section, we mainly introduce the related works to our work.

2.1   Works on RGB-D Based Object Recognition

A number of unsupervised deep learning methods for RGB-D based objects recognition have been introduced in the past few years. Bo et al. [5] introduced a hierarchical matching pursuit (HMP) algorithm, which used sparse coding to learn hierarchical feature representations from raw RGB-D data in an unsupervised way. Socher et al. [6] introduced a new model based on a combination of convolutional and recursive neural networks (CNN and RNN) for learning features. These methods mainly extracted features from RGB and depth images separately and simply concatenated them as a feature vector.

2.2   Sparse Auto-Encoder

Sparse auto-encoder learning algorithm [10] is one unsupervised approach to automatically learn features from unlabeled data. Assume that the raw unlabeled training examples set is $\{x^{(1)},...,x^{(t)}\}$ where $x^{(t)} \in R^n$. Let n_v and n_h represent input layer and hidden layer size respectively. Let $a^{(t)} = \{a_1, a_2,..., a_{n_h}\} \in R^{n_h}$ denotes the value of hidden layer through the following activation function when given the input $x^{(t)}$:

$$a_i^{(n_h)} = \varphi(\sum_{j=1}^{n_v} W_{ij} x_j^{(t)} + b_i) \tag{1}$$

Where $\varphi(z)$ is a activation function and W is the weight matrix, and b denotes the biased weight of input layer. With the output $a^{(t)}$ of hidden layer we can reconstruct the input $x^{(t)}$. To obtain optimal model parameters of $\{W,b,c\}$, we can compute and minimize the loss function by error back propagation as follows:

$$J_{sparse} = \frac{1}{m}\sum_{i=1}^{m}(\frac{1}{2}\|x^{(i)} - \hat{x}^{(i)}\|^2) + \frac{\lambda}{2}\sum_{i=1}^{n_v}\sum_{j=1}^{n_h}(W_{ij})^2 + \beta P(x) \tag{2}$$

where first term of formula is the reconstruction error, and second term a regularization term (L_2 regularization) preventing overfitting, and last term denotes sparsity penalty function In a typical SAE, L_2 regularization (as specified in formula (2)) is common used in training, which sometimes is not very suitable for multimodal data because it always ignores correlated modality information. To change the situation, Lenz et al. [9] introduced a variant of SAE which can learn feature from multimodal data by incorporating a structured penalty term called multimodal regularization into the optimization problem. For structured multimodal regularization, each modality will be used as a regularization group separately for each hidden unit. In the function they use the L_0 norm, it takes a value 0 for an input of 0, and 1 otherwise:

$$f(W) = \sum_{j=1}^{k}\sum_{r=1}^{R}I\{(\max_i {}_iS_{r,i}|W_{i,j}|) > 0\} \tag{3}$$

Where matrix S is an RxN binary matrix and R is distinct modality of input data X. Each element $S_{r,i}$ indicates membership of visible unit x_i in a particular modality r. I is an indicator function, which takes a value of 1 if its argument is true, 0, otherwise. The multimodal regularization function now encodes a direct penalty on the number of modes, successful learning correlated features between multiple input modalities.

2.3   Recursive Neural Networks

RNN [12, 13] is a widely used neural network which can learn hierarchical feature representations by applying the same neural network recursively in a tree structure. Previous RNN works [6, 13, 14] structured the tree. For some task, it is simply to design the input to be balanced to structure a fixed-trees. Previous works only combined pairs of vectors. Socher et al. [6] generalized the RNN architecture to allow each layer to merge blocks of adjacent vectors.

Assume 3D matrix $X \in R^{K\times r\times r}$ denotes the input of the RNN for each image. In [6], Socher defined a block to be a list of adjacent column vectors which were merged into a parent vector $P \in R^K$. If blocks are of size $K\times b\times b$, we get b^2 vectors in each block. The neural network for computing the parent vector is

$$P = f(W[x_1,......, x_{b^2}]) \tag{4}$$

where the parameter matrix $W \in R^{K\times b^2 K}$, f is a nonlinearity such as $\varphi(z)$ which will be applied to all blocks of vectors in X with the same weights W. Generally, there will be $(r/b)^2$ many parent vectors P, forming a new matrix P_1. The vectors in P_1 will again be merged in blocks just as those in matrix X using Eq. 4 with the same tied weights resulting in matrix P_2. This procedure continues until one parent vector remains. For detailed introduced refers to [6].

For recognition task, we used the top vector P^{top} as the feature vector to a classifier. Previous works only use a single RNN. We can actually use the 3D matrix X as input to a number of RNNs, we found that even RNNs with random weights produce high quality feature vectors. Each of N RNNs will output a K-dimensional vector. After we forward propagate through all the RNNs, we concatenate their outputs to a NK-dimensional vector which is then given to a softmax classifier.

3   SAE-RNN Neural Networks

In this section, we describe our new SAE-RNN deep learning model. Firstly, we train a variant of the SAE to learn features from patches extracted from raw data in an unsupervised way. Secondly, we use the trained variant SAE to extract features from raw images in a convolution way. The resulting features are given to RNNs. RNNs compose higher order features that can then be used to classify images. Fig 2 shows the total structure figure of our new SAE-RNN deep learning model.

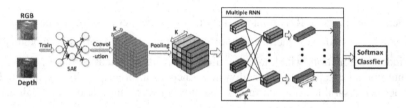

Fig. 2. The total structure figure of our new SAE-RNN deep learning model

3.1   A Two Layers SAE

To extract features from the high resolution images, we train a variant of the sparse auto-encoder (SAE) and use it as a mapping function to extract low level features in a convolutional way. In the multimodal setting, we use a variant of the SAE related above to deal with the RGB-D based object recognition problem. Firstly, we resize all images to size of dr and randomly extract patches with size $d \times d \times 3$ from raw RGB images and size $d \times d$ from raw depth images on the corresponding position separately. Then we combine every RGB patch with depth patch into a single vector with size of $d \times d \times 4$ as input of SAE. We train the first layer and the second layer separately where the first layer has the structured penalty term while the second layer has not this term. We use the first layer's output as the second layer's input, and we use the same size h as the first layer's hidden unit and the second layer's hidden unit.

For each image, we resize the raw input image it to size dr and then extracted the combined patches with size of $d \times d \times 4$ from raw RGB and depth image, so we can get $dl \times dl$ patches for one image. Then we use the $dl \times dl$ patches as the input of the trained SAE, resulting in a $h \times dl \times dl$ dimensional 3D matrix for each image. We then average pool them with square regions of size ds and a stride size of s, to obtain a pooled response with width and height equal to $r = (dl - ds)/s + 1$, resulting in a $h \times r \times r$ dimensional 3D matrix X.

3.2   Multiple Random RNNs

By applying the same neural network recursively in a tree structure, previous works on RNN flexible structure the tree depend on the input, but in our method, by combining with the two layers SAE we can simply limit the input to a fixed size which can structure a fixed-tree. Socher et al. [6] proposed a multiple random RNNs, with which we can actually use the 3D matrix X as input to a number of RNNs. The leaf nodes of the tree are h-dimensional vectors, use each RNN with random weights we can merge a block of vectors to its parent node, recurrent this step layer by layer we finally get a vector size of h dimension for each image with each RNN. If we use N RNN, after we forward propagate through all the RNNs, we concatenate their outputs to a N ×h-dimensional vector for each image which is then given to a softmax classifier.

4   Experiments

In this paper we focus on the problem of RGB-D based object category recognition, we evaluate our method on the recent RGB-D dataset of Lai et al. [1]. This RGB-D dataset contains 41877 RGB-D images of 51 different classes of objects and totally 300 instances of these classes. To compare our work with previous comparable works, we use the same setup as [1] and the 10 random splits they provide. The final accuracy is averaged over the 10 experiments.

For each experiment, we use one split to sample one instance from each class which result in 51 instances from different classes. The rest 249 instances are used for train and the 51 instances are used for test, and each instance contains about 130 images. We randomly extract 600000 RGB and 600000 patches with size 7×7 from raw depth images. All the patches are normalized and ZCA whitening is performed. After the preprocessing, we combine every RGB patch and its corresponding depth patch into a single vector with size of 196. We use these vectors as the input of SAE, we use 400 as the hidden unit size of both the first and second layer of the SAE. We train the two layers separately. With the trained two layers SAE, we extract features from each image in a convolution way pixel by pixel after all the images are resized to size of 146×146, resulting in a 400×140×140 dimensional 3D matrix. Then average pooling with pooling regions of size 10 and stride size 5 is performed which result in a 3D matrix of size 400×27×27 for each image. We use this matrix as the input of the RNN layer. We define 3x3 as the size of each RNN block and this lead to the following matrices at each depth of the tree: 400×27×27→400×9×9→400×3×3 and to the finally vector 400. We use 64 randomly initialized RNNs, so we get the finally features vector for each image of 400×64 = 25600 dimensions. After the features vector of all images obtained, we put these vectors into a softmax classifier to obtain the finally recognition accuracy of our method.

Comparison to Other Methods: Table 1 lists the main accuracy numbers and compares to the published results [1–6]. We can find that our method outperforms all others except that of Bo et al. [5] whose accuracy is about 0.2% greater than ours in recognition accuracy. They use the eight channels: grayscale and RGB for RGB images, and depth and surface normal for depth images, while our method only uses four channels: RGB and Depth. Socher et al. [6] only used four channels same as us, they extracted features from RGB and depth images separately and then simply combined them to a finally features vector. Our recognition accuracy is about 0.5% higher than theirs.

Table 1. Comparison of our SAE–RNN to multiple related methods

Method	Accuracy
SVM [2]	86.2±2.1
CKM [4]	86.4±2.3
SP+HMP [5]	87.5±2.9
CNN-RNN [6]	86.8±3.3
SAE-RNN	87.3±2.6

5   Conclusions

In this work, we introduced a new variant of the SAE which incorporates a structured penalty term into the optimization problem to be solved during learning. With this regularization term, the algorithm can specify how mode-sparse or mode-dense the features should be. It is very useful for RGB-D based object recognition. Combining the two layers SAE with the RNNs, we get very discriminating features and obtain state of the art performance on a standard RGB-D object dataset while being more accurate and faster during training and testing than comparable architectures.

References

1. Lai, K., Bo, L., Ren, X.: A Large-scale Hierarchical Multi-view RGB-D Object Dataset. In: ICRA, pp. 1817–1824 (2011)
2. Bo, L., Ren, X., Fox, D.: Depth Kernel Descriptors for Object Recognition. In: IROS, pp. 821–826 (2011)
3. Lai, K., Bo, L., Ren, X.: Sparse Distance Learning for Object Recognition Combining RGB and Depth Information. In: ICRA, pp. 4007–4013 (2011)
4. Blum, M., Springenberg, J.T., Wulfing, J.: A Learned Feature Descriptor for Object Recognition in RGB-D Data. In: ICRA, pp. 1298–1302 (2012)
5. Bo, L., Ren, X., Fox, D.: Unsupervised Feature Learning for RGB-D Based Object Recognition. In: Desai, J.P., Dudek, G., Khatib, O., Kumar, V. (eds.) Experimental Robotics. STAR, vol. 88, pp. 387–402. Springer, Heidelberg (2013)
6. Socher, R., Huval, B., Bath, B.P., et al.: Convolutional-Recursive Deep Learning for 3D Object Classification. In: NIPS, pp. 665–673 (2012)
7. Cireşan, D.C., Meier, U., Masci, J.: Flexible, High Performance Convolutional Neural Networks for Image Classification. In: IJCAI, pp. 1237–1242 (2011)
8. Ngiam, J., Khosla, A., Kim, M.: Multimodal Deep Learning. In: ICML, pp. 689–696 (2011)
9. Lenz, I., Lee, H., Saxena, A.: Deep Learning for Detecting Robotic Grasps. arXiv preprint arXiv 1301.3592 (2013)
10. Ng, A.: Sparse autoencoder. CS294A Lecture notes, 72 (2011)
11. Jalali, A., Ravikumar, P.D., Sanghavi, S., et al.: A Dirty Model for Multi-task Learning. In: NIPS, pp. 77–105 (2010)
12. Socher, R., Lin, C.C., Manning, C.: Parsing Natural Scenes and Natural Language with Recursive Neural Networks. In: ICML, pp. 129–136 (2011)
13. Socher, R., Pennington, J., Huang, E.H.: Semi-supervised Recursive AutoEncoders for Predicting Sentiment Distributions. In: EMNLP, pp. 151–161 (2011)

Visual Servoing of Robot Manipulator Based on Second Order Sliding Observer and Neural Compensation

Tran Minh Duc[1], Van Mien[1], Hee-Jun Kang[1,*], and Le Tien Dung[2]

[1] School of Electrical Engineering, University of Ulsan, 680-749 Ulsan, Korea
[2] Department of Electrical Engineering, The University of Da Nang - UST, Viet Nam
{ductm.ctme,vanmien1,dung.letien}@gmail.com,
hjkang@ulsan.ac.kr

Abstract. In this paper, we propose a PD-like visual servoing with second order sliding mode observer and neural network compensation algorithm for planar robot manipulators with only position measurement. This controller is designed based on a combining of a PD controller, neural network compensation and a velocity observer. First, a PD controller is designed as a nominal controller to control robot. Then, in order to compensate the uncertainties, an online learning neural network is designed. Furthermore, the controller incorporates a super-twisting second-order sliding mode observer for estimating the joint velocities; therefore, the velocity measurement is not required. The stability of the closed-loop controller-observer is proved based on the Lyapunov method. Finally, a computer simulation results are presented to evaluate the proposed controller.

Keywords: Radial basis function neural network, visual servoing, sliding mode observer, robot control, stability.

1 Introduction

The use of visual data to control the motion of robotic system is referred as visual servoing [1-2]. Visual servoing is based on the visual perception of a robot and workpiece location. The camera may be stationary or held in the robot hand, it can be classified into fixed cameras and camera in hand configurations. For fixed camera configuration, cameras are in world coordinate frame, and they can capture images of robots and environment. For camera in hand configuration, cameras are mounted on the end-effector of the robot. It means that they obtain the only visual information of the environment. In this paper, we use the fixed camera approach. An objective of visual servoing is to move the end effector of robot toward the desired positions.

Various PD-like visual servoing control schemes have been developed [3-4]. However, these PD visual servoing approaches cannot guarantee that the steady state error becomes zero when the friction and gravity forces exist. If the friction and gravity are unknown, neural networks can be applied. Lewis et al, [5] used neural networks to approximate the nonlinearity of robot dynamic. Stable visual servoing

---

* Corresponding author.

D.-S. Huang et al. (Eds.): ICIC 2014, LNCS 8588, pp. 241–247, 2014.
© Springer International Publishing Switzerland 2014

with neural network compensation is presented in [6]. However, this method still has a limitation; the controller requires measurements of joint velocities. It is very important to realize visual servoing with only joint position. One way to alleviate the above difficulty is used a velocity observer. The first Observer based visual servoing was presented in [7]. A linear observer was introduced in [8] by neglecting the nonlinear dynamic of the robot. Other observer based technique, which is so-called sliding mode (SM) observer, has been developed. For example, traditional sliding mode observer has been proposed in [13-14]. PD–based visual servoing combining with traditional sliding mode observer and neural network compensator has been developed in [9]. However, the use of the traditional sliding mode generates a big chattering. The most widely used in practical applications to eliminate chattering are higher-order sliding mode [10-11]. Davila et al, [10], used second-order sliding mode observer to estimate the velocity from the position measurement for mechanical systems. The results had shown the advantages of the observer in terms of their ability to exact velocity estimation.

In this paper, we propose a neural network-based visual servoing combined with second order sliding mode observer for a planar robot manipulator in fixed camera configuration. The proposed algorithm is designed based on a combination of PD controller, second-order sliding mode observer and neural network compensator. First, PD-like visual servoing is used as the main controller. Then, second-order sliding mode observer is designed to estimate the joint velocities from joint position measurement. Finally, a radial basis function neural network is used to compensate for the uncertainties. The stability and convergence of the proposed system is proved by Lyapunov theory. Simulation results are shown to verify the performance of the proposed controller.

2   Background

2.1   Robot System Model

The dynamics of a serial n-links robot manipulator can be written as [12]

$$M(q)\ddot{q} + C(q,\dot{q})\dot{q} + G(q) + F(\dot{q}) = \tau \qquad (1)$$

where $q \in R^n$ is the joint variable vector, $\dot{q} \in R^n$ is the joint velocity vector, $M(q) \in R^{n \times n}$ is the inertial matrix, $C(q,\dot{q}) \in R^{n \times n}$ is the centripetal and Coriolis matrix, $G(q) \in R^n$ is the gravity vector, $F(\dot{q}) \in R^n$ is a positive diagonal matrix of frictional terms, and $\tau \in R^n$ is the control torque.

The inertial matrix $M(q)$ and the Coriolis matrix $C(q,\dot{q})$ in the dynamic model (2) satisfy the following properties;

a) $M(q)$ is a symmetric positive definite matrix.

b) $M(q) - 2C(q,\dot{q})$ is a skew symmetric matrix.

$$x^T[\dot{M}(q) - 2C(q,\dot{q})]x = 0 \qquad (2)$$

2.2   2-D Visual Servoing of Robot

Consider a perspective transformation as an ideal pinhole camera model [4]. The image center is in origin of the screen coordinate frame, the camera is rotated about its optical axis by clockwise θ radian, so the rotation matrix $R(\theta)$ is

$$R(\theta) = \begin{bmatrix} \cos\theta & -\sin\theta \\ \sin\theta & \cos\theta \end{bmatrix} \tag{3}$$

where $[C_x, C_y]^T$ is the image center, the position of optical axis is $[O_{r1}, O_{r2}]^T$, a point P in robot coordinate frame is $[P_{r1}, P_{r2}]^T$, this point in image coordinate frame is $[P_{s1}, P_{s2}]^T$, so the description of a point P_r in the robot coordinate frame is given in terms of the image coordinate frame as.

$$\begin{bmatrix} P_{s1} \\ P_{s1} \end{bmatrix} = \alpha h R(\theta) \left\{ \begin{bmatrix} P_{r1} \\ P_{r2} \end{bmatrix} - \begin{bmatrix} O_{r1} \\ O_{r2} \end{bmatrix} \right\} + \begin{bmatrix} C_x \\ C_y \end{bmatrix} \tag{4}$$

where α is the scale factor of length units, and h is the magnification factor defined as

$$h = \frac{\lambda}{\lambda - z} \tag{5}$$

where λ is the focal length and z is the distance between camera and robot frame. The image position error $\tilde{x}_s$ is defined as

$$\tilde{x}_s = \begin{bmatrix} x_s^* \\ y_s^* \end{bmatrix} - \begin{bmatrix} x_s \\ y_s \end{bmatrix} \tag{6}$$

where (x_s^*, y_s^*) and (x_s, y_s) are the positions of target and end-effector in the image, (6) can be written as

$$\tilde{x}_s = \alpha h R(\theta) \left\{ \begin{bmatrix} x_s^* \\ y_s^* \end{bmatrix} - \begin{bmatrix} x_s \\ y_s \end{bmatrix} \right\} = \alpha h R(\theta)[K(q^*) - K(q)] \tag{7}$$

The Velocity property is

$$\frac{d}{dt}\tilde{x}_s = \alpha h R(\theta)[J(q^*)\dot{q}^* - J(q)\dot{q}] \tag{8}$$

The proposed PD-like visual servoing control is (for case $\dot{q}^* = 0$)

$$\tau = J^T k_p \tilde{x}_s - k_d \dot{q} \tag{9}$$

3   Second-Order Sliding Mode Observer for Velocity Estimation

Using $x_1 = q \in R^n$ and $x_2 = \dot{q} \in R^n$, the robot dynamics as expressed in equation (1) can be written in state space form as

$$\dot{x}_1 = x_2$$
$$\dot{x}_2 = M^{-1}(q)[\tau - C(q,\dot{q})\dot{q} - G(q)] + \Delta(q,\dot{q},t) \tag{10}$$
$$y = x_1$$

where $\Delta(q,\dot{q},t) = M^{-1}(q)F(\dot{q})$, $f(x_1,\hat{x}_2,\tau) = M^{-1}(q)[\tau - C(q,\dot{q})\dot{q} - G(q)]$.

Based on [15], the second-order sliding mode based state observer can be described by

$$\dot{\hat{x}}_1 = \hat{x}_2 + \alpha_1|\tilde{x}_1|^{1/2} sign(\tilde{x}_1)$$
$$\dot{\hat{x}}_2 = f(x_1,\hat{x}_2,\tau) + \alpha_2 sign(\tilde{x}_1) \tag{11}$$

where $\hat{x}_1$ and $\hat{x}_2$ are states estimation of x_1 and x_2 with initial condition $\hat{x}_1 = x_1$, $\hat{x}_2 = 0$. $\tilde{x}_1 = x_1 - \hat{x}_1$ is the state estimation error.

Substituting Eq. (10) into Eq. (11), we obtain the states estimation error:

$$\dot{\tilde{x}}_1 = \tilde{x}_2 - \alpha_1|\tilde{x}_1|^{1/2} sign(\tilde{x}_1)$$
$$\dot{\tilde{x}}_2 = d(x_1,\hat{x}_2,\tilde{x}_2,\tau) + \Delta(q,\dot{q},t) - \alpha_2 sign(\tilde{x}_1) \tag{12}$$

where $d(x_1,\hat{x}_2,\tilde{x}_2,\tau) = f(x_1,x_2,\tau) - f(x_1,\hat{x}_2,\tau)$, If we defined $f(x_1,\hat{x}_2,\tilde{x}_2,\tau,t) = d(x_1,\hat{x}_2,\tilde{x}_2,\tau) + \Delta(q,\dot{q},t)$

Assumption 1: The states of the robot system are bounded for all time.

Assumption 2: The modeling uncertainty is bounded.

$$\left\|M^{-1}(q)F(\dot{q})\right\| \le \overline{\Delta} \tag{13}$$

where $\overline{\Delta}$ is the known constant.

Based on the assumptions 1-2, there exist a constant f^+ such that:

$$f(x_1,\hat{x}_2,\tilde{x}_2,\tau,t) < f^+ \tag{14}$$

Based on the Lyapunov approach in Ref. [13], Theorem 1 gives the convergence of the estimation error to zero in finite time.

Theorem 1: Suppose that condition (14) holds for a system as in Eq. (13); if the sliding mode gains of the observer scheme in Eq. (12) are chosen as

$$\alpha_1 > 0$$
$$\alpha_2 > 3f^+ + 2\frac{f^{+2}}{\alpha_1} \tag{15}$$

Then the observer scheme is stable, and the states of the observer in Eq. (11) ($\hat{x}_1,\hat{x}_2$) converges to the true state (x_1,x_2) in Eq. (10) in finite time.

By similar way with the Ref. [16], we can verify that the observer states converge to the true states in finite time.

4   PD-like Visual Servoing Based on Second-Order Sliding Mode Observer and Neural Network Compensation

The ideal PD-likes visual servoing with neural compensation can be expressed as

$$\tau = J^T k_p \tilde{x}_s - k_d \dot{q} + \hat{W}_t \sigma(x) \tag{16}$$

When $x = \begin{bmatrix} q^T, \dot{q}^T \end{bmatrix}$ is the input vector of the neural networks. The joint velocity $\dot{\hat{q}}$ is estimated by the second order sliding mode observer. Eq. (11), the observer based PD-like visual servoing is

$$\tau = J^T k_p \tilde{x}_s - k_d \dot{\hat{q}} + \hat{W}_t \sigma(x) \tag{17}$$

We define estimation errors for $\dot{q}$ and $\ddot{q}$ as

$$\dot{\hat{q}} = \dot{q} + \varepsilon_q$$
$$\ddot{\hat{q}} = \ddot{q} \tag{18}$$

ε_q is the neural network error, we assume that it is bounded.

$$\varepsilon_q \Lambda_q \varepsilon_q \le \bar{\varepsilon}_q, \Lambda_q^{\ T} = \Lambda_q > 0$$

Now considering the case of neural network cannot approximate the uncertainty exactly

$$G(q) + F(\dot{q}) = W^T \sigma(x) + \varepsilon_\sigma \tag{19}$$

$\varepsilon(x)$ is the neural network error, we assume that it is bounded.

$$\varepsilon_\sigma \Lambda_\sigma \varepsilon_\sigma \le \bar{\varepsilon}_\sigma, \Lambda_\sigma^{\ T} = \Lambda_\sigma > 0 \tag{20}$$

We also consider the case of camera model is not exactly known,… then

$$\dot{\tilde{x}}_s = \alpha h R J \dot{q} + e J \dot{q} \tag{21}$$

Where, the image modeling error is assumed to be bounded as

$$e^T \Lambda_e e \le \bar{e}, \Lambda_e^{\ T} = \Lambda_e > 0 \tag{22}$$

Theorem 2: if k_d in the 2D visual servoing control law is selected big enough

$$k_d > \left[\Lambda_\sigma^{-1} + \dot{M}^T \Lambda_{\varepsilon_p}^{\ -1} \dot{M} + \tilde{x}_s^{\ T} \left(\frac{J^T R^{-1} k_p}{\alpha h} \Lambda_s^{\ -1} \right) \tilde{x}_s \right] \tag{23}$$

Then the weight of the neural compensation is updated as

$$\dot{\hat{W}} = -k_w \sigma(x) \dot{\hat{q}}^T \tag{24}$$

And dead-zone control is used as

If $\left\| \dot{\hat{q}} \right\|^2 > \lambda_{\min}^{\ -1} Q \bar{E}$ then we use the control law (17)

If $\left\| \dot{\hat{q}} \right\|^2 \le \lambda_{\min}^{\ -1} Q \bar{E}$ then stop the control law

here

$$Q = k_d - \left[\Lambda_\sigma^{-1} + \dot{M}^T \Lambda_{\varepsilon_p}^{-1} \dot{M} + \tilde{x}_s^T \left(\frac{J^T R^{-1} k_p}{\alpha h} \Lambda_s^{-1} \right) \tilde{x}_s \right] \quad (25)$$

The closed-loop system is stable.

5   Simulation Results

To verify the effectiveness of the proposed algorithm, a two-link planar robot manipulator is considered. The orientation of the camera with respect to the robot coordinate frame was $\theta = 0.8$ rad. The scale factor was $\alpha = 500$, the magnification factor was $h = 0.01$ m. The initial robot position was $q = (-1.4, 0.6)^T$, the target have been placed at $q^* = (0 - 1, 0.8)^T$. In the RBF neural the centers and its width are assumed to have been fixed. The symmetric positive definite proportional, derivative and k_w matrices were chosen as $k_p = diag(150, 100)$, $k_d = diag(50, 50)$, $k_w = diag(6, 6)$.

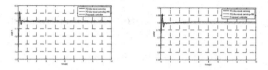

Fig. 1. Comparison among PD-like visual servoing, PD-like visual servoing with neural network compensation and propose controller

In the simulation set, we compare three kinds of visual servoing, PD-likes visual servoing, visual servoing with neural compensation, visual servoing with second-order sliding mode velocity observer and neural compensation proposed in this paper. From the data presented in this figure, it can be seen that the combination of second-order sliding mode observer and neural compensator is a good way to improve the performance of the normal PD-like visual servoing, when gravity, friction are unknown. Moreover, in case joint velocities are unknown, we still get good performance results by using second-order sliding mode observer.

6   Conclusion

This paper presented a stable visual servoing based on second order sliding mode observer and radial basis function neural network compensator for a planar robot manipulator in a fixed camera configuration. The theory proved that the controller could make the closed-loop system stable when the gain of the PD controller was larger than the upper bounds of the uncertainties. The results of computer simulations for 2-DOF robot manipulator were shown to evaluate the controller. Future work will be carried out on the augmentation of acceleration command in the visual servoing.

Acknowledgments. This work was supported by Special Research Fund of Electrical Engineering in University of Ulsan.

References

1. Hutchinson, S., Hager, G., Corke, P.: An Introduction to Inertial and Visual Sensing. IEEE Transactions on Robotics and Automation 12(5), 12–22 (1996)
2. Chaumette, F., Hutchinson, S.: Visual Servo Control. Part I: Basic Approaches. IEEE Robotics & Automation Magazine 12 (2002)
3. Espiau, B., Chaumette, F., Fixot, N.: A new approach to visual servoing in robotic. IEEE Transactions on Robotics and Automation 8(3), 34–44 (1992)
4. Keily, F.: Robust asymptotically stable visual servoing of planar robots. IEEE Transactions on Robotics and Automation 12(5), 53–74 (1996)
5. Lewis, F.: Neural network output feedback control with guaranteed stability. International Journal of Control 70(3), 34–57 (1998)
6. Loreto, G., Yu, W.: Stable visual servoing with neural networks compensation. In: IEEE International Symposium on Intelligent Control (2001)
7. Hashimotom, K.: Visual servoing with nonlinear observer. In: IEEE International Conference on Robotic and Automation (1995)
8. Hashimotom, K.: Visual servoing with linearized observer. In: IEEE International Conference on Robotic and Automation (1999)
9. Yu, W., Li, X.: Visual servoing with velocity observer and neural compensation. In: IEEE International Symposium on Intelligent Control (2004)
10. Davila, J., Fridman, L.: Second-order sliding-mode observer for mechanical systems. IEEE Transactions on Automatic Control 50, 1785–1789 (2005)
11. Van, M., Kang, H.-J.: A robust fault diagnosis and accommodation scheme for robot manipulators. International Journal of Control, Automation, and Systems 11, 377–388 (2013)
12. Craig, J.J.: Introduction to robotics mechanics and control, 3rd edn.
13. Fixot, N.: Adaptive control of robot manipulator via velocity estimated state feedback. IEEE Transactions on Automatic Control 37(8), 1234–1246 (1992)
14. Astrom, K.J.: Trajectory tracking in robot manipulators via nonlinear estimated state feedback. IEEE Transactions on Robotics and Automation 8(1), 1283–1295 (1992)
15. Davila, J., Fridman, L., Levant, A.: Second-order sliding-mode observer for mechanical systems. IEEE Transactions on Automatic Control 50, 1785–1789 (2005)
16. Moreno, J.A., Osorio, M.: A Lyapunov approach to second-order sliding mode controllers and observers. In: 47th IEEE Conference on Decision and Control, pp. 2856–2861 (2008)

An Automatic Image Segmentation Algorithm Based on Spiking Neural Network Model

Xianghong Lin[1], Xiangwen Wang[1], and Wenbo Cui[2]

[1] School of Computer Science and Engineering, Northwest Normal University, Lanzhou 730070, China
[2] Linyi Branch Corporation, China Telecom, Linyi 276000, China
linxh@nwnu.edu.cn

Abstract. Inspired by the structure and behavior of the human visual system, an automatic image segmentation algorithm based on a spiking neural network model is proposed. At first, the image pixel values are encoded into the timing of spikes of neurons using the time-to-first-spike coding strategy. Then the segmentation model of spiking neural networks is applied to generate the matrix of spike timing for the visual image. Finally, using the maximum Shannon entropy as the fitness function of genetic algorithm, the evolved segmentation threshold is obtained to segment the visual image. The experimental results show that the method can obtain the optimum segmentation threshold, and achieve satisfactory segmentation results for different images.

Keywords: spiking neural network, image segmentation, genetic algorithm, segmentation threshold, maximum Shannon entropy.

1 Introduction

Humans show a remarkable ability to extract salient information from visual inputs. This has inspired researchers to develop new computational visual models to help understand the biological mechanisms that underlie image processing in the brain. The process of segmenting images is one of the most critical ones in automatic image analysis whose goal can be regarded as to find what objects are presented in visual images. The project described in this paper discusses image segmentation in the visual system and aims to build a computational model based on spiking neural networks (SNNs). SNNs are artificial abstract models of different parts of the brain or nervous system, featuring essential properties of these systems using biologically realistic models. They are often referred to as the third generation of neural networks in which computation is performed in the temporal domain and relies on the timings between spikes, and have potential to solve problems related to biological stimuli [1], [2]. In general a spiking neuron operates by integrating spike responses from presynaptic neurons and generating an output spike when the membrane potential reaches a threshold value.

In recent years, researchers used SNNs to segment visual images, and achieved better results [3], [4], [5]. Meftah et al. [6] used the SNN based on spike response

D.-S. Huang et al. (Eds.): ICIC 2014, LNCS 8588, pp. 248–258, 2014.
© Springer International Publishing Switzerland 2014

model to segment image and detect edge. The image pixel values are encoded into spike sequences using the population coding strategy in the input layer, then the unsupervised or supervised learning algorithm is applied to adjust the connection weights in the network, and the results of image segmentation or edge detection are obtained in the output layer. Soni et al. [7] proposed a SNN model which is constituted of the leaky integrate-and-fire neurons to segment gray-scale images, and clustering approach is used for the segmentation of visual images. After the selection and the extraction of the image features, the feature samples are grouped together in compact but well-separated clusters corresponding to each class of the image. Similar work was done in [8]. However, the result of image segmentation is encoded by the spike firing rate. Wu et al. [9] used integrate-and-fire neuron network model for color image segmentation. In their work, the network is constructed in the two parts. The first part is a SNN which extract color features through different ON/OFF pathways. The second part is a BP neural network which is trained to recognize the color features and segment the visual image. In essence, the approach is still based on the traditional rate coding neural networks for image segmentation.

In this paper, we propose an automatic image segmentation algorithm based on a SNN model. The characteristics of this image segmentation method are: (1) the SNN is constructed in the three layers using the leaky integrate-and-fire neuron model, the input and output of the entire network adopts the time to first-spike coding strategy. That is, the image pixel value is encoded into the timing of spike in which each neuron fire only one spike, the information encoding of images is more close to the basic physiology of human visual nervous system [10], [11]; (2) the effect of image segmentation depends on the parameters of the network model, especially adaptive selection of the segmentation threshold. So that, the genetic algorithm can be applied for choosing the optimum segmentation threshold to achieve the better accuracy in image segmentation.

2   Spiking Neural Network Model for Image Segmentation

2.1   Spike Coding Method of Image Pixel

It is well known that biological neurons use pulses or spikes to encode information. Neurological research shows that the biological neurons store information in the timing of spikes [12]. SNN cannot directly calculate analog, so we first need to encode the analog value to timing of spike and input it into spiking neuron network, then output the calculated results in the spike mode. The time-to-first-spike coding strategy is an information encoding method of converting the analog value to spike timing, it assumes that the neurons fire only one spike, and the timing of spike is proportional to the analog value. The time-to-first-spike coding stresses that the first spike is generated in the neuron which received the stimulation signal [11].

In this paper, the time-to-first-spike coding strategy is applied to SNN for image segmentation, and use a nonlinear function that realizes the corresponding relationship from pixel values to the spike times of the neurons [13]. The conversion between gray value of image pixels and firing time of neuronal spikes using the Sigmoid function is represented as follow:

$$t = \frac{T_{max}}{1 + \exp\left(\sigma(128 - p)\right)} \tag{1}$$

where p is the gray value of pixels in the interval [0, 255], t is the firing time of spikes, σ is the nonlinear coding parameter, T_{max} is the maximum firing time, and T_{max}=10ms. Fig. 1 shows the diagram of the time-to-first-spike coding strategy as the nonlinear coding parameter σ=0.05. The nonlinear corresponding relationship from pixel values to spike times of the neurons is shown in Fig. 1(a). When the gray values of pixels 25, 128, and 250 are inputted into a neuron, the neuron generate the output spikes at 0.06 ms, 5 ms and 9.98 ms respectively (Fig. 1(b)).

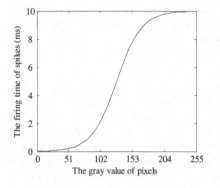
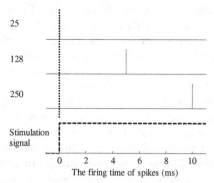

(a) The nonlinear corresponding relationship   (b) The examples of spike coding

Fig. 1. The diagram of the time-to-first-spike coding strategy

2.2   Spiking Neural Network Model

The human visual system can efficiently perform image segmentation and recognition, neuroscientists have found that there are a variety of neuronal receptive fields from simple cells to those cells in the retina and lateral geniculate [14]. On the basis of the receptive field mechanism and the leaky integrate-and-fire neuron which is probably the best-known example of a formal spiking neuron model [15], the three-layer SNN model for image segmentation is shown in Fig. 2. The first layer is light-sensing input layer, all image pixels are converted to spike times using the time-to-first-spike coding strategy. In the middle layer, each neuron corresponds to a receptive field of the input layer and integrating the spike responses from presynaptic neurons in the receptive field. The third layer is the output layer, each neuron stores the spike timing of the corresponding neuron in the middle layer, and the matrix of spike timing in output layer represents the result of image segmentation through the segmentation threshold.

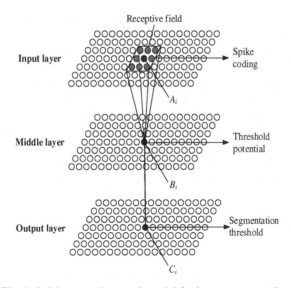

Fig. 2. Spiking neural network model for image segmentation

Firstly, the original image needs to convert the gray values of image pixels to spike times in the input layer, we adopts the nonlinear method of time-to-first-spike coding strategy [13]. Assuming that the size of the original image is $W{\times}H$, and therefore there are $W{\times}H$ neurons in the input layer, where W and H is the width and height of the image respectively. Each neuron A_i ($i=1, 2, ..., W{\times}H$) in this layer as the center forms a rectangular receptive field whose size is $3{\times}3$. If neuron is located in the edges of the input layer, then not constitutes a complete $3{\times}3$ size of the receptive field, and the vacant positions in matrix will not fire spikes.

In the middle layer, each neuron B_i receives the corresponding neuronal inputs in the input layer where the neuron A_i is the center of the receptive field, the connection weight w_{ij} between each neuron R_j ($j=1, 2, ..., 3{\times}3$) in the receptive field and the neuron B_i in the middle layer is calculated as follows:

$$w_{ij} = W_{max} \exp\left(-\left(\frac{\left\|F(A_i)-F(R_j)\right\|^2}{d} + \frac{\left\|X(A_i)-X(R_j)\right\|^2}{d}\right)\right) \qquad (2)$$

where W_{max} is the interaction strength or the weight scale between neurons, in our simulation $W_{max}=10$ mV. $F(A_i)$ and $F(R_j)$ are the pixel gray values of the neuron A_i and R_j respectively, $\|F(A_i)-F(R_j)\|$ is the difference of the pixel gray values. $X(A_i)$ and $X(R_j)$ are the pixel coordinate of the neurons respectively, $\|X(A_i)-X(R_j)\|$ is the Euclidean distance between the pixel coordinates. d is a constant and we set $d=3$. After the firing times of spikes and the connection weights between the neurons are known, the membrane potential of neuron B_i can be calculated. The spike sequence in the receptive field is transmitted to neuron B_i according the order of timing of spikes. When the membrane potential reaches the threshold potential, the first spike is generated, and ignore the rest of spike inputs. If all spikes are inputted to the neuron B_i but the membrane potential of the neuron B_i still have not reached the threshold potential, it means that the neuron do not fire spike, the spike time of this neuron is represented as $+\infty$.

Each neuron C_i in the output layer corresponds to the neuron B_i in the middle layer. In this layer we set a segmentation threshold T_{thresh}, when the spike time of the neuron in the middle layer exceeds the threshold value, the neuron C_i do not fire spike, and labeling the corresponding pixel gray value as 255. When the firing time of the middle layer neuron does not exceed the threshold value, the neuron C_i fire a spike and labeling the corresponding pixel gray value as 0. So that, according to the segmentation threshold T_{thresh}, the segmented binary image can be obtained.

3   Automatic Image Segmentation with Genetic Algorithm

Genetic algorithm [16] is a random algorithm that imitating the biological group evolution. It starts from a population of potential solutions of the representative problems. A population consists of a certain number of genes encoding individuals, each individual is an entity with a characteristic chromosomal. At beginning we need to achieve the mapping of phenotype to genotype, namely genetic coding. In this paper we choose the binary coding, that is chromosome genetic code consisting of binary strings as the operation object of genetic algorithm.

For the original image, we can obtain the spike timing matrix of the image using the SNN image segmentation model. That is, the matrix stores the firing times of spikes of all neurons in the middle layer. Then, a certain number of individuals generated randomly constitute the initial population, decoding each individual in the population can get the segmentation threshold. Using the segmentation threshold for the spike timing matrix, the binary image can be obtained. For the segmented binary image, we evaluate the fitness using the fitness function. Applying the selection operator based on the individual fitness, and crossover and mutation operators, the new population can be reproduced. Therefore, the segmentation threshold is evolved, and ultimately get the ideal segmented image. The SNN image segmentation process with genetic algorithm is shown in Fig. 3.

For each generation, the individuals in the population are decoded to generate segmentation threshold, and they are used to the matrix of firing times to get the segmented binary image. According to the Shannon entropy [17], the fitness of the individual is calculated as follow:

$$Fitness = -P_0 \log_2 P_0 - P_1 \log_2 P_1 \tag{3}$$

where P_0 and P_1 are the probability that the pixel gray values of segmented binary image are 0 and 255 respectively. For most images, Shannon entropy represents the amount of information of the image. If the Shannon entropy of the segmented image is larger, the segmented image will get the greater amount of information from the original image, better result is obtained for high detailed image, and the overall segmentation effect will be better accuracy.

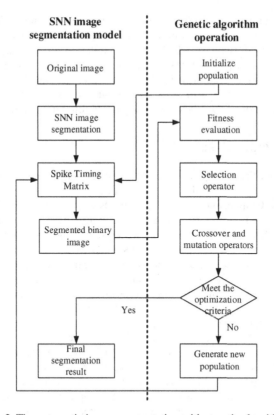

Fig. 3. The automatic image segmentation with genetic algorithm

Table 1. The SNN image segmentation with genetic algorithm

---

(1) Apply the SNN image segmentation model to original image with the size of $W{\times}H$, and obtain the spike timing matrix whose size is $W{\times}H$.

(2) Initialize a segmentation threshold population with a certain size (popsize).

(3) For each segmentation threshold in the population do the follow:

    (a) Divide the spike times in the matrix into two categories by the segmentation threshold that decoding from the individuals in the population and output the binary image.

    (b) Evaluate this binary image and obtain the fitness according to formula (3).

(4) Repeat until the new segmentation threshold population with popsize is generated.

    (a) Select two individuals in the population by the tournament selection.

    (b) Apply single-point crossover operator according to a certain probability to generate new individuals.

    (c) Apply single-point mutation operator according to a certain probability.

(5) Replace the current population with the new generated population.

(6) If the termination condition is satisfied, end the loop; otherwise, go to step 3.

Therefore, generating the next generation individuals under the influence of genetic operators, the specific evolutionary processes are: (1) according to the fitness of the individual, select two individuals in the population by the tournament selection method; (2) for the two parent individuals, using the single-point crossover and the single-point mutation operators [18], the two new individuals are generated. In addition, our evolutionary algorithm uses elitist strategy where the elite ratio is 20%. Table 1 shows the steps of the SNN image segmentation with genetic algorithm.

4   Experimental Results

In our experiments, the resting potential value of the leaky integrate-and-fire neuron model is -70 mV, the threshold potential value is -55 mV, the membrane time constant is 20 ms. Comparing the coding methods in [13], we found that the method of non-linear coding is easier to select optimal parameters, and acquires the better segmentation result of image. Therefore, we used nonlinear coding method and set $\sigma=0.05$, and the gray value of pixels corresponds to the maximum firing time of spikes $T_{max}=10$ ms. To test the effect on the adaptive selection of segmentation threshold, the population size is set to 20, and each evolution iterates 100 generation. The genetic algorithm executes 50 times, crossover rate is 0.5, and mutation rate is 0.05. Reference images use the Lena image with size of 256×256 and the camera man image with size of 256×256.

Firstly we evolve the segmentation threshold of the Lena image, and finally achieve the optimal segmentation threshold $T_{thresh}=7.713077$, and the corresponding Shannon entropy is 0.99997, in [13] where the segmentation threshold is artificially selected, the Shannon entropy is 0.99992. Moreover, we evolve the segmentation threshold of the camera man image, and finally achieve the optimal segmentation threshold $T_{thresh}=6.657836$, and the corresponding Shannon entropy is 0.99998, in [13] where the segmentation threshold is artificially selected, the Shannon entropy is 0.999894. From the experimental results we find that we can achieve the better accuracy in image segmentation by evolving the parameter of segmentation threshold. Compared with the artificial multiple test method for obtaining segmentation threshold, the new algorithm is a high efficient and automatic image segmentation method.

Fig. 4 shows the fitness change curve of the segmented image using genetic algorithm. It can be seen in Fig. 4(a) that the Shannon entropy of segmented Lena image gradually become larger, on average the genetic algorithm can converge and achieve the optimal fitness at around 20th generation, that is to get the maximum Shannon entropy of the segmented image. The average fitness value in general is also showing a rising trend. Fig. 4(b) shows the fitness change curve of the camera man image. The change trend is similar to Fig. 4(a). On average, the genetic algorithm can converge and achieve optimal segmentation threshold at around 17th generation.

Fig. 5 shows the results of Lena image segmentation in an evolution process of genetic algorithm. Fig. 5(a) is the original Lena image. Fig. 5(b) is the optimal segmentation threshold $T_{thresh}=6.475695$ at the 5th generation, the Shannon entropy of the segmented image is 0.9679625. Fig. 5(c) is the optimal segmentation threshold $T_{thresh}=7.148962$ at the 10th generation, the Shannon entropy of the segmented image is 0.982473. Fig. 5(d) is the optimal segmentation threshold $T_{thresh}=7.713077$ at the 50th generation, the Shannon entropy of the segmented image is 0.99997. No matter the Shannon entropy or directly observed from the segmented image, it can be seen clearly that the result of Fig. 5(d) is better than one of Fig. 5(b) and Fig. 5(c).

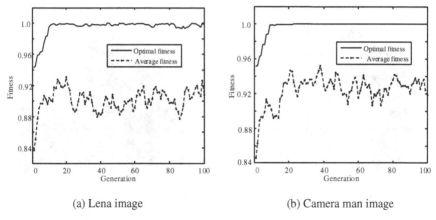

(a) Lena image   (b) Camera man image

Fig. 4. The fitness change curve of the image segmentation

(a) Original image   (b) 5th generation, T_{thresh}=6.475695

(c) 10th generation, T_{thresh}=7.148962   (d) 50th generation, T_{thresh}=7.713077

Fig. 5. Results of Lena image in an evolutionary process

(a) Original image

(b) 5th generation, T_{thresh}=5.013658

(c) 15th generation, T_{thresh}=6.436875

(d) 45th generation, T_{thresh}=6.6578

Fig. 6. Results of camera man image in an evolutionary process

Fig. 6 shows the results of camera man image segmentation in an evolution process of genetic algorithm. Fig. 6(a) is the original camera man image. Fig. 6(b) is the optimal segmentation threshold T_{thresh}=5.013658 at the 5th generation, the Shannon entropy of the segmented image is 0.976646. Fig. 6(c) is the optimal segmentation threshold T_{thresh}=6.436875 at the 15th generation, the Shannon entropy of the segmented image is 0.998953. Fig. 6(d) is the optimal segmentation threshold T_{thresh}=6.657836 at the 45th generation, the Shannon entropy of the segmented image is 0.99998.

5   Conclusions

In this paper, we propose an automatic image segmentation algorithm based on a SNN model, in which the key parameter of segmentation threshold is evolved using genetic algorithm. The experimental results show that the optimum segmentation threshold is obtained rapidly through the evolution process, the maximum Shannon entropy corresponding to the segmented image, namely the ideal segmentation effect is achieved. It should be noted that in this paper we only select segmentation threshold adaptively in the SNN image segmentation model, the main reasons are that: (1) segmentation threshold is the key parameter of the image segmentation model, select the optimal value by its evolution process, the segmented image with the maximum Shannon entropy can be achieved; (2) encoding the image pixel values into spike times, the corresponding spike timing matrix can be obtained by inputting the spikes into the SNN, it needs to be executed only once. Each evolved segmentation threshold in the genetic algorithm is applied only on the spike timing matrix, thus it is to improve the image segmentation efficiency.

Acknowledgement. The work is supported by the National Natural Science Foundation of China under Grants No. 61165002 and No. 61363059.

References

1. Ghosh-Dastidar, S., Adeli, H.: Spiking neural networks. International Journal of Neural Systems 19(4), 295-308 (2009)
2. Paugam-Moisy, H., Bohte, S.M.: Computing with spiking neuron networks. In: Rozenberg, G., Bäck, T., Kok, J.N. (eds.) Handbook of Natural Computing, pp. 335-376. Springer Berlin Heidelberg (2012)
3. Wu, Q.X., McGinnity, M., Maguire, L.P., Belatreche, A., Glackin, B.: Processing visual stimuli using hierarchical spiking neural networks. Neurocomputing 71(10-12), 2055-2068 (2008)
4. Borisyuk, R., Kazanovich, Y., Chik, D., Tikhanoff, V., Cangelosi, A.: A neural model of selective attention and object segmentation in the visual scene: An approach based on partial synchronization and star-like architecture of connections. Neural Networks 22(5-6), 707-719 (2009)
5. Finger, H., König, P.: Phase synchrony facilitates binding and segmentation of natural images in a coupled neural oscillator network. Frontiers in Computational Neuroscience 7, (2013)
6. Meftah, B., Lezoray, O., Benyettou, A.: Segmentation and edge detection based on spiking neural network model. Neural Processing Letters 32(2), 131-146 (2010)
7. Soni, C., Meftah, B., Khurshid, A.A.: Image segmentation using leaky integrate and fire model of spiking neural network. International Journal of Wisdom Based Computing 2(1), 21-28 (2012)
8. Buhmann, J.M., Lange, T., Ramacher, U.: Image segmentation by networks of spiking neurons. Neural Computation 17(5), 1010-1031 (2005)
9. Wu, Q.X., McGinnity, T.M., Maguire, L., Valderrama-Gonzalez, G.D., Dempster, P.: Colour image segmentation based on a spiking neural network model inspired by the visual system. In: Huang, D.S., Zhao, Z., Bevilacqua, V. (eds.) Advanced Intelligent Computing Theories and Applications, pp. 49-57. Springer Berlin Heidelberg (2010)
10. Thorpe, S., Fize, D., Marlot, C.: Speed of processing in the human visual system. Nature 381(6582), 520-522 (1996)
11. Wysoski, S.G., Benuskova, L., Kasabov, N.K.: Brain-like system for audiovisual person authentication based on time-to-first spike coding. Computational Modeling and Simulation of Intellect: Current State and Future Perspectives 384, (2011)
12. Bohte, S.M.: The evidence for neural information processing with precise spike-times: A survey. Natural Computing 3(2), 195-206 (2004)

13. Cui W.B., Lin X.H., Xu M.Y.: Coding method of spiking neural networks for image segmentation. Computer Engineer 38(24), 196-199 (2012)
14. Izhikevich, E.M., Piekniewski, F., Nageswaran, J., Petre, C., Richert, M., Sokol, S., Szatmary, B.: Spontaneous emergence of simple and complex receptive fields in a spiking model of V1. BMC Neuroscience 14(Suppl 1), O6 (2013).
15. Burkitt, A.N.: A review of the integrate-and-fire neuron model: I. Homogeneous synaptic input. Biological Cybernetics 95(1), 1-19 (2006)
16. Sivanandam, S.N., Deepa, S.N.: Introduction to genetic algorithms. Springer Berlin Heidelberg (2008)
17. Barbieri, A.L., De Arruda, G.F., Rodrigues, F.A., Bruno, O.M., Costa, L.D.F.: An entropy-based approach to automatic image segmentation of satellite images. Physica A: Statistical Mechanics and its Applications 390(3), 512-518 (2011)
18. Holland, J.H.: Outline for a logical theory of adaptive systems. Journal of the Association for Computing Machinery 9(3), 297-314 (1962)

A Fast Decoupling Method for Multi-input Multi-output Nonlinear Continuous-Time System

Feng Wang[1,2], Jin-Lin Ding[1], and Guo-Hai Liu[2]

[1] Suzhou Vocational University, Suzhou, Jiangsu Province, China 215104
{wf,djinlin}@jssvc.edu.cn

[2] School of Electrical and Information Engineering, Jiangsu University,
Zhenjiang, Jiangsu Province, China 212013
guohailiu@ujs.edu.cn

Abstract. Neural network inverse (NNI) method has been applied in decoupling control of MIMO nonlinear continuous-time system, but slow learning speed restricts its application. To solve the problem, a new online learning algorithm called modified online regularized extreme learning machine (MO-RELM) is proposed. MO-RELM can learn the training data one by one or chunk by chunk and discard the data for which the training has done. Finally, through connecting NNI based on MO-RELM with the original MIMO system, the constructed pseudo linear composite system (PLCS) can decouple the nonlinear coupled system into a number of independent pseudo linear SISO systems online. Simulation results show that the proposed fast decoupling method realizes online decoupling control of MIMO nonlinear continuous-time system. The method has good application prospect.

Keywords: RELM, online, MO-RELM, inverse system, decoupling control.

1 Introduction

In [1, 2] Dai proposed Neural network $\alpha - th$ order inverse system to decouple nonlinear MIMO continuous-time system with linearization. In which neural network (NN) is used to approximate inverse of the original system. Connecting trained neural network inverse (NNI) with the original nonlinear MIMO system, and the formed pseudo linear composite system (PLCS) can decouple the nonlinear coupled system into a number of independent pseudo linear SISO systems. In [3, 4] NNI is applied in decoupling control of two-motor variable frequency speed-regulating system. The constructed PLCS realizes the decoupling control between speed and tension. In [5] systematic NNI modeling techniques are presented for microwave modeling and filter design. In [6] NNI mapping technique is applied to electromagnetic design. The results illustrate that the method achieves an optimal solution.

Despite the many achievements have been acquired in NNI, the slow training speed of traditional NN restricts its application in control system. Although the ELM model has many advantages, it may lead to over-fitting problem [7, 8]. Compared

D.-S. Huang et al. (Eds.): ICIC 2014, LNCS 8588, pp. 259–268, 2014.
© Springer International Publishing Switzerland 2014

with the original ELM algorithm, regularized ELM (RELM) based on the structural risk minimization principle and weighted least square, has significantly improved generalization performance in most cases without increasing training burden.

In [9] Real-time transient stability assessment model using extreme learning machine is presented. It has significantly improved the learning speed and can enable effective on-line updating. Online sequential learning algorithm called OS-ELM is proposed in [10-12]. The algorithm uses batch learning for ELM and has shown faster training speed and better generalization than other batch training methods. Simulation results show that the OS-ELM is faster than the other sequential algorithms and produces better generalization performance. But OS-ELM needs more learning time as chunk size increases. In order to improve training speed of online learning, modified online regularized extreme learning machine (MO-RELM) is proposed. MO-RELM can not only learn the training data one by one or chunk by chunk, but also improve generalization through updating method than OS-ELM. Finally, we apply MO-RELM in two-input two-output (TITO) nonlinear control system. MO-RELM is used to identify inverse system of TITO online. Connecting identified NNI with the original TITO, PLCS is constructed. Simulation results show that constructed PLCS based on MO-RELM can decouple the nonlinear coupled system into two independent pseudo linear SISO systems online and acquired better performance.

2   Reviews of RELM

In order to improve the generalization ability of the traditional SFLNs based on ELM, Huang et al. proposed the equality constrained optimization-based ELM. The so-called RELM for the proposed constrained optimization can be formulated as

$$\arg\min_{\beta} E(W) = \arg\min_{\beta}\left[\frac{1}{2}\gamma\|\varepsilon\|^2 + \frac{1}{2}\|\beta\|^2\right].$$

$$Subject \quad to \quad H\beta - T = \varepsilon \tag{1}$$

Where $\|\varepsilon\|^2$ is empirical risk, $\|\beta\|^2$ represents structural risk, γ is the proportion parameter of two kinds of risks, and ε is regression error. To solve the above optimization problem Lagrange function is constructed as follows:

$$\xi(\beta,\varepsilon,\alpha) = \frac{1}{2}\gamma\|\varepsilon\|^2 + \frac{1}{2}\|\beta\|^2 - \alpha(H\beta - T - \varepsilon) \tag{2}$$

Where $\alpha = [\alpha_1, \alpha_2, \cdots, \alpha_N]$ is Lagrange multiplier, $\alpha_j \in R^m, (j = 1, 2, \cdots, N)$.

Letting the partial derivatives of $\xi(\beta,\varepsilon,\alpha)$ with respect to β, ε and α be zero, we have

$$\begin{cases} \beta^T = \alpha H \\ \gamma \varepsilon^T + \alpha = 0 \\ H\beta - T - \varepsilon = 0 \end{cases} . \tag{3}$$

In most cases the number of hidden nodes is much less than the number of distinct training samples, $L \prec\prec N$. From equation (3) we can get

$$\beta = \left(\frac{1}{\gamma} I + H^T H \right)^{-1} H^T T . \tag{4}$$

Where I is the identity matrix, and the corresponding regression model of RELM is

$$Y = h(X)\beta = h(X)\left(\frac{1}{\gamma} I + H^T H \right)^{-1} H^T T . \tag{5}$$

3   Proposed MO-RELM

3.1   Modified RELM for MIMO

On the basis of Huang's RELM, modified RELM algorithm for MIMO is proposed by us. In Huang's RELM, the weight vector β is solved by inverse operation of high-order matrix. There exists large amount of calculation, and it reduces modeling efficiency. In order to simplify computation, a new method based on cholesky factorization is introduced.

From Eqn.(3) we obtain

$$\left(\frac{1}{\gamma} I_L + H_L^T H_L \right) \beta_L = H_L^T T . \tag{6}$$

Let

$$A_L = \frac{1}{\gamma} I_L + H_L^T H_L, \qquad B_L = H_L^T T . \tag{7}$$

Then Eqn.(6) can be rewritten as

$$A_L \beta_L = B_L . \tag{8}$$

Since A_L is a symmetric positive definite matrix, it can be uniquely factored as

$$A_L = L_L L_L^T . \tag{9}$$

Where

$$l_{ij} = \begin{cases} \sqrt{a_{ii} - \sum\limits_{k=1}^{i-1} l_{ik}^2}, & i = j \\ a_{ij} - \sum\limits_{k=1}^{j-1} l_{ik} l_{jk} / l_{jj}, & i > j \end{cases} . \tag{10}$$

Substituting Eqn.(9) into Eqn.(8) leads to

$$L_L L_L^T \beta_L = B_L . \tag{11}$$

Denote

$$F_L = L_L^T \beta_L . \tag{12}$$

Substituting Eqn.(12) into Eqn.(11) gives

$$L_L F_L = B_L . \tag{13}$$

Using Eqn.(13), we have

$$f_{ij} = \begin{cases} b_{ij} / l_{ii} & i = 1 \\ \left(b_{ij} - \sum\limits_{n=1}^{i-1} l_{in} f_{nj} \right) / l_{ii} & i > 1 \end{cases} . \tag{14}$$

According to Eqns.(10),(12),(14),we can further obtain

$$\beta_{ij} = \begin{cases} f_{ij} / l_{ii} & i = L \\ \left(f_{ij} - \sum\limits_{n=1}^{m-i} l_{i+n,i} \beta_{i+n,j} \right) / l_{ii} & i < L \end{cases} . \tag{15}$$

3.2   Modified Online RELM

The original training dataset $\aleph_0 = \left\{ (X_i, t_i) \middle| X_i \in R^n, t_i \in R^m \right\}_{i=1}^N$, the corresponding hidden layer output matrix is H_0, the corresponding output is T_0 ,where

$$H_0 = \begin{bmatrix} g(w_1 \cdot X_1 + b_1) & \cdots & g(w_L \cdot X_1 + b_L) \\ \vdots & \ddots & \vdots \\ g(w_1 \cdot X_N + b_1) & \cdots & g(w_L \cdot X_N + b_L) \end{bmatrix}_{N \times L} . \tag{16}$$

And $T_0 = \begin{bmatrix} t_1^T & \cdots & t_L^T \end{bmatrix}_{N \times m}^T$

Suppose now that we are given another training dataset $\aleph_1$, which discards the first old ΔN data of $\aleph_0$ and increases new ΔN data in the rear. We denote

$$\Delta\aleph_1 = \left\{ (X_i, t_i) \middle| X_i \in R^n, t_i \in R^m \right\}_{i=1}^{\Delta N} \text{ and } \Delta\aleph_1^1 = \left\{ (X_i, t_i) \middle| X_i \in R^n, t_i \in R^m \right\}_{i=N+1}^{N+\Delta N}.$$

The corresponding hidden output matrices of $\Delta\aleph_1$ and $\Delta\aleph_1^1$ are ΔH_1, ΔH_1^1, The corresponding outputs are ΔT_1 and ΔT_1^1. Considering both chunks of training data sets $\aleph_0$ and $\aleph_1$, we have

$$\left(\frac{1}{\gamma} I + H_0^T H_0 \right) \beta_0 = H_0^T T_0 \cdot \tag{17}$$

$$\left(\frac{1}{\gamma} I + H_1^T H_1 \right) \beta_1 = H_1^T T_1 \cdot \tag{18}$$

We denote

$$H_0 = \begin{bmatrix} \Delta H_1 \\ H_1^- \end{bmatrix}, H_1 = \begin{bmatrix} H_1^- \\ \Delta H_1^1 \end{bmatrix}, \ T_0 = \begin{bmatrix} \Delta T_1 \\ T_1^- \end{bmatrix}, T_1 = \begin{bmatrix} T_1^- \\ \Delta T_1^1 \end{bmatrix}. \tag{19}$$

Then

$$H_0^T H_0 = \begin{bmatrix} \Delta H_1 \\ H_1^- \end{bmatrix}^T \begin{bmatrix} \Delta H_1 \\ H_1^- \end{bmatrix} = \Delta H_1^T \Delta H_1 + (H_1^-)^T H_1^- \cdot \tag{20}$$

$$H_1^T H_1 = \begin{bmatrix} H_1^- \\ \Delta H_1^1 \end{bmatrix}^T \begin{bmatrix} H_1^- \\ \Delta H_1^1 \end{bmatrix} = (H_1^-)^T H_1^- + (\Delta H_1^1)^T \Delta H_1^1 \cdot \tag{21}$$

$$H_0^T T_0 = \begin{bmatrix} \Delta H_1 \\ H_1^- \end{bmatrix}^T \begin{bmatrix} \Delta T_1 \\ T_1^- \end{bmatrix} = \Delta H_1^T \Delta T_1 + (H_1^-)^T T_1^- \cdot \tag{22}$$

$$H_1^T T_1 = \begin{bmatrix} H_1^- \\ \Delta H_1^1 \end{bmatrix}^T \begin{bmatrix} T_1^- \\ \Delta T_1^1 \end{bmatrix} = (H_1^-)^T T_1^- + (\Delta H_1^1)^T \Delta T_1^1 \cdot \tag{23}$$

Substituting Eqn.(20) into Eqn.(21) leads to

$$H_1^T H_1 = H_0^T H_0 - \Delta H_1^T \Delta H_1 + (\Delta H_1^1)^T \Delta H_1^1 \cdot \tag{24}$$

Substituting Eqn.(22) into Eqn.(23) leads to

$$H_1^T T_1 = H_0^T T_0 - \Delta H_1^T \Delta T_1 + (\Delta H_1^1)^T \Delta T_1^1 . \tag{25}$$

Let

$$K_0 = \frac{1}{\gamma} I + H_0^T H_0 . \tag{26}$$

$$K_1 = \frac{1}{\gamma} I + H_1^T H_1 . \tag{27}$$

$$P_0 = H_0^T T_0 . \tag{28}$$

$$P_1 = H_1^T T_1 . \tag{29}$$

Substituting Eqns.(24)(26) into Eqn.(27) leads to

$$\begin{aligned} K_1 &= \frac{1}{\gamma} I + H_0^T H_0 - \Delta H_1^T \Delta H_1 + (\Delta H_1^1)^T \Delta H_1^1 \\ &= K_0 - \Delta H_1^T \Delta H_1 + (\Delta H_1^1)^T \Delta H_1^1 \end{aligned} . \tag{30}$$

Substituting Eqns.(25)(28) into Eqn.(29) leads to

$$\begin{aligned} P_1 &= H_0^T T_0 - \Delta H_1^T \Delta T_1 + (\Delta H_1^1)^T \Delta T_1^1 \\ &= P_0 - \Delta H_1^T \Delta T_1 + (\Delta H_1^1)^T \Delta T_1^1 \end{aligned} . \tag{31}$$

Generalizing the previous arguments, as new data arrives, a recursive algorithm for updating the least-squares solution can be written as follows. When $(k+1)th$ chunk of data set is received, we have

$$K_{k+1} \beta_{k+1} = P_{k+1} . \tag{32}$$

Where

$$K_{k+1} = \frac{1}{\gamma} I + H_{k+1}^T H_{k+1} = K_k - \Delta H_{k+1}^T \Delta H_{k+1} + (\Delta H_{k+1}^1)^T \Delta H_{k+1}^1 . \tag{33}$$

$$P_{k+1} = H_{k+1}^T T_{1k+1} = P_k - \Delta H_{k+1}^T \Delta T_{k+1} + (\Delta H_{k+1}^1)^T \Delta T_{k+1}^1 . \tag{34}$$

Since K_{k+1} is a symmetric positive definite matrix, we can calculate β_{k+1} according to proposed modified RELM algorithm in section 3.

Obviously MO-RELM learning algorithm can reduce training time and realize on-line learning.

4   Simulation and Analysis

4.1   Decoupling Case

The model of MIMO nonlinear continuous-time system is

$$\begin{cases} \ddot{y}_1 = -6.4y_1 - 1.6\dot{y}_1 + 4.0u_1 / (0.5 + \dot{y}_1^2 + \dot{y}_2^2)^{1/2} \\ \ddot{y}_2 = -5.0y_2 - \dot{y}_2 / e^{y_1} + 2.0u_2(2 - u_1)^{1/3} \end{cases} \tag{35}$$

Initial value : $\begin{cases} y_1(t_0) = \dot{y}_1(t_0) = 0 \\ y_2(t_0) = \dot{y}_2(t_0) = 0 \end{cases}$

The work region is $u_1, u_2 \in (-2, 2)$, when input signals are square waves the response are shown as Fig.1. Fig.1(a) is input signal, Fig.1(b) is response signal. From Fig.1 we can see that the system is strong coupled and nonlinear.

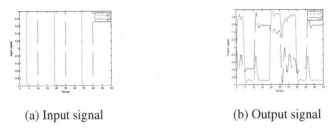

(a) Input signal   (b) Output signal

Fig. 1. Responses when input signals are square waves

In order to realize the decoupling control, the neural network inverse (NNI) is introduced. The relative order of system is $\alpha = (2,2)$, so the inverse model is

$$u = \varphi(y_1, \dot{y}_1, \ddot{y}_1, y_2, \dot{y}_2, \ddot{y}_2) . \tag{36}$$

The NN $\alpha - th$ order inverse system based on MO-RELM is constructed to approximate nonlinear mapping $u = \varphi(y_1, \dot{y}_1, \ddot{y}_1, y_2, \dot{y}_2, \ddot{y}_2)$. The designed NNI based on MO-RELM is connected with original system, then PLCS can be formed as Figure.2. y_1 and y_2 are decoupled into independent linear subsystem, The method can realize the decouple control.

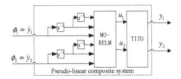

Fig. 2. Diagram of PLCS

4.2   Simulation and Analysis

Network structure is 6-50-2, we training NN to approximate inverse system. Taking $y = \{ y_1, \dot{y}_1, \ddot{y}_1, y_2, \dot{y}_2, \ddot{y}_2 \}$ as the input of SFLNs, and u as the output of SFLNs. The input weights ω_i and the hidden layer biases b_i are arbitrarily chosen. The activation function of hidden-layer is selected as sigmoid function $g(x) = 1/(1 + e^{-x})$. The parameter γ is 4096.

In order to verify online learning performance, OS-ELM and MO-RELM algorithms are adopted. Two algorithms have the same input weights ω_i and hidden layer biases b_i. The training data set N is 20000, Number of initial training data N_0 is 15000 and updated data set is ΔN.

Table.1 shows comparison results of OS-ELM and MO-RELM when online learning is adopted. In training time, OS-RELM is sensible to chunk size ΔN. With increase of chunk size MO-RELM has shorter training time than OS-ELM. MO-RELM algorithm is clearly fitter for online learning than OS-ELM. Although OS-ELM and MO-RELM algorithms have similar training accuracy, testing accuracy of MO-RELM is better than OS-ELM. This shows that MO-RELM has better generalization than OS-ELM. In conclusion, proposed MO-RELM has better generalization and online learning than OS-ELM and can be applied in online learning.

Table 1. Performance comparison of OS-ELM and MO-RELM

Algorithm	Chunk size	Training Time (s)	Training RMSE	Testing RMSE	Nodes
OS-ELM	200	1.1875	0.025	0.0197	50
	100	0.6094	0.025	0.0197	50
MO-RELM	200	0.4969	0.025	0.0194	50
	100	0.7844	0.025	0.0194	50

In order to verify decoupling performance, the identified NNI based on MO-RELM is connected with the original system, PLCS is formed as Fig.2. Where ϕ_1 、 ϕ_2 are input signals, and y_1 、 y_2 are output signals. First, step signal of the amplitude 0.02 is added in ϕ_1 and ϕ_2 is kept zero, the response of system in 13 second is shown as Fig.3. Second, step signal of the amplitude 0.02 is added in ϕ_2 and ϕ_1 is kept zero, the response of system in 13 second is shown as Fig.4.

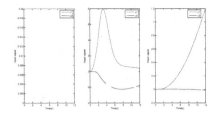

Fig. 3. The response under $\varphi_1 = 0.02$, $\varphi_2 = 0$

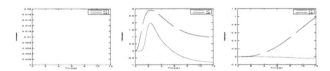

Fig. 4. The response under $\varphi_1 = 0$, $\varphi_2 = 0.02$

We can see that the original system has basically realized linear decouple. The nonlinear composite system has been decoupled into two separate linear subsystems, and the transfer function is $G(s) \approx diag\left(s^{-2}, s^{-2}\right)$.

5   Conclusions

Considering that MIMO nonlinear continuous-time online control, a fast decoupling method has been developed. Firstly, a new online learning algorithm called MO-RELM is proposed to identify inverse model of original system. Compared with OS-ELM, MO-RELM trains data faster and has better generalization through updating learning with increase of chunk size. Then, the identified NN $\alpha - th$ order inverse system based on MO-RELM is connected with the original system, PLCS is formed. Finally, we apply the proposed method in TITO. Simulation results show that the proposed fast decoupling method realizes online decoupling control of TITO nonlinear continuous-time system. The method has good application prospect.

Acknowledgments. This work was supported by the University Graduate Research Innovation Foundation of Jiangsu Province, China (Grant No. CXZZ13_0685), the Natural Science Foundation of Jiangsu Province, China (Grant No. BK2011319), the innovation Foundation of Suzhou vocational university (Grant No.2011SZDCC01).

References

1. Dai, X.Z., He, D., Zhang, X., Zhang, T.: MIMO system invertibility and decoupling control strategies based on ANN ®th-order inversion. IEE Proceedings Control Theory and Applications 148(2), 125–136 (2001)

2. Dai, X.Z., He, D.: Novel method for decoupling nonlinear MIMO system with linearization. Continuous-time System. Control and Decision 14(5), 403–406 (1999)
3. Liu, G., Yang, G., Shen, Y., Chen, Z.: Internal model control of two motor variable frequency systems based on neural network generalized inverse. Transactions of China Electrotechnical Society 25(11), 55–61 (2011)
4. Liu, G., Zhang, J., Zhao, W., Zhang, Y., Jiang, Y.: Internal model control based on support vector machines generalized inverse for two-motor variable frequency system applications. Proceedings of the CSEE 31(6), 85–91 (2011)
5. Kabir, H., Ying, W., Ming, Y.: Neural network inverse modeling and applications to microwave filter design. IEEE Trans. Microwave Theory and Techniques 56(4), 867–879 (2008)
6. Belfore, L.A., Arkadan, A.A., Lenhardt, B.M.: ANN inverse mapping technique applied to electromagnetic design. IEEE Transactions on Magnetics 37(5), 3584–3587 (2001)
7. Han, M., Li, D.: An Norm 1 regularization term elm algorithm based on surrogate function and bayesian framework. Acta Automatica Sinica 37(11), 1344–1350 (2011)
8. Deng, W., Zheng, Q., Chen, L., Xu, X.: Research on extreme learning of neural networks. Chinese Journal of Computers 33(2), 279–287 (2011)
9. Xu, Y., Dong, Z.Y., Meng, K., Zhang, R., Wong, K.P.: Real-time transient stability assessment model using extreme learning machine. IET Gener. Transm. Distrib. 5(3), 314–332 (2011)

A Hand Shape Instruction Recognition and Learning System Using Growing SOM with Asymmetric Neighborhood Function

Takashi Kuremoto[1,*], Takuhiro Otani[1], Masanao Obayashi[1],
Kunikazu Kobayashi[2], and Shingo Mabu[1]

[1] Graduate School of Science and Engineering, Yamaguchi University
Tokiwadai 2-16-1, Ube, Yamaguchi, 755-8611, Japan
{wu,m.obayas,mabu}@yamaguchi-u.ac.jp
[2] School of Information Science and Technology, Aichi Prefectural University
Ibaragabasama 1522-3, Nagakute-Shi, Aichi, 480-1198, Japan
kobayashi@ist.aichi-pu.ac.jp

Abstract. In this paper, we adopt an asymmetric neighborhood function proposed by Aoki and Aoyagi in to a PL-G-SOM to improve the learning performance of the hand shape instruction perspective and learning system. The asymmetric neighborhood function was used in a normal SOM and few applications can be found. The novel PL-G-SOM and its improved version are named as "AGSOM" and "IAGSOM" respectively. The effectiveness of the proposed method was confirmed by the experiments with 8 kinds of instructions.

Keywords: Self-Organizing Map, Neighborhood Function, Human-Machine-Interaction, Instruction Learning System.

1 Introduction

As a Human-Machine-Interaction (HMI) problem, we consider to use the different shapes of a hand as action instructions of partner robots equipped with visual sensors such as pet robots, service robots, entertainment robots, etc. The recognition and understanding of these instructions for robots by image processing have been approached by many mathematical methods such as Self-Organizing Map (SOM), Hidden Markov Model (HMM), Support Vector Machine (SVM) and so on. Specially, Kohonen's SOM is a powerful tool to categorize high-dimensional data as a well-known unsupervised machine learning method. During past decades, there have been thousands publications concerning with the theoretical and practical studies on SOM [7].

To overcome some problems existed in the original SOM such as (a) Exhaustion of units; (b) Disorder of topology of map;(c) Adjustment of parameters, kinds of improved SOM variations are proposed. For example, to deal with the exhaustion of

---

* A part of this work was supported by Grant-in-Aid for Scientific Research (JSPS No. 23500181, No. 25330287).

D.-S. Huang et al. (Eds.): ICIC 2014, LNCS 8588, pp. 269–276, 2014.

units, Growing SOM (GSOM) [4] and Transient-SOM (T-SOM) [8] [9] are proposed. To avoiding defects of map's topology in learning process, asymmetric neighborhood function is used recently [1]-[3]. For the stable learning convergence, Parameterless SOM (PL-SOM) [6] and Parameterless Growing SOM (PL-G-SOM) [9] [10] are given with less parameters such as learning rate and width coefficient in Gaussian neighborhood function.

In this paper, we adopt Aoki & Aoyagi's asymmetric neighborhood function [1]-[3] to PL-G-SOM as novel variations of SOM named "AGSOM" and "IAGSOM" to improve the learning performance at first. And then we apply the proposed SOM to a hand shape instruction recognition and learning system for partner robots. In the remain of this paper, we describe the AGSOM and IAGSOM in Section II, and adopt them in to a hand shape instruction recognition and learning system in Section III. Simulation experiment and real hand image instruction learning experiment were reported in Section IV. Conclusions are in Section V at last.

2   AGSOM

Kohonen's SOM algorithm maps n-dimensional data x(x1, x2 ..., xn) to m units in a low-dimensional space with connections mi (m1, m2 ..., mn) by a winner-takes-all rule: c = arg mini($\| x - mi\|$), where i=1, 2, ..., m means the number of unit on a low-dimension map (1 or 2-dimension grids). c indicates the best-match-unit (BMU) on the map which has the shortest Euclidean distance with the input data x. Initially, connection weight mi is given by a random value, and following learning rule makes input of different class data separately to the different position on the feature map:

Δmi = αhci (x - mi) ,where $\alpha$is a learning rate and hci is a neighborhood function

usually as $h_{ci} = \exp(-\dfrac{\| \mathbf{r}_i - \mathbf{r}_c \|^2}{2\sigma^2})$, where, $\mathbf{r}_i, \mathbf{r}_c$ denotes the positions of an

arbitrary unit i on the output map and BMU, respectively, i=1, 2,...,k $\leq$ m, $\sigma$is a constant. Obviously, $h_{ci}(x) \geq 0$, $h_{ci}(0) = 1$, $h_{ci}(\infty) = 0$. For an arbitrary unit i on the map, the distance between it and the BMU is a scalar, so hci is not a vector.

In the original SOM, the size of the map (for example, $m = N \times M$, N and M are the width and height of a 2-dimensional SOM respectively) is fixed in advance, so it prevents additional learning when the number of categories of data is more than the number of units on the map (i.e. the size of map) m. To overcome this kind of exhaustion of units problem, Growing SOM (GSOM) [4] and Transient SOM (T-SOM) [8] [9] were proposed. In GSOM, few units (one or two) at the beginning of learning process of SOM are set, and then to increase the number of units when a new input cannot find a BMU for its too far distance to all units on the map. Meanwhile T-SOM uses a memory layer to store "matured" units and release the matured units on the map with initial random values. However, according to this forced processing, the topology of the map become to be disordered and this affects the learning performance of T-SOM. So in this study, we concentrate to the former one, GSOM as a solution of unit exhaustion problem.

When a new row/column needs to be insert to the neighbor of a BMU c, the farthest unit f from c is selected among neighbor units of c in the sense of a Euclidean distance $\|\mathbf{m}_c - \mathbf{m}_j\|$, $j = 1,2,\ldots, \leq 8$. Then a new unit r and its neighborhood units (in row or column) are inserted between c and f. The weight of connection between the input and the new unite r is given by average values of connection weights of c and f, mr= 0.5 (mc + mf), and the connection weights of r's neighbors in row or column, mr±l=0.5 (mc±l + mf±l) ,where l=1, 2, ..., N or M. After this process, the map size changes to N×(M+1), or (N+1)×M.

To decide parameters α and σ, Berglund & Sitte proposed a data-driven method in their PL-SOM [6].

Let $\alpha(t) = \dfrac{\|\mathbf{x}(t) - \mathbf{m}_c(t)\|^2}{r(t)})$, $r(t) = \max(\|\mathbf{x}(t) - \mathbf{m}_c(t)\|^2, r(t-1))$,

$\sigma(t) = \sigma_{max}\,\alpha(t) \geq \sigma_{min}$, where $\sigma_{max}, \sigma_{min}$ are positive parameters, for example, the value may be the size of the map and 1.0, respectively. t is the number of training time.

Because the initialization of the original SOM uses random values for connection weights of units, topological defect which shows the disorder (twist) of the map and affects the categorization results. To tackle this problem, Aoki & Aoyagi proposed to use an asymmetric neighborhood function instead of conventional symmetric Gaussian function [1]-[3]:

$$h_\beta(\tilde{r}_{ic}) = 2\left(\frac{1}{\beta_0} + \beta_0\right)^{-1} \exp\left(-\frac{\tilde{r}_{ic}}{2\sigma^2}\right), \tag{1}$$

$$\tilde{r}_{ic} = \begin{cases} \sqrt{\left(\dfrac{r_{ic}\cdot\mathbf{k}}{\beta_0}\right)^2 + \|\mathbf{r}_\perp\|^2}, & \text{if } \mathbf{r}_{ic}\cdot\mathbf{k} \geq 0, \\ \sqrt{(\beta_0\cdot r_{ic}\cdot\mathbf{k})^2 + \|\mathbf{r}_\perp\|^2}, & \text{if } \mathbf{r}_{ic}\cdot\mathbf{k} < 0. \end{cases} \tag{2}$$

$$\mathbf{r}_{ic} = \mathbf{r}_i - \mathbf{r}_c . \tag{3}$$

Where i is an arbitrary unit on the map, c is the BMU, $\mathbf{r}_{ic}$ notes a distance (vector) between c and i, k indicates the asymmetry direction, i.e., horizontal axis or vertical axis in a 2-dimensional SOM, $\mathbf{r}_\perp$ is the component perpendicular to k, and the asymmetric parameter $\beta_0 \geq 1$ gives the degree of asymmetric. The processing means that if a unit locates on the positive direction of BMU c, let the direction is k, then the component parallel to k of the distance between the unit and c is scaled by $1/\beta_0$. Oppositely, if a unit is on the negative location to c, the distance is scaled by β_0.

Furthermore, an improved practical algorithm using asymmetric neighborhood function was also proposed by Aoki et al. [2]. The improved algorithm is during the training process, the direction of asymmetry is inversed in a certain period T. And the disorder of the map's topology is reduced by an asymptotic adjustment of asymmetric parameter β_0:

$$\beta_0(t) \leftarrow \begin{cases} 1 + (\beta_0 - 1)(1 - \dfrac{t}{T_{total}}), & if \quad t < T_{total}, \\ 1, & if \quad t \geq T_{total}, \end{cases} \qquad (4)$$

So the more reasonable topological results can be obtained. Here, we propose to adopt the asymmetric neighborhood function described by Eq. (1)-Eq. (3) into algorithm of PL-G-SOM to give a novel variation of SOM named "AGSOM". When Eq. (4) is also used, a novel variation of SOM-- "IAGSOM (improved AGSOM)" is given.

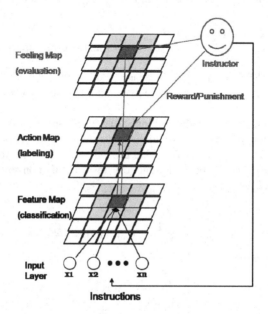

Fig. 1. A hand shape instruction recognition and learning system using AGSOM or IAGSOM

3   A Hand Shape Instruction Recognition and Learning System

To realize human-machine interaction (HMI), the behavior of a user such as pose of body, gesture of hands, voice, face expression, etc need to be recognized and understood by the machine. In this study, a hand shape instruction recognition and learning system is constructed for partner robots, such as pet robot, housework robot, entertainment robot, as an application to HMI of proposed AGSOM and IAGSOM.

Fig. 1 shows the architecture of a hand shape instruction recognition and learning system using AGSOM and IAGSOM proposed in this study.

There are 4 layers in the system: (1) Input layer which accepts input data in n dimension space; (2) Feature map layer, which is an AGSOM or an IAGSOM describe in Section 2, to classify the input data; (3) Action map which labels the input patterns to candidate actions; (4) Feeling map which expresses the success rate of instruction learning.

The instructor presents his/her instructions with the different shapes of his/her hand to a robot, and in the view of the robot, the observed hand shapes mean different states of the environment st, the robot selects a valuable action at (i) adapting to a certain state by a stochastic action policy π. Here, i =1, 2, ..., A is the number of candidate actions, t = 1, 2, ..., Time means the discrete time for the state transitions. Action selection function may be according to Gibbs distribution (Boltzmann distribution) as shown by Eq. (5).

$$\pi_t(a_t(i) \mid s_t) = e^{\frac{Q_t(s_t, a_t(i))}{T}} / \sum_{j=1}^{A} e^{\frac{Q_t(s_t, a_t(j))}{T}}, \qquad (5)$$

where T is a constant named parameter. When an action is selected according to the distribution, and performed by the robot, its instructor evaluates the action by giving a reward/punishment r to robot. The reward is accepted and used to modify the value of Qt by $Q_{t+1}(s_t, a_t(i)) = Q_t(s_t, a_t(i)) + r$, where Qt is called "state-action value function" in reinforcement learning (RL) [12]., r is a reward/punishment (scalar) given by the instructor.

The units on Action map are corresponding to the units on Feature map (i.e., an AGSOM or IAGSOM). According to trial and error learning process, action which yields high reward is selected more easily. The initial value of $Q_{t+1}(s_t, a_t(i))$ is given randomly.

To express the degree of how an instruction is learned by robot, a Feeling map which has the same number of units with Action map is designed as the output layer of the learning system. Feeling map expresses instruction recognition rate, i.e., the feeling of robot: more successful, happier it feels. Feelings of partner robots can be expressed vividly by their face expressions or the gestures. A feeling function is designed by the distance between input pattern and units on Feature map and the reward from instructor. $F_{t+1}(i) = F_t(i) \pm aC - bD_i$, where F(i) notes the feeling value of unit i on the Feeling Map (zero initially), C notes the continue times of reward (+) or punishment (-) accepted from the instructor, Di is the Euclidean distance (squared error) between the unit i on Feature Map and the input data, a, b are constants, and $0 < a < 1$, $0 < b \ll 1$. $F_{t+1}(i)$ is normalized to $-1.0 \le F_{t+1}(i) \le +1.0$, where the highest value +1.0 means 100% success rate of all instructions given by different shapes of a hand of instructor.

4   Experiment and Results

4.1   Skin Area Segmentation

To extract the hand shape from images captured by camera of robot, we used a image processing method proposed in our previous work [8]-[10]. The method use the threshold values of Hue (H), and Saturation (S) [11] of HSV images and Red (R) threshold of RGB images to obtain a binary image of the biggest skin area in images

(HSV and RGB can be transformed to each other by simple mathematical calculations). Noise elimination and holes filling are also effective to segment a hand area from the binary image. To normalize hand area, centering, rotation, expansion are also necessary, and the description of these detail processing is omitted here.

4.2   Feature Data of Hand Shapes

To distinguish the type of a hand shape, feature space definition is important. We discussed the methods of feature space construction of hand shape images in our previous works and proposed a useful feature vector space of hand shapes [8]-[10]. The input images are analyzed by an 80-dimension vector space (See Fig. 2). The longest direction of the hand area is estimated at first. Then the lowest point of the longest line is set as an origin. Start detection from right side 18 degree, the lengths in axes each 1.8-degree increased (80 axes) are the values of the feature vector, i.e., $(x_1, x_2, ..., x_{80})$.

8 kinds of hand shapes, which are supposed to corresponding 8 kinds of instructions such as dancing, sit down, stand up, walk right, etc., were used in experiment (Fig. 3). For each kind of instruction, we recorded 3 samples, so the total input data had 24 samples. In Fig. 3, examples of these hand shapes are shown where original images are shown in the left column, binary images of interesting region are in the center column, and feature vectors are depicted in the right column.

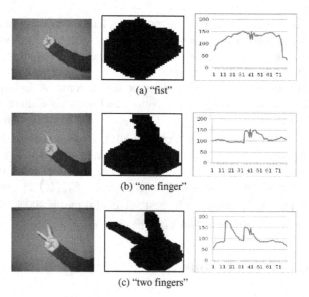

(a) "fist"

(b) "one finger"

(c) "two fingers"

Fig. 2. Examples of hand shapes for instruction learning of robot. 8 kinds of shapes were used in the experiments.

4.3   Results

Table I shows a comparison of learning performance between different SOMs. Recognition rates of 8 instructions were 100% using every method. In the viewpoint of the learning convergence ("MSE"), IAGSOM showed the best performance. And in the sense of success rate, AGSOM had the highest Feeling value. However, the conventional PL-G-SOM marked the smallest size of map and the least calculation time. To overcome this problem of AGSOM and IAGSOM, unit deleting processing can be adopted in to the GSOM algorithm, and we leave it as the future work.

Table 1. Learning performance comparison of different SOMs

Item	PL-G-SOM [9]	AGSOM (proposed)	IAGSOM (proposed)
MSE	0.83	0.39	**0.18**
Feeling value (average)	0.67	**0.74**	0.64
Number of units	**40**	54	81
Recognition rate (%)	100.0	100.0	100.0
Running time (sec.)	**130.91**	163.19	259.3

5   Conclusion

Asymmetric neighborhood function is adopted into PL-G-SOM. The new variation named AGSOM and IAGSOM showed their effectiveness in numeral simulation and real image recognition and learning system. Comparing with the conventional SOM and PL-G-SOM, the proposed SOMs provided better learning convergence and stability. Toward real robot application, the computational cost of the proposed method needs to be reduced in the future.

References

1. Aoki,T., Aoyagi, T.: Self-organizing maps with asymmetric neighborhood function, Neural Computation, Vol. 19, 2515--2535 (2007)
2. Aoki,T., Ota, K., Kurata, K., Aoyagi, T.: Ordering process of self-organizing maps improved by asymmetric neighborhood function, Cognitive Neurodyamics, Vol. 3, 9--15 (2009)
3. Ota, K., Aoki, T., Kurata, K., T. Aoyagi: Asymmetric neighborhood functions accelerate ordering process of self-organizing maps, Physical Review E, Vol. 83, 021903 (2011)
4. Bauer, H. –U., Villmann, Th.: Growing a hypercubical output space in a self-organizing feature map, IEEE Transaction on Neural Networks, Vol. 8, No.2, pp. 218-226 (1997)
5. Baum, L.E., Petrie, T.: Statistical inference for finite state Markov chains, The Annals of Mathematical Statistics, Vol. 37, No.6, 1554--1563 (1966)
6. Berglund, E., Sitte, J.: The parameter-less self-organizing map algorithm, IEEE Transaction on Neural Networks, Vol. 17, No.2, 305--316 (2006)
7. Kohonen, T.:*Self-Organizing Maps*, Berlin; Heidelberg; New-York: Springer, Series in Information Sciences (1995)

8. Kuremoto, T., Hano, T., Kobayashi, K., Obayashi, M.: For partner robots: A hand instruction learning system using transient-SOM. In Proceedings of the 2nd International Conference on Natural Computation and the 3rd International Conference on Fuzzy Systems and Knowledge Discovery (ICNC '06-FSKD'06), 403--414 (2006)
9. Kuremoto, T., Obayashi, M., Kobayashi, K., and Feng, L.-B., "Instruction learning systems for partner robots", *Advances in Robotics-Modeling, Control, and Applications*, iConcept, ch.8 (2012)
10. Kuremoto, T., Otani, T., Feng, L.-B., Kobayashi, K., & Obayashi, M., "A hand image instruction learning system using PL-G-SOM", In *Proceedings of the 12th International Conference on Artificial Intelligence (ICAI 2012)*, CD-ROM (2012)
11. Sherrah, S., Gong, S.,"*Skin Colour Analysis*", online URL: http://homepages.inf.ed.ac.uk/rbf/CVonline/LOCAL_COPIES/GONG1/cvOnline-skinColourAnalysis.html (2001)
12. Sutton, S.S., Barto, A.G., *Reinforcement Learning: An Instruction*, The MIT Press, London (1998)

An Adaptive Tracking Controller for Differential Wheeled Mobile Robots with Unknown Wheel Slips

Ngoc-Bach Hoang[1] and Hee-Jun Kang[2,*]

[1] Graduate School of Electrical Engineering, University of Ulsan
680--749, Ulsan, South Korea
hoangngocbach@gmail.com
[2] School of Electrical Engineering, University of Ulsan,
680-749, Ulsan, South Korea
hjkang@ulsan.ac.kr

Abstract. This paper investigates the tracking control of mobile robot in the presence of wheel slip and external disturbance forces. An adaptive tracking controller is proposed based on a three-layer neural networks. The uncertainties due to the wheel slip and external force are compensated online by neural networks. The stability of the closed-loop system is ensured by using Lyapunov method. The validity of the proposed controller is confirmed by a simulation example of tracking a circle trajectory.

Keywords: Neural networks, Adaptive control, Mobile robot, Wheel slip.

1 Introduction

In most real applications, slip between the ground and wheel often cannot be avoided due to various reasons such as uneven terrain, robot high speed, ground conditions, slippery surface. In [1], a robust controller taking care both slip-kinematic and slip-dynamic models is proposed using the framework of differential flatness. The robust controller requires the measurements of robot acceleration, which are very difficult to obtain in reality. In [2], the wheel slip and external loads are assumed to act as disturbances to the system. Then, the sliding model control method is employed to design a tracking controller of the mobile robot. In [3], longitudinal traction force is included in an omni-directional mobile robot model by externally measuring the magnitude of slip. Another approach to handle the slip relies on estimating the slip in real time using gyros and accelerometers to compensate for it [4,5].

In this paper, we propose a method of using an online tuning neural network to deal with the mobile robot uncertainties due to the wheel slip and external disturbance forces. First, the detail of dynamic model of mobile robot subject to wheel slip and external disturbance load is developed [2]. The friction forces are simply divided into two forces: lateral and longitudinal forces, which are generated due to the slip angle

---

* Corresponding author.

D.-S. Huang et al. (Eds.): ICIC 2014, LNCS 8588, pp. 277–284, 2014.

and the tire slip, respectively. The equation motion of the mobile robot can be obtained by the summation of the external forces and moments in the body centered reference frame based on Newton's Law. In order to reduce the harmful effect of the external loads and wheel slip on the control performance, an online learning neural networks is implemented. The neural network weights are updated based on the backpropagation plus an e-modification term to guarantee its robustness. A simulation example of tracking a circle trajectory is performed to confirm the effectiveness of the proposed algorithm.

2   Kinematic and Dynamic Model of Mobile Robot Subject to Wheel Slip

We derive the kinematic and dynamic model of a differentially driven wheeled mobile robot shown in Fig.1. O-xy is the global coordinate system. $G\text{-}X_cY_c$ is the coordinate system fixed to the mobile platform. G is the center of mass of the mobile robot. u and v are the longitudinal and lateral velocities at the center gravity of the mobile robot, respectively and r denotes the yaw rate. ω_r and ω_l are the wheel angular velocities of right and left wheel, respectively.

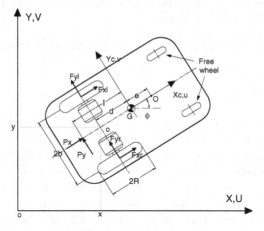

Fig. 1. Mobile Robot

The longitudinal and lateral velocities u , v and the yaw rate r are given as follows

$$u = \frac{1}{2}(u_l + u_r)$$

(1)

$$v = \frac{d}{2b}(u_l - u_r) + v^s$$

(2)

$$r = \frac{1}{2b}(u_l - u_r)$$

(3)

where $u_r = R\omega_r - u_r^s$ and $u_l = R\omega_l - u_l^s$ are the longitudinal speeds of the right and left wheel centers, respectively. v^s is the lateral slip speed of the wheel axle. u_r^s and u_l^s are the longitudinal slip speeds of the right and left wheel centers, respectively. The trajectory of the tracking point O are described by

$$\dot{X}(t) = u.\cos\phi - (v + er).\sin\phi \tag{4}$$

$$\dot{Y}(t) = u.\sin\phi + (v + er).\cos\phi \tag{5}$$

$$\dot{\phi} = r \tag{6}$$

The equation motion of the mobile robot can be obtained based on Newton's Law as

$$F_{xr} + F_{xl} + P_x = m(\dot{u} - vr) \tag{7}$$

$$F_{yr} + F_{yl} + P_y = m(\dot{v} + ur) \tag{8}$$

$$(F_{xl} - F_{xr})b - (F_{yl} + F_{yr})d - P_y l = I_z \dot{r} \tag{9}$$

$$n\tau_r - F_{xr} = I_e \dot{\omega}_r \tag{10}$$

$$n\tau_l - F_{xl} = I_e \dot{\omega}_l \tag{11}$$

where F_{xr}, F_{yr}, F_{xl}, F_{yl} are the tire forces on the driving right and left wheels, respectively. P_x and P_y are the external forces acting on the robot. m is the mass of the robot. I_z is the moment of inertia of the robot platform about the z-axis. n is the gear ratio. $I_e = I_t + n^2 I_m$ is the effective moment of inertia combining the wheel inertia I_t and the driving motor inertia I_m. τ_r and τ_r are the torque input by the servo motor.

By eliminating the tire forces and reformatting the full dynamic equation motion Eqs.(7-11), we obtain the resulting dynamic equations as follows

$$\begin{bmatrix} \dot{X} \\ \dot{Y} \\ \dot{\phi} \\ \dot{u} \\ \dot{r} \end{bmatrix} = \begin{bmatrix} u.\cos(\phi) - (d+e)r.\sin(\phi) \\ u.\sin(\phi) + (d+e)r.\cos(\phi) \\ r \\ mdR^2 r^2 / \Omega_u \\ -dmR^2 ur / \Omega_r \end{bmatrix} + \begin{bmatrix} 0 & 0 \\ 0 & 0 \\ 0 & 0 \\ 2nR / \Omega_u & 0 \\ 0 & 2nbR / \Omega_r \end{bmatrix} \begin{bmatrix} \tau_u \\ \tau_r \end{bmatrix} + \begin{bmatrix} \delta_x \\ \delta_y \\ 0 \\ \delta_u \\ \delta_r \end{bmatrix} \tag{12}$$

where $\Omega_u = mR^2 + 2I_e$; $\Omega_r = 2b^2 I_e + R^2(I_z + md^2)$; $\tau_u = \dfrac{1}{2}(\tau_l + \tau_r)$,

$\tau_r = \dfrac{1}{2}(\tau_l - \tau_r)$ $\delta_u = \dfrac{1}{\Omega_u}\left(R^2 P_x + mR^2 rv^s - I_e(\dot{u}_l^s + \dot{u}_r^s)\right)$ $\delta_x = -v^s \sin\phi$; $\delta_x = v^s \cos\phi$

$$\delta_r = \dfrac{1}{\Omega_r}\left((d-l)R^2 P_y - mdR^2 \dot{v}^s - bl_e(\dot{u}_l^s - \dot{u}_r^s)\right)$$

The vector $\delta_m = \begin{bmatrix} \delta_x & \delta_y & 0 & \delta_r & \delta_r \end{bmatrix}^T$ represents for the uncertainties. We take the world coordinates of tracking point O as the control variable from the posture vector as

$$x = \begin{bmatrix} X & Y \end{bmatrix}^T \text{ and } \dot{x} = \begin{bmatrix} \dot{X} & \dot{Y} \end{bmatrix}^T \tag{13}$$

We obtain the dynamic equation in terms of the position vector as follows

$$\ddot{x} = D(\phi)(\tau + S(\dot{x},\phi,\delta_m) + \zeta(\dot{x},\phi,\delta_m)) \tag{14}$$

$$\dot{\phi} = \alpha(\dot{x},\phi,\delta_m) \tag{15}$$

where

$$D(\phi) = \begin{bmatrix} \dfrac{2nR}{\Omega_u}\cos\phi & -\dfrac{2nRb(d+e)}{\Omega_r}\sin\phi \\ \dfrac{2nR}{\Omega_u}\sin\phi & \dfrac{2nRb(d+e)}{\Omega_r}\cos\phi \end{bmatrix} \quad S = \begin{bmatrix} -\dfrac{emR^2+2(d+e)I_e}{2nR}r^2 \\ \dfrac{2b^2I_e+R^2(I_e-mde)}{2nbR(d+e)}ur \end{bmatrix}$$

$$\zeta = \begin{bmatrix} \dfrac{\Omega_u}{2nR}\left(\delta_u - v^s r\right) \\ \dfrac{\Omega_r}{2nbR(d+e)}\left((b+e)\delta_r - \dot{v}^s\right) \end{bmatrix} \quad r(\phi,\dot{x},\delta_m) = \dfrac{1}{d+e}\left(-\left(\dot{X}-\delta_x\right)\sin\phi + \left(\dot{Y}-\delta_y\right)\cos\phi\right)$$

The nonholonomic model of mobile robot can be described as follows

$$\ddot{x} = D_0(\phi)(\tau + S_0(\dot{x},\phi)) \tag{16}$$

$$\dot{\phi} = \alpha_0(\dot{x},\phi) \tag{17}$$

Where $S_0 = \begin{bmatrix} -\dfrac{emR^2+2(d+e)I_e}{2nR}r^2_{(\phi,\dot{x})} \\ \dfrac{2b^2I_e+R^2(I_e-mde)}{2nbR(d+e)}u_{(\phi,\dot{x})}r_{(\phi,\dot{x})} \end{bmatrix} \quad r(\phi,\dot{x}) = \dfrac{1}{d+e}\left(-\dot{X}\sin\phi + \dot{Y}\cos\phi\right)$

3   Robust Tracking Control Based on Neural Networks

The dynamic model in Eqs.(14)(15) are rewritten in the following form

$$\ddot{x} = D_0(\phi)(\tau + S_0) + \Psi(\tau,\dot{x},\phi,\delta_m) \tag{18}$$

where $\Psi(\tau,\dot{x},\phi,\delta_m) = \Delta D(\tau+S+\zeta) + D_0(\Delta S+\zeta)$. The approximation of the robot unknown uncertainties $\Psi(\tau,\dot{x},\phi,\delta_m)$ can be written in the form of neural network as follows [8]

$$\Psi(\tau,\dot{x},\phi,\delta_m) = W\,\sigma(V\bar{x}) + \varepsilon(\bar{x}) \tag{19}$$

where W and V are the weight matrices of the output and hidden layers, respectively $\|W\|_F \leq W_m$, $\|V\|_F \leq V_m$. $\bar{x} = [\tau, \dot{x}, \phi]^T$ is the neural network input. $\varepsilon(\bar{x})$ is the bounded neural network approximation error. $\sigma(\cdot)$ is the sigmoid activation function $\|\sigma\| \leq \sigma_m$

Given a desired trajectory $x_d, \dot{x}_d, \ddot{x}_d \in R^2$ of the mobile robot, the tracking error is defined as $e = x - x_d$; $\dot{e} = \dot{x} - \dot{x}_d$. The filtered tracking error is defined as [7] $r = \dot{e} + \Lambda e = \dot{x} - \dot{x}_r$ where $\Lambda = \Lambda^T > 0$ is the design parameter matrix.

In this paper, the following adaptive tracking controller for the mobile robot in presence of the wheel slip and the external disturbance forces is proposed

$$\tau = D_0^{-1}(\phi)(\ddot{x}_r - Kr - \hat{W}\,\sigma(\hat{V}\bar{x})) - S_0(\dot{x}, \phi) \tag{20}$$

where K is a positive definite matrix, and $\hat{\Psi}(\tau, \dot{x}, \phi, \delta_m) = \hat{W}\,\sigma(\hat{V}\bar{x})$ is the output of neural network. The filtered tracking error as follows

$$\dot{r} = -Kr + W\,\sigma(V\bar{x}) + \varepsilon(\bar{x}) - \hat{W}\,\sigma(\hat{V}\bar{x}) \tag{21}$$

Lets define the objective function as $J = (1/2)r^T r$, where r is the filtered tracking error. The weight updating laws are obtained as follows [8]

$$\dot{\hat{W}} = -\alpha_1 \left(r^T K^{-1}\right)^T \left(\sigma(\hat{V}\bar{x})\right)^T - \beta_1 \|r\| \hat{W} \tag{22}$$

$$\dot{\hat{V}} = -\alpha_2 \left(r^T K^{-1}\hat{W}(I - diag(\sigma_i^2))\right)^T sgn(\bar{x})^T - \beta_2 \|r\| \hat{V} \tag{23}$$

The following Lyapunov positive definite function is considered

$$L = \frac{1}{2}r^T r + \frac{1}{2}trace(\tilde{W}^T\tilde{W}) + \frac{1}{2}trace(\tilde{V}^T\tilde{V}) \tag{24}$$

The time derivative of Eq.(24) is given by

$$\dot{L} = r^T \dot{r} + tr(\tilde{W}^T\dot{\tilde{W}}) + tr(\tilde{V}^T\dot{\tilde{V}}) \tag{25}$$

Based on the properties of the matrix trace and sigmoidal function, we have

$$\tilde{W}^T(W - \tilde{W}) \leq W_m \|\tilde{W}\| - \|\tilde{W}\|^2 \tag{26}$$

$$\tilde{V}^T(V - \tilde{V}) \leq V_m \|\tilde{V}\| - \|\tilde{V}\|^2 \tag{27}$$

$$\tilde{W}^T \left(r^T K^{-1}\right)^T \left(\sigma(\hat{V}\bar{x})\right)^T \leq \sigma_m \|\tilde{W}\| \|r\| \|K^{-1}\| \tag{28}$$

$$\tilde{V}^T \left(r^T K^{-1}\hat{W}(I - diag(\sigma_i^2))\right)^T sgn(\bar{x})^T \leq \|\tilde{V}\| \|r\| \|K^{-1}\| (W_m + \|\tilde{W}\|) \tag{29}$$

Now we assume that $\beta_1 > \frac{1}{2}\alpha_2 \|K^{-1}\|$ and $\beta_2 > \frac{1}{2}\alpha_2 \|K^{-1}\|$, and let's define

$$A = \frac{\left(\sigma_m + \alpha_1 \sigma_m \|K^{-1}\| + \beta_1 W_m\right)}{2\left(\beta_1 - \frac{1}{2}\alpha_2 \|K^{-1}\|\right)} \; ; B = \frac{\left(\alpha_2 \|K^{-1}\| W_m + \beta_2 V_m\right)}{2\left(\beta_2 - \frac{1}{2}\alpha_2 \|K^{-1}\|\right)} \tag{30}$$

The Eq.(25) becomes

$$\dot{L} \leq -\lambda_{min}(K)\|r\|^2 + \|r\|\left(\varepsilon_m + \left(\beta_1 - \frac{1}{2}\alpha_2 \|K^{-1}\|\right)A^2 + \left(\beta_2 - \frac{1}{2}\alpha_2 \|K^{-1}\|\right)B^2\right) \tag{31}$$

The above analysis shows the ultimate bounded of the filtered tracking error r.

4   Simulation Results

The robot parameter values used for simulations are $R = 0.16$ (m) , $b = 0.23$ (m), $d = 0.2$ (m) , $l = 0.25$ (m) , $e = 0.05$ (m), $n = 100$, $I_m = 1e^{-6}$ (kgm^2) , $I_t = 0.1$ (kgm^2) , $I_z = 4$ (kgm^2) , $m = 40$ (kg). The sampling time: 1(ms). The stable matrix K and Λ $K = diag([6\ 6])$, $\Lambda = diag([4\ 4])$. The learning rate: $\alpha_{1,2} = 30$ and the damping factor $\beta_{1,2} = 3$ .The initial value of the weight matrices: $\hat{W}_0 = rand(0,1)$, $\hat{V}_0 = rand(0,1)$. The number of neurons in the hidden layer are 30 neurons. The lateral slip speed of the wheel axle, the longitudinal slip speeds of the right and left wheel centers $v^s = 0.1 + 0.05\sin(2t)$ (m/s); $u_r^s = 0.1 + 0.1\sin(t)$ (m/s); $u_l^s = 0.1 + 0.1\sin(t)$ (m/s). The external forces at point P : $P_x = 10$ (N), $P_y = 10$ (N). To illustrate the improvement in tracking performance, we compare the controller Eq.(20) with the following conventional computed torque controller

$$\tau = D_0^{-1}(\phi)(\ddot{x}_d - K_D\dot{e} - K_P e) - S_0(\dot{x},\phi) \tag{32}$$

The mobile robot is controlled to track a circle trajectory as shown in Fig. 2. The desired trajectory are

$$\begin{cases} x_d = 3.25 + 2.25\cos(\frac{7\pi}{40}t - \frac{\pi}{2}) \\ y_d = 3.25 + 2.25\sin(\frac{7\pi}{40}t - \frac{\pi}{2}) \end{cases} (m) \tag{33}$$

The initial positions of the mobile robot are: $[x\ y\ \phi]^T = [3.25\ 0.75\ \pi/2]^T$. The simulation results are shown in Figs. 2-3. The tracking performance and tracking error in the case of circle tracking control are shown in Fig. 3.a-d. It can be seen that the tracking accuracy of the proposed controller in this case is also better if the neural network uncertainties compensation is used. Fig. 3.e and 3.f show the results of tuning the output weights and hidden weight of the neural network. It can be seen that the convergence of the online tuning algorithms can also be achieved.

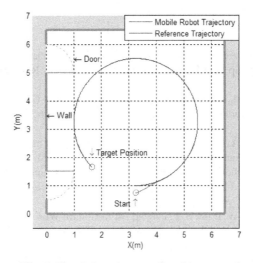

Fig. 2. The circle trajectory of tracking control

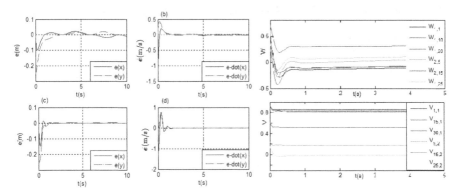

Fig. 3. Circle trajectory tracking errors: (a),(b) Without NN, (c),(d) With NN And Tuning results of NN weights (e) Output weights , (f) Hidden layer weights

5   Conclusions

In this paper, a novel adaptive tracking control of the mobile robot subject to the wheel slip and the external disturbance forces have been presented. The neural networks with an online weight updating laws is used to compensate the effect of wheel slip and external forces. Detailed simulation results of tracking a circle trajectory has been given to confirm the effectiveness of our proposed algorithm.

Acknowledgments. This work was supported by Special Research Fund of Electrical Engineering in University of Ulsan

References

1. Ryu, J.-C. and S.K. Agrawal.: Differential flatness-based robust control of mobile robots in the presence of slip. The International Journal of Robotics Research, vol. 30, pp. 463-475 (2011)
2. Zhang, Y., J. Chung, and S. Velinsky.: Variable Structure Control of a Differentially Steered Wheeled Mobile Robot. Journal of Intelligent and Robotic Systems, vol. 36, pp. 301-314 (2003)
3. Balakrishna, R. and A. Ghosal.: Modeling of slip for wheeled mobile robots. Robotics and Automation, IEEE Transactions on, vol. 11, pp. 126-132 (1995)
4. Seyr, M. and S. Jakubek.: Proprioceptive Navigation, Slip Estimation and Slip Control for Autonomous Wheeled Mobile Robots. In: Robotics, Automation and Mechatronics, 2006 IEEE Conference on. vol. 1, pp 1-6 (2006).
5. Chang Boon, L. and W. Danwei. Integrated Estimation for Wheeled Mobile Robot posture, velocities, and wheel skidding perturbations. in: Robotics and Automation, 2007 IEEE International Conference on. vol 1 pp. 2355-2360 (2007).
6. Lateral and Longitudinal Tire Forces, in Vehicle Dynamics and Control. Springer US. pp. 387-432 (2006)
7. Huang, S.N. and T. Kok Kiang.: Fault Detection, Isolation, and Accommodation Control in Robotic Systems. IEEE Transactions on Automation Science and Engineering, vol. 5, pp. 480-489 (2008)
8. Abdollahi, F., H.A. Talebi, and R.V. Patel.: A stable neural network-based observer with application to flexible-joint manipulators. Neural Networks, IEEE Transactions on, vol. 17 pp. 118-129 (2006)

Functional Link Neural Network
with Modified Cuckoo Search Training Algorithm
for Physical Time Series Forecasting

Rozaida Ghazali, Zulaika Abu Bakar, Yana Mazwin Mohmad Hassim,
Tutut Herawan, and Noorhaniza Wahid

Faculty of Computer Science and Information Technology,
Universiti Tun Hussein Onn Malaysia
rozaida@uthm.edu.my

Abstract. Functional link neural network (FLNN) naturally extends the family of theoretical feedforward network structure by introducing nonlinearities in inputs patterns enhancements. It has emerged as an important tool used for function approximation, pattern recognition and time series prediction. The standard learning algorithm used for the training of FLNN is the Backpropagation (BP) learning algorithm. However, one of the crucial problems with BP algorithm is it tends to easily get trapped in local minima, resulting the degrading performance of FLNN. To overcome this problem, this work proposed an alternative learning scheme for FLNN by using a Modified Cuckoo Search algorithm (MCS), and the model is called FLNN-MCS. The performance of FLNN-MCS is evaluated based on the prediction error, testing on two physical time series data; relative humidity and temperature. Simulation results have shown that the prediction performed by FLNN-MCS is much superior compared to Multilayer Perceptron and FLNN trained with BP, and FLNN trained with Artificial Bee Colony algorithm. The significant performance has proven that FLNN-MCS is capable in mapping the input-output function for the next-day ahead forecasting.

Keywords: Artificial Bee Colony, Cuckoo Search Algorithm, Modified Cuckoo Search Algorithm, Functional Link Neural Network, physical time series prediction.

1 Introduction

Time series forecasting is one of the most important and interesting problems when studying natural occurring phenomena. Traditional methods for time series forecasting are statistical-based, including moving average (MA), autoregressive (AR), autoregressive moving average (ARMA) models, linear regression and exponential smoothing. These approaches do not produce fully satisfactory results, due to the nonlinear behavior of most of the natural occurring time series. Other more advanced techniques such as neural networks, fuzzy logic, and genetic algorithm have been successfully used in time series prediction.

D.-S. Huang et al. (Eds.): ICIC 2014, LNCS 8588, pp. 285–291, 2014.

The application of neural networks in time series prediction has shown better performance in comparison to statistical methods because of their nonlinear nature and training capability. In addition, it has been shown that neural networks are universal approximators and have the ability to produce complex nonlinear mappings. However, when the number of inputs to the model and the number of training examples becomes extremely large, the training procedure for ordinary neural network architectures becomes tremendously slow. To overcome such time-consuming operations, this research work focuses on using Functional Link Neural Network (FLNN) [1] which has a single layer of learnable weights, therefore reducing the networks' complexity.

FLNN can capture non-linear input-output mapping, by utilizing higher combination of it inputs. The most common method for tuning the weight in FLNN is using a Backpropagation (BP) algorithm. However, one of the crucial problems with BP algorithm is that it can easily get trapped in local minima. To recover this drawback, the Modified Cuckoo Search (MCS) algorithm is used in this work to optimize the connection weights in FLNN instead of the BP.

2   Background Works

In this section, the properties of FLNN and the learning procedure of MCS are briefly discussed..

2.1   Functional Link Neural Network

FLNN is a class of Higher Order Neural Networks [1] and has been successfully used in many applications such as system identification [2], [3], classification [4], pattern recognition [5], and prediction [6]. The FLNN architecture is basically a flat network without any existence of hidden layers which provides less complexity to the training of the network. In FLNN, the input vector is extended with a suitably enhanced representation of the input nodes, thereby artificially increase the dimension of the input space.

Figure 1 shows an example of a third order FLNN with 3 external inputs x_1, x_2, and x_3 and four high order inputs which are x_1x_2, x_1x_3, x_2x_3, and $x_1x_2x_3$ act as supplementary inputs to the network, making the total inputs of the network seven. The basic equation for calculating FLNN's output of 3rd degree is given in Equation 1.

$$Y = \sigma\left(W_o + \sum_i W_i X_i + \sum_i \sum_j W_{ij} X_i X_j + \sum_i \sum_j \sum_k W_{ijk} X_i X_j X_k\right) \tag{1}$$

where 'σ' is a nonlinear transfer function, w_0 is the adjusted threshold,   and w_i, w_{ij}, w_{ijk} are adjustable weights that link the external inputs x_i, x_j, x_k,   and also the high order inputs, x_ix_j, x_ix_k, x_jx_k, and $x_ix_jx_k$ to the output node.

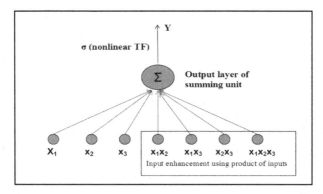

Fig. 1. 3rd order FLNN structure with three inputs

2.2 The Learning Algorithm of FLNN

Most researches have reported that the standard algorithm used for training the FLNN is the BP algorithm [4], [6]. The BP algorithm however, has several limitations which can affect the performance of FLNN. The network prone to get trapped in local minima which caused by the gradient descent method used by BP learning. Another limitation of BP algorithm is that the network became very dependent on the choices of initial weight values as well as the parameters in the algorithm such as the learning rate and momentum. The learning rate and momentum need to be finely adjusted to meet the desired convergence criterion, resulting the difficulty in training of the network. For these reasons, a further investigation to improve the procedure of adjusting the learnable weights in FLNN is desired.

3   MCS: The Proposed Learning Algorithm for FLNN

Cuckoo Search (CS) is one of metaheuristic searching algorithm based on cuckoo bird reproduction behavior [7]. This algorithm is stimulated by the obligate brood parasitism of some cuckoo species by laying their eggs in the nests of other host birds. If an egg is discovered by the host bird as not its own, it will either throw the unknown egg away or simply abandon its nest and build a completely new nest elsewhere. The function of cuckoo eggs is to swap the finest solution in the other bird's nest. The main reason of exchanging the eggs of host nest is because the cuckoo egg carries a better solution to the existing eggs in the nest.

The number of available host nests is fixed. The probability of the egg laid by a cuckoo discovered by the host bird can be approximated by a fraction $pa \in [0,1]$ of the n nests that are replaced by new nests (with new random solutions). When generating new solutions x^{t+1} for a cuckoo j, a Levy flight is performed:

$$x_j^{t+1} = x_j^t + \alpha \oplus levy(\lambda) \,, \tag{2}$$

where $\alpha > 0$ is the step size, which should be related to the scales of the problem of interest. The Levy flight essentially provides a random walk while the random step length is drawn from a Levy distribution as shown in the Equation 3:

$$\text{Lavy} \sim u = t^{-\lambda}, 1 < \lambda \leq 3 \tag{3}$$

This has an infinite variance with an infinite mean. Here the steps essentially construct a random walk process with a power-law step-length distribution and a heavy tail. Notably, based on adequate computation, the result of CS always gives the best in finding the optimum weight. However, the problem of finding the whole region via random walks still gives a slow convergence. As a result, two modifications have been made to the original CS whilst keeping the various algorithmic parts for a wider region of application. The modified and improved CS is called Modified Cuckoo Search (MCS) algorithm [8].

In the first modification, changes had been made to the size of the Levy Flight step α. In CS, the value of α is a constant of 1. There is an increase in the number of generations when α decreases for MCS. This same technique is also implemented in Particle Swarm Optimization (PSO) where the inertia constant is reduced with the aim of enhancing the localized search of individuals, or eggs to the nearest solution.

The second modification, on the other hand rests solely with the addition of information exchanged between the eggs to adequately minimize the convergence to its lowest. This is credited to MCS while in CS, there exist no such exchange of information and the search is performed by individual self.

4   FLNN-MCS for Physical Time Series Prediction

Time series is a sequence of observations of a random variable. A common goal of time series data analysis is to extrapolate past behavior into the future called forecasting or prediction. On another side, physical time series such as astrophysical, geophysical, and meteorological may appear as an output of an experiment, or it may come out as a signal from a dynamical system, or it may contain some sociological, economic or biological information. This research focuses on physical time series which inclusive of 2 univariate datasets namely temperature and relative humidity.

In this work, the FLNN trained with MCS (FLNN-MCS), is used to predict future values based on previously observed data. In the initial process, the FLNN structure (weight and bias) is transformed into objective function along with the training dataset. This objective function is then fed to the MCS algorithm in order to search for the optimal weight parameters. The weight changes are then adjusted by the MCS based on the error calculation (difference between actual and expected results). Based on this algorithm, each egg represents the solutions with a particular weight vector.

5   Experimental Design

In this work, a univariate physical time series data of relative humidity and temperature index for Batu Pahat region, Johor ranging from the year 2005 to 2009 are used. These data were obtained from Malaysian Meteorological Department, Malaysia. Both data series are partitioned into three parts; 50% for training set, and 25% for each validation and test set.

Table 1. Parameters setting

Network model	MLP-BP	FLNN-BP	FLNN-ABC	FLNN-MCS
Hidden nodes/network order	3–8 nodes	2-5	2-5	2-5
Learning rate/probability	0.1–0.5	0.1–0.5	-	0.7
Momentum	0.1–0.5	0.1–0.5	-	-
Colony size/no.nest	-	-	50	25

All network models were trained for 3,000 epochs/cycles on the training set. Training is stopped when the minimum error of 0.001 reached; or when the maximum of 3,000 epochs/cycles reached. We also implemented an early stopping procedure to avoid over-fitting. With this procedure, the networks were trained by observing the point at which the validation error began to rise, and then restored the weights at the iteration cycle where the validation error was minimum.

The parameters setting for the experiment are presented in Table 1. The number of input nodes for all network models was experimentally selected between four to eight. The same trial-and-error procedure was also performed on other parameters. As for forecasting horizon, a one-step-ahead prediction is chosen since the main target is to predict the upcoming measure of daily forecasting. The average results of ten simulations were determined for all networks. The Mean Squared Error (MSE), Normalized Mean Squared Error (NMSE), Signal to Noise Ratio (SNR) were used to evaluate each network performance.

6   Simulation Results

The comparison of simulation results for MLP-BP, FLNN-BP, FLNN-ABC and FLNN-MCS on physical time series prediction data is presented in Tables 2 and 3. All the results discussed in this section were based on the out-of-sample performance of all network models.

Results in Table 2 compare the networks in terms of their complexity. 'Complexity' is defined as the number of adaptive weights used by the neural network. While there are other measures for complexity, such as the training algorithm, weights are a good complexity measure. As in the case when all other properties of two neural networks are equal, a network which has less weights will usually execute faster and require less time for training. Such properties are especially desirable in real-time systems. On this issue, the comparison of MLP and FLNN network in terms of network complexity showed that the FLNN network required less numbers of free parameters (trainable weights + biases) than MLP. The less numbers of parameters indicates that the network required less computational load as there are small numbers of tunable weights to be updated at every epoch or cycle.

Table 2. Complexity of all network models

Datasets	Network model	MLP-BP	FLNN-BP	FLNN-ABC	FLNN-MCS
Relative	Network structure	5-6-1	2nd order	2nd order	3rd order
Humidity	No. tunable weight	43	16	16	22
Temperature	Network structure	5-3-1	2nd order	2nd order	3rd order
	No. tunable weight	22	16	16	22

Meanwhile, Table 3 presents the performance of all network models based on their NMSE, MSE and SNR. It can be seen that the FLNN-MCS outperformed very well on both physical time series datasets. The FLNN-MCS exhibited the least prediction errors in both NMSE and MSE, which indicates that the predicted and the actual values obtained by the network are better in terms of overall deviations of scatter between the prediction and the actual values. This shows that FLNN-MCS was able to produce close forecast to the actual time series data. In terms of SNR, FLNN-MCS gained higher value which shows that the model can track time series signal better than the benchmarked models. Clearly it can be seen that the performance of the FLNN trained with the proposed MCS algorithm have shown a better prediction result on the unseen data when compared to the MLP and FLNN trained with BP and ABC algorithms.

Table 3. Performance evaluations from all network models

Datasets	Network model	MLP-BP	FLNN-BP	FLNN-ABC	FLNN-MCS
Relative	NMSE	0.9872	0.7363	0.6416	**0.006485**
Humidity	MSE	0.0092	0.0069	0.0060	**0.000061**
	SNR (dB)	17.7145	18.9882	19.5862	**23.1105**
Temperature	NMSE	0.9942	0.8023	0.7731	**0.000585**
	MSE	0.0083	0.0067	0.0065	**0.001905**
	SNR(dB)	17.6518	18.5832	18.7439	**29.9842**

On the whole, the FLNN trained with MCS algorithm has the ability to make a good forecast with minimum prediction error as compared to the standard FLNN-BP, FLNN-ABC and MLP-BP. This indicates that FLNN-MCS is capable of representing non-linear function better than MLP-BP, FLNN-BP and FLNN-ABC. Hence it shows that MCS training scheme facilitates the FLNN with better learning by providing a good exploration and exploitation as compared to other learning algorithms.

7   Conclusion

FLNN-MCS has shown its significant advantages in forecasting both relative humidity and temperature datasets. It is suggested that the FLNN network trained with MCS provides better results compared to the standard MLP and FLNN trained with the BP, and FLNN trained with ABC. As MCS algorithm combines the exploration and exploitation properties in its search strategy, it can successfully avoid trapping in local minima, resulting in a better searching for optimal weights set during the training phase. Thus, MCS algorithm can be considered as an alternative learning algorithm for training the FLNN instead of standard BP and ABC learning algorithm. Simulation results have provided additional evidence supporting the application of this model to physical time series, thus suggesting that the model is good at the prediction of relative humidity and temperature index.

Acknowledgments. The authors would like to thank Universiti Tun Hussein Onn Malaysia and Ministry oh Higher Education Malaysia for financially supporting this research under the Exploratory Research Grant Scheme, Vote No. 0882.

References

1. Giles, C.L., Maxwell, T.: Learning, invariance and generalization in high-order neural networks. Applied Optics 26(23) (1987)
2. Emrani, S., Salehizadeh, S.M.A., Dirafzoon, A., Menhaj, M.B.: Individual particle optimized functional link neural network for real time identification of nonlinear dynamic systems. In: 2010 the 5th IEEE Conference on Industrial Electronics and Applications, ICIEA (2010)
3. Patra, J.C., Bornand, C.: Nonlinear dynamic system identification using Legendre neural network. In: The 2010 International Joint Conference on Neural Networks, IJCNN (2010)
4. Abu-Mahfouz, I.-A.: A comparative study of three artificial neural networks for the detection and classification of gear faults. International Journal of General Systems 34(3), 261–277 (2005)
5. Klaseen, M., Pao, Y.H.: The functional link net in structural pattern recognition. In: 1990 IEEE Region 10 Conference on Computer and Communication Systems, TENCON 1990 (1990)
6. Ghazali, R., Hussain, A.J., Liatsis, P.: Dynamic ridge polynomial neural network: forecasting the univariate non-stationary and stationary trading signals. Expert Systems with Applications 38(4), 3765–3776 (2011)
7. Yang, X.S., Deb, S.: Cuckoo search via Lévy flights. In: World Congress on Nature & Biologically Inspired Computing, Coimbatore, India, pp. 210–214 (2009)
8. Walton, S., Hassan, O., Morgan, K., Brown, M.R.: Modified cuckoo search: A new gradient free optimisation algorithm. Chaos, Solitons & Fractals 44, 710–718 (2011)

The Research of Codeword
Based on Fuzzy Hopfield Neural Network[*]

Xian Fu and Yi Ding

Department of Computer Science and Technology,
Hubei Normal University, Huangshi 435002, China
{teacher.fu,teacher.dingyi}@live.com

Abstract. Hopfield neural networks are known to cluster analysis and unsupervised learning programs. In this paper, a new Hopfield-model net called Corrected Fuzzy Hopfield Neural Network (CFHNN) is proposed for vector quantization in image correction. In CFHNN, the Correct Fuzzy C-means (CFC) algorithm, modified from penalized fuzzy c-means, is embedded into Hopfield neural network so that the parallel implementation for codeword design is feasible. The proposed network also proposed to avoid the network to determine the weight factors of the energy function value. In addition, it plans to make e-learning training more quickly than FC and PFC, more effectively. In the experimental results, CFHNN method shows the FC and PFC method is relatively encouraging results.

Keywords: Fuzzy Hopfield, Corrected Fuzzy Hopfield Neural Network, Corrected Fuzzy c-means.

1   Introduction

Today, communication technology and promote the importance of image correction increases. Limited hardware and budget is also important to quickly send data in. The amount of data associated with visual information is so large that its storage requires a huge storage capacity [1]. Image data correction techniques to reduce the storage or transmission of image information of interest without significant loss of any desired number of bits. Because of its wide range of applications in digital image processing, data correction is very important [2].During the past decade, the Hopfield neural network has been studied extensively with its feature of simple architecture and potential for parallel implementation [3].The Hopfield neural network is a well-known technique used for solving optimization problems based on Lyapunov energy function. In Refs[4-8]. Fuzzy Hopfield neural network has been widely studied and successfully applied in image corrected.

Image correction is the coding of transformed image using a code of fixed or variable length. Vector quantization is a significant methodology in image correct, in

---

[*] Xian Fu work is partly supported by the S&T plan projects of Hubei Provincial Education Department of China (No.Q20122207).

D.-S. Huang et al. (Eds.): ICIC 2014, LNCS 8588, pp. 292–298, 2014.

which blocks of divided pixels are formed as training vectors rather than individual scales. Such a method results in the massive reduction of the image information in image transmission. The image is reconstructed by replacing each image block with its nearest code vector. Codeword design can be considered as a clustering process in which the training vectors are classified into the specific classes based on the minimization of average distortion between the training vectors and codeword vectors [9-13].

The remainder of this paper is organized as follows. Section 2 discusses fuzzy c-means, penalized fuzzy c-means; Section 3 proposes the corrected fuzzy c-means techniques and vector quantization by the CFHNN; Section 4 presents several experimental results; Section 5 gives the discussion and conclusions.

2   Fuzzy Hopfield Neural Network

2.1   Fuzzy C-means Clustering Algorithm

Cluster analysis is based on the partition of the set of data points is divided into several sub-groups, a group or a certain extent close to the performance of similar objects (sub-group). It played an important role in dealing with many issues need to be addressed other aspects of pattern recognition and image. The algorithm is an iterative clustering methyl-OD, produced by minimizing the sum of squared error weighted objective function best group C partition:

$$OF_{FC} = \sum_{x=1}^{n} \sum_{i=1}^{c} \frac{\left\| Z_x - \sigma_i \right\|^2 \alpha_{x,i}^m}{2} \tag{1}$$

Where $|\cdot|$ is the Euclidean distance between the training sample and cluster center, c is the number of clusters with $2 \le c < n$, The component $\sigma_{x,i}$ denotes the degree of possibility that a training sample Z_x belongs to an i th fuzzy cluster.

It is composed of (1), points out that the objective function FC algorithm does not consider any spatial information. This limitation makes the FC is very sensitive to noise. The general principles of the proposed approach are to neighborhood information to the classification of the FC algorithm. Spatial context in order to make changes to the FC algorithm is inspired by and based on the criteria above FC algorithms. The new objective function is defined as follows in PFC:

$$OF_{PFC} = \sum_{x=1}^{n} \sum_{i=1}^{c} \frac{\left(\left\| Z_x - \sigma_i \right\|^2 - \zeta \ln \left(\varphi_i \right) \right) \alpha_{x,i}^m}{2} \tag{2}$$

Where φ_i is a proportional constant of class i and $\zeta \ge 0$ is also a constant. When $\zeta = 0$, OF_{PFC} equals the energy function of fuzzy c-means algorithm.

2.2   Penalized FC Algorithm

The PFC algorithm is presented for the clustering problem as follows:

1.Randomly initialize class centers $\sigma_i \left(2 \leq i \leq c\right)$ and fuzzy c-partition $A^{(0)}$.Give fuzzification parameter m $m\left(1 \leq m < \infty \right)$, constant ζ, and the value $\varepsilon > 0$;

2.Compute

$$\varphi_i = \sum_{x=1}^{n} \frac{\alpha_{x,i}^m}{\sum_{i=1}^{c} \alpha_{x,i}^m} \quad , \quad \sigma_i = \sum_{x=1}^{n} \frac{z_x \sum_{i=1}^{c} \alpha_{x,i}^m}{\alpha_{x,i}^m} \quad \text{and} \quad \text{with}$$

$$\alpha_{x,i} = \sum_{x=1}^{n} \sum_{i=1}^{c} \left(\sum_{j=1}^{c} \frac{\left\| Z_x - \sigma_i \right\|^2 - \zeta \ln(\varphi_i)}{\left\| Z_x - \sigma_j \right\|^2 - \zeta \ln(\varphi_j)} \right)^{\frac{1}{(m-1)}}$$,calculate the membership matrix

$A = \left[\alpha_{x,i} \right];$

3.Compute $\Delta = \max \left(\left\| A^{(t+1)} - A^{(t)} \right\| \right)$. If $\Delta > \varepsilon$, then go to Step 2; otherwise go to Step 4;

4.Find the results for the final class centers.

The PFC penalty degree is too heavy to rapidly converge. The penalty term $\sum_{x=1}^{n} \sum_{i=1}^{c} \frac{-\zeta \ln\left(\varphi_i\right) \alpha_{x,i}^m}{2}$, the corrected term $\sum_{x=1}^{n} \sum_{i=1}^{c} \frac{\zeta \tanh\left(\varphi_i\right) \alpha_{x,i}^m}{2}$,then the energy function and membership function in a so-called Corrected Fuzzy C-Means (CFC) algorithm is defined as:

$$OF_{CFC} = \sum_{x=1}^{n} \sum_{i=1}^{c} \frac{\left(\left\| Z_x - \sigma_i \right\|^2 + \zeta \tanh\left(\varphi_i\right) \right) \alpha_{x,i}^m}{2} \tag{3}$$

Since $0 < \varphi_i < 1$, we can find $\tanh(-\varphi_i) \subset \ln(\varphi_i)$ which implies that OF_{CFC} can also be convergent.

3   The Codeword for Corrected Fuzzy Hopfield Neural Network

3.1   Correct Fuzzy C-means

Hopfield neural network, has a simple structure and parallel potential, has been applied in many fields. Hopfield neural networks with the structure and strategy of CFC, CFHNN, image segmentation classification training vectors to produce a viable codeword. In order to improve the ability of the method, the energy function class scatter matrix formulation, is widely used in basic pattern classification concepts. The total input to neuron (x, i) is computed as:

$$N_{x,i} = \sum_{x=1}^{n} \sum_{i=1}^{c} \left(I_{x,i} + \left| Z_x - \sum_{y=1}^{n} \sum_{j=1}^{c} T_{x,i;y,j} \left(\alpha_{y,j} \right)^m \right|^2 \right) \tag{4}$$

The modified Lyapunov energy function of the two dimensional Hopfield neural network using CFC strategy is given by:

$$E = \sum_{x=1}^{n} \sum_{i=1}^{c} \frac{\left(\alpha_{x,i} \right)^m \left(I_{x,i} + \left| Z_x - \sum_{y=1}^{n} \sum_{j=1}^{c} T_{x,i;y,j} \left(\alpha_{y,j} \right)^m \right|^2 \right)}{2} \tag{5}$$

Where $|\cdot|$ is the average distortion between the training vector to cluster center on the divided image, $\sum_{y=1}^{n} \sum_{j=1}^{c} T_{x,i;y,j}$ is the total weighted input received from neuron (y,i) in row $i, \mu_{x,i}$ is the output state at neuron (x,i), and m is the fuzzification parameter.

3.2   CFHNN Algorithm

In order to reduce the quality of the classification result is a very sensitive weighted factor to a 2D Hopfield neural networks and fuzzy c-means clustering correction strategy, called CFHNN proposal to allow constraints can be handled more efficiently. Then, CFHNN scattered energy can be further simplified to:

$$E = \sum_{x=1}^{n} \sum_{i=1}^{c} \frac{\left(\alpha_{x,i} \right)^m \left(\zeta \tanh(\varphi_i) + \left| Z_x - \sum_{y=1}^{n} \sum_{j=1}^{c} Z_{x,i;y,j} \left(\alpha_{x,i} \right)^m \right|^2 \right)}{2} \tag{6}$$

Each column of this modified Hopfield network represents a class and each row represents a training vector. The network reaches a stable state when the modified Lyapunov energy function is minimized. The CFHNN can classify C clusters in a parallel manner that is described as follows:

1.Input a set of training vector $Z = z_1, z_2, ..., z_n$, fuzzification parameter $m(1 \leq m < \infty)$, the number of clusters c, constant ζ and initialize the states for all neurons $A = \left| \alpha_{x,i} \right|$ (membership matrix);

2.Compute $\displaystyle \varphi_i = \sum_{x=1}^{n} \frac{\alpha_{x,i}^{m}}{\displaystyle\sum_{i=1}^{c} \alpha_{x,i}^{m}}$ and weighted matrix $T_{x,i;y,j} = \dfrac{Z_x}{\left(\alpha_{x,i}\right)^{m}}$;

3.Use $\quad I_{x,i;} = \zeta \tanh(\varphi_i)$,calculate   the   input   to   each   neuron

$$N_{x,i} = \zeta \tanh(\varphi_i) + Z_x \left| 1 - \sum_{y=1}^{n} \left(\frac{\alpha_{y,j}}{\alpha_{x,i}} \right)^{m} \right| ;$$

4.Apply (6) to update the neurons' membership values in a synchronous manner:

$$\alpha_{x,i} = \sum_{j=1}^{c} \left(\frac{N_{x,i}}{N_{y,j}} \right)^{-\frac{1}{(m-1)}} ;$$

5.Compute $\Delta = \max \left| A^{(t+1)-} A^{(t)} \right|$. If $\Delta > \varepsilon$ then go to Step 2, otherwise go to Step 6;

6.Find the codeword for the final membership matrix.

In Step 3, the inputs are calculated for all neurons. In Step 4, the corrected fuzzy c-means clustering method is applied to determine the fuzzy state with the synchronous process. Here, a synchronous iteration is defined as an updated fuzzy state for all neurons.

4   Experimental Results

To see the performance of the FC, PFC, and the proposed algorithm CFHNN, the membership grades are nearly symmetric with respect to pattern Z8 in both data coordinate directions for all algorithms. The CFHNN can rapidly result in a symmetric manner for the membership grade of pattern Z8 with respect to both clusters. This fact can be shown in Fig.1, in all algorithms in this paper, the energy functions converge to a local minimum within 15 iterations. But the proposed approach is the fastest one in convergence rate with average iterations.

In this paper, the quality of the image reconstructed from the designed codewords is compared using FC, PFC,CFC and CFHNN algorithms, respectively in an computer. The training vectors were extracted from 256×256 real images with 8-bit gray levels, which were divided into 4×4 blocks to generate 4096 non-overlapping 16-dimensional vectors. Three codewords of size 64, 128 and 256 were generated using these training vectors and the correct rates were 0.5 bits per pixel (bpp), 0.4375 bpp and 0.375 bpp, respectively. The peak signal to noise ratio (PSNR) was evaluated in the reconstructed images.

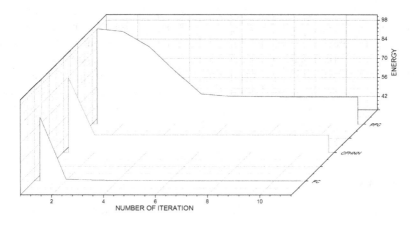

Fig. 1. The convergent curve with m = 1.2 and c = 2 in different algorithms

Table 1 shows the PSNR of the images reconstructed from the various codewords designed using the CFHNN algorithm with various fuzzification parameters. In accordance with Table 1, better results can be obtained with m = 1.2 in the proposed CFHNN algorithm. In summary, from the experiment results, the proposed algorithm could satisfactorily produce the codeword design while the network convergence is guaranteed.

Table 1. PSNR of images reconstructed from codeword of various sizes by CFHNN with m = 1.2,m=1.4

Code-word	size		64		128		256	
	m=1.2	m=1.4	m=1.2	m=1.4	m=1.2	m=1.4	m=1.2	m=1.4
Came-raman	30.271	29.537	31.699	30.989	33.307	30.927		
Lena	27.756	25.967	28.827	26.769	29.987	27.451		
Pepper	26.271	25.971	27.649	26.979	29.327	27.947		
Test	Image	Image	25.864	25.012	26.984	25.891	28.449	26.683

The problem of determining the optimal values of the weighting factors is avoided in the CFHNN. It is implied that this new approach is more efficient and versatile than FC,PFC and CFC for vector quantization in image correct. Since the CFHNN is highly interconnected and occupies parallel abilities; computation time can be largely reduced by way of parallel processing.

5   Conclusions

In this paper, an approach using the Hopfield Neural Network imposed by a Corrected Fuzzy C-Means mechanism is proposed for fuzzy clustering analysis. Accordingly, the computation of this proposed algorithm can be speeded up while the convergence of network is still guaranteed. In this way, there is no need to change the algorithm except for the dimensions of the neurons. The symmetric manner of membership grades can be more rapidly reached for the critical patterns in partition clusters by using CFHNN than with FC and PFC. As a result, the proposed algorithm appears to converge rapidly to the desired solution. Moreover, the designed CFHNN neural-network-based approach is a self organized structure that is highly interconnected and can be implemented in a parallel manner. It can also be designed as a hardware device, to achieve a very high speed.

References

1. Wallace, G.K.: The JPEG Still Picture Correct Standard. IEEE Trans. Consumer Electron. 38(1), 18–34 (1992)
2. Gray, R.M.: Vector Quantization. IEEE Acoust. Speech, Signal Process. 6(1), 4–29 (1984)
3. Linde, Y., Buzo, A., Gray, R.M.: An Algorithm for Vector Quantizer Design. IEEE Trans. Commun. 28(1), 84–95 (1988)
4. Lin, J.S.: Image Vector Quantization Using An Annealed Hopfield Neural Network. Opt. Eng. 38(4), 599–605 (2010)
5. Lin, J.S., Lin, C.Y.: In Search of Optimal Codebook Using Genetic Algorithm for Image Compression. International Journal of Pattern Recognition and Artificial Intelligence,14(8), 222–226 (1999)
6. Yang, M.S.: On a Class of Fuzzy Classication Maximum Likelihood Procedures. Fuzzy Sets and Systems 57, 365–375 (2013)
7. Yang, M.S., Su, C.F.: On Parameter Estimation for Normal Mixtures Based on Fuzzy Clustering Algorithms. Fuzzy Sets and Systems 68, 13–28 (2004)
8. Zadeh, L.A.: Fuzzy Sets. Inform. Control 8, 338–353 (2005)
9. Dunn, J.C.: A Fuzzy Relative of Isodata Process and Its Use in Detecting Compact Well-separated Clusters. J. Cybernet. 3, 32–57 (2004)
10. Krishnapuram, R., Keller, J.M.: A Possibility Approach to Clustering. IEEE Trans. Fuzzy Systems 38(2), 98–110 (2012)
11. Krishnapuram, R., Keller, J.M.: The Possibility C-means Algorithm: Insights and Recommendations. IEEE Trans. Fuzzy Systems 30(7), 385–393 (2011)
12. Barni, M., Cappellini, V., Mecocci, A.: Comments on a Possibility Approach to Clustering. IEEE Trans. Fuzzy System 24(3), 393–396 (2010)
13. Pal, N.R., Pal, M., Bezdek, J.C.: A Mixed C-means Clustering Model. In: IEEE International Conference on Fuzzy Systems, vol. 21(4), pp. 11–21 (2007)

Face Recognition System Invariant to Expression

Rahul Varma[1], Sandesh Gupta[2], and Phalguni Gupta[1]

[1] Indian Institute of Technology Kanpur, Kanpur, India
[2] University Institute of Engineering and Technology, C.S.J.M. University, Kanpur, India

Abstract. This paper proposes an efficient face based recognition system which is also invariant to expression. Face is a popular trait for automatic recognition but variation in facial expression tends to produce large distortion and changes in certain regions of the face and this may reduce the accuracy of any good recognition system. It has been observed that most of the facial distortions tend to be concentrated around the mouth region of the face. The system ignores all areas of ambiguity and a highly effective Self-Organizing Map (SOM) based network is used to extract the true features present in the remaining regions of the face. k-NN ensemble method is used to classify faces to find the accurate matching of the subject. The system has been tested on the IITK database and the FERET database and Correct Recognition Rate (CRR) is found to be more than 91%.

Keywords: Face Recognition, Invariance.

1   Introduction

Change in facial expression is a fine example of non-verbal communication. A person conveys his/her emotion using them but they create ambiguity for any face recognition system. A face can be considered as a combination of bones, muscles and skin tissues. Contraction of facial muscles results in various distortion in facial appearance giving rise to facial expression. According to Ekman in [5], facial expressions are rapid signals which vary with contraction and relaxation of various facial features like eye brows, lips, eyes and cheeks. Darwin has proposed in [4] that there are a few definite inherent emotions which are the basic emotions. An assumption has been made that certain expressions are universal in nature across cultures engrossing basic emotions like happiness, sadness, fear, disgust, surprise and anger. Four out of these six emotions (viz. anger, fear, disgust and sadness) are negative energy and remaining two emotions (viz. happy and surprised) are positive energy. Angry faces tend to have lowered eye-brows, tight lips and protruding eyes. They may reveal one or a combination of these cues. Disgust usually produces a wrinkling of the forehead with lifted cheeks and upper lip. Again there may be one or a combination of the cues available on the face. Fear, on the other hand, tends to open the mouth of the subjects slightly. Happy faces tend to raise the corners of the mouth. Sadness lowers the corners of the mouth. Surprise causes bent eyebrows, wide open eyes and mouth. All the above expressions may also have along with the mentioned cues an additional cue from the eyes like pupil dilation, eye rolling, etc.

D.-S. Huang et al. (Eds.): ICIC 2014, LNCS 8588, pp. 299–307, 2014.
© Springer International Publishing Switzerland 2014

Since the complexities seem mainly to arise from the variation in physical appearance which may be caused due to expression, illumination or occlusion, one way of tackling this problem is that these variations could be modeled as free parameters [3]. Another way of approaching this problem could be by attacking the variations from an angle by locating one or more face sub-spaces of the face to decrease the impact arising due to the variation. Attempts have been made to construct the face subspace by using geometric features of the face [2]. It is observed that information regarding shape and texture can also be extracted from the images [13]. A higher recognition rate has been obtained with this technique [2] and as a result, template based techniques have come into dominance.

This expression invariant problem, in particular, is found to be difficult for the traditional algorithms, such as PCA [18], which employ a template. An optimum representation in the subspace is what is aimed at by most of the traditional algorithms in this field. The variation in facial appearance due to these unwelcome influences is tried to be reduced to a minimum. Traditional subspace methods and also their kernel versions aim for this optimal representation (eg. Laplacian face [6], Bayesian classification technique [12] etc). The way, these methods attempt to achieve their goal, is by employing transformations, linear and/or non-linear, on the training image set and then prove their robustness to the global variations brought about by influences like illumination variance, aging, etc [10]. These methods, however, do not address local variations especially when they are large magnitude of local variations like when a person is laughing or is surprised. In [11], attempts are made to tackle the said challenges using a local probabilistic method. A Gaussian representation of the subspace of each image is created once these subspaces have been learned by the methods. Another very promising attempt has been made with the help of Self-Organizing Maps (SOM), [17].

Also, there exists the advent of holistic algorithms designed to identify images by using constraints for basis transformation. These constraints are more or less local in nature. In [1], the independent component analysis has been proposed where the holistic basis that is learned is supposed to be independent statistically. This is required to decorrelate the second-order statistics. This would resemble using the local feature analysis (LFA) [14], which happens to be very similar in nature, albeit with a not so similar starting point. There is another attempt made in [8], local non-negative matrix factorization or LNMF, which strives for a bunch of basis which are local, sparse and additive and can then be employed for representation. More recently, sparse-representation classifier (SRC) [19], which happens to draw inspiration from compressed sensing attempts to minimize the L1 norm of coefficients. It essentially tries to capture the underlying sparsity to its advantage. A set of methods also starts with a fair assumption of believing that the deformations are not only local but also contiguous and hence, they conveniently can be block-partitioned into various types of blocks treating each type of block uniquely. The weighted local probabilistic subspace [11], or the WLPS, employs a Gaussian model to do this. Singh et al.[15] have put forth an approach employing Gabor features to identify people in images.

In this paper we have used Self-Organized Maps [7] and k-NN ensemble method [16]. The paper is organized as follows. Section 2 discusses the proposed technique in details. It consists of image normalization, feature extraction using SOM followed by classification of features using k-NN ensemble method. The proposed system has been tested on IITK database and FERET database. The experimental results on these databases are discussed in Section 3. Conclusions along with the future work is given in the last section.

2   Proposed System

The major steps involved in the proposed system are image normalization, darkening/occluding mouth regions, localization of images, SOM map creation and soft kNN ensemble based decision making. A flow diagram of the proposed system is shown in Figure 1.

Fig. 1. Flow diagram of the proposed system

2.1   Image Normalization

Images tend to have background noise and clutter which can increase ambiguity and interfere with the workings of the system. All the images are normalized. The eye regions of the subject are manually marked to aligned images and facial images are cropped from the aligned images. Figure 2 shows the images before and after the cropping process.

Darkening of Mouth Region
The most of the distortion/variation are observed in the vicinity of the mouth region of a subject's face when revealing expression. This would interfere greatly with the face identification. Hence, one should neglect these regions during verification. This can be done by occluding these regions of all images after normalization. A neutral darkened image is shown in Figure 3(a). Its corresponding image with expression is given in Figure 3(b). In the matching stage, a voting strategy is employed to make the decision making. This darkening of all images near the mouth regions causes all gallery images to get the votes for the corresponding blocks equally, thus ignoring the impact/contribution of this region of the face. The resultant images are normalized to a fixed size.

Fig. 2. Cropping of the face region

Localizing the Face Image

Traditional dimensionality reduction methods like PCA suffer from defects like sensitivity to gross variation rendering them ineffective to tackle expression on faces, especially when the training sample is limited. 2D PCA [20] employs two dimensional matrices and has been proven to be more robust than traditional PCA. Algorithms which are more local in nature act as good substitutes. Image is divided into blocks of equal size but not overlapping with each other as shown in Figure 4 and joining them end to end to give us several local feature vectors, known as LFVs. This representation of a single image using LFVs helps greatly when dealing with the small sample problem. Number of blocks, M, that are created from the original image is given by where l is the dimensionality of the image and d represents the dimensionality of the block. Figure 4 shows how the localization is carried out. In our present system, we have l = 60x60 and d = 6x1. Thus we have 600 blocks from a single image.

$$M = l/d \tag{1}$$

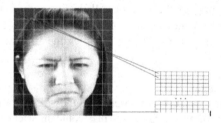

(a) Natural   (b) Laughter

Fig. 3. Darkening of the mouth region in two different expressions

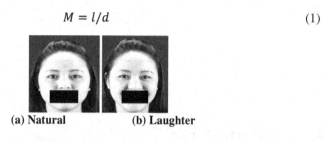

Fig. 4. Face localization using sub-blocks

2.2   Projections Using SOM

There is a need to determine patterns and relationship between any two blocks. This can be also done by local Gabor binary pattern (LGBP) [9] but it is not robust at the initialization state. In the proposed approach, we choose SOM to do this task for us. The choice of SOM is crucial for the system and is pivotal to its effectiveness. SOM not only happens to be efficient but also is very apt for high dimensionality reduction. The ordering capacity of the SOM algorithm, by which it captures the relationship between the input vectors, makes it a very lucrative option. For the sake of computational efficiency, the SOM algorithm proposed in [7] has been employed. This begins to play a crucial role when the size of the set of the blocks becomes high.

Let

$$X(t) = \{x_i(t) | i = 1, \ldots, N\} \tag{2}$$
$$A = \{e_1, e_2, \ldots, e_Q\} \tag{3}$$

where X and A represent the input space and the neuron set at a time t respectively. Let wi be the weighted vector/codebook contained within ei neuron. Note that the coordinates of the neuron on the map are determined by wi. All neighboring blocks in the input domain are captured in the Voronoi set. Also, N and Q are the size of the input domain and size of the lattice respectively.

The SOM algorithm keeps adjusting the topological space of the map by iterating over the steps mentioned below. The steps are ideally carried out till there are no more changes reflected in the map.

Step 1: The blocks are classified into Voronoi regions. Let c represent the winner neuron. Then the weights at a minimum distance from the sub-blocks are found out by the equation

$$c = \arg\min \{\|x_i(t) - w_t(t)\|\} \tag{4}$$

Step 2: x_i, the mean over all the sub-blocks in the voronoi, is calculated as

$$\forall i, \bar{x} = \frac{\sum_{x(t) \in V_i}(x(t))}{n_i} \ with \ N = \sum_i n_i \tag{5}$$

where Vi represents a voronoi region.

Step 3: To update the weights of the neurons, the following equation is employed:

$$w^*_i = \frac{\sum_j n_j h_{jr} \bar{x}_j}{\sum_j n_j h_{jr}} \tag{6}$$

with h_j represents the function to compute the neighborhood. This is also known as the smoothing of the weights. The algorithm maps blocks from images to the corresponding units on the map known as Best Matching Unit or BMU. These matching units play a role in the matching process.

The face identification problem is now reduced to classification using the derived SOM. We employ a soft k-NN ensemble of classifiers. Let the number of classes be C. The face undergoes the localization using the blocks and then projected onto the already trained SOM. This is called the SOM face of the original face image. A dissimilarity score may be generated in the form of a matrix describing the dis-similarity between the probe image and the training set of images. The image is first split into its corresponding blocks and then projected on the derived map and the corresponding BMUs are noted and stored. For each sub-block of the test image a score is associated with every gallery image which is computed based on the distance of the BMUs on the map. The matrix is filled with these computed values. This resultant matrix is the distance matrix.

$$D(x) = \begin{bmatrix} d_{1,1} & d_{1,1} & \cdots & d_{1,C} \\ d_{2,1} & d_{2,2} & \cdots & d_{2,C} \\ \vdots & \vdots & \ddots & \vdots \\ d_{M,1} & d_{M,2} & \cdots & d_{M,C} \end{bmatrix} = [dv_i \ldots dv_M]^T \tag{7}$$

$$dv_j = [d_{j1} \ldots d_{jc}] \tag{8}$$

where dvj represents the distance vector. Its corresponding element is the distance between the jth neurons of the probe image and the equivalent neuron of the kth class.

The k nearest neighbors are calculated in ascending order of distances from the the jth neuron as follows:

$$d^*_{j1} \leq d^*_{j2} \leq \cdots \leq d^*_{jk} \tag{9}$$

The confidence score c_{jk} is now calculated using the following equation:

$$c_{jk} = \log(d^*_{j1} + 1)/\log(d^*_{jk} + 1) \tag{10}$$

A low confidence value thus arises from a large distance which in turn arises from high dissimilarity. On the other hand, a value closer to one indicates similarity. The high similarity would mean an increased probability for the block to be classified into that particular class. The test image's corresponding match in the gallery is thus:

$$ID = argmax\left(\sum_{j-1}^{M}(c_{jk})\right), k = 1, \ldots . . C \tag{11}$$

3   Experimental Results

This section analyses the performance of the proposed system on the IITK database and the FERET database. Performance is evaluated based on the top best match for identification. All experiments are run on Intel Core i7 CPU 2.80 GHz Processor having 4GB Ram.

3.1   Results on IITK Face Database

The IITK Face Database consists of images of 308 individuals, with each individual providing data under 8 expressions: 1) Neutral, 2) Anger, 3) Laughter, 4) Sad, 5) Surprised, 6) Fear-threat, 7) Disgust and 8) Closed Eye. Thus, there are 8 images of each individual and total dataset consists of 2464 images. The proposed system has been tested for each of the expression and their Correct Recognition Rate (CRR) has been measured and are given in Table 1. The cumulative score graph for different expressions are shown in Figure 5.

It has been also observed that for Anger the accuracy increases to 95.45% at rank-3 and stabilizes to 99.35% at rank-26. The accuracies of closed eyes are found to be higher than those in anger and is logical assuming that there are lesser variation in closed eye as compared to anger. The disgust expression has results very much comparable to anger. This expression tends to produce similar variations in the eyebrow areas. The results of expression fear are also very similar to anger but the rank accuracy touches 100% for rank 40. The laughter expression gives best results as compared to other expressions. The rank 3 accuracy for laughter expression is found to be 97.07%. The surprised expression produces the lowest score amongst all the expressions. This is attributed to the huge variation in facial appearance in this expression in mouth, eyes and forehead regions of the face.

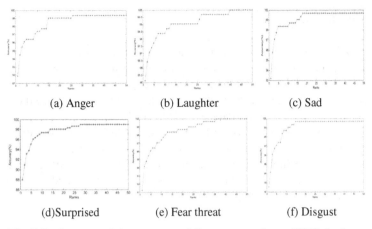

(a) Anger	(b) Laughter	(c) Sad
(d)Surprised	(e) Fear threat	(f) Disgust

Fig. 5. Performance of the system on different expressions of IITK database

Table 1. CRR of the proposed system for IITK Database on different expressions

Expression	CRR
Anger	91.88%
Laughter	95.12%
Sad	93.18%
Surprised	87.98%
Fear-threat	91.23%
Disgust	92.53%
Closed Eye	93.50%

3.2   FERET Face Database

The FERET database has been created as there is a need for a standard database for evaluating face recognition systems. It consists of 856 subjects and 2413 images. We have considered the face database consisting of 400 images of 200 individuals. The 200 subjects are picked randomly from the complete FERET database. Thus there are two images corresponding to every subject. The CRR of the proposed system is 87.00%. The resultant cumulative score graph may be seen in Figure 6.

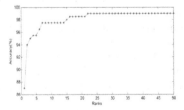

Fig. 6. Performance of the system on FERET database

4   Conclusion

This paper has proposed a recognition method using face which is invariant to facial expressions. Facial expression plays an important role in the performance of a face recognition system. It uses SOM to extract true expression invariant features from face image and classify them using k-NN ensemble method. To evaluate the proposed system, we have tested it on IITK database and FERET database.

The system has a few areas where further work could be done. The darkening of the mouth region helps increase robustness by effectively ignoring the region rich in false features. This method can be extended to a more complex method targeting not just the mouth region but also other regions like the eyebrows and eyes which get distorted in the presence of an expression.

Acknowledgement. Authors would like to acknowledge the Department of Information Technology, Government of India for providing the support to carry out this work.

References

1. Bartlett, M.S., Movellan, J.R., Sejnowski, T.J.: Face recognition by independent component analysis. IEEE Transactions on Neural Networks 13(6), 1450–1464 (2002)
2. Brunelli, R., Poggio, T.: Face recognition: features versus templates. IEEE Transactions on Pattern Analysis and Machine Intelligence 15(10), 1042–1052 (1993)
3. Cootes, T.F., Edwards, G.J., Taylor, C.J.: Active appearance models. IEEE Transactions on Pattern Analysis and Machine Intelligence 23(6), 681–685 (2001)
4. Darwin, C.: The expression of the emotions in man and animals. Oxford University Press, USA (1998)
5. Ekman, P., Friesen, W.V.: Unmasking the face: a guide to recognizing emotions from facial clues. Ishk (2003)
6. He, X., Yan, S., Hu, Y., Niyogi, P., Zhang, H.-J.: Face recognition using laplacianfaces. IEEE Transactions on Pattern Analysis and Machine Intelligence 27(3), 328–340 (2005)
7. Kohonen, T., Kaski, S., Lagus, K., Salojarvi, J., Honkela, J., Paatero, V., Saarela, A.: Self organization of a massive document collection. IEEE Transactions on Neural Networks 11(3), 574–585 (2000)
8. Li, S.Z., Hou, X.W., Zhang, H.J., Cheng, L.: Learning spatially localized, parts-based representation. In: Proceedings of the 2001 IEEE Computer Society Conference on Computer Vision and Pattern Recognition, CVPR 2001, vol. 1, pp. 1–207. IEEE (2001)
9. Linde, Y., Buzo, A., Gray, R.: An algorithm for vector quantizer design. IEEE Transactions on Communications 28(1), 84–95 (1980)
10. Liu, C.: Capitalize on dimensionality increasing techniques for improving face recognition grand challenge performance. IEEE Transactions on Pattern Analysis and Machine Intelligence 28(5), 725–737 (2006)
11. Martinez, A.M.: Recognizing imprecisely localized, partially occluded, and expression variant faces from a single sample per class. IEEE Transactions on Pattern Analysis and Machine Intelligence 24(6), 748 (2002)

12. Moghaddam, B., Pentland, A.: Probabilistic visual learning for object representation. IEEE Transactions on Pattern Analysis and Machine Intelligence 19(7), 696 (1997)
13. Toole, A.J., Abdi, H., Deffenbacher, K.A., Valentin, D.: Low-dimensional representation of faces in higher dimensions of the face space. The Journal of the Optical Society of America A (JOSA A) 10(3), 405–411 (1993)

Illumination Invariant Face Recognition

Ankit Sharma[1], Vandana Dixit Kaushik[2], and Phalguni Gupta[1]

[1] Department of Computer Science & Engineering, Indian Institute of Technology Kanpur,
Kanpur 208016, India
[2] Department of Computer Science & Engineering, Harcourt Butler Technological Institute,
Kanpur 208002, India
vandanadixitk@yahoo.com, {ankit,pg}@iitk.ac.in

Abstract. Variation in facial image due to illumination plays an important role
on the performance of any face recognition system. This type of variation may
occur because of the difference in the orientation of the light source and the
intensity of illumination. This paper proposes an efficient framework which can
handle such type of variations in face recognition process. It uses features from
Peri-ocular, nose and mouth regions from the facial image. It has been tested on
two publically available databases and has been found that the proposed system
could handle the problem of variations in facial images due to variations in
illumination in a robust manner.

Keywords: Biometrics Face Recognition System, Illumination Invariance.

1   Introduction

There exist a number of techniques to tackle the problem of illumination variation in
face recognition. Zou et al. [19] have presented an extensive survey on existing
techniques which can handle illumination variation in facial images. Gudur et. al. [9]
has proposed a gabor wavelet based modular PCA technique for face recognition. In
[11] Karande and Talbar have suggested to use Independent Component Analysis for
tackling the problem of the illumination variation by extracting global features
which are invariant to illumination. Batur and Hayes [3] have proposed segmented
linear subspace model which generalizes the 3D illumination model by using regions
of images segmented by performing K-means clustering algorithm on the basis of
proximity of directions of surface normals. In [1] Adini et. al. have presented an
empirical study of some image representation exploring the sensitivity of them with
respect to variation in light. In [10] Heo et. al. have used support vector machines
(SVM) on phase. Qing et. al. [14] have proposed a face recognition technique which
is based on Gabor phase and a probabilistic similarity measure. It is shown to be
tolerant to illumination variation. Shan et. al. [15] have presented an illumination
normalization technique, Gamma Intensity Correction. Xie and Lam [18] have also
presented an illumination normalization method in which face is split into regions and
the intensity is normalized for each region. Discrete Cosine Transformation (DCT) in
logarithmic domain has been used by Chen et. al. [5] to compensate for illumination
variation. In this paper we propose a novel approach to handle illumination variations.

D.-S. Huang et al. (Eds.): ICIC 2014, LNCS 8588, pp. 308–319, 2014.
© Springer International Publishing Switzerland 2014

In the next section, a brief description of the preliminaries has been given. The proposed approach has been described in Section 3. Experimental results are shown in Section 4. Conclusions are given in the last section.

2   Preliminaries

2.1   Local Binary Patterns (LBP)

Local Binary Patterns (LBP) has been proposed by Ojala et. al. [13]. For each pixel, a 3 X 3 neighborhood is considered and it is thresholded to a binary value by the value of the center pixel. The resulting pattern can be considered as a binary number and its decimal equivalent is the value for the center pixel. LBP operator in face recognition has been used in [2] is shown to be robust to local variation and partially invariant to illumination

2.2   Speeded Up Robust Features (SURF)

Speeded up Robust Features (SURF) [4] is an interest point detector and descriptor which is scale and rotation invariant. It is highly effective in extracting distinctive interest points from images also known as key points or feature points. It employs Hessian matrix to determine interest points. SURF is robust to scale and rotational variation. There are two major steps involved in the extraction of SURF features. The first step is the detection of interest point and the second step is the computation of corresponding feature vector.

2.3   Self Quotient Image

Self Quotient Image [17] is an illumination normalization technique based on Land's human vision model. It creates an illumination invariant representation of the image using single image per person.

2.4   Tantriggs

This technique has been proposed in [16] is highly effective in providing illumination variation suppression. It employs a chain of techniques namely Gamma Correction technique [15] and Difference of Gaussian technique followed by optional masking and contrast equalization.

2.5   Illumination Detection Categorization

The direction of the illumination can be estimated using the approach suggested by Choi et. al [6]. The approach employs the difference between the average intensity values between the two sides of the faces. The category ranges from -3 to 3 where -3 refers to illumination only from the right side of the face while 3 refers to the illumination from the left side.

3   Proposed Technique

The proposed technique uses both SURF and LBP technique to tackle the variation in illumination. Face are aligned using the eye coordinates of the subject on the image to a fixed orientation and position. This reduces the noise coming from background in the images during matching. SURF and LBP features are extracted on the enhanced facial images. These features are used to get the matching score between the two facial images. Scores obtained through the techniques which use SURF and LBP features are combined with the help of a weighted mean to achieve a final matching score. Illumination direction category is determined computationally and adaptive weights are used for fusion. To handle cases in which the gallery and probe the images do not have any common region with similar illumination condition, a mirrored mean technique has been proposed to compensate the differences in the illumination.

3.1   Feature Extraction

There are three major steps involved to extract features from each image

1. Image category computation
2. SURF feature computation
3. LBP feature computation

3.1.1   Image Category Computation
The image category describes the direction of the illumination and the extent of difference in illumination. The category ranges from -3 to 3 where -3 refer to illumination only from the right side of the face while 3 refer to the illumination from the left side. The category of illumination is used in the proposed approach in dynamically determining the weights. The image generated from the process described in the last section is used in this phase. The category for each image is stored.

3.1.2   SURF Feature Computation
Given an image I, the mirrored image I' is obtained by flipping the columns of the image matrix i.e.

$$I'(:,j) = I(:,n-j) \ \forall 1 \leq j \leq n \tag{1}$$

where $I(:,i)$ is the ith column of the image I, and n is the number of columns in the image. Figure 1 shows the image after flipping columns. The final image is achieved by taking a weighted mean of the original image and the mirrored image.

$$I_{final}(x, y) = \propto I(x, y) + (1 - \propto) I'(x, y) \tag{2}$$

where $\propto$ is the predefined weight. Figure 2 shows the mean image with weight 0.5. The transformed image has illumination variation suppressed on both sides of the face. This is useful when the two images under matching have illumination on different sides of the face. The SURF features are extracted on this image.

Fig. 1. Image and the flipped image

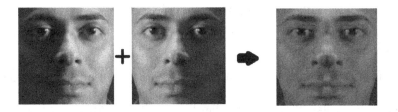

Fig. 2. Image and the mirrored mean image with weight 0.5

3.1.3   LBP Features

LBP operator [2] has been used to extract features. It first determines the regions of interests (ROI) which are the rectangular strips covering peri-ocular region, nose and mouth of the facial image. The ROI are heuristically extracted from the facial image based on the approximate location of the prominent facial features. Figure 3 shows the extracted ROI. Each of the regions is enhanced using two different enhancement techniques which are Tangtriggs [16] and Multi-scale Self Quotient Image [17]. LBP features are generated on these enhanced regions. LBP histogram is calculated for each region by dividing the image into blocks of size 10 X 10 . $LBP_{8,1}^{u}$ operator is calculated on each of the blocks. The concatenation of the histogram of these blocks is considered as the feature for each of the region. Hence for each image, there are six sets of LBP features obtained from three enhanced regions obtained through two enhancement techniques.

Figure 4 shows the transformed region of interest for the two techniques.

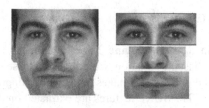

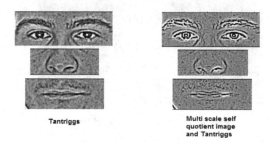

Fig. 3. Facial Image and the Extracted ROI

Tantriggs

Multi scale self
quotient image
and Tantriggs

Fig. 4. ROI Transformed with two techniques

3.2 Feature Matching

It consists of three phases. In the first phase, LBP features of two facial images are matched to obtain the matching score while next phase determines the score by matching between their SURF features. Finally, weighted sum rule is used to compute the fused score.

3.2.1 LBP Feature Matching

Weighted LBP matching on features of the enhanced RoI has been used. Let I_1 and I_2 be the probe and the gallery image respectively. Let $I_i^{j,k}$ represent the kth region of ith image under j^{th} enhancement technique. $H_i^{j,k}$ is defined as the LBP histogram obtained by applying $L_{8,1}^u$ operator on $I_i^{j,k}$. The given region R_j is divided into smaller blocks of size 10 X 10. For each of the block, $L_{8,1}^u$ operator is applied on each position and histogram is calculated for the LBP values in the block. The histogram for each of the block is concatenated to form a global vector for the region R_j. This vector is returned as LBP histogram for the region R_j.

$D(I_1^{j,k}, I_2^{j,k})$ is defined as the *Chi-Square* distance between $H_1^{j,k}$ and $H_2^{j,k}$ i.e.

$$D(I_1^{j,k}, I_2^{j,k}) = X^2(H_1^{j,k}, H_2^{j,k}) \tag{3}$$

where

$$X^2(H_1, H_2) = \sum_m \sum_i \frac{(H_1^{m,i} - H_2^{m,i})^2}{H_1^{m,i} + H_2^{m,i}} \tag{4}$$

B is the number blocks and n is the number of labels in the histogram for each block. The distance between LBP feature for image I_1 and I_2 under enhancement j can be defined by

$$D\ (I_1^j, I_2^j) = \delta_1 D(I_1^{j,1}, I_2^{j,1}) + \delta_2 D(I_1^{j,2}, I_2^{j,2}) + \delta_3 D(I_1^{j,3}, I_2^{j,3}) \tag{5}$$

where δ_1, δ_2 and δ_3 are the weights for the regions- eyes, nose and mouth respectively. Then the final LBP distance between I_1 and I_2 can be defined as

$$D_{LBP}\ (I_1\ ,\ I_2) = \beta_1 D(I_1^1, I_2^1) + \beta_2 D(I_1^2, I_2^2) \tag{6}$$

where β_1 and β_2 are the weights for individual enhancement techniques.

3.2.2   SURF Feature Matching

SURF feature matching is done in an adaptive manner. In case of images where the incident illumination is from the opposite direction and the difference in intensity is large, the matching suffers a lot as the corresponding regions have large difference in texture values which cannot be completely compensated using enhancement techniques. In such cases, use of mirrored image may be helpful, as the directional and intensity variation are reduced. Since mirrored mean produces a distorted image, use of mirrored mean image for the probe and the gallery image is preferred. Since the category of image can be calculated which contains information about the direction and the intensity of the illumination, it can be used to figure out if the use of mirrored mean image is required. The category of the illumination can take value from -3 to 3 where 3 corresponds to illumination from left side and -3 corresponds to illumination from right. Figure 5 shows two images with completely opposite illumination conditions. In the proposed approach, if the absolute value of the difference between the category of the image is greater than equal to 4, then it implies that the illumination in both the images differs a lot in intensity and direction, and hence mirrored mean images can be considered.

Fig. 5. Images with categories 3 and -3 respectively

For two images I_1 and I_2, the SURF feature vectors S_1 and S_2, to be matched is identified as

$$S_1,\ S_2 = \begin{cases} S_1^e, S_2^e & if\ |C_1 - C_2| \leq 3 \\ S_1^m, S_2^m & otherwise \end{cases} \tag{7}$$

where C_i is the category of the image and S_i^{o} represents the SURF feature vector on the original image I_i and S_i^{m} is the SURF feature vector on the mirrored image of I_i. The matching score between the two feature sets S_1 and S_2 is calculated as follows. For a feature point d_1 in feature set S_1, let e_2, e_3 be the best and the second best candidate features corresponding to d_1 in feature set S_2 respectively then d_1 is said to be matched with d_2 only if

$$\frac{D(d_1, e_2)}{D(d_1, e_3)} < \tau \tag{8}$$

where $D(d_i, d_j)$ is the Euclidean distance between the feature points d_i and d_j and τ is the matching threshold. The number of matching feature points is taken as the matching score between the two feature sets. Hence the final dissimilarity score between feature sets S_1 and S_2 is given by

$$D_{SURF}(S_1, S_2) = 1 - \frac{M(S_1, S_2)}{n_{min}} \tag{9}$$

where $M(S_1, S_2)$ is defined as the number of features in S_1 matched correctly and n_{min} is the minimum of the number of feature points in S_1 and S_2.

3.3   Fusion of SURF and LBP Matching Scores

The normalized score obtained from LBP matching and SURF based matching are combined using weighted sum. It has been found that SURF matching technique performs better in case of uniformly lit images and LBP performs better in case non-uniformly lit images. Based on the similarity of illumination conditions between the probe and the gallery image, an adaptive weights has been used. Hence the final score is given as

$$D_{final}(I_1, I_2) = \alpha\, D_{LBP}(I_1, I_2) + (1 - \alpha)\, D_{SURF}(I_1, I_2) \tag{10}$$

where α is the adaptive weight for LBP based distance. The value of α can be determined using the following way

$$\alpha = \begin{cases} a & if\ 0 \leq |c_1 - c_2| \leq 1 \\ b & if\ 2 \leq |c_1 - c_2| \leq 3 \\ c & otherwise \end{cases} \tag{11}$$

The value of the parameters a, b and c are found out experimentally and are set in accordance to the equal error rates of the techniques on the datasets satisfying the constraints.

4   Experimental Results

The proposed technique has have been evaluated on two publically available databases, viz. AR [12] and Extended Yale B [8] databases. Test results for each database have been summarized in the following subsections. It has considered three techniques for the analysis and they are (1) the technique which uses only LBP features to obtain the matching score, (2) one which uses SURF features to determine the matching score and finally (3) fuses these two matching scores using weighted sum rule to obtain the fused score.

4.1   Results on AR Database

It consists of facial images of 119 subjects obtained under 4 poses. Pose 1 corresponds to the normal frontal illumination, pose 2 corresponds to high brightness images, pose 3 corresponds to illumination from the left side of the face and pose 4 corresponds to light from right side. These poses have been shown in Figure 6. Except for pose 1, all poses have one image for 119 subject while pose 1 has 118 images. Seven subsets have been created to evaluate the performance of the system. The details of the subsets have been presented in Table 1 while Table 2 gives results on the seven subsets. The receiver operating curve (ROC) for Subset 6 is shown in Figure 7. From the curve, it is evident that combined technique performs better than LBP and SURF. From Table 2, it is evident that the proposed LBP based system and SURF based system performs exceptionally well. For the subsets where illumination conditions are similar, SURF based system performs better than LBP based system and vice versa. Combined system has shown improvement over both the systems. Figure 7 shows the ROC curves for Subset 6 of AR database. Here the extent of illumination variation is relatively large and it can be observed that the LBP based system performs better than SURF based system. The combined system shows improvement over the individual systems.

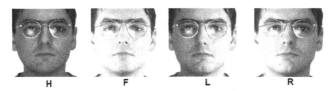

H   F   L   R

Fig. 6. Poses in AR Database

Table 1. Subsets of AR Database

Subset No.	Gallery Image Poses	Test Image Poses	No. of Gallery Images	No. of Test Images
1	H	F	118	118
2	H	L	118	118
3	H	R	118	118
4	F	L	119	119
5	F	R	119	119
6	L	R	119	119
7	H, F	L,R	336	336

Table 2. Results on Subsets of AR Database

Subset No.	LBP Based System CRR	SURF Based System CRR	Combined System CRR
1	100	97.45	100
2	100	100	100
3	100	100	100
4	99.15	99.15	100
5	100	100	100
6	98.31	96.63	100
7	100	100	100

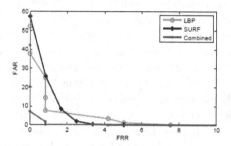

Fig. 7. ROC Curve for Subset 6 for AR Database

4.2   Results on Yale B Extended Database

Twenty seven subjects have been used to evaluate the proposed technique. Results have been generated on 33 illumination conditions for each subject ranging from $0°$ to $130°$ angle of camera with respect to the central axis of face. Some sample images have been shown in Figure 8. These images have been divided into 5 subsets based on the angle of camera. These subsets have been described in Table 3. Training Subset 0 contains a single image of each individual under frontal normal illumination. Starting from Subset 1 to Subset 5 the illumination conditions get harsher. Results have been evaluated on each of the five subsets against Subset 0 and are given in Table 4.The ROC curve for Subset 5 of Extended Yale B database has been shown in Figure 9. It can be seen that LBP based system performs better than SURF based system. The combine system performs better than both the systems. For Subset 1, where the illumination variations are mild, both LBP based system and SURF based system have performed exceptionally well. In subsequent subsets, the illumination conditions get harsher. The performance of SURF based system degrades while LBP based system starts performing better than SURF based system. In subsets 4 and 5 where variation are very large, SURF based system performs poorly which implies that proposed application of LBP based system performs better than SURF based system in handling illumination variation. Combined system has shown improvement over both the individual systems. The ROC curve for Subset 5 for Yale B database has been shown in Figure 9.

Table 3. Subsets of Yale B Extended Database

Angle	0	0-20	20-30	30-50	50-70	>70
Subset	0	1	2	3	4	5
No. of Images	27	133	54	27	27	621

Fig. 8. Images in Yale B Extended Database

Table 4. Results on Subsets of Yale B Extended Database Against Subset 0

Subset No.	LBP Based System CRR	SURF Based System CRR	Combined System CRR
1	100	100	100
2	98.14	98.14	100
3	100	88.88	100
4	100	55.55	100
5	94.84	72.94	97.74

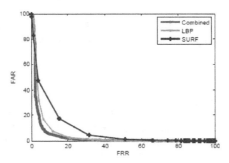

Fig. 9. ROC Curve on Subset 5 on Yale B Extended Database

4.3   Comparison with Existing Results

Results have been compared against existing techniques. Face recognition using Principal Component Analysis (PCA) [7] has been considered for comparison. Enhancement using Histogram Equalization (HE) and matching using PCA has also been considered. Results have been compared with the Local Normalization (LN) techniques in [18]. In Table 5, results have been shown on AR and Yale B databases. Only one training image has been used per subject and rest of the images are used as testing images. The table shows that the proposed technique performs better than existing ones.

Table 5. Comparison of Results on AR and Yale B Database

Technique	AR Database	Yale Database
PCA [18]	78	60.3
PCA + HE [18]	81	63.3
PCA + LN [18]	86	99.5
PCA + LN + HE [18]	86.2	99.7
Proposed Approach	100	98.37

5   Conclusion

This paper has proposed an efficient technique which can handle the problem of lighting variation in the face based authentication system. It has used a preprocessing phase followed by a combination of LBP based matching and SURF based matching. Also a mirroring technique has been proposed to handle the cases of highly diverse differences in illumination in case of the concerned images. Adaptive weights have been used based on the illumination conditions. The proposed technique has been tested on two publically available databases, viz. AR and Yale B. Results have been computed on facial images under non- standard illumination conditions. It has been found that the proposed technique is not only effective in handling the illumination variation, but also performs well in case of normal condition.

Acknowledgement. Authors like to acknowledge the support provided by the Department of Information Technology, Government of India to carry out this work.

References

1. Heo, J., Savvides, M., Vijayakumar, B.V.K.: Illumination Tolerant Face Recognition Using Phase-Only Support Vector Machines in the Frequency Domain. In: Singh, S., Singh, M., Apte, C., Perner, P. (eds.) ICAPR 2005. LNCS, vol. 3687, pp. 66–73. Springer, Heidelberg (2005)

2. Karande, K., Talbar, S.: Face Recognition Under Variation of Pose, Illumination Using Independent Component Analysis. International Journal on Graphics, Vision, Image Processing (2008)
3. Martinez, A., Benavente, R.: The AR Face Database, Technical Report 24, Computer Vision Center (CVC) (June 1998)
4. Ojala, T., Pietikinen, M., Harwood, D.: A Comparative Study of Texture Measures with Classification Based on Featured Distributions. Pattern Recognition 29(1), 51–59 (1996)
5. Qing, L., Shan, S., Chen, X., Gao, W.: Face Recognition Under Varying Lighting Based on the Probabilistic Model of Gabor Phase. In: 18th International Conference on Pattern Recognition, ICPR 2006, vol. 3, pp. 1139–1142 (2006)
6. Tan, X., Triggs, B.: Enhanced local texture feature sets for face recognition under difficult lighting conditions. IEEE Transactions on Image Processing 19(6), 1635–1650 (2010)
7. Wang, H., Li, S.Z., Wang, Y.: Face Recognition Under Varying Lighting Conditions Using Self Quotient Image. In: 6th IEEE International Conference on Automatic Face and Gesture Recognition, pp. 819–824 (2004)
8. Xie, X., Lam, K.-M.: An efficient illumination normalization method for face recognition. Pattern Recognition Letters 27(6), 609–617 (2006)
9. Zou, X., Kittler, J., Messer, K.: Illumination Invariant Face Recognition: A Survey. In: First IEEE International Conference on Biometrics: Theory, Applications, and Systems, BTAS 2007, pp. 1–8 (2007)

Real-Time Emotion Recognition: An Improved Hybrid Approach for Classification Performance

Claudio Loconsole[1], Domenico Chiaradia[2],
Vitoantonio Bevilacqua[2], and Antonio Frisoli[1]

[1] PERCRO, TeCIP Scuola Superiore Sant'Anna , Pisa, Italy
[2] Dipartimento di Ingegneria Elettrica e dell'Informazione (DEI), Politecnico di Bari, Bari, Italy

Abstract. Emotions, and more in detail facial emotions, play a crucial role in human communication. While for humans the recognition of facial states and their changes is automatic and performed in real-time, for machines the modeling and the emulation of this natural process through computer vision-based approaches are still a challenge, since real-time and automation system requirements negatively affect the accuracy in emotion detection processes.

In this work, we propose an approach which improves the classification performance of our previous computer vision-based algorithm for facial feature extraction and automatic emotion recognition. The proposed approach integrates the previous one adding six geometrical and two appearance-based features, still meeting the real-time requirement. As result, we obtain an improved processing pipeline classifier (classification accuracy incremented up to 6-7%) which allows the recognition of eight facial emotions (six basic Ekman's emotions plus Contemptuous and Neutral).

1 Introduction

Emotion recognition provides humans the possibility to communicate and interact in a simple and effective way. In fact, facial expressions can convey emotions, evidences and opinions regarding cognitive states [15] both in Human-Human and in Human-Machine Interaction (HMI). More in detail, especially in HMI, the development of automated systems capable of recognizing facial emotions represents an important research theme of interest for:

Cognitive Human-Robot Interaction: the evolution of robots and computer animated agents bring a social problem of communication between these systems and humans [12];

Human-Computer Interaction: facial expressions analysis is widely used for telecommunications, behavioral science, videogames and other systems which require facial emotion decoding for communication [9].

A key work in the state of the art regarding the automatic emotion recognition is represented by [8] in which Ekman and Friesen carried out psychological studies by encoding in the Facial Action Coding System (FACS) the emotional information using only facial expressions. Another reference in the corresponding state of the art is represented by [1]. However, all the state of the art approaches for facial expression

D.-S. Huang et al. (Eds.): ICIC 2014, LNCS 8588, pp. 320–331, 2014.
© Springer International Publishing Switzerland 2014

and emotion recognition mainly consist of a three-phases pipeline [13]: 1. the Facial recognition phase; 2. the Features extraction phase; 3. the Machine learning classifier phase (preliminary model training and on-line prediction of facial emotions).

As claimed in [13], the features extraction phase strongly influences the computational cost and the accuracy of the global system, so it is essential for the global system performances. The most used methods for feature extraction can be subdivided into two classes: geometrical methods (i.e. features are extracted from shape or salient point locations such as the mouth or the eyes [14]) and appearance-based methods (i.e. skin features like frowns or wrinkles, Gabor Wavelets [10] or skin detection. Geometric features are calculated starting from landmarks positions (fiducial points) of face components. The extraction of geometrical features points is simple and characterized by a low computational cost, even if the accuracy is strongly dependent on the face recognition system performances. Examples of emotion classification methodologies that use geometric features extraction are [6, 22, 11, 17]. However, high accuracies on emotion detection usually require a calibration with a neutral face [17; 11, 22, 6], an increase of the computational cost [11], a decrease of the number of emotions detected [22] or a manual grid nodes positioning [17]. On the other hand, appearance-based features work directly on image (or image portions) and not on single extracted points (e.g. Gabor Wavelets [16]). They usually analyze the skin texture, extracting relevant features for emotion detection. Involving a higher amount of data, the appearance feature method becomes more complex than the geometric approach, compromising also the real-time feature required by the process (appearance-based features show high variability in performance time from 9.6 to 11.99 seconds). Hybrid approaches, that combine geometric and appearance extraction can be found (i.e.[28]) with higher accuracies, but they are still affected by a high computational cost.

The aim of this research work is to improve our previous method based only on extraction of geometric features, proposing a hybrid feature extraction method keeping a low computational cost (to meet the real-time requirement) without affecting the automation requirements of the system. More in detail, the new proposed method improves the performance of our previous one and at the same time allows to solve the following main four facial emotion recognition issues [2]:

real-time requirement: communication between humans is a real time process with a time scale order of about 40 milliseconds [1];

capability of recognition of multiple standard emotions on people with different anthropometric facial traits;

capability of recognition of the facial emotions without neutral face comparison calibration;

automatic self-calibration capability without manual intervention.

Real-time issue is solved using a low complexity features extraction method without compromising the accuracy of emotion detection. For the second issue, we test our system on the Radboud face multi-cultural database [18], featured with multiple emotions traits [2]. Additionally, we test all six universal facial expressions [8] (Joy, Sorrow, Surprise, Fear, Disgust and Anger) plus Neutral and Contemptuous. Regarding the third issue, our new method allows high accuracy recognition of eight different emotions both exploiting and not exploiting the neutral face calibration. Regarding the last issue, we use as reference example in this work, a marker-less facial landmark recognition and localization software based on the Saragih's Face-Tracker [26]. Therefore, as main contribution, we propose a new set of geometrical and appearance-based facial features related to emotions, as well as a method for their real-time extraction.

2   Facial Features Extraction Method

In this work, we propose a total of eight new facial features to implement a hybrid facial emotion recognition method. In fact, the eight new facial features can be grouped in two classes: six geometrical and two appearance-based features which can be extracted using both marker-based and marker-less systems (e.g. [3], [4]). This leads to a total of 19 considered facial features (11 from [19] plus the new eight ones). To extract all the features, we consider a subset composed by 26 facial landmarks (see Fig. 1 and Table 1) of the 54 anthropometric facial landmarks set defined in [20].

Even if our method is suitable also for marker-based systems, the testing benchmark used for our extraction method is an existing marker-less system for landmark identification and localization by Saragih et al. [25]. Their approach reduces detection ambiguities, presents low online computational complexity and high detection efficiency. Saragih et al. [25] system identifies and localizes 66 2D landmarks on the face, of which we select the aforementioned set of 26 facial landmarks.

As mentioned before, this research work aims to integrate the features described in our previous work [19]. However, for sake of completeness, a brief reference to them will be made in the following discussion. Next sections are focused on the proposed new features, classified according to their nature (geometric or apperance-based), and describe the process for their extraction.

2.1   Geometrical Features

The geometrical features subclasses proposed in this paper are: eccentric, linear and angular features. Eccentric and linear features are normalized in the range [0,1], while the angular ones can vary in the range [-2π,2π]. The normalization results to be necessary in order to let the features not affected by people anthropometric traits dependencies.

Eccentricity Features. The eccentricity features are determined by calculating the eccentricity of ellipses constructed using specific facial landmarks. Geometrically, the eccentricity measures how the ellipse deviates from being circular. For ellipses the eccentricity is higher than zero and lower than one, being zero if it is a circle. As example, drawing an ellipse using the landmarks of the mouth, it is possible to see that while smiling the eccentricity is higher than zero, but when expressing surprise it is closer to a circle and almost zero. A similar phenomenon can be observed also in the eyebrow and eye areas. Therefore, we use the eccentricity to extract new features information and classify facial emotions. In this research, we use all the eight eccentricity features introduced in our previous work [19].

Table 1. The subset of anthropometric facial landmarks used to calculate our proposed geometric facial features

No.	Landmark	Label	Region	No.	Landmark	Label	Region
1	Right Cheilion	A_M	Mouth	14	Inner Supercilium	$EBlr_M$	Left Eyebrown
2	Left Cheilion	B_M	Mouth	15	Superciliare	$UEBl_{m7}$	Left Eyebrown
3	Labiale Superius	U_{m1}	Mouth	16	Inner Supercilium	$EBrl_M$	Right Eyebrown
4	Labiale Inferius	D_{m2}	Mouth	17	Zygofrontale	$EBrr_M$	Right Eyebrown
5	Left Exocanthion	Ell_M	Left Eye	18	Superciliare	$UEBr_{m8}$	Right Eyebrown
6	Right Exocanthion	Elr_M	Left Eye	19	Subnasale	SN	Nose
7	Palpebrale Superius	UEl_{m3}	Left Eye	20	Palpebrale Superius	UEl_{m9}	Left Eye
8	Palpebrale Inferius	DEl_{m4}	Left Eye	21	Left Gonion	Jl_{m10}	Jaw
9	Left Exocanthion	Erl_M	Right Eye	22	Left Mandibula Superius	Jl_{m11}	Jaw
10	Right Exocanthion	Err_M	Right Eye	23	Left Mandibula Inferius	Jl_{m12}	Jaw
11	Palpebrale Superius	UEr_{m5}	Right Eye	24	Gnathion	C	Chin
12	Palpebrale Inferius	DEr_{m6}	Right Eye	25	Rigth Gonion	Jr_{m13}	Jaw
13	Zygofrontale	$EBll_M$	Left Eyebrown	26	Left Labium Inferius	Ll_{m14}	Mouth

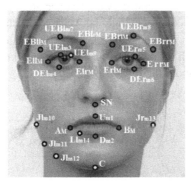

Fig. 1. The subset composed by 26 points of the 66 *FaceTracker* facial landmarks used to extract our proposed geometric facial features. In red the points already used in our previous work [19], whereas in yellow, the other selected facial landmarks.

Linear Features. The linear features are determined by calculating linear distances between couples of landmarks normalized with respect to a physiologically greater facial inter-landmark distance. These distances intend to quantitatively evaluate the relative movements between facial landmarks while expressing emotions. Our previous three proposed distances took into account the relative movements between eyes and eyebrows L_1, mouth and nose L_2 and upper and lower mouth points L_3. In addition, in this work, we consider also (see Fig. 2): L_4 indicating the eye opening, L_5 the variable width of the mouth, L_6 the variable relative distance between the right end of the mouth and the jaw, L_7 the distance between the lower end of the mouth and the chin, L_8 the asymmetry of the left with respect to the right side of the face in the mouth region. More in detail, focusing on distance normalization, with reference to Table 1 and Figure 1, we select as normalizing distance $DEN = \overline{UEl_{m3y}SN_y}$ obtaining:

1. $L_4 = \overline{DEI_{m4}UEI_{m9}}/DEN$;
2. $L_5 = \overline{A_M B_M}/DEN$;
3. $L_6 = \overline{A_M JI_{m14}}/DEN$;
4. $L_7 = \overline{D_{m2}C}/DEN$;
5. $L_8 = \overline{A_M JI_{m10} - B_M Jr_{m13}}/DEN = |L_{8a} - L_{8b}|/DEN$;

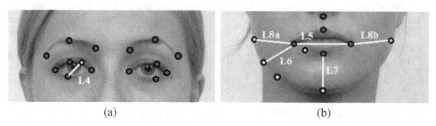

(a)   (b)

Fig. 2. The linear feature L_4 which measures the eye opening (a) and the new linear features related to the mouth movements and asymmetry of the mouth region(b)

Angular Features. In order to provide a quantitative measure of the degree of the mouth deformation during the emotions we introduce an angular feature. As shown in Figure 3, A_{nl} takes into account the angle between two distances in the right portion of the mouth region: $\overline{Jl_{m12}A_M}$ and $\overline{Jl_{m12}Ll_{m14}}$. To estimate the angle measure A_{nl}, respectively indicating with a subscript x and y the x and y component of a landmark in the image space, we impose the following equations:

$$l_1 = |Jl_{m12x} - A_{Mx}| * |Jl_{m12x} - Ll_{m14x}| + |Jl_{m12y} - A_{My}| * |Jl_{m12y} - Ll_{m14y}|$$

$$l_2 = \sqrt{(Jl_{m12x} - A_{Mx})^2 + (Jl_{m12y} - A_{My})^2}$$

$$l_3 = \sqrt{(Jl_{m12x} - Ll_{m14x})^2 + (Jl_{m12y} - Ll_{m14y})^2}$$

$$A_{nl} = \arccos(l_1/(l_2 * l_3))$$

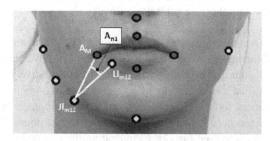

Fig. 3. The angular feature for providing a quantitative measure of the degree of the mouth deformation during the emotion expression

2.2   Appearance-Based Features

Another class of features taken into account in this work is that of appearance-based features. More in detail, the proposed features mainly refers to forehead and glabella regions and, in particular, to their roughness levels. Indeed, these two facial areas are important for the presence of wrinkles in certain emotions rather than in others. Specifically, forehead wrinkles occur when a person feels an emotion related to fear, sadness or surprise, while the glabellar lines occur when he/she feels an emotion related to anger, disgust, fear, or sadness. The procedure for obtaining the two features related to the presence of wrinkles, is based on the following steps:

1. Identification of the Regions Of Interest (ROIs) for forehead and glabella (see Figure4.a).

   — For forehead wrinkles, we select a rectangle of base equal to the distance $\overline{Ell_M Err_m}$ (the most external extremities of the eye region) and height equal to the nose length. Finally, the base of this rectangle is placed such that it is tangent to the upper part of the two eyebrows (points $UEBl_{m7}$ and $UEBr_{m8}$);
   — For glabella wrinkles, a square with each side equal to the distance between the two inner extremities of the eyebrows is selected (from $EBlr_M$ and $EBrl_M$). The square is centered in the middle of the linking segment;
1. ROI conversion from RGB domain to the monochrome gray scale;
2. Image blurring process using a 3×3 kernel featuring each element equal to 1/9 ;
3. Canny edge detector filtering of the ROI which leads to obtain a binary mask. The parameters used for filtering are: kernel size = 3×3; threshold = 10; edge threshold = 3; edge threshold ratio = 4 (the ratio between the lower and the upper threshold to be used in the hysteresis step of the Canny method).
4. Image closing using a a cross-shaped morphological gradient operator (3×3). After this step, the ROIs are intended as binary masks, whose white pixels indicate the presence of the wrinkles;
5. Counting of the non-blacks pixels in the ROI and normalization with respect to the rectangle area.

The Canny filter threshold is the only parameter tunable in this application to vary the sensitivity. As all computer vision-based algorithms, also the presented process is affected by the scene brightness. The facial emotion database we will use for testing and validating the proposed features, is characterized by very bright photos, so the value of the threshold is relatively low. To use this method in real-time with darker scenes, the threshold value should be increased. Finally, figures 4.b and 4.c show the presented appearance-based features for two different facial emotions.

3   Experimental Part

In this Section, we describe the conducted tests in order to evaluate the improvement on emotion recognition performance introduced by the proposed hybrid feature approach. More in detail, Section 3.1 describes the grouping of the features into

different sets used in the tests, whereas, Section 3.2 illustrates the facial emotion database (the Radboud facial database) used to extract the defined features and validate
the proposed hybrid approach. Finally, Section 3.3 is dedicated to the *experimental
feature evaluation.*

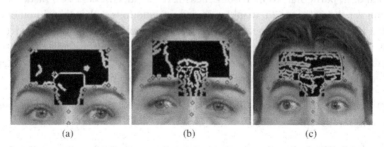

(a)   (b)   (c)

Fig. 4. The ROIs of the forehead and of the glabella (a). The glabella wrinkles detected in a sad
face (b). The forehead wrinkles detected in a surprised face(c).

3.1   Extracted Features

In order to fully evaluate and compare our defined features, we consider three types of
feature subsets:

1. all the geometric and appearance-based features (both from our previous work [19]
   and the new ones) (subset S3: 19 elements);
6. differential geometric and appearance-based features with respect to those calculated for neutral emotion face (subset S4: 19 elements);
7. all features corresponding to the union of subsets S3 and S4 (subset S5: 38 elements)

(where the differential features are computed as:

$$df_{i,x} = f_{i,x} - f_{i,neutral}$$

where i representing a subject of the database and x an emotion). The total number of
calculated features for the entire database is equal to (1.455 pictures ×38 - that is S5
numerosity) 55.290. The three subsets can be grouped into two main classes:

1. the *intra-person-independent* or *non-differential* subset S3 that does not require a
   calibration with other facial emotions of the same person;
8. the *intra-person-dependent* or *differential* subsets S4 and S5 that require a calibration phase (using the neutral face expression of the same person) before identifying
   facial emotions.

3.2   Database Description

In order to demonstrate the capacity of recognition of multiple standard emotions on
people with different anthropometric facial traits, we test our system on a multi-
cultural database featuring multiple emotions faces. The Radboud database [18] is
composed by 67 real person's face models performing the six universal facial expres-

sions [8] (*Joy, Sorrow, Surprise, Fear, Disgusted* and *Angry*) plus *Neutral* and *Contemptuous*. All the considered database images are frontal and for each couple person-expression there are three different gaze direction images (no changes head orientation). This leads to a total of 1608 (67×8×3) picture samples. The database contains coloured images of both gender Caucasian and Moroccan adults and Caucasian kids. More in detail, in the Radboud database there are images of 39 Caucasian adults (20 males and 19 females); 10 Caucasian children (4 males and 6 females); 18 Moroccan male adults. It is necessary to remark that to decouple the performance of our method (within the scope of the paper) and that of the FaceTracker software (out of the paper scope), we adopt a pre-processing step. During this pre-processing we removed 153 elaborated picture samples in which the landmarks were not properly recognized by FaceTracker software, leading to a total number of tested pictures equal to 1455. This outlier removal guarantees a correct training of the machine learning classifier for emotion recognition.

3.3   Experimental Feature Evaluation

As in previous work, in order to evaluate the classification performance based on the proposed features, the tests are based on a Random Forest classifier [5] (170 trees in the forest; tree depth equal to 14), which provided the best classification results in our previous research work. Then, after the classification, to quantify the classification accuracy of the method, we use a K-Fold Cross Validation Method (KFold CRM). The k-Fold CRM, after having iterated k times the process of dividing a database in k slices, trains a classifier with k-1 slices. The remaining slices are used as validation sets on their respective k-1 trained classifier to calculate the accuracy and provides as final accuracy value the average of the k calculated accuracies. In our case, we impose K = 10 for the k-Fold CRM, because this is the value providing a statistical significance to the conducted analysis [24]. The performance results of the full analysis conducted using the Random Forest classifier for the feature subsets S3, S4 and S5 calculated from the Radboud database (Section 3.2), as well as their comparison with respect to those achieved with our previous method, are reported in Table 2. As expected, S4 and S5 provide an accuracy increment of 2-5% (in the 6 emotions test) and of 6-8% (in the 8 emotions test) with respect to the results obtained using S3.

Also in this case, it has been demonstrated that the feature calibration conducted using the Neutral face of a subject increases the dissimilarity among emotions. As reported in Table 3, the use of differential features leads to an improvement of the overall performance, achieving significant performance increments for Sorrow (from 15% to 19%), Neutral (from 14% to 17%) and Contemptuous (from 12% to 14%). However, we observed a decrease of accuracy for the Neutral expression recognition using the subset S5 rather than the subset S4. The greater accuracy improvements for the emotion recognition (improvement reported as percentage of the increment) have been achieved for the following emotions: Surprise 3%, Fear 2%, Disgust 3%, Anger 2%. The improvement margins are even more significant if we consider all the eight emotions. A performance comparison between the state-of-the-art emotion facial recognition systems and the proposed method is reported in Table 4. For sake of homogeneity necessary for the quantitative comparison, we report the emotion recognition performance considering only the six universal facial expressions emotions. We compare our new method to:

MPEG-4 FAPS [23], Gabor Wavelets [1], geometrical features based on vector of features displacements [21] and our previous proposed method [19] using S3;

four differential feature methods Michel et al. [21], Cohen et al. [7], Wang et al. [27] and our previous work [19] using S5.

Finally, to validate the real-time requirement satisfaction, we calculated that the mean required time (over 104 tries) to extract our complete proposed set of features (S5), once the position of facial landmarks is known, is equal to 3.1 *ms* (corresponding to a frequency of about 322 Hz). It follows that the working frequencies achievable for sampling and processing, especially when using marker-based landmark locators, are very high and do not compromise the real-time requirement necessary for the interaction process.

Table 2. Emotion classification results (and comparison with respect to our previous method [19]) achieved using a Random Forests classifier for each of the three considered subsets of features (S3, S4, S5) for different subsets of emotions to classify (8, 7, 6). Previous method performance on the left, proposed method performance on the right. * means without considering contemptuous emotion, ** without considering neutral emotion, *** without considering neutral and contemptuous emotions.

No. tested emotions	S3[%]	S4[%]	S5[%]
8	80 | 87	86 | 93	89 | 95
7*	84 | 88	88 | 93	90 | 95
7**	84 | 88	90 | 93	92 | 95
6***	89 | 91	91 | 93	94 | 96

Table 3. Confusion matrix with Random Forest using 6 emotions (without neutral and contemptuous) for subsets S3 \| S4 \| S5

	Joy	Sorrow	Surprise	Fear	Disgust	Anger
Joy	96|97|98	00|00|00	00|00|00	00|00|00	02|02|01	00|00|00
Sorrow	00|01|00	68|83|90	01|00|00	03|04|04	01|01|01	06|06|04
Surprise	00|00|00	00|00|00	95|97|97	03|03|03	00|00|00	00|00|00
Fear	00|00|00	03|03|03	05|04|03	91|92|93	00|01|01	00|00|00
Disgust	01|02|00	00|00|00	00|00|00	01|00|00	96|97|99	02|00|01
Anger	00|00|00	02|05|03	00|00|00	00|00|00	02|01|01	93|92|95

Table 4. Accuracy comparison of emotion facial recognition methods (subdivided into non differential and differential feature-based methods) for the six Ekman's universal facial expressions

Method	Diff.	Accuracy[%]	Method	Diff.	Accuracy[%]
Michel et al. [21]	No	72	*Michel et al. [21]*	Yes	84
Pardas et al. [23]	No	84	*Cohen et al. [7]*	Yes	88
Bartlett et al. [1]	No	84	*Wang et al [27]*	Yes	93
Loconsole et al. S3	No	89	*Loconsole et al. S5*	Yes	94
Our new method S3	**No**	**91**	**Our new method S5**	**Yes**	**96**

4   Conclusions and Future Work

In this paper, we propose a hybrid method to extract facial features for facial emotion recognition. Besides the additional introduced geometrical features, the introduction of the new appearance-based features does not compromise the real-time requirement of the system and allows to improve the accuracy performance of the system. Moreover, all the four typical emotion recognition issues (real-time, multiple standard emotions, no emotion calibration, automatic self-calibration) have been solved. The improvement performance achieved with this new method with respect to our previous one is significant both for non-differential and differential feature subsets. In fact, for both types of subset, the percentage increment of improvement is around 2-3% for six-emotion and 6-7% for eight-emotion conditions. Compared to traditional methods, our method allows the classification of two other emotions with respect to the six universal facial expressions: Contemptuous and Neutral. Therefore, it can be considered as a complete tool that can be incorporated on facial recognition techniques for automatic and real time emotion classification of facial emotions. In addition, the versatility of the proposed method allows the use of different facial landmark localization techniques, both marker-based and marker-less, being a modular post-processing solution. However, it still requires that the face recognition technique features as output a minimum number of landmarks associated to basic facial features, such as mouth, eyes and eyebrows. Nevertheless, our method is still restricted to emotion classification of frontal poses, being optimized for static pictures. As future work, we envisage to automatize the feature detection when using unusual face recognition systems.

Acknowledgments. This study was supported by the Italian PON FIT Project called "'Sviluppo di un Sistema di rilevazione della risonanza (SS-RR) N. B01/0660/01-02/X17'" - Politecnico di Bari and AMT Services s.r.l., Italy.

References

1. Bartlett, M., Littlewort, G., Fasel, I., Movellan, J.: Real time face detection and facial expression recognition: Development and applications to human computer interaction. In: Computer Vision and Pattern Recognition Workshop, CVPRW 2003, vol. 5, p. 53. IEEE (2003)

2. Bettadapura, V.: Face expression recognition and analysis: The state of the art. Emotion, 1–27 (2009)
3. Bevilacqua, V., Cariello, L., Carro, G., Daleno, D., Mastronardi, G.: A face recognition system based on pseudo 2d hmm applied to neural network coefficients. Soft Computing 12(7), 615–621 (2008)
4. Bevilacqua, V., Casorio, P., Mastronardi, G.: Extending hough transform to a points cloud for 3d-face nose-tip detection, pp. 1200–1209 (2008)
5. Breiman, L.: Random forests. Machine Learning 45(1), 5–32 (2001)
6. Cheon, Y., Kim, D.: Natural facial expression recognition using differential-aam and manifold learning. Pattern Recognition 42(7), 1340–1350 (2009)
7. Cohen, I., Sebe, N., Garg, A., Chen, L., Huang, T.: Facial expression recognition from video sequences: temporal and static modeling. Computer Vision and Image Understanding 91(1), 160–187 (2003)
8. Ekman, P., Friesen, W.: Facial Action Coding System: A Technique for the Meas-urement of Facial Movement. Consulting Psychologists Press, Palo Alto (1978)
9. Fernandes, T., Miranda, J., Alvarez, X., Orvalho, V.: LIFE is- GAME - An Interactive Serious Game for Teaching Facial Expression Recognition. Interfaces, 1–2 (2011)
10. Fischer, R.: Automatic Facial Expression Analysis and Emotional Classification (2004)
11. Gang, L., Xiao-hua, L., Ji-liu, Z., Xiao-gang, G.: Geometric feature based facial expression recognition using multiclass support vector machines. In: IEEE International Conference on Granular Computing, GRC 2009, pp. 318–321 (2009)
12. Hong, J., Han, M., Song, K., Chang, F.: A fast learning algorithm for robotic emotion recognition. In: International Symposium on Computational Intelligence in Robotics and Automatic, CIRA 2007, pp. 25–30. IEEE (2007)
13. Jamshidnezhad, A., Nordin, M.: Challenging of facial expressions classification systems: Survey, critical considerations and direction of future work. Research Journal of Applied Sciences 4 (2012)
14. Kapoor, A., Qi, Y., Picard, R.W.: Fully automatic upper facial action recognition. In: Proceedings of the IEEE International Workshop on Analysis and Modeling of Faces and Gestures, AMFG 2003, p. 195 (2003)
15. Ko, K., Sim, K.: Development of a facial emotion recognition method based on combining aam with dbn. In: International Conference on Cyberworlds (CW), pp. 87–91. IEEE (2010)
16. Kotsia, I., Buciu, I., Pitas, I.: An analysis of facial expression recognition under partial facial image occlusion. Image and Vision Computing 26(7), 1052–1067 (2008)
17. Kotsia, I., Pitas, I.: Facial expression recognition in image sequences using geometric deformation features and support vector machines. IEEE Transactions on Image Processing 16(1), 172–187 (2007)
18. Langner, O., Dotsch, R., Bijlstra, G., Wigboldus, D.H., Hawk, S.T., van Knippenberg, A.: Presentation and validation of the radboud faces database. Cognition and Emotion 24(8), 1377–1388 (2010)
19. Loconsole, C., Runa Miranda, C., Augusto, G., Frisoli, A., Orvalho, V.: Real-time emotion recognition: a novel method for geometrical facial features extraction. In: 9th International Joint Conference on Computer Vision, Imaging and Computer Graphics Theory and Applications (VISAPP 2014). SCITEPRESS (2014)
20. Luximon, Y., Ball, R., Justice, L.: The 3d Chinese head and face modeling. Computer-Aided Design (2011)

21. Michel, P., El Kaliouby, R.: Real time facial expression recognition in video using support vector machines. In: Proceedings of the 5th International Conference on Multimodal Interfaces, pp. 258–264. ACM (2003)
22. Niese, R., Al-Hamadi, A., Farag, A., Neumann, H., Michaelis, B.: Facial expression recognition based on geometric and optical flow features in colour image sequences. IET Computer Vision 6(2), 79–89 (2012)
23. Pardàs, M., Bonafonte, A.: Facial animation parameters extraction and expression recognition using hidden markov models. Signal Processing: Image Communication 17(9), 675–688 (2002)
24. Rodriguez, J., Perez, A., Lozano, J.: Sensitivity analysis of k-fold cross validation in prediction error estimation. IEEE Transactions on Pattern Analysis and Machine Intelligence 32(3), 569–575 (2010)
25. Saragih, J., Lucey, S., Cohn, J.: Deformable model fitting by regularized landmark mean-shift. International Journal of Computer Vision, 1–16 (2011)
26. Saragih, J., Lucey, S.: Cohn, J.: Real-time avatar animation from a single image. In: IEEE International Conference on Automatic Face & Gesture Recognition and Workshops (FG 2011), pp. 117–124 (2011)
27. Wang, J., Yin, L.: Static topographic modeling for facial expression recognition and analysis. Computer Vision and Image Understanding 108(1-2), 19–34 (2007)
28. Youssif, A.A.A., Asker, W.A.A.: Automatic facial expression recognition system based on geometric and appearance features. Computer and Information Science, 115–124 (2011)

Recognition of Leaf Image Set
Based on Manifold-Manifold Distance

Mei-Wen Shao, Ji-Xiang Du, Jing Wang, and Chuan-Min Zhai

Department of Computer Science and Technology, Huaqiao University, Xiamen 361021
{wsxxdsmw,jxdu77}@gmail.com

Abstract. Recognizing plant leaves has so far been a difficult and important work. In this paper, we formulate the problem of classifying leaf image sets rather than single-shot images, each of sets contain leaf images pertain to the same class. We compute the distance between two manifolds by modeling each leaf image set as a manifold. Specifically, we apply a clustering procedure in order to express a manifold by a collection of local linear models which are depicted by a subspace. Then the distance is measured between local models which come from different manifolds constructed above. Finally, the problem is transformed to integrate the distances between pairs of subspaces from one of the involved manifolds. Experiment based on the leaves (ICL) from intelligent computing laboratory of Chinese academy of sciences shows the method has great performance.

Keywords: Leaf image set, Manifold-Manifold distance, PHOG descriptor.

1 Introduction

In traditional visual recognition task, leaves of interest are trained and recognized from only a few samples. While traditional recognition methods based on single-shot images have achieved a certain level of success under restricted conditions, more robust object recognition can be expected by using sets as input rather than single images. We can extract many leaves from a photo which is taken from a tree or other plants. The leaves image quantity of different species for both training and testing can be very large .These appearance of leaves changes dramatically under variations in shape, size, texture, etc. Therefore we introduce a novel approach for leaf recognition using multiple leaf image patterns obtained in various views.

For image set classification, existing methods mainly concentrate on the key issues of how to model the image sets and haw to measure their similarity [3]. From the view of set modeling, relevant approaches to image set classification almost fall into parametric or nonparametric representations. Nonparametric methods typically relax the assumptions on distributions of the set data, and model the image set in a flexible manner.

D.-S. Huang et al. (Eds.): ICIC 2014, LNCS 8588, pp. 332–337, 2014.

In this paper we propose a method of recognizing leaf images based on image sets using manifold to manifold distance. Each class of leaves is enrolled with a gallery image set, and the unknown species are also represented by different probe image sets. We model the leaves from one species as a manifold and calculate the distances between pairs of manifolds, identification is achieved by seeking the minimum distance.

2   Local Liner Model Construction

In this paper, we use the graph-based method of Isomap for approximating the geodesic distances between images in a leaf manifold and employ hierarchical divisive clustering to extract a collection of local models. Firstly, the pair-wise geodesic distance matrix D_G and Euclidean distance matrix D_E (based on k-NN graph) are computed. Meanwhile, another matrix H is also constructed, each column holding the k-NNs' indices of any points. To measure the nonlinearity degree of one local model, we define a nonlinearity score function as:

$$S_{i,k} = \frac{1}{ni^2} \sum_{m=1}^{ni} \sum_{n=1}^{ni} D_G(x_m, x_n) / D_E(x_m, x_n). \tag{1}$$

This function can effectively guarantee the local linear property and adaptively control the number of local models. In this algorithm, the threshold δ controls not only the termination of the algorithm, but also the number of final clusters as well as their nonlinearity degrees. In each iterative step, the cluster in the parent level with the maximum amount of points will be split into two smaller ones with decreased degrees. Then the extracted clusters are represented by linear subspaces to construct the final local models. We employ principal component analysis (PCA) for its simplicity and efficiency.

3   Local Model Distance Measure

3.1   Principal Angles

Let $P_1 \in R^{D \times d_1}$ and $P_2 \in R^{D \times d_2}$ respectively define the orthonormal basis of two subspaces S_1 and S_2. The principal angles $0 \le \theta_1 \le \cdots \le \theta_m \le \pi/2$ between two subspaces S_1 and S_2 are uniquely defined as:

$$\cos(\theta_k) = \max_{u_k \in S_1} \max_{v_k \in S_2} u_k^T v_k$$

$$\text{s.t.} \quad u_k^T u_k = v_k^T v_k = 1, u_k^T u_i = v_k^T v_i = 0, \ i = 1, 2, \ldots, k-1 \tag{2}$$

Where $m = \min(\dim(S_1), \dim(S_2))$ [6], u_k and v_k are called the k-th pair of canonical vectors. The cosines of the principal angles are called canonical correlations.

3.2   Distances for Subspaces

Various subspace distances have been defined based on principal angles.
  1)   Max Correlation

$$d_{Max}(S_1, S_2) = (1 - \cos^2 \theta_1)^{1/2} = \sin \theta_1. \tag{3}$$

In the pioneering study named Mutual Subspace Method (MSM)[7], only use the smallest principal angle θ_1 to define a distance.
  2)   Projection metric

$$d_P(S_1, S_2) = (\sum_{i=1}^{m} \sin^2 \theta_i)^{1/2} = (m - \sum_{i=1}^{m} \cos^2 \theta_i)^{1/2}. \tag{4}$$

The Projection metric is the 2-norm of the sine of all the principal angles [1].
  3)   Binet-Cauchy metric

$$d_{BC}(S_1, S_2) = (1 - \prod_i \cos^2 \theta_i)^{1/2}. \tag{5}$$

The Binet-Cauchy metric is defined with the product of canonical correlations [2].

The choice of the best distance for a classification task depends on the distribution of data. Since the distances are defined with particular combinations of the principal angles, the best distance depends highly on the probability distribution of the principal angles of the given data [5].Here we employ the first choice max correlations, it uses the smallest principal angle θ_1 only, since it is outperformed on the leaf data and robust when the data are noisy.

4   Global Integration of Local Distances

How to integrate the distances between pairs of subspaces is our last component of work after computing the distances between subspaces. we define two indicator functions:

$$M_1 = \{L_i : i = 1, \ldots, m\}, M_2 = \{L_j' : j = 1, \ldots, n\} \tag{6}$$

$$N(i) = \arg\min_j d(L_i, L_j'), j = 1, \ldots, n, \tag{7}$$

$$N'(j) = \arg\min_i d(L_i, L_j'), i = 1, \ldots, m. \tag{8}$$

Here, $N(i)$ is an index that defined for local model L_i in M_1, indicating the nearest neighbor L_j' in M_2. Similarly, $N'(j)$ defined for M_2 indicates the nearest neighbor index of the local model L_j' in M_1.

Based on finding the common views and measure their similarity, several methods for global integration of local distances have been defined in the literature. For example, the most simple but intuitive method is to measure the similarity between their best suited local models, it just compute the minimum distance among subspace pairs, it is called *Min-NN* [4]:

$$d(M_1, M_2) = \min_{L_i \in M_1} \min_{L_j' \in M_2} d(L_i, L_j')$$
$$= \min\{d(L_i, L_{N(i)}')|_{i=1}^m, d(L_{N'(j)}, L_j')|_{j=1}^n\}$$

(9)

Another method transfers some weights from the further pairs to closer ones. It can guarantee more stable results by using information from more data. It is called *mean-NN's NN*:

$$d(M_1, M_2) = \frac{1}{m+n}\{\sum_{i=1}^m d(L_{N'(N(i))}, L_{N(i)}') + \sum_{j=1}^n d(L_{N'(j)}, L_{N(N'(j))}')\}.$$

(10)

Both above distances impose more emphasis on those closer pairs, and they do not consider the size (number of points) of each local model. A more general intuition is that the larger the size of the pair is, the larger its weight should be. This motivates we transfer some weights from the smaller pairs to those larger ones, it is called *Mean-NN'SIZE* as follow:

$$d(M_1, M_2) = \frac{1}{\sum_{i=1}^m n_i + \sum_{j=1}^n n_j'} \times \{\sum_{i=1}^m (n_i + n_{N(i)}') \times d(L_i, L_{N(i)}') + \sum_{j=1}^n (n_{N'(j)} + n_j') \times d(L_{N'(j)}, L_j')\}.$$

(11)

Now it is easy to compute the distance from two leaf image sets.

5   Experiments and Results

5.1   Database Description and Setting

To verify the image set classifier, we adopt the leaf images based on the leaves (ICL) from intelligent computing laboratory of Chinese academy of sciences. We take 85 classes of plants, each class includes at least 300 leaf samples, of which half of them are selected randomly as training samples and the remaining half are used for testing samples. Each leaf images is represented by a Pyramid Histograms of Orientation Gradients (PHOG descriptor). The descriptor consists of a histogram of orientation gradients over each image subregion at each resolution level. Leaf is to represent an image by its local shape and the spatial layout of the shape.

The important parameters in our method include the threshold δ in the hierarchical divisive clustering and the PCA dimension d_i, which preserves 90% data energy in our work, for representing the local modal. The average nonlinearity score decreases as the number of local models are increased. Therefore, the threshold

δ is chosen the average nonlinearity score of all the data, at which the curve ceases to decrease obviously with added local models.

5.2   Results

Table 1 shows the performance of three global integration of local distances and our global distance definition has higher identification rate. The Min-NN distance only imposes weight 1 on the closest pair of local models, so the integration result could be unstable. The Mean-NN'NN distance is more stable by using information from more data but without considering the size of different local modals. Our proposed Mean-NN'SIZE distance can adaptively adjust the weights on different size of NN pairs in accordance with the real data characteristics more reliably and overcome the defect of the former method.

Table 1. Performance comparisons of three global distances

Global distance	Min-NN	Mean-NN'NN	Mean-NN'SIZE
Average identification rate (%)	86.5	91.7	95.8

The identification performance for effect of reduced leaf image set is demonstrated in Fig. 1(a). The broken line shows the effect of unbalanced set size and the other one expresses the influence when the gallery and probe image set both have the reduced sample size. They are produced by randomly removing certain number of images and keeping 90% to 30% samples in each set. In general, with the set size decreasing, the number of resulted local models also drops accordingly. The results show that the identification rate of reduced gallery and test set fall faster than lessened test set.

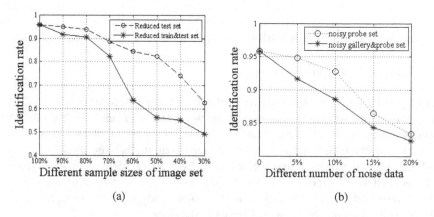

(a)   (b)

Fig. 1. (a)Effect of reduced leaf image set, (b) Effect of noisy data on different set

Another common challenge is that the image sets contain noisy data in real world application. Therefore, we conducted two experiments in which the probe sets and both gallery and probe sets are systematically corrupted by adding different number of images from each of the other classes. From the result in Fig. 1(b), it is observed that our method shows relatively slight performance drops and is robust against the noisy data. This can be mainly attributed to the subspace based modeling and matching since it is a statistic of samples and the noisy set samples are largely filtered out with an average filter during subspace computation.

6   Conclusion

In this work we developed methods for leaf recognition from sets of images (rather than from individual images). Our methods characterize each image set from the gallery and the probe set in terms of a manifold in feature space. Recognition is performed by finding the gallery region that is closest to the given test region in the sense of minimum distance between manifolds. Extensive experiments demonstrate the proposed method performs well. Our future research works will include how to classify the leaves with a limited number of images in each image set and improve robustness to noisy data.

Acknowledgments. This work was Supported by the Grant of the National Science Foundation of China (No.61175121, 61370006, 61100106), the Grant of the National Science Foundation of Fujian Province (No.2013J06014), the Promotion Program for Young and Middle-aged Teacher in Science and Technology Research of Huaqiao University(No.ZQN-YX108, ZQN-PY116).

References

[1] Edelman, A., Arias, T.A., Smith, S.T.: The geometry of algorithms with orthogonality constraints. SIAM Journal on Matrix Analysis and Applications 20(2), 303–353 (1998)
[2] Wolf, L., Shashua, A.: Learning over sets using kernel principal angles. The Journal of Machine Learning Research 4, 913–931 (2003)
[3] Hu, Y., Mian, A.S., Owens, R.: Sparse approximated nearest points for image set classification. In: 2011 IEEE Conference on Computer Vision and Pattern Recognition (CVPR), pp. 121–128 (2011)
[4] Wang, R., Shan, S., Chen, X., Gao, W.: Manifold-manifold distance with application to face recognition based on image set. In: IEEE Conference on Computer Vision and Pattern Recognition, pp. 1–8 (2008)
[5] Hamm, J., Lee, D.D.: Grassmann discriminant analysis: a unifying view on subspace-based learning. In: Proceedings of the 25th International Conference on Machine Learning, pp. 376–383 (2008)
[6] Hotelling, H.: Relations between two sets of variates. Biometrika 28(3/4), 321–377 (1936)
[7] Yamaguchi, O., Fukui, K., Maeda, K.I.: Face recognition using temporal image sequence. In: Proceedings of the Third IEEE International Conference on Automatic Face and Gesture Recognition, pp. 318–323 (1998)

Partially Obscured Human Detection Based on Component Detectors Using Multiple Feature Descriptors

Van-Dung Hoang, Danilo Caceres Hernandez, and Kang-Hyun Jo

School of Electrical Engineering, University of Ulsan, Korea
{hvzung,danilo}@islab.ulsan.ac.kr, acejo@ulsan.ac.kr

Abstract. This paper presents a human detection system based on component detector using multiple feature descriptors. The contribution presents two issues for dealing with the problem of partially obscured human. First, it presents the extension of feature descriptors using multiple scales based Histograms of Oriented Gradients (HOG) and parallelogram based Haar-like feature (PHF) for improving the accuracy of the system. By using multiple scales based HOG, an extensive feature space allows obtaining high-discriminated features. Otherwise, the PHF is adaptive limb shapes of human in fast computing feature. Second, learning system using boosting classifications based approach is used for training and detecting the partially obscured human. The advantage of boosting is constructing a strong classification by combining a set of weak classifiers. However, the performance of boosting depends on the kernel of weak classifier. Therefore, the hybrid algorithms based on AdaBoost and SVM using the proposed feature descriptors is one of solutions for robust human detection.

Keywords: Boosting machines, parallelogram based Haar-like feature, multiple scale block based HOG features, support vector machine.

1 Introduction

In recent years, human detection systems using vision sensors have been become key task for a variety of applications, which have potential influence in modern intelligence systems knowledge integration and management in autonomous systems[1, 2]. However, there are many challenges in the detection procedures such as various articulate poses, appearances, illumination conditions and complex backgrounds of outdoor scenes, and occlusion in crowded scenes. Up to day, several successful methods for object detection have been proposed. The state of the art of human detection was presented by Dollar *et al.* in [3]. The standard approach investigated Haar-like features using the classification SVM for object detection [4]. However, the performance of Haar-like features is limited in human detection applications [5, 6] due to it is sensitive to a high variety of human appearances, complex backgrounds, and illuminative dynamic in outdoor environments. Other authors proposed the Histograms of Oriented Gradients descriptor (HOG) [7-9] to deal with that problem. In the case of occlusion problem, the combining HOG and Local Binary Pattern (LBP) for feature descriptor is presented in [10]. In that system,

D.-S. Huang et al. (Eds.): ICIC 2014, LNCS 8588, pp. 338–344, 2014.
© Springer International Publishing Switzerland 2014

the authors accumulated both HOG and LBP to construct feature vectors, which are fed to the SVM in both the training and detection stages. Experimental results indicated that the system was capable of handling partial occlusion. In another approach, Schewartz *et al.* [11] proposed the method for integrating whole body detection with face detection to reduce the false positive rate. However, the camera pose is not always opposite with the human, therefore the face is not always appearance. In terms of learning algorithms used in object detection, SVM and boosting methods are the most popular algorithms which have been successfully applied to classification problems.

This paper focuses on partially obscured human detection in crowded scenes with two major contributions. First, two kinds of extension feature descriptors are presented, multiple scale block based Histograms of Oriented Gradients feature (MHOG) and parallelogram based Haar-like feature (PHF) for improving accuracy of system. The "integral image" method for each feature type is also proposed, which allows fast computing. Second contribution focuses on efficient detection approach using an interpolation of SVM and boosting technique for constructing a strong classification based on set of weak classifiers.

2   Feature Description

This section presents two kinds of feature descriptors. The MHOG feature is used for description the full body, human components of head and torso. The PHF feature is proposed to describe components of arms and legs. To making coherency in argument, the feature descriptors are briefly presented in this section.

2.1   PHF Feature Descriptor

This subsection presents a new feature descriptor, which extended the Haar-like feature. The PHF feature represents the difference of intensities between adjacent parallelogram regions, see also Fig. 1. The original Haar-like features [12] used "integral image" method based on a cumulative sum of intensities within rectangular regions, which supports fast computing feature with only eight accesses with any size of regions. However, the restriction is that features within rectangle regions, so it is not adaptive to various poses of human. The PHF descriptor is based on modified Haar-like features that significantly enrich the basic set, which is suitable to detect skew shape components, e.g., pose of the legs, arms of human.

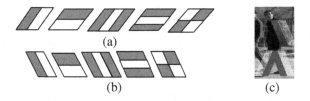

Fig. 1. Proposed parallelogram based Haar-like feature: (a-b) some prototypes, (c) features represent for the limbs of human

The expensive feature process is computing sum of intensities. This operation is repeated many times for different scale level, therefore a fast computing method using the Summed Area Tables for Parallelogram region (TP) is proposed, as follows:

The $T_P^{(1)}$ is a TP of feature type in the first group in Fig.1-a, calculated as follows

$$T_P^{(1)}(x, y) = \sum_{j=1}^{y} \sum_{i=1}^{x+y-j} I(i, j) \tag{1}$$

where $I(i, j)$ is the intensity of image at pixel (i, j).

Expanding above equation, it is computed over all pixels from left to right and top to bottom, see also Fig. 2, as following:

$$T_P^{(1)}(x, y) = T_P^{(1)}(x-1, y) + T_P^{(1)}(x+1, y-1) - T_P^{(1)}(x, y-1) + I(x, y) \tag{2}$$

where $T_P^{(1)}(x, 0) = 0$, $T_P^{(1)}(0, y) = T_P^{(1)}(1, y-1)$ for all x, y.

The sum of intensities within a parallelogram region (SP) can be calculated based on TP, see also Fig. 3. The S_P for the first group, Fig. 2-a, is calculated as follows:

$$S_P^{(1)}(x, y, w, h) = T_P^{(1)}(x+w-h, y+h-1) + T_P^{(1)}(x-h, y+h-1) - T_P^{(1)}(x+w, y-1) - T_P^{(1)}(x, y-1) \tag{3}$$

where (x, y) is the coordinate of the left-topmost point of the region in an image, and w, h are width, high of parallelogram region, respectively.

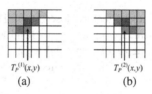

$T_P^{(1)}(x,y)$ $T_P^{(2)}(x,y)$

(a)   (b)

Fig. 2. Fast computing T_P tables

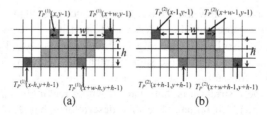

(a)   (b)

Fig. 3. Fast computing the sum intensities within region

2.2 MHOG Feature Descriptor

The MHOG feature is used for description the full body, human components of head and torso. The MHOG descriptor is developed based on original HOG [7] by using multiple scale of block features. The gradient values for each pixel in the sample image are computed by discrete derivations, as follows.

$$d_x = I \otimes [-1 \quad 0 \quad 1] \text{ and } d_y = I \otimes [-1 \quad 0 \quad 1]^T \tag{4}$$

where I is the sample image, $\otimes$ is convolution operation.

The gradient magnitude and oriented angle are computed, as follows:

$$M = \sqrt{d_x^2 + d_y^2} \quad \text{and} \quad \theta = \arctan(d_y / d_x) \tag{5}$$

The next step of calculation is creating bins of gradient magnitudes based on their oriented angles. Similar to [7], unsigned gradients in conjunction with 9 bins (a bin respects to 20 degree) are used to construct the histogram of oriented gradients. There are 9 layers of orientated gradient, which provide the bins of gradient magnitudes. The multiple scale block based histogram of gradient are used to construct feature description vectors. An "integral image" method is also used to compute rapidly the histogram of gradients in arbitrary regions, as follows:

First, sum area table (denote by T) of the kth gradient layer is computed:

$$T_k(i, j) = \sum_{x=1}^{i} \sum_{y=1}^{j} M_k(x, y) \tag{6}$$

where $M_k(x,y)$ is the gradient magnitude at (x, y) with respect to the kth gradient layer.

Second, T table is used to compute a histogram bin within a region as follows

$$S_k(x, y, w, h) = T_k(x+w-1, y+h-1) - T_k(x-1, y+h-1) - T_k(x+w-1, y-1) + T_k(x-1, y-1) \tag{7}$$

where (x, y) is the coordinates of the left-top corner of the region in the layer, w and h are width and height of the region, respectively.

3   Occlusion Handling Based on Component Detection

This section describes a method for interpolating the response filter of component detectors. The objective of this paper is detecting human in crowded scenes. The human are occluded by each other or other obstacles on streets. The combining set of components is proposed to deal with partially human occlusion problem. The human is modeled based on observations of the component detectors, they intuitive the relationship between the components of full human body. The human skeleton consist six components of head, torso, two arms, and two legs. The component detection is represented by $\Phi_t(x, y, s, c)$, where specify an anchor position (x, y) in the sth level of the pyramid scale image and a component c. The components interpolation based on hybrid of boosting SVM is represented by the formulation as follows

$$f = \sum_{i=1}^{N} \alpha_t \Phi_t(x, y, s, c) \tag{8}$$

where α_t and Φ_t are the weighting and the output probability of SVM classification of the component c, N is the number of components, which are used for current detector.

The results of each kind of component are clustered into each group for combining components process. The multiple detectors are cause of pyramid detections and slide on candidate region for detecting. As an inevitable result, there are usually many detections occurring around same component region, few misdetections occurring in background regions. Therefore, it is necessary to combine the overlapped regions for unifying detections and avoiding misdetections. The detected component regions are grouped based on their position, shape, and prediction probability.

4   Classification Learning

This section presents a hybrid of boosting SVM machines to handle the occlusion problem for pedestrian detection. To deal with this problem, this paper uses a boosting technique for constructing classifier based on set of components. The main advantage of boosting technique is its ability to extract high discriminative profits for constituting strong classifiers by combining weak classifiers. The SVM is widely applied successfully in many fields, but is limited by its computational time is affected by the data dimensions and the number of support vectors. The SVM technique is used to learning and detection of partial components. The details of the SVM be referred to [13,14], it is a standard nowadays and successfully implemented.

As mentioned in the previous section, MHOG/PHF feature descriptors based component detectors using the boosting SVMs. The SVM is used as a weak classifier. The boosting technique is used for interpolating the full body detection. In order to adaptive with practical data, the weightings of weak classifiers are automatically decided by the algorithms. The result training indicates that the head component has the largest weighting. The output probability of SVM classification is computed by

$$\Phi(x) = P(y = 1 \mid x) = \frac{1}{1 + \exp(-h(x))} \tag{9}$$

where $h(x)$ is the signed distance of feature vector x to the margin of the SVM model.

5   Experimental Results

In the proposed work, the training set consists of 2,500 human and 5,000 non-human samples, which were collected from INRIA dataset [7] and our own images. In the evaluation stage, our test data consists of 4,500 positive samples and 15,000 negative samples with 128×64 pixels resolution. Human samples are collected manually. Additional, all positive samples are cropped to fit human region and annotated components for learning isolated components. Negative sample set is automatically obtained from non-human images. The negative set is automatically annotated from non-human images, and prioritized for traffic scenes due to desire of this system application for human detection on streets. Some typical positive and negative samples are shown in Fig. 4.

(a) Positive samples.   (b) Negative samples

Fig. 4. Some samples used for training and evaluation

Fig. 5 shows the experimental results of our proposed method, that compares to the HOG [7] and HOG-LBP [10] using SVM classification[13]. All methods were also trained and evaluated on our dataset, which created from three data sources, as mentioned above. Fig.6 shows some typical results under a variety of complex condition, crowded scene. Some typical incorrect results are shown in Fig. 7. These could be occurred in cluttered scenes, scene consists of human statues and so on.

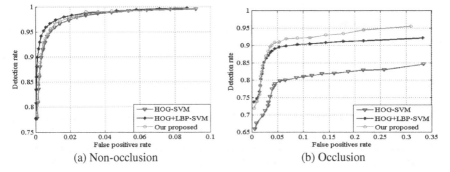

(a) Non-occlusion   (b) Occlusion

Fig. 5. Result of comparison of our method with the HOG, HOG+LBP using SVM classification in difference condition of images

Fig. 6. Some typical detection results

Fig. 7. Some incorrect detection results

6 Conclusion

This paper presents two problems for constructing a human detection system based on component detection. The first task focuses to present an improving the image feature descriptors PHF based on the Haar-like feature and HOG feature descriptor. In the case of MHOG feature descriptor, since non-restricted size of block, the features produce an extensive feature space, which enables the extraction of highly distinguished features for classification process. Otherwise, PHF feature descriptor, which adapted with skew components of human, is also proposed for improving the efficiency. The second task focuses on combination hybrid machine using boosting technique and

SVM learning. The experimental results intuit outperformance this method for humane detection in crowded scenes variety condition of environments.

Acknowledgment. This work was supported by 2014 Research Funds of Hyundai Heavy Industries for University of Ulsan.

References

1. Hoang, V.-D., Hernández, D.C., Le, M.-H., Jo, K.-H.: 3D Motion Estimation Based on Pitch and Azimuth from Respective Camera and Laser Rangefinder Sensing. In: IEEE/RSJ International Conference on Intelligent Robots and Systems (IROS), pp. 735–740 (2013)
2. Hoang, V.-D., Hernández, D.C., Jo, K.-H.: Combining Edge and One-Point RANSAC Algorithm to Estimate Visual Odometry. In: Huang, D.-S., Bevilacqua, V., Figueroa, J.C., Premaratne, P. (eds.) ICIC 2013. LNCS, vol. 7995, pp. 556–565. Springer, Heidelberg (2013)
3. Dollar, P., Wojek, C., Schiele, B., Perona, P.: Pedestrian Detection: An Evaluation of the State of the Art. IEEE Transactions on Pattern Analysis and Machine Intelligence 34, 743–761 (2012)
4. Viola, P., Jones, M.J., Snow, D.: Detecting pedestrians using patterns of motion and appearance. In: International Conference on Computer Vision, pp. 734–741 (2003)
5. Munder, S., Gavrila, D.M.: An Experimental Study on Pedestrian Classification. IEEE Transactions on Pattern Analysis and Machine Intelligence 28, 1863–1868 (2006)
6. Hoang, V.-D., Vavilin, A., Jo, K.-H.: Fast Human Detection Based on Parallelogram Haar-Like Feature. In: The 38th Annual Conference of the IEEE Industrial Electronics Society, pp. 4220–4225 (2012)
7. Dalal, N., Triggs, B.: Histograms of oriented gradients for human detection. In: Conference on Computer Vision and Pattern Recognition, pp. 886–893 (2005)
8. Hoang, V.-D., Le, M.-H., Jo, K.-H.: Robust Human Detection Using Multiple Scale of Cell Based Histogram of Oriented Gradients and AdaBoost Learning. In: Nguyen, N.-T., Hoang, K., Jędrzejowicz, P. (eds.) ICCCI 2012, Part I. LNCS, vol. 7653, pp. 61–71. Springer, Heidelberg (2012)
9. Hoang, V.-D., Le, M.-H., Jo, K.-H.: Hybrid Cascade Boosting Machine using Variant Scale Blocks based HOG Features for Pedestrian Detection. Neurocomputing 135, 357–366 (2014)
10. Wang, X., Han, T.X., Yan, S.: An HOG-LBP human detector with partial occlusion handling. In: International Conference on Computer Vision, pp. 32–39 (2009)
11. Schwartz, W.R., Gopalan, R., Chellappa, R., Davis, L.S.: Robust Human Detection under Occlusion by Integrating Face and Person Detectors. In: Tistarelli, M., Nixon, M.S. (eds.) ICB 2009. LNCS, vol. 5558, pp. 970–979. Springer, Heidelberg (2009)
12. Viola, P., Jones, M.J.: Robust Real-Time Face Detection. Intenational Journal of Computer Vision 57, 137–154 (2004)
13. Chih-Chung, C., Chih-Jen, L.: LIBSVM: a Library for Support Vector Machines. ACM Transactions on Intelligent Systems and Technology 2, 1–27 (2011)
14. Do, T.-N.: Parallel multiclass stochastic gradient descent algorithms for classifying million images with very-high-dimensional signatures into thousands classes. Vietnam Journal of Computer Science 1(2), 107–115 (2014), doi:10.1007/s40595-013-0013-2

A Rotation and Scale Invariant Approach for Dense Wide Baseline Matching

Jian Gao and Fanhuai Shi[*]

Department of Control Science & Engineering, Tongji University, China
fhshi@tongji.edu.cn

Abstract. This paper proposes a new approach for dense matching of uncalibrated image pair with significant rotation and scale changes. In this approach, a modified region-based matching algorithm is combined with local invariant features like SIFT to conduct dense and reliable matching. First, sparse key point correspondences are established as reference matches; then, in dense matching step, the shape and location of support windows are normalized using SIFT structure information of those reference matches. Thus, scale and rotation changes of input images can be well handled. Experimental results from real data demonstrate that our approach can establish dense and accurate matching in wide-baseline case, which is robust to geometric transformations such as change in scale and rotation, as well as some extent of viewpoint change.

Keywords: dense matching, wide baseline, rotation and scale invariant, SIFT.

1 Introduction

Finding dense and reliable match is a challenging task in computer vision, especially when the input images have significant rotation and scale changes. Dense and accurate matching plays an essential role in many applications such as 3D reconstruction, image-based modeling and image mosaicing. Many researches have been conducted on this topic, but dense and reliable matching of two wide baseline images taken with uncalibrated cameras still remains a tough problem.

Traditional dense two-frame matching algorithms aim at computing disparity maps from image pairs with short baseline [1]. In this situation, the images are almost identical, and the search space of disparities can be reduced to one dimension. These algorithms can be classified into two classes: global and local. In the local framework, dense matching is obtained by searching for correspondences pixel by pixel, thus the similarity measure of candidate pairs is quite important. In the window-based methods, matching cost is the aggregation of pixel-level dissimilarity within a support window; they assume that pixels in the matching window have similar disparities [2].

---

[*] Corresponding author.

D.-S. Huang et al. (Eds.): ICIC 2014, LNCS 8588, pp. 345–356, 2014.

When the baseline is wide, for example, with the images obtained by a handheld camera, dense matching becomes quite tough. In this situation, images may be quite different, so fixed window becomes ineffective to capture the same image content.

Several approaches have been proposed to handle the wide-baseline case. Plane-sweep-based algorithms test a family of plane hypotheses and choose the best one for each pixel [3]. These approaches can simultaneously perform dense matching and compute the depth map, but the cameras are required to be fully calibrated. In the uncalibrated case, quasi-dense approaches [4, 5] based on match propagation were proposed. In these methods, new matches are propagated from reliable matches obtained in the last iteration. The idea of propagation is to use reliable match at one pixel to guide new matches of its nearby pixels. However, the denseness cannot be guaranteed, which largely depends on the image content and propagation strategies.

In this paper, we propose a new approach for dense matching. Inspired by the great success of SIFT-based descriptors [6], which was reported to have best performance [7], we combine it with the window-based method in short-baseline matching to handle the problem of rotation and scale changes. Our framework contains the following steps: First, local invariant features are used for sparse matching. Second, the support windows are normalized based on reliable sparse matching result. Then, matching score of the warped patches is calculated. Finally, the best correspondence is obtained by searching for the highest matching score, and dense matching result is achieved. Sampson error from known homography is used for evaluation.

The main contribution of the proposed method is to use normalized support windows to deal with the problem of rotation and scale changes in wide baseline matching. Our approach can establish dense and reliable pixel correspondences.

This paper is organized as follows: In section 2, our novel dense matching scheme is presented. Establishment of reliable sparse key point matching is also introduced. In section 3, the detail of dense matching is described, including support window normalization, cost aggregation and searching strategy. Some experimental results on real data are presented in section 4. Finally, this paper is concluded in section 5.

2   Algorithm Framework and Sparse Matching

In this section, we first give a brief overview of the proposed method, and then discuss the initial sparse matching step.

2.1   Framework

The proposed dense matching approach starts with the detection of sparse key points, such as SIFT key-points and image corners, which are invariant to rotation and scale changes. These robust matches are used as reference points to guide the following dense matching. In the window-based method, correspondences are established pixel by pixel. For each pixel, we assign one nearest SIFT match and several corner matches to guide the matching. Firstly, its local geometric transformation from one image to the other is estimated based on these reference matches. Secondly, its hypothesis corresponding location is directly set using this transformation. Then, pixels near this location are searched. Finally, the one with highest matching score is accepted. To deal with the rotation and scale changes, the support windows are normalized, so that it is possible to capture the same image content during searching. The flowchart of the entire algorithm is presented in Figure 1.

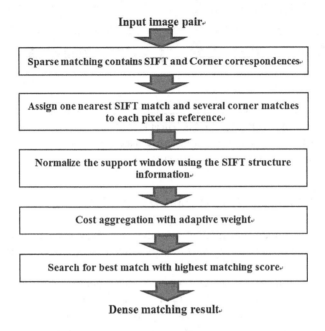

Fig. 1. Flowchart of our approach

2.2 Sparse Key Point Matching

Our goal in this step is to obtain enough reliable reference matches, which need to be invariant to rotation and scale changes. The local invariant feature detectors [8] and descriptors [7] have shown good performance in sparse wide baseline matching. Hence, we use DoG detector and SIFT descriptor [6] to generate initial matches. For matching, we adopt a modified nearest neighbor strategy: both the nearest distance and the first-to-second nearest distance ratio are required to be greater than a threshold; and another requirement is that both corresponding descriptors need to be the nearest neighbor of each other. Besides SIFT matches, we extract image corners and use a robust corner matching method in [9, 10] to produce new matches, which is also invariant to scale and rotation changes. In addition, we use the epipolar constraint to refine those matches. During refinement, both the fundamental matrix F and the true correspondences are estimated iteratively to minimize the Sampson cost function [11]; a RANSAC [12] scheme is adopted to eliminate outliers.

In the test we found that our sparse matching strategy is effective to increase the number of reference matches as well as improve the reliability. An illustrating example of sparse wide baseline matching is presented in Figure 2.

(a)   (b)

Fig. 2. Sparse matching result for an indoor image pair. The images are 816×612 pixels. (a): the initial SIFT match without refinement, using Vlfeat [15] library, 168 matches shown by the line linking dot; note the clear mismatches. (b): sparse matching result using our approach, 187 matches in total; note that mismatches in (a) is eliminated and new reliable matches are found.

3   Rotation and Scale Invariant Dense Matching

In window-based methods, it is implicitly assumed that pixels in the matching windows should have similar disparities [2]. When the baseline is short, this assumption is easy to meet, as the image pairs are almost identical except for a little translation in one dimension, so it is effective to search for correct correspondences by sliding the window along this dimension. However, in the wide baseline case, when geometric transformations such as rotation and scaling exist, this assumption becomes invalid. Therefore, the support windows need to be adjusted to capture the image of the same object surface. In section 3.1, we will discuss how to normalize the support windows using the sparse matching result. The definition and calculation of matching score are introduced in section 3.2. The dense matching process is introduced in section 3.3.

3.1   Normalize the Support Window

In this subsection, we discuss how to construct rotation and scale covariant support windows based on sparse matching result. First, a transformation matrix is estimated using the reference matches, and then it is used to normalize the support window.

Use SIFT Match to Estimate a Transformation Matrix

The goal of this step is to find optimal support window with adaptive shape. For an arbitrary candidate match, the support window is adjusted refer to the blob structure nearby. As each SIFT feature reliably represents the scale and orientation information of the blob, it is reasonable to take use of the SIFT matches as spatial prior to guide the adjustment. Give two image pairs I_1 and I_2. Let $p_1 = I_1(x_1, y_1)$, $p_2 = I_2(x_2, y_2)$ be an candidate match, where x_1, y_1, x_2, y_2 are image coordinates. The SIFT feature points $s_1 = \{I_1(x_{s1}, y_{s1}), \sigma_1, \theta_1\}$ and $s_2 = \{I_2(x_{s2}, y_{s2}), \sigma_2, \theta_2\}$ are the reference matches assigned to p_1 and p_2 respectively, where $x_{s1}, y_{s1}, x_{s2}, y_{s2}$ are image coordinates, and $\sigma_1, \sigma_2, \theta_1, \theta_2$ are the characteristic scale and dominant orientation of the SIFT feature. In order to achieve scale and orientation invariance, the coordinates of the candidate match and its support region should be transformed refer to the SIFT keypoint scale and orientation. We introduce a new SIFT coordinate system, in which the origin is at the SIFT keypoint (x_{s1}, y_{s1}), (x_{s2}, y_{s2}), and the horizontal axis is oriented to the dominant orientation. The relative position of p_1 in the local SIFT coordinate system is presented in Figure 3. Then, we normalize the putative correspondence p_1 and p_2 into the SIFT coordinate systems centered at (x_{s1}, y_{s1}), (x_{s2}, y_{s2}) respectively, which can be expressed as follows:

$$\tilde{p}_1 = \begin{bmatrix} \tilde{x}_1 \\ \tilde{y}_1 \end{bmatrix} = \frac{1}{\sigma_1} \begin{bmatrix} \cos\theta_1 & -\sin\theta_1 \\ \sin\theta_1 & \cos\theta_1 \end{bmatrix}^T \begin{bmatrix} x_1 - x_{s1} \\ y_1 - y_{s1} \end{bmatrix} \tag{1}$$

$$\tilde{p}_2 = \begin{bmatrix} \tilde{x}_2 \\ \tilde{y}_2 \end{bmatrix} = \frac{1}{\sigma_2} \begin{bmatrix} \cos\theta_2 & -\sin\theta_2 \\ \sin\theta_2 & \cos\theta_2 \end{bmatrix}^T \begin{bmatrix} x_2 - x_{s2} \\ y_2 - y_{s2} \end{bmatrix} \tag{2}$$

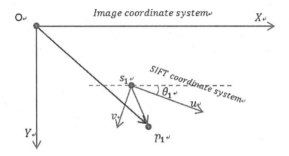

Fig. 3. Relationship between point p_1 and the local SIFT coordinate system centered at s_1; note that s_1 is the nearest SIFT feature point of p_1. $\overrightarrow{Op_1} = (x_1, y_1)^T$ and $\overrightarrow{s_1p_1} = (\tilde{x}_1, \tilde{y}_1)^T$

Suppose p_1 and p_2 are true matches, and this match is consistent with the blob structure represented by SIFT, the normalized coordinates $(\tilde{x}_1, \tilde{y}_1)^T, (\tilde{x}_2, \tilde{y}_2)^T$ are expected to be identical. Combine (1), (2) with the assumption that $(\tilde{x}_1, \tilde{y}_1)^T = (\tilde{x}_2, \tilde{y}_2)^T$, then p_2, the corresponding point of p_1 in I_2, can be predicted with a transformation matrix H. Rewrite p_1, p_2 in homogeneous coordinates $p_1 = (x_1, y_1, 1)^T$ and $p_2 = (x_2, y_2, 1)^T$; H is a local geometric transformation matrix. Then, H can be estimated as follow:

$$p_2 = H * p_1; \text{ and } H = \begin{bmatrix} S * R & t \\ 0^T & 1 \end{bmatrix} \tag{3}$$

Let $\sigma = \sigma_2 / \sigma_1$ and $\theta = \theta_2 - \theta_1$, thus,

$$S = \sigma, \ R = \begin{bmatrix} \cos\theta & -\sin\theta \\ \sin\theta & \cos\theta \end{bmatrix}$$

$$t = \begin{bmatrix} t_x \\ t_y \end{bmatrix} = -S * R * \begin{bmatrix} x_{s1} \\ y_{s1} \end{bmatrix} + \begin{bmatrix} x_{s2} \\ y_{s2} \end{bmatrix} \tag{4}$$

Use Corner Match to Refine the Transformation Matrix

If a pixel p_1 is far from the nearest blob, the estimated local transformation H may be inaccurate. In our approach, we refine the transformation matrix with the help of corner matches. First, several corner matches near p_1 are selected: $c_{k1} \leftrightarrow c_{k2}$, $k = 1, 2, \ldots M$. For each corner c_{k1} in I_1, we use a candidate transformation matrix H to predict its corresponding point c_{k2}' in I_2. Transfer error of this set of corner correspondences is defined as follow:

$$\text{transfer\_error}(c_{k1}, c_{k2}, H) = \sum_k d(H * c_{k1}, c_{k2})^2, \ k = 1, 2, \ldots M \quad (5)$$

Then, the transformation matrix H can be optimized by minimizing this transfer error. As described above, the calculation of H is based on SIFT feature information, which includes keypoint position, characteristic scale and dominant orientation. The scale ratio $\sigma = \sigma_2/\sigma_1$ and orientation difference $\theta = \theta_2 - \theta_1$ may be not accurate due to non-rigid distortion of the local structure. So, we perturb these two parameters in a discrete set around their original value and obtain a set of transformation matrix H, and use each H to construct several pairs of corresponding patches W_1, W_2. If $p_1 \leftrightarrow p_2$ is a correct match, W_1, W_2 are expected to be identical when H is well estimated. So, the H with lowest transfer error can be accepted. Some experimental results are demonstrated in Figure 4. It is obviously shown that the image content covered by the corresponding support windows tend to be similar after normalization.

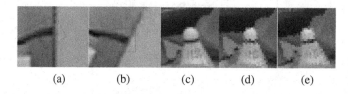

(a)   (b)   (c)   (d)   (e)

Fig. 4. Normalized support window. (a)(b) are two image patches with rotation change. The support windows can cover the same image content after normalization. (c) is a local patch in I_1, and (d),(e) are the normalized corresponding patch in I_2. The support window of (d) has not been refined using corner matches, while (e) has. Note that (c)(d) have some error in orientation. WZNCC score of (c)(d) is 0.8682, and (c)(e) is 0.9424, which is an obvious improvement.

In this way, the modified window-based method is invariant to rotation and scale changes and thus effective in wide baseline case. Following, we will discuss the similarity measure and cost aggregation strategy in the normalized window.

3.2   Matching Score

Dense matching depends on a cost function to measure the similarity of image locations. The normalized cross-correlation (NCC) and its improved version ZNCC are reported to have the best performance [13].

Suppose a pixel p_1 in the left image I_1, and its correspondence in the right image I_2 is p_2; N_{p1}, N_{p1} are pixels within the support windows, $q_1 \in N_{p1}, q_2 \in N_{p2}$ and $q_2 = q_1 - d$, $d = [d_x \quad d_y]^T$ for the disparity. Then, the similarity of two image patches centered at p_1 and p_2 can be calculated by ZNCC score:

$$Score_{ZNCC}(p_1, p_2) = \frac{\sum_{q_1 \in N_{p1}, q_2 \in N_{p2}} (I_1(q_1) - \bar{I}_1(q_1))(I_2(q_2) - \bar{I}_2(q_2))}{\sqrt{\sum_{q_1 \in N_{p1}} (I_1(q_1) - \bar{I}_1(q_1))^2 \sum_{q_2 \in N_{p2}} (I_2(q_2) - \bar{I}_2(q_2))^2}} \qquad (6)$$

However, in the window-based method, pixels in the support window have expected to have similar disparities, so that they can largely support the central pixel. In fact, this local planarity assumption is difficult to achieve, as finding the optimal support region not straddle object boundary is very difficult. Yoon proposed a method called adaptive support weighted window to solve this problem [14]. It can effectively cope with occlusions and depth discontinuities near object boundaries. In our approach, we add support weight in ZNCC to measure the similarity of corresponding image patches. The weight is decided according to intensity similarity $\Delta_{c_{pq}}$ and geometric proximity $\Delta_{g_{pq}}$; γ_c and γ_p are two coefficients to keep the balance.

$$w(p, q) = \exp\left(-\left(\frac{\Delta_{c_{pq}}}{\gamma_c} + \frac{\Delta_{g_{pq}}}{\gamma_p}\right)\right) \qquad (7)$$

Where $\Delta_{c_{pq}}$ represents the Euclidean distance between two colors c_p and c_q in the CIELab color space, and $\Delta_{g_{pq}}$ is the distance between p and q in the image domain:

$$\Delta_{c_{pq}} = |c_p - c_q| \quad , \quad \Delta_{g_{pq}} = |p - q| \qquad (8)$$

Thus, our matching score WZNCC (weighted ZNCC) is:

$Score_{WZNCC}(p_1, p_2) =$

$$\frac{\sum_{q_1 \in N_{p1}, q_2 \in N_{p2}} w_1(p_1, q_1) w_2(p_2, q_2)(I_1(q_1) - \bar{I}_1(q_1))(I_2(q_2) - \bar{I}_2(q_2))}{\sqrt{\sum_{q_1 \in N_{p1}} (w_1(p_1, q_1)(I_1(q_1) - \bar{I}_1(q_1)))^2 \sum_{q_2 \in N_{p2}} (w_2(p_2, q_2)(I_2(q_2) - \bar{I}_2(q_2))^2)}} \qquad (9)$$

The matching score is normalized in [0, 1]. In this way, we can search for the best match according to this score. Some results are presented in Figure 5.

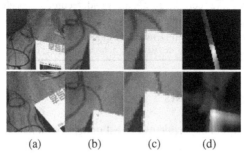

(a)   (b)   (c)   (d)

Fig. 5. Adaptive weighted support window. (a) original image; (b) corresponding point; (c) the normalized support window; (d) WZNCC score during search, light means high score; in the upper image, search range is limited to the region near the epipolar line, with threshold 3.

3.3   Procedure of Dense Matching

In dense matching, the disparity map is obtained by searching for correspondences pixel by pixel. This is the most time consuming step, so we need to limit the search range to reduce redundancy. For each pixel on a selected reference image, the correct match in the other view is obtained by searching within a limited range. In our method, the normalized support windows need to be computed for every pixel in the neighborhood from which new matches are searched. An effective way to reduce redundancy is to limit the search space. In multiple view geometry, there is a map from a point in one image to its corresponding epipolar line in the other image, which is represented by the fundamental matrix. In other word, the confidence of a putative location is low when it is far away from the epipolar line. During search, if the distance from a candidate point to this line exceeds a threshold, it is rejected, as it violates the epipolar constraint. Then, the search range is limited to a strip region, see Figure 5(d). The dense matching procedure is presented in Algorithm 1.

Algorithm 1. Dense Matching Procedure

Input: Color images I_1, I_2; Sparse matching result
         Search range N, Window size W
Output: Dense matching result
1. Compute the fundamental matrix, refer to [11]
2. For each pixel in I_1, calculate its local geometric transformation matrix H, refer to section 3.1.
3. Initialize the corresponding position in I_2, form a putative match $p_1 \leftrightarrow p_{2\_initial}$
4. If $p_{2\_initial}$ is out of the image range, recorded as occlusion; return
5. Limit the search space with epipolar constraint
For each pixel within the search range in I_2,
6. Extract corresponding image patches Patch1 and Patch2, warp the window refer to section 3.1.
7. Calculate matching score according to section 3.2,
8. If matching score is below a threshold, recorded as occlusion; return
9. Select the one with highest score as correct match, denoted by p_2
10. Add $\{p_1 \leftrightarrow p_2\}$ to Dense correspondences.
End for

---

4   Experimental Results

In this section, we present experiments on real images, which are collected from public dataset (e.g. INRIA on the web) or from our lab dataset. The latter are obtained by handheld cameras at arbitrary positions. The geometric distortion of the image pairs is apparent, such as rotation and scale changes. Firstly, we use images with significant rotation and scale changes to show the robustness of our algorithm. Dense matching results on several wide baseline image pairs are demonstrated. Then, we discuss the details of the experiment on a couple of indoor image pairs, and use the Sampson error from known homography for evaluation. We implemented the proposed method in MATLAB, and tested it on a laptop with Intel Core i5 CPU and 6GB RAM.

4.1   Test on Images with Significant Rotation and Scale Changes

The top row in Figure 6 is captured by a mobile phone, and the bottom is the image pair New York (from INRIA). We choose the first image as target, and put the corresponding pixel in the second image on it. Dense matching result is shown on the right column. The black region represents inaccurate matches, that is, the matching score is below 0.8. Scale change is apparent in the top row images, while the bottom row contains images with significant rotation change (about 100 degrees). The dense matching result is good on both cases, which demonstrates that our method is robust against rotation and scale changes.

Fig. 6. Images with significant rotation and scale changes: The left two columns are original images, and the right column is dense matching results

4.2   Evaluation and Analysis

The 'office' data consists of two images with resolution 640×480 pixels taken with two uncalibrated cameras. First, we conduct sparse wide baseline matching on the image pair. Outliers have been removed using the epipolar constraint with RANSAC. Before dense matching, we use bilateral filter to remove noise. Then, for a current pixel, the support window is normalized using the sparse matching result. The window size is set to 15x15, empirically, as too small windows lead to low distinctiveness in homogenous regions, and too large windows tend to straddle depth discontinuities. In Figure 4 (c)-(e), we see that, the image content captured in the normalized support windows tend to be identical. After the refinement of the transformation matrix, the matching score

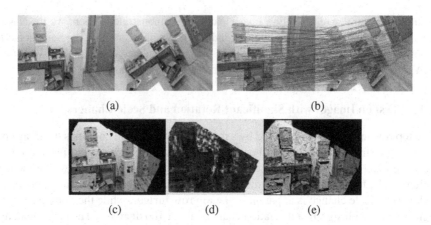

Fig. 7. Dense matching result on image data "office". (a): original images; (b): sparse matching result; (c): our dense matching result; (d): Sampson error image, bright area means large error; (e): quasi-dense matching result [5], black region means no match.

Fig. 8. Depth map from stereo-pair images. (a), (b): original images, (c): depth map.

Table 1. Percentage of good matches

Sampson distance (pixel)	<1	<2	<3	<4
%	67.57%	87.49%	95.21%	98.06%

is improved from 0.8682 to 0.9424. Search range for a candidate pixel is set to N=10 pixels. The computing time is 8.21 milliseconds per match. Correspondences with matching score less than 0.8 are rejected. Dense matching result is shown in Figure 7(c). Compared with the quasi-dense method [5], our matching result is much denser. The Sampson error [11] to the ground truth homography is shown in Table 1. Finally, in order to demonstrate the quality of the matches, we calibrated the cameras and computed a depth map of the scene, as shown in Figure 8.

5   Conclusion

In this paper, we proposed a modified window-based method for dense matching between two uncalibrated images with significant rotation and scale changes. With help of the powerful local invariant features, the support window is normalized before matching. Thus, we achieve robust matching against rotation and scale changes. In this way, our approach can deal with images obtained from handheld cameras without calibration. Experimental results on real images show that our approach is effective to obtain accurate and dense correspondences. However, our method is originally designed to handle scale and rotation changes, which may suffer from significant geometric transformations like affine changes or perspective changes. A major direction of future work is to develop an affine invariant approach to handle this problem.

Acknowledgement. This work is supported in part by National Natural Science Foundation of China under Grant No. 61175014, and the Fundamental Research Funds for the Central Universities of China.

References

1. Scharstein, D., Szeliski, R.: A Taxonomy and Evaluation of Dense Two-Frame Stereo Correspondence Algorithms. IJCV 47(1-3), 7–42 (2002)
2. Min, D.B., Lu, J.B., Do, M.N.: A Revisit to Cost Aggregation in Stereo Matching: How Far Can We Reduce Its Computational Redundancy? In: ICCV, pp. 1567–1574 (2011)
3. Gallup, D., Frahm, J.-M., et al.: Real-time plane-sweeping stereo with multiple sweeping directions. In: CVPR, pp. 2110–2117 (2007)
4. Lhuillier, M., Quan, L.: A quasi-dense approach to surface reconstruction from uncalibrated images. TPAMI 27(3), 418–433 (2005)
5. Kannala, J., Brandt, S.S.: Quasi-dense wide baseline matching using match propagation. In: CVPR, pp. 2126–2133 (2007)
6. Lowe, D.G.: Distinctive image features from scale-invariant keypoints. IJCV 60(2), 91–110 (2004)
7. Mikolajczyk, K., Schmid, C.: A performance evaluation of local descriptors. TPAMI 27(10), 1615–1630 (2005)
8. Mikolajczyk, K., Tuytelaars, T., Schmid, C., et al.: A comparison of affine region detectors. IJCV 65(1-2), 43–72 (2005)
9. Shi, F., Huang, X., Duan, Y.: A Hybrid Approach for Robust Corner Matching. In: Tarn, T.-J., Chen, S.-B., Fang, G. (eds.) Robotic Welding, Intelligence and Automation. LNEE, vol. 88, pp. 169–177. Springer, Heidelberg (2011)
10. Yan, B., Shi, F., Yue, J.: An Improved Image Corner Matching Approach. In: Huang, D.-S., Bevilacqua, V., Figueroa, J.C., Premaratne, P. (eds.) ICIC 2013. LNCS, vol. 7995, pp. 472–481. Springer, Heidelberg (2013)
11. Hartley, R., Zisserman, A.: Multiple View Geometry in Computer Vision. Cambridge University Press, Cambridge (2004)

12. Fischler, M., Bolles, R.: Random Sample Consensus: A Paradigm for Model Fitting with Applications to Image Analysis and Automated Cartography. Comm. ACM 24(6), 381–395 (1981)
13. Hirschmueller, H., Scharstein, D.: Evaluation of Stereo Matching Costs on Images with Radiometric Differences. TPAMI 31(9), 1582–1599 (2009)
14. Yoon, K.J., Kweon, I.S.: Adaptive support-weight approach for correspondence search. TPAMI 28(4), 650–656 (2006)
15. Vedaldi, A., Fulkerson, B.: Vlfeat: An open and portable library of computer vision algorithms. In: Proc. Int. Conf. on Multimedia, pp. 1469–1472 (2010)

Computational Aesthetic Measurement of Photographs Based on Multi-features with Saliency

Yimin Zhou[1], Yunlan Tan[1,2], and Guangyao Li[1]

[1] College of Electronics and Information
Tongji University
Shanghai, 201804, China
[2] School of Electronics and Information
Jinggangshan University
Ji'an, Jiangxi, 201804, China
{1010080053,2011tan,lgy}@tongji.edu.cn

Abstract. Based on the existent computational aesthetic measurements, we present a new approach that combining both saliency region detection and extraction with a feature set in line with the principle of human vision. We first extract the saliency region using frequency-based method, then extract 53 features from both local and global regions, and select top 15 features which can determine the best aesthetic value. We run both SVM classification & regression and CART as well as linear regression on the filtered dataset. The experiments show a meaningful result of an accuracy above 70%.

Keywords: saliency features, computational aesthetic measurement, aesthetics of photography, SVM classification and regression, CART.

1 Introduction

With the emergence of computer, internet and digital camera, photographs acquisition, storage and sharing become even more convenient. Aesthetics of photography is how people assess the beauty of one photograph. A professional photographer may decompose it into parts, then judge each in detail, even evaluate brightness, color combination and contrast etc., which are integrated factors in addition to some standards of the photography industry. All these kinds of assessment are much largely dominated by the particular individual who engaged in the job. According to Flickr's statistics[1], users upload about 6.5 million photographs every day. It is a heavy challenge to make aesthetics measurement of such huge amount of photographs. Nevertheless, computer, again, plays the central role in dealing with the situation. To tackle this knotty problem, in this paper, we present a computational and experimental approach to handle it in a statistical learning way. This allows us to exclude those irrelevant factors of photographs and focus on only those certain key features that can identify ones of high aesthetics from the vast amount of the collection of albums on

D.-S. Huang et al. (Eds.): ICIC 2014, LNCS 8588, pp. 357–366, 2014.

the internet in a statistical sense. In the age of information explosion as well as the digital image information, to develop intelligent systems for automatic evaluation of digital photographs is much critically demanded.

In our work, we present a process of images' related features detection, calculation, training, classification and prediction which is a variant of Datta's[2]. Unlike Datta's, we consider both image's saliency region and traditional aesthetic feature including Kolmogorov complexity, improved wavelet feature and NSCT wavelet decomposition etc. We achieve our experiment results with promising aesthetic measurement' values based on authoritative online photograph galleries. The accuracy close to those of Datta's[2] and other former researchers proves our method's availability and validity. The remainder of this article is organized as follows. In the next section, we review related work; In section 3, we illustrate the method of image feature extraction, training and classification; In section 4, we present the experimental results and some insight from them; In section 5, a conclusion of our work has been drawn.

2   Related Works

Measurement of the aesthetics value can be retraced to the writing of American Mathematician Birkhoff in 1933, the "Aesthetic Measure"[3]. In this work, he presented the formula of aesthetics measurement,

$$M = \frac{O}{C} \tag{1}$$

Where M means measurement, O represents order and C denotes complexity. Machado et al.[4] treated aesthetics measurement as proportional image complexity and inversely proportional to processing complexity of the image in brain. They gave the formula of aesthetic measurement of the image,

$$Measure = \frac{IC}{PC} \tag{2}$$

IC is the ratio of JPEG compressed image's error and the compression ratio, while PC is fractal image compressibility. Rigau et al.[5-6] combined information theory, which extended Birkhoff's theory by applying colors' distribution, Shannon's Entropy and Kolmogorov complexity. According to their framework, Juan Romero and Penousal Machado[7] explored the use of image compression estimates to predict the aesthetic merit of the images.

Although a theoretical measure of computing the aesthetics value is a fast and accurate way, classical assessments of features such as similarity, balance, combination, orientation, hue, harmony and contrast, still dominate most observers and connoisseurs' sensory. Wang and Datta et al.[2]from Pennsylvania were the first to realize the quantization of aesthetics measurement of image features. They extracted image features from the images, including the brightness, color distribution, wavelet, region composition and depth of field, and then applied SVM or linear regression to classify the high from the low quality photographs. They achieved an accuracy of 70.12%. The ACQUINE[8] aesthetics value measurement system developed by them is a typical aesthetics evaluation and search engine. Wu et al.[9] extended the SVM classification method utilized by Datta and Wang to predict aesthetics measure values. Ke et al.[10] distinguished professional high quality images from low quality snapshots by evaluating the aesthetics value from images' visual features. But, they ignored the differentiation between the visual attention regions and the remains. Wong et al.[11] presented saliency-enhanced image classification method but with a partial-automatic saliency detection approach and a relatively stale feature sets. Luo et al.[12] exploited the blur detection to estimate roughly focused on main area's features. Subhabrata et al.[13] discussed the rules of thirds and visual weight balance in detail and developed an interactive application that enabled users to improve the visual aesthetics of their digital photographs using spatial re-composition.

3   Aesthetic Classification and Sorting

3.1   Frequency-Tuned Saliency Region Detection

Frequency-tuned saliency-region detection was presented by Achanta[14]. Figure 1 illustrates one of the source images and its detected saliency region. Every RGB image is converted to HSV color space, generating the two-dimensional matrices I_H, I_S, I_V, each size of $X \times Y$. In comparison to our saliency method with the former largest patches selection method, we extract both features of saliency and largest patches to demonstrate that saliency features are dominant than the largest patches features.

Fig. 1. (a)Source image (ID. 65200)   (b) the corresponding result of saliency detection (c) The related top 5 patches detected

Altogether we extract 53 features of local and global features for every selected image. The first selected features are referred as candidate ones for further refinement and are denoted as F={ f_i |1≤i≤53} that are described as follows.

3.2   Global Features Extraction

- **Aspect Ratio.** As is known to all, the value approximating 4:3 and 16:9, which are the golden ratio and chosen as television screens, are much related to the viewing pleasure. The aspect ratio features is

$$f_1 = Width/Height$$

Brightness. We use the average pixel intensity as the brightness of the image, which is defined as f_2. The saliency brightness is defined as f_3, which represents the average pixel intensity of the saliency region.

$$f_2 = \frac{1}{XY} \sum_{x=o}^{X-1} \sum_{y=0}^{Y-1} I_V(x, y), \quad f_3 = \frac{1}{XY} \sum_{(x,y) \in saliency} I_V(x, y)$$

- **Saturation.** Saturation or as its definition, subjective experience of vividness or richness of color indicates chromatic purity. We compute the average saturation as

$$f_4 = \frac{1}{XY} \sum_{x=0}^{X-1} \sum_{y=0}^{Y-1} I_s(x, y)$$

$$f_5 = \frac{1}{XY} \sum_{(x,y) \in saliency} I_S(x, y)$$

- **Dark Channel .** Dark channel was introduced by He et al.[15] for haze removal. The value of dark channel[16] of an image I is defined as

$$I_{dark}(i) = \min_{c \in R,G,B} (\min_{i' \in \Omega(i)} I_c(i'))$$

Where I_c is the color channel of I and $\Omega(i)$ is the neighborhood of pixel i. The $\Omega(i)$ is chosen as 11×11 local patch. The dark channel feature of the photograph I is computed as the average of the normalized dark channel values in the area of the whole image

$$f_6 = \frac{1}{\|S\|} \sum_{(i) \in S} \frac{I_{dark}(i)}{\sum_{c \in R,G,B} I_C(i)}$$

Where S is the subject area of the photograph.

- **Kolmogorov Complexity.** Kolmogorov[17] [5] complexity is defined as the normalization of the order

$$f_7 = M_k = \frac{NH_{max} - K}{NH_{max}}$$

M_k takes the values within [0,1] and expresses the photograph's degree of order without any prior knowledge of its size. Here NH_{max} is the original size of the photograph and K is the Kolmogorov complexity of the compressed size. JPEG compressor is selected because of its ability to discover patterns while losing little information which is imperceptible to the human eyes. We also compute the Kolmogorov complexity of the saliency region and the corresponding feature f_8 in a similar way.

- **NSCT Texture Features.** We apply a non-subsampled contourlet transform (NSCT)[18] to the HSV color space of the image and use these samples as our selected features. The Non-subsampled Directional Filter Bank(NSDFB) is constructed by combining critically-sampled two-channel fan filter banks and resample operations. The result is a tree-structured filter bank that splits the 2-D frequency plane into directional wedges. Figure 2 illustrates a three-level decomposition.

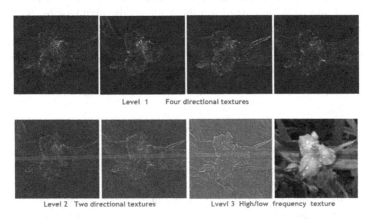

Level 1   Four directional textures

Level 2   Two directional textures   Lvevl 3 High/low frequency texture

Fig. 2. The different level textures of NSCT on ID. 217257

We take the average NSCT wavelet transform of level 1 to 3 of HSV color space of each image as our texture features.

$$f_{8+i}= \frac{1}{XY} \sum_{W_H,S=1} (x, y)$$

$$f_{9+i}= \frac{1}{XY} \sum_{W_H,S=2,D_1} (x, y) + \sum_{W_H,S=2,D_2} (x, y)$$

$$f_{10+i}= \frac{1}{XY} \sum_{W_H,S=2,D_1} (x, y) + \sum_{W_H,S=2,D_2} (x, y) + \sum_{W_H,S=2,D_3} (x, y)$$

where i=1,11,15. So are S and V. We also define similar features for saliency region.

3.3   Local Features Extraction

- **Rule of Thirds.**   Rule of Thirds is a very popular rule of thumb in photography. We consider the rule of thirds inside saliency region, so we do not omit the rule of thirds features. We compute the average hue of the inner thirds region as,

$$f_{27} = \tfrac{9}{XY} \sum_{x=X/3}^{2X/3} \sum_{y=Y/3}^{2Y/3} I_H(x, y)$$

f_{28}, f_{29} are computed similarly for I_s and I_v respectively.

- **The top5-Patches Features.** We still present the old top 5 patches related features for comparison with the saliency related features. The image is transformed in the LUV space because the perceived color changes well in this locally Euclidean distance model space. With a fixed threshold for all the photographs, we use a K-means algorithm after a K-center algorithm computes clusters. Following a connected component analysis, color-based segments are obtained. The largest segments formed are denoted as {s1,...,s5}. One sample and its top 5 patches are shown in Figure 1(c). Then we calculate the average hue, saturation and value for each of the top 5 patches as features 30-32, 34-36, 38-40, 42-44, 46-48. That is,

$$f_{i+30} = \tfrac{|S_i|}{(XY)}$$

i=1,5,9,13,17. And finally, the rough positions of each segment are stored as features $f_{33}, f_{37}, f_{41}, f_{44}, f_{49}$. The image is divided into 3 equal parts along horizontal and vertical directions, locate the centroid of each patch s_i in the block. $f_{32+i}=(10r+c)$ where (r,c) $\in \{(1,1),...,(3,3)\}$ indicates the corresponding block starting with left and top.

- **Visual Attention Center.** The stress points[13] are the four points in an image which are the intersections of horizontal and vertical 1/3 and 2/3 dividing lines. We include four features of visual attention centers, which are represented by f_{50} through f_{53}.

$$f_{50} = \tfrac{1}{XY} \| v - s_1 \|_2^2$$

3.4   Training and Testing

SVM Classification and regression. All 53 features in $\mathcal{F}$ were extracted and normalized to the range of [0, 1] to form the data of experiment. We make a filter of aesthetics scores where greater than or equal to 5.8 being treated as high values and less than or equal to 4.2 being treated as low values. The scores between 4.2 and 5.8 are considered as ambiguous. People often don't fully distinguish high from low or vice versa. We then replicate the data to generate equal number of samples of the high and the low to ensure equal priors. We use the standard RBF kernel with the parameter γ=18.5 and cost=1.0 to perform our classification job. We chose livSVM as our algorithm package. SVM is run 10 times per feature by m-fold cross-validation. Then top 27 features were filtered out. We then use the greedy algorithm to stop after found the top 15 features and used them to build the SVM-based classifier. After doing this, we try to predict the aesthetic values by using the SVM based linear regression by setting $\gamma=200.0$, $\varepsilon=0.1$ and cost=5.0 to have a moderate result.

CART Classification and regression. Unlike SVM, the benefit of CART is we construct a regression tree to predict aesthetic values in a fast and effective way. We chose the recursive partitioning implementation(RPART) as our CART building utility, and built a two-class classification tree model for the same training set in SVM turn.

Linear Regression. We perform linear regression in polynomial terms of the feature values to see whether it can directly predict the aesthetics scores from the feature vector or not. All data were used. For each feature f_i, the polynomial terms f_i^2, f_i^3, $f_i^{1/3}$, $f_i^{2/3}$ are used as independent variables.

4   Experimental Results and Analysis

We test the proposed filter algorithm on hundreds of color images depicting typical scenes of natural landscapes by a PC with Duo CPU 2.8GHz. Our experiments are programmed in c++ and MATlab.

4.1   Dataset

We use dpchallenge.com as our dataset source. The dataset contains user scores of 304,073 images in a grade from 1 to 10. In the collection, each photo has been evaluated by at least one hundred users. The collection contains several evaluation indices for each image, including the number of aesthetics ratings received, the means of ratings and a distribution of quality ratings on a 1-10 scale. We download 28,896 photographs from a miscellaneous contents set from its portal and chose 5,967 from them containing all ranges of aesthetic values every of which were voted by no less than two hundred viewers. Figure 3 shows the distribution of samples according to their aesthetic value.

From our experiences, only about one-fifth to one-fourth results by the frequency-tuned saliency detection algorithm can be considered as very satisfactory. The complete samples set consist of 1860 images. We chose two classes of data, high ones containing samples with aesthetics scores above 5.8 and low ones containing samples with aesthetics scores below 4.2. Then the high set contains 434 samples and the low contains 503. As to give equal prior of the two classes, we replicate some of the high set randomly to increase to 503 samples. So the whole training set becomes 1006. Similarly, the amount of all samples goes to 1929.

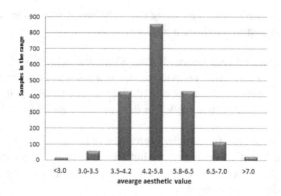

Fig. 3. The distribution of images according to their aesthetic value

4.2 Experimental Results

For the SVM performed on the individual features, the 27 features are obtained by directly extracted from the images. In decreasing prediction accuracy order, the features are { f_7, f_{50}, f_{21}, f_{12}, f_{53}, f_{24}, f_{15}, f_{51}, f_{30}, f_{26}, f_{17}, f_{34}, f_2, f_{49}, f_{22}, f_{13}, f_{52}, f_{46}, f_{18}, f_9, f_6, f_{25}, f_{23}, f_{16}, f_{20}, f_{11}, f_3 }. The correct classification rate achieved by single feature is f_7 with 62.8%. After the run of wrapper based greedy algorithm, features are clearer, they are { f_7, f_{15}, f_6, f_9, f_{11}, f_{20}, f_{12}, f_{18}, f_{46}, f_3, f_{52}, f_{17}, f_2, f_{30}, f_{53} }. The accuracy achieved with the 15 features is 72.0%, with the precision of detecting high class being 85.9% and low class 63.8%. Then we use the same features set and run SVM linear regression prediction onto the training set to have a moderate result of variance of 0.43 with all the samples.

Figure 4 shows part of the CART decision tree with all 53 features. In this figure, the decision nodes are denoted by squares while leaf nodes are denoted by circles. The number of the observations in each node and the splits is shown in the figure. Low nodes with low classification are dark and high nodes with high classification are light. We use 5-fold cross validation to increase the accuracy. The complexity parameter is set to 0.0072 yielding 39 splits. We achieve 73% model accuracy and 60% 5-CV accuracy.

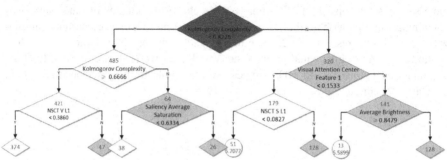

Fig. 4. The CART decision tree obtained

For linear regression, the variance σ^2 of the aesthetics score over 1929 samples is 0.96. With 5 polynomial terms for each of the 53 features, we achieve a residual sum-of-squares $R_{res}^2 =0.63$, which is a 34% reduction of the variance σ^2.

4.3   Discussion

We guess the classification model in the SVM and CART can explain some hidden phenomenon behind the aesthetic evaluation of the viewers. First of all, human eyes and comprehensive system are very vulnerable to complex visual elements. This perhaps because one cannot understand too complex information during a short time or repulsive to incompressible random information as noise. So Kolmogorov complexity largely dominates the classification process from both SVM features filtering process and CART. Next we can see the saliency region of an image plays an important role in determining the image's aesthetic value because half of the features in the SVM selected ones are saliency related. Another essence, we can get from the classification process is that the basic features as brightness, saturation of the images are still prior to human evaluation.

5   Conclusion

We present an automatic approach for machine determining aesthetic value of photographs. We use SVM, CART and linear regression to construct the model. The SVM features filtering is a tedious work, which exploited both filter based and wrapper based method to distill final 15 features for SVM training. The CART is a fast and convenient method (certainly all short enough to within tens of minutes), but it is not so accurate as the former. We also find that the complexity of an image to a great extent dominates the classification process. We also can conclude that human eyes and comprehensive system tend to accept simple and ordered image information. For images with an obvious salient region, viewers are inclined to focus on this region and its features in evaluating the aesthetic value of an image. Furthermore, brightness, saturation etc. are fundamental, but still important factors in aesthetic evaluation.

Acknowledgements. This work is supported by Key Laboratory of Advanced Engineering Surveying of National Administration of Surveying, Mapping and Geoinformation (No.TJES1205). We are grateful to the anonymous referees for useful comments and suggestions.

References

1. Flickr, http://www.flickr.com
2. Datta, R., Joshi, D., Li, J., Wang, J.Z.: Studying aesthetics in photographic images using a computational approach. In: Leonardis, A., Bischof, H., Pinz, A. (eds.) ECCV 2006. LNCS, vol. 3953, pp. 288–301. Springer, Heidelberg (2006)

3. Birkhoff, D.G.: Aesthetic measure, Cambridge, Mass. (1933)
4. Machado, P., Cardoso, A.: Computing aesthetics. In: de Oliveira, F.M. (ed.) SBIA 1998. LNCS (LNAI), vol. 1515, pp. 219–228. Springer, Heidelberg (1998)
5. Rigau, J., Feixas, M., Sbert, M.: Conceptualizing Birkhoff's aesthetic measure using Shannon entropy and Kolmogorov complexity. In: Proceedings of the Third Eurographics Conference on Computational Aesthetics in Graphics, Visualization and Imaging, pp. 105–112. Eurographics Association (2007)
6. Rigau, J., Feixas, M., Sbert, M.: Informational aesthetics measures. IEEE Computer Graphics and Applications 28(2), 24–34 (2008)
7. Romero, J., Machado, P., Carballal, A., Osorio, O.: Aesthetic classification and sorting based on image compression. In: Di Chio, C., et al. (eds.) EvoApplications 2011, Part II. LNCS, vol. 6625, pp. 394–403. Springer, Heidelberg (2011)
8. Datta, R., Wang, J.Z.: ACQUINE: aesthetic quality inference engine-real-time automatic rating of photo aesthetics. In: Proceedings of the International Conference on Multimedia Information Retrieval, pp. 421–424. ACM (2010)
9. Wu, Y., Bauckhage, C., Thurau, C.: The good, the bad, and the ugly: Predicting aesthetic image labels. In: 2010 20th International Conference on Pattern Recognition (ICPR), pp. 1586–1589. IEEE (2010)
10. Ke, Y., Tang, X., Jing, F.: The design of high-level features for photo quality assessment. In: 2006 IEEE Computer Society Conference on Computer Vision and Pattern Recognition, vol. 1, pp. 419–426. IEEE (2006)
11. Wong, L.K., Low, K.L.: Saliency-enhanced image aesthetics class prediction. In: 2009 15th IEEE International Conference on Image Processing (ICIP), pp. 997–1000. IEEE (2009)
12. Luo, Y., Tang, X.: Photo and video quality evaluation: Focusing on the subject. In: Forsyth, D., Torr, P., Zisserman, A. (eds.) ECCV 2008, Part III. LNCS, vol. 5304, pp. 386–399. Springer, Heidelberg (2008)
13. Bhattacharya, S., Sukthankar, R., Shah, M.: A framework for photo-quality assessment and enhancement based on visual aesthetics. In: Proceedings of the International Conference on Multimedia, pp. 271–280. ACM (2010)
14. Achanta, R., Hemami, S., Estrada, F.: Frequency-tuned salient region detection. In: IEEE Conference on Computer Vision and Pattern Recognition (CVPR), pp. 1597–1604. IEEE (2009)
15. He, K., Sun, J., Tang, X.: Single image haze removal using dark channel prior. IEEE Transactions on Pattern Analysis and Machine Intelligence 33(12), 2341–2353 (2011)
16. Luo, W., Wang, X., Tang, X.: Content-based photo quality assessment. In: 2011 IEEE International Conference on Computer Vision (ICCV), pp. 2206–2213. IEEE (2011)
17. Cover, T.M., Thomas, J.A.: Elements of information theory. John Wiley & Sons (2012)
18. Da Cunha, A.L., Zhou, J., Do, M.N.: The nonsubsampled contourlet transform: theory, design, and applications. IEEE Transactions on Image Processing 15(10), 3089–3101 (2006)

Multi-layers Segmentation Based Adaptive Binarization for Text Extraction in Scanned Card Images

Chunmei Liu

Department of Computer Science and Technology, Tongji University, Shanghai, China
chunmei.liu@tongji.edu.cn

Abstract. This paper proposes an adaptive text binarization algorithm based on multi-layers to improve the binarization performance of scanned card images. It combines gray information and position information to divide a scanned card image into multiple layers, and proposes a division rate to identify whether to continue layer division. On each text layer, the dual-threshold is applied to eliminate the disturbance of noise and background pattern. Experimental results demonstrate that this approach is robust to various situations and can achieve a good performance in a scanned card image binarization system.

Keywords: Card image binarization, Text extraction.

1 Introduction

In automatic scanned card processing system, binarization algorithm is a crucial step to the accuracy of overall card processing system. However, in scanned card processing system, some difficulties are involved in text extraction due to noise characteristics, background patterns, text quality, and histogram modalities in the scanned card images.

The binarization methods can be categorized into two types: region-based methods and threshold-based methods. The region-based methods extract some regions and identify text regions according to text properties, such as approximately consistent color and stroke width. This kind of approaches is suitable for the texts with wide strokes. It is unstable for the texts with narrow strokes as it is hard to acquire the accurate stroke width. The threshold-based methods binarize an image to extract texts by global, local or multiple thresholds. In scanned card images, texts generally have narrow strokes. So for this kind of images the threshold-based method is attractive as it is a simple and effective tool. The threshold-based techniques can be categorized into two classes: single stage and multi-stages.

The single stage thresholding methods try to find good values at once using either global or local thresholding techniques [1]. In global thresholding algorithms, a single threshold is estimated and applied to separate objects from background. The typical global thresholding algorithms are Otsu's method [2], Kittler and Illingworth's method [3], and others. The global methods are simple and runtime saving, and can obtain satisfying results when a card image is relative consistent. In complex situations, the

D.-S. Huang et al. (Eds.): ICIC 2014, LNCS 8588, pp. 367–375, 2014.

local thresholding techniques might outperform the global method. The local thresholding algorithms calculate threshold values which are determined locally based on the neighborhood of a pixel. The popular local thresholding algorithms are Niblack's method [4], Bernsen's method [5], Sauvola method [6], White and Rohrer's method [7] and so on. One of the limitations of local thresholding method is that it expends a high amount of processing time.

The multi-stages thresholding techniques can utilize more information to extract texts from background [8-10]. Sue Wu [8] utilizes two stages to separate texts from image. In the first stage, the global thresholding is used to get some children images. The second stage is performed on the children region by looking for the first minimum of histogram as thresholding value. This method made a success in the postal envelopes. Solihin and Leedham [9] propose a two-stages thresholding approach for the handwriting image binarization which classifies a image into three classes: foreground, background, and fuzzy area. Chia-Shaud Hong et al. propose a 2-D intelligent block detection algorithm with more accurate region segmentation for document image binarization [10]. H.B. Song and D.J. He split the image into several layers including text layers and background which were processed using different techniques to pick up the text document successfully [11]. In Seop Na et al. propose three thresholding methods for binarizing the address region of Korean Identity card images [12].

The problem we address in this paper is to binarize texts from a scanned card image with complex pattern background, such as passport, ID card, name card, mail image, bank check, bill and so on. We present an adaptive binarization method based on multi-layers to improve binarization performance, in which the multiple layers are obtained by combination of gray information and position information. On each text layer it binarizes texts from a scanned card image to eliminate background pattern. The second contribution of our paper is to propose the division rate to identify whether there is a text layer in the scanned card image. If there were a text layer, the image would continue to be divided into more text layers. Thus it can reduce the disturbance of noise and background pattern on text layer and improve text binarization performance. In Fig. 1, we can see that the proposed approach is mainly performed by three steps: the first binarization, adaptive binarization, the last binarization. In contrast to other approaches, we use the first binarization to divide a scanned card image into several initial text layers. Then the adaptive binarization is proposed to classify these layers into the multiple text layers. The division rate is computed to decide whether the layer division is continued or not. Finally, the last binarization is processed on each text layer to obtain the final results. Experimental results demonstrate that our approach could efficiently be used as an automatic binarization in scanned card processing system. It is robust for complicated background, pattern background, noise and various image qualities.

The paper is organized as follows. In section 2 the adaptive binarization based on multi-layers is described. The experimental results are shown in section 3, and prove the advantages of the proposed algorithm. Finally, we draw conclusions and discuss the open issues for future research in section 4.

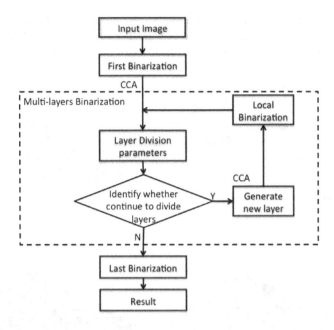

Fig. 1. Flow chart of the proposed approach

2 Adaptive Binarization Based on Multi-layers Segmentation

In this paper, the card images that we research include passport, ID cards and other card images. They are more complicated than the plain text document images as they include both text and some kinds of pattern background. However these kinds of pattern background in card image is relatively simpler than the images from nature. So we can use multiple gray thresholds to extract texts from a card image. An example is shown in Fig. 2(a). The card image is mainly composed of texts. In such kind of image, the gray values of texts are generally darker than the gray values of background, as shown in Fig. 2 (b). According to the gray information we can set multiple thresholds to simply divide one image into the multiple layers. These layers may include texts, or background, or both texts and background, as shown in Fig. 2 (c). If we only use the multi-thresholds to binarize these layers, we still cannot obtain the good results. The reason is that the texts that belong to the different sentences may have the different gray values and background gray values are not always different with texts gray values. It will lead to the failure of text extraction only by several gray thresholds. So it is not enough just only to use the gray information to acquire text layers. We combine gray information and position information to obtain the multiple text layers. The texts that belong to the same sentence generally have the similar gray value. At the same time they have the adjacent position. So the text layer division is not only related with gray information, but also with position information. The proposed method adaptively binarizes the image into multiple layers by combination of gray information and position information. So our aim is to find the text layers and do binarization on each text layer. In the following discussion, we provide details on how to adaptively binarize the scanned card image based on multi-layers.

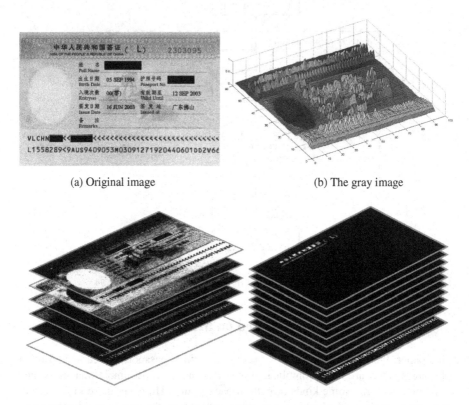

(a) Original image

(b) The gray image

(c) Multi-layers based on gray thresholding (d) Multi-layers based on gray information and position information

Fig. 2. Multi-layers analysis

2.1   First Binazarition

This stage aims to obtain initial layers and remove simple background by binarization. For a color image, it is necessary to convert it to the gray-level image. Then we begin with a threshold method to obtain the preliminary binary image. Comparing with local adaptive threshold method, we focused on the global thresholding method which is parameter independent and with low computation cost. Here we chose Otsu's method because it is simple and has been cited an effective scheme.

After we obtain the binary image, we consider that it is background part that has the same binary value with peripheral border region and the other parts are text candidates because generally there are few texts in the peripheral image region. Then the connected component analysis (CCA) is applied on text candidates to obtain n_0

connected component regions. Each connected component is extracted as a single layer. The position of each connected component is projected on the original image. Then we get n_0 layers: $L_1^0, L_2^0, \cdots L_{n_0}^0$, which include text candidate regions, and the non connected components (non-CC) are eliminated as background region or noise.

2.2 Adaptive Binarization

According to context information, texts to the same phrase or sentence generally have similar gray value and adjacent position. The aim in this stage is to acquire the multiple layers. There are texts to the same phrase or sentence on each layer.

After the first binarization, we acquire the initial n_0 layers $L_1^0, L_2^0, \cdots L_{n_0}^0$, which can be classified into two classes: texts to the same phrase or sentence and texts to the different phrases or sentences. For the layer with texts to the same phrase or sentence, we may use any threshold technique to binarize it directly. For the layer with texts to the different sentences, we continue to divide it into several layers until the layer with texts only to the same phrase or sentence.

As the disturbance of background pattern and noise, it is hard to obtain the layer with texts only to the same phrase or sentence. Here, we propose the division ratio r_d to identify whether the layer need to be divided. For example, we apply dual-threshold method on one layer L_i^0. The dual-threshold method binarizes the image twice to obtain the first coarse foreground and the second refined foreground [13]. Then we obtain the first binary foreground I_{sw1}^0 and the second binary foreground I_{sw2}^0 of layer L_i^0. The division ratio r_i^0 of the layer L_i^0 is computed as Equation (1).

$$r_i^0 \quad \text{sum}\left(\left|I_{sw1}^0 \quad I_{sw2}^0\right|\right)/\text{sum}\left(I_{sw1}^0\right) \tag{1}$$

If r_i^0 is more than the division ratio limit r_d, it means the texts on the current layer L_i^0 belong to the different sentences. The layer L_i^0 need to be divided into new layers. Thus it can avoid the disturbance of background pattern and noise.

The CCA is applied on the second binary foreground I_{sw2}^0 of layer L_i^0 to obtain n_i^0 connected component regions. Each connected component is extracted as a single layer. The position of each connected component is projected on original image. Then for the layer L_i^0, we get n_i^0 new layers: $L_1^0, L_2^0, \cdots L_{n_i^0}^0$. For all initial layers $L_1^0, L_2^0, \cdots L_{n_0}^0$, after the first layer division we can obtain the new multiple layers $L_1^1, L_2^1, \cdots L_{n_1}^1$.

It is uncertain that the texts on the layer L_i^1 belong to the same phrase or sentence. It needs to be identified by division ratio r_i^1. So the next layer division continues until there is not any layer which texts belong to the different sentences. The whole adaptive binarization is an iterative process. The detailed binarization process is implemented as Table 1.

When the adaptive binarization process is accomplished after m iteration, the text layers $L_1^m, L_2^m, \cdots L_{n_m}^m$ are generated. In order to avoid falling into the dead loop, the program set the iteration limit n_t.

Table 1. Adaptive binarization procedure

Algorithm: Adaptive binarization
Input: initial layers $L_1^0, L_2^0, \cdots L_{n_0}^0$
Output: text layer $L_1^m, L_2^m, \cdots L_{n_m}^m$
1. Initialize two layer sets: ELS, DLS. ELS stores the layers that need not be divided. It is initially set empty. DLS stores the layers that need to be divided. Initially put initial layers $L_1^0, L_2^0, \cdots L_{n_0}^0$ to DLS.
2. For each layer L_i^k in k iteration in DLS, compute the division ratio r_i^k and obtain the foreground of layer L_i^k. If r_i^k is less than the division ratio limit r_d, move L_i^k to ELS. Otherwise, go to step 3.
3. Obtain the foreground's bounding box of layer L_i^k. Project the bounding box onto the original gray image to obtain the gray foreground I_{gi}^k. Apply dual-thresholding on I_{gi}^k to obtain the binary foreground I_{bwi}^k. Then apply CCA to obtain n_i^k new connected component regions. Each connected component is extracted as a single layer. The position of each connected component is projected on original image. Then we get n_i^k layers: $L_1^k, L_2^k, \cdots L_{n_i^k}^k$. Move $L_1^k, L_2^k, \cdots L_{n_i^k}^k$ to DLS.
4. Obtain the final text layers $L_1^m, L_2^m, \cdots L_{n_m}^m$ repeating step 2-3 until DLS is empty or iteration number k is more than the top iteration limit n_t.

2.3   Last Binarization

This stage includes two parts: text layer binarization and refined binarization. Any thresholding technique can be used in text layer binarization. For the comparison of experiment we apply Otsu's method on each text layer. After the binarization on all text layers, the corresponding results merge together to form the result of text extraction.

After text layer binarization there are still some non-text parts in the card image, such as portrait in passport and ID card, stamp on envelop. The size of non-text part is generally larger than character's. Here we compute the average height of character hei_{avg} to eliminate non-text part if its height is much larger than average height of all foregrounds, $r_{hh} \times hei_{avg} < hei < r_{ht} \times hei_{avg}$. r_{hh} and r_{ht} is a bottom limit and a top limit of height ratio. The small part is removed as noise if its height is much smaller than the average height. Thus the final result is acquired, as shown in Fig. 3 (b).

3   Experiment

We evaluate the performance of the proposed approach on a total of 26 scanned card images, which include passports, name cards, ID cards. These images contain variety of layouts, complicated background, different color texts, and different text sizes. Our experiment was implemented in Matlab. In the experiment, the relative parameters are empirically set as τ_d 0.8, n_t 4 according to the binarized results on the test set.

A comparative experiment is implemented by the proposed method and Otsu's method. Fig. 2(a) shows the original image of a scanned passport image. It has complicated background and text characters with different size and different gray scale. Fig. 3(a) is the result by Otsu's method which is not satisfactory because of the disturbance of complicated background and noise. Fig. 3(b) shows the result by the proposed method. It can be shown that the proposed method can effectively remove background pattern that the Ostu method cannot deal with. And it can provide a good result with the clear texts.

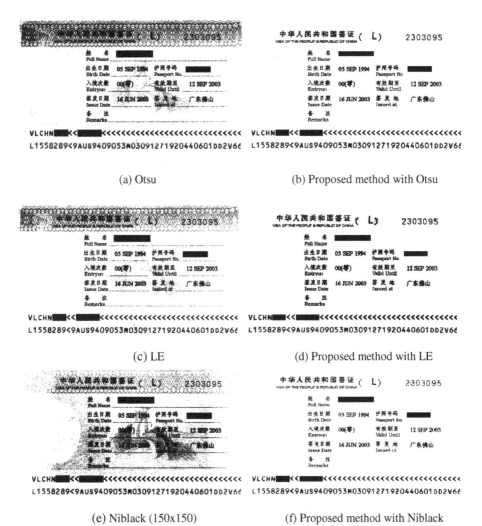

(a) Otsu

(b) Proposed method with Otsu

(c) LE

(d) Proposed method with LE

(e) Niblack (150x150)

(f) Proposed method with Niblack

Fig. 3. Comparative experiment results

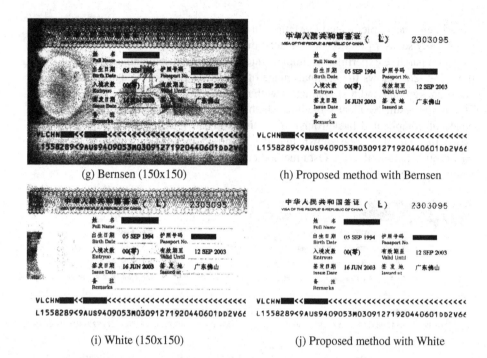

(g) Bernsen (150x150)   (h) Proposed method with Bernsen

(i) White (150x150)   (j) Proposed method with White

Fig. 3. (*continued*)

In order to evaluate the effectiveness of the proposed method, several comparative experiments are carried out with other methods, which are local entropy (LE), Niblack, Bernsen, and White method. In the proposed method, any thresholding technique can be used for the last binarization. Fig. 3 (c,e,g,i) respectively show the results with Local Entropy method (LE), Niblack, Bernsen and White method. Fig. 3 (d,f,h,j) show the corresponding results by the proposed method which the last binarization is respectively implemented by LE, Niblack, Bernsen and White method. It is very obvious that the proposed algorithm can improve the performance of the scanned card binarization system.

4   Conclusion

In this paper, an adaptive binarization algorithm is proposed which is based on multi-layers to improve the performance of scanned card processing system. The multiple layers are obtained by combination of gray information and position information. Furthermore, we propose the division rate to identify whether there is text layer in the scanned card image. It can help to obtain good text layer and reduce the disturbance of noise and background on the text layer. Experimental results demonstrate that our approach could improve the binarization performance of the scanned card processing system.

Currently, our system is able to binarize text from the scanned card images with complicated background, noise and various image qualities. However, the proposed method has difficulties in binarizing the texts which have the approximate gray value with their surrounding background. In this case it is a good way to add color information. These problems need to be tackled in the future research.

Acknowledgement. This work is supported by the National Natural Science Foundation of China (Grant No. 61003102 & No. 61103072 & 61272271) and the natural science foundation program of shanghai (Grant No.12ZR1434000).

References

1. Golder, S.A., Huberman, B.A.: Usage patterns of collaborative tagging systems. J. Inf. Sci. 32(2), 198–208 (2006)
2. Quintarelli, E.: Folksonomies: power to the people. ISKO Italy-UniMIB Meeting (2005)
3. Bao, S., Xue, G., Wu, X., Yu, Y., Fei, B., Su, Z.: Optimizing web search using social annotations. In: WWW 2007, pp. 501–510 (2007)
4. Zhou, D., Bian, J., Zheng, S., Zha, H., Giles, C.L.: Exploring social annotations for information retrieval. In: WWW 2008, pp. 715–724 (2008)
5. Xu, S., Bao, S., Fei, B., Su, Z., Yu, Y.: Exploring folksonomy for personalized search. In: Porc. of International ACM SIGIR Conference on Research and Development in Information Retrieval, pp. 155–162 (2008)
6. Conde, J.M., Vallet, D., Castells, P.: Inferring user intent in web search by exploiting social annotations. In: Proc. of International ACM SIGIR Conference on Research and Development in Information Retrieval, pp. 827–828 (2010)
7. Vallet, D., Cantador, I., Jose, J.M.: Personalizing web search with folksonomy-based user and document profiles. In: Gurrin, C., He, Y., Kazai, G., Kruschwitz, U., Little, S., Roelleke, T., Rüger, S., van Rijsbergen, K. (eds.) ECIR 2010. LNCS, vol. 5993, pp. 420–431. Springer, Heidelberg (2010)
8. Guo, J., Cheng, X., Xu, G., Shen, H.: A structured approach to query recommendation with social annotation data. In: Proc. of the ACM Conference on Information and Knowledge Management, pp. 619–628 (2010)
9. Heymann, P., Ramage, D., Garcia-Molina, H.: Social tag prediction. In: SIGIR 2008: Proceedings of the 31st Annual International ACM SIGIR Conference on Research and Development in Information Retrieval, pp. 531–538 (2008)
10. Lu, Y.-T., Yu, S.-I., Chang, T.-C., Hsu, J.Y.: A content-based method to enhance tag recommendation. In: Proc. of IJCAI 2009, pp. 2064–2069 (2009)

Integrating Visual Saliency Information into Objective Quality Assessment of Tone-Mapped Images

Xueyabo Liu[1], Lin Zhang[1,*], Hongyu Li[1], and Jianwei Lu[1,2]

[1] School of Software Engineering, Tongji University, Shanghai, China
[2] Institute of Advanced Translational Medicine, Tongji University, Shanghai, China
{8281xyb,cslinzhang,hyli}@tongji.edu.cn

Abstract. Tone-mapped images are the low dynamic range (LDR) images converted from high dynamic range (HDR) images. Recently, the objective quality assessment of tone-mapped images is becoming a challenging problem. However, there is no mature algorithm to deal with this issue until the tone-mapped image quality index (TMQI) was proposed recently, which is tone-mapped image quality index (TMQI). Unfortunately, the pooling method of the structural fidelity map in TMQI is the simple "mean", which makes the result unsatisfying. On the other hand, recent studies have found that different locations of an image may have different contributions to the quality perception of the human visual system. The significance of a local image region can be well characterized by a visual saliency (VS) model. Inspired by this insight, in this paper, we propose a VS-based pooling strategy for the objective quality assessment of tone-mapped images. The experimental results clearly demonstrate the efficacy of our proposed method.

Keywords: Tone-mapped images, high dynamic range images, objective quality assessment, visual saliency.

1 Introduction

Recently, researchers have shown a growing interest in high dynamic range (HDR) images. Compared to the low dynamic range (LDR) images, the range of intensity levels of HDR images is largely wider, thus enjoying advantages. Specifically, the range of intensity level of HDR images could be on the order of 10000 to 1, which allows for accurate representations of luminance variations in real scenes, ranging from direct sunlight to faint starlight [1], while the LDR images only have 256 intensity levels. However, General displays are designed for displaying LDR images, so we must use tone-mapping operators (TMOs) to create the corresponding LDR images for the visualization of HDR images. As the reduction in dynamic range, the LDR images converted from HDR images cannot preserve all the information. With an increasing number of TMOs being developed, a question appears, that is which TMO has a better performance? This question drives us to develop an index to compare them, and then utilize the index for optimizing parameters in TMOs.

---

* Corresponding author.

D.-S. Huang et al. (Eds.): ICIC 2014, LNCS 8588, pp. 376–386, 2014.

In the past few years, the quantitative evaluation of image perceptual quality has become mature, and there has been a lot of image quality assessment (IQA) indices of full reference (FR) methods, some of which show good consistency with subjective ratings when measure the image quality. However, all of the above are only for measuring images when the reference image and the test image have the same dynamic range, so we cannot use them for our purpose.

Until Yeganeh and Wang proposed tone-mapped image quality index (TMQI) [2], TMO assessment relied almost entirely on human subjective evaluations. However, there are many limitations which make human subjective evaluations difficult to be popularized: 1) it is expensive and inefficient; 2) it is tough to optimize; 3) human observers may ignore some missing information. What must be acknowledged is that TMQI has made a great contribution to the objective quality assessment of tone-mapped images.

TMQI proposed by Yeganeh and Wang [2] is quite efficient. In this method, a multi-scale structural fidelity measure and a statistical naturalness measure are combined. However, it has deficiency in pooling strategy it uses. Recent studies have found that different locations of an image may have different contributions to the quality perception of the human visual system. And the significance of a local image region can be well characterized by a visual saliency model. Consequently, in this paper, we attempt to extent TMQI by proposing a novel VS-based pooling strategy. Specifically, in our method, we use VS as a weight map, which is inspired by its successful use on IQA [4, 5]. The experimental results demonstrate that the VS-based pooling strategy can achieve significantly better performance than the simple "mean" that used in TMQI.

The rest of the paper is organized as follows. Section 2 provides some background knowledge about two visual saliency models which have outstanding performance when used in IQA. Section 3 presents how visual saliency information is integrated into TMQI. Section 4 reports the experimental results. Finally, Section 5 concludes the paper with a summary.

2   Two Visual Saliency Models

Building effective computational models to simulate human visual attention has been studied by scholars for a long time. Most of the existing visual attention models are bottom-up VS models, which suppose visual attention is driven by low-level stimulus in the scene, such as orientation, edges and intensities. Here, we only introduce two visual saliency models of this kind. In [3], dozens of various VS models have been proposed and several of them could predict the human visual attention accurately.

Recently, researchers have gradually found that VS is closely relevant to the image's perceptual quality. Various methods have been attempted to integrate VS information into IQA metrics [4]. These methods demonstrate that a VS-weighted pooling strategy could perform better than the simple "mean" scheme. However, no one tries to use VS-based pooling strategy in the objective quality assessment of tone-mapped images. In this paper, we tried 5 kinds of VS models, but in this section we only give a brief introduction about two of them which we found have outstanding performance when used in IQA [5] or saliency detection [7], namely SR and SDSP.

2.1   Spectral Residual Visual Saliency

Spectral residual visual saliency (SRVS) model was proposed in [6]. The computation of SRVS consists of two steps. First, we calculate the spectral residual from the log spectrum of the examined image, and then the VS map is obtained by transforming the spectral residual to the spatial domain. According to the information theory, the information in an image consists of innovation part and prior knowledge part, and the former is much easier to catch the attention of a person. In this model, spectral residual approximately represents the innovation part of an image by removing the statistical redundant components. Suppose I is the examined image. According to [6], SRVS can be computed as the following:

$$M(u,v) = abs\left(F\{I(x,y)\}(u,v)\right)$$
(1)

$$A(u,v) = angle\left(F\{I(x,y)\}(u,v)\right)$$
(2)

$$L(u,v) = \log\left(M(u,v)\right)$$
(3)

$$R(u,v) = L(u,v) - h_n(u,v) * L(u,v)$$
(4)

$$SRVS(x,y) = g(x,y) * \left(F^{-1}\{\exp(R+jA)\}(x,y)\right)^2$$
(5)

where F and F^{-1} denote the Fourier Transform and Inverse Fourier Transform, respectively. (x, y) denotes the pixel point in the spatial domain while (u, v) denotes the pixel point in the frequency domain. $abs(\cdot)$ returns the magnitude of a complex number, $angle(\cdot)$ returns the argument of the complex number, $h_n(u, v)$ is a $n \times n$ mean filter, $g(x, y)$ is a Gaussian function, which smoothes the saliency map for better visual effects, and $*$ represents the convolution. The equations (1-4) are for the computation of spectral residual, and equation (5) is for transforming from frequency domain to spatial domain.

2.2   SDSP: Saliency Detection by Combining Simple Priors

Saliency detection by combining simple priors (SDSP) was proposed in [7]. This method was inspired by three simple priors. Firstly, the behavior that the human visual system detects salient objects in a visual scene can be well modeled by band-pass filtering. Secondly, people would pay more attention on the center of an image. Thirdly, warm colors are much easier to catch people's attention than cold colors. The SDSP can be obtained as follows:

$$SDSP(\mathbf{x}) = S_F(\mathbf{x}) \cdot S_D(\mathbf{x}) \cdot S_C(\mathbf{x})$$
(6)

where $S_F(\mathbf{x})$, $S_D(\mathbf{x})$, $S_C(\mathbf{x})$ are the maps using the three simple priors, corresponding to frequency prior, location prior and color prior, respectively. They can be computed in equations (7), (9) and (10), respectively.

As for frequency prior maps,

$$S_F(\mathbf{x}) = \left((I_L * g)^2 + (I_a * g)^2 + (I_b * g)^2 \right)^{\frac{1}{2}} \tag{7}$$

where I_L, I_a, and I_b are the three resulting channels. For a given image I in the RGB color space, at first, it will be converted to the $CIEL*a*b*$ color space. $g(\mathbf{x})$ ($\mathbf{x} = (x, y) \in R^2$) is the transfer function of a log-Gabor filter in the frequency domain, which can be expressed aş:

$$G(\mathbf{u}) = \exp\left(-\left(\log\frac{\|\mathbf{u}\|_2}{\omega_0} \right)^2 \Bigg/ 2\sigma_F^2 \right) \tag{8}$$

where ($\mathbf{u} = (u, v) \in R^2$) is the coordinate in the frequency domain, ω_0 is the filter's center frequency, and σ_F controls the filter's bandwidth. $g(\mathbf{x})$ could be approximately obtained by performing a numerical inverse Fourier transformation to $G(\mathbf{u})$.

As for location prior maps,

$$S_D(\mathbf{x}) = \exp\left(-\frac{\|\mathbf{x} - \mathbf{c}\|_2^2}{\sigma_D^2} \right) \tag{9}$$

this prior can be simply and effectively modeled as a Gaussian map. $\mathbf{c}$ denotes the center of the input image I and σ_D is a parameter.

As for color prior maps,

$$S_C(\mathbf{x}) = 1 - \exp\left(-\frac{I_{an}^2(\mathbf{x}) + I_{bn}^2(\mathbf{x})}{\sigma_C^2} \right) \tag{10}$$

where σ_C is a parameter and I_{an} and I_{bn} can be computed as:

$$I_{an}(\mathbf{x}) = \frac{I_a(\mathbf{x}) - mina}{maxa - mina}, \quad I_{bn}(\mathbf{x}) = \frac{I_b(\mathbf{x}) - minb}{maxb - minb} \tag{11}$$

where I_a and I_b have the same definitions as above, $mina$ ($maxa$) is the minimum (maximum) value of I_a, and $minb$ ($maxb$) is the minimum (maximum) value of I_b.

3   Integrating Visual Saliency into TMQI

TMQI proposed by Yeganeh and Wang [2] was inspired by two successful design principles in IQA literature. The first is the structural similarity (SSIM) approach [8] and its multi-scale derivations [9, 10], and the second is the natural scene statistics (NSS) approach [11]. SSIM advocates that the main purpose of the human visual system is to extract structural information from the visual scene, which makes the structural fidelity a good method to predict the perceptual quality of an image. NSS maintains that the human visual system is highly adaptive to viewing field and uses the departure from natural image statistics as a measure of perceptual quality of an image. Based on these two principles, Yeganeh and Wang proposed the Tone Mapped image Quality Index (TMQI), which combines a multi-scale structural fidelity measure and a statistical naturalness measure [2].

3.1   Structural Fidelity

Same as the original SSIM, the structural fidelity is applied locally, but it contains only two components, which was defined as:

$$S_{local}(x,y) = \frac{2\sigma'_x\sigma'_y + C_1}{\sigma'^2_x + \sigma'^2_x + C_1} \cdot \frac{\sigma_{xy} + C_2}{\sigma_x\sigma_y + C_2} \tag{12}$$

where x, y are two local patches extracted from the HDR image and the tone-mapped LDR one. σ_x, σ_y and σ_{xy} are the local standard deviations and cross correlation between the two corresponding patches in HDR and LDR images. C_1 and C_2 are two positive stabilizing constants, used to control the result between 0 and 1. From the equation, we can see the second item is the same as the structure comparison component in the original SSIM, while the luminance comparison part is absolutely ignored since TMOs are mainly to change local intensity and contrast. The first item is the improved version of the corresponding one in the SSIM. It will work if one of HDR and LDR images is significant and the other one is insignificant. Otherwise, it makes little contributions.

To determine the significance of the signal strength, we pass the local standard deviation σ through a nonlinear mapping, and obtain the σ' value. The nonlinear mapping has to highly meet the visual sensitivity, and be designed so that significant signal strength is mapped to 1 and insignificant signal strength to 0, with a smooth transition in-between, which has been extensively studied in [12]. The computation of σ' is as the following:

$$\sigma' = \frac{1}{\sqrt{2\pi}\theta_\sigma} \int_{-\infty}^{\sigma} \exp\left[-\frac{(x-\tau_\sigma)^2}{2\theta_\sigma^2}\right]dx \tag{13}$$

$$\theta_\sigma(f) = \frac{\tau_\sigma(f)}{k} \tag{14}$$

$$\tau_\sigma(f) = \frac{\bar{\mu}}{\sqrt{2}\lambda A(f)} \tag{15}$$

$$A(f) \approx 2.6[0.0192 + 0.114f]\exp\left[-(0.114f)^{1.1}\right] \tag{16}$$

where λ is a constant to fit psychological data, f is spatial frequency and $k = 3$. Using these equations, we could obtain σ_x, σ_y, and then the local structural fidelity is computed. Then we use a sliding window running across the image to obtain the local structural fidelity of all the patches. Then we need to pool all the locals to get a single score of the image. The pooling method in TMQI is the simple "mean", which does not take the characters of human visual system into account. To overcome this deficiency, we propose a new method to use VS-based pooling strategy in the objective quality assessment of tone-mapped images. It has been widely accepted that a good quality score pooling strategy should correlate well with human visual fixation. We use VS models to characterize the visual importance of a local region. If the LDR image has a high VS value, it implies that this position $\mathbf{x}$ will have a high impact on human visual system (HVS), so we use $V_m(\mathbf{x})$ to represent the SRVS value or the SDSP value of each patch. Therefore, the structural fidelity at each scale can be computed as:

$$S_l = \frac{\sum_{\mathbf{x} \in \Omega} S_{\text{local}}(\mathbf{x}) \cdot V_m(\mathbf{x})}{\sum_{\mathbf{x} \in \Omega} V_m(\mathbf{x})} \tag{17}$$

where Ω means the whole image spatial domain. Following the idea used in multi-scale [9] and information-weighted SSIM [10], we also adopt a multi-scale approach, where the images are iteratively low-pass filtered and down-sampled to create an image pyramid structure [13]. So far, the overall structural fidelity could be calculated by combining scale level structural fidelity scores using the method in [9]:

$$S = \prod_{l=1}^{L} S_l^{\beta_l} \tag{18}$$

where L is the number of scales and β_l is the weight of the l-th scale. The equations (12-18) demonstrate the computation of the structural fidelity, and we discover that there are some parameters in the procedure. The same with TMQI, we set $C_1=0.01$ and $C_2=10$ in (12), $\bar{\mu}=128$ in (15) and $L=5$ in (18), then spatial frequency are 16, 8, 4, 2, 1 cycles/degree and the values of β_l are 0.0448, 0.2856, 0.3001, 0.2363, 0.1333 for each scale. Finally, if the input image is color image, we have to convert it from RGB color space to Yxy space and then apply the proposed structural fidelity measure on the Y component only.

3.2   Statistical Naturalness

Except faithfully preserving the structural fidelity of the HDR image, the tone-mapped LDR image should also look natural. The statistical naturalness model used here is also from [2], which is built upon brightness and contrast, because the brightness and contrast of an image could be reflected by the mean and standard deviation. The histograms of the means and standard deviations of natural images could be well fitted using a Gaussian and a Beta probability density functions given by:

$$P_m(m) = \frac{1}{\sqrt{2\pi}\sigma_m} \exp\left[-\frac{m-\mu_m}{2\sigma_m^2}\right] \tag{19}$$

$$P_d(d) = \frac{(1-d)^{\beta_d-1} d^{\alpha_d-1}}{\mathrm{B}(\alpha_d, \beta_d)} \tag{20}$$

where μ_m =115.94 and σ_m =27.99, α_d =4.4 and β_d =10.1, and B($\cdot,\cdot$) is the Beta function. Recently, studies suggested that brightness and contrast are highly independent [14], which means their joint probability density function would be the product of them. Therefore, the statistical naturalness measure could be computed as:

$$N = \frac{1}{K}P_m P_d \tag{21}$$

where K= max$\{P_m, P_d\}$, is to constrain the value between 0 and 1.

3.3   Quality Assessment Model

The model is the same as TMQI, which is a three-parameter function combining the structural fidelity measurement S and the statistical naturalness N:

$$Q = aS^\alpha + (1-a)N^\beta \tag{22}$$

where a adjusts the relative importance of the two components, and α and β determine their sensitivities, respectively. Based on experience, we set a =0.8800, α=0.3046, and β=0.7088. Until now, we could get a single score of a tone-mapped image. Figure 1 illustrates the scheme of our method for computing Q. The part circled by red frame is the difference between our method and TMQI.

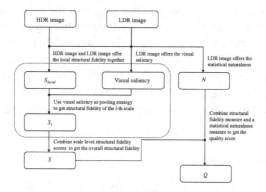

Fig. 1. Illustration for the computing scheme of Q

4   Experiments and Results

4.1   Test Protocol

In our experiment, we use a database introduced in [2], in which there are 15 data sets, and each data set has 1 HDR image and 8 LDR images created by 8 TMOs from the HDR image. The first five LDR images generated by the TMOs developed by Reinhard *et al.* [15], Drago *et al.* [16], Durand & Dorsey [17], Mantiuk *et al.* [18] and Pattanaik *et al.* [19] are computed using the publicly available software Qtpfsgui [20]. In addition, the last three LDR images were created using the built-in TMOs in Adobe Photoshop, namely "Exposure and Gamma," "Equalize Histogram," and "Local Adaptation," respectively. The parameters used in 8 TMOs are their default values. The reference HDR images include different indoor and outdoor scenes and are all available online [21-24]. In the subjective test, each subject is required to give the ranking scores from 1 to 8 representing the best and the worst quality, respectively. The subjective rankings for each image set are the average ranking scores given by 20 subjects.

The metric employed to evaluate our index is Spearman rank-order correlation coefficient (SRCC) [25], which is defined as:

$$SRCC = 1 - \frac{6\sum_{i=1}^{N} d_i^2}{N(N^2 - 1)} \tag{22}$$

where d_i is the difference between the i-th image's ranks in subjective and objective evaluations. There is no parameter in this metric which is rank-order based correlation and independent of any monotonic nonlinear mapping between subjective and objective scores. From the equation we can find that a smaller d_i value can lead to a higher SRCC value, and SRCC value will be more close to 1, especially when there are no differences between all of the image's ranks in subjective and objective evaluations, the SRCC value will be 1.

4.2   Performance Evaluation

The experiments have been carried out in our validation process, using the same database which is from [2], and all the parameters we use are the ones we discussed in section 3.3. In the first experiment, we calculate the SRCC values between the mean ranking scores and 6 sets of objective quality assessment scores obtained from 6 pooling strategies: simple "mean", 2 visual saliency models we have mentioned in section 2, and 3 typical visual saliency models, respectively. The result is given in Table 1. We use the SRCC metric to compare the scores obtained from the objective assessment index using 6 different kinds of pooling methods with the behavior of an average subject on each image set. The row called "MSP" means the mean subjective performance, which is calculated by comparing the mean ranking scores and the ranking scores given by each individual subject for each image set, and the range of ± 1 standard deviation from SRCC values of the mean over all subjects indicates that TMQI behaves quite similarly to an average subject. From Table 1, we observed that except the Itti [26] and GBVS [27] visual saliency models, which are developed a few years ago, the remaining three models show better performance than the simple "mean" pooling strategy used in TMQI, especially SDSP.

The running speed of each selected pooling methods was also evaluated. Experiments were performed on a Lenovo Qitian M7150 PC with a Pentium (R) Dual-Core CPU. The software platform was Matlab R2013a. The total time cost consumed by each IQA pooling methods from reading images to obtain the mean SRCC of 15 images sets is listed in Table 2. From Table 2 we can find that since there is no complex computation involved in our index, it has low computational cost, and what is worth mentioning is that the best SDSP is only 55 seconds slower than the simple "mean" pooling strategy, while the negligible extended time leads to a much better performance.

Table 1. Performance evaluation using 15 image sets

Image set	SRCC							
	MSP	STD of MSP	mean	Itti	GBVS	SR	LRK	SDSP
1	0.9071	0.0650	0.9048	0.9048	0.8810	0.9048	0.8810	0.9524
2	0.8251	0.1709	0.7857	0.7857	0.8095	0.7857	0.7619	0.8095
3	0.8797	0.0758	0.8095	0.8333	0.8333	0.8333	0.8333	0.7619
4	0.9130	0.0746	0.8810	0.8571	0.7857	0.7143	0.8810	0.9048
5	0.6000	0.2030	0.7381	0.7619	0.7619	0.7381	0.7381	0.6905
6	0.7630	0.1707	0.9762	0.9762	1.0000	0.9762	0.9762	1.0000
7	0.8285	0.1006	0.6905	0.8333	0.8810	0.8333	0.8810	0.7857
8	0.8023	0.1813	0.7143	0.7381	0.6667	0.7143	0.7143	0.7857
9	0.7857	0.1625	0.6905	0.5000	0.6905	0.6905	0.6190	0.7857
10	0.9276	0.0581	0.9286	0.9286	0.9286	0.9286	0.9286	0.9048
11	0.8523	0.1352	0.8810	0.7143	0.9048	0.8571	0.8571	0.8810
12	0.7595	0.2055	0.7143	0.6190	0.7143	0.7143	0.7143	0.7143
13	0.6970	0.2343	0.6826	0.6467	0.3952	0.7545	0.7904	0.8503
14	0.7702	0.1474	0.7381	0.7857	0.8333	0.8333	0.8333	0.8333
15	0.9035	0.0705	0.9524	0.9768	0.8333	0.9762	0.9762	0.9524
Average	**0.8143**	**0.1368**	**0.8058**	**0.7907**	**0.7946**	**0.8170**	**0.8257**	**0.8408**

Table 2. Total time cost of each pooling method

Pooling strategy	Time (seconds)
mean	114.051
Itti	204.347
GBVS	391.510
SR	160.513
LRK	213.780
SDSP	169.468

5   Conclusion

In this paper, we propose a new method to use VS-based pooling strategy in the objective quality assessment of tone-mapped images. This pooling strategy is designed based on the assumption that an image's VS map has a close relationship with its perceptual quality. Experimental results indicate that VS-based pooling schemes could lead to better performance than just simple "mean". Moreover, VS models used in our paper have very low computational complexity, similar as simple "mean". In conclusion, VS can be a better candidate of pooling strategy for the objective quality assessment of tone-mapped images for real-time applications.

Acknowledgement. This work is supported by the Natural Science Foundation of China under grant nos. 61201394 and 61303112, the Shanghai Pujiang Program under grant no. 13PJ1408700, and the Innovation Program of Shanghai Municipal Education Commission under grant no. 12ZZ029.

References

1. Reinhard, E., Ward, G., Pattanaik, S., Debevec, P., Heidrich, W., Myszkowski, K.: High Dynamic Range Imaging: Acquisition, Display, and Image-Based Lighting. Morgan Kaufmann, San Mateo (2010)
2. Yeganeh, H., Wang, Z.: Objective quality assessment of tone-mapped images. IEEE Trans. IP 22, 657–667 (2013)
3. Toet, A.: Computational versus psychophysical bottom-up image saliency: A comparative evaluation study. IEEE Trans. PAMI 33, 2131–2146 (2011)
4. Engelke, U., Kaprykowsky, H., Zepernick, H., Ndjiki-Nya, P.: Visual attention in quality assessment. IEEE Signal Processing Magazine 28, 50–59 (2011)
5. Zhang, L., Li, H.: SR-SIM: A fast and high performance IQA index based on spectral residual. In: Proc. ICIP, pp. 1473–1476 (2012)
6. Hou, X., Zhang, L.: Saliency detection: A spectral residual approach. In: Proc. CVPR, pp. 1–8 (2007)
7. Zhang, L., Gu, Z., Li, H.: SDSP: A novel saliency detection method by combining simple priors. In: Proc. ICIP, pp. 171–175 (2013)
8. Wang, Z., Bovik, A.C., Sheikh, H.R., Simoncelli, E.P.: Image quality assessment: From error visibility to structural similarity. IEEE Trans. IP 13, 600–612 (2004)
9. Wang, Z., Simoncelli, E.P., Bovik, A.C.: Multi-scale structural similarity for image quality assessment. In: Proc. Signals, Systems and Computers, pp. 1398–1402 (2003)
10. Wang, Z., Li, Q.: Information content weighting for perceptual image quality assessment. IEEE Trans. IP 20, 1185–1198 (2011)
11. Cadík, M., Slavík, P.: The naturalness of reproduced high dynamic range images. In: Proc. Information Visualisation, pp. 920–925 (2005)
12. Barten, P.G.J.: Contrast Sensitivity of the Human Eye and Its Effects on Image Quality. SPIE, Washington, DC (1999)
13. Burt, P.J., Adelson, E.H.: The Laplacian pyramid as a compact image code. IEEE Trans. Communications 31, 532–540 (1983)
14. Mante, V., Frazor, R., Bonin, V., Geisler, W., Carandini, M.: Independence of luminance and contrast in natural scenes and in the early visual system. Nature Neuroscience 8, 1690–1697 (2005)
15. Reinhard, E., Stark, M., Shirley, P., Ferwerda, J.: Photographic tone reproduction for digital images. ACM Transactions on Graphics 21, 267–276 (2002)
16. Drago, F., Myszkowski, K., Annen, T., Chiba, N.: Adaptive logarithmic mapping for displaying high contrast scenes. Computer Graphics Forum 22, 419–426 (2003)
17. Durand, F., Dorsey, J.: Fast bilateral filtering for the display of high dynamic-range images. ACM Transactions on Graphics 21, 257–266 (2002)
18. Mantiuk, R., Myszkowski, K., Seidel, H.: A perceptual framework for contrast processing of high dynamic range images. ACM Transactions on Applied Perception 3, 286–308 (2006)
19. Pattanaik, S.N., Tumblin, J., Yee, H., Greenberg, D.P.: Time dependent visual adaptation for fast realistic image display. In: Proc. Computer Graphics and Interactive Techniques, pp. 47–54 (2000)
20. Open Source Community (2007),
   http://qtpfsgui.sourceforge.net/index.php
21. Cadík, M., Wimmer, M., Neumann, L., Artusi, A.: Image attributes and quality for evaluation of tone mapping operators. In: Proc. Computer Graphics and Applications, pp. 35–44 (2006)

22. Reinhard, E.: High Dynamic Range Data (2009),
    http://www.cs.utah.edu/~reinhard/cdrom/hdr/
23. Ward, G.: High Dynamic Range Data (2008),
    http://www.anyhere.com/gward/pixformat/tiffluvimg.html
24. Debevec, P.: High Dynamic Range Data (2010),
    http://www.debevec.org/Research/HDR/
25. Wang, Z., Li, Q.: Information content weighting for perceptual image quality assessment. IEEE Trans. IP 20, 1185–1198 (2011)
26. Itti, L., Koch, C., Niebur, E.: A model of saliency based visual attention for rapid scene analysis. IEEE Trans. PAMI 20, 1254–1259 (1998)
27. Harel, J., Koch, C., Perona, P.: Graph-based visual saliency. In: Advances in Neural Information Processing Systems, vol. 19, pp. 545–552 (2007)

Moving Object Detection Method with Temporal and Spatial Variation Based on Multi-info Fusion

Yaochi Zhao, Zhuhua Hu[*], Xiong Yang, and Yong Bai

College of Information Science & Technology, Hainan University,
Haikou, Hainan 570228, China
{yaochizi,eagler_hu}@163.com

Abstract. Aiming at the problem that precision of frame difference for moving object detection is weak, a multi-info fusion model and a novel moving object detection algorithm based on the model are presented in this paper. Firstly, the temporal difference image of two frames in a motion sequence is reconstructed with morphologic operator to obtain target area. Then the spatial-temporal information in the target area is integrated into a fusion image by using the multi-info fusion model. Finally the accurate moving object is detected with automatic threshold segmentation method. The methods of fusion in fusion model are discussed, and a static linear fusion and dynamic self-adaptive fusion based on temporal entropy are presented. The experimental results show that the edge of the obtained moving target with the multi-info fusion method proposed is more accurate than the existing method, and the time complexity is low, which meets the requirement of real-time detection.

Keywords: Moving object detection, Multi-info fusion, Morphological reconstruction.

1   Introduction

Moving object detection is a fundamental task in computer vision and pattern recognition. Detecting moving object efficiently and real-timely can provide evidence for further object tracking, behavior understanding and analysis. At present, the conventional methods for moving object detection are frame difference, optical flow and background subtraction [1-4]. Optical flow is high time complexity and cannot meet the need of real time detection. Background subtraction is efficient in the condition that the background model is accurate; otherwise, large detection error will be generated when the background is changed or it is interfered by light or weather. Frame difference use temporal difference between consecutive frames to get the motion object, this method is simple to implement and has strong adaptability to dynamic environment, But it is not efficient to get accurate and continuous edge of objection, especially for the edge that have the same direction with moving direction.

---

[*] Corresponding author.

D.-S. Huang et al. (Eds.): ICIC 2014, LNCS 8588, pp. 387–397, 2014.
© Springer International Publishing Switzerland 2014

However, edge carries a lot of information and it always is the base on which image is analyzed and understood. So obtaining accurate and continuous edge is an important task in the stage of image segmentation. Aiming at the precision problem that exists in frame difference, many improved methods have been investigated. One improved idea is to increase the number of frames joined in the calculation [5-8], which improves the continuity of objection edge to some extent, but the obtained edges is the addition of multiple frames' target edges. Another improved idea is to use the edge difference [9, 10,11], with which other method combine for moving object detection [12], or on which subsequent processing will be made [13,14]. Edge difference has some robustness, but it still is not efficient to get accurate and continuous edge of objection when the edge direction has the same with moving direction [6].

To improve the integrity and continuity of target edge, a multi-info fusion model is constructed and a new method for moving objection detection based on the multi-info fusion is presented in this paper. Firstly, the area of moving object is obtained using frame difference and morphologic reconstruction. Then the temporal and spatial variation information is integrated into a fused image according to the multi-info fusion model. Lastly, the accurate moving objection is obtained with automatic threshold segmentation method. The fusion methods in the multi-info fusion model are considered with emphasis, and a static linear fusion and a dynamic self-adaptive fusion based on temporal entropy are presented for reference. The experimental results show that the edges of the moving target obtained with the multi-info fusion method are more accurate and continuous than the traditional and the existing improved methods and the time complexity of the algorithms are low, which meet the requirement of real-time detection.

2   Brief Introduction to the Multi-info Fusion Algorithm

The algorithm process is shown in Fig.1.

Firstly, temporal difference image where the absolute difference is calculated at each pixel between two consecutive frames and a threshold by Otsu [15] is applied to get the binary image. Secondly, the morphologic open-reconstruction operation is applied to remove noise keeping the characteristics of edge of moving objection, and then the region of interesting (ROI) is obtained. Thirdly, the spatial-temporal variation information, such as temporal difference in ROI, the spatial differences (edge) of the current frame and the former frame in ROI, the edge difference in ROI, is integrated into a fused image based on the multi-information fusion model. Finally, a threshold by Otsu also applied to segment the fused image, and a more precise moving object edge can be obtained with a common edge detection operator.

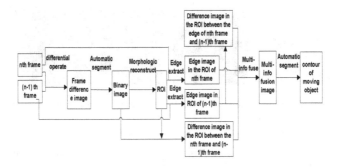

Fig. 1. Flowchart of multi-info fusion algorithm for moving objection detection

3   Motion Region Extraction

3.1   Frame Difference

Ideally, every pixel value of the background in the difference image is zero, so moving target can be obtained by analyzing the difference image. Detailed description is as follows.

Let $G(x, y, t_{i-1})$ and $G(x, y, t_i)$ are successive images of two adjacent frames, $G_{dif_{(i-1,i)}}(x, y)$ is difference for two successive frames, the equation is as follow:

$$G_{dif_{(i-1,i)}}(x, y) = \left| G(x, y, t_i) - G(x, y, t_{i-1}) \right| \tag{1}$$

Using a threshold T to segment the difference image, the binary image $BW\_G_{dif_{(i-1,i)}}(x, y)$ can be obtained, as shown in equation (2).

$$BW\_G_{dif_{(i-1,i)}}(x, y) = \begin{cases} 1 & if \ G_{dif_{(i-1,i)}}(x, y) > T \\ 0 & otherwise \end{cases} \tag{2}$$

3.2   Morphologic Reconstruction

The segmented binary image $BW\_G_{dif_{(i-1,i)}}(x, y)$ above-mentioned, is always affected by noise and interference, and exists some isolate region that is not moving objection. So some filter algorithm is needed to remove the non motion region. The traditional method is to use morphologic open operation to filter noise or small interference [16, 17]. Morphologic open operation [18] is efficient to remove noise and some small isolate interference, but the moving target always destructed partly, as shown in Fig.2(c).

Fig. 2. Example of (a) Video sequence, (b) difference image, (c) morphologic open filter, (d) morphologic open reconstruction filter

In order to solve the problem of contour destructed, morphologic open reconstruction operate is introduced as a lossless filter. Detailed description is as follows:

$$O_R(F) = R_F^D \left[(F \ominus B) \right] \qquad (3)$$

In equation (3), F is the binary image to be filtered. B is the structural element. $F \ominus B$ is morphologic erode operation. R_F^D is morphologic reconstruction operation, the equation is as follow [19].

$$R_F^D(I) = \bigcup_{F \cap I_K \neq \phi} I_K \qquad (4)$$

In equation (4), I is the image to be reconstructed, $I_1, I_2, \dots I_K$ are the connected components of the image I.

As shown in Fig.2(d), the contour of edge preserves better with morphologic open reconstruction filter than morphologic open filter.

4   Multi-info Fusion

Frame difference is widely used because of small calculating amount and strong adaptability for scene change, but it exists essential defect. If the moving object is of region homogeneity, the obtained moving object would exist cavity. As shown in Fig.3, if the gray rectangular object in Fig.3(a) moves horizontally, the obtained object with frame difference in Fig.3(b) is fractured with a big cavity, and the horizontal edge cannot be obtained. However the moving target with region homogeneity is common and the most topical target is moving vehicles. As shown in Fig.2, the bottom edge of the car cannot be obtained with frame difference. In addition frame difference is not effective for non-rigid motion.

(a)   (b)

Fig. 3. Disadvantage schematic diagram of traditional frame difference method

To improve the precision of frame difference and expand its application range, a multi-info fusion model, as shown in equation (5), is proposed, with which frame difference image, edge image and edge difference image are integrated into a multi-info fusion image, and then a more precise moving object edge can be obtained by the subsequent treatment.

4.1   Multi-info Fusion Model

$$\varepsilon(I_{IR}(x, y)) = f(P(G_{dif}(x, y)), P(E_{dif}(x, y)), P(E_{cur}(x, y)), P(E_{for}(x, y)))$$
(5)

In equation (5), $\varepsilon(I_{IR}(x, y))$ is the multi-info fusion value in the position of (x, y) at the current time, G_{dif} is the temporal difference value between the current time and the former time, E_{dif} is the edge difference value between the current time and the former time, E_{cur} is the edge intensity at the current time, E_{for} is the edge intensity at the former time, P is the normalized function with which the all information are normalized into the range from 0 to 255, f is the multi-info fusion method.

4.2   Multi-info Fusion Method

In equation (5), the choice of f is the key for the fusion model. In consideration of precision, real-time performance and the characteristic of the scene, a static linear fusion method and a dynamic self-adaptive fusion method based on temporal entropy are proposed.

Static Linear Fusion

$$f(P(G_{dif}(x, y)), P(E_{dif}(x, y)), P(E_{cur}(x, y)), P(E_{for}(x, y))) =$$
$$k * P(G_{dif}(x, y)) + l * P(E_{dif}(x, y)) + m * P(E_{cur}(x, y)) + n * P(E_{for}(x, y))$$
(6)

In equation (6), k, l, m, n are const as the weight coefficient of all information. The key of static linear fusion method is the value of the weight coefficient. If the weight coefficients are suitable, the intensity of background edge will be low in the fusion image, and then will be considered as background in the subsequent stage. According to simple background, a relatively large value for m and n in the fusion model is suitable, and then smooth and continuous edge will be obtained. To complex background, the value for m and n would be smaller, because the contours of the background influence greatly in the fusion model. There are some special situations for the value of the weight coefficient.

If $m = 0$, and $n = 0$, and $l = 0$, the model is simplified as the current frame difference.

If $m = 0$, and $n = 0$, and $k = 0$, the model is simplified as the edge sub difference, as introduced in the reference [9].

If $m = 0$, and $n = 0$, only the time difference and edge difference are considered in the model.

Dynamic Self-Adaptive Fusion Based on Temporal Entropy

In equation (6), the weight coefficients represent the scene's prior knowledge, however obtaining the best parameter need debug many times. Considering the question of parameter's self-adaptive, entropy is introduced to the multi-info fusion model.

In moving sequence, the pixel value at a particular location could change due to noises or motion object. Along with time, the pixel's state would change from one to another, pixel state's change brought by noises would be in a small range, but those brought by motion would be large. So temporal entropy, as a measure of the diversity of state can be used to obtain moving object. In the reference [20] spatial-temporal entropy image (STEI) was proposed to detect the moving object, and the spatial-temporal entropy of frame difference image (DSTEI) was introduced to get more precise moving target in the reference [21].However, by using the method of spatial-temporal entropy the obtained moving object is the superposition of multi-frames', and the time-complexity is high.

In this paper, temporal entropy is introduced as the weight coefficient of E_{cur} and E_{for}. The temporal entropy of every pixel is calculated as the probability of that the pixel is background, and the coefficient of E_{cur} and E_{for} are decided.

$$
\begin{aligned}
f(P(G_{dif}(x, y)), P(E_{dif}(x, y)), P(E_{cur}(x, y)), P(E_{for}(x, y))) = \\
P(G_{dif}(x, y))^r + 0 * P(E_{dif}(x, y))^r \\
+ entr(N, x, y) * P(E_{cur}(x, y))^r + entr(N, x, y) * P(E_{for}(x, y))
\end{aligned}
\tag{7}
$$

In equation (7), $entr(N, x, y)$ is the temporal entropy in the position of (x, y) at the time N, detailed description is as follows.

$$
entr(N, x, y) = -\sum_{i=0}^{n-1} p_{N-i}(x, y) * \log p_{N-i}(x, y)
\tag{8}
$$

In equation (8), n is the length of time window, $p_{N-i}(x, y)$ is the probability of the intensity value in the position of (x, y) at the time $N - i$, detailed description is as follows.

$$p_{N-i}(x, y) = \frac{H_{N-i}(x, y)}{n} \qquad (9)$$

In equation (9), $H_{N-i}(x, y)$ is the number of the intensity value in the position of (x, y) at the time $N - i$ in the temporal entropy histogram.

It is worth noting that the coefficient of l is 0 in the method of dynamic self-adaptive fusion based on temporal entropy, it is because that the coefficient of l is used to inhibit background edge, and $entr(N, x, y)$ as the coefficient of $E_{cur}(x, y)$ and $E_{for}(x, y)$ is the probability of that the pixel is background and can inhibit background edge too.

5   Experimental Results

In this section, we applied the proposed algorithms to three different disturbed scenes to detect moving car or person. The part of experimental results is demonstrated in Fig.4.

In Fig.4(a) one frame in experiment's sequences is illustrated. The main interference in scene 1 is that moving object is occluded. In scene 2 swing leaves are the source of disturbance. In scene 3 the light reflection of ground surface affects accuracy of object detection.

Fig.4(b) shows the results where the global DSTEI in [21] is automatically segmented in every scene. In Fig.4(c) the DSTEI in ROI obtained by using the algorithm showed in Fig.1 is automatically segmented. Fig.4(d) shows the results where the multi-info fusion method based on temporal entropy is applied in the ROI. From the experiment effects it is clear that the global spatial-temporal method in Fig.4(b) works in failure, and the local spatial-temporal method in Fig.4(c) works better than the global, yet the moving object obtained by using local spatial-temporal method is the addition of multi-frame's, and the multi-info fusion method based on temporal entropy in ROI in Fig.4(d) has a significantly better effect than the former two methods, especially in scene 3.

In Fig.4(e) we applied edge difference method presented in [9] on the three scenes. In Fig.4(f) the multi-info linear fusion method in ROI proposed in this paper is used, the fusion parameter settings are shown as Table 1, the edge coefficient m and n was set at zero in the scene 2 because there is the great leaves swing interference, in which case the multi-info fusion is actually the integration of frame difference and edge difference together. Comparing the effects between the multi-info linear fusion method in ROI in this paper and the edge difference method in [9], the details of pedestrian legs in scene 1 and the edge of car roof in scene 2 in the former are significantly better than the later.

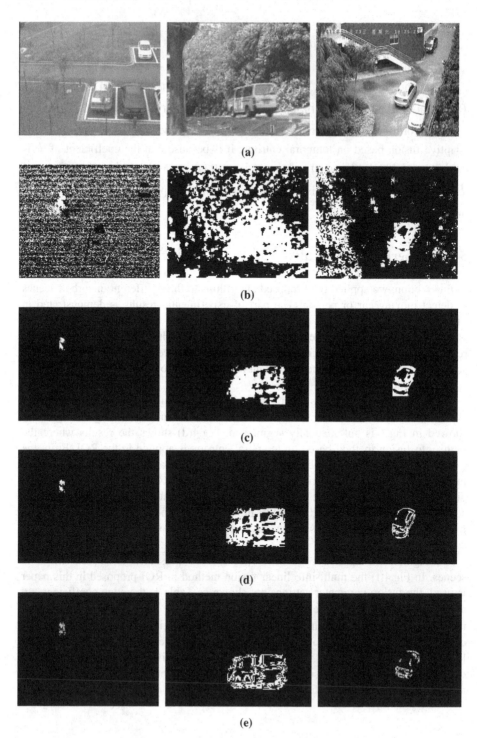

Fig. 4. Comparison of experimental effects

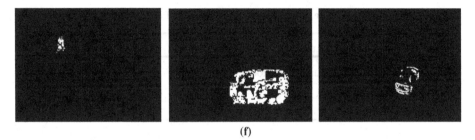

(f)

Fig. 4. (*continued*)

Table 1. Parameters setting

Parameters setting	Scene 1	Scene 2	Scene 3
k	1	1	1
l	1	1	1
m	1	0	1
n	1	0	1

Comparing the proposed multi-info dynamic self-adaptive fusion method based on temporal entropy in ROI in fourth line with the proposed multi-info static linear fusion method in ROI in Fig.4(f), in the scene 2 and the scene 3, there are non-rigid motion of pedestrian or swing leaves, which have influence on the entropy value, so the experiment effects by the former method are inferior to the latter. As for the scene 3, the motion object has significant region homogeneity, and the edge obtained is more continuous and with more details by using the former method than the latter.

As far as the time complexity of algorithms is concerned, the proposed algorithms in this paper are simple and easy to implement. The Table 2 shows the consumed average time of detecting the moving object in each scene. The detection time is related to the size of fame and the size of motion object. The size of the frame in scene 1, scene 2 and scene 3 are 704*576, 352*288 and 160*128 respectively. In general case, comparing the multi-info static linear fusion method with the multi-info dynamic self-adaptive fusion method based on temporal entropy, the time complexity of the latter is 2 to 3 times higher than that of the former. If the code can be optimized, even for the high resolution frame such as scene 1, the both methods can meet completely the demand of real-time detection.

All experiments were conducted on an Inter(R) Core(TM) 3.30GHZ CPU, and 3.24GB RAM running 32-bit Windows XP operating system. The proposed algorithms were implemented in Matlab 7.1 and visual C++ 6.0 cross-compiler environments.

Table 2. Comparison of experimental time

Fusion method	Scene 1	Scene 2	Scene 3
multi-info static linear fusion method/s	0.0677	0.0221	0.0213
multi-info dynamic self-adaptive fusion method based on temporal entropy/s	0.2206	0.0637	0.0525

6   Conclusion

The traditional frame difference method for moving target detection can't obtain the continuous edge. In this paper, we propose a novel multi-info fusion method, which first gets the ROI and then fuses the information of frame difference, edge and edge difference together, finally obtains the relatively continuous edge of motion object by automatic threshold method. In the stage of getting ROI, to filter noise and guarantee the continuity of edge simultaneously, morphological open-reconstruction operation was used to obtain the accurate ROI. In the stage of multi-info fusion, we built the multi-info fusion model, and the static linear fusion method based on experience and dynamic self-adaptive fusion method based on temporal entropy is proposed for reference. The static fusion has lower time complexity than the dynamic fusion, yet both methods can meet the demand of real-time detection. On the basis of proposed multi-info fusion model in this paper, to seek multi-info fusion method from a statistical point of view is a topic worthy of our further study.

Acknowledgements. The authors thank for the supports by the Youth Found in Hainan University (No. qnjj1245, No. qnjj1185), the scientific research fund of Hainan provincial education department (No. hjkj2013-14), and National Natural Science Foundation of China (No. 61261024).

References

1. Zhan, W., Yang, J.: Moving Object Detection from Video with Optical Flow Computation. International Information Institute (Tokyo). Information 15(10), 4157–4164 (2012)
2. Subudhi, B.N., Ghosh, S., Ghosh, A.: Change Detection for Moving Object Segmentation with Robust Background Construction under Wronskian Framework. Machine Vision and Applications 24(4), 795–809 (2013)
3. Ince, S., Konrad, J.: Occlusion-Aware Optical Flow Estimation. IEEE Transactions on Image Processing 17(8), 1443–1451 (2008)
4. Barbu, T.: Multiple Object Detection and Tracking in Sonar Movies using an Improved Temporal Differencing Approach and Texture Analysis. UPB Scientific Bulletin, Series A 74, 27–40 (2012)
5. Zhang, Y.J.: Image analysis. Tsinghua University Press, Beijing (2005) (in Chinese)
6. Sonka, M., Hlavac, V., Boyle, R.: Image Processing, Analysis, and Machine Vision. Thomson, Toronto (2008) (in Chinese)

7. Li, J., Lan, J.H., Li, J.: A Novel Fast Moving Target Detection Method. Journal of Central South University (Science and Technology) 44(3), 978–984 (2013)
8. Hao, H.G., Chen, J.Q.: Moving Object Detection Algorithm Based on Five Frame Difference and Background Difference. Computer Engineering 38(4), 146–148 (2012)
9. Gan, M.G., Chen, J., Liu, J.: Moving Object Detection Algorithm Based on Three-Frame-Differencing and Edge Information. Journal of Electronics & Information Technology 32(4), 894–897 (2010)
10. Luan, Q.L., Zhao, W.S.: Moving Object Detection Algorithm Based on Three-Frame-Difference of Moving Background and Edge Information. Opto-Electronic Engineering 38(10), 77–83 (2011)
11. Song, N., Shang, Z.H., Liu, H.: Moving Object Detection Algorithm Based on Color Image Edge-Differencing Extraction. Microcomputer & Its Applications 30(24), 36–42 (2011)
12. Gao, K.L., Qin, T.F., Wang, Y.Z.: A Novel Approach for Moving Objects Detection Based on Frames Subtraction and Background Subtraction. Telecommunication Engineering 51(10), 86–91 (2011)
13. Zhan, C.H., Duan, X.H., Xu, S.Z.: An Improved Moving Object Detection Algorithm Based on Frame Difference and Edge Detection. In: 4th International Conference on Image and Graphics, pp. 519–523. IEEE Press, Chengdu (2007)
14. Cui, Y.Y., Zeng, Z.Y., Cui, W.H., Fu, B.T., Liu, W.: Moving Object Detection Based on Edge Pair Difference. Advanced Materials Research 204, 1407–1410 (2011)
15. Otsu, N.: A Threshold Selection Method from Gray-Level Histogram. Automatica 9(1), 62–66 (1979)
16. Xu, W.W., Duan, X.H.: Infrared Moving Target Detection Based on Transition Region Extraction. Laser & Infrared 40(7), 775–778 (2010)
17. Li, J.C., Liu, X.M., Pan, W.Q., Xue, F.L., Liu, J.: A Method of Infrared Image Moving Object Detection on Dynamic Background. Opto-Electronic Engineering 40(3), 1–6 (2013)
18. Gonzalez, R.C., Woods, R.E.: Digital Image Processing, 3rd edn. Electronics Industry Publish House, Beijing (2011) (in Chinese)
19. Liu, J.Q., Ruan, Q.Q.: The Research and Application of Morphological Filter by Reconstruction. Journal on Communications 23(1), 116–121 (2002)
20. Ma, Y.F., Zhang, H.J.: Detecting Motion Object by Spatio-Temporal Entropy. In: IEEE International Conference on Multimedia and Expo, Tokyo, pp. 265–268 (2001)
21. Guo, J., Chng, E.S., Deepu, R.: Foreground Motion Detection by Difference-Based Spatial Temporal Entropy Image. In: 2004 IEEE Region 10 Conference TENCON 2004, pp. 379–382 (2004)

Images Denoising with Feature Extraction for Patch Matching in Block Matching and 3D Filtering

Guangyi Chen[1], Wenfang Xie[2], and Shu-Ling Dai[3]

[1] Department of Computer Science and Software Engineering, Concordia University,
Montreal, Quebec, Canada H3G 1M8
guang_c@cse.concordia.ca
[2] Department of Mechanical and Industrial Engineering, Concordia University,
Montreal, Quebec, Canada H3G 1M8
wfxie@encs.concordia.ca
[3] State Key Lab. of Virtual Reality Technology and Systems, Beihang University, ZipCode
100191, No 37, Xueyuan Rd., Haidian District, Beijing, P.R. China
sldai@yeah.net

Abstract. In this paper, we propose a new method for grey scale image denoising. Our method takes advantage of the fact that the mean of the Gaussian white noise is zero. For every patch in the noisy image, we use a line to divide the image into two regions with equal area, and then take the mean of one of the two regions. We select lines with different slopes in order to extract a number of features. We use these extracted features to match the patches in the noisy image. All other steps in our method are the same as those in the standard BM3D. Our experimental results show that our new method outperforms the standard BM3D for $\sigma_n > 120$, and they are identical, otherwise.

Keywords: Image denoising, block matching and 3D filtering (BM3D), Gaussian white noise.

1 Introduction

Noise reduction of an image is a very important problem in a number of real-life applications. Gaussian white noise is the most popular topic in the literature. We formulate this kind of noise in image B as [1]:

$$B = A + \sigma_n Z \tag{1}$$

where A is the noise-free image, σ_n is the noise standard deviation, and Z is the Gaussian white noise with $N(0,1)$ distribution.

There are many methods in the literature for reducing this kind of noise. Sendur and Selesnick [1] proposed a bivariate denoising method by employing parent-child relationship in the wavelet domain. Dabov et al. [2] developed the block matching and 3D filtering (BM3D) for image denoising. Chen and Wu [3] investigated the so-called bounded BM3D (BBM3D) method for image denoising. Luisier et al. [4] proposed a SURE-based denoising method for image denoising. Chen and Kegl [6] proposed an

D.-S. Huang et al. (Eds.): ICIC 2014, LNCS 8588, pp. 398–406, 2014.
© Springer International Publishing Switzerland 2014

image denoising method by using complex ridgelets. Chen et al. ([7], [8]) developed two image denoising methods by considering coefficient dependency in the wavelet domain. Cho et al. [9] also proposed several denoising techniques for images. Cho and Bui [10] studied multivariate statistical modeling for image denoising using wavelet transforms. Recently, there are a few new image denoising methods appeared in the literature. Fathi and Naghsh-Nilchi [11] proposed an efficient image denoising method based on a new adaptive wavelet packet thresholding function. Chatterjee and Milanfar [12] studied patch-based near-optimal image denoising. Rajwade et al. [13] worked on image denoising using the higher order singular value decomposition. Motta et al. [14] proposed the iDUDE framework for gray scale image denoising. Miller and Kingsburg [15] studied image denoising using derotated complex wavelet coefficients.

In this paper, we propose a new method for reducing Gaussian white noise in images. Our method is based on block matching and 3D filtering (BM3D) [2], which is the state-of-the-art in image denoising. We use lines with different slopes to divide each patch into two regions and then calculate the means of these regions because we can take advantage of the zero mean of the Gaussian white noise. We use these extracted features to align the image patches. All other steps in our method are identical to those in the BM3D method. Experimental results show that our method outperforms existing denoising methods in heavy noisy environment.

The organization of this paper is as follows. Section 2 proposes a new method for reducing the Gaussian white noise in an image. Section 3 conducts some experiments in order to show the superiority of our new method. Finally, Section 4 concludes the paper and proposes future research directions.

2   Proposed Method

In this section, we propose a new method to reduce the Gaussian white noise in a noisy image. The noise standard deviation can be approximated as [5]:

$$\sigma_n = \frac{median(|\, y_{1i}\, |)}{0.6745}, \quad y_{1i} \in \text{subband } HH_1. \tag{2}$$

where HH_1 is the finest scale of wavelet coefficient subband. Note that we only need to perform the wavelet transform on the noisy image for one decomposition scale in order to estimate σ_n.

We know that the mean of Gaussian white noise is zero, so we divide each image patch into two regions with equal area and then calculate the mean of one of these two regions. We choose a number of lines with different slopes so that a moderate number of features can be extracted from the image patches. Let $k \in [1, K]$, where K×K is the size of the image patch. We choose the lines passing through the following two points as in Cases 1 and 2:

Case 1:   (1,k) and (k,K-k)
Case 2:   (k,1) and (K-k,k)

The total number of features for each patch is then $2K+1$ whereas the patch size is K^2. It is easy to know that these lines divide the image patches into two regions with equal area. We then take the mean of one of the two regions for each line and use the extracted mean features to match the patches, where the nearest neighbour classifier is utilized. The closest patches should have the smallest distance in these features. All other procedures in our denoising method are the same as those of the BM3D. Figs. 1 and 2 show the noise-free patch with 8×8 and 16×16 pixels, its noisy patch (σ_n =140), the extracted features from both the noise-free and noisy patches, and the difference between the noise-free and noisy features. The horizontal axes of the two lower sub-figures are the feature size, and the vertical axes represent the extracted features (lower-left) and difference of features between the noise-free and noisy patches. It can be seen that our extracted features are very robust to Gaussian white noise.

In summary, we list the steps of our new method as follows:

Step 1. Given the noisy grey scale image B, estimate the noise standard deviation σ_n from B according to equation (2).

Step 2. If $\sigma_n \leq 120$, then perform BM3D to B as $B_1 = BM3D(B,\sigma_n)$. Set $\tilde{B} = 255 \times B_1$ since BM3D scales the output image to the range of [0,1]. Stop.

Step 3. If $\sigma_n > 120$, then use a number of lines with different slopes to divide each patch into two regions with equal area. Calculate the means of one of the two regions.

Step 4. Use these extracted features to match the patches in the noisy image.

Step 5. Denoise the 3D patches and then put back the denoised patches, just like the standard BM3D. Stop.

The major contribution of this paper is the following. Our proposed method falls back the standard BM3D when the noise level is not high. In addition, it outperforms the BM3D when $\sigma_n > 120$ in terms of PSNR. Furthermore, the feature length of our extracted features from patches is shorter than the patch size. This means that our method should be fast as well. However, the calculation of the mean features from each patch is time consuming. Our experiments show that our new denoising method is a bit slower than the standard BM3D for denoising images.

3   Experimental Results

We conducted a number of experiments in order to demonstrate the power of our proposed method in this paper. We tested our method with grey scale images: Lena, Boat, and Barbara. We generate the noisy images by using equation (1). Fig. 3 displays these three images, which are frequently used in other denoising papers in the literature. We compared our method with the standard BM3D [2], the bounded BM3D (BBM3D [3]), and the SURELET [4] for image denoising. Tables 1-3 tabulate the peak signal to noise ratio (PSNR) of the BBM3D, SURELET, BM3D and our proposed method in this paper for these three images, respectively. The PSNR is defined as

$$PSNR = 10 \log_{10} \left(\frac{M \times N \times 255^2}{\sum_{i,j} (B(i, j) - A(i, j))^2} \right) \tag{3}$$

where $M \times N$ is the number of pixels in the image, and A and B are the noise-free and denoised images. It can be seen that our proposed method is identical to BM3D for noise standard deviation $\sigma_n \leq 120$, and our method outperforms the SURELET, BBM3D, and BM3D methods for nearly all other cases. The only exception is in Table 2, where the SURELET is the best for $\sigma_n = 220$ and $\sigma_n = 240$. However, the SURELET is only a bit better than our proposed method for these two cases. Figs. 4-6 show the noise-free, noisy ($\sigma_n = 220$), and the denoised images by SURELET, BBM3D, BM3D, and the proposed methods for the Lena, Boat, and Barbara images. It can be seen that our denoised images are closer to the noise-free images than images generated by all other three methods in the experiments. The images obtained by BBM3D and BM3D do not have smooth regions as the noise-free image, but our new method does.

Table 1. The PSNR of different denoising methods for image Lena with Gaussian white noise. The best results are highlighted in bold font.

σ_N	NOISY	SURELET	BBM3D	BM3D	PROPOSED
20	22.10	31.36	**33.03**	**33.03**	**33.03**
40	16.08	28.29	**29.82**	**29.82**	**29.82**
60	12.56	26.59	26.63	**28.15**	**28.15**
80	10.06	25.45	24.97	**26.82**	**26.82**
100	8.12	24.59	23.67	**25.76**	**25.76**
120	6.53	23.90	22.60	**24.89**	**24.89**
140	5.20	23.32	21.59	23.62	**24.10**
160	4.04	22.81	20.65	21.95	**23.47**
180	3.01	22.37	19.77	20.86	**22.91**
200	2.10	21.96	18.88	19.98	**22.37**
220	1.27	21.59	17.99	19.15	**21.87**
240	0.51	21.25	17.09	18.45	**21.41**

Table 2. The PSNR of different denoising methods for image Boat with Gaussian white noise. The best results are highlighted in bold font.

σ_N	NOISY	SURELET	BBM3D	BM3D	PROPOSED
20	22.10	29.40	30.76	**30.79**	**30.79**
40	16.08	26.39	**27.63**	**27.63**	**27.63**
60	12.56	24.80	25.03	**25.90**	**25.90**
80	10.06	23.79	23.63	**24.74**	**24.74**
100	8.12	23.06	22.48	**23.88**	**23.88**
120	6.53	22.49	21.53	**23.16**	**23.16**
140	5.20	22.03	20.73	22.21	**22.43**
160	4.04	21.63	19.96	20.92	**21.93**
180	3.01	21.29	19.23	20.12	**21.49**
200	2.10	20.99	18.56	19.46	**21.06**
220	1.27	**20.71**	17.91	18.81	20.67
240	0.51	**20.46**	17.21	18.22	20.30

Table 3. The PSNR of different denoising methods for image Barbara with Gaussian white noise. The best results are highlighted in bold font.

σ_N	NOISY	SURELET	BBM3D	BM3D	PROPOSED
20	22.10	27.81	**31.73**	**31.73**	**31.73**
40	16.08	24.53	**28.00**	**28.00**	**28.00**
60	12.56	23.14	23.58	**26.33**	**26.33**
80	10.06	22.35	22.36	**24.84**	**24.84**
100	8.12	21.79	21.46	**23.66**	**23.66**
120	6.53	21.32	20.70	**22.69**	**22.69**
140	5.20	20.93	19.97	21.31	**21.64**
160	4.04	20.58	19.29	20.14	**21.20**
180	3.01	20.28	18.62	19.45	**20.79**
200	2.10	20.00	17.91	18.78	**20.44**
220	1.27	19.74	17.20	18.16	**20.09**
240	0.51	19.51	16.51	17.63	**19.77**

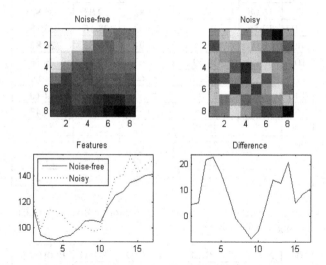

Fig. 1. The noise-free patch with 8×8 pixels, its noisy patch (σ_n =140), the extracted features from both patches, and the difference between the noise-free and noisy features. The horizontal axes of the two lower sub-figures are the feature size, and the vertical axes represent the extracted features (lower-left) and difference of features between the noise-free and noisy patches. It can be seen that our extracted features are robust to Gaussian white noise.

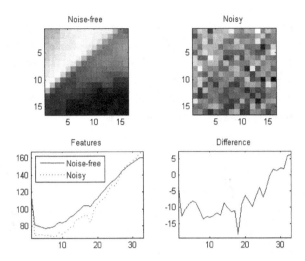

Fig. 2. The noise-free patch with 16×16 pixels, its noisy patch (σ_n =140), the extracted features from both patches, and the difference between the noise-free and noisy features. The horizontal axes of the two lower sub-figures are the feature size, and the vertical axes represent the extracted features (lower-left) and difference of features between the noise-free and noisy patches. It can be seen that our extracted features are robust to Gaussian white noise.

Fig. 3. The three images used in our experiments: Lena (left), Boat (middle), and Barbara (right)

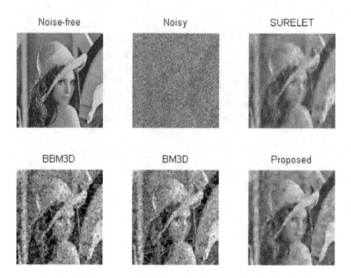

Fig. 4. The noise-free, noisy (σ_n=220), and the denoised images by SURELET, BBM3D, BM3D, and the proposed methods for the Lena image

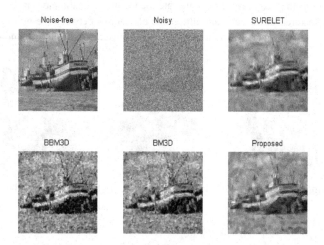

Fig. 5. The noise-free, noisy (σ_n=220), and the denoised images by SURELET, BBM3D, BM3D, and the proposed methods for the Boat image

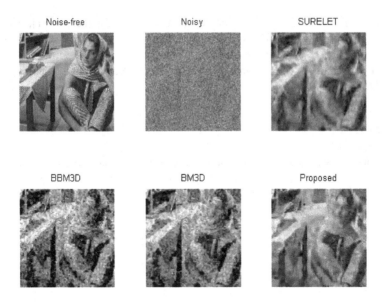

Fig. 6. The noise-free, noisy (σ_n=220), and the denoised images by SURELET, BBM3D, BM3D, and the proposed methods for the Barbara image

In standard BM3D, the noise variance σ_n is a known parameter for the noisy image. We estimate it by using equation (2) in this paper. Since we only need to perform the wavelet transform for one decomposition scale, the time to estimate σ_n is fast.

4   Conclusions and Future Works

In this paper, we have proposed a new method for reducing the Gaussian white noise by extracting features from each patch and align the patches by using these features. The closest patches should have the smallest distance in these features. All other steps in our new method are identical to the BM3D. Our experiments show that our new method outperforms the standard BM3D under heavy noise environment and it is identical to the BM3D method for other noisy conditions. Our new method is nearly always better than the BBM3D and SURELET methods for image denoising except two cases in Table 2, where the SURELET method is the best among all three compared methods.

Future research will be conducted in order to deal with other types of noise in the noisy 1D signals, 2D images, and 3D videos. We believe that our proposed method in this paper may also be applied to multi-spectral or hyper-spectral satellite imagery as well. Furthermore, we would like to align the image patches by affine transformation so that better denoising results can be obtained.

Acknowledgements. The authors would like to thank the authors of [1], [2], [3], [4] and [16] for posting their denoising software on their websites. This research was supported by the research grant from the Natural Science and Engineering Research Council of Canada (NSERC) and Beijing Municipal Science and Technology Plan: Z111100074811001.

References

1. Sendur, L., Selesnick, I.W.: Bivariate Shrinkage With Local Variance Estimation. IEEE Signal Processing Letters 9(12), 438–441 (2002)
2. Dabov, K., Foi, A., Katkovnik, V., Egiazarian, K.: Image Denoising By Sparse 3d Transform-Domain Collaborative Filtering. IEEE Transactions on Image Processing 16(8), 2080–2095 (2007)
3. Chen, Q., Wu, D.: Image Denoising By Bounded Block Matching and 3d Filtering. Signal Processing 90, 2778–2783 (2010)
4. Luisier, F., Blu, T., Unser, M.: New Sure Approach to Image Denoising: Interscale Orthogonal Wavelet Thresholding. IEEE Transactions on Image Processing 16(3), 593–606 (2007)
5. Donoho, D.L., Johnstone, I.M.: Ideal Spatial Adaptation By Wavelet Shrinkage. Biometrika 81(3), 425–455 (1994)
6. Chen, G.Y., Kegl, B.: Image Denoising With Complex Ridgelets. Pattern Recognition 40(2), 578–585 (2007)
7. Chen, G.Y., Bui, T.D., Krzyzak, A.: Image Denoising Using Neighbouring Wavelet Coefficients. Integrated Computer-Aided Engineering 12(1), 99–107 (2005)
8. Chen, G.Y., Bui, T.D., Krzyzak, A.: Image Denoising With Neighbour Dependency And Customized Wavelet and Threshold. Pattern Recognition 38(1), 115–124 (2005)
9. Cho, D., Bui, T.D., Chen, G.Y.: Image Denoising Based on Wavelet Shrinkage Using Neighbour and Level Dependency. International Journal Of Wavelets, Multiresolution and Information Processing 7(3), 299–311 (2009)
10. Cho, D., Bui, T.D.: Multivariate Statistical Modeling for Image Denoising Using Wavelet Transforms. Signal Processing: Image Communication 20(1), 77–89 (2005)
11. Fathi, A., Naghsh-Nilchi, A.R.: Efficient Image Denoising Method Based on a New Adaptive Wavelet Packet Thresholding Function. IEEE Transactions on Image Processing 21(9), 3981–3990 (2012)
12. Chatterjee, P., Milanfar, P.: Patch-Based Near-Optimal Image Denoising. IEEE Transactions on Image Processing 21(9), 1635–1649 (2012)
13. Rajwade, A., Rangarajan, A., Banerjee, A.: Image Denoising Using the Higher Order Singular Value Decomposition. IEEE Transactions on Pattern Analysis and Machine Intelligence 35(4), 849–862 (2013)
14. Motta, G., Ordentlich, E., Ramirez, I., Seroussi, G., Weinberger, M.J.: The Idude Framework for Grayscale Image Denoising. IEEE Transactions on Image Processing 20(1) (2011)
15. Miller, M., Kingsburg, N.: Image Denoising Using Derotated Complex Wavelet Coefficients. IEEE Transactions on Image Processing 17(9), 1500–1511 (2008)
16. Lebrun, M.: An Analysis and Implementation of the Bm3d Image Denoising Method. Image Processing on Line (2012), http://Dx.Doi.Org/10.5201/Ipol.2012.L-Bm3d

A Novel Detail-Enhanced Exposure Fusion Method Based on Local Feature

Mali Yu[*] and Hai Zhang

School of Information Science & Technology, Jiujiang University, Jiujiang 332005, China
mary8011@163.com

Abstract. Digital imaging system has limited capability of capturing images for the scenes with very High Dynamic Range(HDR). Exposure fusion is a useful method to form an HDR-like Low Dynamic Range(LDR) image by directly combining differently exposed LDR images. The common methods have problems in dealing with weak edges and textures especially when they are not clear in all input images. To address this problem, a novel gradient adjustment method is proposed based on the local feature. Then, the gradient-magnitude difference between the small detail and strong edge is significantly decreased. Experimental results demonstrate that the proposed method is robust and effective in local detail enhancement and meanwhile preserving the global consistency.

Keywords: exposure fusion, adjusted gradient, local feature detail enhancement.

1 Introduction

Real scenes usually span a large range in luminance. Due to the hardware limitation of digital imaging system, the captured images have too limited range to represent the scenes [1, 2], in which there exist the over-exposed/under-exposed areas. High Dynamic Range (HDR) imaging devices have been developed but cannot be applied widely because of their high cost. In contrast, HDR imaging method is indeed an attractive candidate, which combines differently exposed images with the same scene into a realistic image. There are two kinds of HDR imaging methods: HDR tone mapping and exposure fusion. Compared with HDR tone mapping, exposure fusion does not need the exposure time of images and does not estimate the camera response function. And the result image produced by exposure fusion method can be directly displayed on the common devices without additional HDR compression.

Exposure fusion belongs to one kind of image fusion methods, but the common image fusion methods cannot be directly applied for fusing differently exposed images. So far, there are three kinds of exposure fusion methods, including the block-based method, multi-resolution-decomposition-based method and pixel-based method. The block-based method was first formally proposed by Goshtasby etc. [3] and was improved later [4, 5]. This kind of method splitted images into blocks, selected the

---

[*] Corresponding author.

D.-S. Huang et al. (Eds.): ICIC 2014, LNCS 8588, pp. 407–414, 2014.

block having the most information for each block and blended all selected blocks. The block-based method had difficulty in dealing with the boundaries between blocks. Multi-resolution decomposition technologies, such as gradient pyramids [6-8] and discrete wavelets [9, 10], had been widely used to extract the base-layers and the detail-layers under different resolutions. After that, the same or different strategies were applied on blending each layer. The pixel-based method directly determined the weight of each pixel based on the probabilistic models [11, 12]. When the input images provide enough information for every pixel, the multi-resolution-decomposition-based method and pixel-based method can achieve good results. However, small details not clear in all input images are still missing in their results. To address this problem, we propose a novel method to adjust the gradient based on the local feature, by applying which on the exposure fusion method the global consistency is preserved and the local detail is enhanced. The rest parts of this paper are organized as the following. Section 2 describes in detail the proposed method and its implementation. Section 3 analyses the experimental results, followed by conclusions in Section 4.

2   Proposed Method

Our fundamental goal is to effectively enhance the small details in the over-exposed/under-exposed area, and meanwhile preserve the global consistency between differently exposed regions. The proposed method consists of three steps. First, the initial gradient is generated by the weighted sum of the gradients of all input images. Then, the gradient is adjusted based on the local feature. Finally, the solution of Poisson equation based on the adjusted gradient by using the multigrid method is the result image.

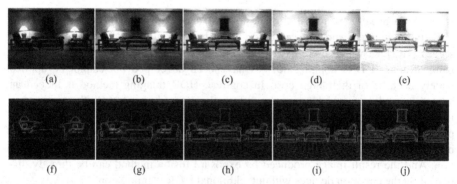

Fig. 1. Luminance images and corresponding gradient magnitudes. (a)-(e) are luminance images. The exposure time gets longer from (a) to (e). (a) is the shortest-exposure image and (e) is the longest-exposure image. (f)-(j) are the gradient magnitudes of (a)-(e).

2.1   Initial Gradient Generation

Let L_k, $k=1,2,\ldots,N$, denote the luminance image corresponding to the kth input image, where N refers to the number of input images. Gradient varies with different exposure parameters. Commonly, gradient in the well exposed region is more reliable than one in the over-exposed/under-exposed region. As shown in Fig. 1, the gradient near the lamp is obvious and reliable when the exposure time is short; on the contrary, the gradient far away the lamp is obvious and reliable when the exposure time becomes long. At a certain pixel, the gradient in the well exposed image must be assigned to a big weight and the gradient in other images must be assigned to a small weight.

Goshtasby etc. proposed the entropy to evaluate the block quality [3], which cannot effectively evaluate the quality of the uniform region. Mertens etc. proposed the contrast, saturation, and well-exposedness to evaluate the pixel quality [7]. The information of single pixel is easily affected by noises especially in the dark region. Thus, we propose to combine the luminance stand deviation in the neighborhood and the luminance value to determine the weights. The weight for the gradient, $w^k(x,y)$, is given by,

$$w^k(x,y) = \exp\left(-\alpha \frac{\left(S^k(x,y)-0.5\right)^2}{2*0.2^2} - (1-\alpha)\frac{\left(L^k(x,y)-0.5\right)^2}{2*0.2^2}\right) \tag{1}$$

$$w^k(x,y) = \frac{w^k(x,y)}{\sum_{k=1}^{N} w^k(x,y)} \tag{2}$$

where $S^k(x,y)$ denotes the luminance stand deviation of the pixel (x,y) and its surrounding pixels in the kth image, and α is the parameter making a trade-off between the stand deviation and the luminance. In all experiments, α is set to 0.9. For the non-uniform region, the pixel with the big standard deviation will be assigned to a big weight. And for the uniform region, the pixel with the luminance near to 0.5 will be assigned to a big weight. Equation (2) is used to normalize the weights for each pixel. Then, the initial gradient, $(G_x(x,y), G_y(x,y))$, is the weighted sum of gradients in all images and is given by,

$$G_x(x,y) = \sum_{k=1}^{N} w^k(x,y)G_x^k(x,y), G_y(x,y) = \sum_{k=1}^{N} w^k(x,y)G_y^k(x,y) \tag{3}$$

where $(G_x^k(x,y), G_y^k(x,y))$ denote the gradient in the kth images.

2.2   Gradient Adjustment Based on the Local Feature

By applying the section 2.1, the best gradient among all input images is preserved for each pixel, but the gradients maybe vary with regions because they come from different images. As shown in Fig.2 (a), the normalized gradient magnitude generated by section 2.1, the boundary of sofa and the texture on the ground come from differently exposed images, so their gradient magnitudes are different. By contrast, the adjusted gradient magnitude of the weak edges or textures in Fig.2 (b) is more visible.

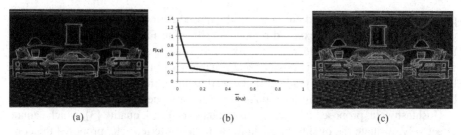

Fig. 2. Normalized gradient magnitudes and their relationship. (a) is the normalized gradient magnitude before adjusted; (b) is the function relationship between the standard deviation and the adjustment factor where k_1, k_2, μ_2 and μ_3 are 0.3, 1, 0.2, 1; (c) is the normalized gradient magnitude after adjusted.

To enhance the weak edges or textures and suppress the strong edges, we propose an adjustment factor. A certain pixel at the strong edges, having the big standard deviation in the mean luminance image, must be assigned to a small adjustment factor; and a certain pixel at the weak edges or textures, having the small standard deviation in the mean luminance image, must be assigned to a big adjustment factor. Then, we determine the adjustment factor $F(x,y)$ by using

$$F(x,y) = \begin{cases} k_1 + k_2 \cot\left(\dfrac{\pi\left(\overline{S}(x,y)+2\mu_2-3\mu_1\right)}{4(\mu_2-\mu_1)}\right), & \overline{S}(x,y) < \mu_1 \\[4mm] k_1 + k_2 \cot\left(\dfrac{\pi\left(\overline{S}(x,y)+2\mu_3-3\mu_1\right)}{4(\mu_3-\mu_1)}\right), & \overline{S}(x,y) \geq \mu_1 \end{cases} \tag{4}$$

where $\overline{S}(x,y)$ denotes the mean luminance, and k_1, k_2 are always set to 0.3 and 1, which are determined experimentally. μ_1, μ_2 and μ_3 are determined by,

$$\mu_1 = \frac{\displaystyle\sum_{(x,y)\in\Omega\cap\overline{S}(x,y)>3/255} \overline{S}(x,y)}{\displaystyle\sum_{(x,y)\in\Omega\cap\overline{S}(x,y)>3/255} 1}, \mu_2 = 3\mu_1, \mu_3 = 40\mu_1 \tag{5}$$

where Ω denotes the entire image domain. Fig. 2 (b) shows the function relationship between the standard deviation and the adjustment factor where μ_2 and μ_3 are 0.2, 1. The adjustment factor decreases with the increase of the standard deviation and the decrease speed is different. When the standard deviation comes close to 0, the adjustment factor changes rapidly; whereas, when the standard deviation comes close to 1, the adjustment factor changes stably. Then, the difference between gradients decreases. The final gradient is given by,

$$G'_x(x,y) = G_x(x,y)F(x,y), G'_y(x,y) = G_y(x,y)F(x,y) \tag{6}$$

As shown in Fig. 2(c), the gradient magnitudes of the weak edges or textures are enhanced and the gradient magnitudes of the strong edges are suppressed.

2.3   Implementation

The proposed method is applied on the single channel, so we first convert all input images from RGB (Red, Green, Blue) space to HSV (Hue, Saturation, Value) space and the luminance image is represented by the V channel in HSV space. The following steps are applied on the luminance image.

Step 1: Compute the gradients and the standard deviation for all luminance images;

Step 2: According to Equation (1)-(2), calculate the weights and get the initial gradient;

Step 3: According to Equation (4), compute the adjustment factor and get the adjusted gradient;

Step 4: Based on the adjusted gradient, the result luminance image I is the solution of the following Poisson equation.

$$\arg \min_{I} \left\{ \left\| \nabla I - G' \right\|^2 \right\} \tag{7}$$

where G' refers to the adjusted gradient $(G'_x(x,y), G'_y(x,y))$. Equation (10) can be solved by using the multigrid method[1]. In the code, the width and the length of image are required to be 2^n+1, where n is an integer. Thus, the image is padded to the specified size with zero when implemented.

Step 5: Let R_1, R_2, R_3 denote the ratios of R, G and B channels to the mean luminance, respectively. The fused color image I_f is computed by [13],

$$I_f = \left(I \bullet R_1^\beta, I \bullet R_2^\beta, I \bullet R_3^\beta \right) \tag{8}$$

where β is set to 0.8.

3   Experiments and Discussions

We test the performance of the proposed method and compare to the classical exposure fusion method proposed by Mertens etc. [7] and another detail-enhanced exposure fusion method proposed by Singh etc. [8].

Fig. 3 shows the result images of the proposed method, Mertens's method and Singh's method. The brightness difference between the bright regions and the dark regions in (a) is smaller than the brightness difference in (b) and (c), which means that the brightness difference of differently exposed regions is compressed. Whereas, the dark regions become clear. (d)-(f) are the close-ups of (a)-(c). In (d), the color of the tower is rich, while in (e) and (f), the tower is so dark that the detail is weakened.

---

[1] http://www.eecs.berkeley.edu/~demmel/

Compared with Mertens's method and Singh's method, the proposed method can suppress the brightness difference of differently exposed regions and meanwhile enhance the weak edges and textures. Although the proposed method can effectively enhance the detail, the halo effect cannot occur, so the transition in the edge of the tower is smooth. The halo effect can be seen in (f) because the contrast is enhanced excessively in Singh's method. To objectively assess the proposed method, the entropy is used, where the big entropy means rich detail. The entropies of Fig. 3(a)-(c) are 7.08, 6.84 and 6.83, respectively. Thus, the proposed method can enhance the weak edges and textures.

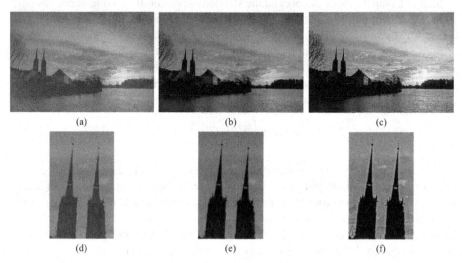

Fig. 3. Comparison of the results of the proposed method, Mertens's method and Singh's method. (a) is the result image of the proposed method; (b) is the result image of Mertens's method; (c) is the result image of Singh's method; (d)-(f) are the close-ups of (a)-(c).

Fig. 4 shows other comparison results, where the first column is produced by the proposed method, the second column is produced by Mertens's method, and the third column is produced by Singh's method. The color is rich is Fig. 4 (a) and (d), and the brightness difference between the differently exposed regions is also suppressed by the proposed method. The trees on the bank show more detail in Fig. 4 (a), while the bank in Fig. 4 (b) and (c) is too dark to weaken the detail.

The region near the window in Fig. 4 (d) shows less detail than that in Fig. 4 (e), because the indoor and outdoor difference is big and the proposed method takes consideration of the smooth transition between differently exposed regions. And far away from the transition area, the details on the shelf are enhanced by the proposed method.

(a)   (b)   (c)

(d)   (e)   (f)

Fig. 4. More comparison results. The images in the first column are produced by the proposed method, the images in the second column are produced by Mertens's method, and the other images are produced by Singh's method.

Fig. 5 shows the result images with different numbers of input images, where the images from (a) to (c) are the results produced with 2, 3 and 4 input images, respectively. All these images look similar to one another. From the experimental results, we see that the number of input images has a small impact on the performance of the proposed method.

(a)   (b)   (c)

Fig. 5. Comparison of the results of the proposed method with different numbers of input images. (a)-(c) are the results produced with 2,3 and 4 input images, respectively.

4   Conclusions

Exposure fusion can blend differently exposed images into an HDR-like LDR image. The common exposure fusion method is difficult in dealing with weak edges or textures especially when they are not clear in all input images. The proposed method proposes a novel gradient adjustment method. First, the initial gradient is generated by the weighted sum of gradients of all input images. Then, the adjustment factor is constructed based on the standard deviation to decrease the difference between the gradients of the strong edges and the weak edges or textures. Finally, the multigrid method is used to solve the Poisson equation based on the adjusted gradient, the solution of which is the result image. Experiments have been carried out and compared to Mertens's method and Singh's method. Experimental results demonstrate that more weak details have been enhanced effectively and the global consistency has been preserved.

Acknowledgement. This work has been partially supported by the National Natural Science Foundation of China (Grant No. 61362032), by the Natural Science Foundation of Jiangxi,China(Grant No. 20132BAB211025) and by the Educational Commission of Jiangxi Province, China(Grant No.GJJ13716). Finally we sincerely thank the anonymous reviewers for their valuable suggestions and comments, which help to improve our manuscript.

References

1. Debevec, P., Reinhard, E., Ward, G., Pattanaik, S.: High Dynamic Range Imaging. In: ACM SIGGRAPH 2004 Course Notes, p. 14. ACM (2004)
2. Reinhard, E., Heidrich, W., Debevec, P., Pattanaik, S., Ward, G., Myszkowski, K.: High Dynamic Range Imaging: Acquisition, Display, and Image-Based Lighting. Morgan Kaufmann (2010)
3. Goshtasby, A.A.: Fusion of Multi-Exposure Images. Image and Vision Computing 23, 611–618 (2005)
4. Bachoo, A.K.: Real-Time Exposure Fusion on a Mobile Computer (2009)
5. Cho, W.-H., Hong, K.-S.: Extending Dynamic Range of Two Color Images under Different Exposures. In: Proceedings of the 17th International Conference on Pattern Recognition, ICPR 2004, pp. 853–856. IEEE (2004)
6. Kartalov, T., Ivanovski, Z., Panovski, L.: Optimization of the Pyramid Height in the Pyramid-Based Exposure Fusion Algorithms
7. Mertens, T., Kautz, J., Reeth, F., Van, E.F.: A Simple and Practical Alternative to High Dynamic Range Photography. Computer Graphics Forum, Wiley Online Library, 161–171 (2009)
8. Singh, H., Kumar, V., Bhooshan, S.: A Novel Approach for Detail-Enhanced Exposure Fusion Using Guided Filter. The Scientific World Journal, Article ID 659217, 1- 8 (2014)
9. Ray, L.A., Adhami, R.R.: Dual Tree Discrete Wavelet Transform with Application to Image Fusion. In: Proceeding of the Thirty-Eighth Southeastern Symposium on System Theory, SSST 2006, pp. 430–433. IEEE (2006)
10. Wang, J., Xu, D., Lang, C., Li, B.: Exposure Fusion Based on Shift-Invariant Discrete Wavelet Transform. J. Inf. Sci. Eng. 27, 197–211 (2011)
11. Shen, R., Cheng, I., Shi, J., Basu, A.: Generalized Random Walks for Fusion of Multi-Exposure Images. IEEE Transactions on Image Processing 20, 3634–3646 (2011)
12. Song, M., Tao, D., Chen, C., Bu, J., Luo, J., Zhang, C.: Probabilistic Exposure Fusion. IEEE Transactions on Image Processing 21, 341–357 (2012)
13. Hall, R.: Illumination and Color in computer Generated Imagery. Springer, New York (1989)

Frequency Domain Directional Filtering Based Rain Streaks Removal from a Single Color Image

Changbo Liu[1], Yanwei Pang[1], Jian Wang[1], Aiping Yang[1], and Jing Pan[1,2]

[1] School of Electronic Information Engineering, Tianjin University, Tianjin 300072, China
[2] School of Electronic Engineering, Tianjin University of Technology and Education,
Tianjin 300222, China

Abstract. Bad weather conditions, such as rain or snow, degrade outdoor vision system performance. Rain removal from a single image has been investigated extensively. However, existing built rain streak models are greatly influenced by inaccurate parameter estimation or non-stationary background due to camera motions. In this paper, we propose a rain streak removal algorithm from a single color image based on frequency-domain directional filter. Experiment results demonstrate that the proposed algorithm can deal with rain steaks of any directions, and maintains more details than existing algorithms.

Keywords: rain removal, directional filter, image enhancement.

1 Introduction

Adverse weather condition, such as rain or snow, severely degrades outdoor vision system's performance. Vision system can't offer reliable target detection [1], object recognition and tracing, feature extraction [2] results due to image degradation by rain. As a result, to improve the robustness of outdoor surveillance system, we have to restore the corrupted image and remove the bad effects of bad weather. Meanwhile, in the field of computer vision, adverse weather condition will increase the noise of the image, reduce the robustness of feature extraction and the generalization ability of classifier, degrade the performance of image segmentation.

Removal of rain streaks from images or images or videos has recently received much attention in the past few years. Many algorithms have been proposed for the removal of rain streaks. These algorithms can be classified into three categories [3], spatial-temporal domain based, frequency domain based and matrix factorization based. Traditional methods on detecting and removing rain streaks usually focus on the temporal domain. Garg and Nayar[4] proposed a method on detecting and removing rain streaks in a video. They developed a correlation model demonstrating the dynamics of rain and a motion blur model characterizing the photometry of rain. Subsequently, they found that some camera parameters can be adjusted to reduce the effects of rain without altering the appearance of the scene [5]. Moreover, Zhang [6] proposed an improved video rain streak removal algorithm incorporating both temporal and chromatic properties. Brewer [7] further utilized the shape characteristics of

D.-S. Huang et al. (Eds.): ICIC 2014, LNCS 8588, pp. 415–424, 2014.

rain streaks for identifying and removing rain streaks from videos. Furthermore, for the frequency based circumstances, Barnum [8] built a model of the shape and appearance of a single rain or snow streak in the image space, and remove rain or snow streaks with its frequency spectrum. Bossu [9] proposed selection rules based on photometry and size to select the potential rain streaks in a video. They computed a histogram of orientations of rain streaks estimated with geometric moments. Moreover, Kang [10] used the method of dictionary leaning. A rain image was first separated into geometry layer and texture layer by bilateral filtering. Then they divided the dictionary online-trained by texture layer into rain-dictionary and no-rain dictionary, and restored the rain-removed image with the no-rain dictionary only.

By analyzing existing algorithms, we found that the existing built rain streak models are greatly influenced by inaccurate parameter estimation or non-stationary background due to camera motions. Consequently, their performances are badly degraded. In this paper, we propose a framework for a single color image rain streaks removal, which utilizes both the uniform spread feature of rain streaks in spatial domain and directional feature in frequency domain. The major contribution of this paper is twofold: 1) we propose a new rain streak direction determination approach, which utilizes the uniform spread feature in spatial domain and directional feature in frequency domain, to build a block-wise HOE feature to determine the direction of rain streaks. 2) our method is a new rain streaks removal method using directional filter in frequency domain. On the basis of rain streaks' direction, we design a corresponding directional filter and remove rain streaks in frequency domain.

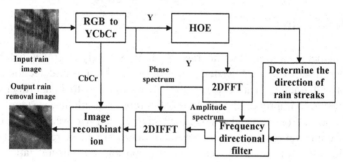

Fig. 1. Block diagram of proposed algorithm

2 Proposed Algorithm

The block diagram in Fig.1 shows the proposed rain streak removal framework, which includes 3 major steps, pre-processing, determination of the direction of rain steaks, frequency domain directional filtering based rain streaks removal. First, the input color image is converted into YCbCr color space, and the Y-component image is divided into blocks. Then we compute the dominant direction of the global edge based on the local Histogram of Oriented Edge to determine the direction of the rain streaks. After that, we design a corresponding 2D frequency domain directional filter, and then utilize the filter to remove rain streaks in frequency domain of Y-component.

At last, we reconstruct the color image by combining the filtered Y-component image with original Cb and Cr component image.

2.1   Pre-processing

Fig. 2 shows an input color image and its corresponding RGB channel images. It is observed that rain steaks are obvious in all 3 channels. However, rain steaks can only influence the Y-channel image [3], and hardly influence the CbCr color channels.

The above observations are fit for most rain images. Considering the reduction of computation cost and making the algorithm more robust, the raised algorithm convert the input color image into YCbCr color space, and then deals with the Y-component image only. The RGB-to-YCbCr relation is shown in (1).

$$Y = 16 + 0.257 \times R + 0.564 \times G + 0.098 \times B$$
$$C_b = 128 - 0.148 \times R - 0.291 \times G + 0.439 \times B \qquad (1)$$
$$C_r = 128 + 0.439 \times R - 0.368 \times G - 0.071 \times B$$

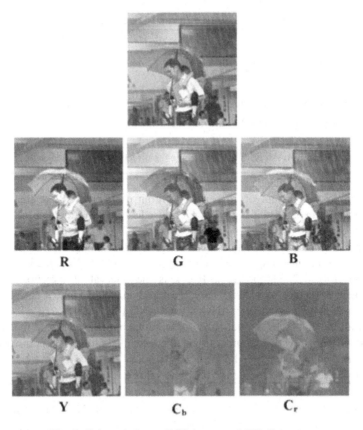

Fig. 2. Color rain image RGB image and YCbCr image

The proposed algorithm determines the rain steaks' direction based on the local Histogram of Oriented Edge. Considering that many rain images are in poor light conditions, rain steaks are not so distinct from background. Classic Canny operator is used to detect the Y-component edges. Comparing with other edge detection methods, Canny operator is exactly fit for detecting the weak edges. There is better connectivity between the detected edges.

2.2 Determination of the Direction of Rain Steaks

The directions of rain steaks are almost the same in the original image and the edge image. Based on the former edge binary image, we compute the horizontal gradient amplitude G_x, vertical gradient amplitude G_y and the gradient orientation ϕ for each edge point in Y-component image. G_x, G_y and ϕ are defined as (2)~(4),

$$G_x(x, y) = L(x+1, y-1) + L(x+1, y) + L(x+1, y+1) \\ - L(x-1, y-1) - L(x-1, y) - L(x-1, y+1) \tag{2}$$

$$G_x(x, y) = L(x-1, y+1) + L(x, y+1) + L(x+1, y+1) \\ - L(x-1, y-1) - L(x, y-1) - L(x+1, y-1) \tag{3}$$

$$\phi(x, y) = \arctan(G_y / G_x) \tag{4}$$

where $-90° \leq \phi \leq 90°$. Considering the edge orientation θ is perpendicular to ϕ, we have the relations $\theta = \phi + 90°$ and $0° \leq \theta \leq 180°$.

Table 1. Algorithm to determine the direction of rain streaks

1. Divide the edge image into 4*4=16 blocks
2. Compute the amount of edge points within each block. Take the block as clean(no rain steaks) if the proportion of edge points in the sum of points in each block is too low.
3. For the rest blocks, compute the histogram of oriented edge in each block based on the inner edge information within each block. Note the histogram as $HIST_b(k)(k=1,2,...,8; b=1,2,..., 16)$, k is the interval of direction, b is the sequence number of blocks. Take the top 3 maximal value's corresponding interval, noted as $ANG_b(1)$, $ANG_b(2)$ and $ANG_b(3)$.
4. For all the 16 blocks, we get 16*3 = 48 interval numbers. Take all the numbers and we get a new histogram of oriented edge called global salient histogram of oriented edge, noted as GHIST(k).
5. With GHIST(k), we choose the maximal value as the direction of rain streaks.

We divide 180° into 8 intervals, noted as interval 1: [0, 22.5°), interval 2:[22.5°, 45°), interval 3:[45°, 67.5°), interval 4:[67.5°, 90°), interval 5:[90°, 112.5°), interval 6:[112.5°, 135°), interval 7:[135°, 157.5°), interval 8:[157.5°, 180°). Considering the background of image corresponds to edge point in all directions, it tends to be imposs-ible to detect the direction of rain streaks with the global histogram of oriented edge. As a result, we propose to deal with the edge in blocks.

The global histogram of oriented edge shows the distribution of edge information in the image. Since the rain streaks are uniform spread within the image, and the edge of the background is usually intensive, the uniform spread of the edge of rain streaks can be well shown with the use of blocks. Also, the local background edge informa-tion can be suppressed effectively.

2.3   Frequency Domain Directional Filtering Based Rain Streaks Removal

Most algorithms remove the rain streaks with the spatial domain, which is easily per-formed. However, there tends to be obvious profile between rain streaks and the background, which causes bad visual effect. To overcome that, we propose a new frequency domain directional filtering based rain streaks removal method.

A frequency wedge directional filter is needed for the removal. Wedge directional filter is a wedge-pass 2D digital filter in frequency domain. There are 3 ways to de-sign a wedge filter, direct optimization method, down-sample method and transform method. Direct optimization is complicated and the filtering structure can't be well designed. The down-sample method designs the 2D filter with 2D down-sample. Horizontal and vertical wedge directional filter can be obtained by shift and down-sample. However, the filter is restricted by sample-factor and only fits for the condi-tion of a single direction filtering. The McClellan-transform [11] based method is most widely used to design a 2D directional filter and works best. Shyu [12] extended the McClellan transform and applied it to all kinds of 2D FIR filters. The pass band can be designed by adjusting the parameters. The proposed method designs the filter with Shyu's method as below.

1D FIR filter impulse response $h(n)$ can be noted as

$$H(\omega) = \sum_{n=0}^{N} a_p(n)\cos(\omega n) \tag{5}$$

The coefficient $a_p(n)$ can be represented by a p-polynomial,

$$a_p(n) = \sum_{m=0}^{M} a(n,m)p^m \tag{6}$$

Use Chebyshev polynomial substituting the $\cos(n\omega)$,

$$H(\omega) = \sum_{n=0}^{N} a_p(n)T_n\left(\cos(\omega)\right) \tag{7}$$

Table 2. Procedure of designing an 8-direction frequency filter

1. Design a 1D prototype filter. Using the Kaiser Window to design a 1D low-pass filter with pass frequency $w_p = 0.45\pi$, stop frequency $w_s = 0.55\pi$, pass band amplitude is 1, peak passband ripple $\alpha_p = 0.5\text{dB}$, minimum stopband attenuation $\alpha_s = 40\text{dB}$.

2. With the transform function (9) and different value of {A, B, C, D, E}, we can get different 90° wedge filters and parallelogram filters. The wedge filters can be used to design four 8-directional sub-band filters by multiplying the 4-directional filters and the parallelogram filters. Then with the 4-directional filters subtracting the four 8-directional filters, we get the other four 8-directional sub-band filters. Subtracting by 1, we can get the all eight 8-directional masking filters.

Bringing in the transform function $\Phi(\omega_1, \omega_2)$, the transform is realized by tiny impulse transform. According to McClellan transform, design the 2D filter with 1D ones, the frequency response is

$$H(\omega_1, \omega_2) = \sum_{n=0}^{N} a(n) T_n \left(\Phi(\omega_1, \omega_2) \right) \tag{8}$$

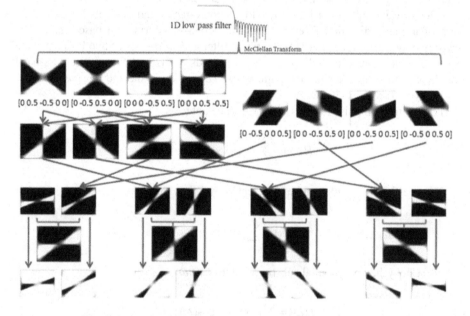

Fig. 3. 8-directional filters design based on McClellan Transform

The transform function is

$$\Phi(\omega_1, \omega_2) = A + B\cos(\omega_1) + C\cos(\omega_2)$$
$$+ D\cos(\omega_1 - \omega_2) + E\cos(\omega_1 + \omega_2) \tag{9}$$

In the function, the choice of {A, B, C, D, E} depends on the shape of transfer function.

Table 3. Algorithm procedure of frequency domain directional filter based rain removal

1. Note 2D-FFT of Y-component as **I**, amplitude spectrum as **F**, phase spectrum as **P**.
2. Choose a corresponding directional filter from the eight pre-designed 8-directional filters based on the rain streaks direction we determined before. We denote the **F**-filtering result as **F′**. Then we combine **F′** with the phase spectrum **P** to conduct inverse-FFT to return to the spatial domain. The result is denoted as **I′**.
3. With the original Cb-component, Cr-component images and **I′**, we convert the three components from YCbCr to RGB image, and then reconstruct the rain-removed color image.

3   Experiments and Results

Our experiments were implemented in MATLAB 2013b on an Intel I7 CPU (Core 3.4 GHz), 16G RAM PC. To testify the validity of our method, we chose 20 images to do the experiment, among which 8 images come from [10], the others were from the internet or taken by us in real life.

Fig. 4 shows part of the results. The images in the left column are original rain color images, the images in the left column are rain streaks removal result images. Though the proposed algorithm is not very complicated, the results in the first three rows are satisfying.

The last 2 rows of Fig. 4 show the circumstance of unsatisfying results. In the second row from bottom, even the direction of rain streaks can be correctly determined, since the rain streaks have a large influence on the background, the rain removal results are not so well as the previous ones. The image in the last row is in a complicated background. There are a lot of bush in the image, which have rich edge information. Though the rain streaks are obvious, the background has a bad influence on the edges, which makes the determination of the direction of rain streaks impossible. For the first circumstance, the bilateral technique can be applied first to divide the image into geometric and texture layers. Deal with the texture layer only. For the latter circumstance, it's suggested that the image content be analyzed as the image being divided into blocks. If the image block contains rich texture information, its edge direction information could be ignored while analyzing the rain streaks direction.

Fig. 5 shows the comparison results between our algorithm and Kang's method. It is obvious that in the result of our method, the texture is better preserved, especially the parts of face and hand. The reason is that, since Kang first separate the image into geometric layer and texture layer, most high-frequency details are separated into the high-frequency layer, which is used to train the dictionary. After that, using HOG feature and K-means to classify the atoms will misclassify the high frequency texture atoms into the rain dictionary, which will result in the loss of original texture information.

Fig. 4. Rain removal results: Left: original image. Right: rain-removed image.

Besides that, in our algorithm, the look-up-table could be used to accelerate the processing speed. Specifically, before (4), we set up the relations among G_x, G_y and ϕ, stored that as a file. During the processing, after we get G_x and G_y, the look-up-table file is loaded into RAM. By look up the table, we can easily get ϕ of each edge pixel. Using the optimized algorithm, it costs only 280ms to process a color image of size 512*512 while Kang's algorithm needs to train dictionary online, which would cost more than 10s to process a color image of the same size.

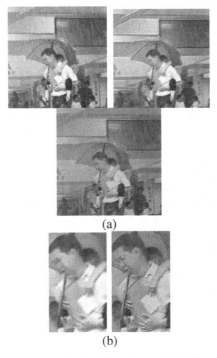

(a)

(b)

Fig. 5. (a) Comparison between our method and Kang's method: Left top: original image. Right top: Kang's result. Bottom: our result. (b) Left: Kang's result. Right: our result.

4   Conclusion

Considering the characteristics of rain in spatial domain and frequency domain, we propose a frequency domain directional filtering based rain streaks removal algorithm for a single color image. Combining with the thought of dividing image into blocks, we propose a salient global histogram of oriented edge feature to determine the direction of rain streaks. Based on that, we choose the corresponding directional filter to remove the rain in frequency domain. The experiments show that comparing with the existing algorithms, the proposed algorithm can correctly determine the direction of rain streaks, and provide significantly improvement on the artifacts brought by removal process. For the future work, the performance of our work may be further improved by combining features in the time domain, spatial domain, and frequency domain and further realize the time-spatial-frequency three-domain-joint rain streaks removal.

Acknowledgement. This work was supported in part by the National Basic Research Program of China 973 Program (Grant No. 2014CB340400), the National Natural Science Foundation of China (Grant Nos. 61372145, 61172121, 61271412, and 61222109), and the Open Funding Project of State Key Laboratory of Virtual Reality Technology and Systems, Beihang University (Grant No. BUAA-VR-13KF).

References

1. Dalal, N., Triggs, B.: Histograms of Oriented Gradients for Human Detection. In: Proceedings of International Conference on Computer Vision and Pattern Recognition, San Diego, CA, vol. 1, pp. 886–893 (2005)
2. Lowe, D.G.: Distinctive Image Features from Scale-invariant Keypoints. International Journal of Computer Vision 60(2), 91–110 (2004)
3. Tripathi, A.K., Mukhopadhyay, S.: Removal of Rain from Videos: a Review. Signal, Image and Video Processing (2012)
4. Garg, K., Nayar, S.K.: Detection and Removal of Rain from Videos. In: Proceedings of International Conference on Computer Vision and Pattern Recognition, vol. 1, pp. 528–535 (2004)
5. Garg, K., Nayar, S.K.: When does a camera see rain? International Journal of Computer Vision 2, 1067–1074 (2005)
6. Zhang, X., Li, H., Qi, Y., Leow, W.K., Ng, T.K.: Rain Removal in Video by Combining Temporal and Chromatic Properties. In: Proceedings of International Conference on Multimedia & Expo, Toronto, ON, Canada, pp. 461–464 (2006)
7. Brewer, N., Liu, N.: Using the Shape Characteristics of Rain to Identify and Remove Rain from Video. In: da Vitoria Lobo, N., Kasparis, T., Roli, F., Kwok, J.T., Georgiopoulos, M., Anagnostopoulos, G.C., Loog, M. (eds.) S+SSPR 2008. LNCS, vol. 5342, pp. 451–458. Springer, Heidelberg (2008)
8. Barnum, P.C., Narasimhan, S., Kanade, T.: Analysis of rain and snow in frequency space. International Journal of Computer Vision 86(2/3), 256–274 (2010)
9. Bossu, J., Hautière, N., Tarel, J.P.: Rain or Snow Detection in Image Sequences through Use of a Histogram of Orientation of Streaks. International Journal of Computer Vision 93(3), 348–367 (2011)
10. Kang, L., Lin, C., Fu, Y.: Automatic Single-Image-Based Rain Streaks Removal via Image Decomposition. IEEE Transactions on Image Processing 21(4), 1743–1755 (2012)
11. McClellan, J.M.: The Design of 2-D Digital Filters by Transformations. In: Proceedings of 7th Princeton Conference on Information and Systems, pp. 247–251 (1973)
12. Shyu, J.J., Pei, S.C., Huang, Y.D.: Design of Variable Two-dimensional FIR Digital Filters by McClellan Transformation. IEEE Transactions on Circuits and Systems 56(3), 574–582 (2009)

Multi-scale Fractal Coding
for Single Image Super-Resolution

Wei Xie, Jiwei Liu, Lizhen Shao, and Fengwei Jing

School of Automation and Electrical Engineering
University of Science and Technology Beijing, Beijing, China
xwei2013@163.com, liujiwei@ustb.edu.cn

Abstract. Fractal image coding can specially utilize spatial information and self-similarity structural information of an image to achieve image super-resolution. However, fractal image coding based on single scale brings the problems of block effect and loss of details. In this paper we propose a multi-scale fractal coding method for single image super-resolution. The proposed method integrates the fractal results of different scales and uses back-projection to further optimize the result. Experimental results show that the proposed method can remove the block effect, improve the loss of details and keep smooth of flat area and sharpness of edges in the reconstructed image. Compared with conventional fractal coding and cubic splines interpolation, our method is superior to both of them subjectively and objectively.

Keywords: fractal image coding, multi-scale, back-projection.

1   Introduction

Image super-resolution (SR) is a technique which can apply appropriate algorithms to reconstruct a clear high-resolution image, using one or multiple low-resolution images. The super-resolution reconstruction can increase image resolution effectively without increasing hardware burden, so it has great prospects in the medical image processing, satellite imaging, video surveillance, and many other fields [1].

In general, image super-resolution can be mainly classified into three categories: interpolation-based methods, reconstruction-based methods and learning-based methods. Interpolation-based methods [2,3] estimate the value of new pixels using neighbor information. Reconstruction-based methods [4,5] estimate a high-resolution image by adding prior knowledge and constraints. Learning-based methods [6,7,8] estimate the high-frequency information depending on training samples. However, the above mentioned methods do not fully utilize spatial information and self-similarity structural information of an image.

Fractal image coding can specially utilize spatial information and self-similarity structural information of an image to achieve image super-resolution. With the application of fractal set in different technical fields, the fractal technology has become a research hot topic in the field of image processing. Fractal image coding was

D.-S. Huang et al. (Eds.): ICIC 2014, LNCS 8588, pp. 425–434, 2014.

proposed by Barnsley [9] in the late 1980s, and its theoretical foundation is the iterative function system and collage theorem. Because the decoding process of fractal image coding is resolution independent, and the fractal code contains not only spatial information, but also self-similarity structural information of an image, fractal image coding can achieve image super-resolution reconstruction [10,11].

However, natural images do not have strictly self-similarity, and the fractal codes are only the approximate estimation for original low-resolution image. Using fractal codes to achieve image super-resolution restoration will introduce errors inevitably, and cause block effect and loss of parts of details.

To overcome the above shortcomings, Chung et al. [12] has proposed to use an additional enhancement layer and Lai et al. [13] has proposed to use adaptive overlapped range block partitioning to achieve a magnified image. However, those methods cause the loss of details and cannot enhance the high frequency components efficiently. In this paper, we propose a multi-scale fractal coding method to overcome the above defects effectively.

The rest of this paper is organized as follows. In section 2, we briefly introduce the encoding and decoding of the conventional fractal technology. The proposed method is presented in section 3. The experimental results are shown in section 4. Section 5 concludes this paper.

2   Conventional Fractal Image Coding for SR

In the image fractal coding, an image I of size $N \times N$ (N is usually a power of 2 in practice), firstly, is partitioned into a set of non-overlapping blocks called range blocks with the fixed size $b_r \times b_r$ each. Each range block is denoted as R_i. Thus the image I is the combination of the range blocks, i.e.

$$I = \bigcup_{i=1}^{N_r=(N/b_r)^2} R_i \; . \tag{1}$$

Moreover, the image I is also partitioned into a set of domain blocks, denoted as $\{D_i\}_{i=1}^{N_d}$, which can overlap in neighborhood and are not necessary to cover the whole image I. The domain blocks can be obtained by sliding a window of size $b_d \times b_d$ with a step δ across image I according to the order of raster scan. Generally, the size of the domain blocks is two times that of the range blocks, namely $b_d = 2b_r$.

At the encoding phase, we need to search a best matching block D_k in the domain pool for each R_i. We also have to obtain the optimal adjustment parameters of contrast and luminance, respectively denoted as s_i and o_i. Thus we need to solve the problem

$$\{k, j, s_i, o_i\} = \arg\min_{k,j,s_i,o_i} \left\| R_i - \left(s_i \cdot \left(\psi_k \left(T(D_j) \right) \right) + o_i \cdot 1 \right) \right\|^2 \; , \tag{2}$$

where ψ_k is one of the eight isometric transformations [14], 1 is the constant block whose luminance value is one, and T is the spatial contractive transformation with the use of transforming the size of D_j to be the same as that of R_i. Generally, T is implemented by averaging four surrounding pixels.

Let $\hat{D}_j = \psi_k\left(T(D_j)\right)$, and ignore the inner minimization problem. Equation (2) can be simplified as

$$\{s_i, o_i\} = \arg\min_{s_i, o_i} \left\| R_i - \left(s_i \cdot \hat{D}_j + o_i \cdot 1 \right) \right\|^2 . \tag{3}$$

Using the least square methods to solve (3), we can obtain

$$s_i = \frac{\frac{1}{m}\sum_{j=1}^{m} r_j d_j - \left(\frac{1}{m}\sum_{j=1}^{m} r_j\right)\left(\frac{1}{m}\sum_{j=1}^{m} d_j\right)}{\frac{1}{m}\sum_{j=1}^{m} d_j^2 - \left(\frac{1}{m}\sum_{j=1}^{m} d_j\right)^2}, \tag{4}$$

$$o_i = \frac{1}{m}\sum_{j=1}^{m} r_j - s_i\left(\frac{1}{m}\sum_{j=1}^{m} d_j\right), \tag{5}$$

where r_j and d_j denote the j-th pixel value of R_i and D_i, respectively, and m is the total number of pixels in a range block.

To guarantee the reconstruction accuracy, exhaustive search method is used to search the best matching domain block. Finally, we will obtain the fractal code $\{s_i, o_i, k, j\}$ of each range block R_i, where k denotes the k-th isometric transformation and j denotes the position of the best matching domain block of the range block R_i.

At the decoding phase, the fractal code can be applied to an arbitrary initial image, regardless of the size of the encoded image. Then we recursively apply the fractal code to the intermediate result, and finally the result will converge to an image similar to the original after ten iterations or less. When we want to enlarge the image I of size $N \times N$ by the scale factor q, we need to scale the size of the initial image to $qN \times qN$. Notably, the size of range blocks, domain blocks and the sliding step δ should be enlarged by the same scale factor q.

3   Proposed Algorithm

In this paper, we propose a multi-scale fractal coding method for single image super-resolution. The process of our proposed method is described as follows. Firstly, we partition the original low-resolution image into blocks with different scales, and get the corresponding reconstructed result using the conventional fractal coding for each scale. Then we integrate the reconstructed images to generate one result. Finally, we update the final result using back-projection to enforce global optimization.

3.1   Multi-scale Fractal Coding

Generally, the size of range blocks in conventional fractal coding is 4×4 [12,13]. However, fractal coding based on single scale introduces block effect and loss of details [11]. Consider that fractal results under different scales contain different information. Our proposed method is proposed based on multiple scales, not just single scale. Our experimental results prove that the reconstructed results introduce unwanted noises when the scale is less than 4×4 and we cannot figure out the result due to the serious noises when the scale is 2×2, block effect becomes obvious when the scale is larger than 4×4 and the larger the scale, the more obvious the block effect. For example, Fig.1 shows the reconstructed results of image "Lena" using the conventional fractal coding for different scales.

Based on the fractal reconstruction effect, the multiple scales as mentioned in this paper are selected as four scales, namely 3×3, 4×4, 5×5 and 6×6. For each scale the reconstruction process is the same as the conventional fractal coding described in section 2. Then we can obtain one fractal reconstruction result for each scale.

The range blocks is non-overlapping and should cover the whole image, thus N needs to be divisible by b_r. The proposed method is based on multiple scales, so that would introduce one problem, which is the N generally cannot be divisible by each b_r. Thus we need to preprocess the initial image in the size. In this paper, we apply the nearest neighbor interpolation to extend the initial image around. For example, for an image of 256×256, $N=256$ cannot be divided by $b_r = 3$. We extend the image to 258×258 by the nearest neighbor interpolation, while the image needs to be partitioned into 3×3 range blocks. Enlarge the extended image by the fractal coding, then remove the extended part to obtain the reconstructed result of the initial image.

3.2   Image Integration Based on Reconstruction Term

The results reconstructed under different scales contain different structure information and high-frequency information, thus the purpose of this section is to integrate the fractal reconstruction images to improve the reconstruction quality.

Let f_1, f_2, f_3, f_4 represent the reconstructed results when $b_r = 3,4,5,6$ respectively. Y represents the initial low-resolution image. We formulate the image integration problem as

$$\{a_i\}_{i=1}^4 = \arg\min_{a_1,a_2,a_3,a_4} \left\| Y - \sum_{i=1}^4 a_i \cdot f_i \downarrow_2 \right\|^2 , \quad s.t. \sum_{i=1}^4 a_i = 1 , \qquad (6)$$

where $\downarrow_2$ represents the averaging of four neighbor pixels or down-sampling. An reconstructed image close to the initial image is generated through training the weights to minimize the error. Training the weights is a constrained least squares problem. We define a gram matrix H for $\vec{Y}$ as

$$H = \left(\vec{Y}1^T - F\right)^T \left(\vec{Y}1^T - F\right) ,$$ (7)

where $\vec{Y}$ is the column vector form of Y, 1 is a column vector of ones. Let $f_i^d = f_i \downarrow 2$, then $F = \left[\vec{f}_1^d, \vec{f}_2^d, \vec{f}_3^d, \vec{f}_4^d\right]$. We group the weights of the fractal results to form a 4-dimensional vector a, namely $a = \left[a_1, a_2, a_3, a_4\right]^T$. For the constrained least squares problem, we can get the following solution

$$a = \frac{H^{-1}1}{1^T H^{-1}1} ,$$ (8)

In order to avoid inverting H, we solve the linear system of equation $Ha = 1$, and then normalize the weights so that $\sum_{i=1}^{4} a_i = 1$. Finally, we can get the integration image X_0 by

$$X_0 = \sum_{i=1}^{4} a_i \cdot f_i .$$ (9)

3.3   Enforcing Global Optimization through Back-Projection

While the integration image X_0 does not exactly meet equation $Y = X_0 \downarrow_2$, coupled with the impact of the noises, we need to further modify the reconstructed image by back-projection [1,8]. Map X_0 to the low-resolution space, then get the residual which contains the parts of high-frequency information and update the result iteratively. The process is formulated as

$$X_{t+1} = X_t + \lambda\left(Y - X_t \downarrow_2\right)\uparrow_2, t = 0,1,2,\cdots ,$$ (10)

where $\uparrow_2$ represents the up-sampling by a interpolation method, and λ is a parameter used to balance the reconstructed result and the reconstruction error. When the reconstruction error $\left\|Y - X_t \downarrow_2\right\|^2$ is less than the threshold or the update reaches the preset iteration times, stop the iteration.

4   Experimental Results

In the experiment, we choose image "Lena", "Cameraman" and "Peppers" to verify the effectiveness of the proposed algorithm. The original size of image "Lena", "Cameraman" and "Peppers" are 512×512, 256×256, 512×512, respectively. The images are first down-sampled 2 times as the test images. Then they will be restored to the original size with the proposed algorithm. Let $\lambda = 1$. In order to compare our method with conventional fractal coding and cubic splines interpolation, we also implemented the simulation experiments of the both. All the experimental results were implemented in Matlab 2012a, and ran on a laptop of Core(TM) i3 with 2G memory.

Fig. 1. The reconstructed results by the conventional fractal coding: a). $b_r = 3$; b). $b_r = 4$; c). $b_r = 5$; d). $b_r = 6$

Fig.1 shows the reconstructed results for the "Lena" image applying the conventional fractal coding based upon multiple scales of the range blocks. We found that the reconstructed result is better than that of the others when the scale is equal to

4×4 . We concluded that, the larger the scale, the more obvious the block effect and when the scale is less than   4×4 , the conventional fractal coding generates unwanted noisy results. So a median filtering needs to be applied to the reconstructed result of the scale being   3×3   before we are going to integrate the images. Although the restored results of the other three scales do not have so good effect as that of the scale being   4×4 , the results reconstructed by different scales contain different information. So we can integrate the different results to enhance the quality of the reconstructed image.

Fig. 2. Comparison of super-resolution results on "Lena" image by different methods: a). Initial high-resolution image; b). Cubic spline interpolation, PSNR=33.5386; c). Conventional fractal coding, PSNR=30.6624; d). The proposed method, PSNR=35.2340

Fig.2, Fig.3 and Fig.4 compare the result of our method with those of the cubic spline interpolation and the conventional fractal method. Objectively, our method can achieve higher PSNR than both the cubic spline interpolation and the conventional fractal. Subjectively, our method performs well on reconstructing sharp edges and restoring high-resolution details while the others introduce a certain amount of blurring. Our proposed method exceeds the others in keeping smooth in flat area and sharpness of the whole image. Compared with the conventional fractal coding, we can see that the proposed method can effectively eliminate the block effect and improve the loss of details.

Comparing the reconstructed results of different test images shown in Fig.2, Fig.3 and Fig.4, we notice that the quality of the reconstructed image "Peppers" and "Lena" are better. That is because they contain many simple edges and flat regions. That is to say fractal coding is suitable for the images that possess high self-similarity.

Fig. 3. Comparison of super-resolution results on "Cameraman" image by different methods: a). Initial high-resolution image; b). Cubic spline interpolation, PSNR=26.5226; c). Conventional fractal coding, PSNR=24.8891; d). The proposed method, PSNR=27.2482

Fig. 4. Comparison of super-resolution results on "Pepper" image by different methods: a). Initial high-resolution image; b). Cubic spline interpolation, PSNR=32.0085; c). Conventional fractal coding, PSNR=30.6253; d). The proposed method, PSNR=34.1000

5   Conclusion

In this paper, we proposed a multi-scale fractal coding method to achieve single image super-resolution. The proposed method can remove the block effect and improve the situation of the loss of details through the image integration and back-projection. Because the proposed method specially utilizes spatial information and self-similarity structural information of an image to achieve super-resolution, it can be combined with some previously proposed methods such as reconstruction-based methods and learning-based methods. However, the proposed method exists the problem of time-consuming, we will focus on how to speed up the encoding process while maintaining comparable magnified image quality in future work.

References

1. Zhang, Z.-H., Su, H., Zhou, J.: Survey of super-resolution image reconstruction methods. Acta Automatica Sinica 39(8), 1202–1213 (2013)
2. Yusheng, D.: Interpolated algorithm research in digital image processing. Computer Knowledge and Technology: Academic Exchange 6(6), 4502–4503 (2010)
3. Zhang, T., Xu, X.: Construction and realization of cubic spline interpolation function. Ordnance Industry Automation 25(11), 76–78 (2007)
4. Schultz, R.R., Stevenson, R.L.: Improved definition video frame enhancement. In: 1995 International Conference on Acoustics, Speech, and Signal Processing, ICASSP 1995, vol. 4, pp. 2169–2172. IEEE (1995)
5. Banham, M.R., Katsaggelos, A.K.: Digital image restoration. IEEE Signal Processing Magazine 14(2), 24–41 (1997)
6. Freeman, W.T., Jones, T.R., Pasztor, E.C.: Example-based super-resolution. IEEE Computer Graphics and Applications 22(2), 56–65 (2002)
7. Chang, H., Yeung, D.-Y., Xiong, Y.: Super-resolution through neighbor embedding. In: Proceedings of the 2004 IEEE Computer Society Conference on Computer Vision and Pattern Recognition, CVPR 2004, vol. 1, p. I. IEEE (2004)
8. Yang, J., Wright, J., Huang, T.S., Ma, Y.: Image super-resolution via sparse representation. IEEE Transactions on Image Processing 19(11), 2861–2873 (2010)
9. Barnsley, M.: Fractals everywhere. Academic Press (1988); Besicovitch, A.S.: On the existence of tangent to rectifiable curves. J. London Math. Soc. 19, pp. 205–207 (1944)
10. Li, X., Zhuo, L., Wang, S.: Image/Video Super-resolution. Posts & Telecom Press (2011) (in Chinese)
11. Fisher, Y.: Fractal Image Compression: Theory and Application. Springer, New York (1995)
12. Chung, K.-H., Fung, Y.-H., Chan, Y.-H.: Image enlargement using fractal. In: Proceedings of the 2003 IEEE International Conference on Acoustics, Speech, and Signal Processing (ICASSP 2003), vol. 6, pp. VI–273. IEEE (2003)
13. Lai, C.-M., Lam, K.-M., Chan, Y.-H., Siu, W.-C.: An efficient fractal-based algorithm for image magnification. In: Intelligent Multimedia, Video and Speech Processing (2004)
14. Chuanjiang, H.: Study on algorithms for fractal image coding technology. Ph.D. dissertation, Chongqing University (2004)

Image Denoising with BEMD
and Edge-Preserving Self-Snake Model

Ting-qin Yan[1,2], Min Qu[1], and Chang-xiong Zhou[1]

[1] Department of Electronic and Informational Engineering,
Suzhou Vocational University, Suzhou 215104, China
[2] Suzhou High-Tech Key Laboratory of Cloud Computing & Intelligent Information Processing,
Suzhou, 215104, China
ytqmax@gmail.com

Abstract. Image denoising is important in digital image processing. In this paper, an image denoising method using edge-preserving self-snake model(ESM) and bidimensional empirical mode decomposition(BEMD) is presented. The ESM includes an edge stopping function which is constructed with nonlocal gradient having maximum peak only at edges and good tolerance for noise. This model can preserve edge information while removing noise from digital images. The BEMD transforms the image into intrinsic mode functions(IMFs) and residue. Different components of IMFs present different frequency of the image. we use ESM of the IMFs to filter noise. Finally, we reconstruct the image with the filtered IMFs and residue. Experiments show that this algorithm has a better result than ESM.

Keywords: Image denoising, BEMD, Nonlocal gradient, Edge stopping function, BEMD_ESM.

1 Introduction

Noisy images are harmful to image processing. And image denoising is important for other aspects of image processing. Many denoising algorithms had been proposed, such as algorithms based on nonlocal transform-domain filter[1], algorithms based on wavelet[2] [3] [4] and algorithm based on Combining independent component analysis and hierarchical fusion of classifiers theory [5]. In [6] and [7] the authors used the method of multi-diagonal matrix filter and laplacian regularization respectively. Lately, some researchers proposed an algorithm using image fusion [8] and compressive sensing theory [9].

Empirical mode decomposition (EMD) was first proposed by Huang [10]. At the beginning, EMD is used on one dimension signals analysis, such as sound signals. Later, BEMD was introduced into image processing field[11].

In general, edges are fundamental elements which underlie more complicated features in images. The latter can be maintained as long as edges are preserved during

D.-S. Huang et al. (Eds.): ICIC 2014, LNCS 8588, pp. 435–442, 2014.

denoising. The goal of image denoising is to remove noise from images without blurring sharp edges. Nonlocal gradient[12] has maximum peak only at edges and good tolerance for noise. A new edge stopping function[13] is constructed with the nonlocal gradient. The edge stopping function is approximated to zero around fine structures and is close to one in the interior of homogeneous regions. To keep edges better while denoising, the edge stopping function with nonlocal gradient is introduced into ESM. In this paper, we proposed an algorithm using BEMD and ESM. Firstly, we execute BEMD on original image and get the IMFs and residue. Secondly, we filter noise of the IMFs with ESM. Lastly, we reconstruct the image with the filtered IMFs and residue.

2   Bidimensional Empirical Mode Decomposition

In data analysis, EMD is an important method, which can decompose a complex signal into a set of IMFs and residue, so that more detailed information of signal can be extracted. EMD has been used for image denoising, where an image is viewed as one dimensional signal. But the relationship between adjacent pixels is ignored. In 2003, BEMD was proposed[11]. BEMD is used to decompose bidimensional image into IMFs and residue. The detailed process is presented as follows,

  a) Look for the extrema (both minima and maxima) of the original image $I(x, y)$ and form the envelopments by connecting extrema.

  b) Determine the average $ave_1(x, y)$ of the top envelopment and bottom envelopment, and subtract the average from the image.

$$h_1(x, y) = I(x, y) - ave_1(x, y) \tag{1}$$

  c) Replace $I(x, y)$ with $h_1(x, y)$ and execute the above three steps. Then $h_i(x, y)$ can be obtained, until the standard deviation SD is smaller than the threshold predefined. $h_i(x, y)$ is viewed as IMF_1. Where

$$SD = \sum_{x=0}^{X} \sum_{y=0}^{Y} \left[\frac{|h_{i+1}(x, y) - h_i(x, y)|^2}{h_i^2(x, y)} \right] \tag{2}$$

  d) Replace $I(x, y)$ with $I(x, y) - IMF_1$ and repeat step a)-c), we can get IMF_i. If we have N IMFs, the decomposition can be expressed as,

$$I(x, y) = \sum_{i=1}^{N} IMF_i(x, y) + R_N(x, y) \tag{3}$$

3   Nonlocal Gradient

For one dimensional signal $x = [x_1, x_2, ..., x_i, ..., x_N]$,   the neighbourhood region P_{x_i} of x_i comprising W pixels centred around x_i.

$$P_{x_i} = [x_{i-(W-1)/2}, x_{i-(W-1)/2+1}, ...x_i, ...x_{i+(W-1)/2}] \tag{4}$$

The nonlocal difference between two signal samples P_{x_i} and P_{xj} can be measured as a Gaussian weighted Euclidean difference,

$$d_{NL}(x_i, x_j) = \left\| P_{x_i} - P_{x_j} \right\|_{2,\sigma} \tag{5}$$

where σ is the standard deviation (Std) of a normalized Gaussian kernel. first-order nonlocal derivative at pixel can be defined as

$$|\nabla_{NL} x_i| = \left\| P_{x_{i-(W+1)/2}} - P_{x_{i+(W+1)/2}} \right\|_{2,\sigma}$$

$$= \left(\sum_{k=-(W-1)/2}^{k=(W-1)/2} G_\sigma(k) \left(x_{i-(W+1)/2+k} - x_{i+(W+1)/2+k} \right)^2 \right)^{1/2} \tag{6}$$

where G_σ is a Gaussian kernel with Std σ. In a two-dimensional image $u(i, j)$, nonlocal derivative of pixel (i, j) can be defined as the maximum value of level and vertical nonlocal derivatives.

$$|\nabla_{NL} u(i, j)| = \max(|\nabla_{NLrow} u(i, j)|, |\nabla_{NLcol} u(i, j)|) \tag{7}$$

The nonlocal derivative is calculated in the neighborhood window of the given pixel, therefore, it gets the maximum value at edges of the image, and is not sensitive to the noise[14].

4   BEMD Edge-Preserving Self-Snake Model(BEMD_ESM)

4.1   Edge-Preserving Self-Snake Model

Image filtering equations based on self-snake model is as follows

$$\frac{\partial u}{\partial t} = div(g(|\nabla u|)\nabla u / |\nabla u|)|\nabla u| \tag{8}$$

Where u is the gray value of the image, div is the divergence. The equation (8) can be expressed as:

$$\frac{\partial u}{\partial t} = g(|\nabla u|)div(\frac{\nabla u}{|\nabla u|})|\nabla u| + \nabla g(|\nabla u|) \cdot \nabla u \tag{9}$$

The first item is direction diffusionwith edge stopping function $g(|\nabla u|)$, and the second is shock filter function with image enhancement. The edge stopping function in general is:

$$g(|\nabla u|) = 1 \Big/ \left(\left(\frac{|\nabla u|}{k1} \right)^2 + 1 \right) \tag{10}$$

where K1 is a constant, the self-snake model with edge stopping function (10) is a standard self snake model. In the smooth homogeneous regions of image, the derivative is small, edge stopping function is approximatly 1, while the derivative is significant at the edge, edge stopping function is approximatly 0, so the edge stopping function can better preserve the edges. The equation (10) is a function of the derivative, and it is sensitive to the gray change. When the noise gradient is large, or the image resolution is low, The equation (10) can not effectively eliminate noise, the image edge blur. In view of the above situation, the nonlocal gradient is taken as the edge stopping function, i.e.

$$g(|\nabla_{NL} u(i,j)|) = 1 \big/ (|\nabla_{NL} u(i,j)| \big/ k2 + 1) \tag{11}$$

where K2 is a constant, ESM is presented as follows.

$$\partial u / \partial t = div(g(|\nabla_{NL} u(i,j)|) \nabla u / |\nabla u|) |\nabla u| \tag{12}$$

where $g(|\nabla_{NL} u(i,j)|)$ is edge stopping function.

For quantitative evaluation of algorithm performance, the Equivalent Number of Looks (ENL) and the edge Definition(DEF) are introduced.

$$ENL = \bar{u}^2(i,j) / var(i,j) \tag{13}$$

$$DEF = \sum_{i,j=m1,n1}^{m2,n2} \{[u(i,j) - u(i-1,j)]^2 \tag{14}$$
$$+ [u(i,j) - u(i,j-1)]^2\}^{1/2} / (m_2 - m_1)(n_2 - n_1)$$

Where $\bar{u}$ is image gray mean, var is gray variance. The larger the ENL, the better the denoising; Conversely, the algorithm denoises badly with smaller ENL. Where $\bar{u}$ is the denoised image, and m1,n1,m2,n2 are the coordinate values of the window. The larger DEF indicates edges are preserved better.

ESM algorithm is used to remove the noise of an original image. But this may filter some image information. To solve this problem, we proposed to transform the image into IMFs and the residue, and remove noise from IMFs respectively, then reconstruct image with the denoised IMFs and the residue. The detail of the BEMD_ESM algorithm can be expressed as follows,

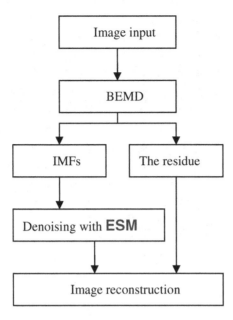

Fig. 1. Process of BEMD_ESM Denoising

5   Experiments

We test the BEMD_ESM algorithm on the Lenna image and compare the results with existing popular denoising methods, including ESM method and block matching and Feature-preserving Nonlinear Anisotropic Diffusion Method (FP-NAD) [14], the visual and numerical results obtained by using BEMD_ESM algorithm are presented in this section.

We undertake experiments on a classical image: Lenna (512x512). Fig. 2 (a) shows the noisy-free and Fig. 3 (a) shows the noisy image with AWGN of a Std $\sigma n = 20$. Fig. 2(b)-(d) show the edge stopping functions of FP-NAD, ESM and BEMD_ESM, respectively. The edge stopping function of BEMD_ESM has higher brightness at the edges than FP-NAD and ESM, which can accurately highlight the regional edge position. and lower brightness of edge stopping function in the background area shows stronger noise suppressing ability.

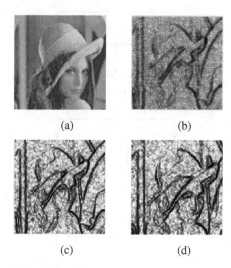

Fig. 2. (a) Original image, (b)-(d) are edge stopping functions of FP-NAD, ESM and BEMD_ESM, respectively

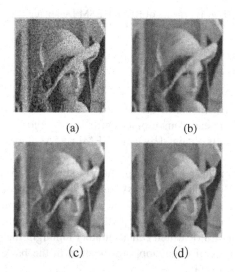

Fig. 3. (a) The noisy image with AWGN of a Std = 20, (b)-(d) Denoised results by FP-NAD, ESM and BEMD_ESM, respectively

The denoising results are shown in Fig.3 (b)-(d). Compared with algorithm FP-NAD and ESM, algorithm BEMD_ESM has the best visual effects, denoising better and preserving image edges.

We report in Table 1 the ENL and DEF values of row 55 and 56 for the Lenna image shown in Figure 3(a) by BEMD_ESM, ESM and FP-NAD. As seen, the noisy image has larger niose, the variance is larger, so the ENL is smaller and the DEF is larger. As for the three algorithms(FP-NAD, ESM, BEMD-ESM), denoised images have larger

ENL than the original one. and the ENL in denoised image by BEMD-ESM is the largest, shows the best denoising ability. The DEF of the three algorithms(FP-NAD, ESM, BEMD-ESM) become larger, and the DEF of BEMD-ESM algorithm is the largest shows the best edge preserving ability. The ENL of ESM is larger than FP-NAD algorithm shows better denoising ability, the DEF of FP-NAD is larger than ESM shows better edge preserving ability. The ENL and DEF by BEMD-ESM are largest in these three algorithms show best denoising ability and best edge preserving ability. The data in Table 1 are corresponding to the visual effects in Fig.3, our algorithm can effectively reduce the background noise and provide improved detection performance on edges.

Table 1. Comparison of Vars, Aves, ENL and DEF by FP-NAD, ESM and our Algorithm (BEMD-ESM)

	Noisy image	FP-NAD	ESM	BEMD-ESM
vars	2136.8656	1592.8503	1427.0201	1487.2397
aves	108.8933	108.6634	108.5479	111.9567
ENL	5.5491	8.2519	8.2568	8.4279
DEF	41.5094	18.1965	13.6934	18.4214

6   Conclusions and Discussion

Noise is inevitable and harmful. Just as edges, Noise belongs to high frequency information. Denoising by traditional technology will often blur edges. Nonlocal gradient has the maximum value at edges of images, and is not sensitive to noise. So we intruduce nonlocal gradient to construt edge stopping function. The edge stopping function with nonlocal gradient is close to 0 at image edges, and close to 1 in the flat regions. Self snake model is useful to enhance image, so ESM is introduced into our algorithm. In contrast to the other algorithms, we introduce both ESM and BEMD to image denoising. The BEMD transforms the image into IMFs which have different frequencies. Then we use ESM to filter the IMFs with different threshold parameters. The final image is composed of the filtered IMFs and the residue.

We have tested the new algorithm on the Lenna image, the experimental result shows that, our algorithm is significantly better than FP-NAD and ESM algorithm . It demonstrates good performance in preserving edges. It can also effectively reduce the background noise.

Acknowledgements . This research was sponsored by the grants of National Science Foundation of China (No.61373098), Project of Suzhou Vocational University (No.2012SZDYY04), the Opening Project of Suzhou High-tech Key Laboratory of Cloud Computing & Intelligent Information Processing(No. SXZ201304).

References

1. Maggioni, M., Katkovnik, V., Egiazarian, K., et al.: Nonlocal Transform-Domain Filter for Volumetric Data Denoising and Reconstruction. IEEE Transactions on Image Processing 22(1), 119–133 (2013)
2. Xie, Z.J., Song, B.Y., Zhang, Y., et al.: Application of an Improved Wavelet Threshold Denoising Method for Vibration Signal Processing. Advanced Materials Research 889, 799–806 (2014)
3. Yan, R., Shao, L., Liu, Y.: Nonlocal Hierarchical Dictionary Learning Using Wavelets for Image Denoising. IEEE Transactions on Image Processing 22(12), 4689–4698 (2013)
4. Zhong, J.J., Fang, S.N., Linghu, C.Y.: Research on Application of Wavelet Denoising Method Based on Signal to Noise Ratio in the Bench Test. Applied Mechanics and Materials 457, 1156–1162 (2014)
5. Salimi-Khorshidi, G., Douaud, G., Beckmann, C.F., et al.: Automatic denoising of functional MRI data: Combining independent component analysis and hierarchical fusion of classifiers. NeuroImage 90(15), 449–468 (2014)
6. Anbarjafari, G., Demirel, H., Gokus, A.E.: A Novel Multi-diagonal Matrix Filter for Binary Image Denoising. Journal of Advanced Electrical and Computer Engineering 1(1), 14–21 (2014)
7. Liu, X., Zhai, D., Zhao, D., et al.: Progressive image denoising through hybrid graph laplacian regularization: a unified framework. IEEE Transactions on Image Processing: A Publication of the IEEE Signal Processing Society 23(4), 1491–1503 (2014)
8. Raj, V., Venkateswarlu, T.: Denoising of 3D Magnetic Resonance Images Using Image Fusion. In: 2014 International Conference on Electronic Systems, Signal Processing and Computing Technologies (ICESC), pp. 295–299. IEEE (2014)
9. Jin, J., Yang, B., Liang, K., et al.: General image denoising framework based on compressive sensing theory. Computers & Graphics 38, 382–391 (2014)
10. Huang, N.E., Shen, Z., Long, S.R., et al.: The empirical mode decomposition and the Hilbert spectrum for nonlinear and non-stationary time series analysis. Proceedings of the Royal Society of London. Series A: Mathematical, Physical and Engineering Sciences 454(1971), 903–995 (1998)
11. Nunes, J.C., Bouaoune, Y., Delechelle, E., et al.: Image analysis by bidimensional empirical mode decomposition. Image and Vision Computing 21(12), 1019–1026 (2003)
12. Yang, L., Peng, J.S.: Scale Effect on Dynamic Analysis of Electrostatically Actuated Nano Beams Using the Nonlocal-Gradient Elasticity Theory. Applied Mechanics and Materials 411, 1859–1862 (2013)
13. Zhao, J., Qi, Y.M., Pei, J.Y.: A New Enhancement Model Combined Anisotropic Diffusion and Shock Filter. Advanced Materials Research 889, 1089–1092 (2014)
14. Qiu, Z., Yang, L., Lu, W.: A new feature-preserving nonlinear anisotropic diffusion for denoising images containing blobs and ridges. Pattern Recognition Letters 33(3), 319–330 (2012)

Enhanced Local Ternary Pattern
for Texture Classification

Jing-Hua Yuan[1], Hao-Dong Zhu[2], Yong Gan[2], and Li Shang[3]

[1] College of Electronics and Information Engineering, Tongji University, Shanghai, China
[2] School of Computer and Communication Engineering,
Zhengzhou University of Light Industry, Zhengzhou, Henan, China
[3] Department of Communication Technology, College of Electronic Information Engineering,
Suzhou Vocational University, Suzhou, Jiangsu, China

Abstract. The Local Ternary Pattern (LTP) extends the conventional LBP to ternary codes and makes a significant improvement. LTP is more resistant to noise, but no longer strictly invariant to gray-level transformations. To improve the performance of LTP, this paper proposes the Enhanced Local Ternary Pattern (ELTP) by adopting the Average Local Gray Level (ALG) to take place of the traditional gray value of the center pixel, taking an auto-adaptive strategy on the selection of the threshold and introducing a novel coding process. Finally, the Completed Enhanced Ternary Pattern (CELTP) is also presented.

Keywords: Local Binary Pattern (LBP), Local Ternary Pattern (LTP), Enhanced Local Ternary Pattern (ELTP).

1 Introduction

In [1], Ojala *et al.* proposed the Local Binary Pattern (LBP) to address rotation invariant texture classification. LBP is efficient to represent local texture and invariant to monotonic gray scale transformations. Heikkila *et al.* [2] presented center-symmetric LBP (CS-LBP) by comparing center-symmetric pairs of pixels. Liao *et al.* [3] proposed Dominant LBP (DLBP) for texture classification. Tan and Triggs [4] presented Local Ternary Pattern (LTP), extending the conventional LBP to 3-valued codes. However LTP is no longer strictly invariant to gray-level transformations due to the simple strategy on the selection of the threshold. Zhang *et al.* [5] proposed Local Derivative Pattern (LDP) to capture more detailed information by introducing high order derivatives. However, if the order is greater than three, LDP is more sensitive to noise than LBP. Guo *et al.* [6] proposed Completed LBP (CLBP) by combining the original LBP with the measures of local intensity difference and central pixel gray level. Zhao *et al.* [7] proposed the Local Binary Count (LBC), to address rotation invariant texture classification by totally discarding the micro-structure which is not absolutely invariant to rotation under the huge illumination changes. However, LBC (or LBP) is sensitive to random noise and quantization noise in the near-uniform regions, as the

D.-S. Huang et al. (Eds.): ICIC 2014, LNCS 8588, pp. 443–448, 2014.

LBC (or LBP) threshold is the value of the central pixel as mentioned in [4]. Zhao *et al.* [8] presented Completed Robust Local Binary Pattern (CRLBP) by modifying the center pixel gray level to improve the LBP, but a more parameter has to be tuned.

Motivated by [4], [7], and [8],, this paper tries to address these potential difficulties by proposing the Enhanced Local Ternary Pattern (ELTP). An auto-adaptive strategy for threshold selection is adopted and a novel coding process is introduced. Furthermore, the Completed Enhanced Local Ternary Pattern (CELTP) is presented.

The remainder of this paper is organized as follows. Section 2 presents the ELTP and CELTP. Section 3 reports experimental results and Section 4 concludes the paper.

2   Enhanced Local Ternary Pattern

In order to address the demerits of LBPs, we propose Enhanced Local Ternary Pattern (ELTP). Gray-levels in a tolerance zone of width $\pm t^e$ around g_c^e are quantized to zero, ones above this are quantized to +1 and ones below it to -1, i.e., the indicator $s(x)$ used in LBP is replaced with a 3-valued function:

$$s^e\left(g_p, g_c^e, t^e\right) = \begin{cases} 1, & g_p - g_c^e \geq t^e \\ 0, & \left|g_p - g_c^e\right| < t^e, \quad p = 0, 1, ..., P-1 \\ -1, & g_p - g_c^e \leq -t^e \end{cases} \tag{1}$$

$$g_c^e = mean(G), \quad t^e = mad(G), \quad G = \{g_i \mid i = 0, 1, ..., 8\}$$

where P is the size of the neighbor set of pixels, g_p ($p=0, 1,..., P-1$) denotes the gray value of the neighbor, G is the set of the gray-level values in a 3×3 local region, $mean(G)$ is the mean of the set G, $mad(G)$ is the median absolute deviation of the set G.

g_c is replaced by g_c^e, which make the ELTP more robust to noise. On the other hand, instead of using the simple user-specified threshold, we adopt new strategy to set the threshold t^e. To make the selection of threshold auto-adaptive, the threshold is set as the median absolute deviation (MAD), which not only reflect derivation of the local region but make the ELTP code invariant to gray-level transformations and insensitive to noise. In addition, the use of MAD does not affect the complexity of the model.

For simplicity, the experiments use a coding scheme that splits each ternary pattern into its positive half (ELTP_P) and negative half (ELTP_N) as illustrated in **Fig. 1**, subsequently combining these two separate channels of LBC descriptors to form the final ELTP descriptor and finally computing the histogram and similarity metric. Obviously, ELTP is also rotation invariant. Using the same notations as in (1), the ELTP descriptor used in the experiment is defined as follows:

$$\text{ELTP}_{P,R} = \text{ELTP\_P}_{P,R} * (P+2) - (\text{ELTP\_P}_{P,R} * (\text{ELTP\_P}_{P,R} + 1))/2 + \text{ELTP\_N}_{P,R}$$

$$\text{ELTP\_P}_{P,R} = \sum_{p=0}^{P-1} e(s^e\left(g_p, g_c^e, t^e\right), 1), \quad \text{ELTP\_N}_{P,R} = \sum_{p=0}^{P-1} e(s^e\left(g_p, g_c^e, t^e\right), -1), \quad e(x, y) = \begin{cases} 1, & x = y \\ 0, & x \neq y \end{cases} \tag{2}$$

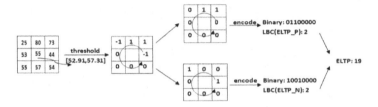

Fig. 1. The process of forming the ELTP code

Aiming to improve the discriminative capability of the local structure, the enhanced image local differences d_p^e are decomposed into two complementary components [6], i.e., the signs (s_p^e) and the magnitudes (m_p^e), respectively

$$d_p^e = s_p^e * m_p^e, \quad s_p^e = s^e\left(g_p, g_c^e, t^e\right), \quad m_p^e = ((g_p - g_c^e) - s_p^e * t^e) * s_p^e, \; p = 0, 1, ..., P. \quad (3)$$

Similar to the CLBP, we also proposed Completed Enhanced Local Ternary Pattern (CELTP) which contain three operators: CELTP_S, CELTP_M and CELTP_C. Generally, the CELTP_S equals to the original ELTP described above. And the CELTP_M and CELTP_C can be defined as:

$$\text{CELTP\_M}_{P,R} = \text{CELTP\_MP}_{P,R} * (P+2) - (\text{CELTP\_MP}_{P,R} * (\text{CELTP\_MP}_{P,R} + 1)) / 2 + \text{CELTP\_MN}_{P,R}$$

$$\text{CELTP\_MP}_{P,R} = \sum_{p=0}^{P-1} s(m_p^e - m_I^e) * e(s_p^e, 1), \quad \text{CELTP\_MN}_{P,R} = \sum_{p=0}^{P-1} s(m_p^e - m_I^e) * e(s_p^e, -1) \qquad (4)$$

$$\text{CELTP\_C}_{P,R} = s(g_c^e - c_I)$$

where the threshold m_I^e is the mean value of m_p^e of the whole image and the threshold c_I is set as the mean gray level of the whole image.

3   Experimental Results

To evaluate the proposed method, we carried out experiments on two benchmark texture databases: the Outex database [9], which includes 24 classes of textures collected under three illuminations and at nine angles, and the CUReT database [10], which contains 61 classes of real-world textures captured at different viewpoints and illumination orientations. And we assume that the nearest neighborhood classifier and the χ^2 statistics [11][12,13,14,15,16,17] are used. For two histograms, $H=\{h_i\}$ and $K=\{k_i\}$, their dissimilarity can be calculated as: $d_{\chi^2}(H, K) = \sum_{i=1}^{N} (h_i - k_i)^2 / h_i + k_i$.

Table 1. Classification Rate (%) on TC10 and TC12 Using Different Methods

| | TC10 | R=1, P=8 TC12 | | Average | TC10 | R=2, P=16 TC12 | | Average | TC10 | R=3, P=24 TC12 | | Average |
		t184	horizon			t184	horizon			t184	horizon	
LBP[15]	84.82	65.46	63.68	71.32	89.40	82.27	75.21	82.29	95.08	85.05	80.79	86.97
LTP[15]	76.06	62.56	63.42	67.34	96.11	85.20	85.87	89.06	98.64	92.59	91.52	94.25
CLBP_M[15]	81.74	59.31	62.78	67.94	93.67	73.80	72.41	79.96	95.52	81.18	78.66	85.12
CLBP_S/M/C[15]	96.56	90.30	92.29	93.05	98.72	93.54	93.91	95.39	98.93	95.32	94.54	96.26
CLBC_S/M/C[16]	97.16	89.79	92.92	93.29	98.54	93.26	94.07	95.29	98.78	94.00	93.24	95.67
CRLBP_S/M/C(α=1)[17]	96.54	91.16	92.06	93.25	**98.85**	96.67	96.97	97.50	**99.48**	97.57	97.34	98.13
CRLBP_S/M/C(α=8)[17]	**97.55**	**91.94**	**92.45**	**93.98**	98.59	95.88	96.16	96.96	99.35	96.83	96.16	97.45
ELTP	79.06	67.82	76.34	74.41	97.14	90.95	91.00	93.03	96.54	93.96	93.22	94.57
CELTP_M	80.83	70.81	74.63	75.42	97.06	90.88	91.50	93.15	98.59	96.46	96.25	97.10
CELTP_M/C	87.99	78.31	83.17	83.16	92.27	86.92	89.58	89.59	94.51	92.08	93.54	93.38
CELTP_S_M/C	87.81	76.67	82.94	82.47	93.67	90.56	91.34	91.86	95.55	94.65	95.76	95.32
CELTP_S/C	87.58	76.64	86.32	83.51	98.18	94.93	95.76	96.29	98.46	93.63	95.72	95.94
CELTP_M_S/C	87.89	77.50	86.20	83.86	98.83	**96.71**	**97.29**	**97.61**	99.40	**98.56**	**98.61**	**98.86**
CELTP_S_M	82.68	71.25	77.94	77.29	98.39	96.00	95.97	96.79	99.40	98.45	98.08	98.64
CELTP_S_M_C	87.50	76.39	84.56	82.82	98.54	95.58	95.70	96.64	99.43	98.17	97.92	98.51

Table 2. Classification Rate (%) on CUReT Using Different Methods

| | R=1, P=8 | | | | R=2, P=16 | | | | R=3, P=24 | | | |
	46	23	12	6	46	23	12	6	46	23	12	6
LBP[15]	80.63	74.81	67.84	58.70	86.37	81.05	74.62	66.17	86.37	81.21	74.71	66.55
LTP[15]	85.13	79.25	72.04	63.09	92.66	87.30	80.22	70.50	91.81	85.78	77.88	67.77
CLBP_M[15]	75.20	67.96	60.27	51.49	85.48	79.01	71.24	61.59	82.16	76.23	69.22	60.45
CLBP_S/M/C[15]	**95.59**	**91.35**	84.92	74.80	95.86	92.13	86.15	77.04	94.74	90.33	83.82	74.46
CLBC_S/M/C[16]	94.78	90.12	82.92	72.85	95.39	91.30	85.91	75.14	95.26	90.55	84.07	73.18
CRLBP_S/M/C(α=1)[17]	94.55	89.47	82.72	73.22	94.78	91.10	85.47	76.39	95.35	90.73	85.05	76.34
CRLBP_S/M/C(α=8)[17]	95.39	91.33	**85.40**	**76.56**	95.88	91.85	86.44	**77.79**	96.27	91.83	86.06	**78.43**
ELTP	84.98	78.73	70.85	60.57	90.00	85.20	78.01	67.90	90.45	85.70	78.92	69.28
CELTP_M	84.83	78.46	70.36	60.25	92.31	87.44	80.25	70.23	91.24	86.01	78.55	68.32
CELTP_M/C	85.62	77.23	67.15	54.77	90.47	83.59	74.68	63.40	92.48	86.84	78.72	67.75
CELTP_S_M/C	87.36	79.53	69.62	57.05	92.03	85.99	77.86	66.84	94.25	89.51	82.32	72.00
CELTP_S/C	87.36	79.70	69.98	57.70	95.02	90.85	84.49	74.67	94.95	91.14	85.02	75.58
CELTP_M_S/C	88.43	80.65	70.90	58.66	**96.46**	**92.89**	**86.97**	77.62	**96.30**	**92.68**	**86.59**	77.03
CELTP_S_M	88.89	83.06	75.09	64.79	94.50	90.21	83.59	73.84	94.66	90.34	83.67	73.79
CELTP_S_M_C	91.49	85.67	77.57	66.99	95.34	91.08	84.42	74.57	95.18	90.95	84.28	74.35

Table 1 and Table 2 list the experimental results by different schemes on the two databases, from which we could make the following findings. Firstly, on both databases, the proposed CELTP method is inferior to CLBP_S/M/C, CLBC_S/M/C and CRLBP_S/M/C on the condition ($R=1$, $P=8$) .This mainly because we use the mean of the gray-level values in a 3×3 region to replace the value of the center pixel. Secondly, on the Outex database, ELTP performs much better than LBP except for the experiment ($R=1$, $P=8$) on TC10 dataset and outperform LTP except for the experiment($R=3$, $P=24$) on TC10 dataset LTP; on the CUReT database, ELTP outperforms LBP and achieve similar accurate classification rate compared with LTP. It should be noticed that ELTP is much more robust to viewpoint and illumination variations. Thirdly, CELTP_M achieves impressive performance as compared with CLBP_M. This means that CELTP_M could capture more discriminative gray-level information than CLBP_M. Finally, except for the experiment ($R=1$, $P=8$), CELTP_M_S/C and CELTP_S_M almost always outperform other methods on TC12 dataset and on the CUReT database, which is impressive when the feature dimensionality is concerned.

In one word, CELTP, with low feature dimensionality, achieves better results than other methods and is less sensitive to viewpoint and illumination variations.

4   Conclusion

The Enhanced Local Ternary Pattern (ELTP) is proposed by introducing an average local gray-level strategy and an auto-adaptive threshold selection scheme, which make the proposed ELTP not only resistant to noise but also strictly invariant to gray-level transformations. To use the 3-valued codes, we also gave a novel coding scheme. Finally, we proposed Completed Enhanced Local Ternary Pattern (CELTP). Experimental results obtained from two representative databases clearly demonstrate that the proposed CELTP can obtain impressive texture classification accuracy.

Acknowledgments. This work is supported by the Science and Technology Innovation Outstanding Talent Plan Project of Henan Province of China under Professor De-Shuang Huang (No.134200510025), the National Natural Science Foundation of China (Nos. 61373105, 61373098 & 61133010).

References

1. Ojala, T., Pietikainen, M., Maenpaa, T.: Multiresolution Gray-Scale And Rotation Invariant Texture Classification with Local Binary Patterns. IEEE Trans. Pattern Anal. Mach. intell. 24(7), 971–987 (2002)
2. Heikkila, M., Pietikainen, M., Schmid, C.: Description of interest Regions with Center-Symmetric Local Binary Patterns. Proc. of Computer Vision, Graphics and Image Processing 43(38), 58–69 (2006)
3. Liao, S., Law, M.W.K., Chung, A.C.S.: Dominant Local Binary Patterns for Texture Classification. IEEE Trans. Image Process. 18(5), 1107–1118 (2009)
4. Tan, X.Y., Triggs, B.: Enhanced Local Texture Feature Sets for Face Recognition Under Difficult Lighting Conditions. IEEE Trans. Image Process. 19(6), 1635–1650 (2010)
5. Zhang, B., Gao, Y., Zhao, S., Liu, J.: Local Derivative Pattern Versus Local Binary Pattern: Face Recognition with High-Order Local Pattern Descriptor. IEEE Trans. Image Process. 19(2), 533–544 (2010)
6. Guo, Z.H., Zhang, L., Zhang, D.: A Completed Modeling of Local Binary Pattern Operator for Texture Classification. IEEE Trans. Image Process. 19(6), 1657–1663 (2010)
7. Zhao, Y., Huang, D.-S., Jia, W.: Completed Local Binary Count for Rotation invariant Texture. IEEE Trans. on Image Process. 21(10), 4492–4497 (2012)
8. Zhao, Y., Jia, W., Hu, R.-X., Min, H.: Completed Robust Local Binary Pattern for Texture Classification. Neurocomputing 10(6), 68–76 (2013)
9. Ojala, T., Maenpaa, T., Pietikainen, M., Viertola, J., Kyllönen, J., Huovinen, S.: Outex - New Framework for Empirical Evaluation of Texture Analysis Algorithms. In: Proc. 16th int. Conf. Pattern Recognit., vol. 1, pp. 701–706 (2002)

10. Dana, K.J., van Ginneken, B., Nayar, S.K., Koenderink, J.J.: Reflectance and Texture of Real World Surfaces. ACM Trans. Graph. 18(1), 1–34 (1999)
11. Varma, M., Zisserman, A.: A Statistical Approach to Texture Classification from Single Images. Int. J. Comput. Vision 62(1-2), 61–81 (2005)
12. Wang, X.F., Huang, D.S., Xu, H.: An Efficient Local Chan-Vese Model for Image Segmentation. Pattern Recognition 43(3), 603–618 (2010)
13. Li, B., Huang, D.S.: Locally Linear Discriminant Embedding: An Efficient Method for Face Recognition. Pattern Recognition 41(12), 3813–3821 (2008)
14. Huang, D.S.: Radial Basis Probabilistic Neural Networks: Model and Application. Int. Journal of Pattern Recognit., and Artificial Intell. 13(7), 1083–1101 (1999)
15. Huang, D.S., Ma, S.D.: Linear And Nonlinear Feedforward Neural Network Classifiers: A Comprehensive Understanding. Journal of Intelligent Systems 9(1), 1–38 (1999)
16. Huang, D.S.: A Constructive Approach for Finding Arbitrary Roots of Polynomials by Neural Networks. IEEE Trans. on Neural Networks 15(2), 477–491 (2004)
17. Huang, D.S., Ip, H.H.S., Chi, Z.-R.: A Neural Root Finder of Polynomials Based on Root Moments. Neural Computation 16(8), 1721–1762 (2004)

Feature Points Selection Algorithm of DSA Images Based on Adaptive Multi-scale Vascular Enhancement and Mean Shift

Xinhong Zhang[1,2] and Fan Zhang[1,3]

[1] Institute of Image Processing and Pattern Recognition, Henan University,
Kaifeng 475001, China
[2] Software School, Henan University, Kaifeng 475001, China
[3] School of Computer and Information Engineering, Henan University, Kaifeng 475001, China
{zhangfan,zxh}@henu.edu.cn

Abstract. In the Digital Subtraction Angiography (DSA) image registration algorithm, the precision of the control points as well as their number and the distribution in image determines the accuracy of geometric correction and registration. Control points usually adopt the grid points; however, a more effective method is to extract control points adaptively according to the image feature. In this paper, a control point's selection algorithm of DSA images is proposed based on adaptive multi-Scale vascular enhancement, error diffusion and means shift algorithms. This paper introduce error diffusion algorithm is to take advantage of the image intrinsic characteristics to adaptively select control points. In the edge and texture complex area, more control points (grid intensive) will be selected. On the contrary in the flat area less control points will be selected. Introduction of the mean shift is to reduce the number of control points, and achieve the optimal image registration. Experimental results show that the proposed algorithm can adaptively put the control points to blood vessels and other key image characteristics, and can optimize the number of control points according to practical needs, which will ensure the accuracy of DSA image registration.

Keywords: Digital subtraction angiography, Error diffusion, Mean shift.

1 Introduction

Digital Subtraction Angiography (DSA) is an X-ray imaging. DSA is the technology combined conventional angiography with computer image processing technology. In traditional angiography images are acquired by exposing an area of interest with time-controlled x-rays while injecting contrast medium into the blood vessels. The image obtained would also include all overlying structure besides the blood vessels in this area. This is useful for determining anatomical position and variations but unhelpful for visualizing blood vessels accurately.

In order to remove these distracting structures to see the vessels better, first a mask image is acquired. The mask image is simply an image of the same area before the

D.-S. Huang et al. (Eds.): ICIC 2014, LNCS 8588, pp. 449–458, 2014.
© Springer International Publishing Switzerland 2014

contrast is administered. The radiological equipment used to capture this is usually an image intensifier, which will then keep producing images of the same area at a set rate, taking all subsequent images away from the original 'mask' image. The radiologist controls how much contrast media is injected and for how long. Smaller structures require less contrast to fill the vessel than others. Images produced appear with a very pale grey background, which produces a high contrast to the blood vessels, which appear a very dark grey. Tissues and blood vessels on the first image are digitally subtracted from the second image, leaving a clear picture of the artery which can then be studied independently and in isolation from the rest of the body [1,2].

However, DSA image may exist in a variety of artifacts which will affect the image quality. The motion artifact is the most common form of artifacts. Motion artifact is inevitable, but we can reduce, eliminate motion artifacts through the image processing methods and significantly improve image quality, and providing doctors with more effective diagnostic information. Recent studies in the elimination of motion artifacts are mainly based on image registration methods. Image registration is the process of transforming different sets of data into one coordinate system. Data may be multiple photographs, data from different sensors, times, depths, or viewpoints [3-6]. Since multiple images that reflect the same or partly same characteristics, so a part of the image pixels should be represented the same point on the corresponding points in the other image. They are registration control points (RCP). Control point selection is the key for image registration. Control points can be selected in three ways: manual, semi-automated search and computer automatically search. Regardless of what kind of matching algorithm, the final accuracy, quantity and distribution of the selected RCP largely determines the accuracy of geometric correction and registration.

Control point selection generally has the following requirements. (1) Control point should generally be more clear and stable markers, and be easily identifiable on both the reference image and the corrected image; (2) Control point should be distributed as evenly as possible in the image, otherwise registration accuracy is better in the area where registration control points are dense, and the accuracy of registration is worse in the area where registration control points distributed are more sparse; (3) The number of control points should be appropriate, too much RCP will impact on computer processing speed, too less RCP is not conducive to accurate registration.

In most image registration algorithm, control points usually exploit regular grid points. Defined on a Cartesian coordinate image system, if equidistant parallel lines are drawn along the X, Y direction, respectively, then the point of intersection of the parallel lines will be selected as the control point. This control point selection method is simple, but it does not take full advantage of the feature information of the image itself. If using such regular grid points, there are little influence with the X-ray dose changes and target geometry changes area of absence of significant image structure. And subsequent registration algorithm computes these points do not make sense, only spend a lot of computing time.

A more effective method is using image intrinsic characteristics to extract control points. This approach not only allows more efficient registration, but also reduce the number of control points, furthermore, further reduces the computing time. This paper presents a novel feature selection algorithm of DSA based on error diffusion and mean shift. The proposed feature point selection algorithm introduces error diffusion to take advantage of the image intrinsic characteristics and select control points adaptively. More control points are selected in the area of the edge and complex texture region (with density grid). And less control points are selected in the area of flat region. The purpose of the introduction of mean shift is to reduce the number of control points to achieve optimal image registration.

2   Adaptive Multi-scale Vessel Enhancement

In order to improve accuracy of feature points selection in DSA image, multi-scale adaptive filtering algorithm is used for DSA vascular enhancement processing. This paper proposes an adaptive multi-scale vessel enhancement filtering algorithm that the filter parameters can be determined adaptively according to image intrinsic characteristics. The main purpose of this algorithm is to improve the visual quality of blood vessel in DSA images.

The purpose of vessel enhancement is to emphasize vascular structures in image, while restrain the non-blood vessel characteristics. So the image recognition will be strengthened. Multi-scale vessel enhancement algorithm is based on multi-scale theory, using eigenvalues characteristics of Hessian matrix. The filter parameters and the scale factor are adapted the vascular structures is enhanced by image filtering. In digital image processing, Hessian matrix is used to extract image orientation feature and to analyze particular shape, while also be used in the curve structure segmentation and reconstruction. Eigenvalues of the Hessian matrix can be used to determine the corner where the density changes rapidly. Therefore the direction and intensity of blood vessels in DSA image can be characterized by Hessian matrix eigenvalues and eigenvectors.

Since the differences of blood vessel diameter and the differences of concentrations of contrast agent, gray value is higher in thick blood vessel of DSA image, and gray value is lower in small blood vessels. Although gray-scale transformation can highlight the great vessels, but it is difficult to distinguish small blood vessels with surrounding tissue, which will affect DSA vascular feature point extraction. In this paper, a Hessian matrix eigenvalue algorithm is used to enhance the blood vessels further. The principle of this algorithm is extracting blood vessels by use of vascular tubular structure. The vessel enhancement is viewed as a filtering process to look for two-dimensional graphical data which is similar as tubular structure [7,8].

On the basis of reference [7], this paper improved filter design of vessel enhancement. The parameters of filter can be adjusted adaptively according to the characteristics of image. In the multi-scale iterative process, the noise tends to be amplified, thus we increase a noise elimination filter design. In the proposed adaptive multi-scale vessel enhancement algorithms, the vascular similarity filter is:

$$V(x,y;\sigma) = \begin{cases} 0 & \text{if } \lambda_2 > 0, \\ \exp\left(-\frac{\left(\arctan\left(\lambda_2/\lambda_1\right)\right)^2}{2\beta^2}\right)\left(1-\exp\left(-\frac{S^2}{2w^2}\right)\right)\left(1-\exp\left(-\frac{|\lambda_1|}{\gamma}\right)\right) \end{cases} \quad (1)$$

According to scale space theory, for the linear structural elements, when the scale factor σ best match the actual width of vascular, the filter output is the maximum. Multi-scale vessel enhancement is to obtain the V value at different scales according to iterative scaling factor σ at each point in two-dimensional image, and take the maximum V value as the actual output of this point.

In the Eq. (1), λ_1, λ_2 is two eigenvalues of two-dimensional Hessian matrix ($|\lambda_1| \leqslant |\lambda_2|$). The eigenvector corresponding to the large eigenvalue represents the direction of maximum curvature in this point, and the small eigenvector corresponding to the small eigenvalue represents the direction of minimum curvature in this point. β is the scale affect factor. $S = \sqrt{\lambda_1^2 + \lambda_2^2}$, where S is the norm of Hessian matrix in current scale. That is to say, the larger the scale factor is, the greater the radius of Gaussian filters is; the smaller the scale factor is, the smaller the radius of Gaussian filters is. This means that the smaller diameter blood vessel has smaller response. In the Eq. (1) the initial parameter w determines the radius of the Gaussian filter directly. The greater the initial parameter w is, the greater the radius of Gaussian filters is, and therefore the greater response the vascular with greater diameter is. The smaller the initial parameter w is, the smaller the radius of Gaussian filters is, and therefore the smaller response the vascular with greater diameter is.

3   Control Point Selection Based on the Error Diffusion

In DSA images, we can use the image edge, corners and other information to complete the control point extraction. Firstly, detecting image edge, and then picking out specific points according to a threshold value. This gradient threshold approach procedure is as follows:

(1) Edge detection processing for the mask image.

(2) Calculate the gradient magnitude by $\frac{\partial f}{\partial x} = f(x+1, y) - f(x, y)$ and $\frac{\partial f}{\partial y} = f(x, y+1) - f(x, y)$, and then compared with a threshold value.

(3) Set the minimum and maximum distances between points to avoid too little or too many points.

To take advantage of image intrinsic characteristics selecting control points adaptively, the proposed feature selection algorithm introduced the error diffusion method. Error diffusion is an important half-tone image algorithm. Because of its efficient and good quality and special expression, error diffusion has been used widely in the image printing, display, as well as non-realistic rendering graphics, and other areas of computer graphics. In the image color conversion process, due to the different color range, the conversion process may introduce some error. Error diffusion algorithm can pass error to surrounding pixels and mitigate their visual error. Error diffusion algorithm is based on human visual characteristics and colorimetric properties of images. We can achieve the optimal binary image reproduction by this algorithm. Digital half-tone technology includes: Dot Diffusion, Error Diffusion, Ordered Dither and so on.

Error diffusion algorithm was proposed firstly by Floyed and Steinberg. This algorithm requires neighborhood processing. Its produces rich tones image, while the distribution of pixel anisotropy. The basic idea is to scan and quantitate image pixel according to certain path with a threshold, then the quantization error is spread to adjacent unprocessed pixels in a certain way. From another angle, error diffusion algorithm can achieve optimal image reproduction because the dots in the image space are proportional to the density and the local image gray.

The proposed error diffusion based algorithm of feature point selection can select control point adaptively according to image intrinsic characteristics. The algorithm is designed to distribute control points in image and make the control point's space density is proportional to the local image gray. This paper placed grid points according to Floyd-Steinberg algorithm in the DSA vessel image, so we can as much as possible make the grid points distributing on the vessel, and use these grid points on blood vessels as control points.

For the input image $f(x, y)$, DSA feature point selection algorithm based on the error diffusion is described below:

(1) Beginning with the first point (i_0, j_0), processing pixels according to raster scan order.

(2) $f(i, j)$ represents the value at point (i, j) of input image. For each point (i, j), $f(i, j)$ is compared with the threshold value q to obtain the indicator function $b(i, j)$, and then to obtain the position of the control point, as shown in the Eq. (2).

$$b(i, j) \triangleq \begin{cases} 1, & \text{if } f(i, j) \geq q \\ 0, & \text{otherwise.} \end{cases}, \tag{2}$$

where q is an experienced threshold value. $b(i, j) = 1$ means that there is a control point in the point (i, j).

Then at point (i, j), the quantization error $e(i, j)$ is calculated, as shown in the Eq. (3).

$$e(i, j) = f(i, j) - (2q)b(i, j), \tag{3}$$

At four direct adjacent to the points of point (i, j), $(i, j+1)$, $(i+1, j-1)$, $(i+1, j)$, $(i+1, j+1)$, we diffuse error $e(i, j)$. The values of these adjacent points are modified. The modified method as Eq. (4):

$$f(i,j+1) \Leftarrow f(i,j+1) + w_1 e(i,j),$$
$$f(i+1,j-1) \Leftarrow f(i+1,j-1) + w_2 e(i,j),$$
$$f(i+1,j) \Leftarrow f(i+1,j) + w_3 e(i,j), \qquad (4)$$
$$f(i+1,j+1) \Leftarrow f(i+1,j+1) + w_4 e(i,j),$$

Where $\Leftarrow$ represents that the value of right side in equation is assigned to the left.

(4) Repeat step 2 and 3 until all pixels in the image traversal.

(5) Extracting control point from the points which gray value is 1 in the result binary image.

In the diffusion process, in order to avoid possible omission of error diffusion occurs in the image boundary, the error diffusion pattern should be adjusted in the image border. For example, for a point on the right boundary, the weight w_1 and w_4 should be set to 0; however, the remainder weight should be adjusted so as to their sum is equal to a fixed value. Of course, if no significant features in image boundary, this adjustment can simply be ignored. For the classic Floyd-Steinberg error diffusion algorithm, to produce vivid digital halftone image, weight w_i, $i=1,\cdots,4$ are assigned respectively to 7/16, 3/16, 5/16, 1/16. For the control point setting, a more reasonable approach should adjust the value of weights w_i based on the geometric distance of adjacent points.

In the conventional raster scan order, the pixel will be processed in scan order from left to right. As an alternative, in a so-called zigzag raster scan order, the odd-numbered lines are processed in order from left to right, and the even-numbered rows from right to left. This approach is intended to enable error diffusion from left to right and right to left in the direction with a more balanced spread. Experimental results show that the zigzag raster scan order can generate slightly more accurate results.

In the proposed algorithm, the error diffusion threshold determines the number of control points. This is very important because it allows the error diffusion algorithm predetermined number of grid points, their corresponding number as follows:

$$K \approx \frac{1}{2q} \sum_{i=1}^{M} \sum_{j=1}^{N} f(i,j), \qquad (5)$$

Where M, N is ranks number of image $f(i,j)$, q is a threshold.

4   Control Points Simplify Based on Mean Shift

Due to the large number of control points based on above error diffusion method, it cannot adapt to the subsequent image registration. So the proposed algorithm simplifies these control points further using mean shift algorithm.

Mean shift clustering algorithm is a technique of nonparametric kernel density estimation based on Parzen window, which is a process to find local maximum point of density function (also called pattern dot). Starting from any sample point in the data set as the initial point, mean shift procedure converges to the same pattern where pixels can be considered as belonging to the same cluster, and determine the number of clusters automatically, and then describe the clusters boundaries. This algorithm originally developed by Fukunaga *et al* [9], and later Yizong Cheng applied it to clustering algorithm [10]. Dorin Comaniciu *et al* proposed an adaptive mean shift algorithm based on estimated bandwidth [11], and applied it to image segmentation and clustering [12].

Given n points $X_i \in R^d$, $i=1,\cdots,n$, in d-dimensional space, multivariate kernel density estimation function in point X can be expressed as [11]:

$$\hat{f}_K(x) = \frac{1}{n}\sum_{i=1}^{n}\frac{1}{h_i^d}k\left(\left\|\frac{x-X_i}{h_i}\right\|^2\right) \tag{6}$$

Where the function $k(x)$ is the profile function of kernel function $K(x)$. Kernel function $K(x)$ is a defined function. It is usually a class of radial symmetry function with special kernel, as follows:

$$K(x) = c_{k,d}k\left(\|x\|^2\right) > 0 \quad \|x\| \le 1 \tag{7}$$

Where $c_{k,d}$ is a normalization constant, to guarantee $K(x)$ integral as 1. In the Eq. (6), h_i is called the bandwidth or the window size of kernel function, which determines the kernel scope of X_i. In the proposed algorithm, the clustering algorithm uses Epanechnikov kernel function, which is defined as follows:

$$K(x) = \begin{cases} \frac{1}{2}c_d^{-1}(d+2)(1-x^Tx) & \text{if } x^Tx<1 \\ 0 & \text{otherwise} \end{cases} \tag{8}$$

Epanechnikov kernel function enables the integral mean square error minimized, while the standard deviation is probability density function and kernel density estimation based on the potential integration of data.

The gradient of multivariate kernel density estimation function in Eq. (6) is expressed as:

$$\hat{\nabla}f_K(x) \equiv \nabla\hat{f}_K(x) = \frac{2}{n}\sum_{i=1}^{n}\frac{x-X_i}{h_i^d}k'\left(\left\|\frac{x-X_i}{h_i}\right\|^2\right)$$

$$= \frac{2}{n}\sum_{i=1}^{n}\frac{X_i-x}{h_i^{d+2}}g\left(\left\|\frac{x-X_i}{h_i}\right\|^2\right) \tag{9}$$

$$= \frac{2}{n} \left[\sum_{i=1}^{n} \frac{1}{h_i^{d+2}} g\left(\left\| \frac{\mathbf{x} - \mathbf{x}_i}{h_i} \right\|^2 \right) \right] \times \left[\frac{\sum_{i=1}^{n} \frac{\mathbf{x}_i}{h_i^{d+2}} g\left(\left\| \frac{\mathbf{x} - \mathbf{x}_i}{h_i} \right\|^2 \right)}{\sum_{i=1}^{n} \frac{1}{h_i^{d+2}} g\left(\left\| \frac{\mathbf{x} - \mathbf{x}_i}{h_i} \right\|^2 \right)} - \mathbf{x} \right]$$

Where $g(x) = - k'(x)$. The last section of Eq. (9) contains the mean shift vector:

$$M(\mathbf{x}) = \left[\frac{\sum_{i=1}^{n} \frac{\mathbf{x}_i}{h_i^{d+2}} g\left(\left\| \frac{\mathbf{x} - \mathbf{x}_i}{h_i} \right\|^2 \right)}{\sum_{i=1}^{n} \frac{1}{h_i^{d+2}} g\left(\left\| \frac{\mathbf{x} - \mathbf{x}_i}{h_i} \right\|^2 \right)} - \mathbf{x} \right]_{,} \tag{10}$$

Assume $G(\mathbf{x}) = c_{g,d} g(\|\mathbf{x}\|^2)$, Eq. (9) can be expressed as [11]:

$$M(\mathbf{x}) = C \frac{\hat{\nabla} f_K(\mathbf{x})}{\hat{f}_G(\mathbf{x})}, \tag{11}$$

Eq. (11) shows that mean shift vector is proportional to gradient estimation of density function based on kernel function K. So mean shift vector pointing in the direction of increasing the maximum density. This is basis of mean shift clustering algorithm.

In the proposed algorithm, firstly, control points are grouped by mean-shift algorithm. Secondly, searching the nearest neighbor of each group pattern point, and taking them as simplify control point in each group. Experimental results

5   Conclusions

This paper proposes a DSA feature point selection algorithm based on error diffusion and mean shift. The error diffusion algorithm is introduced in order to take advantage of image intrinsic characteristics and select control point adaptively. The purpose is to control the locations of control points in the image, and make the spatial density of control point is proportional to the local image gray level. The proposed algorithm using Floyd-Steinberg algorithm to place grid points in DSA images, so we can make as many as possible grid points distributed on the vessel, and use these grid points in blood vessels as control points. We arrange more complex control points (concentrated grid) in the region of edge and complex texture. On the contrary, we arrange less control points in the flat region. The purpose of introduction of mean shift is to reduce the number of control points, and to achieve optimal image registration. Mean shift vector is based on kernel function K, which is proportional to the density function of gradient estimation. So mean shift vector pointing the direction of increase of the maximum density. In the proposed algorithm, firstly, control points are grouped by mean-shift algorithm. Secondly, searching the nearest neighbor of each group pattern point, and taking them as simplify control point in each group. Experimental results show that the adaptive multi-scale vessel enhancement algorithm of DSA has good blood enhanced performance, can effectively filter image background and non-vascular structure, and to avoid the increase of blood deformation. This algorithm can set control points adaptively at blood vessels and other key image characteristics, and can optimize the number of control points if needed to ensure the accuracy of DSA image registration.

Acknowledgements. This research was supported by the Foundation of Education Bureau of Henan Province, China grants No. 2010B520003, Key Science and Technology Program of Henan Province, China grants No. 132102210133 and 132102210034, and the Key Science and Technology Projects of Public Health Department of Henan Province, China grants No. 2011020114.

References

1. Schuldhaus, D., Spiegel, M., Redel, T., Polyanskaya, M., Struffert, T., Hornegger, J., Doerfler, A.: Classification-based Summation of Cerebral Digital Subtraction Angiography Series for Image Post-processing Algorithms. Physics in Medicine and Biology 56(1), 1791–1802 (2011)
2. Sang, N., Li, H., Peng, W., Zhang, T.: Knowledge-based Adaptive Threshold Segmentation of Digital Subtraction Angiography Images. Image and Vision Computing 25(8), 1263–1270 (2007)
3. Cebral, R., Castro, A., Appanaboyina, S., Putman, M., Millan, D., Frangi, F.: Efficient Pipeline for Image-Based Patient-Specific Analysis of Cerebral Aneurysm Hemodynamics: Technique and Sensitivity. IEEE Transactions on Medical Image 24(4), 457–467 (2005)
4. Zwiggelaar, R., Astley, M., Boggis, R., Taylor, J.: Linear Structures in Mammographic Images: Detection and Classification. IEEE Transactions on Medical Image 23(9), 1077–1086 (2004)
5. Vukadinovic, D., Walsum, T., Manniesing, R., Rozie, S., Hameeteman, R., Weert, T., Lugt, A., Niessen, W.: Segmentation of the Outer Vessel Wall of the Common Carotid Artery in CTA. IEEE Transactions on Medical Image 29(1), 65–76 (2010)
6. Shang, Y., Deklerck, R., Nyssen, E., Markova, A., Mey, J., Yang, X., Sun, K.: Vascular Active Contour for Vessel Tree Segmentation. IEEE Transactions on Biomedical Engineering 58(4), 1023–1032 (2011)
7. Frangi, A.F., Niessen, W.J., Vincken, K.L., Viergever, M.A.: Multiscale Vessel Enhancement Filtering. In: Wells, W.M., Colchester, A.C.F., Delp, S.L. (eds.) MICCAI 1998. LNCS, vol. 1496, pp. 130–137. Springer, Heidelberg (1998)
8. Frangi, A.F., Niessen, W.J., Hoogeveen, R.M., Walsum, T., Viergever, M.A.: Model-based Quantization of 3D Magnetic Resonance Angiographic Images. IEEE Transactions on Medical Imaging 18(10), 946–956 (1999)

9. Fukunaga, K., Hostetler, L.: The Estimation of the Gradient of a Density Function, With Applications in Pattern Recognition. IEEE Trans. Information Theory 21(1), 32–40 (1975)

10. Cheng, Y.: Mean Shift, Mode Seeking, and Clustering. IEEE Trans. Pattern Analysis and Machine Intelligence 17(8), 790–799 (1995)

11. Comaniciu, D., Ramesh, V., Meer, P.: The Variable Bandwidth Mean Shift and Data-driven Scale Selection. In: IEEE Int. Conf. Computer Vision, vol. 1, pp. 438–445 (2001)

12. Jimenez-Alaniz, J., Pohl-Alfaro, M., Medina-Banuelos, V., Yanez-Suarez, O.: Segmenting Brain MRI Using Adaptive Mean Shift. In: Proceedings of 28th International IEEE EMBS Annual Conference, pp. 3114–3117 (2006)

Parallel Image Texture Feature Extraction
under Hadoop Cloud Platform

Hao-Dong Zhu[1,*], Zhen Shen[1], Li Shang[2] and Xiao-Ping Zhang[3]

[1] School of Computer and Communication Engineering
ZhengZhou University of Light Industry
Henan Zhengzhou, 450002, China
zhuhaodong80@163.com
[2] Department of Communication Technology, College of Electronic Information Engineering,
Suzhou Vocational University, Suzhou 215104, Jiangsu, China
[3] College of Electronics and Information Engineering
Tongji University Shanghai, China

Abstract. With the increasing amount of digital image data, massive image process and feature extraction process have become a time-consuming process. As an excellent mass data processing and storage capacity of the open source cloud platform, Hadoop provides a parallel computing model MapReduce, HDFS distributed file system module. Firstly, we introduced Hadoop platform programming framework and Tamura texture features. And then, the image processing and feature texture feature extraction calculations involved in the process to achieve Hadoop platform. The results which comparison with Matlab platform shows it is less obvious advantage of Hadoop platform in image processing and feature extraction of lower-resolution images, but for image processing and feature extraction of high-resolution images, the time spent in Hadoop platform is greatly reducing, data processing capability the advantages is obvious.

Keywords: Hadoop, Tamura Texture Feature, Image Processing, Feature Extraction

1 Introduction

As one of the three underlying features of the image, Texture Features is an important visual feature, does not depend on the color or brightness, and includes the alignment and organizational order of the things surface structure. It shows the connection with context content and reflects the homogeneous phenomenon which is recurring in visual feature. So the Texture Feature has been widely used in image classification and retrieval.

Base on the psychology study of human visual perception about texture, Tamura [1]-[6] proposed a new theory to expression texture feature. The six components of

---

* Corresponding author.

D.-S. Huang et al. (Eds.): ICIC 2014, LNCS 8588, pp. 459–465, 2014.
© Springer International Publishing Switzerland 2014

texture correspond to the texture's six properties from psychological perspective, namely coarseness, contrast, directionality, lifelikeness, regularity, roughness. The most important of these characteristics is coarseness、 directionality and contrast. Extracted four kinds of leaf image from three different image databases which are ImageClef, Flavia and ICL. The image resolutions of three databases are 500×800,800×600 and 300×400.Through the experiment, we know the average time on Flavia and ImageClef is about 30s, and ICL is 6s. Higher resolution means the detail information will be better reflected and we can get better texture feature, but the consequent is computation time growth.

Image preprocessing, feature extraction, classification is three main steps in Image retrieval or identification. In pattern recognition field, some advanced methods or algorithms [23]-[28] will be applied to our plant identification based on cloud platform. In order to achieve high efficiency, we shall combine related algorithms with Hadoop, especially make full use of Hadoop ability in the facet of parallel computing. The next step is how to use related algorithms [29]-[36] to processing the features from plant image data. At last, an image retrieval system based on plant leaves will be implemented in the near future.

References Paper [1]-[6] describe the basic and improved Tamura algorithm, [7]-[22] describe the basic of Hadoop Platform and related application on it. Due to limited space, we have no describes about Tamura and Hadoop in this paper and introduce the system directly.

2   System Design and Implementation

This system treats an image file as a split which would be processed as a job record. Each Map task corresponds to an image file, and then parallel computes multiple image texture features. The system uses a single Reduce task to written the result to the specified location in the specified format. In order to achieve the above functions, first: need to implement a new data type Image, it was used to image processing and storage. Second, the function of Input Format and RecordReader need redefined to transforming image files to specific data types; Last, the calculation of image texture features in the Map function.

2.1   Image Type

When a class achieved writable interface, it was able to act as a value in the Hadoop, but hadoop has no class to act as an alternative type of key and value. To solve this problem, we define the Image type, it was not only implements the basic functionality required by hadoop, but also can processing image to become gray, get the size of the target image, get pixel data and other functions.

2.2 Image Processing and Feature Calculation

About k value set problem in coarseness calculation, Tamura described two cases: ①
k=0,1,2,3: image have no noise, within this range Sbest is always maximum, small
amount of calculation;② k=0,1,...,5: Image noise is present, Sbest results is unstable,
not only have an impact on the calculation results, and the amount of calculate
become larger. Therefore, image must be pretreatment before calculation of texture
feature to eliminate noise.

3   Experiments

We used two experiment platform In this experiment: ① Common platform: OS is
Windows 7, configured as eight nuclear IntelCorei7 processor, 4GB memory, 1TB
hard drive, Matlab version R2012a; ②Hadoop platform: which have one master node
and four slave node, OS are CentOS-6.4 64bit, configured as eight nuclear IntelCorei7
processor, 4GB memory, 1TB hard drive, Hadoop version 1.1.2, Java version 1.7.25,
each node via 100Mb / s LAN connection.

3.1   Experiments Result and Analysis

In order to verify the efficiency of algorithm with different resolutions, we selected
three data sets: Flavia, ICL and ImageClef, extracted 2000 images from three data sets
and divided into 100, 200, 500, 1000, and 2000 five groups. We used two platforms to
calculation and comparison. The time shown in Figure 5, Figure 6 and Figure 7,
Figure 8, Figure 9 shows the algorithm's speedup.

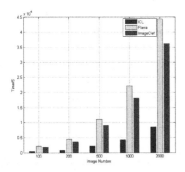
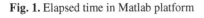
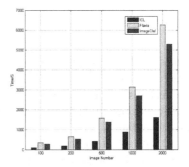

Fig. 1. Elapsed time in Matlab platform   **Fig. 2.** Elapsed time in hadoop platform
with 3 nodes

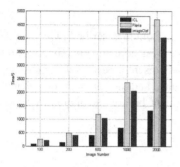

Fig. 3. Elapsed time in hadoop platform with 4 nodes

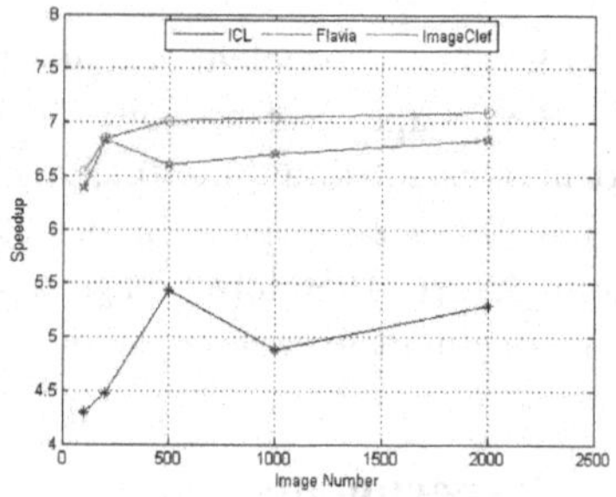

Fig. 4. Speedup Ratio with 3 nodes

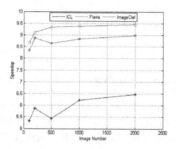

Fig. 5. Speedup Ratio with 4 nodes

From Figure 1, Figure 2, Figure 3, we can see that with the increase of image number, the time used between two platforms growth exponential. There are two reasons: ①The image resolution in Flavia are 800 × 600, ImageClef is about 500 × 800, ICL is about 300 × 400, because the amount of calculation about Tamura's Coarseness is closely related to the image resolution, Therefore, the computation time grew more obvious; ② Matlab platform uses a serial approach to calculate the image texture features, while Hadoop platform uses multiple Map parallel computing method to perform, so the computation time compared with the Matlab platform has significantly reduces.

As we can see from the five charts, Tamura run on hadoop platform is more efficiency than Matlab platform, the algorithm speedup is increasing with the number of image; with the increase number of nodes in cluster, and the algorithm is more efficiency when processing the larger datasets. Compared with the traditional method, Tamura in Hadoop have higher efficiency and scalability, it is available to extracted features with larger image datasets.

4   Conclusion and Future Work

We carried out the Tamura algorithm on Hadoop to extract texture features from massive image files. In order to solve the problem that Hadoop cannot load image files directionality, we especially designed a new Inputformat ImageInputFormat and new data type image to meet the needs of image input and processing. The algorithm makes full use of the parallel processing capabilities of larger data on hadoop, ensures the data's accuracy and shortens the computation time. Through experimental analysis, we show the efficiency of the algorithm. However, In the course of the experiment, because hadoop block size is 64MB and the size of image used in experiment does not exceed 1MB, the storage space was wasting a lot. Limited to scheduling policy on platform, the algorithm timeliness was affected. Improve the utilization of system memory when storing large number of small file and designs better scheduling strategy is the focus point of our future research.

Acknowledgment. The authors would like to thank the editors and anonymous reviewers for their valuable comments. This work is supported by the Science and Technology Innovation Outstanding talent Plan Project of Henan Province of China (No.134200510025), the National Natural Science Foundation of China (No. 61201447) and the Science and Technology Innovation Fund project of postgraduate of Zhengzhou University of Light Industry.In addition, this work also received guidance from Huang De-Shuang who is a distinguished professor in Henan Province.

References

1. Tamura, H., Mori, S., Yamawaki, T.: Textural features corresponding to visual perception. IEEE Transactions on Systems, Man and Cybernetics 8(6), 460–473 (1978)
2. Hao, Y.B., Wang, R.L., Ma, J., Su, B., Zheng, J.H.: Image retrival based on improved Tamura texture features. Science of Surveying and Mapping 35(4), 136–138 (2010)
3. Lu, X.Q., Guo, J.G., Zhao, Y.H., Ren, X.Y.: Research and realization of Tamura texture feature extraction method based on image segmentation. Chinese Journal of Tissue Engineering Research 16(17), 3160–3163 (2012)
4. Wang, S.J., Qi, C., Cheng, Y.S.: Application of Tamura texture feature to classify underwater targets. Applied Acoustics 31(2), 135–139 (2012)
5. Jing, J.F., Zhang, H.H., Li, P.F., Wang, J.: Fabric defect classification based on local binary patterns and Tamura texture method. Computer Engineering and Applications 48(23), 155–160 (2012)
6. Tomas, M., David, S.: Extension of Tamura Texture Features for 3D Fluorescence Microscopy. In: Proceedings of 2012 Second International Conference on 3D Imaging, Modeling, Processing, Visualization and Transmission, pp. 301–307 (2012)
7. http://lucene.apache.org/hadoop/2006
8. http://en.wikipedia.org/wiki/Apache_Hadoop
9. White, T.: Hadoop: the definitive guide, 2nd edn. O'Reilly Media, Sebastopol (2011)
10. Chang, F.: BigTable: A Distributed Storage System for Structured Data. In: OSDI 2006 Proceedings of the 7th Symposium on Operating Systems Design and Implementation, pp. 205–218 (2006)

11. Dean, J., Ghemawat, S.: MapReducer: Simplified Data Processing on large Clusters. In: OSDI 2004 Proceedings of the 6th Symposium on Operating Systems Design and Implementation, pp. 137–150 (2004)
12. Zhu, Y.M.: Image Classification Based on Hadoop Platform. Journal of Southwest University of Science and Technology 26(2), 70–73 (2011)
13. Zhang, L.J., Huan, F., Wang, Y.D.: Parallel Image Processing Implementation under Hadoop Cloud Platform. Information Security and Communication Privacy 10, 59–62 (2012)
14. Li, Q., Shi, X.P.: Data type design based on Hadoop MapReduce Image Processing. Software Guide 11(4), 182–183 (2012)
15. Ranajoy, M., Naga, V.: A CUDA-enabled Hadoop Cluster for Fast Distributed Image Processing. In: Proceedings of the 2013 National Conference on Parallel Computing Technologies, pp. 1–5 (2013)
16. Liu, Y.H., Chen, B., He, W.X., Fang, Y.: Massive Image Data Management using HBase and MapReduce. In: Proceedings of the 2013 21st International Conference on Geoinformatics, pp. 1–5 (2013)
17. Li, W.W., Zhao, H., Zhang, Y., Wang, Y.: Research on massive data mining based on MapReduce. Computer Engineering and Applications 49(20), 112–117 (2013)
18. Chi, Z.W., Zhang, F., Du, Z.H., Liu, R.Y.: Cloud Storage of Massive Remote Sensing Data Based on Distributed File System. In: Proceedings of 2013 IEEE International Conference on Communication and Computing, Signal Processing, pp. 1–4 (2013)
19. Yang, C.T., Huang, K.L., Chu, W.C., Lai, K.C., Chang, C.H., Lu, C.W.: Implementation of Video and Medical Image Services in Cloud. In: Proceedings of 2013 IEEE 37th Annual Computer Software and Application Conference Workshops, pp. 451–456 (2013)
20. Wichian, P., Anucha, T., Sarayut, I., Nucharee, P.: Improving Performance of Content-Based Image Retrieval Schemes using Hadoop MapReduce. In: Proceedings of 2013 International Conference on High Performance Computing and Simulation, pp. 615–620 (2013)
21. Muneto, Y., Kunihiko, K.: Parallel Image Database Processing with MapReduce and Performance Evaluation in Pseudo Distributed Mode. International Journal of Electronic Commerce Studies 3(2), 211–228 (2012)
22. Sitalakshmi, V., Siddhivinayak, K.: MapReduce Neural Network Framework for Efficient Content Based Image Retrieval from Large Datasets in the Cloud. In: Proceedings of 2012 12th International Conference on Hybrid Intelligent Systems, pp. 63–68 (2012)
23. Wang, X.F., Huang, D.S., Xu, H.: An efficient local Chan-Vese model for image segmentation. Pattern Recognition 43(3), 603–618 (2010)
24. Wang, X.F., Huang, D.S., Xu, H.: An efficient local Chan-Vese model for image segmentation. Pattern Recognition 43(3), 603–618 (2010)
25. Li, B., Huang, D.S.: Locally linear discriminant embedding: An efficient method for face recognition. Pattern Recognition 41(12), 3813–3821 (2008)
26. Huang, D.S., Du, J.X.: A constructive hybrid structure optimization methodology for radial basis probabilistic neural networks. IEEE Trans. Neural Networks 19(12), 2099–2115 (2008)
27. Huang, D.S., Zhao, W.B.: Determining the centers of radial basis probabilistic neural networks by recursive orthogonal least square algorithms. Applied Mathematics and Computation 162(1), 461–473 (2005)
28. Huang, D.S.: Radial basis probabilistic neural networks: Model and application. Int. Journal of Pattern Recognit., and Artificial Intell. 13(7), 1083–1101 (1999)
29. Wang, X.F., Huang, D.S.: A novel density-based clustering framework by using level set method. IEEE Trans. Knowledge and Data Engineering 1(11), 1515–1531 (2009)

30. Shang, L., Huang, D.S., Du, J.X., Zheng, C.H.: Palmprint recognition using Fast ICA algorithm and radial basis probabilistic neural network. Neurocomputing 69(13-15), 1782–1786 (2006)
31. Zhao, Z.Q., Huang, D.S., Sun, B.Y.: Human face recognition based on multiple features using neural networks committee. Pattern Recognition Letters 25(12), 1351–1358 (2004)
32. Huang, D.S., Ip, H.H.S., Chi, Z.R.: A neural root finder of polynomials based on root moments. Neural Computation 16(8), 1721–1762 (2004)
33. Huang, D.S.: A constructive approach for finding arbitrary roots of polynomials by neural networks. IEEE Trans. on Neural Networks 15(2), 477–491 (2004)
34. Huang, D.S., Chi, Z.R., Siu, W.C.: A case study for constrained learning neural root finders. Applied Mathematics and Computation 165(3), 699–718 (2005)
35. Huang, D.S., Ip, H.H.S., Law, K.C.K., Chi, Z.R.: Zeroing polynomials using modified constrained neural network approach. IEEE Trans. on Neural Networks 16(3), 721–732 (2005)
36. Huang, D.S., Ip, H.H.S, Law, K.C.K., Chi, Z.R., Wong, H.S.: A new partitioning neural network model for recursively finding arbitrary roots of higher order ar-bitrary polynomials. Applied Mathematics and Computation 162(3), 1183–1200 (2005)

Image Super-resolution Reconstruction Utilizing the Combined Method of K-SVD and RAMP

Li Shang[1], Tao Liu[1], and Zhan-Li Sun[2]

[1] Department of Communication Technology, College of Electronic Information Engineering,
Suzhou Vocational University, Suzhou 215104, Jiangsu, China
[2] Department of Automation, College of Electrical Engineering and Automation,
Anhui University, Hefei 230601, Anhui, China
`{sl0930,lt}@jssvc.edu.cn, zhlsun2006@yahoo.com.cn`

Abstract. A new image super-resolution reconstruction (SRR) method, combined a modified K-means based Singular Value Decomposition (K-SVD) and Regularized Adaptive Matching Pursuit (RAMP) algorithm is proposed in this paper. In the modified K-SVD algorithm, the maximum sparsity is considered. First, the K-SVD denoising model is first to preprocess the Low Resolution (LR) images. And then, the high-resolution (HR) and LR images are both trained by the RAMP based K-SVD algorithm. The LR and HR dictionaries are also classed by the K-mean method. In test, a human-made and real LR image, namely millimeter wave (MMW) image are respectively used to testify our method proposed. Further, compared our image SRR method with methods of the basic K-SVD and RAMP, experimental results testified the validity of our method proposed.

Keywords: Image super-resolution reconstruction,Sparse representation, High resolution (HR), Low resolution (LR), K- Singular Value Decomposition (K-SVD), Regularized adaptive matching pursuit (RAMP).

1   Introduction

At present, learning based image super-resolution reconstruction (SRR) methods are condemned as the hot research topics [1]. Among these published methods, the sparse representation based methods are the most popular [2-4]. The K-means Based Singular Value Decomposition (K-SVD) algorithm used widely in image processing is just such a sparse representation based method [5-6]. In the common K-SVD algorithm, the over-complete dictionary is trained by utilizing Orthogonal Matching Pursuit (OMP) or Regularized OMP (ROMP) [3] or Stage-wise OMP (StOMP) [7], which require more measurements for perfect reconstruction, especially require the sparse measurement of original data to be known, at the same time, they also lack provable reconstruction quality [8]. To avoid these defects, in this paper, a modified K-SVD method is proposed, which considers the maximum sparsity of image feature coefficients, and uses a Regularized Adaptive Matching Pursuit (RAMP) [3] to respectively train LR dictionary and HR dictionary. At the same time, in order to

D.-S. Huang et al. (Eds.): ICIC 2014, LNCS 8588, pp. 466–472, 2014.
© Springer International Publishing Switzerland 2014

reduce the iteration time, LR dictionary and HR dictionary are further classed by K-mean method. In test, a simulation LR image and a real LR image called Millimeter Wave (MMW) image are respectively used to testify our method. The validity of our method is also proved by comparing it with other methods of the basic K-SVD and RAMP.

2   RAMP Algorithm

Let ε_1 and ε_2 denote respectively the iteration times and the threshold of stage switch. x and y are the measurement vector and the reconstructed vector. Then, RAMP algorithm is generalized in main as follows:

Step 1. Initialized setting. Let the initial residue $r_0 = y$, the step size $\eta \neq 0$, the stage index $j = 1$, the iteration time $t = 1$, and the index set $J \neq \varnothing$, $\Lambda = \varnothing$.

Step 2. If the residue $\|r\|_2 \leq \varepsilon_1$, stopping the iteration. The reconstruction can be implemented by atoms trained. Else, go to Step 3.

Step 3. Using Eqn. (1) to calculate relative coefficients u, and the L biggest indexes in u are saved in J.

$$u = \left\{ u_j \,\middle|\, u_j = \left| \langle r, \varphi_j \rangle \right|, j = 1, 2, \cdots, N \right\} \tag{1}$$

Step 4. Regularized relative coefficients of atoms responding to the index set J, and saved the regularized results in set J_0. In J_0, all coefficients satisfy $|u(i)| \leq 2|u(j)|$ ($i, j \in J_0$), and choose J_0 with the maximal energy $\|u|J_0\|_2$.

Step 5. Updating the support set ϕ_Λ, where $\Lambda = \Lambda \bigcup J_0$;

Step 6. Using Eqn. (2) to obtain $\hat{x}$, and updating the residual using Eqn. (3):

$$u = \hat{x} = \arg \min_{i \in R^\Lambda} \left\| y - \phi_\Lambda x \right\|_2 \tag{2}$$

$$r_{new} = y - \phi_\Lambda \hat{x} \tag{3}$$

Step 7. If $\|r_{new} - r\|_2 \leq \varepsilon_2$, let $j = j + 1$, $\eta = \eta * j$. Then go to Step 3. Else, let $r = r_{new}$, $n = n + 1$, go to Step 2.

According to the above steps, it is clear to see that by the setting threshold ε_1 and ε_2, RAMP algorithm can define self-adaptively to adjust the current step size to define whether the next stage or iteration is implemented or not.

3   The Modified K-SVD Denoising Model

3.1   The Cost Function

Assume that x is an image patch, and consider its noisy version (namely $y=x+\sigma$), contaminated by an additive zero-mean white Gaussian noise with standard deviation σ, the maximum a posteriori (MAP) estimator for denoising y is built by solving the following form

$$\hat{s} = \arg\min_{s} \|s\|_0 \quad \text{subject to} \quad \hat{s} = \|Ds-y\|_2^2 \leq T \tag{4}$$

where D denotes the code dictionary, s denotes the sparse coefficient vector, and $T>0$ is a small threshold, dictated by σ and the number of atoms K. Thus, the denoised image is thus given by $\hat{x} = D\hat{s}$. In practice, the optimization task of K-SVD is changed to be

$$\hat{s} = \arg\min_{s} \|y-Ds\|_2^2 + \mu\|s\|_0 \tag{5}$$

However, the maximum sparsity can not be ensured in Eqn. (5). To solve this question, a modified K-SVD denoising model is proposed by us. This K-SVD model is described as in the following formula:

$$J\left(\hat{D},\hat{S}\right) = \arg\min_{s_{ij},D}\left[\lambda\|X-Y\|_2^2 + \sum_{i,j}\mu_{ij}\|S_{ij}\|_0 + \gamma\sum_{i,j}\left(D_{ij}^T D_{ij}\right) + \sum_{i,j}\|DS_{ij}-R_{ij}X\|_2^2\right] \tag{6}$$

In Eqn. (6), the first term is the log-likelihood global force that demands the proximity between the measured image Y, and its denoised version X. The second and the third terms are parts of the image prior that makes sure that in the constructed image X, every patch $X_{ij} = R_{ij}X$ of size $p \times p$ in every location has a sparse representation with bounded error. R is an $p \times N$ matrix that extracts the (i,j) block from an image with $N \times N$ size. The coefficients μ_{ij} must be location dependent so as to comply with a set of constraints of the formula $\|DS_{ij}-X_{ij}\|_2^2 \leq T$.

3.2   Image Denoising by K-SVD Model

In training the dictionary, each LR image is sampled randomly image patches with $p \times p$ size, thus, the image patch set $Y = \{y_i\}_{i=1}^{M}$ is obtained, where y_i is the ith overlap image patch, and M is the number of overlap image patches, which can be calculated by $(N-p+1)^2$. For the set Y, again sampling randomly $M'(M'<M)$ times to obtain training set $Y' = \{y_i\}_{i=1}^{M'}$,. Thus, the training sample set Y' can be obtained, and using the following steps, Eqn.(6) can be minimized.

4   Experimental Results and Analysis

4.1   Dictionary Learned by K-SVD Method

In test, a degenerated version of Lena image and a real MMW image shown in Fig.1 were used. In the task of image reconstruction, the modified K-SVD denoising model was first used to denoise LR images. In fact, the denoised images were used as the LR images. Then, each HR and LR image were both sampled 5000 times with 8×8 image patch with overlap of three pixels between adjacent patches, thus the LR and HR image patch sets could be obtained and denoted X_L and X_H respectively.

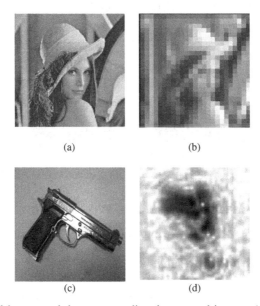

(a)   (b)

(c)   (d)

Fig. 1. The original images and the corresponding degenerated images. (a) The original Lena image; (b) Noise version of Lena image with noise level 0.01; (c) The imaging target ; (d) MMW image.

Next, using the K-SVD model based on RAMP method to train X_L and X_H data set, the corresponding HR dictionary D_H and the LR dictionary D_L were obtained. In dictionary training, the dimension of X_L and X_H was selected as 64. The length of atoms of D_H and D_L was chosen as 256. For HR Lena images, the D_H dictionary of our method and the basic K-SVD method were shown in Fig.2 (a) and Fig.2 (b). For MMW image, the D_L dictionary of our method and the basic K-SVD method was shown in Fig.3(a) and Fig.3(b).

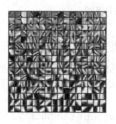 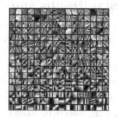

(a) modified K-SVD   (b) The basic K-SVD algorithm
    algorithm

Fig. 2. Lena image's HR dictionary with 256 atoms of K-SVD model

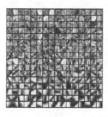 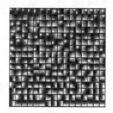

(a) Modified K-SVD   (b) Basic-K-SVD dictionary
    dictionary

Fig. 3. MMW Dictionary with 256 atoms of K-SVD model. (a) Dictionary obtained by modified K-SVD model. (b) Dictionary obtained by basic K-SVD model.

4.2   Image Reconstruction Results

Using the trained HR dictionary D_H and the optimal sparse coefficients s^* of LR images learned by K-SVD, the HR image patches can be reconstructed as $x = D_H s^*$, and then the image patch x is put into the HR image X_0. Using this method, the image SRR results corresponding to Lena LR image and MMW image were shown in Fig.4(a) and Fig.5(a) respectively. At the same time, Restored results of Lena LR image and MMW image obtained by the basic K-SVD and RAMP algorithms were also shown in Fig.4(b) to Fig.4 (c) and Fig.5(b) to Fig.5 (c). Furthermore, the reconstructed results of Lena LR image were evaluated by SNR values, and for the MMW image, the Relative SNR (RSNR) criterion is used, and the RSNR values are calculated by using the following formulas:

$$\text{RSNR} = \frac{1}{\sqrt{NM}} \left[\sum_{i=1}^{N} \sum_{j=1}^{M} \hat{I}(i,j) \right] \bigg/ \sqrt{\sum_{i=1}^{N} \sum_{j=1}^{M} \left[\hat{I}(i,j) - \tilde{I}(i,j) \right]^2} \tag{7}$$

where $I(i,j)$ denotes the input image with the size of $N \times M$, $\hat{I}(i,j)$ denotes the restored image and $\tilde{I}(i,j)$ denotes the mean of this image. The SNR values of Lena image restored were listed in Table 1, and those of RSNR of the MMW image were listed in Table 2.

(a) Modified K-SVD method   (b) RAMP method   (c) The basic K-SVD method

Fig. 4. Reconstructed Lena images by different algorithms

(a) Modified K-SVD method   (b) RAMP method   (c) The basic K-SVD method

Fig. 5. Reconstructed MMW image by different algorithms

Table 1. SNR values of different algorithms of restored Lena images

Algorithms	Restored Lena images	Lena LR image
(our method)	24.58	
RAMP	22.79	14.03
The basic K-SVD model	16.43	

Table 2. RSNR values of different algorithms of restored Lena images

Algorithms	Restored Lena images	Lena LR image
(our method)	17.36	
RAMP	15.68	12.37
The basic K-SVD model	13.23	

5   Conclusions

The image super resolution reconstruction (SRR) method using the combined method of RAMP and K-SVD is proposed in this paper. Here, the artificial and real LR images are considered. At first, LR images are preprocessed by the basic K-SVD denoising model. Then, using the RAMP based K-SVD method, the HR and LR dictionary pairs are learning respectively. Further, using HR dictionart and the sparse coefficients of LR images, the SRR task is implemented well. Compared with SRR methods of the basic K-SVD, RAMP and SAMP, experimental results show that our method behaves in indeed the best effect in the image SRR task.

Acknowledgement. This work was supported by the National Nature Science Foundation of China (Grant No. 61373098, 60970058 and 61370109), the Innovative Team Foundation of Suzhou Vocational University (No. 3100125), and the "333 Project" of Jiangsu Province.

References

1. Liu, Y.X., Zhao, R.Z., Hu, S.H., Jiang, C.H.: Regularized adaptive matching pursuit algorithm for signal reconstruction based on compressive sensinig. Journal of Electronics & Information Technology 32(11), 2713–2717 (2010)
2. Zeyde, R., Elad, M., Protter, M.: On single image scale-up using sparse-representations. In: Boissonnat, J.-D., Chenin, P., Cohen, A., Gout, C., Lyche, T., Mazure, M.-L., Schumaker, L. (eds.) Curves and Surfaces 2011. LNCS, vol. 6920, pp. 711–730. Springer, Heidelberg (2012)
3. Dong, W.S., Zhang, L., Shi, G.: Image deblurring and super-resolution by adaptive sparse domain selection and adaptive regularization. IEEE Transaction on Image Process. 20(7), 1838–1857 (2011)
4. Yang, J.C., Wright, J., Huang, T.S.: Image super-resolution via sparse representation of raw image patchs. IEEE Transactions on Image Processing 19(11), 1–8 (2010)
5. Aharon, M., Elad, M., Bruckstein, A.: K-SVD: an algorithm for designing overcomplete dictionaries for sparse representation. IEEE Transaction on Image Processing 54(11), 4311–4322 (2006)
6. Elad, M., Aharon, M.: Image denoising via learned dictionaries and sparse representation. Computer Vison and Pattern Recogniton 1, 895–900 (2006)
7. Sun, J., Xu, Z.B., Shum, H.Y.: Gradient profile prior and its applications in image super-resolution and enhancement. IEEE Transactions on Image Processing 20(6), 1529–1542 (2011)
8. Do, T.T., Lu, G., Nam, N., Trac, D.T.: Sparsity adaptive matching pursuit algorithm for practical compressed sensing. In: The 42nd Asilomar Conference on Signals, Systems, and Computers, Pacific Grove, California, pp. 234–240 (2008)

Dispersion Constraint Based Non-Negative Sparse Coding Neural Network Model

Li Shang[1], Yan Zhou[1], and Zhan-Li Sun[2]

[1] Department of Communication Technology, College of Electronic Information Engineering,
Suzhou Vocational University, Suzhou 215104, Jiangsu, China
[2] Department of Automation, College of Electrical Engineering and Automation,
Anhui University, Hefei 230601, Anhui, China
{sl0930,zhy}@jssvc.edu.cn, zhlsun2006@yahoo.com.cn

Abstract. A dispersion constraint based non-negative sparse coding (DCB-NNSC) model is discussed in this paper. To ensure the sparsity in self-adaptive, the kurtosis criterion is used to measure the sparse priori knowledge of feature coefficients. And to enhance the capability of feature separability, the dispersion ratio of within-class and between-class of sparse coefficient vectors is utilized. Simulation results show that, just as those sparse coding (SC) and NNSC published, DCB-NNSC model can also simulate successfully the respective field of V1 in the primary visual system of human beings. Moreover, compared with common NNSC models, DCB-NNSC model behave clearer sparsity. Using DCB-NNSC features to test image reconstruction task, the results obtained prove further that the DCB-NNSC model is indeed effective in extracting image features and modeling the V1 mechanism of the primary visual system of human in application.

Keywords: Non-negative sparse coding (NNSC), Dispersion constraint, Sparse coefficients, Feature extraction, Within-class dispersion, Between-class dispersion; Image reconstruction.

1 Introduction

As sparse coding (SC) model [1], non-negative sparse coding (NNSC) model can also find an efficient representation for high dimension data. Nowdays, it has found applications in numerous domains, especially in image processing field [2]. The early NNSC model was proposed by Hoyer P.O. in 2002 [3]. At present, Hoyer's model has been proved to be efficient to simulate the respective field of V1 in the primary visual system of human [3-5] and used widely in the image processing field [2, 6]. However, Hoyer's model only considers two terms. The first one is the image reconstruction error and another is the sparse priori distribution of sparse coefficients. In training Hoyer's model [3-5], the combination of gradient projection and auxiliary function-based multiplicative updating method is used, so its performance is influenced hardly by the iterative step size in gradient projection, and the convergence precision can not be very high, at the same time, the sparsity of coefficients can not be ensured in

D.-S. Huang et al. (Eds.): ICIC 2014, LNCS 8588, pp. 473–479, 2014.

self-adaptive. Moreover, when being used to feature recognition task, Hoyer's model can't obtain high classification precision. To solve these faults above-mentationed, considered class priori information and the maximum sparsity, a novel dispersion constraint based NNSC model (denoted by DCB-NNSC here), is proposed by us here. In the DCB-NNSC model, the kurtosis measure is selected as the priori distribution of coefficients to ensure the sparsity. And the dispersion ratio of within-class and between-class of feature coefficients is used to improve the feature reparability in the feature extraction task. Further, using the PolyU palm images to test the DCB-NNSC model proposed by us, simultaneously, compared with Hoyer's model in the same experimental condition, simulation results both testify that it is efficient indeed in image feature extraction task.

2   The Dispersion Constraint of Feature Coefficients

Assumed that $S = [s_1, s_2, s_3, \cdots, s_k, \cdots, s_C]$ denotes the sparse feature coefficient set, where s_k is the kth class feature coefficient sample vector ($k = 1, 2, \cdots, C$), and C is the class number of training samples. Let n_k denote the number of elements in s_k, and m_k denote the mean of the kth class sample, then the within-class dispersion matrix S_W and between-class dispersion matrix S_B are defined respective as follows:

$$
\begin{cases}
S_W = \sum_{k=1}^{C} \sum_{s \in S_k} (s - m_k)(s - m_k)^T \\
S_B = \sum_{k=1}^{C} n_k (m_k - m)(m_k - m)^T
\end{cases}
\tag{1}
$$

where $m_k = \dfrac{1}{n_k} \sum_{s \in S_k} s$, and $m = \dfrac{1}{n} \sum_{i=1}^{n} s_i$. Here the variable s represents each element in s_k, and parameters of m and n are respectively the mean and the total number of all class samples. After S_W and S_B are computed, the dispersion ratio is obtained by the following formula:

$$
\frac{S_W}{S_B} = \frac{\sum_{k=1}^{C} \sum_{s \in S_k} (s - m_k)(s - m_k)^T}{\sum_{k=1}^{C} n_k (m_k - m)(m_k - m)^T} V_k \cdot
\tag{2}
$$

Noted that the within-class dispersion S_W and the between-class dispersion S_B are both a global description of data set, and they both represent the second order statistical information of data. The smaller S_W is, the larger the within-class aggregation degree is. And the larger S_B is, the larger the dispersion between different classes is. Therefore, the smaller the dispersion ratio S_W / S_B, the better the within-class aggregation is. In this paper, in order to derivation operation, we utilize the logarithm of S_W/S_B, namely $\ln(S_W/S_B)$, to be the class constrain term added in the NNSC model and this makes the features trained to be better in implementing feature recognition task.

3   The DCB-NNSC Model AMP Algorithm

3.1   Hoyer's NNSC Model

The early NNSC model, proposed by Hoyer P. O. in 2002, combines the sparseness with the algorithm of non-negative matrix factorization (NMF) [7], and the cost function was formulated as follows [3-5]:

$$\text{min} \qquad L(A,S) = \frac{1}{2}\left\| X - AS \right\|^2 + \lambda \sum_{ij} S_{ij} \; . \tag{3}$$

subject to the non- negativity of X , A , S , and $\left\| A \right\|_1 = 1$. Here, $X = [x_1, x_2, \cdots, x_N]^T$ is the N dimension input matrix, A is the feature basis matrix with the size of $N \times M$, and S is the M dimension sparse coefficient matrix. In training Hoyer's NNSC, matrix A and S are updated in turn. Fixed S , A is updated by using the gradient descent algorithm, and computed by the iteration form $\hat{A} = A^t - \mu(A^t S - X)S^T$. As such, fixed A , S is updated by using a multiplicative method, and computed by using $S^{t+1} = S^t \cdot \left[((A^{t+1})^T X) \cdot / ((A^{t+1})^T A^{t+1} S^t + \lambda)\right]$. The detailed updated process can be found in the Ref. [5]. Because of using the gradient projection, the property of Hoyer's model is influenced hardly by the size of step length, and its precision can not be high in practice.

3.2   The DCB-NNSC Model

On the basis of Hoyer's NNSC model, to ensure the sparsity of feature coefficients and improve the feature reparability, considered the adaptive-self sparsity of data and class priori knowledge, a new NNSC model is proposed in this paper and the cost function is defined as follows:

$$\text{min} \qquad J(\mathbf{A},\mathbf{S}) = \frac{1}{2}\left\| \mathbf{X} - \mathbf{AS} \right\|^2 + \lambda_1 \sum_i f\left(\frac{s_i}{\sigma_i}\right) + \lambda_2 \ln\left(\frac{\mathbf{S}_W}{\mathbf{S}_B}\right) \; . \tag{4}$$

where matrixes of X , A , S have the same meaning as Eqn. (3). Vector s_i is the *ith* row vector of the sparse coefficient matrix S . Parameter σ_i is estimated noise variance calculated by the equation $\sigma_i^2 = \langle s_i^2 \rangle$ and symbol $\langle \cdot \rangle$ means the mean value operation. S_W and S_B are respectively the within-class and between-class dispersion matrix, which is described in the subsection 2. Here, the sparsity measure function $f(\cdot)$ is chosen to be the absolute value of kurtosis of a sparse coefficient vector, which is defined as:

$$f(s_i) = \left| kurt(s_i) \right| = \left| E\{s_i^4\} - 3\left(E\{s_i^2\} \right)^2 \right| .$$ (5)

and maximizing the term of $\left| kurt(s_i) \right|$ (i.e., minimizing $-\left| kurt(s_i) \right|$) is equivalent to maximizing the sparseness of coefficient vectors. Thus, the object function described in Eqn. (5) is rewritten as

$$J(\mathbf{A},\mathbf{S}) = \frac{1}{2}\sum_i \left[\mathbf{X}_i(x,y) - \sum_{j=1}^{M} S_{ij}(x,y) A_j \right]^2 - \lambda_1 \sum_{ij} \left| kurt(S_{ij}) \right| +$$

$$\lambda_2 \ln\left(\frac{\sum_{k=1}^{C}\sum_i (s_i - m_k)(s_i - m_k)^T}{\sum_{k=1}^{C} n_k (m_k - m)(m_k - m)^T} \right)$$ (6)

Here, the feature matrix A and coefficient matrix S are still updated in turn. To reduce the convergence time, the matrix A and matrix S are first updated by the gradient descent algorithm, and then they are updated by using a multiplication factor method. Combined the partial derivative of the ith row vector a_i of the matrix A and the multiplication factor rule, the updating process of A is deduced as follows:

$$\begin{cases} \nabla a_i = -\left[X(x,y) - \sum_{i=1}^{n} a_i(x,y) s_i \right] s_i^T \\ a_i^{(t+1)} = \left(\frac{a_i^{(t)} \sum_i x_i \dfrac{s_k^{(t)}}{\sum_k a_i^{(t)} s_k^{(t)}}}{\sum_k s_k^{(t)}} \right) \Big/ \left\| a_i^{(t)} \right\| \end{cases} .$$ (7)

And fixed matrix A, the updating process of matrix S is written as follows:

$$\begin{cases} \nabla_{S_i} = -a_i^T \left[X(x,y) - \sum_{i=1}^n a_i(x,y) s_i \right] - \frac{\lambda_1}{\sigma_i} kurt' \left(\frac{S_i}{\sigma_i} \right) + 2\lambda_2 \left[\frac{(S_i - m_k)}{S_W} - \frac{(m_k - m)}{S_B} \right] \\ S_i^{(t+1)} = \sqrt{S_k^{(t)} \left(\frac{\left[a_i^{(t)} \right]^T x_i}{a_i^{(t)} s_k^{(t)}} \right)} \end{cases} \tag{8}$$

By using the above updating methods described, A and S can be ensured to be positive, at the same time, the sparse coefficient vectors are guided to approach the true class center of samples.

4   Experimental Result Analysis

4.1   Feature Basis Learning

In test, the Hong Kong Polytechnic University (PolyU) palmprint database was used. This database includes 600 images from 100 individuals with 6 images from each [8]. Here, only 10 images of different individuals are chosen from this database. For an original image, it was first processed by the region of interesting (ROI) extraction method. For each ROI image patch, it was randomly sampled 5000 times with the 12×12 pixel patch, thus, a training set with the size of 144×5, 000 was obtain for each ROI image patch. Therefore, for 10 ROI patches, the non-negative matrix X with 144×50, 000 pixels could be obtained. Further, the dimension of X was reduced by 2-dimension principal component analysis (2D-PCA) method. The 64 feature basis vectors of 10 palmprint images learned by the DCB-NNSC model are shown in Fig.1. From Fig.1, it is easy to see that basis vectors behave clearer locality and are sparser.

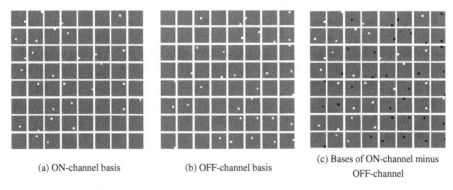

(a) ON-channel basis (b) OFF-channel basis (c) Bases of ON-channel minus OFF-channel

Fig. 1. Basis vectors DCB-NNSC of ON-channel and OFF-channel image data

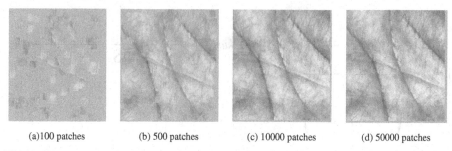

(a)100 patches   (b) 500 patches   (c) 10000 patches   (d) 50000 patches

Fig. 2. The original image and restored results of DCB-NNSC model. (a) to (d): Restored images of 100, 500, 10000 and 50000 image patches.

4.2   Testifying Features

To testify the features extracted, at first, the image reconstruction task was implemented. A test image shown in Fig. 2(a) was randomly selected from the PolyU palmprint database. This image was sampled different image patch with 8×8 image patch. Using features of DCB-NNSC, some of reconstruction images, corresponding to 500, 10000, and 50000 image patches, were shown in Fig. 2(b) to Fig. 2(d). Observing the restored images, it is clear to see that the more the image patches are, the clearer the restored results are, and the better the visual efficiency is.

Moreover, when the number of image patches is 50000, it is difficult to tell it from the original images. Further, the signal to noise ratio (SNR) criterion was used to measure the quality of restored images. And SNR values corresponding to different image patches were listed in Table 1. At the same time, those SNR values of restored images obtained by Hoyer's NNSC model were also listed in Table 1. And some of restored images obtained by Hoyer's NNSC model, corresponding to the same image patches, were shown in Fig. 3. According to experimental results, it is clear to see that the largest the number of image patches is, the clearer the visual effect of restored images is, and the largest the SNR value is despite of algorithms. Especially, when the number of image patches is 50000, only in term of the visual effect of restored images, it is very difficult to tell them from the original image in different algorithms. Otherwise, from Table 1, it is also easy to see that in the same number of image patches, the SNR values of our method are larger than those of Hoyer's model. Therefore, in a way, the experimental results can prove that our model is indeed effective in modeling the V1 mechanism of the primary visual system of human and extracting image features.

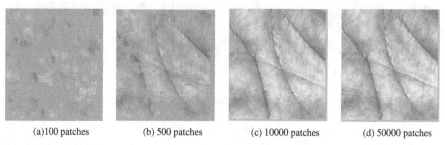

(a)100 patches   (b) 500 patches   (c) 10000 patches   (d) 50000 patches

Fig. 3. Restored results of NNSC model. (a) to (d): Restored images of 100, 500, 10000 and 50000 image patches.

Table 1. SNR values of different image patches obtained by different algorithms

Algorithms Image patches	SNR	
	DCB-NNSC	Hoyer's NNSC
100	19.99	19.92
500	24.02	23.44
10000	31.11	27.46
30000	31.25	30.42
50000	31.27	30.73

5   Conclusions

On the early NNSC model, a novel NNSC model with dispersion constraint, denoted DCB-NNSC model, is proposed in this paper. In this model, to ensure the sparsity of feature coefficients, the sparse penalty function is selected as the kurtosis function. At the same time, in order to enhance the capability of feature separability, the dispersion ratio of within-class and between-class of sparse coefficient vectors is utilized. Simulation results show that image features extracted by DCB-NNSC are much sparser and behave clearer locality than those of Hoyer's NNSC model, and our DCB-NNSC model is indeed efficient in image feature extraction and modeling the respective field of V1 in the primary visual system.

Acknowledgement. This work was supported by the National Nature Science Foundation of China (Grant No. 61373098 and 61370109), the Innovative Team Foundation of Suzhou Vocational University (No. 3100125), and the "333 Project" of Jiangsu Province.

References

1. Olshausen, B.A., Field, D.J.: Emergence of simple-cell receptive field properties by learning a sparse code for natural images. Nature 381(6583), 607–609 (1996)
2. Li, L., Zhang, Y.J.: Sense: a stable and efficient algorithm for nonnegative sparse coding. Acta Automatica Sinica 35(10), 439–443 (2009)
3. Hoyer, P.O.: Non-negative sparse coding. In: Proceedings of IEEE Workshop on Neural Networks for Signal Processing, pp. 557–565. IEEE, Martigny (2002)
4. Hoyer, P.O., Hyvärinen, A.: A multi-layer sparse coding network learns contour coding from nature images. Vision Research 42(12), 1593–1605 (2002)
5. Hoyer, P.O.: Modeling receptive fields with non-negative sparse coding. Neurocomputing 52(54), 547–552 (2003)
6. Shang, L.: Non-negative sparse coding shrinkage for image denoising using normal inverse Gaussian density model. Image and Vision Computing 26(8), 1137–1147 (2008)
7. Lee, D.D., Seng, H.S.: Learning the parts of objects by non-negative matrix factorization. Nature 401, 788–891 (1999)
8. Kong, W.K., Zhang, D., Li, W.: Palmprint recognition using eigenpalm features. Pattern Recognition Letters 24(9), 1473–1477 (2003)

Moving People Recognition Algorithm
Based on the Visual Color Processing Mechanism

XiaoJin Lin, QingXiang Wu*, Xuan Wang, ZhiQiang Zhuo, and GongRong Zhang

Key Laboratory of OptoElectronic Science and Technology for Medicine
of Ministry of Education, College of Photonic and Electronic Engineering
Fujian Normal University, Fujian, Fuzhou, 350007, China
qxwu@fjnu.edu.cn, xiaojin900912@163.com

Abstract. In this paper a target people recognition algorithm is proposed, in which the color processing mechanism is inspired by the biological visual system. The algorithm is constructed in two parts. In the first part a spiking neural network is proposed to extract the color features of the objects which are captured from videos. In the second part, after a feature reduction, a Support Vector Machine is used to fuse the color features and recognize the target. The algorithm has been successfully applied to recognize target people appeared in video sequence with a high recognition rate and suitable for generic recognition domain.

Keywords: visual color processing, moving people recognition, spiking neural networks, support vector machine.

1   Introduction

Vision-based human motion recognition plays a vital role in intelligent monitoring system, the system will be a timely warning to protect the safety of person and property when unusual circumstances occur. With the development of computer networks and image processing technologies, a number of issues such as human detection and face recognition technology attract attentions over the world. Present portrait recognition methods mainly rely on face recognition, but in real life, the actual pixels of human face images are often inadequate because of the camera insufficient, long distances or undesirable illumination and result in difficulty for utilization of face recognition. Therefore, this paper focuses on identification of the overall portrait, and it makes important significance in tracking people's whereabouts between multiple cameras. We study the people recognition in details in two aspects: color feature extraction by spiking neural network and recognition algorithm by color features. The highest recognition rate in this paper reach at 99.3435% in the test using the CASIA Database A.

In this paper a simplified conductance-based integrate-and-fire model [1] is used for each neuron in SNN. There are various receptive fields in the striate cortex, retina

---

* Corresponding author.

D.-S. Huang et al. (Eds.): ICIC 2014, LNCS 8588, pp. 480–487, 2014.

and lateral geniculate nucleus [2-3] in human visual system. The visual images are transferred among these neurons in the form of spiking trains through ON or OFF pathways [4]. Different ON/OFF pathways are used to construct the specific networks forcolor feature extracting in a biological manner. Thus a people recognition algorithm based on visual color processing is proposed in this paper and it obtains a good result in the mainstream pedestrian gait databases.

2   Pretreatment

2.1 Moving People Extraction

This article adopts the CASIA Database A[5] of the Chinese academy of sciences. The binary images of the moving people region are provided by Chinese academy of sciences, and they are used to do the mask operation with video image sequences to obtain the moving targets. Finally all the extracted target pictures are normalized to 150*220 pixels.

2.1 Body Segmentation

To enhance the recognition rate, in this work the human body is divided into three parts : head-shoulder part, torso part and leg part. Body segmentation is significant to capture information of color distribution over people's body. The human anatomy proportion of normal people is roughly the same. Head and shoulder part occupies roughly 20% of body proportions, torso part is roughly 30% and leg part is roughly 50%. Figure 1 shows the definition of the body parts. Figure 2 shows the segment results of our targets.

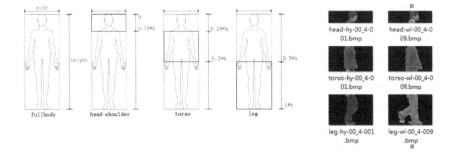

Fig. 1. The definition of body parts   **Fig. 2.** The segment results

3   SNN Model for Color Feature Extraction

In this paper a SNN model for color feature extraction proposed by Q.Wu et al. [6] is selected. Inspired by the roles of rods and cones in retina, the spiking neural network is constructed using four types of receptive fields. The structure of the network is shown in Figure 3. Each pixel of the input image corresponds to a receptor. RCx,y is a 5*5 receptive field simulates three types of cones corresponding to color components R, G, and B. Each receptive field forms two ON/OFF pathways. The 7*7 RNx,y receptive field simulates rod receptive field corresponding to gray scale image and also forms two ON/OFF pathways.

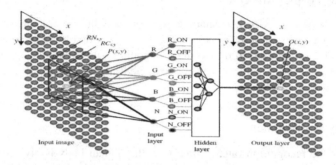

Fig. 3. Spiking neural network for eight ON/OFF pathways

Photonic receptors are represented in the first layer. Each neuron of the spiking neural network connected to each pixel of the input image, and convert the pixels into pulse sequences according to RGB values and gray value. When each of the neurons in an input layer fires, a firing rate can be obtained for corresponding pixel value in the final output layer. For color red, we can have the firing rate for the neuron $F_{R\_ON(x,y)}$ by the following expression:

$$F_{R\_ON\,(x,y)}(t) = \frac{1}{T}\sum_{t}^{t+T} S_{R\_ON\,(x,y)}(t) \tag{1}$$

$S_{R\_ON(x,y)}(t)$ represents the spike train generated by the neuron as follow:

$$S_{R\_ON(x,y)}(t) = \begin{cases} 1 & \text{if neuron } R\_ON(x,y) \text{ fires at time } t. \\ 0 & \text{if neuron } R\_ON(x,y) \text{ dose not fire at time } t. \end{cases} \tag{2}$$

By analogy, a firing rate of $F_{R\_OFF(x,y)}$ neuron is obtained:

$$F_{R\_OFF\,(x,y)}(t) = \frac{1}{T}\sum_{t}^{t+T} S_{R\_OFF\,(x,y)}(t) \tag{3}$$

For color green, color blue and gray scale:

$$F_{G\_ON\,(x,y)}(t) = \frac{1}{T}\sum_{t}^{t+T} S_{G\_ON\,(x,y)}(t) \,, \qquad F_{G\_OFF\,(x,y)}(t) = \frac{1}{T}\sum_{t}^{t+T} S_{G\_OFF\,(x,y)}(t) \tag{4}$$

$$F_{B\_ON(x,y)}(t) = \frac{1}{T}\sum_{t}^{t+T} S_{B\_ON(x,y)}(t), \qquad F_{B\_OFF(x,y)}(t) = \frac{1}{T}\sum_{t}^{t+T} S_{B\_OFF(x,y)}(t) \quad (5)$$

$$F_{N\_ON(x,y)}(t) = \frac{1}{T}\sum_{t}^{t+T} S_{N\_ON(x,y)}(t), \qquad F_{N\_OFF(x,y)}(t) = \frac{1}{T}\sum_{t}^{t+T} S_{N\_OFF(x,y)}(t) \quad (6)$$

Color feature is extracted through different ON/OFF pathways represent as F_s nn:

$$F_{snn} = (F_{R\_ON(x,y)}, F_{R\_OFF(x,y)}, F_{G\_ON(x,y)}, F_{G\_OFF(x,y)}, F_{B\_ON(x,y)}, F_{B\_OFF(x,y)},$$
$$F_{N\_ON(x,y)}, F_{N\_OFF(x,y)}) \qquad (7)$$

Eight images are obtained corresponding to the eight ON/OFF pathways of a sam ple as shown in the Figure 4.

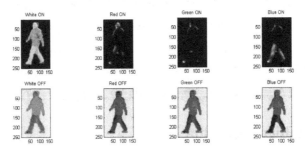

Fig. 4. Eight images corresponding to the eight ON/OFF pathways

4   Feature Statistics

The dimension of one input image is 150*220, so that one ON/OFF pathway feature's dimension is 150*220, then eight ON/OFF pathways features' dimension would reach 8*150*220 equal to 264000. So feature reduction is required; the computational cost can be greatly reduced and classification will become much efficient because of low-dimensional space.

In this paper a feature statistics method which is similar to color moment [7] is used to reduce features' dimension. Color distribution information focused on lower moments, so three low-order moments E_i, σ_i and t_i of the images could express the color distribution adequately. In this paper, the first central moment E_i denotes the average color of each ON/OFF pathway feature, the second central momentσ_i denotes the variance, and the third central moment t_i denotes the skewness. Definitions are as follows:

$$E_i = \frac{1}{N}\sum_{j=1}^{N} p_{ij}, \qquad \sigma_i = (\frac{1}{N}\sum_{j=1}^{N}(p_{ij}-E_i)^2)^{1/2}, \qquad t_i = (\frac{1}{N}\sum_{j=1}^{N}(p_{ij}-E_i)^3)^{1/3} \qquad (8)$$

Where p_{ij} represents the pixel of j-th firing rate of the i-th ON/OFF pathway component. N is the number of pixels. Three low-order moments E_i, σ_i and t_i of eight color pathways constitute a 24-dimensional histogram vector F_{8color}:

$$F_{8color} = (E_{R_{ON}}, E_{R_{OFF}}, E_{G_{ON}}, E_{G_{OFF}}, E_{B_{ON}}, E_{B_{OFF}}, E_{N_{ON}}, E_{N_{OFF}}, \sigma_{R_{ON}}, \sigma_{R_{OFF}}, \sigma_{G_{ON}}, \sigma_{G_{OFF}},$$
$$\sigma_{B_{ON}}, \sigma_{B_{OFF}}, \sigma_{N_{ON}}, \sigma_{N_{OFF}}, t_{R_{ON}}, t_{R_{OFF}}, t_{G_{ON}}, t_{G_{OFF}}, t_{B_{ON}}, t_{B_{OFF}}, t_{N_{ON}}, t_{N_{OFF}}) \qquad (9)$$

A 264000-dimensional color feature through the eight ON/OFF pathways of each image as a sample matrix , after dimensionality reduction through the above steps ,the size of the feature vector reduce to 24 dimension for each sample.

The eight ON/OFF pathways represent all the visual color components of the image, some components maybe not use for the recognition algorithm or even be the noise components. So in this paper, in order to test and verify whether the SNN using eight ON/OFF pathways is effective or not, we analyze different recognition rates corresponding to less components including three ON/OFF pathways and four ON/OFF pathways. Three low-order moments E_i, σ_i and t_i of R_{ON}, G_{ON}, B_{ON} color pathways constitute a 9-dimensional histogram vector F_{3color}:

$$F_{3color} = (E_{R_{ON}}, E_{G_{ON}}, E_{B_{ON}}, \sigma_{R_{ON}}, \sigma_{G_{ON}}, \sigma_{B_{ON}}, t_{R_{ON}}, t_{G_{ON}}, t_{B_{ON}}) \qquad (10)$$

Three low-order moments E_i, σ_i and t_i of R_{ON}, G_{ON}, B_{ON}, N_{ON} color pathways constitute a 12-dimensional histogram vector F_{4color}:

$$F_{4color} = (E_{R_{ON}}, E_{G_{ON}}, E_{B_{ON}}, E_{N_{ON}}, \sigma_{R_{ON}}, \sigma_{G_{ON}}, \sigma_{B_{ON}}, \sigma_{N_{ON}}, t_{R_{ON}}, t_{G_{ON}}, t_{B_{ON}}, t_{N_{ON}}) \qquad (11)$$

5   Recognition Algorithm

LibSVM [8] is used to verify the algorithm and In this paper SVM use the polynomial kernel function, parameters g and c are selected in range $[2^{-10}, 2^{10}]$, c is the penalty factor of nuclear function which is determined according to different kernel functions, g is the control parameter of nuclear function. As c increases while the number of support vectors reduces, the number of support vectors and recognition rates are stabilizing so that the recognition rate can reach most excellent result.

Five types of feature patterns are extracted by the proposed algorithm, and they are putting into the SVM for identification: Full-body color feature, Head-shoulder color feature, Torso color feature, Leg color feature, and Three-part-combination color feature which is fused orderly using the Head-shoulder, Torso and Leg color feature.

6   Experimental Results

CASIA Database A is composed of 20 people corresponding to three angles (0-angle, 45-angle, 90-angle), each angle includes four image sequences. This article adopts the 0-angle image sequences. After cleaning some noise data we obtain 4606 images which are divided into 10 groups randomly. One group is chosen as a test set and the remaining nine as training set in turn. Ten times tests are conducted and the average of recognition rate is obtained. Finally the trained LibSVM classifier can be used to identify moving peoples. In order to analyze the recognition rate vs different feature pathways, in this paper the SNNs corresponding to different pathways are used to test the recognition rate for five types of features. Results are shown in Table 1.

Table 1. The correct recognition rate of different pathways vs different part-combinations

Feature \ Algorithm	SNN+eight pathways	SNN+four pathways	SNN+three pathways
Full-body	98.2495%	92.5602%	89.2779%
Head-shoulder	76.1488%	71.7724%	62.8009%
Torso	95.4048%	88.8403%	79.2123%
Leg	91.2473%	81.6193%	76.3676%
Three-part-combination	99.3435%	96.2801%	94.3107%

Obviously the algorithm based on eight pathways gets the highest recognition rate. It is shown that eight pathways contain the key features of color information more completely than fewer pathways and demonstrates the effectiveness of the algorithm proposed in this paper. On the other hand, the algorithms used Three-part-combination feature get the highest recognition rate. This result demonstrates the effectiveness of our segment algorithm. On the contrary, the algorithms used Head-shoulder feature get the lowest recognition rate, which shows that it is difficult for face recognition to obtain so high recognition rate for this data base.

To prove the effectiveness of the proposed algorithms, we also compare with some other methods which are from Mass Vectors[9], GEI+LBP+DCV[10], BenAbdelkader [11], Collins [12], Lee [13]and Phillips [14] respectively, as they both use the CASIA Database A 0-angle image sequences to verify algorithm.

Table 2. Comparison of proposed algorithms and other algorithms on people recognition

Algorithms	Recognition Rate		
SNN+eight pathways	99.3435%		
SNN+four pathways	96.2801%		
SNN+three pathways	94.3107%		
Mass Vector[9]	96.25%		
GEI+LBP+DCV[10]	93.08%(rank1)	100%(rank5)	
BenAbdelkader[11]	72.50%(rank1)	88.75%(rank5)	96.25%(rank10)
Collins[12]	71.25%(rank1)	78.75%(rank5)	87.50%(rank10)
Lee [13]	87.50%(rank1)	98.75%(rank5)	100% (rank10)
Phillips[14]	78.75%(rank1)	91.25%(rank5)	98.75%(rank10)

In [11-14], three norms are used to evaluate their gait recognition algorithms: Rank 1 shows the proportion of correct category to be ranked first in the test samples, rank 5 shows the proportion of correct classes to be ranked in the top five of the test sample, and rank 10 shows the top ten. Seen from Table 1 and Table 2, the proposed algorithm has a high recognition rate compared with the other algorithms, it can handle massively parallel compute, distributed processing and possess powerful self-learning and self-organization ability. But each algorithm has unique advantages and disadvantages under different testing conditions, so further evaluations and comparisons on more realistic and challenging databases can be considered to further work.

7   Conclusion

This paper gives a recognition algorithm based on SNN, different from the traditional color feature extraction methods, SNN is imitating biological visual mechanisms to extract the color feature. Then a feature statistics method which is similar to color moment is proposed. Experimental results base on CASIA Data A show that the results is quite satisfactory. But most of the time, the walking directions of people are not fixed and sometimes person may be occlusion, so overcoming the impacts will be a challenging issue. Future research will focus on exploring more essential features while using multi-feature fusion technology to enhance the accuracy further. The processing speed is another vital factor, in future we can take advantage of the modern Graphic Processing Units to speed up the simulation.

Acknowledgement. The authors gratefully acknowledge the fund from the Natural Science Foundation of China (Grant No.61179011 & No. 61070062) and Science and Technology Major Projects for Industry-academic Cooperation of Universities in Fujian Province (Grant No.2013H6008&2011H6010).

References

1. Muller, E.: Simulation of High-Conductance States in Cortical Neural Networks. In: Masters thesis, University of Heidelberg (2003)
2. Kandel, E.R., Schwartz, J.H., Jessell, T.M. (eds.): Principles of Neural Science. McGraw-Hill, New York (2000)
3. Masland, R.H.: The Fundamental Plan of the Retina. Nature Neuroscience 4(9), 877–886 (2001)
4. Nelson, R., Kolb, H.: ON and OFF Pathways in the Vertebrate Retina and Visual System. The Visual Neurosciences 1, 260–278 (2004)
5. CASIA Database, http://www.cbsr.ia.cn
6. Wu, Q., McGinnity, T.M., Maguire, L., Valderrama-Gonzalez, G.D., Dempster, P.: Colour Image Segmentation based on a Spiking Neural Network Model Inspired by the Visual System. In: Huang, D.-S., Zhao, Z., Bevilacqua, V., Figueroa, J.C. (eds.) ICIC 2010. LNCS, vol. 6215, pp. 49–57. Springer, Heidelberg (2010)
7. Stricker, M.A., Orengo, M.: Similarity of color images. In: IS&T/SPIE's Symposium on Electronic Imaging: Science & Technology, pp. 381–392. International Society for Optics and Photonics (1995)
8. Chang, C.C., Lin, C.J.: LIBSVM: A Library for Support Vector Machines. ACM Transactions on Intelligent Systems and Technology (TIST) 2(3), 27 (2011)
9. Hong, S., Lee, H., Nizami, I.F., Kim, E.: A New Gait Representation for Human Identification: Mass Vector. In: 2nd IEEE Conference on Industrial Electronics and Applications, ICIEA 2007, pp. 669–673. IEEE (2007)
10. Liu, Z., Feng, G., Chen, W.: Gait Recognition Based on Local Binary Patterns and Identify Common Vector. Computer Science 40(9), 262–265 (2013)
11. BenAbdelkader, C., Cutler, R., Nanda, H., Davis, L.: EigenGait: Motion-Based Recognition of People Using Image Self-Similarity. In: Bigun, J., Smeraldi, F. (eds.) AVBPA 2001. LNCS, vol. 2091, pp. 284–294. Springer, Heidelberg (2001)

12. Collins, R.T., Gross, R., Shi, J.: Silhouette-based Human Identification From Body Shape and Gait. In: Proceedings of the Fifth IEEE International Conference on Automatic Face and Gesture Recognition, pp. 366–371. IEEE (2002)
13. Lee, L., Grimson, W.E.L.: Gait Analysis for Recognition and Classification. In: Proceedings of the Fifth IEEE International Conference on Automatic Face and Gesture Recognition, pp. 148–155. IEEE (2002)
14. Phillips, P.J., Sarkar, S., Robledo, I., Grother, P.: Baseline Results for the Challenge Problem of Human ID using Gait Analysis. In: Proceedings of the Fifth IEEE International Conference on Automatic Face and Gesture Recognition, pp. 130–135. IEEE (2002)

Finding Specific Person Using Spiking Neural Network Based on Texture Features

Xuan Wang[1,2], QingXiang Wu[1,*], XiaoJin Lin[1], and ZhiQiang Zhuo[1]

[1] Key Laboratory of Optoelectronic Science and Technology for Medicine of Ministry
of Education, College of Photonic and Electronic Engineering
Fujian Normal University, Fuzhou, 350007, P.R. China
[2] College of Information Science and Engineering
Fujian University of Technology, Fuzhou, 350007, P.R. China
xwang1978068sina.com, qxwu@fjnu.edu.cn

Abstract. This paper has proposed a new texture-based method to find a specific person. A spike neural network is constructed to extract texture features. The network is constructed using integrate-and-fire neuron model, and it has behaviors similar to the Gabor filters. The person image is divided into three parts: head, torso and leg. Three parts of texture features are extracted by means of this network. BP neural network and multi-class Support Vector Machine are used as classifiers. CUDA technology is applied in the experiment, which greatly reduces the computation time.

Keywords: Texture feature, Gabor filters, Spike neural network, Support Vector Machine, BP neural network.

1 Introduction

In video surveillance, it is pretty common to identify a specific person. Color feature and texture feature of a person are distributed in whole body for a person.The distribution of color and texture of whole body cannot be the same for different persons because different persons have different appearances. Therefore, a special person can be discriminated by color features and texture features of the appearances.

In recent decades, the visual nervous system has become a hot research domain. Spiking neuron model was firstly proposed by Alan Lloyd Hodgkin and Andrew Huxley in [1]. The model describes how the neurons exchange information encoded by pulse. Since the complexity, in practice, a lot of simplified models have been derived. The most popular model is the conductance based integrate-and-fire model. Therefore, this neuron model is used to construct spiking neural network to extract texture features in this paper.

Ganglion cells, which are a type of neurons in the visual system, have on-center and off-center receptive fields. Based on the conductance based integrate-and-fire model and receptive field model, Wu extracted the feature of the edge of object in [2].

---

* Corresponding author.

D.-S. Huang et al. (Eds.): ICIC 2014, LNCS 8588, pp. 488–494, 2014.

Xie in [3] described how the process of taking advantage of the parallel execution of CUDA to segment color image. Wu applied hierarchical spiking neural networks to simulate visual attention in [4].

The receptive field can be modeled by Gabor function.[5]. Spike neural network can perform functions similar to Gabor filters[6,7]. Adjusting the firing threshold value of spike neuron, the noise and some unimportant information can be ignored.

In this paper a Spike Neural Network model is proposed to extract texture features from gray images and play the roles of Gabor filters.The texture features are used to classify different persons.The output is encoded by the fire rate of the neuron in the network. Experimental results show that spike neural network could reach good effect. In order to speed up the process of feature extraction, CUDA technology is also employed.

2   Spiking Neural Network Model for Texture Feature Extraction

The structure of the spiking neural network is showed in Figure 1.

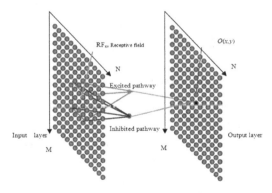

Fig. 1. Structure of Spike Neural Network Model

The dimension of input layer is M×N, where M is the height of image, and N is the width of image. The neurons in input matrix correspond to image pixels. The dimension of intermediate layer is also M×N. Two intermediate layers correspond to parallel excitatory and inhibitory pathways respectively. Each neuron in intermediate layer is connected to the neurons within its receptive field RF_{xy} in input layer. The neurons receive spikes from the neurons in the receptive field RF_{xy} and the membrane potentials of neurons in two intermediate layers are convoluted with $w_{x,y}^{gabor\_ex}$ / $w_{x,y}^{gabor\_ih}$. According to [8], the value of $w_{x,y}^{gabor\_ex}$ and $w_{x,y}^{gabor\_ih}$ can be computed as follows:

$$w_{x,y}^{gabor\_ex} = \frac{\left\|\vec{k}_j\right\|^2}{\sigma^2}\exp\left(-\frac{\left\|\vec{k}_j\right\|^2((x-x_c)^2+(y-y_c)^2)}{2\sigma^2}\right)\left[\exp(i\vec{k}_j((x-x_c)+(y-y_c)))-\exp(-\frac{\sigma^2}{2})\right] \tag{1}$$

$$\vec{k}_j = k_v(\cos\theta,\sin\theta)^T \quad , \quad k_v = 2^{\frac{-(v+2)}{2}}\pi \tag{2}$$

$$\theta = 8*u/\pi \,, w_{x,y}^{gabor\_ih} = -w_{x,y}^{gabor\_ex} \tag{3}$$

Where (x_c, y_c) is the center of receptive field RF_{xy}, (x, y) is the coordinate of the pixels in the receptive field RF_{xy}, u is the orientation , v is the level scale, σ is the standard deviation of Gaussian function. In this paper, the scale number is 5, and the orientation number is set to 8.

According to [2, 9], the membrane potential of the neurons of intermediate layer can be inferred by the following equations

$$\frac{dI_{x,y}(t)}{dt} = \alpha\left\|\frac{w_{x,y}^{gabor\_ex}\otimes G_{x,y}}{A_{ex}} - \frac{w_{x,y}^{gabor\_ih}\otimes G_{x,y}}{A_{ih}}\right\| \tag{4}$$

$$c_m\frac{dV_N(t)}{dt} = g_l\left(E_l - V_N(t)\right) + I_{x,y}(t) \tag{5}$$

Where $I_{x,y}(t)$ is synapse current, α is a constant, c_m is membrane capacity, $V_N(t)$ is the membrane potential of the neuron, g_l is membrane conductance, E_l is reverse potential, and $G_{x,y}$ is the gray value of the pixel at (x, y), A_{ex} is the contacting area between excitatory synapses and dendrites, and A_{ih} is the contacting area between inhibitory synapses and dendrites, $\otimes$ is convolution symbol.

When the membrane potential of the neuron of intermediate-layer reaches the threshold value, the neuron will fire, $S_{x,y}(t)$ will be set a value of 1.The generated pulse is transmitted to output layer. The dimension of output layer is the same as that of input layer. Unlike the neuron of intermediate layer, the neuron of output layer at (x, y) is only connected to the neuron of intermediate layer at (x, y). The membrane potential of the neuron of output layer can be inferred by the following equations:

$$\frac{g_{x',y'}(t)}{dt} = -\frac{1}{\tau}g_{x',y'}(t) + \beta S_{x,y}(t) \tag{6}$$

$$c_m\frac{dV_{x',y'}^{out}(t)}{dt} = g_l(E_l - V_{x',y'}^{out}(t)) + \frac{g_{x',y'}(t)}{A}\left(E_r - V_{x',y'}^{out}(t)\right) \tag{7}$$

Where $V_{x',y'}^{out}(t)$ is the membrane potential of the neuron of output layer, $g_{x',y'}(t)$ is synapse conductance, β is a constant, τ is time constant, E_r is reverse potential, and A is the contacting area between synapses and dendrites. The neuron will fire if the threshold voltage is reached. The firing rate can be calculated by the following equation:

$$fRate_{x',y'}^{u,v} = \frac{1}{T}\sum_{t}^{t+T} S_{x',y'}^{u,v}(t) \tag{8}$$

The calculation of fire rate of each neuron of output layer is similar to (10). The extraction of texture feature requires forty Gabor filters, which have five level scale and eight orientations. The fire rate matrices are regarded as the gray images derived from different SNN-Gabor filters. According to [10], the mean and the variance of different firing rate matrices of a sample form a 1×80 vector F representing its texture feature. The mean $\mu_{u,v}$ and the variance $\sigma_{u,v}$ are calculated by the following equations:

$$\mu_{u,v} = \frac{1}{MN}\sum_{y'=1}^{M}\sum_{x'=1}^{N}\left|fRate_{x',y'}^{u,v}\right|, \sigma_{u,v} = \sqrt{\frac{1}{MN}\sum_{y'=1}^{M}\sum_{x'=1}^{N}\left[fRate_{x',y'}^{u,v} - \mu_{u,v}\right]^2}, \tag{9}$$

$$F = \left\{\mu_{0,0}, \sigma_{0,0}, ..., \mu_{4,7}, \sigma_{4,7}\right\} \tag{10}$$

The extraction of texture feature requires forty fire rate matrices. It is time consuming. CUDA technology can be applied to assign one GPU thread to each neuron. CUDA is a parallel computing architecture proposed by NVIDIA in 2006. The architecture partitions all the GPU threads into different blocks. Different blocks are carried in parallel on different stream processors. In a block, 32 threads are assigned to the same warp. The threads in the same warp are implemented in the same time. The experiment uses NVIDIA GeForce GTS 450 which has 192 CUDA cores. In this experiment, the block number is 150, so 150×32 GPU threads can be executed concurrently. This greatly shortens the time of the calculation. Each GPU thread calls the same kernel function that discriminate different neurons through blockIDx.x and threadIDx.x.

3   Experimental Results

Experimental data are from the silhouettes and the color images from gait dataset A provided by CASIA[11]. The dataset contains walking pictures of 20 pedestrians. The walking pictures of 0 degree are selected for test of the algorithm. In the experiments the size of the pedestrian image is normalized to 220×150. The data set is split into 10 parts, 9 parts of which will be used as training data in turn and 1 part as the test data in turn.

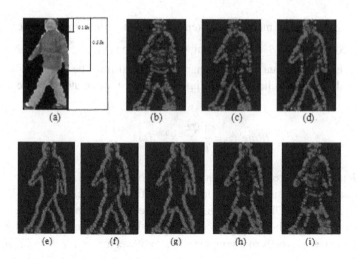

Fig. 2. (a) The Human Body Divided into Three Parts. (b-i) Images Filtered by SNN-Gabor Filters of 0 Level Scale and 8 Different Orientations.

According to the proportion of human body, the body can be roughly divided into three parts: head, torso and leg.Using the Spike Neural Network architecture mentioned above, different parts of texture features can be extracted. The texture features of more than 4000 pictures of 20 persons are classified by BP neural network and multi-class Support Vector Machine. Results are showed in the following table:

Table 1. Average Recognition Rate Obtained From Different Parts of Texture Features

Classifier	Recognition Rate (head texture)	Recognition Rate (torso texture)	Recognition Rate (leg texture)
BP	74.224%	91.25%	86.552%
SVM	74.6336%	94.806%	92.3491%

It is clear that the accuracy based on torso texture features is the highest. Table 2 indicates GPU implementation of texture feature extraction speed up a lot.

Table 2. The Time of the Calculation of the Texture Feature of a Sample

Body Part	Calculation Time (C)	Calculation Time (CUDA)
Head	21203ms	3812ms
Torso	41797ms	7157ms
Leg	49578ms	8716ms

Based on three parts of texture features, the following fusion methods are used to determine the class to achieve the identification results:

(1)Three parts of texture features $F_{head}, F_{torso}, F_{leg}$ are fused into F_{fusion} in series. It is clear that F_{fusion} is a 1×240 vector. Then all the feature vectors are normalized and training samples are entered into multi-class Support Vector Machine to train the classifiers.

(2)Voting mechanism. There are three classifiers that are based on head texture feature, torso texture feature and leg texture feature respectively. They vote the result to classify the samples. A test sample is identified as the class which has the most votes. If the outputs of three classifiers are different, the classifier based on torso texture feature has the highest priority.

(3)Three parts of texture features are classified by BP neural network respectively. The number of output layer of neural network is set as the same as the number of classes. The value of output in the neural network is normalized to between 0-1. The output value of neuron j for head classifier, torso classifier or leg classifier is labeled as $P_{head,j}$, $P_{torso,j}$ and $P_{leg,j}$ respectively, and is regarded as the probability P_j that a sample belongs to class j based on head texture feature, torso texture feature and leg feature. The class number i of a sample is determined by the following:

$$P_j = P_{head,j} + P_{torso,j} + P_{leg,j} \tag{12}$$

$$i = \text{argmax } P_j \quad 1 \le j \le 20 \tag{13}$$

The experimental results are showed in Table 3:

Table 3. Average Recognition Rate Obtained From Different Methods

Method	Average Recognition Rate
SNN-Gabor+GPU+SVM+Voting	97.802%
SNN-Gabor+GPU+BP+Probability	98.168%
SNN-Gabor+GPU+ Serial Fusion +SVM	98.7716%
Gabor+ Serial Fusion +SVM	98.7069%

As the results show, the proposed model works efficiently. The output value of BP neural network is converted to the probability and three outputs of the neural networks are integrated to make decision. Recognition rate is improved clearly compared with approach SNN-Gabor+GPU+SVM+Voting. SNN could extract good features because some nonsignificant information and noise have been removed through fire threshold adjustment. For the classifiers used in the experiments, multi-class Support Vector Machine is relatively better than Bp neural network. Therefore, approach SNN-Gabor+GPU+Serial Fusion+SVM gets the highest rate of recognition.

4   Discussion

This paper presents a set of algorithms based on spiking neural network model to detect specific person. Experimental results show that it is useful for texture feature extracted using SNN-Gabor filters to identify a specific person. The method also can be applied to video surveillance systems to find a specific person. Some issues can also be studied in further.For example, other useful information is neglected. Future work includes that the non-classical receptive field model is applied, texture feature is fused with the features of color, hog and gait based on the model inspired by biological vision. If those problems will be solved effectively and more accurate result will be acquired.

Acknowledgments. The authors gratefully acknowledge the fund from the Natural Science Foundation of China (Grant No.61179011 & No. 61070062) and Science and Technology Major Projects for Industry-academic Cooperation of Universities in Fujian Province (Grant No.2013H6008&2011H6010).

References

1. Hodgkin, A.L., Huxley, A.F.: A Quantitative Description of Membrane Current and Its Application to Conduction and Excitation in Nerve. The Journal of Physiology 117(4), 500–544 (1952)
2. Wu, Q., McGinnity, M., Maguire, L., Belatreche, A., Glackin, B.: Edge Detection Based on Spiking Neural Network Model. In: Huang, D.-S., Heutte, L., Loog, M. (eds.) ICIC 2007. LNCS (LNAI), vol. 4682, pp. 26–34. Springer, Heidelberg (2007)
3. Xie, E., McGinnity, M., Wu, Q., Cai, J., Cai, R.: GPU Implementation of Spiking Neural Networks for Color Image Segmentation. In: 2011 4th International Congress on Image and Signal Processing (CISP), vol. 3, pp. 1246–1250. IEEE (2011)
4. Wu, Q.X., McGinnity, T.M., Maguire, L., Cai, R.T., et al.: A Visual Attention Model Based on Hierarchical Spiking Neural Networks. Neurocomputing 116, 3–12 (2013)
5. Daugman, J.G.: Uncertainty Relation for Resolution in Space, Spatial Frequency, and Orientation Optimized by Two-Dimensional Visual Cortical Filters. JOSA A 2(7), 1160–1169 (1985)
6. Van Rullen, R., Gautrais, J., Delorme, A., Thorpe, S.: Face Processing Using One Spike per Neuron. Biosystems 48(1), 229–239 (1998)
7. Elmir, Y., Benyettou, M.: Gabor Filters Based Fingerprint Identification Using Spike Neural Networks. In: Proc. of the 5th Sciences of Electronic, Technologies of Information and Telecommunication Conference (2009)
8. Wiskott, L., Fellous, J.M., Krüger, N., von der Malsburg, C.: Face Recognition and Gender Determination (1995)
9. Destexhe, A.: Conductance-Based Integrate-and-Fire Models. Neural Computation 9(3), 503–514 (1997)
10. Manjunath, B.S., Ma, W.Y.: Texture Features for Browsing and Retrieval of Image Data. IEEE Transactions on Pattern Analysis and Machine Intelligence 18(8), 837–842 (1996)
11. Wang, L., Tan, T., Ning, H., Hu, W.: Silhouette Analysis-Based Gait Recognition for Human Identification. IEEE Transactions on Pattern Analysis and Machine Intelligence 25(12), 1505–1518 (2003)

Unsupervised Image Segmentation Based on Contourlet Texture Features and BYY Harmony Learning of t-Mixtures

Chenglin Liu and Jinwen Ma[*]

Department of Information Science, School of Mathematical Sciences and LMAM,
Peking University, Peking University, Beijing, 100871, China
jwma@math.pku.edu.cn.

Abstract. This paper proposes an unsupervised color image segmentation approach excellent in multi-texture image segmentation. Actually, it employs a novel texture feature extraction mechanism through the contourlet subband coefficient clustering, which is more effective in image segmentation than the discrete cosine transform based normalization technique (DCT). In addition, it adopts the gradient Bayesian Ying-Yang harmony learning of t-mixtures (BYY-t) for automatic image objects detection so that the image segmentation is in an unsupervised mode. The experiments on the images in Berkeley Segmentation Database and Benchmark (BSDB) database demonstrate the improved performances of this approach in varied and complex color image segmentation. Additional experiments on multi-texture color images further demonstrate its better performances   in comparison with those of the state-of-art algorithms.

Keywords: image segmentation, Bayesian Ying Yang (BYY), texture feature.

1   Introduction

In the field of color image segmentation, clustering analysis is one of the most powerful methods   due to its complicity and computational efficiency [1]. In this way, the pixels of an image are treated as feature vectors in a special feature space. The clustering method is performed on these pixels to group the ones with similar features. Many well-known clustering algorithms such as k-means [2], EM algorithm [3] have been widely applied to image segmentation.

There are, however, two challenges to the clustering methods which critically influence the segmentation performances: (1). the selection of a proper feature space; (2). the determination of an appropriate number of clusters. In fact, many supervised classification methods for image segmentation cannot guarantee the performances on the general images due to the complexity and high variation of the actual images. Thus, a proper feature space as well as   a good clustering or unsupervised learning method with automatic cluster number selection   are desired for image segmentation.

---

[*] Corresponding author.

D.-S. Huang et al. (Eds.): ICIC 2014, LNCS 8588, pp. 495–501, 2014.

In color image segmentation, the feature space is often composed of the position, color and texture features. The color feature can be based on one of many popular color spaces, such as RGB, HIS, CIE and their variants. As to the texture features, many researches are focusing on modeling the distributions of discrete wavelet coefficients to generate a convenient measurement between different textures. In this paper, we evaluate the texture variation of an image based on a texture classification method contourlet subband clustering. This method excels the state-of-art methods due to fully employment of the information of all contourlet subband coefficients without the assumption of their distributions. A texture similarity matrix is then obtained, and the spectral clustering method is performed to represent features by the eigenvectors of the similarity matrix.

In addition, we address the second challenge by using an unsupervised BYY-t algorithm, which can realize the model selection during parameter estimation, leading to the clustering analysis with automatic cluster number selection [4]. The segmentation performances are evaluated on a set of 100 color images in Berkeley segmentation database test dataset, and are quantized by a similarity measurement, the probability rand (PR) index.

The rest of this paper is organized as follows. Section 2 describes the contourlet subband clustering method for texture classification and presents the steps of the proposed algorithm for image segmentation including the feature space construction and BYY-t for similar pixel clustering analysis. Section 3 presents the experimental results of the proposed algorithm on different kind of images. Finally, a brief conclusion is given in Section 4.

2   Methods

In this section, the feature extraction method is firstly introduced on the three aspects of texture, color and position features. Then, BYY-t clustering method is described. Combining them together, the image segmentation procedures are finally summarized.

2.1   Contourlet Texture Feature Space Construction

The texture feature space is constructed based on the following steps: cutting image into windows, extracting texture features using contourlet subband coefficient clustering, building texture similarity matrix and determining texture eigenvectors.

First of all, it is meaningless talking about the texture of a single pixel, because textures are patterns presented by a window of pixels. For each analyzed window, the distribution of the histograms of the wavelet coefficients can only be obtained based on enough number of pixels. Here, we resize the images and cut them to windows of size no less than 50×50 pixels.

The texture features of each window are based on the contourlet subband coefficient clustering method. In fact, it is a new texture feature extraction approach that adopts k-means algorithm for contourlet subband coefficients modeling [5].

Actually, it fully employs the information of the wavelet coefficients without any hypothesis of their distributions. The extracted features consists of the sorted cluster centers of the contourlet subband coefficients based on k-means algorithm $F = (f_1, \ldots, f_J)$ as well as the dispersion degrees of cluster centers:

$$f_{J+1} = \frac{1}{N}\sum_{n=1}^{N}(x_n - \frac{1}{N}\sum_{n=1}^{N}x_n)^2, f_{J+2} = \frac{1}{N}\sum_{n=1}^{N}x_n^2. \tag{1}$$

We firstly extract the texture features of each window using the above method, and calculate their similarities by

$$SL(F^1, F^2) = \frac{1}{\tau + RL(F^1, F^2)}, \tag{2}$$

where $F^1 = (f_1^1, \ldots, f_{J+2}^1)$ and $F^2 = (f_1^2, \ldots, f_{J+2}^2)$ are the texture features of two windows, RL is the relative-L1 distance calculated by

$$RL(F^1, F^2) = \sum_{j=1}^{\tau} \frac{|f_j^1 - f_j^2|}{1 + |f_j^1| + |f_j^2|}, \tag{3}$$

where a small value τ is to avoid the zero value divisor when $RL(F^1, F^2) = 0$.

At last, we extract the texture eigenvectors based on spectral clustering. Given a similarity matrix S, a stochastic transition matrix P is obtained by

$$P = D^{-1}S, \tag{4}$$

where D is a diagonal matrix, and $D_{i,i} = \sum_j S_{i,j}$, $P_{i,j}$ is interpreted as a transition probability in a Markov random walk. Based on the eigenvector selection method [6], we choose the important eigenvectors of P to represent the texture features of the pixels of each window.

2.2   Color and Position Feature Space Construction

We further adopt the CIELUV representation for color feature space construction. The CIE color representation is demonstrated as the best space for image segmentation [7]. In the CIELUV, L^* is the initialism of lightness, and u^*, v^* are color measurements of images. For typical images, the ranges of u^* and v^* are ± 100, and $0 \leq L^* \leq 100$. In consideration of the robustness to noises, the color feature of pixel (i, j) is the mean value of $\{L^*, u^*, v^*\}$ among its surrounded 3×3 region unit.

The position features of (i, j) is simply its corresponding coordinate values, that is, its position features are i and j.

2.3   BYY-t for Image Segmentation

BYY-t is an unsupervised learning algorithm with the favorite feature of automated model selection during parameter learning for t-mixtures. Indeed, it is an application of Bayesian Ying-Yang (BYY) harmony learning system on the t-mixture learning problem.

Consider an observation $x \in \mathcal{X} \subset \mathcal{R}^n$ and its inner representation $y \in \mathcal{Y} \subset \mathcal{R}^m$, their relationship can be described via two types of Bayesian decomposition of the joint density: $p(x, y) = p(x)p(y \mid x)$ and $q(x, y) = q(y)q(x \mid y)$, which are called Yang machine and Ying machine in the BYY learning system [12, 13]. In BYY-t algorithm, x is assumed to follow a t-mixture distribution, and y is the discrete variable with $y \in \{1, 2, \ldots, k\}$. BYY-t estimates the model by maximizing the harmony function:

$$J(\Theta_k) = \frac{1}{N} \sum_{t=1}^{N} \sum_{j=1}^{k} \frac{\alpha_j q(x_t \mid \theta_j)}{\sum_{i=1}^{k} \alpha_i q(x_t \mid \theta_i)} \ln[\alpha_j q(x_t \mid \theta_j)] \tag{5}$$

where $q(x_t \mid \theta_j)$ is a d-dimensional t distribution with parameters $\theta_j = \{\mu_j, \Sigma_j, \nu_j\}$.

Assuming that the image features follow the t-mixture distribution, we perform BYY-t algorithm to obtain the t-mixture model. After that, we can obtain an eigenvector as well as the posterior probability of each of its components as to each pixel. The index of the component with the maximum probability value of its eigenvector identifies the class of its corresponding pixel.

2.4   Image Segmentation Procedures

Given a color image I of size $M \times N$, the color image segmentation algorithm is performed as follows.

– Construct feature space.

• Subdivide the image I into smaller windows. Extract texture features via contourlet subband clustering, and obtain the texture feature values of the pixels based on eigenvectors of the texture similarity matrix using spectral clustering.

• Determine the position and color features. The color feature of each pixel is based on its surrounded 3×3 region unit using the rules in Section 2.2.

• Normalize these three types of features to [0, 8].

– Perform BYY-t algorithm to generate the t-mixture model.

– Class determination and image segmentation.

• Obtain the eigenvector of each image pixel, and calculate the posterior probability of each of its components. The index of the component with the maximum probability value of its eigenvector identifies the class of its corresponding pixel.

• Embed the pixels of each class with a unique color.

3   Experimental Results

3.1   Average Segmentation Performance Evaluation

Firstly, we evaluate the average performance of our algorithm on a set of 100 color images in the BSDB test dataset [8]. These images cover a variety of complex scenarios, and are accompanied with their corresponding manually produced ground-truth segmentation datasets. The performances based on these images exhibit the average segmentation levels and stabilities.

The probability rand (PR) index is adopted to evaluate the segmentation performance [9]. PR index evaluates the difference of segmented images and a set of ground-truth segmentation results. The range of PR index is [0, 1], and larger value indicates a better segmentation.

We perform our algorithm on these 100 test images and obtain the mean and standard deviation of PR index (Table 1). Moreover, our algorithm is compared with two state-of-art algorithms: the rival penalized competitive learning (RPCL) and competitive agglomeration clustering (CAC) algorithm. We also include the comparisons with our algorithm performed only using the DCT normalization texture feature extraction method [10]. Experimental results show that our algorithm is better than PRCL and CAC algorithm, and the adoption of these new texture features improves the segmentation performance.

Table 1. The mean and standard deviation of PR index based on 100 image segmentation

Segmentation Algorithm	Mean of PR index	Standard Deviation of PR
RPCL	0.727	0.132
CAC	0.732	0.154
DCT	0.737	0.108
Contourlet	0.741	0.112

Contourlet: the algorithm in this paper.

DCT: our algorithm but based on DCT normalization texture feature exaction method.

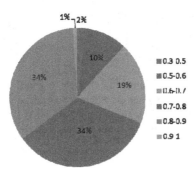

Fig. 1. The PR index of the 100 test images. The segmentation PR indexes of 69% of images are above 0.7, which indicate good segmentation results. Only 2% of them are less than 0.5.

Furthermore, we present the number of images of different performance levels, shown in Figure 1. About seventy percentage of images obtain relatively good segmentation results, with PR index larger than 0.7. Two percent of images have very bad segmentation results, with PR index less than 0.5. This case is acceptable considering the image content complexity.

3.2   Experiments on Color Images with Multiple Textures

Next, in order to demonstrate the advantage of our algorithm in the texture image segmentation, we focus on the images with multiple textures adopted from Vision Texture Database. Our algorithm is compared with the Gaussian-EM algorithm [3], k-means algorithm and the DCT based segmentation method. The initial number of clusters is set as 3. Two examples are shown in Figure 2. For convenience, we embed the pixels corresponding to different clusters with different colors.

As shown in Figure 2 (a), the walls in the corridor have consistent texture patterns but varied colors and brightness degrees. Gaussian-EM and k-means algorithms are too sensitive to image color and brightness, and assign the walls to different clusters. In comparison, our algorithm captures the texture features of the image, and correctly segments the components of images. Furthermore, our algorithm automatically determines the number of components in this image as 2. Figure 2 (b) shows the scenery of a city around the sea. Since the color of sea is more similar with that of building than that of the sky, the algorithms are more potential to group the sea and buildings together. In contract, our algorithm is more sensitive to texture variation, and correctly recognizes the city region from the image. However, this segmentation method is not accurate as to the pixels around the boundary due to the window-based texture feature extraction strategy. Further work will explore this point.

Fig. 2. Multi-texture image segmentation.(a) Original corridor image.(b) Original city-sea image. (*-1) Image segmentation by our algorithm; (*-2) image segmentation based on DCT normalization method; (*-3) image segmentation based on Gaussian- EM algorithm; (*-4) image segmentation based on k-means algorithm. * refers to (a) or (b).

4   Conclusions

In this paper, we proposed a new image segmentation algorithm by incorporating a new texture feature extraction method and the BYY-t unsupervised algorithm together. The framework has addressed the problems of proper feature space construction and proper cluster number selection in image segmentation. The experiments show that the proposed algorithm is more insensitive to texture variations, and have a better segmentation performance on general color images even with varied brightness degrees. Since texture is often more consistent than color and brightness in image objects, our algorithm excels in complex and varied content image segmentation. The improved performances based on BSDB image segmentation further demonstrate this point.

Since the texture feature extraction is based on a window-based strategy, our framework cannot obtain accurate boundary in the image segmentation. However, by correctly selecting the objects from the backgrounds, our algorithm is sufficient in image tracking and cataloging. Future work will focus on clear boundary detection.

Acknowledgments. This work was supported by the Natural Science Foundation of China for Grant 61171138.

References

1. Ilea, D.E., Whelan, P.F.: Image Segmentation Based on the Integration of Colour–Texture Descriptors—a Review. Pattern Recognition 44, 2479–2501 (2011)
2. Yusoff, I.A., Mat Isa, N.A., Hasikin, K.: Automated Two-Dimensional K-Means Clustering Algorithm for Unsupervised Image Segmentation. Computers & Electrical Engineering 39, 907–917 (2013)
3. Fu, Z., Wang, L.: Multimedia and Signal Processing, pp. 61–66. Springer (2012)
4. Liu, C., Ren, Z., Ma, J.: Byy Harmony Learning of T Mixtures and Its Application to Unsupervised Image Segmentation. In: IPCV 2013, vol. II, pp. 721–727 (2013)
5. Dong, Y., Ma, J.: Feature Extraction through Contourlet Subband Clustering for Texture Classification. Neurocomputing 116, 157–164 (2013)
6. Xiang, T., Gong, S.: Spectral Clustering with Eigenvector Selection. Pattern Recognition 41, 1012–1029 (2008)
7. Ooi, W., Lim, C.P.: Fusion of Colour and Texture Features in Image Segmentation: An Empirical Study. The Imaging Science Journal 57, 8–18 (2009)
8. Martin, D., Fowlkes, C., Tal, D., Malik, J.: Proceedings of the Eighth IEEE International Conference on Computer Vision, ICCV 2001, vol. 2, pp. 416–423. IEEE (2001)
9. Unnikrishnan, R., Pantofaru, C., Hebert, M.: Toward Objective Evaluation of Image Segmentation Algorithms. IEEE Transactions on Pattern Analysis and Machine Intelligence 29, 929–944 (2007)
10. Chen, W., Er, M.J., Wu, S.: Illumination Compensation and Normalization for Robust Face Recognition Using Discrete Cosine Transform in Logarithm Domain. IEEE Transactions on Systems, Man, and Cybernetics, Part B: Cybernetics 36, 458–466 (2006)

Using Spectral Feature for Face Recognition of One-Sample-Per-Person Problem[*]

Xing-Zhu Wang[1], Zhan-Li Sun[1,**] and Li Shang[2]

[1] School of Electrical Engineering and Automation,
Anhui University, Hefei, China
zhlsun2006@126.com
[2] Department of Communication Technology, Electronic Information Engineering
College, Suzhou Vocational University, Suzhou, China

Abstract. In this paper, we propose a more accurate local spectral feature based face recognition approach for the one-sample-per-person problem. In the proposed algorithm, multi-resolution local spectral features are first extracted to represent the face images to enlarge the training set. A weaker classifier is then constructed based on the spectral features of each local region. A strategy of classifier committee learning is proposed further to combine the results obtained from different local spectral features. Experimental results on the standard databases demonstrate the feasibility and effectiveness of the proposed method.

Keywords: Spectral Feature, Face Recognition, Classifier Committee Learning.

1 Introduction

In recent years, face recognition for the one-sample-per-person problem has received increasing attentions owing to its wide range of potential applications, e.g., law enforcement, surveillance identification, forensic identification and access control, etc [1, 2, 3]. So far, many approaches have been developed to address the one-sample-per-person face recognition problem. Generally speaking, the common strategy to deal with the problem is to enlarge the training set by extracting the various discriminative features [4], generating the virtual samples [5], or constructing the discriminant model by means of the generic set [6]. A typical strategy of the discriminative feature extraction method is to employ the features from the local region [7]. A prominent advantage of using local representations is its fair robustness to variations in lighting, expression and occlusion.

---

[*] The work was supported by two grants from National Natural Science Foundation of China (Nos. 61370109, 61373098), a grant from Natural Science Foundation of Anhui Province (No. 1308085MF85), and 2013 Zhan-Li Sun's Technology Foundation for Selected Overseas Chinese Scholars from department of human resources and social security of Anhui Province (Project name: Research on structure from motion and its application on 3D face reconstruction).
[**] Corresponding author.

D.-S. Huang et al. (Eds.): ICIC 2014, LNCS 8588, pp. 502–506, 2014.
© Springer International Publishing Switzerland 2014

Among these methods, the information in the frequency domain is frequently utilized to strengthen the recognition performance [8]. A combination of the frequency invariant features and the moment invariant features [9], and a fusion of the directionality of edges and the intensity facial features [10], are proposed for face recognition with a single training sample. Recently, we presented a very effective recognition method for the one-sample-per-person problem, named MR_2DLDA [11]. In the proposed method, multi-resolution spectral feature images are constructed to represent the face images, which greatly enlarge the training set. 2DLDA (two-dimensional linear discriminant analysis) is then applied on the spectral representations.

Inspired by the robustness of the local feature, and the superiority of the spectral feature representation, in this paper, we propose a more accurate local spectral feature based face recognition approach for the one-sample-per-person problem. In the proposed method, multi-resolution spectral features are first extracted and used as the representations of training face images by means of a method similar to [12]. Further, the spectral feature images are divided into patches with the same sizes. Then, the patches of the same position are collected, and used as the training samples of one weak classifier. In order to determine the classes of the testing images, a strategy of classifier committee learning (CCL) is designed further to combine the results obtained from different local spectral feature images.

The remainder of the paper is organized as follows. In Section 2, we present our proposed algorithm. Experimental results and related discussions are given in Section 3, and concluding remarks are presented in Section 4.

2   Methodology

2.1   Local Spectral Feature Representation

Assume that there are L training images I_i $(i = 1, ..., L)$ with size $m \times n$, and that each belongs to one subject. We first extract the spectral feature images of each training image. The image is first pre-filtered to reduce the effect of illumination by using a local normalization method of intensity variance. Next, a set of Gabor filters with n_s scales and n_o orientations is applied on the Fourier transform of the pre-filtered image [12].Then, the amplitude of the resulting image is computed as the spectral feature image. As a result, for the given N_f (i.e. $n_s \times n_0$) filters, N_f spectral feature images can be obtained for each training sample. In the same way, N_f spectral feature images can also be extracted for each test sample.

Subsequently, the spectral images of the training images and the testing images are divided into blocks with the same sizes. Take the kth testing image for example, we then compute the Euclidean distance $D(T_{ij}^k, N_{ij}^l), (i = 1, 2, ..., m, j = 1, 2, ..., n, l = 1, 2..., L)$ of the pixel values between the blocks T_{ij}^k and N_{ij}^l .

2.2   Combining the Weaker Classifiers

After computing the distances $D(T_{ij}^k, N_{ij}^l)$, we arrange the distances $D(T_{ij}^k, N_{ij}^l)$, $(l = 1,...,L)$ of each block as one row of the matrix D^k. If we consider to construct a nearest-neighbor classifier based on each block, the label of the testing image I_k derived from the block T_{ij} can be assigned by

$$C_{ij}^k = \arg \min_l (D(T_{ij}^k, N_{ij}^l)), 1 \le l \le L. \tag{1}$$

Further, we can get the label vector C^k of the testing image I_k, which is obtained by collecting the labels predicted from all the filters.

Finally, a classifier-combination strategy is adopted to determine the class label of the test image. As the outputs of the weaker classifiers are the class labels of the test images, the majority-vote rule is the most suitable strategy to combine these outputs. To count the votes of the qth class, a binary-valued vector b_q^k is defined according to C^k,

$$b_q^k = (b_1^k, b_2^k, ..., b_R^k)^T. \tag{2}$$

where $R = m \times n \times N_f$, the element b_i^k can be given by

$$b_i^k = \begin{cases} 1, C_i^k = q \\ 0, C_i^k \ne q \end{cases}, i = 1,...,R. \tag{3}$$

Further, we sum these vectors to obtain the number of votes for each class as follows:

$$v_l^k = \sum_{i=1}^R b_i^k, l = 1,...,L. \tag{4}$$

Finally, the label of I_k can be assigned by

$$c^k = \arg \min_l (v_l^k), 1 \le l \le L. \tag{5}$$

3   Experimental Results

The performance of our proposed method is evaluated on 2 standard face image databases: Extended Yale Face database B and PIE database, which are relatively large and widely used databases for face recognition. The Extended Yale Face database B has 38 individuals and around 64 near frontal images under different illuminations per individual. The PIE database contains images of 68 individuals [13].

To verify the performance of our proposed method (denoted as LSF-PC), we compare it to three other face recognition methods, i.e. the 2DLDA method [14], the spectral feature method (denoted as SF) [12], and the MR_2DLDA method [11]. For the SF method, the global spectral features are directly used to determine the class labels of the testing images.

Table 1. The mean and the standard deviation ($\mu \pm \sigma$) of the recognition rates (%) for four methods when different face images are used as the training samples

	2DLDA	MR_2DLDA	SF	LSF-PC
YaleB 32x32	18.64±8.83	67.93±11.45	65.29±12.31	73.88±11.16
PIE 32x32	10.75±3.73	43.40±7.67	40.94±7.84	47.26±8.77

In the experiments, to investigate the robustness of these algorithms, one face image is randomly selected from every class and used as the training sample. The trials are performed for ten times. Table 1 depicts the mean and the standard deviation ($\mu \pm \sigma$) of the recognition rates (%) for four methods when different face images are used as the training samples. Note that the experimental results of 2DLDA and MR_2DLDA are given by [11]. It can be seen from Table 1 that the performances of MR_2DLDA and LSF-PC are obviously better than other two methods. Further, we can see from Table 1 that the mean recognition rates of LSF-PC are higher than those of MR_2DLDA. Therefore, the proposed method is effective and feasible for the one-sample-per-person problem.

4   Conclusions

In this paper, we propose a more accurate local spectral feature based face recognition algorithm for the one-sample-image-per-person problem. The experimental results have demonstrated that our proposed method is more accurate than some recently reported methods.

References

1. Wang, Z.Q., Sun, X.: Multiple Kernel Local Fisher Discriminant Analysis for Face Recognition. Signal Processing 93(6), 1496–1509 (2013)
2. Tan, X.Y., Chen, S.C., Zhou, Z.H., Zhang, F.Y.: Face Recognition from A Single Image per Person. A Survey. Pattern Recognition A Survey 39(9), 1725–1745 (2006)
3. Levine, M.D., Yu, Y.F.: State-of-the-Art of 3D Facial Reconstruction Methods for Face Recognition based on A Single 2D Training Image per Person. Pattern Recognition Letters 30(10), 908–913 (2009)
4. Koc, N., Barkana, A.: A New Solution to One Sample Problem in Face Recognition Using FLDA. Applied Mathematics and Computation 217(24), 10368–10376 (2011)

5. Li, Q., Wang, H.J., You, J., Li, Z.M., Li, J.X.: Enlarge the Training Set based on inter-Class Relationship for Face Recognition from One Image per Person. PLOS ONE (2013), 10.1371/journal.pone.0068539

6. Kan, M.N., Shan, S.G., Sun, Y., Xu, D., Chen, X.L.: Adaptive Discriminant Learning for Face Recognition. Pattern Recognition 46(9), 2497–2509 (2013)

7. Kanan, H.R., Faez, K., Gao, Y.S.: Face Recognition Using Adaptively Weighted Patch PZM Array from A Single Exemplar Image per Person. Pattern Recognition 41(12), 3799–3812 (2008)

8. Sharma, P., Arya, K.V., Yadav, R.N.: Efficient Face Recognition Using Wavelet-based Generalized Neural Network. Signal Processing 96(3), 1557–1565 (2013)

9. Chen, Y.M., Chiang, J.H.: Face Recognition Using Combined Multiple Feature Extraction based on Fourier-Mellin Approach for Single Example Image per Person. Pattern Recognition Letters 31(13), 1833–1841 (2010)

10. Chen, Y.M., Chiang, J.H.: Fusing Multiple Features for Fourier Mellin-based Face recognition with Single Example Image per Person. Neurocomputing 73(16-18), 3089–3096 (2010)

11. Sun, Z.L., Lam, K.M., Dong, Z.Y., Wang, H., Gao, Q.W., Zheng, C.H.: Face Recognition with Multi-Resolution Spectral Feature Images. PLOS ONE 8(2), e55700 (2013), doi:10.1371/journal.pone.0055700

12. Oliva, A., Torralba, A.: Modeling the Shape of the Scene: A Holistic Representation of the Spatial Envelope. International Journal of Computer Vision 42(3), 145–175 (2001)

13. Cai, D., He, X.F., Han, J.W., Zhang, H.J.: Orthogonal Laplacianfaces for Face Recognition. IEEE Transactions on Image Processing 15(21), 3608–3614 (2006)

14. Gao, Q.X., Zhang, L., Zhang, D.: Face Recognition Using FLDA with Single Training Image per Person. Applied Mathematics and Computation 205, 726–734 (2008)

Analyzing Feasibility for Deploying Very Fast Decision Tree for DDoS Attack Detection in Cloud-Assisted WBAN

Rabia Latif[1], Haider Abbas[1,2], Saïd Assar[3], and Seemab Latif[1]

[1] National University of Sciences and Technology, Islamabad, Pakistan
{rabia.msis6,seemab}@mcs.edu.pk
[2] Centre of Excellence in Information Assurance (COEIA),
King Saud University, Saudi Arabia
hsiddiqui@ksu.edu.sa, haidera@kth.se
[3] Institut Mines-Télécom, Telecom Ecole de Management
Information System Department, France
said.assar@it-sudparis.eu

Abstract. In cloud-assisted wireless body area networks (WBAN), the data gathered by sensor nodes are delivered to a gateway node that collects and aggregates data and transfer it to cloud storage; making it vulnerable to numerous security attacks. Among these, Distributed Denial of Service (DDoS) attack could be considered as one of the major security threats against cloud-assisted WBAN security. To overcome the effects of DDoS attack in cloud-assisted WBAN environment various techniques have been explored during this research. Among these, data mining classification techniques have proven itself as a valuable tool to identify misbehaving nodes and thus for detecting DDoS attacks. Further classifying data mining techniques, Very Fast Decision Tree (VFDT) is considered as the most promising solution for real-time data mining of high speed and non- stationary data streams gathered from WBAN sensors and therefore is selected, studied and explored for efficiently analyzing and detecting DDoS attack in cloud-assisted WBAN environment.

Keywords: Cloud-assisted WBAN, Distributed denial of service (DDoS) attack, Data mining (DM), Decision trees, Very fast decision trees (VFDT).

1 Introduction

The research into cloud-assisted WBAN environment is still in its initial stage. Numerous security issues and challenges need to be addressed in order to provide better health care services to patients.

In cloud-assisted WBAN, the aim of cloud computing is to provide a powerful, flexible and cost effective infrastructure in order to provide healthcare services to people anywhere and at any time with improved quality of treatment. Therefore, a high- level of security mechanism is required to ensure the service availability of cloud provider at critical times especially in case of DDoS attack [1].

D.-S. Huang et al. (Eds.): ICIC 2014, LNCS 8588, pp. 507–519, 2014.
© Springer International Publishing Switzerland 2014

In WBAN, both security and system performance is of equal importance, therefore integrating same high- level security mechanism in a less power and resource constrained WBAN sensors increases the computational and communication cost [2][3].

Cloud-assisted WBANs are vulnerable to various types of attacks. Among these DDoS is an attack that directly affects the availability of e-healthcare service provider in sensor cloud and often a source of cloud service disruptions [4].

Distributed denial of service (DDoS) attack poses a serious threat to network security. A number of methodologies and tools have been proposed to detect DDoS attacks in order to reduce the damage they cause. Still, most of the existing methods fail to provide efficient detection with low false alarms and real- time transfer of packets simultaneously [5]. The DDoS attack can be classified into bandwidth attacks, protocols attacks and software vulnerability attacks. Taking into account the cloud-assisted WBAN scenario our focus is on bandwidth attacks which are intended to overflow and consume network resources available to victim e.g., TCP SYN Flood, ICMP Flood and UDP Flood [5].

2   Existing Techniques for DDoS Attack Detection

Currently, a number of solutions have been available for handling DDoS attack in traditional wired and wireless networks, but they are not appropriate for cloud-assisted WBAN due to the resource constrained nature of WBAN. Similarly, few traditional WSN security schemes can be embedded in a WBAN because WBAN devices are limited in terms of power efficiency, computation and communication.

One of the most efficient solution for detecting DDoS attack is to use Intrusion detection systems (IDS) to ensure the continuity of cloud services for e-healthcare. However, the major drawback of using IDS sensor is they yield excessive amount of alerts and produce high false positive and negative rates, which limits their use in cloud-assisted WBAN [6].

To detect DDoS attack, Wu et al. [7] proposed a technique based on decision tree and grey relational analysis. For tree classification, fifteen different network packet attributes are selected which monitor the in/ out packet/ byte rate and compile the TCP, SYN and ACK flag rates to show the traffic flow pattern. C4.5 decision tree technique is applied by picking selected attributes as tests to detect abnormal traffic flow. Finally, the traffic pattern matching technique is used to identify an attack source and traceback the origin of an attack.

Zhang et al. [8] proposed a real- time DDoS attack detection and prevention system based on per-IP traffic behavioural analysis. According to authors, for each IP user, the system will create records for every single IP user's sending and receiving traffic and judge whether its behavior meets the normal principles. A specific packet identification technique is utilized to reach real-time flooding attack detection goal. A non-parameter CUSUM (Cumulative Sum) algorithm is applied to detect the abnormal behavior of each IP. Based on a decision algorithm, each IP user will be classified as attacker, victim or legitimate user. After the attack identification, the system will block its traffic and forward the normal user packets. As the approach needs less computation and memory, the system could be deployed for on-line DDoS detection. The system can detect attacks accurately at the earlier attack stage and quickly filter the attack traffics and forward the normal traffics. The system has high DDoS detection accuracy and short detection time. One of the major drawback of the system is that it does not immediately take defensive measures to stop the attack, but keep observing the suspected IP record.

3   Data Mining in Cloud-assisted WBAN Applications

Data mining also called knowledge discovery is a recent topic in the field of computer science but employ a number of existing computational techniques from statistics, information retrieval, machine learning and pattern recognition [9]. According to Patcha et al. [10], *"Data mining is concerned with learning patterns, association, changes anomalies and statistically significant structures and events from large quantities of data"*. Fig.1 depicts the process of data mining that transforms the raw data to valuable knowledge.

Fig. 1. Data mining process

Data miners are trained at using specialized automated software to discover regularities and irregularities in large and complex data sets. Data mining techniques are widely used by security specialists for the detection of fraudulent activities such as network attacks and recently it became an important component for the detection and prevention of DoS and DDoS attacks [11].

The aim of deploying WBAN for e-health applications is to make the real-time decision which seems to be very challenging due to the highly resource constrained computing, communication capacities and high speed of non-stationary data generated by WBAN sensors. This challenge is the source of motivation to select and explore the data mining techniques that is light-weight and deals with dicovering patterns from large continuous stream of WBAN sensors data. Few specific tasks that data mining might contribute to a DDoS attack dectection in Cloud-assisted WBAN are as follows:

i. Help to mine the sensor data for uncovering patterns in order to make intelligent decisions immediately after an attack occurs

ii. Detect anomalous activity that expose a real attack

iii. Identify large continous patterns in ongoing streams of sensor data i.e., different IP address, same activity etc.

iv. Identify bad sensors signatures

v. Detect previously unknown network anomalies

To fulfill these specific tasks, data miners utilizes a single or a combination of data mining techniques. These includes: statistical techniques/data summarization, visualization, clustering, association rule and classification techniques [12]. The effectiveness of each techniques depends upon the application scenrio on which it is applied. Table 1 shows the detail of these techniques alongwith their pros and cons [13],[14]. Finally, we analyze the consequence of choosing data mining techniques in cloud-assisted WBAN environemnt.

Table 1. Data mining techniques

	Description	Pros	Cons	Consequences in Cloud-assisted WBAN
Association rule Mining	These are mainly focussed on the discovery of patterns and dependencies in data sets. It is an expression of the form X=>Y, where X and Y are sets of items. Make use of two measures: support and confidence i.e., supp(X=>Y)= supp(XuY) and conf(X=>Y)= supp(XuY)/ supp(X) [13]	Formulated to look for sequential patterns [13].	Complex algorithm. Does not show reasonable patterns with dependent variables and cannot reduce the number of independent variables by removing. Cannot be useful if the information do not provide correct support and confidence of rules [14].	As WBAN sensors has less computational capacity, therefore building complex algorithms is not a good choice.
Clustering	Grouping together objects that are similar to each other but different to the object belonging to other cluster. These algorithms are used for the detection of underlying structures within the data [14].	Provides end users with abstract view of database operations. Very fast computation on databases [14].	The effectiveness of clustering techniques depends on the applications. Once a merge or a split is committed, it cannot be rollbacked or refined [14].	As WBAN data is high speed continuous data streams, which leads to missing data values within the input data.
Visualization	Visualization is a way to transform poor data into meaningful form by using a wide variety of data mining techniques in order to discover hidden patterns [14].	Useful for discovery of patterns in data sets. Transforms poor data into meaningful form [14].	Difficult to understand when buit hierarchically . As it is difficult for humans to understand numbers, so the summarization is required to put the data into graphical form [14].	Computational intensive for WBAN because sumarization is required to put the data into human understandable form.

Table 1. (*continued*)

Statistics	Statistics is a branch of mathematics dealing with collecting data and counting it [13].	Gives a high-level view of the database. Provides a useful and important informatiom about the database [13].	These techniques make certain assumptions about data [13].	As we are dealing with human health, making assumptions about data is not a good idea.
Classification	Classification Techniques predicts the category to which a particular record belongs [14].	Simplicity and interpretability of their rules. Better performance and understanding then other DM techniques [14].	Data prepartion procedures are not restricted and imposed by any requirements [14].	Appropriate for WBAN due to their simplicity & interpretability of their rules, derived easily from the organization of the tree in case of decision trees

From the perspective of network attack detection, classification algorithms can efficiently distinguish network data as malicious, benign, scanning or any other class based on the attributes chosen e.g., IP addresses, source and destination addresses, ports etc.

4   Decision Tree: A Classification Technique

At the core of classification technique is decision tree which is a predictive modeling technique formed by a learning algorithm that can be viewd as a simple tree like structure to design the underlying   patterns   of attributes. Other classification techniques include Bayesian network algorithm, neural networks, support vector machine can also be used to design non-linear relations, but their efficacy depends on the application in which they are used [15].

4.1   Significance of Decision Tree Techniques

For Cloud-assisted WBAN

As we are dealing with an environment in which a flexible stack of massive computing and storage is integrated with a resource constrained sensor network (WBANs) therefore, there is a need for a solution that can be used effectively and change the existing capabilities of WBAN. Decision trees have been considered as an excellent solution for WSN and WBAN because of the following reasons [15]:

i. Simplicity and explicability of their rules, which can be simply obtained from the structure of decision tree algorithms

ii. Decision tree gives better performance and understandability as compared to other classification techniques whose internal working are difficult to understand and implement in WBAN environment.

iii. Rules can be derived from decision tree routes which is a sequence of conditions and used in WBAN environment to determine an outcome based on the attributes of the WBAN sensor data.

For DDoS Detection

Decision tree proves to be a viable tool for detecting network attacks both in wired and wireless networks because they produce real results from the sensed data. The accuracy of the results depends upon the environment. The main advantage of using decision trees for network attack detection is its efficiency in both generalization and new attack detection [16].

4.2   An Overview of Decision Tree Techniques

Taking into account the scarcity of resources and computational power in WBANs, exploring an appropriate solution to secure cloud-assisted WBAN environment from DDoS attack has gained significant importance. Table 2 discusses the decision tree techniques present in literature along with their limitations and conclusion [15]. After briefly discussing each technique, the discussion will turn to a specific technique that would better suit the cloud-assisted WBAN environment and efficiently detects DDoS attack.

Table 2. Decision tree techniques

	Description	Limitations	Concluding Remarks
ID3	ID3 is a very simple DT algorithm that uses information gain as a splitting criteria [17].	Doesn't handles numeric attributes or missing values. Doesn't give accurate result in noisy training data sets [17].	ID3, C4.5 and CART are not appropriate for handling DDoS attack in cloud-assisted WBAN environment because: — Appropriate only for small datasets. — Required that all or a portion of the entire dataset remains permanently in memory. — Low classification accuracy when training data is large.
C4.5	An extension of ID3. Uses gain ratio as a splitting criteria. It handles numeric attributes and missing values [15], [17].	Appropriate for small Data sets. Slower to classify [15]	
CART	CART (Classification and Regression Trees) constructs binary Trees. It uses Gini Index for splitting procedure [18].	Not accurate when training data is large [18].	These issues limits their suitability for mining over large datasets.

Table 2. (*continued*)

SLIQ	SLIQ (Supervised Learning in Quest) algorithm is used for decision tree induction. It doesn't require loading entire dataset into memory. It creates a single decision tree from a large dataset [19].	SLIQ uses a class list data structure that resides in memory causing memory restrictions on the data [15], [19].	Both SLIQ and SPRINT are suitable for large datasets thus removing the memory restrictions on data. Also they can handle both continuous and categorial attributes. These techniques are expensive to implement in Cloud-assisted WBAN and also they are not appropriate to process high speed continuous data streams of sensor.
SPRINT	SPRINT (Scalable Parallelizable Induction of decision Tree algorithm) is an efficient version of SLIQ in terms of memory consumption. It is a fast, scalable decision tree classifier. SPRINT uses breadth-first greedy technique as a splitting criterion [20].	Expensive [20] in terms of computation.	
VFDT	VFDT (Very Fast Decision Tree) algorithm reduces training time for large incremental data sets by subsampling the incoming data stream. It does not require reading full dataset as part of the learning process [15].		For detecting DDoS attack in cloud-assisted WBAN, VFDT is appropriate because: — Of being leight weight. — It handles huge amount of streaming data obtained from WBAN — Fast in detecting DDOS attack because it learns pattern by simply reading an incoming stream of data and evaluated either its tree nodes should further expand or not.

5   Proposed Technique

After an extensive study of machine learning techniques, we came across the point that data mining techniques have been considered as one of the most promising solutions for resource constrained WBAN. Among the data mining techniques [12], data classification techniques is the most prevalent due to the simplicity and interpretability of their rules and thus considered as more appropriate for low- power sensor networks. Data classification is used to provide information about the changing environment covered by a dense sensor network [21].

At the core of classification techniques is decision tree built by learning algorithms. Among the Data classification techniques discussed in Table 2, VFDT technique have been assumed as an efficient solution to detect DDoS attack in cloud-assisted WBAN environment. Fig 2 shows the hierarchy of data mining techniques. The shaded boxes show the flow of our selected techniques towards detecting DDoS attack in cloud-assisted WBAN environment.

Fig. 2. Hierarchy of Data mining techniques

"VFDT is stream based data mining classification algorithm built incrementally by splitting nodes into two based on the incoming data streams. The tree is evaluated and expands as more data arrives" [22].

Following are the reasons for the selection of VFDT algorithm:

1. Light weight making it suitable for fast detection of DDoS attack.
2. Can progressively build decision tree from scratch which helps in detecting DDoS attack at any stage.
3. Each time a new segment of sensor data arrives, a test and train process is performed over it keeping the stored data up to date.
4. Doesn't require reading full dataset yet adjusts decision tree according to the newly incoming and gathered statistical attributes thus consuming less memory space.
5. Appropriate for huge amount of non-stationary and streaming data obtained from WBAN sensors.
6. Provides a transparent learning process.

These aforementioned features make VFDT a suitable candidate for implementing an autonomous decision maker for detecting DDoS attack in cloud-assisted WBAN.

5.1   Very Fast Decision Tree Process for Detecting DDoS Attack

Fig 3 shows the process of implementing decision tree in WBAN environment. The same process is applied on e-Health care service provider to detect the occurrence of DDoS attack in cloud-assisted WBAN environment. Implementing a decision tree process for detecting DDoS attack includes [23]:

1. Two pre- requisites for analysis are data collection and tool acquisition.
2. For DDoS attack detection, to differentiate between normal data and attack data requires both known normal dataset and known attack dataset to train the VFDT algorithm. The next important consideration is where to collect data required training the decision tree; storing the data at base station is an optimal solution.
3. The whole gathered data requires a pre- processing phase to extract the sensor attributes necessary for VFDT.
4. The processed data is then entered the VFDT for instant testing and training to calculate the accuracy of a learnt model based on how well it classifies the testing set.

A final classification output will be generated by VFDT for further decision making to detect the existence and non- existence of DDoS attack at Base Station.

Fig. 3. Decision tree process for detecting DDoS attack

6   Simulation Experiments and Result Analysis

6.1   Employing VFDT for Cloud-assisted WBAN

The wireless body area networks poses new challenges: First, a single WBAN consists of a number of sensor nodes, each of them having a limited storage capacity making it difficult to process and cache the training data of a whole body network. Second, the sensor nodes are powered by non-rechargeable batteries, making energy proficiency an important consideration. These limitations give rise to the straight forward solution in which a data mining process can be carried out only at base station that fulfills the memory and computational requirements. To address the above challenges, we present a decision tree based solution deploying VFDT for DDoS attack detection. Fig 4 shows the gateway sensor node employing VFDT for the detection of DDoS attack.   The same decision center can be deployed at cloud Side for DDoS attack detection at both e-Health service provider and health cloud storage.

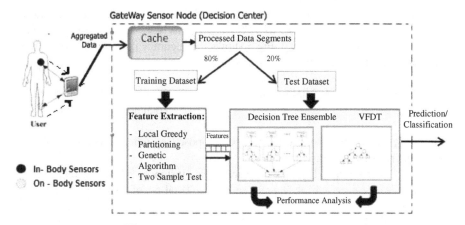

Fig. 4. The workflow of VFDT at base station

While analyzing the feasibility of VFDT in detecting DDoS attack, we came across the following benefits:

- The high performance of VFDT in handling large amount of data streams makes it a useful candidate in real-time detection of DDoS attack.
- Due to the ever increasing volume of data flow across network (specially under DDoS attack), VFDT have the ability to construct simple and easily interpretable model that helps security experts to inspect and assist in the analysis if malicious data flow across the network in less time.
- Generalization accuracy of VFDT in detection of new attacks is another useful property that makes them attractive to deploy in cloud-assisted WBAN environment.

6.2   Experimentation

The preliminary simulation experiment was carried out using Massive Online Analysis (MOA) stream mining software. The data used in the experiment is downloaded from the KDDCUP 1999 dataset provided by paralyzed veterans of America [24]. The data comprise information concerning human localities and activities measured by monitoring sensors attached to patients. Our experiment is divided into three main steps: first, the data is divided into two subsets such as training dataset and testing dataset. In the second step, the training dataset is further classified to handle different types of DDoS attacks (TCP SYN, ICMP,UDP) and normal data which are then given to the proposed system for identifying the appropriate attributes. Third, we train our VFDT classifier to differentiate between normal and attack traffic. The acquired result is then used to compute the overall accuracy of the proposed system.

6.3   Results and Analysis

The overall accuracy of the proposed system was evaluated based on the correct classified attack instances; similarly the error rate is the misclassified attack instances using 41 attributes of KDD dataset to VFDT algorithms (e.g., in case DDoS, there are 1000 instances in 40K of dataset). Fig 5 shows the accuracy rate of proposed system compared with C4.5 classifier using the following formula:

$$\text{Accuracy Rate} = \frac{\text{Total no. of correct classified instances} * 100}{\text{Total no. of instances}} = \mathbf{93.67\%}$$

Similarly, Fig 6 shows the error rate of proposed system compared with C4.5 classifier using the following formula.

$$\text{Error rate} = \frac{\text{Total no. of misclassified instances} * 100}{\text{Total no. of instance}} = \mathbf{2.95\%}$$

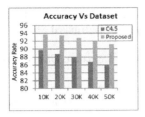

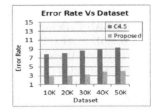

Fig. 5. Accuracy rate of the proposed system   **Fig. 6.** Error rate of the proposed system

In the future work, the performance matrix of proposed system for DDoS attack detection will be calculated using the data in a confusion matrix [25]. Table 3 shows the confusion matrix.

Table 3. Confusion matrix

Predicted / Actual	Attack	Normal
Attack	True Positive (TP)	False Negative (FN)
Normal	False Positive (FP)	True Negative (TN)

7 Conclusion and Future Work

Data mining techniques have proven themselves as a valuable tool in many research areas. In sensor WBAN, traditional data mining techniques works on relatively organized and stationary data stored historically in database and need to be updated and retrained periodically which consumes time and increase computational complexities. VFDT replaces the traditional data mining techniques employed in WBAN by embracing huge amount of incoming data streams. VFDT is light weight and does not require storage capacity making it suitable for resource constrained WBAN.

There is very limited prior work found that uses VFDT for analyzing and detecting DDoS attack in cloud-assisted WBAN environment. The future intensions of this research would be to investigate VFDT classifier for detecting DDoS attack by collecting a reasonable volume of data set from cloud-assisted WBAN sensors and then prove the efficiency of the proposed model by implementing a simulation prototype in a particular environment.

References

1. Perumal, B., Pallikonda, R.M., Ramalingam, H.M.: WSN Integrated Cloud for Automated Telemedicine (ATM) based e-Healthcare Applications. In: 2012 4th International Conference on Bioinformatics and Biomedical Technology, IPCBEE, vol. 29 (2012)
2. Bhargavi, B.D., Smita, J.: Security Requirements and Solution in Wireless Body Area Network (WBAN). In: VESIT, International Technological Conference-2014 (I-TechCON), January 03-04 (2014)

3. Irum, S., Ali, A., Khan, F.A., Abbas, H.: A Hybrid Security Mechanism for Intra-WBAN and Inter-WBAN Communications. International Journal of Distributed Sensor Networks 2013, Article ID 842608, 11 pages (2013)
4. Markey, J.: Using Decision Tree Analysis for Intrusion Detection. Global Information Assurance Certification (2011)
5. Katkamwar, N.S., Puranik, A.G., Deshpande, P.: Securing Cloud Servers against Flooding Based DDoS Attacks. International Journal of Application or Innovation in Engineering & Management (IJAIEM) 1(3) (November 2012)
6. Lonea, A.M., Popescu, D.E., Tianfield, H.: Detecting DDoS Attacks in Cloud Computing Environment. Int. J. Comput. Commun. 8(1), 70–78 (2013)
7. Wu, Y.C., Tseng, H.R., Yang, W., Jan, R.H.: DDoS detection and traceback with decision tree and grey relational analysis. International Journal of Ad-hoc and Ubiquitous Computing, IJAHUC 7(2) (2011)
8. Zhang, Y., Qiang, L., Guofeng, Z.: A Real-Time DDoS Attack Detection and Prevention System Based on per-IP Traffic Behavioral Analysis. IEEE (2010)
9. Manohar, T.B., Jyothi, E.V.N., Rajani, B.: Traceback of DDoS Attacks Based on Decision Trees Model Using Intrusion Detection System. International Journal of Computer Science and Management Research 1(4) (2012)
10. Patcha, A., Park, J.: An overview of anomaly detection techniques: Existing solutions and latest technological trends. The International Journal of Computer and Telecommunications Networking 51(12) (2007)
11. Jain, N., Sharma, S.: The Role of Decision Tree Techniques for Automating Intrusion Detection System. International Journal of Computational Engineering Research (ijceronline.com) 2(4) (2012)
12. Mahmood, A., Shi, K., Shaheen, K.: Data Mining Techniques for Wireless Sensor Networks: A Survey. International Journal of Distributed Sensor Networks 2013(2013)
13. Sarat, M.K., Christopher, G.H.: Summarization Techniques for Visualization of Large Multidimensional Datasets. Technical Report
14. Moawia, E.Y., El-mukashfi, M.: A New Approach for Evaluation of Data Mining Techniques. IJCSI International Journal of Computer Science Issues 7(5) (September 2010)
15. Venkatadri, M., Lokanatha, C.R.: A Comparative Study on Decision Tree Classification Algorithms in Data Mining. International Journal of Computer Applications in Engineering, Technology and Sciences 2(2) (2010)
16. Kanwal, G., Rshma, C.: Detection of DDOS Attacks using Data Mining. International Journal of Computing and Business Research (IJCBR) 2(1) (2011)
17. Anyanwu, M.N., Shiva, S.G.: Comparative Analysis of Serial DecisionTree Classification Algorithms. International Journal of Computer Science and Security (IJCSS) 3(3) (2009)
18. Breiman, L., Friedman, J.H., Olshen, R., Stone, C.J.: Classification and regression trees. Wadsworth & Brooks/Cole Advanced Books & Software, Monterey (1984)
19. Mehta, M., Agarwal, R., Rissanen, J.: SLIQ: A fast scalable classifier for data mining. In: Apers, P.M.G., Bouzeghoub, M., Gardarin, G. (eds.) EDBT 1996. LNCS, vol. 1057, pp. 18–32. Springer, Heidelberg (1996)
20. Shafer, J.C., Agrawal, R., Mehta, M.: SPRINT: A scalable parallel classifier for data mining. In: Proceedings of 22nd International Conference in Very Large databases, VLDB, pp. 544–555 (1996)

21. Hang, Y., Simon, F., Guangmin, S., Raymond, W.: A Very Fast Decision Tree Algorithm for Real-Time Data Mining of Imperfect Data Streams in a Distributed Wireless Sensor Networks. International Journal of Distributed Sensor Networks (2012)
22. Subrahmanyam, M.V.V.S., Venkateswara, R.V.: VFDT Algorithm for Decision Tree Generation. International Journal for Development of Computer Science and Technology (IJDCST) 1(7) (November 2013)
23. Markey, J.: Using Decision Tree Analysis for Intrusion Detection. Global Information Assurance Certification (2011)
24. The Third International Knowledge Discovery and Data Mining Tools Competition sponsored by the American Association for Artificial Intelligence (AAAI), http://kdd.ics.uci.edu/databases/kddcup99/kddcup99.html
25. Radhika, G., Anjali, S., Ramesh, C.J.: Parallel Misuse and Anomaly Detection Model. International Journal of Network Security 14(4), 211–222 (2012)

A Data Obfuscation Based on State Transition Graph of Mealy Automata

Xin Xie, Fenlin Liu, and Bin Lu

Zhengzhou Information Science and Technology Institute,
and the State Key Laboratory of Mathematical Engineering and Advanced Computing,
Zhengzhou, Henan 450002, China
{xiexin0011,stoneclever}@gmail.com, liufenlin@vip.sina.com

Abstract. Mealy automata model can be used to obfuscate constants and strings in programs, as to the obfuscation, the structure of state transition graph of mealy machine is simple and easy to test. To solve this problem, a data obfuscation based on state transition graph of mealy machine is proposed. With iteration of state transition graph of mealy machine, redundant states, transition functions and output functions based on probability are added into the graph, and then constants and strings in programs are obfuscated by the mealy machine. Analysis and experiment validated that redundant states and transition functions can increase the complexity of the structure of state transition graph. Output functions based on probability can increase the randomization of output obfuscated data. Obfuscation can be effective to improve the performance of mealy machine to resist static and dynamic reverse analysis.

Keywords: data obfuscation, reverse analysis, mealy automata, state transition graph, iteration.

1 Introduction

With the rapid development of network, internet has been great effective on people's daily life. Various software are widely employed in different fields such as mathematical computation, entertainment, and business. Consequently, malicious reverse attackers are analyzing the key algorithms and acquiring information of these software to obtain illegal profits. Common reverse analysis methods used by attackers are static disassembling, static decompiling, dynamic debugging and dynamic tracing. The illegal distribution and dissemination of cracked software are serious violations of software intellectual property rights. Therefore, more and more attentions are being paid to the effective protection of the software intellectual property in the software security fields.

In order to solve the security problem of malicious reverse attacking and illegal distribution, code obfuscation, tamper-proofing, software watermarking are developed to protect the software intellectual property [1]. Code obfuscation is to confuse the code of programs in order to make programs harder to analyze and understand by reverse engineering, while keeping the original semantics [2]. By different abstract representation of software, code obfuscation can be classified into: source code

D.-S. Huang et al. (Eds.): ICIC 2014, LNCS 8588, pp. 520–531, 2014.

obfuscation, medial language obfuscation, binary obfuscation. By different abstract properties of software, code obfuscation can also be classified into: layout obfuscation, structure obfuscation, control-flow obfuscation and data obfuscation. Layout obfuscation is to obfuscate key layout information of programs, such as identifier renaming, whitespace and comment deleting, variable and functional name modification [3,4]. Structure obfuscation is to transform design model and abstract structure 5, such as class splitting, module combination and function type hiding. Control obfuscation is to make the control logic more complicate [6,7], such as opaque predicate inserting, control-flow flattening, trap hiding. Data obfuscation transforms data and data structures of a program to obscure the original data form, it can make programs leak less sensitive semantics information. To implement the obfuscation of Boolean variables and operations, true and false of Boolean variables are multiplied different number or split into two parts [8,9]. Based on pointer, compound function [10] and homomorphic function [11], array obfuscation is implemented by the method of splitting, combination, folding and flattening. With extending the GCC compiler, data structure obfuscation is automatically implemented by field reordering [12]. Based on state transition graph, data obfuscation of constants and strings is proposed [13].

The integer numbers, strings and initial arrays in programs are always containing some valuable information, such as secret key, register code and copy right information. Aiming at obfuscation for constants and strings, they are transformed into codes that can generate obfuscated data in the process of program runtime. Mealy machine codes are used to complete obfuscation of constants and strings with strings and state transition graph as input, and constants and strings are output. To use mealy machine to obfuscate strings and constant directly, it can be effectively resist from the search attack. However, the structure of state transition graph is simple, it can be easily understood by static and dynamic reverse analysis. Obfuscation should have high potency against static and dynamic analysis. To solve this problem, a data obfuscation based on state transition graph of mealy machine is proposed.

2   Obfuscation Based on Mealy Machine Iteration

2.1   Formal Description

In order to effectively resist from search attack, sensitive and core data is transformed into the code of mealy machine [14]. During the program execution, constants or strings are output by using automata, the principle is shown as Fig. 1. To use mealy machine to transform constant data, after transformation, the difficulty of program analysis is transformed from sensitive and core data to the code and memory data of mealy machine. To transform sensitive and core data directly, the state transition graph corresponded to the mealy machine is simple; it can easy be analyzed and understood by static and dynamic reverse analysis. With the iteration of state transition graph of mealy machine, redundant states, transition functions, output functions based on probability are added into it, in this way, it can increase the potency of obfuscation against reverse analysis.

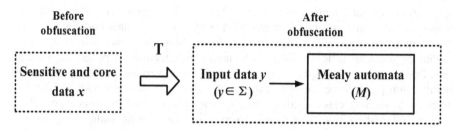

Fig. 1. Data obfuscation based on mealy machine

Let mealy machine be $M = (Q, \Sigma, \Delta, \delta, \lambda, q_0)$. In the expression, Q is the set of non-empty state, Σ denotes finite alphabet, Δ indicates output alphabet, δ is state transition function, λ denotes output function, q_0 indicates initial state. The formal description for obfuscation of mealy machine is shown as equal (1):

$$O(M_{mealy}) = M'_{mealy} \tag{1}$$

, where M_{mealy} and M'_{mealy} are before and after obfuscation of mealy machine respectively.

Let sensitive data be $S = \{a_1 a_2 ... a_n\}$, where a_i is a character. The state transition of mealy machine M is $G = (V, E)$, where V is the set of state in Q, E includes two parts of state transition function δ and output function λ. Based on the idea of randomization iteration, set of finite states Q, transition function δ and output function λ in the state transition graph of mealy machine are obfuscated, the formal description of obfuscation for state transition are equal (2) and (3), while T is transformation. G_{mealy}, G_n are original and n times iteration state transition of mealy machine, q_{new} denotes new state node, Δ^+ denotes set of new output characters, V^*, E^* and Δ^* and represent set of states after obfuscation, set of functions and alphabet respectively, $\Pr(\{V^* \times \Sigma \to \Delta^*\}, \phi)$ denotes output function based on probability, namely character Δ^* is output by probability $\Pr$.

$$O(G_{mealy}) = G_n = \begin{cases} T(G_{n-1}), n \geq 1 \\ G_{mealy}, n = 1 \end{cases} \tag{2}$$

$$T = \begin{cases} V^* = Q \cup \{q_{new}\} \\ E^* = \begin{cases} \delta \cup \{Q \times \Sigma \to q_{new}\} \cup \{q_{new} \times \Sigma \to Q\} \\ \lambda \cup Pr(\{V^* \times \Sigma \to \Delta^*\}, \phi) \end{cases} \\ \Delta^* = \Delta \cup \Delta^+ \end{cases} \tag{3}$$

Due to the equal (2) and (3) for the formal description of state transition diagram obfuscation, let state number be 5, iteration number be 3, the probability value of output function be 1/2, omit the input and output characters of new transition function, the obfuscation procedure of state transition graph iteration of mealy machine is shown as Fig. 2, if "0000" is input into the mealy machine, "a1a2a3a4" will be output, while input is other "01" data, "a1a2a3a4" will not be output.

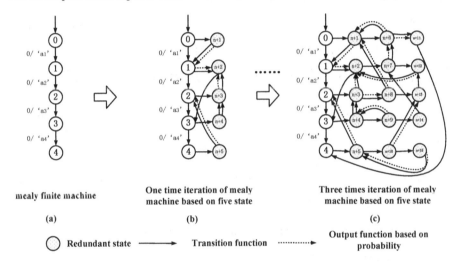

mealy finite machine

One time iteration of mealy machine based on five state

Three times iteration of mealy machine based on five state

(a)   (b)   (c)

◯ Redundant state   ⟶ Transition function   ┈┈┈▶ Output function based on probability

Fig. 2. Iteration obfuscation based on state transition graph of mealy machine

2.2   Obfuscation Algorithm

With redundant states and transition functions of mealy machine increased, the structure of state transition graph will be complicated. It can increase the difficulty of static reverse analysis. By adding the output function based on probability, the output characters will be uncertainty, it can increase the difficulty of dynamic test and analysis. Based on the idea of random iteration, complexity of state transition diagram will be increased to defend reverse analysis. According to formalization of section 2.1, the steps of mealy machine obfuscation are as follows:

1. To obtain sensitive data $S = \{a_1 a_2 ... a_n\}$ that need to obfuscate in program, where a_i is a character.

2. Initialize a mealy machine M due to S, where let set of finite state be $Q = \{0,1,2,...,n\}$, input alphabet $\Sigma = \{0,1\}$ initial state $q_0 = \{0\}$, output function $\lambda = \{\{(0,0) \rightarrow a_1\}, \{(1,0) \rightarrow a_2\},...,\{(n-1,0) \rightarrow a_n\}\}$, output alphabet $\Delta = \{a_i \mid i \in \{1,2,...,n\}\}$, state transition function $\delta = \{\{(0,0) \rightarrow 1\}, \{(1,0) \rightarrow 2\},...,\{(n-1,0) \rightarrow n\}\}$

3. Iteration obfuscation of state transition graph of mealy machine

   (a) Extend the output alphabet $\Delta^* = \Delta \cup \Delta^+$, where Δ^+ is the new set of characters.

   (b) Let number of obfuscation iteration be m, control variables be $i = 0$, $j = 1$

   (c) Add new redundant state $Q = Q \cup q_j$, where $q_j = \{i + j * (n+1)\}$

   (d) Add new transfer function $\delta_1 = \delta \cup e_{\delta i}$ and $\delta_2 = \delta_1 \cup e_{\delta j}$, where

   $$e_{\delta i} = \{(i+(j-1)*(n+1),b) \rightarrow i+j*(n+1) \mid b \in \{0,1\}\}$$,

   $$e_{\delta j} = \{(i+j*(n+1),b) \rightarrow \{0,1...,i+j*(n+1)\}_{random} \mid b \in \{0,1\}\}$$

   (e) Set output function based on probability $Pr(\alpha_1,\alpha_2) = \begin{cases} \alpha_1, Pr \\ \alpha_2, 1-Pr \end{cases}$ where

   Pr is probability value, α_1, α_2 are output characters.

   (f) Add new output function $\lambda_1 = \lambda \cup Pr(e_{\lambda i}, \phi)$ and $\lambda_2 = \lambda_1 \cup Pr(e_{\lambda j}, \phi)$,while

   $$e_{\lambda i} = \{(i+(j-1)*(n+1),b) \rightarrow \Delta^+_{random} \mid b \in \{0,1\}\}$$,

   $$e_{\lambda j} = \{(i+j*(n+1),b) \rightarrow \Delta^+_{random} \mid b \in \{0,1\}\}$$

   (g) $i++$, if $i <= n$, go to the step c)

   (h) $j++$, if $j <= m$, go to the step c), else go to the end to complete the obfuscation of mealy machine.

3   Theory Analysis

Theory analysis of mealy machine obfuscation includes algorithm cost analysis, potency analysis and stealth analysis. While cost analysis contains time complexity of obfuscation, time and space overhead by using obfuscation. The potency analysis is the strength to resist static and dynamic reverse. Stealth analysis is to measure whether program is obfuscated nor not.

3.1 Cost Analysis

Cost measures the extra overhead of obfuscation brought to program, on the one hand, it is the cost of obfuscating a program, and on the other hand, it is the time and space overhead that is increased by original program compared with obfuscated program.

Cost of data obfuscation is mainly the time overhead of obfuscation algorithm. Iteration obfuscation based on state transition graph includes determining iteration variable, creating iteration formula of redundant state and probability function, and setting mode and times of iteration. Obfuscation time cost is based on redundant states, transition functions and probability output functions that are new added by iteration formula. The number of new redundant states and functions are determined by the number of initial states and times, if initial state and iteration number are n and m respectively, the complexity of iteration is $O(nm)$.

Obfuscated program compared with original program has extra cost of memory space and execution time. Iteration obfuscation based on state transition graph implements the transformation of sensitive and core data with adding state transition data and automata code into programs. The overhead space cost is determined by the memory space and code of mealy machine, the overhead time cost is determined by the execution time of mealy machine code. Extra space and time cost are shown in (4) and (5), where S_{code_ori} and S_{code_obf} denote code size of original and obfuscated programs. S_{data_ori} and S_{data_obf} denote memory space of original and obfuscated programs, T_{ori} and T_{obf} represent the execution of original and obfuscated program.

$$Space_{ovh} = \frac{S_{code\_obf} + S_{data\_obf}}{S_{code\_ori} + S_{data\_ori}} \tag{4}$$

$$Time_{ovh} = \frac{T_{obf}}{T_{ori}} \tag{5}$$

With the iteration number increased, state transition graph will be more and more complexity by obfuscation, the memory space occupied will continue to increase. It will produce automata state explosion problem. In practice, the balance of memory overhead and effectiveness should be considered; the iteration number of obfuscation needs to be controlled.

3.2 Potency Analysis

Potency analysis is obfuscation to resist reverse analysis, including the strength against static reverse analysis and dynamic analysis.

1) Resist Static Reverse Analysis

The effectiveness against static reverse analysis is mainly due to the complexity of code and memory data of mealy machine, mealy machine code can be obfuscated by instruction reorder, junk insertion and control structure transformation. In the way, it can increase the difficulty of static reverse analyzing for the understanding of mealy machine code. State transition graph represented by memory data will enhance its complexity with the iteration number increased.

Let original state transition graph of mealy machine be $G = (V, E)$, where V denotes state node, E denotes transition function. Set number of V be $n+1$, number of E be n iteration number be k. After iteration obfuscation by mealy machine, V will be $n+1+kn$, E will be $n+2k(n+1)$ complexity measure of cyclomatic complexity [15] for mealy machine is shown in equals (6) and (7). Complexity of state transition graph obfuscated will increase $kn+2k$.

$$S_m = \frac{|C(G_m') - C(G_m)|}{C(G_m)} \tag{6}$$

Where,

$$C(G) = E - V + 2 \tag{7}$$

2) Resist Dynamic Reverse Analysis

The effectiveness against dynamic reverse analysis is mainly due to the diversity of codes, output function number, and probability of output characters. To diverse control flow of mealy transformation codes [16], it can increase the difficult for dynamic reverse analysis, output function number and probability of output characters determine the difficulty of dynamic testing for data.

To input data I for original state transition graph of mealy machine G, the transition function sequence is $T = (e_1, e_2, ..., e_n)$, where e_i denotes transition function, output character sequence is $t_a = (a_1, a_2, ..., a_n)$, where a_i is output character. To the iteration obfuscation of state transition graph of mealy machine, all the new output functions are probability, the output characters and probabilities are $((a_1, p_1), (a_2, p_2), ..., (a_n, p_n))$ To dynamic input test for original mealy, output character sequence t_a is obtained. After obfuscation, mealy should test $\dfrac{1}{p_1 p_2 p_3 \cdots p_n}$ times for input data I, it can obtain the complete output character sequence. With the same condition, test cost of original and obfuscated mealy is shown in (8).

$$D_m = \frac{|TestTime(G_m', I)|}{|TestTime(G_m, I)|} = \frac{1}{p_1 p_2 p_3 \cdots p_n} \qquad (8)$$

3.3   Stealth Analysis

Stealth is used to measure the difficulty of recognition whether program is obfuscated or not. It can measure by the similarity between original and obfuscated program code, data, and transition graph. The graph structure of state transition of mealy machine that is directly converted is shown in figure 2(a). Through iteration, obfuscation will continue to make state transition graph of mealy machine more complicated, the obfuscated graph structure is shown in figure 2(c).

With the similarity calculation of sequence, set, and graph [17], the similarity of original and obfuscated state diagram and source code can be calculated, Transform state transition graph into sequence or set form to calculate similarity, it has low computational complexity, but it will lose some structure information of mealy machine. By the similarity calculation of subgraph isomorphism [18], similarity of state transition graph can be more precise, However subgraph isomorphism is NP-complete problem, it has a high computational cost.

4   Experiment Result and Analysis

In this paper, obfuscation algorithm is implemented on the environment of Intel (R) Core (TM) 2 CPU 1.86 GHz, Microsoft Windows XP Professional 5.1.2600 Service Pack 3, Visual Studio 2008. Benchmark programs for string operations in the boost library of 1.53.0 version are obfuscated by the algorithm. Based on the obfuscation theory, static disassembly tool IDA and dynamic debugging tool Ollydbg are used to analyze and test the original and obfuscated programs.

Table 1. Benchmark programs

Num	Name	Size (kb)	Number of obfuscated strings	Memory (kb)	Execution Time (ms)
(1)	conv_example	104	5	1520	14
(2)	find_example	102	6	1512	15
(3)	predicate_example	99	6	1500	16.3
(4)	replace_example	136	9	1540	28.4
(5)	rle_example	136	6	1560	17.9
(6)	split_example	173	4	1588	14
(7)	trim_example	88	8	1496	16.7

4.1   Cost Validation of Algorithm

Test run time and memory space of all benchmark programs, in order to reduce the effects of other factors that can affect program time measure in the computer, the average of 10 times of program running time is measured. Mealy machine is used to obfuscate programs which size kept between 85~180kb, information of benchmark programs are shown in Table. 1.

Based on mealy machine data obfuscation, state transition graph is iterated. Let iteration number be k, when k is 0, data in programs is obfuscated by original mealy machine, else according to k values, state transition graph of mealy machine is iterated, and then data in programs is obfuscated by this mealy machine, where runtime space and time overhead of obfuscation iteration are shown in figure 3(a) and (b), abscissa represents iteration number, ordinate represents run time size (kb) and obfuscation iteration time (ms) respectively. As shown in the figure, when iteration number is 0, original mealy machine is used to obfuscate constants and strings data. With the iteration number increased, state transition graph occupied space will also be increased linearly. The more strings obfuscated in program, the more occupied space of state transition graph and iteration time overhead will be. As shown in Table 1 the most and the least number of obfuscated strings of benchmark program are replace_example.exe and split_example.exe, they are the two lines in picture most up and down respectively. When the number of iterations is 0, run time space of programs remain at around 1500kb, when iterations number is 2048, the run time space of replace_example.exe and split_example.exe are around 19500kb and 5000kb respectively. In figure 3(b) iteration obfuscation time are gained by executing programs for the average of 10 times. They increase correspondingly to 48 and 270ms. As the code size of mealy machine does not change, execution time of original and obfuscated program will be consistent.

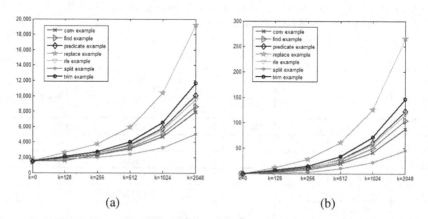

(a)   (b)

Fig. 3. Programs runtime space and iteration time of mealy machine

4.2   Potency Validation of Algorithm

Resist static reverse analysis: using static disassembly tool IDA[1] to analyze the state transition graph, it can analyze the nodes, edges and complexity of state transition graph. The size of state transition graph with one node and two state edges is shown in Figure 4. Abscissa represents iteration number, coordinate represents the size of state transition graph. While iteration number is 0, size of state transition graph is about 2k, number of nodes and edges of state transition graph will increase with iteration number increased. When iteration number is 2048, sizes of state transition graph of replace_example.exe and split_example.exe are 17400kb and 2860kb respectively. With the order of state transition graph from big to small, benchmark programs are replace_example.exe, trim_example.exe, predicate_example.exe, rle_example.exe, find_example.exe, conv_example.exe and split_example.exe.

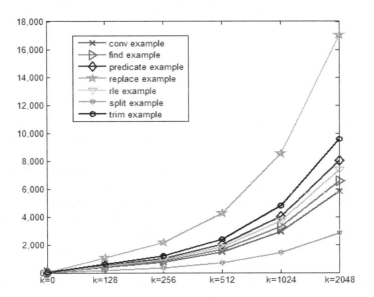

Fig. 4. Size of state transition graph of mealy machine

Resist the dynamic test and analysis: benchmark programs are tested many times by using dynamic debugging tool OllyDBG[2]. Length of obfuscated strings and transition probability are set, and mealy machine is tested and analyzed. To insert redundant nodes into the state transition graph, where dynamic test input data is (11111), output data obfuscated by mealy machine is "abcde". If all output function in the state transition graph are based on the probability, probability are 1/2, 1/3, 2/3, 1/4 respectively, mealy machine will be dynamic tested 100 times, dynamic test results are shown in Table. 2. The proportion of output times of "abcde" to the total number

---

[1] IDA Multi-Processor Disassembler and Debugger. http://www.hex-rays.com
[2] OllyDbg 32-bit assembler level analyzing debugger for Windows. http://www.ollydbg.de

of test are 4%, 1%, 12% and 0%. It suggests that a lot of input should be tested in order to output the full string. Based on the probability of output function, the difficulty of dynamic inverse analysis has increased greatly.

Table 2. String dynamic testing based on mealy machine (n=100)

Output strings	The sum of strings			
	prob=1/2	prob=1/3	prob=2/3	prob=1/4
"abcde"	4	1	12	0
"abcd", "abce", "acde", "abde", "bcde"	15	5	33	1
"abc", "abd", "abe", "acd", "ace" "ade", "bcd", "bce", "bde", "cde"	29	16	34	9
"ab", "ac", "ad", "ae", "bc" "bd", "be", "cd", "ce", "de"	34	33	17	26
"a", "b", "c", "d", "e"	16	32	4	40
null	2	13	0	24

5   Conclusions

At present, a typical way to protect sensitive and core data in program is to use mealy finite state machine to convert the data to a different kind of expression in a program. Directly use mealy automata for data transformation, attackers can be easy to analyze, locate and seize the data by static and dynamic reverse analysis. Through the iteration obfuscation of mealy machine, redundant state and state transition function are added into the state transition graph, and based on the graph, probability output functions are added. Thus it can improve the ability of mealy machine to resist the static and dynamic analysis. Analysis and experiment validated obfuscation algorithm from three aspects that are potency, cost and stealth respectively. Experiments show that method can effectively increase the complexity of state transition graph of mealy machine, it can enhance the performance for mealy automata to against static analysis and dynamic testing.

Acknowledgement. This work was supported by the grants of National Natural Science Foundation of China No. 61379151, 61274189 and 61302159. Excellent Youth Foundation of Henan Province of China No. 144100510001.

References

1. Collberg, C., Thomborson, C.: Watermarking, Tamper-Proofing, and Obfuscation-Tools for Software Protection. IEEE Transactions on Software Engineering 28(8), 735–746 (2002)

2. Collberg, C., Thomborson, C., Low, D.: A taxonomy of obfuscating transformations. Technical Report 148, Department of Computer Science, the University of Auckland, New Zealand (1997)
3. Chan, J., Yang, W.: Advanced obfuscation techniques for Java bytecode. Journal of Systems and Software 71(1/2), 1–10 (2004)
4. Xu, W., Zhang, F., Zhu, S.: The power of obfuscation techniques in malicious JavaScript code: A measurement study. In: Proceedings of 7th IEEE International Conference on Malicious and Unwanted Software, pp. 9–16 (2012)
5. Sosonkin, M., Naumovich, G., Memon, N.: Obfuscation of design intent in object-oriented applications. In: Proceedings of 3rd ACM Workshop on Digital Rights Management, pp. 142–153 (2003)
6. Popov, I.V., Debray, S.K., Andrews, G.R.: Binary obfuscation using signals. In: Proceedings of 16th USENIX Security Symposium, pp. 275–290 (2007)
7. Balachandran, V., Emmanuel, S.: Potent and Stealthy Control Flow Obfuscation by Stack Based Self-Modifying Code. IEEE Transactions on Information Forensics and Security 8(4), 669–681 (2013)
8. Badger, L., D'Anna, L., Kilpatrick, D., et al.: Self-protecting mobile agents obfuscation techniques evaluation report. Technical Report 01-036, NAI Labs (2001)
9. Collberg, C., Thomborson, C., Low, D.: Breaking abstractions and unstructuring data structures. In: Proceedings of IEEE International Conference on Computer Languages, pp. 28–38 (1998)
10. Praveen, S., Sojan, P.: Array Data Transformation for Source Code Obfuscation. Proceedings of World Academy of Science, Engineering and Technology 1, 83–87 (2007)
11. Zhu, W., Thomborson, C.D., Wang, F.Y.: Obfuscate arrays by homomorphic functions. In: Proceedings of IEEE International Conference on Granular Computing, pp. 770–773 (2006)
12. Xin, Z., Chen, H., Han, H., Mao, B., Xie, L.: Misleading malware similarities analysis by automatic data structure obfuscation. In: Burmester, M., Tsudik, G., Magliveras, S., Ilić, I. (eds.) ISC 2010. LNCS, vol. 6531, pp. 181–195. Springer, Heidelberg (2011)
13. Collberg, C., Nagra, J.: Surreptitious software: obfuscation, watermarking, and tamper-proofing for software protection, pp. 269–271. Addison-Wesley (2009)
14. George, H.M.: A method for synthesizing sequential circuits. Bell System Technical Journal 34(5), 1045–1079 (1955)
15. McCabe, T.: A complexity measure. IEEE Transaction on Software Engineering SE-2(4), 308–320 (1976)
16. Schrittwieser, S., Katzenbeisser, S.: Code Obfuscation against Static and Dynamic Reverse Engineering. In: Filler, T., Pevný, T., Craver, S., Ker, A. (eds.) IH 2011. LNCS, vol. 6958, pp. 270–284. Springer, Heidelberg (2011)
17. Cesare, S., Xiang, Y.: Software Similarity and Classification, pp. 63–70 (2012)
18. Patrick, P.F., Lucas, C.K., Yiu, S.M.: Heap Graph Based Software Theft Detection. IEEE Transactions on Information Forensics and Security 8(1), 101–110 (2013)

Synchronization of Chen's Attractor and Lorenz Chaotic Systems by Nonlinear Coupling Function

Hongjie Yu and Hui Qian

State Key Laboratory of Ocean Engineering (Shanghai Jiao Tong University),
200240 Shanghai, China
yuhongjie@sjtu.edu.cn

Abstract. The chaotic synchronizations of Chen's attractor and Lorenz chaotic systems linked by the difference of the nonlinear coupling function of the two systems are discussed. The method is an expansion of SC method (the method of chaotic synchronization based on the stability criterion). The special nonlinear-coupled functions are constructed by suitable separation between the linear term and the nonlinear term of chaotic systems. The state variable of the response system can be synchronized easily and completely onto that of the driver system without calculation of the maximum conditional Lyapunov exponent when the coupling strength is taken as constant one, and the zero point of synchronization error is asymptotically stable. The proposed method is effective for chaotic synchronization of Chen's and Lorenz system by numerical simulations.

Keywords: Chaotic synchronization, nonlinear coupling function, stability criterion, Chen's attractor, Lorenz attractor.

1 Introduction

Pecora and Carroll first observed the phenomenon of chaotic synchronization in 1990[1]. They showed that chaotic systems could be made to synchronize by linking them with common signals. Hence the synchronization problem of chaotic and hyper-chaotic dynamical system has been widely studied and expanded [2-9], such as identical synchronization [1,2], phase synchronization [3], generalized synchronization [4] and projective synchronization [7-9]. Practical applications of chaotic synchronization to secure communication systems have been reported in Ref. [10].

We have noticed that synchronization of chaos and hyperchaos are principally studied by using the same linear function of variables from each chaotic oscillator to couple other chaotic oscillators in departed research. Pecora and Carroll gave master stability function for any linear coupling of oscillators [6]. Recently, there are some authors used nonlinear coupling feedback functions for synchronization of chaotic and hyperchaotic systems [5].

The methods in early stage were mainly used to realize the synchronization between identical chaotic systems that are coupled properly. They can achieve

D.-S. Huang et al. (Eds.): ICIC 2014, LNCS 8588, pp. 532–540, 2014.

complete synchronization by following the same chaotic trajectory. However, experimental and even more real systems are often not fully identical; Especially there are mismatches in parameters of the systems. It is thus important and also interesting to investigate synchronization behavior between non-identical systems. Different types of synchronization behavior in coupled non-identical low-dimensional chaotic systems have been investigated. In literature [12], authors studied the chaotic synchronization of third-order systems and second-order driven oscillator. Such a problem is related to synchronization of strictly different chaotic systems[13-16].

Yu and Liu [17] proposed The SC method based on the stability criterion of linear systems in 2003 for chaotic synchronization the two different systems, and the synchronization of two chaotic systems is achieved by a unidirectional nonlinear coupling from driving system to response system. The SC method is based on the stability criterion of linear systems, and the synchronization of constructed systems is ensured by that all eigenvalues of the matrix **A** have negative real parts. The synchronization can be robustly achieved without the requirement to calculate the conditional Lyapunov exponents.

The new approach suggested in this letter is an expansion of the SC method. We pay attention to the synchronization problems of the two essentially different (topologically nonequivalent) chaotic systems with identical dimensions such as Lorenz system and Chen's system that derived from the classical Lorenz system. We unidirectionally couple the two non-identical systems using the nonlinear-coupled function constructed specially by inspiration of SC method based on stability criterion. Although there does not exist such an invariant manifold $\mathbf{x} = \mathbf{y}$ between the two topologically nonequivalent systems. However, in this paper, the chaotic trajectories of the response system can be easily and completely synchronized onto that of the driver system with the developing of time by influence of nonlinear coupled function when coupling strength is taken as critical value 1. The complete synchronization of systems is ensured by that error equation is asymptotically stable due to all eigenvalues of the Jacobian matrix **A** have negative real parts and without the requirement to calculate the conditional Lyapunov exponents. This show that the state variables of one chaotic attractor can be completely transformed into that of another chaotic attractor by proper unidirectionally coupling while they are topologically nonequivalent before coupling. This is important and interesting for potential applications. We also show that two unidirectionally coupled non-identical chaotic systems undergo transitions from non-synchronization to phase then to complete synchronization with increasing coupling strengths which is belong in the range [0,1], and from complete synchronization to non-synchronization with increasing coupling strengths when which is larger than 1.

2   Chaotic Synchronization of Two Different Systems

We consider the two chaotic continuous systems described by

$$\dot{X}(t) = F(X(t)) \tag{1-1}$$

$$\dot{Y}(t) = \overline{F}(Y(t)) \tag{1-2}$$

where $X(t) \in \mathbf{R}^n, Y(t) \in \mathbf{R}^n$ are n-dimensional state vector of the two chaotic systems respectively, and $F : \mathbf{R}^n \to \mathbf{R}^n, \overline{F} : \mathbf{R}^n \to \mathbf{R}^n$ define the vector field in n-dimensional space respectively. We first suitably decompose the function $F(X(t))$ as

$$\dot{X}(t) = G(X(t)) + D(X(t)) \tag{2}$$

Where function $G(X(t)) = AX(t)$ is specially disposed as the linear part of $F(X(t))$, and is required that A is a full rank constant matrix, all eigenvalues of which have negative real parts. Function $D(X(t)) = F(X(t)) - G(X(t))$ is the nonlinear part of $F(X(t))$, and is called as the nonlinear feedback function. Then the system (1-1) can be rewritten as

$$\dot{X}(t) = AX(t) + D(X(t)). \tag{3}$$

Similarly, we also decompose the function $\overline{F}(Y(t))$, and then the system (1-2) can be rewritten as

$$\dot{Y}(t) = AY(t) + \overline{D}(Y(t)) \tag{4}$$

where the matrix A is same with that in Eq.(3), Function $\overline{D}(Y(t)) = \overline{F}(Y(t)) - AY(t)$ is the nonlinear part of $\overline{F}(Y(t))$. We construct a nonlinear coupling function $u(t)$ between two chaotic systems by the difference of feedback functions $D(X(t))$ and $\overline{D}(Y(t))$ as follows

$$\dot{X}(t) = AX(t) + D(X(t)) \tag{5}$$

$$\dot{Y}(t) = AY(t) + \overline{D}(Y(t)) + \alpha u(t) \tag{6}$$

where $u(t) = D(X(t) - \overline{D}(Y(t))$, α is the coupling strength. The systems (5) and (6) are the driver and the response systems respectively. The synchronization error between system (5) and system (6) is defined as $e(t) = X(t) - Y(t)$. The evolutional equation of the difference $e(t)$ is determined as follows

$$\dot{e}(t) = Ae(t) + (1 - \alpha)[D(X) - \overline{D}(Y)] \tag{7}$$

Obviously the zero point of $e(t)$ corresponds to the equilibrium point when the coupling strength α is taken as $\alpha = 1$. Since all eigenvalues of the matrix A have negative real parts mentioned as before, according to the stability criterion of linear system $\dot{e} = Ae$, the zero point of synchronization error is asymptotically stable and $e(t)$ tends to zero when $t \to \infty$. Then the state vector $Y(t)$ of system (6) is synchronized completely onto the state vector $X(t)$ of systems (5). We notice that the calculation of conditional Lyapunov exponents is not necessary in stability analysis of synchronization as $\alpha = 1$.

3   Application to the Numerical Example

Chen found a new chaotic attractor (Chen's chaotic attractor) in 1999 [18]. The nonlinear differential equations of Chen attractor are

$$
\begin{aligned}
\dot{x}_1 &= a(x_2 - x_1) \\
\dot{x}_2 &= (c - a)x_1 - x_1 x_3 + c x_2 \\
\dot{x}_3 &= x_1 x_2 - b x_3
\end{aligned}
\tag{8}
$$

The system (8) has chaotic behavior at the parameter values $a = 35, b = 3$, and $c = 28$. According to Eq. (3), we decompose Eq.(8) into AX and $D(X)$ as

$$
A\begin{pmatrix} x_1 \\ x_2 \\ x_3 \end{pmatrix} = \begin{pmatrix} -a & a & 0 \\ c-a & 1 & 0 \\ 0 & 0 & -b \end{pmatrix}\begin{pmatrix} x_1 \\ x_2 \\ x_3 \end{pmatrix}, \quad D(x_1, x_2, x_3) = \begin{pmatrix} 0 \\ (c-1)x_2 - x_1 x_3 \\ x_1 x_2 \end{pmatrix}
\tag{9}
$$

the matrix A has negative real eigenvalues (-25.882, -8.118, -3.0).

As another chaotic system, we consider the Lorenz system described by

$$
\begin{aligned}
\dot{y}_1 &= \sigma(y_2 - y_1) \\
\dot{y}_2 &= r y_1 - y_2 - y_1 y_3 \\
\dot{y}_3 &= y_1 y_2 - \overline{b} y_3
\end{aligned}
\tag{10}
$$

Eqs.(10) give rise to a chaos when $\sigma = 10$, $r = 28$ and $\overline{b} = 8/3$. We can decompose Eq.(10) with functions AY and $\overline{D}(Y)$ as

$$
A\begin{pmatrix} y_1 \\ y_2 \\ y_3 \end{pmatrix} = \begin{pmatrix} -35 & 35 & 0 \\ -7 & 1 & 0 \\ 0 & 0 & -3 \end{pmatrix}\begin{pmatrix} y_1 \\ y_2 \\ y_3 \end{pmatrix}, \overline{D}(y_1, y_2, y_3) = \begin{pmatrix} 25y_1 - 25y_2 \\ 35y_1 - 2y_2 - y_1 y_3 \\ \dfrac{1}{3}y_3 + y_1 y_2 \end{pmatrix}
\tag{11}
$$

Using the difference between nonlinear feedback functions $D(X)$ and $\overline{D}(Y)$, Chen's attractor and Lorenz attractor are coupled nonlinearly according to Eq.(5) and Eq. (6)

$$\begin{cases} \dot{x}_1 = a(x_2 - x_1) \\ \dot{x}_2 = (c-a)x_1 - x_1 x_3 + cx_2 \\ \dot{x}_3 = x_1 x_2 - bx_3 \end{cases} \tag{12}$$

$$\begin{cases} \dot{y}_1 = \sigma(y_2 - y_1) + \alpha u_1(t) \\ \dot{y}_2 = ry_1 - y_2 - y_1 y_3 + \alpha u_2(t) \\ \dot{y}_3 = y_1 y_2 - \overline{b} y_3 + \alpha u_3(t) \end{cases} \tag{13}$$

where the controllable nonlinear coupling function

$$u(t) = \begin{pmatrix} -25 y_1 + 25 y_2 \\ 27 x_2 - x_1 x_3 - (35 y_1 - 2 y_2 - y_1 y_3) \\ x_1 x_2 - (\frac{1}{3} y_3 + y_1 y_2) \end{pmatrix} \tag{14}$$

The initial condition are given as $x_1 = 3.0, x_2 = 4.0, x_3 = 3.5$ and $y_1 = 5.1$, $y_2 = 4.2$, $y_3 = 4.5$. The Fig.1 shows that the time histories of the synchronization error $e(t) = \sqrt{(e_1^2 + e_2^2 + e_3^2)}$, $e_i = (x_i - y_i), i = 1,2,3$, converges to zero rapidly. The synchronization of chaotic systems (13) and (12) is achieved eventually.

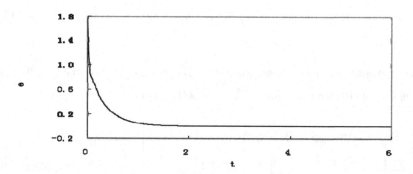

Fig. 1. The time history of the synchron

Notice that the Chen' system (12) is the driver system and the Lorenz system is the response system here. Thus the state variables (y_1, y_2, y_3) of the Lorenz system are completely synchronized onto the state variables (x_1, x_2, x_3) of the Chen's system. Fig.2(a)-(b) shows that the Lorenz's attractor has been changed as the Chen's attractor in plane (y_1, y_2) and (y_1, y_3) after achieving synchronization.

In addition, we also consider the effects of the coupling strength α to the synchronization of the two systems. We find that the state variables (y_1, y_2, y_3) of the Lorenz system can completely be synchronized onto the state variables (x_1, x_2, x_3) of the Chen's system when $\alpha \in [0.995, 1.005]$ in which the synchronization error $|e| \le 0.1$, while this synchronizing process can't be achieved completely when $\alpha \in [0, 0.994]$ and $\alpha \in [1.006, 5.1]$.

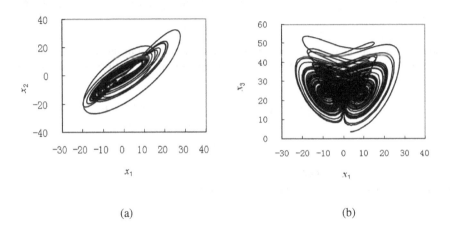

(a)   (b)

Fig. 2. The Lorenz's attractor has been synchronized onto the Chen's attractor in plane(y_1, y_2) and (y_1, y_3)

In Fig.3 (a)-(b), the Lorenz's attractors in plane (y_1, y_2) in synchronizing process are given with different values of α. We see that the Lorenz's attractor be changed onto Chen's attractor gradually with increasing value of α.

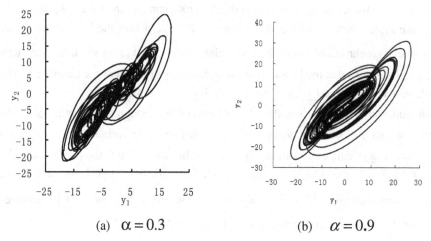

(a) $\alpha = 0.3$ (b) $\alpha = 0.9$

Fig. 3. Metamorphic Lorenz's attractor in synchronizing process with different α in plane(y_1,y_2)

Besides, we can also set the Lorenz system as driver system but Chen's system as the response system. Now, we linked two unidirectional coupling chaotic systems similar to (5) and (6) as follows

$$\dot{X}(t) = AX(t) + D(X(t)) + \alpha[\overline{D}(Y(t)) - D(X(t))] \qquad (14)$$

$$\dot{Y}(t) = AY(t) + \overline{D}(Y(t)) \qquad (15)$$

Setting $\alpha = 1.0$, the Chen's system and Lorenz system coupled according above formulas (14) and (15) are described as follows

$$\begin{cases} \dot{x}_1 = a(x_2 - x_1) + (25y_1 - 25y_2) \\ \dot{x}_2 = (c - a)x_1 - x_1x_3 + cx_2 + [(35y_1 - 2y_2 - y_1y_3) - (27x_2 - x_1x_3)] \\ \dot{x}_3 = x_1x_2 - bx_3 + [(-\overline{b} + b)y_3 + y_1y_2 - x_1x_2] \end{cases} \qquad (16)$$

$$\begin{cases} \dot{y}_1 = \sigma(y_2 - y_1) \\ \dot{y}_2 = ry_1 - y_2 - y_1y_3 \\ \dot{y}_3 = y_1y_2 - \overline{b}y_3 \end{cases} \qquad (17)$$

The parameter $\overline{b}$ is taken as $\overline{b} = 3$, another parameters a, c, b, σ, r are taken as before.

Fig.4 shows the time histories of ratio x / y of the synchronization state between Lorenz and Chen's system. It converges to constant one rapidly. We find that the state variables (x_1, x_2, x_3) of the Chen's system are completely synchronized onto the state variables (y_1, y_2, y_3) of the Lorenz system here.

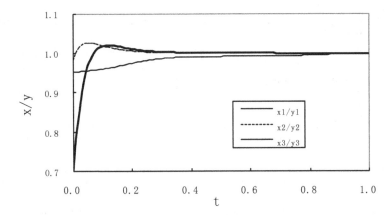

Fig. 4. Time histories of ratio x/y of the synchronization state between Lorenz and Chen's system

4 Conclusion

In this letter, a new approach of chaotic synchronization for Chen' and Lorenz attractor that have the same dimension linked through a nonlinear coupling function is proposed. The method expands SC method based on the stability criterion. A special nonlinear coupling function is constructed by suitable separation for function of chaotic systems. The evolutional equations of difference are given with diversified coupling strength. The state variable of the response system can be synchronized easily and completely onto that of the driver system without calculation of the maximum conditional Lyapunov exponent when the coupling strength is taken as constant one, and then the zero point of synchronization error is asymptotically stable. The proposed method is effective for chaotic synchronization of Chen's and Lorenz system by numerical simulations.

Acknowledgment. The research is supported by the National Natural Science Foundation of China (Project No. 10802030)

References

1. Pecora, L.M., Carroll, T.L.: Synchronization in Chaotic System. Phys. Rev. Lett. 64, 821–824 (1990)
2. Carroll, T.L., Pecora, L.M.: System for producing synchronized signals. Physica D 67, 126–140 (1993)

3. Rosenblum, M.G., Pikovsky, A.S., Kurths, J.: Phase Synchronization of Chaotic Oscillators. Phys. Rev. Lett. 76, 1804–1807 (1996)
4. Kocarev, L., Parlitz, U.: Generalized synchronization and equivalence of unidirectionally coupled dynamical systems. Phys. Rev. Lett. 76, 1816–1819 (1996)
5. Ali, M.K., Fang, J.Q.: Synchronization of chaos and hyperchaos using linear and nonlinear feedback functions. Phys. Rev. E 55, 5285–5290 (1997)
6. Pecora, L.M., Carroll, T.L.: Synchronization in chaotic system. Phys. Rev. Lett. 80, 2109–2113 (1998)
7. Xu, D., Chee, C.Y.: Secure digital communication using controlled projective synchronisation of chaos. Phys. Rev. E 66, 046218-1–046218-5 (2002)
8. Fink, K.S., Johnson, C., Carroll, T., Mar, D., Pecora, L.: Three coupled oscillators as a universal probe of synchronization stability in coupled oscillator arrays. Phys. Rev. E 61, 5080–5090 (2000)
9. Yu, H.J., Peng, J.H., Liu, Y.Z.: Projective synchronization of unidentical chaotic systems based on stability criterion. International Journal of Bifurcation and Chaos 16, 1049–1056 (2006)
10. Cai, J., Lin, M., Yuan, Z.: Secure communication using practical synchronization between two different chaotic systems with uncertainties. Mathematical and Computational Applications 15, 166–175 (2010)
11. Rosenblum, M.G., Pikovsky, A.S., Kurths, J.: From phase to lag synchronization in coupled chaotic oscillators. Phys. Rev. Lett. 78, 4193–4196 (1997)
12. Femat, R., Perales, G.S.: Synchronization of chaotic systems with different order. Phys. Rev. E 65, 036226 (2002)
13. Guan, S., Lai, C.H., Wel, G.W.: Phase synchronization between two essentially different chaotic systems. Phys. Rev. E 72016205, 1–5 (2005)
14. Park, J.H.: Chaos synchronization between two different chaotic dynamical systems. Chaos, Solitons and Fractals 27, 549–554 (2006)
15. Chen, X., Lu, J.: Adaptive synchronization of different chaotic systems with fully unknown parameters. Phys. Lett. A 364, 123–128 (2007)
16. Roopaei, M., Jahromi, M.Z.: Synchronization of two different chaotic systems using novel adaptive fuzzy sliding mode control. CHAOS 18, 033133 (2008)
17. Yu, H.J., Liu, Y.Z.: Chaotic synchronization based on stability criterion of linear systems. Phys. Lett. A 314, 292–298 (2003)
18. Chen, G., Ueta, T.: Yet another chaotic attractor. Int. J. Bifurcation Chaos 9, 1465–1466 (1999)
19. Shuai, J.W., Durand, D.M.: Phase Synchronization in two coupled Chaotic Neurons. Phys. Lett. A 264, 289–296 (1999)

A Data Hiding Algorithm for 3D Videos Based on Inter-MBs

Guanghua Song[1], Zhitang Li[1,2], Juan Zhao[1], and Mingguang Zou[1]

[1] Department of Computer Science and Technology,
Huazhong University of Science and Technology,Wuhan, China
[2] Network and Computing Center, Huazhong University of Science and Technology,
Wuhan, China
ghsong@hust.edu.cn

Abstract. Distortion drift, including intra-frame and inter-frame, is a big problem of data hiding in 3D video streams. With the analysis of the coding structure of MVC videos, we propose a new data hiding algorithm without inter-frame and intra-frame distortion drift. Firstly, an ideal that modify the 4×4 QDCT coefficients of the inter-MBs except the rightmost column and the bottom row is proposed, and it can completely prevent intra-frame distortion drift. Then, we point out that hiding data into b4-frames can completely prevent inter-frame distortion drift. The algorithm has lower complexity and a greater embedding capacity than the current algorithms. Experimental results confirm the embedding capacity and complexity superiority of the proposed algorithm while keeping the similar human visual effect.

Keywords: Data hiding, Multi-view coding, Distortion drift.

1 Introduction

Data hiding techniques are being used in a variety of application domains, such as the authentication, identification, and secret communication. With the gradual rise of 3D videos, traditional data hiding algorithm researchers begin to turn their attention to the new areas. In 2010, Joint Video Team (JVT) released the Multi-view coding (MVC) [1] standard based on H.264/AVC, and wrote them as an appendix in the form of the latest H.264/AVC. JVT adopted the structure of hierarchical B-frames proposed in [2] which combined intra prediction, inter prediction and inter-view prediction. The proposed hierarchical B-frames prediction structure with two viewpoints is shown in Fig. 1.

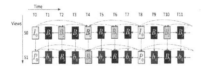

Fig. 1. Hierarchical B frame structure with two viewpoints [2]

D.-S. Huang et al. (Eds.): ICIC 2014, LNCS 8588, pp. 541–552, 2014.
© Springer International Publishing Switzerland 2014

Currently, the research on data hiding algorithms for H.264/AVC mainly focuses on 2D fields, most of the data hiding algorithms have been developed in [3]-[7]. The existing H.264 video data hiding schemes can be classified into two types according to their extraction approaches: detectable algorithm and readable algorithm. The former inserts a code that can only be detected, and the latter embeds a message that can be read. However, they all suffer from the error propagation problem in including intra-frame and inter-frame, which degrades the video quality.

To avoid this error propagation problem, Ma et al. [8] presented an efficient discrete cosine transform-based (DCT-based) data hiding algorithm. They exploited several paired-coefficients of a 4×4 DCT block to accumulate the embedding induced distortion. The algorithm can achieve a high embedding capacity and low visual distortion. But this algorithm would bring inter-frame distortion drift, and the prediction mode of all MBs must be achieved before embedding and extracting, which greatly increases the complexity of the algorithm. Lin et al. [9] developed an improved error propagation-free DCT-based perturbation scheme that fully exploited the remaining 54% of luma blocks and thereby doubled the data hiding capacity of Ma et al.'s algorithm. Further, in order to preserve the visual quality and increase the embedding capacity of the embedded video sequences, a new set of sifted 4×4 luma blocks was considered in the proposed DCT-based perturbation scheme. But this algorithm would also bring inter-frame distortion drift.

Based on this, combining MVC videos coding features, we propose a new data hiding algorithm without inter-frame and intra-frame distortion drift. Instead of perturbing QDCT coefficient-pairs to embed data only into intra-frame 4×4 luma blocks, this paper allows all the inter-MBs, whose transformation matrix is 4×4, to embed data. In contrast to Ma et al.'s and Lin et al.'s approaches, we can fully utilize the large number of existing inter-MBs in the 3D videos to obtain a larger embedding capacity. At the same time, the proposed algorithm does not require to obtain the data of each MB's MB-type and prediction mode in advance, which greatly reduces the complexity of the algorithm, as well as improves the embedding efficiency. Another contribution of this paper is that we can embed data into the b4-frames to prevent the inter-frame distortion drift. The proposed algorithm has lower complexity and a greater embedding capacity than the current algorithms.

The remainder of this paper is organized as follows. In the next section, we analyze the prediction structure of the 3D videos with two viewpoints, and point out that Ma et al.'s and Lin et al.'s algorithms are not suitable for the coding structure of 3D videos. Further, we propose the methods to avoid the distortion drift for 3D videos. In Section 3, two proposed algorithms are presented and the related theoretical analysis is given. In Section 4, the experiment results are presented and the conclusion is given in section 5.

2   The Prediction Structure of 3D Videos

In this section, we briefly explain the inter-frame and intra-frame prediction structure under H.264/MVC, and point out that Ma et al.'s and Lin et al.'s algorithms are not suitable for the coding structure of 3D videos. Further, we propose the methods to protect against the distortion drift for 3D videos.

2.1   The Inter Prediction Structure of 3D Videos

JVT adopt the structure of hierarchical B-frames proposed in [2] which combines intra prediction, inter prediction and inter-view prediction. There are only one I-frame and one P-frame in a GOP in 3D videos with two viewpoints, as shown in Fig. 1. As a key frame, I-frame will be referenced by all the remaining frames. If we modify the I-frame, the cause of the error will be passed to all the rest of frames, and it will result in serious inter-frame distortion drift. Therefore, we generally do not take I-frame as the embedded frames. In addition to the I-frame, there is one P-frame in a GOP, and the remaining are all B-frames.   Through the study of the prediction structure of MVC videos, we find that the prediction direction is from S0 to S1 as shown in Fig. 2. So, we can embed data into the S1 viewpoint to prevent drift between the viewpoints, which is proved by the following Conclusion 1.

C1. The inter-view-prediction direction of 3D videos with two viewpoints is from S0 to S1. So if we modify the frames in S0, the distortion will be passed to not only the frames in S0 but also in S1 which will cause serious distortion drift. However, if we modify the frame in S1, the distortion will be passed only to the frames in S1, and the inter-view distortion drift is prevented.

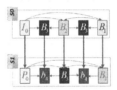

Fig. 2. The prediction direction of 3D videos with two viewpoints

2.2   The Intra Prediction Structure of 3D Videos

It is well known that intra prediction plays an important role in reducing spatial redundancy among H.264/AVC intra frames. The existing algorithms, included Ma et al.'s and Lin et al.'s algorithms, mainly target intra prediction mode. Due to the requirement for using the prediction relationship of the intra-MBs, these algorithms must obtain the data of each MB's MB-type and prediction mode before hiding data. It increases the complexity of the algorithm, and is not appealing to real-time processing. Therefore, we turn our attention to the inter-MBs. Through the study of the prediction structure among H.264/MVC, we find that the predictive mode with the rules described in the following Conclusion 2.

C2. The residual coefficients will be referenced by its previously encoded blocks (Pro-blocks) in intra-MBs, and the motion vectors will be referenced by its Pro-blocks in inter-MBs.

As shown in Fig. 3(a), since the current MB is an inter-MB, if its Pro-blocks are also inter-MBs, we can reach the conclusion that the residual coefficients of the current MB's will not be referenced by its Pro-blocks because they only take the motion vector as reference. Then, if we modify the residual coefficients of the current MB's, the cause of the error will not be passed to the Pro-MBs, and the intra-frame distortion drift is prevented.

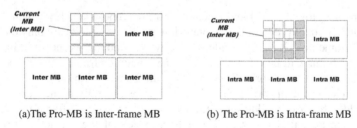

(a)The Pro-MB is Inter-frame MB   (b) The Pro-MB is Intra-frame MB

Fig. 3. The relationship between the current MB and the Pro-MBs

On the other hand, as shown in Fig. 3(b), if its Pro blocks are intra-MBs, we know that the residual coefficients of the rightmost column vector and the bottom row vector of current MB's will be referenced by its Pro-blocks. And the rest of the sub-MBs will not be referenced by its Pro sub-MBs because they also only take the motion vector as reference. Hence, if we modify the residual coefficients of the sub-MBs except the rightmost column and the bottom row of current MB's, the cause of the error will not be passed to the Pro-MBs, and the intra-frame distortion drift is prevented.

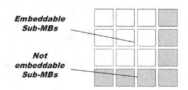

Fig. 4. The selection of embeddable sub-MBs in inter-frame-MBs

The error propagation-free property of the perturbation technique used in inter-MBs is proved by the following Conclusion 3.

C3. For inter-MBs, if we just modify the residual coefficients of embeddable sub-MBs, the cause of the error will not be passed to the Pro-MBs, and the intra-frame distortion drift is prevented.

As shown in Fig. 4, there are nine sub-MBs located in upper left corner that can be used to embed data without intra-frame distortion drift. And we also do not need to obtain the data of the MB type and the prediction mode of the Pro-MBs.

2.3   B4-frame and Inter Distortion Drift Prevention

Though the study of the MVC video coding standard and the prediction structure, as shown in Fig. 1, we have found the method to prevent the inter-frame distortion drift, which lead to the following **Conclusion 4.**

C4. *There is a non-reference frame in the MVC videos with the structure of hierarchical B-frames, in which hiding data will not cause the inter-frame distortion drift.*

Fig. 5. The first b4-frame in a GOP

As shown in Fig. 1, with examining the b4-frames (e.g. the first b4-frame in a GOP as shown in Fig.5.), we draw the following two conclusions. Firstly, as the frame is located in viewpoint **S1**, it will not be as a reference frame for inter-view prediction according to **C1**. Secondly, the b4-frame also will not be as a reference frame for the inter prediction in viewpoint **S1**. So the b4-frames are the non-reference frame in the MVC videos with the structure of hierarchical B-frames. The b4-frames are neither used as the reference frame for inter prediction nor used as the reference for inter-view prediction.   Therefore, it will not result in any inter-frame distortion drift when we hide data into b4-frames.

2.4   The Integer DCT in 3D MVC Videos

As we all know, the integer discrete cosine transform (IDCT), which is developed from the DCT, is used in H.264/MVC standard. There are two types of the transformation matrix including 4×4 and 8×8. Generally, the 4×4 matrix is applied to the sub-MB partition and the 8×8 matrix is applied to the MB partition. So, each MB can be transformed by either 8×8 or 4×4. Here we only consider embedding data into MBs transformed by 4×4 matrix because MBs transformed by 8×8 matrix are homogeneous, and the human eyes are very sensitive to the change of luma values in these MBs.

3   The Proposed Algorithm

In this section, based on the analysis of the structure of MVC, we propose the way to prevent inter-frame and intra-frame distortion drift with the theoretical derivation, respectively. Then two steganography algorithms without distortion drift for MVC are given.

Ag.1 Data Hiding in GOP based on inter-MBs

Input: Message M ,Original GOP, Threshold,   Random sequence RS

Output: Data embedded in the decoded GOP

encrypt: $M \rightarrow M' = (m_i)$

  foreach   decoded frame in the GOP do

  if the frame is b4-frame

      if the MB is inter-MBs && Luma Trans= 4×4

          if the sub-MB is neither the rightmost column nor the bottom row

              then   Obtain QDCT coefficients $Y_{i,j}^{QDCT}$

              if $|Y_{00}| > Threshold$

              then

              Select coefficients with RS

              $if ((Y_{kl}^i \% 2) \oplus m_i == 1 \& \& Y_{kl}^i \geq 0)$

                  $Y_{kl}^{i'} \rightarrow Y_{kl}^i + 1$

              $if ((Y_{kl}^i \% 2) \oplus m_i == 1 \& \& Y_{kl}^i < 0)$

                  $Y_{kl}^{i'} \rightarrow Y_{kl}^i - 1$

              $else$ $Y_{kl}^{i'} \rightarrow Y_{kl}^i$

    end

    until M is completely embedded

end

3.1   Data Hiding Algorithm Based on Inter-MBs

According to the analysis of **C3** and **C4**, a data hiding algorithm without distortion drift is presented. Firstly, in order to prevent the inter-frame distortion drift, we hide data into the b4-frames. Secondly, in order to avoid the intra-frame distortion drift, we modify the QDCT coefficients of inter-MBs with 4×4 transformation matrix using the method of **C3**. The block is selected according to its quantized DCT coefficients, such as whether there exist $|Y_{00}| > Threshold(Threshold = 0,1,2\cdots)$. And the embedded data M is encrypted into M'. We assume that $M' = m_1 m_2 m_3 \cdots, m_i \in \{0,1\}$. To improve security, the embedded data M is encrypted into M'. More importantly, we introduce the idea of a random sequence. The proposed data hidden operates are shown in Algorithm 1.

The data extractor operates to extract the hidden message as a special decoder and our proposal is straightforward without the original video, as shown in **Algorithm 2**. After data extraction from the consecutive GOPs, the hidden message is reconstructed back by concatenation of the extracted bitstream.

Compared with the algorithms of [8] and [9], the contribution of this paper is twofold. First, our algorithm can directly hide data into the MBs without the data of predictable pattern and MB type. This greatly reduces the complexity of the embedding operation, and it is more appealing to real-time processing. Second, our algorithms take full advantage of all the MBs, including not only intra-MBs but also inter-MBs, hence there is a greater embedding capacity than the other algorithms'.

Ag.2 Extraction in GOP based on inter-MBs

Input: Hidden GOP, Threshold, Random sequence RS

Output: Message bitstream

foreach   decoded frame in the GOP do

  if the frame is b4-frame

    if the MB is inter-MBs && Luma Trans= 4×4

      if the sub-MB is neither the rightmost column nor the bottom row

      then Obtain QDCT coefficients $Y_{i,j}^{QDCT'}$

        if $|Y_{00}'| > Threshold$

        Select coefficients with RS

        then

$$m_i' = \begin{cases} 1, if\,(Y_{kl}^{i'}\%2 == 1) \\ 0, otherwise \end{cases}$$

    end

    until M' is completely extracted

    decrypt $M' \rightarrow M$

  end

---

4   Experimental Results

We implement the hiding and extraction **Algorithm 1** and **Algorithm 2,** and integrate them to the MVC video with two viewpoints encoder and decoder operation. We conduct a number of experiments to compare the performance between Ma et al.'s algorithm and the proposed algorithm on the nine test sequences including Akko&Kayo, Ballroom, Crowd, Exit, Flamenco, Objects, Race, Rena and Vassar. The size of GOP is chosen to be 8 and its frame structure is set as shown in Fig 1. All implementations are realized on a Lenovo compatible computer with an Intel Core 2 Due E5500 CPU 2.8 GHz and 2 GB RAM. The operating system is Microsoft Windows 7 SP1, the program development environment is Visual C++2010, and the implementation platform is JM 18.4. All the test sequences have the frame size of 640×480 which corresponds to 1200 MBs per frame. Just like most of the literature, we use PSNR, SSIM, embedding capacity and decoding time [10] as the indicators to evaluation the proposed algorithm.

4.1   The Visual Quality Performance

Selecting the appropriate threshold, such as $Threshold = 2$, we implement the proposed algorithms to nine standard test sequences. The size of GOP is chosen to be 8 and its frame structure is set as shown in Fig 1. Then, Figs. 6–7 are two images set, namely, Ballroom and Crowd sequences, which are used to evaluate the visual quality performance. We compare the visual effect between Fig. 6-7(a) and each of Fig. 6-7(b), and observe that both concerned algorithms slightly degrade the visual quality when compared to the Original videos. It is obtained that our algorithms do not lead to a large decline in video quality.

(a)   (b)

Fig. 6. Visual effect of the resultant b4-frames in Ballroom sequence by (a) Original, (b) algorithm

(a)   (b)

Fig. 7. Visual effect of the resultant b4-frames in Crowd sequence by (a) Original, (b) algorithm 1

4.2   Algorithm Comparison and Discussion

The quality comparison with PSNR for difference algorithms is illustrated in Table 1. When compared to Ma et al.'s and Lin et al.'s algorithms, it is clear that the PSNR degradation of the proposed algorithm is slightly larger than that of their algorithms. The main reason for this is that our algorithm has a larger embedding capacity. However, the overall difference is very small, and our algorithms do not lead to a large decline in video quality. In addition to using PSNR to measure the quality of the embedded video frames, the well-known SSIM index [10] is employed to measure the visual quality as well.

Table 1. The DPSNR comparison between the algorithms in [8, 9] and the proposed algorithm for QP = 28

Sequence	PSNR	Ma et al.	Lin et al.	Proposed
	PSNR1	39.6790	39.6790	39.6790
Akko&Kayo	PSNR2	39.6784	39.6783	39.6782
	DPSNR	0.0006	0.0007	0.0008
	PSNR1	36.9550	36.9550	36.9550
Ballroom	PSNR2	36.9544	36.9542	36.9533
	DPSNR	0.0006	0.0008	0.0017
	PSNR1	35.8218	35.8218	35.8218
Crowd	PSNR2	35.8216	35.8213	35.8191
	DPSNR	0.0002	0.0005	0.0027
	PSNR1	38.4256	38.4256	38.4256
Exit	PSNR2	38.4255	38.4254	38.4253
	DPSNR	0.0001	0.0002	0.0003
	PSNR1	39.3892	39.3892	39.3892
Flamenco	PSNR2	39.3889	39.3886	39.3877
	DPSNR	0.0003	0.0006	0.0015
	PSNR1	37.8566	37.8566	37.8566
Objects	PSNR2	37.8565	37.8564	37.8562
	DPSNR	0.0001	0.0002	0.0004
	PSNR1	37.5640	37.5640	37.5640
Race	PSNR2	37.5635	37.5632	37.5631
	DPSNR	0.0005	0.0008	0.0009
	PSNR1	41.4660	41.4660	41.4660
Rena	PSNR2	41.4659	41.4658	41.4656
	DPSNR	0.0001	0.0002	0.0004
	PSNR1	36.6493	36.6493	36.6493
Vassar	PSNR2	36.6491	36.6489	36.6485
	DPSNR	0.0002	0.0004	0.0008

As shown in Table 2, when compared to Ma et al.'s and Lin et al.'s algorithms, although the PSNR degradation of the proposed algorithm is being, the SSIM index of the proposed algorithm is similar to that of Ma et al.'s and Lin et al.'s algorithms. The similar SSIM indexes of the three concerned algorithms imply that the proposed algorithm improves the embedding capacity of Ma et al.'s and Lin et al.'s algorithms without visual quality degradation. For QP = 28, when compared to Ma et al.'s algorithm, Table 2 demonstrates that PSNR degradation of the proposed algorithm is ranged from 0.007 dBs to 0.031 dBs, but the SSIM index is almost the same with each other.

Table 2. PSNR and SSIM comparison between the algorithms in [8, 9] and the proposed algorithm for QP = 28

Sequence	Item	Ma et al.	Lin et al.	Proposed
Akko&Kayo	PSNR	39.6784	39.6783	39.6782
	SSIM	0.9941	0.9932	0.9931
Ballroom	PSNR	36.9544	36.9542	36.9533
	SSIM	0.9848	0.9846	0.9843
Crowd	PSNR	35.8216	35.8213	35.8191
	SSIM	0.9735	0.9734	0.9733
Exit	PSNR	38.4255	38.4254	38.4253
	SSIM	0.9936	0.9934	0.9932
Flamenco	PSNR	39.3889	39.3886	39.3877
	SSIM	0.9926	0.9925	0.9923
Objects	PSNR	37.8565	37.8564	37.8562
	SSIM	0.9941	0.9932	0.9931
Race	PSNR	37.5635	37.5632	37.5631
	SSIM	0.9768	0.9765	0.9763
Rena	PSNR	41.4659	41.4658	41.4656
	SSIM	0.9956	0.9954	0.9953
Vassar	PSNR	36.6491	36.6489	36.6485
	SSIM	0.9739	0.9736	0.9734
Average	PSNR	38.2004	38.2002	38.1997
	SSIM	0.9866	0.9862	0.9860

Table 3 gives the embedding capacity for the nine video sequences with different algorithms. It is observed that the embedding capacity in Ma et al.'s and Lin et al.'s algorithms are very small since there are exactly one I-frame in a GOP and only a few intra-MBs in b4-frames in hierarchical prediction structure. This indicates that the two algorithms are not suitable for 3D videos with hierarchical B-frames prediction structure. Moreover, compared with Ma et al.'s and Lin et al.'s algorithms, the average capacity improvement ratio of the proposed algorithm is increased by about 131%. The embedding capacity with our algorithms can be improved because we take full advantage of the inter-MBs.

The decoding time performance of the concerned two algorithms is shown in Table 4. Compared with [8] and [9], the proposed algorithms only require about 19.8ms to decode each frame from the compressed video sequence. The reason is that the proposed algorithm can directly hide data into MBs without the data of MB-type and prediction mode. This greatly reduces the complexity of the embedding operation, and more appealing to real-time processing.

Table 3. Embedding capacity comparison between the algorithms in [8, 9] and the proposed algorithm for QP = 28

Sequence	Ma et al.	Lin et al.	Proposed
Akko&Kayo	32	45	106
Ballroom	65	108	219
Crowd	128	190	296
Exit	95	159	253
Flamenco	111	196	305
Objects	67	132	263
Race	157	231	321
Rena	56	106	406
Vassar	201	306	329

Finally, we analyze the security and the robustness of the algorithm. The proposed algorithms are all blind detection algorithm which can retrieve the embedded message exactly without the knowledge of the original videos. And the proposed algorithms are readable data hiding algorithm which aims to retrieve the embedded message exactly without the knowledge of the embedded contents. The security of the algorithm is mainly reflected in the following three aspects. Firstly, the embeddable blocks are selected based on the absolute value of their DC coefficients. So, we can take the threshold as a key. Secondly, the proposed algorithm maintains a great randomness by using the three random sequences, which also improves the security of the algorithm. Thirdly, we can also improve the security by cryptographic operations before hiding data. And with regard to the robustness, it is the same with Ma et al,' algorithm. If the video processing operation changes the either absolute value of their DC coefficients or the value of their AC coefficients, which makes an embedded block un-extractable, then the embedded bit will not be retrieved.

Table 4. Decoding time comparison between the algorithms in [8, 9] and the proposed algorithm for QP = 28. (ms/frame).

Sequence	Ma et al.	Lin et al.	Proposed
Akko&Kayo	36.928	37.662	18.926
Ballroom	37.659	38.026	19.305
Crowd	37.862	38.236	19.415
Exit	37.713	38.215	19.403
Flamenco	37.807	38.225	19.413
Objects	37.509	37.987	19.216
Race	38.219	38.563	19.689
Rena	37.556	37.952	20.299
Vassar	38.401	38.964	20.311

5   Conclusion

We have presented a new data hiding algorithm to resolve the distortion drift problem for MVC videos. The contribution of this work is threefold. First, unlike [8] and [9], the proposed scheme can take full advantage of inter-MBs and all MBs with 4×4 IDCT to hide data. Moreover, the proposed algorithm can directly hide data into MBs without the data of MB-type and prediction mode. This greatly reduces the complexity of the embedding operation, and is more appealing to real-time processing. Second, embedding data into b4-frames can completely prevent the inter-frame distortion drift. Third, the proposed algorithm is oriented carrier for the new field of 3D videos.   Based on nine test video sequences, experimental results confirm the embedding capacity and complexity superiority of the proposed algorithm whilst keeping the similar human visual effect in terms of SSIM index.

Acknowledgments. The work described in this paper was supported by the National Natural Science Foundation of China under Grant (61272407).

References

1. Ho, Y., Oh, K.: Overview of multi-view video coding. In: Proc. 14th Int. Workshop Syst. Signals Image Process., pp. 5–12 (2007)
2. Merkle, P., Smolic, A., Mueller, K., Wiegand, T.: Efficient prediction structures for multiview video coding. IEEE Trans. Circuits Syst. Video Technol. 17(11), 1461–1473 (2007)
3. Noorkami, M., Mersereau, R.: A framework for robust watermarking of H.264-encoded video with controllable detection performance. IEEE Trans. Inform. Forensics Security 2(1), 14–23 (2007)
4. Chen, M.J., Chen, C.S., Chi, M.C.: Temporal error concealment algorithm by recursive block-matching principle. IEEE Transactions on Circuits and Systems for Video Technology 15(11), 1385–1393 (2005)
5. Zhang, J.: Robust video watermarking of H.264/AVC. IEEE Trans. Circuits Syst. II: Express Briefs 54(2), 205–209 (2007)
6. Wong, K.S., Tanaka, K., Takagi, K., Nakajima, Y.: Complete video quality preserving data hiding. IEEE Transactions of the Circuits and Systems for Video Technology 19(10), 1499–1512 (2009)
7. Kim, S.-M., Kim, S.B., Hong, Y., Won, C.S.: Data hiding on H.264/AVC compressed video. In: Kamel, M.S., Campilho, A. (eds.) ICIAR 2007. LNCS, vol. 4633, pp. 698–707. Springer, Heidelberg (2007)
8. Hao, T., Bochao, Z.: A data hiding algorithm for H.264/AVC video streams without intra-frame distortion drift. IEEE Trans. Circuits Syst. Video Technol. 20(10), 1320–1330 (2010)
9. Lin, T.J., Chung, K.: An Improved DCT-based Perturbation Scheme for High Capacity Data Hiding in H.264/AVC Intra-frame Frames. Journal of Systems and Software 86(3), 604–614 (2013)
10. Wang, Z., Bovik, A.C., Sheikh, H.R., Simoncelli, E.P.: Image quality assessment: from error visibility to structural similarity. IEEE Transactions on Image Processing 13(4), 600–612 (2004)

A Robust Reversible Data Hiding Scheme for H.264 Based on Secret Sharing

Yunxia Liu[1,2,*], Suimin Jia[1], Mingsheng Hu[1,*], Zhijuan Jia[1], Liang Chen[3], and Hongguo Zhao[2]

[1] Zhengzhou Normal University, Zhengzhou 450044, China
[2] Huazhong University of Science and Technology, Wuhan 430074, China
[3] National Computer Network Emergency Response Technical Team/Coordination Center of China, Beijing 100029, China
liuyunxia0110@mail.hust.edu.cn, hero_jack@163.com

Abstract. This paper proposes a robust reversible data hiding scheme in H.264. We divide the original video into several groups by using the secret sharing technique to improve the robustness of the embedded data. Then we embed the embedded data into the coefficients of the 4 × 4 discrete cosine transform (DCT) block of the selected frames which meet our conditions to avert the distortion drift. The experimental results show that this new robust reversible data hiding algorithm can get more robustness, effectively avert intra-frame distortion drift.

Keywords: Reversible Data hiding, H.264/advanced video coding (AVC), Secret Sharing, Intra-frame distortion drift.

1 Introduction

Data hiding embeds data into cover media contents and can be used in many applications such as law enforcement, perceptual transparency and medical systems, user identification, copyright protection, video indexing, and access control, etc. There exists a drawback in many data hiding methods that the cover media is permanently distorted since irreversible operations such as, quantization, truncation, etc, which makes the application of data hiding prohibited in the fields of law enforcement, medical systems and perceptual transparency, etc. Then it is desired to reverse the marked media back to the original cover media without any distortions after the hidden data are retrieved. The earliest reversible data hiding method is presented in [1]. After that, many reversible data hiding schemes are proposed [2]-[8]. H.264/advanced video coding (AVC) is the latest standard for video compression with high compression efficiency. It is well-adapted for network transmission which is poised to replace the existing video coding standards [9]. However, a thorough investigation of reversible data hiding scheme shows only in [10]-[13] for H.264 (H.264/AVC) video. Therefore, research on H.264 reversible data hiding methods is very valuable.

---

* Corresponding author.

D.-S. Huang et al. (Eds.): ICIC 2014, LNCS 8588, pp. 553–559, 2014.

The robust data hiding algorithm in video data is proposed since the embedded bits sometimes cannot survive from network transmission, packet loss, video-processing operations, various attacks, and so on. The embedded message can be retrieved correctly because the embedded bits can be recovered without any error. Secret Sharing has also been implemented in the literature [14] [15], but it was not built on H.264. The infinite video sequences of H.264 just meet the redundancy properties of Secret Sharing. To the best of our knowledge, the robust reversible data hiding method based on secret sharing has not been investigated to be compatible to H.264 standard. Consequently, further study and investigation are required.

Intra-frame distortion drift is a huge problem of data hiding in H.264 because this technique increases the dependence of the neighbouring blocks. When one frame is changed by data hiding, the reconstructed pixels of related frames will be influenced, i.e., intra-frame distortion drift happens. The distortion drift problem is first discussed in data hiding for compressed videos and a drift compensation method to counteract this problem is provided by Hartung and Girod [16]. Although many data hiding studies for compressed videos have employed H&G's method to restrain distortion in stego-video [17]-[20], the original video cannot be obtained in the schemes and those schemes are not applicable to the reversible data hiding. [10]-[12] cannot avert the distortion drift, and although [13] that is reversible data hiding method for H.264 can avert the distortion drift, it is not robust enough. Till now, there is no specific robust reversible data hiding algorithm to avert distortion drift for H.264. To deal with this problem, a robust reversible data hiding based on H.264 without intra-frame distortion drift is proposed in this paper. In order to correct the error bits caused by network transmission, packet loss, video-processing operations, various attacks, and so on, we first use secret sharing before data hiding. In order to avert the distortion drift, we use the directions of intra-frame prediction to choose 4×4 blocks which meet our conditions and embed the encoded data into the DCT coefficients of the blocks.

The rest of this paper is organized as follows. Section 2 describes the data hiding algorithm. Experimental results are presented in Section 3 and we draw the conclusions of this paper in Section4.

2   The Proposed Method

In order to improve the robustness of the embedded data, the original video is firstly encoded by the secret sharing before data hiding. The original video is entropy decoded to get the intra-frame prediction modes and DCT coefficients. Then, the 4 × 4 luminance DCT blocks with large residuals and the appropriate coefficients for embedding are selected. The embedded data is embedded into appropriate coefficients based on modulo modulation. All the quantized DCT coefficients are entropy encoded to get the target embedded video. Fig.4 depicts the scheme of our proposed algorithm.

2.1   Embedding

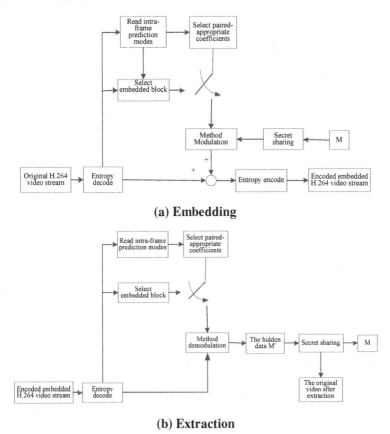

(a) Embedding

(b) Extraction

Fig. 1. Proposed data hiding scheme diagram: (a) Embedding. (b) Extraction

Fig.1 (a) describes the embedding operation. To make the embedding procedure more clear, the block which have a positive integer N and coefficient $\tilde{Y}_{i,j}$ (i,j=0,1,2,3) is exploited for data hiding as an example.

1. Divide the original video.

We divide the original video into several groups by using the secret sharing technique to improve the robustness of the embedded data.

2. Select embeddable blocks.

If the current block meets Condition 1, Condition 2 and Condition 3, the 4 × 4 luminance DCT block for embedding is selected.

Modulation Method

If $\left| \tilde{Y}_{i,j} \right| \neq N$, $\tilde{Y}_{i,j}$ is modified according to the formula (1);

If the embedded bit is 1 and $\left|\tilde{Y}_{i,j}\right|=N$, $\tilde{Y}_{i,j}$ is modified according to the formula (2);

If the embedded bit is 0 an $\left|\tilde{Y}_{i,j}\right|=N$, $\tilde{Y}_{i,j}$ is not modified.

$$\tilde{Y}_{i,j} = \begin{cases} \tilde{Y}_{i,j}+1, & \text{if } \tilde{Y}_{i,j} \geq 0 \text{ and } \left|\tilde{Y}_{i,j}\right| > N, \\ \tilde{Y}_{i,j}-1, & \text{if } \tilde{Y}_{i,j} < 0 \text{ and } \left|\tilde{Y}_{i,j}\right| > N, \\ \tilde{Y}_{i,j}, & \text{if } \left|\tilde{Y}_{ij}\right| < N. \end{cases} \tag{1}$$

$$\tilde{Y}_{i,j} = \begin{cases} \tilde{Y}_{i,j}+1, & \text{if } \tilde{Y}_{i,j} \geq 0 \text{ and } \left|\tilde{Y}_{i,j}\right|=N, \\ \tilde{Y}_{i,j}-1, & \text{if } \tilde{Y}_{i,j} < 0 \text{ and } \left|\tilde{Y}_{i,j}\right| = N. \end{cases} \tag{2}$$

2.2   Data Extraction

Fig. 2 (b) describes the extraction operation. After the entropy decoding of the H.264 video, we choose the embeddable blocks and extract the hidden data M ($M=m_1 m_2,..., m_i, m_i \in \{0,1\}$).Then we extract the hidden data M as follows:

If $\left|\tilde{Y}_{ij}\right| \neq N+1$, then $\tilde{Y}_{ij}$ is modified according to the formula (3);

If $\left|\tilde{Y}_{ij}\right|=N+1$, then the extracted bit is 1 and $\tilde{Y}_{ij}$ is modified according to the formula (4);

If $\left|\tilde{Y}_{ij}\right|=N$, then the extracted bit is 0 and $\tilde{Y}_{ij}$ is not modified.

$$\tilde{Y}_{i,j} = \begin{cases} \tilde{Y}_{i,j}-1, & \text{if } \tilde{Y}_{i,j} \geq 0 \text{ and } \left|\tilde{Y}_{i,j}\right|>N+1, \\ \tilde{Y}_{i,j}-1, & \text{if } \tilde{Y}_{i,j} < 0 \text{ and } \left|\tilde{Y}_{i,j}\right| > N+1, \\ \tilde{Y}_{i,j}, & \text{if } \left|\tilde{Y}_{i,j}\right| < N. \end{cases} \tag{3}$$

$$\tilde{Y}_{i,j} = \begin{cases} \tilde{Y}_{i,j}-1, & \text{if } \tilde{Y}_{i,j} \geq 0 \text{ and } \left|\tilde{Y}_{i,j}\right|=N+1, \\ \tilde{Y}_{i,j}+1, & \text{if } \tilde{Y}_{i,j} < 0 \text{ and } \left|\tilde{Y}_{i,j}\right| = N+1, \end{cases} \tag{4}$$

3   Experiment Result

The proposed method has been implemented in the H.264 reference software version JM16.0. We have successfully applied our method to the commonly used video sequences, such as Container, News, Akiyo and Coastguard, etc. In this paper the block is selected based on Y_{00} since the un-quantized coefficient Y_{00} is the average of the residual block according to the 4×4 DCT, and the usage of criterion $Y_{00} \neq 0$ since embedding data into a block of all zero coefficients would bring significant visual distortion. In order to improve the robustness of the embedded message, we use the secret sharing technique before data hiding. We encoded 300 frames of standard video sequences at 30frame/sand use an intra-period of 15. The survival rate in this paper is that the number of all the embedded bits is divided by the number of the embedded bits retrieved without error. The frame loss rate in this paper is that the number of experimental frame is divided by that of random discarded frames. The embedding capacity of a video sequence in this paper is the number of bits embedded in the20 I frames of that sequence.

Table 1 shows the performance comparison between our algorithm and [13] when using different frame loss rates. The average survival rate of [13] and our method is 24.25 % and 82.42.14% respectively. To all the sequences, the survival rate of the method has increased 57.97 % compared with [13]. Although the embedding capacity of [13] is higher than our method, we can improve the number of the embedded bits by using more H.264 since the infinite video sequences of H.264 just meet the redundancy properties of secret sharing. Thus, we can embed the bits what we want.

Table 1. Embedding performance comparison between our algorithm and [13] when using different frame loss rates

Sequences	Frame loss rate(%)	Survival rate of [13] (%)	Survival rate of our scheme (%)
Container	10	9.37	86.66
	20	14.06	90.00
News	10	26.37	87.12
	20	23.07	78.78
Mobile	10	21.24	77.32
	20	23.53	60.42
Akiyo	10	54.77	99.19
	20	21.62	79.84

4   Conclusion

In this paper, we focus on the robust reversible data hiding scheme of data hiding in H.264 video. This scheme does not conflict with the algorithms presented in [13], and the experiment results show that our method can get good visual quality and more robustness than [13] .In our future work, we will try to combine the proposed algorithm with the techniques in [22]-[28] to improve our method.

Acknowledgments. This paper is sponsored by the National Natural Science Foundation of China (NSFC, Grant U1204703, U1304614, 61272407), the Key Scientific and Technological Project of Henan Province (122102310004), the Innovation Scientists and Technicians Troop Construction Projects of Zhengzhou City (10LJRC190, 131PCXTD597) and the Key Scientific and Technological Project of The Education Department of Henan Province (13A413355, 13A790352, ITE12001, 14A630057).

References

1. Barton, J.: Method and apparatus for embedding authentication information within digital data. U.S. Patent 5 646 997 (1997)
2. Tian, J.: Reversible data embedding using a difference expansion. IEEE Trans. Circuits Syst. Video Technol. 13(8), 890–896 (2003)
3. Ni, Z.C., Shi, Y.Q., Ansari, N., Su, W.: Reversible data hiding. IEEE Trans. Circuits Syst. Video Technol. 16(3), 354–362 (2006)
4. Hu, Y.J., Heung-K., L., Li, J.W.: DE-based reversible data hiding with improved overflow location map. IEEE Trans. Circuits Syst. Video Technol. 19(2), 250–260 (2009)
5. Fridrich, J., Goljan, M., Du, R.: Invertible authentication watermark for JPEG images. In: Proc. of the International Conference on Information Technology: Coding and Computing, pp. 223–227 (2001)
6. Rui, D., Fridrich, J.: Lossless authentication of MPEG-2 video. In: Proc. of the Internationa Conference on Image Processing, vol. 2, pp. 893–896 (2002)
7. Chen, H., Chen, Z.Y., Zeng, X., Fan, W., Xiong, Z.: A novel reversible semi-fragile watermarking algorithm of MPEG-4 video for content authentication. In: Proc. of the Second International Symposium on Intelligent Information Technology Application, vol. 3, pp. 37–41 (2008)
8. Zeng, X., Chen, Z.Y., Xiong, Z.: Issues and solution on distortion drift in reversible video data hiding. Multimed. Tools Appl. 52, 465–484 (2011)
9. Richardson, I.E.G.: H.264 and MPEG-4 video compression: video coding for next-generation multimedia. Wiley, U.K. (2003)
10. Piva, A., Caldelli, R., Filippini, F.: Data hiding for error concealment in H.264/AVC. In: IEEE. MMSP 2004, Siena, Italy, pp. 199–202 (2004)
11. Lie, W.N., Lin, T.C., Tsai, D.C., et al.: Error Resilient Coding Based on Reversible Data Embedding Technique for H.264/AVC Video. In: IEEE International Conference on Multimedia and Expo (2005)
12. Lin, S.D., Meng, H.C., Su, Y.L.: A novel error resilience using reversible data embedding in H.264/AVC. In: International Conference on Information, Communications and Signal Processing, ICICS 2007, Singapore, pp. 1–5 (2007)

13. Liu, Y.X., Ma, X.J., Li, Z.T.: A Reversible Data Hiding Scheme Based on H.264/AVC without Distortion Drift. Journal of Software 7(5), 1059–1065 (2012)
14. Wu, Y.S., Thien, C.C., Lin, J.C.: Sharing and hiding secret images with size constraint. Pattern Recognition 37(7), 1377–1385 (2004)
15. Chang, C.C., Lin, C.Y., Tseng, C.S.: Secret image hiding and sharing based on the (t, n)-threshold. Fundamenta Informaticae 73, 1–13 (2006)
16. Hartung, F., Girod, B.: Watermarking of uncompressed and compressed video. Signal Process. 66(3), 283–301 (1998)
17. Winne, D., Knowles, H., Bull, D., Canagarajah, C.: Spatial digital watermark for MPEG-2 video authentication and tamper detection. In: Proc. of International Conference on Acoustics, Speech, and Signal Processing (ICASSP), vol. 1(4), pp. 3457–3460 (2002)
18. Alattar, A., Lin, E., Celik, M.: Digital watermarking of low bit-rate advanced simple profile MPEG-4 compressed video. IEEE Trans. Circuits Syst. Video Technol. 13(8), 787–800(2003)
19. Biswas, S., Das, S., Petriu, E.: An adaptive compressed MPEG-2 video watermarking scheme. IEEE Trans. Instrum. Meas. 54(5), 1853–1861 (2005)
20. Chen, C., Gao, T.G., Li, L.Z.: A compressed video watermarking scheme with temporal synchronization. In: Proc. of the Congress on Image and Signal Processing (CISP), pp. 605–612 (2008)
21. Ma, X.J., Li, Z.T., Tu, H., Zhang, B.C.: A data hiding algorithm for H.264/AVC video streams Without Intraframe Distortion Drift. IEEE Trans. Circuits and Systems for Video Technology 20(10), 1320–1330 (2010)
22. Wang, X.F., Huang, D.S., Xu, H.: An efficient local Chan-Vese model for image segmentation. Pattern Recognition 43(3), 603–618 (2010)
23. Huang, D.S., Du, J.X.: A constructive hybrid structure optimization methodology for radial basis probabilistic neural networks. IEEE Transactions on Neural Networks 19(12), 2099–2115 (2008)
24. Huang, D.S.: Radial basis probabilistic neural networks: Model and application. International Journal of Pattern Recognition and Artificial Intelligence 13(7), 1083–1101 (1999)
25. Wang, X.F., Huang, D.S.: A novel density-based clustering framework by using level set method. IEEE Transactions on Knowledge and Data Engineering 21(11), 1515–1531 (2009)
26. Huang, D.S., Mi, J.X.: A new constrained independent component analysis method. IEEE Trans. on Neural Networks 18(5), 1532–1535 (2007)
27. Li, B., Huang, D.S., Wang, C., Liu, K.H.: Feature extraction using constrained maximum variance mapping. Pattern Recognition 41(11), 3287–3294 (2008)
28. Du, J.X., Huang, D.S., Wang, X.F., Gu, X.: Shape recognition based on neural networks trained by differential evolution algorithm. Neurocomputing 70(4-6), 896–903 (2007)

A High-Dynamic Invocation Load Balancing Algorithm for Distributed Servers in the Cloud

Zhaoyang Qu, Jiannan Zang, Ling Wang[*], Huiyu Sun, and Yongwen Wang

Department of Computer Science and Technology, School of Information Engineering,
Northeast Dianli University, Jilin, China
qzywww@mail.nedu.edu.cn,
{907331523,690011768,995802183}@qq.com,
smile2867ling@gmail.com

Abstract. Nowadays, cloud storage has received widespread attention for sharing of resources to achieve coherence and economies of scale. Focus on maximizing the effectiveness of the shared resources, how to allocate tasks reasonably and enhance the load balance are critical challenges that enhancing the overall performance of cloud service platform. In this paper, we proposed a high-dynamic invocation load balancing algorithm (LY-Cluster) for distributed servers in the cloud. There are three main contents that automatically allocate services' IDs, multi-level capacity manager, and dynamically reallocated per demand based on sudden tasks. The experimental results show that our method performs well in terms of load balancing across the service replicas and improves the system scalability and response time.

Keywords: cloud storage, load balancing, cluster scheduling, cloud clusters, LY-Cluster.

1 Introduction

With the rapid development of Internet, digitalization and informationization are influencing our lives everywhere and bringing us convenience as well as great risks that exponential growth in the server traffic. The statistics of IDC (International Data Corporation) shows that the global data volume has reached 2.8ZB with an anticipation of 40ZB in 2020, which also means the revolution driven by information technology, is accelerating. Facing the surging tide of incoming data, it is difficult for traditional single web server to meet the demand of actual processing. Accordingly, the parallel processing architecture of cloud storage system has become the effective solution to server processing power shortage situation with its characteristics of high performance, low price and flexible scalability [1]. The mainstream of cloud storage server cluster structure is shown in Fig.1.

In the cluster of cloud storage servers, front desk directory server is mainly responsible for storing data, receiving them and assigning different access request to the corresponding background in the server for processing according to the data

---

[*] Corresponding author.

D.-S. Huang et al. (Eds.): ICIC 2014, LNCS 8588, pp. 560–571, 2014.

directory. The cloud platform, known as the third IT wave after personal computer and Internet revolution for its excellent characteristics, has become an effective approach to solve processing problem due to 4V characteristics of data in current big data era [2]. However, in the cloud storage server cluster environment, reasonable allocation of tasks and improvements of load balance degree are the key to enhance the overall performance of cloud storage system due to the differences among servers in processing and loading capacity.

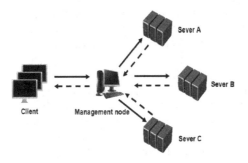

Fig. 1. Server cluster structure

In order to solve the defects of load balancing in cloud storage server cluster, we proposed a new load balancing algorithm, ripple-type load balancing algorithm for cloud storage services (LY-Cluster). When a node in the cluster server overloaded, the high load capacity will be allocated rationally to the lower call server cluster to realize the balancing server cluster. LY-Cluster is a high-dynamic invocation load balancing algorithm for distributed servers in the cloud. There are three main contents that automatically allocate services' IDs, multi-level capacity manager, and dynamically reallocated per demand based on sudden tasks.

2   Related Work

At present, the cluster scheduling algorithm has been widely applied in various fields, how to realize the optimal task allocation is still a NP problem. In the early process of practical application, Heuristic algorithm is often used to solve the scheduling problem [3]. Many scholars have done a lot of research work in this respect: Branu has been tested almost over eleven kinds of regular heuristic scheduling algorithm. These scheduling algorithms include OLB, UDA, Fast Greedy, Min-Min and Max-Min, Greedy, based Algorithm (GA) and Simulated Annealing (SA), GSA, Tabu, and A*, the experimental results show that the Min-Min, the GA and A* has good performance, but the Min-Min Algorithm of load balance is relatively poor; Aimed at this shortage, Sufferage algorithm was proposed in Maheswaran, the algorithm has better comprehensive performance than Min-Min algorithm [4].

As various types of computer application develop rapidly, higher requirements are required on the performance of the server cluster. However, limited by the size of the server cluster, processing ability, the performance of the server cluster does not meet the actual demand. Load balancing technology as an effective means to improve overall system performance have become the focus in the current server cluster. A cost model and incremental adjustment strategy was proposed in Overman, Prins, and Miller to facilitate more efficient use of AFMM (Adaptive Fast Multipole Method), implementation in the heterogeneous system using AFMM achieve load balance [5]; Chen et al., in view of the system in the treatment of complicated GPU (Graphics Processing Unit) when the phenomenon of poor performance, put forward a kind of Multi - GPU system based on task cycle dynamic load balancing solutions [6].

Since Google CEO Eric Schmidt first proposed the concept of "Cloud Computing" in the Search Engine Strategies, Cloud Computing has become the hottest topic in computer field [7]. Cloud computing task distribution in a large number of cheap computer constitute a pool of resources, enabling users to on-demand access to computing power, storage space and information services [8]. Cloud computing with its characteristics of high performance and low price, scalability, become a doubling of the current solution due to the amount of data of an effective method for computing needs serious shortage phenomenon. Cloud computing, with its characteristics of high performance and low price, scalability, has become an effective means to address the current doubling the amount of data computing needs arising from a serious shortage of phenomena.

Given the load balancing strategy cloud cluster environment is fundamentally different, in order to give full play to the advantages of cloud computing, improve system resource utilization, scholars at home and abroad have conducted a lot of research work: A kind of accurate cost estimation based adaptive load balancing algorithm was proposed in Gufler, Augsten, Reiser, and Kemper for MapReduce system in dealing with the science highly skewed data set when the load imbalance. The algorithm effectively solve the load balancing problem brought by the data skew[9]. A AFLBAK (Adaptive Feedback Load Balancing Algorithm in HDFS) algorithm was proposed in Fan and others to improve the shortages of insufficient load balancing in storage server, the algorithm greatly improve the degree of the HDFS load balance [10]; Fang and others are discussed based on cloud computing load balancing two hierarchical task scheduling mechanism, not only can meet the requirements of users, but also can obtain higher resource utilization [11]; A HBB-LB (Honey Bee Behavior inspired Load Balancing) algorithm was proposed in Venkata to achieve a balanced load on the virtual machine throughput maximization [12]; A scheduling strategy on load balancing of VM (Virtual Machine) resources based on genetic algorithm in Hu et al., in view of the load balancing problem in VM resources scheduling, the algorithm is based on historical data, and through the genetic algorithm and the current state of the integrated system for load balancing, has obtained the good effect of load [13].

3   Ripple-Type Load Balancing Algorithm for Cloud Storage Services

The current server cluster load balancing algorithm is mainly through real-time feedback of server status information of each node to adjust the task assignment. According to the different application needs of customers, the algorithm design different load balancing strategy. This kind of algorithm on the server cluster overall invocation pattern is still in the solid form. However, in the practical process, it's hard to guarantee the server of the cluster load balancing, and these algorithms cannot make full use of the resources of the server cluster, processing capacity restricted by the number of servers have some limitations. Compared with the existing load balancing algorithm, ripple-type load balancing algorithm for cloud services(LY-Cluster) can carry on the dynamic deployment to the cloud storage server cluster, and can provide unlimited increase based on ripples mechanism for cluster can ensure load balancing.

3.1   LY-Cluster Algorithm

The core idea of LY-Cluster algorithm is that when a node in the cluster server overloaded, the high load capacity would be allocated rationally to the lower call server cluster to realize the balancing server cluster, so as to realize the load balancing server cluster. In this process, the dynamic adjustment of each node in the cluster server ID, facilitating resource for a hot spot of high access to take corresponding processing mechanism, make server cluster can automatically adjust to adapt to the actual demand.

3.2   Algorithm Process

LY-Cluster algorithm mainly includes the following steps:

Step 1: the server in the cluster all n nodes encode $N_{1...n}$.

Step 2: the data stored in the server cluster according to the URL ascending sort, and the average data stored in each server node, each node storage server cluster all the data of 1/n.

Step 3: when a node N overloads, the node will be saved storage resources stored in the ripples mechanism in the form of partial data replication mechanism identified at a lower level node, and will access request for the data according to the predetermined value of load distribution, F assigned to the lower node for processing.

Step 4: when the monitor to the server on the actual traffic under load, reduce the level of call, release the subordinate relations, and delete the data will be copied to the lower nodes, reduce the data redundancy of server cluster. The general flows as shown in the Fig.2.

The loading condition of server node state is divided into 3 states: from state 1 to state 3 sequentially showing the idle, normal and overload condition. Every 10 seconds each server node using UDP packets (which contains a time stamp and load state) server report forward again. When the load jumps from one state to another, the server would be reported the forward one immediately.

Algorithm is to divide all data which shall be carried out in accordance with the URL, and respectively for storage, can improve the scalability of server cluster and Cache hit ratio; when the monitor to the overload node load returns to normal state, will be copied to the data in a call at a lower level node deletion, release the lower node resources, and reduce the redundancy of the system and to the server cluster can timely deal with other overload condition, improve the reliability of the server cluster.

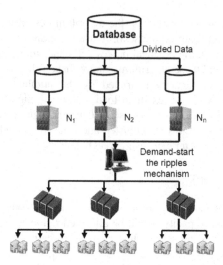

Fig. 2. Ripple server cluster load balancing algorithm flows

4   Ripple-Type Server IDs Allocation Mechanism

Ripple-Type Server IDs Allocation Mechanism, hereinafter referred to as the ripples mechanism, inspired by the lake ripples, when spend a small stones at the calm water of the lake, stir up layers of waves, and a wave of the impact of the water carrying stones together, until the calm.This mechanism will be overloaded server definition is given priority to the server, when the primary server overload occurs, lower nodes activated ripples mechanism determine its call.According to the actual needs of the primary server, which in turn calls primary, secondary and tertiary call node, assisted the master server to deal with the overload. Experiments of cloud storage server cluster of the communist party of China has 30 server nodes, respectively is $N_{1...30}$. Assume that three of them O, R, M server nodes appear overload phenomenon. Three overloaded server node graph shows as follows:

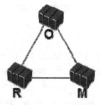

Fig. 3. Schematic diagram of the three overloaded nodes

When the cloud storage servers in the cluster server nodes appear overload phenomenon, ripples mechanism will redistribute server to server in the cluster server node ID, give priority to select the appropriate server nodes form its call cluster at a lower level, in order to help the master server complete access request processing. Server ID in the form of regulation for X_YX_Y, X to invoke the node master server name, Y is the call level.

4.1   The First Call Level

Invoked when the primary server is in level, with O, R, M as the center of the circle, draw three radius is r and a circle circle, determines the level of call server node ID for O_1R_1, O_1M_1, R_1M_1, these three server nodes will provide corresponding master server processing power at the same time support. Level 1 call diagram as shown in the figure below:

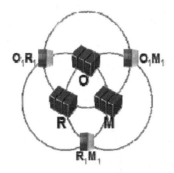

Fig. 4. The first call level schematic

Through above you can see, each node level call to serve two primary server at the same time, assuming that each master server assigned to the lower nodes of maximum load F, guarantees the maximum load are called node 2F, protect the load balance of nodes.

4.2   The Second Call Level

When level 1 call cannot meet the actual demand, ripples mechanism will start the second call.Secondary call, with three main server as the center of the circle, the radius 2r circle, will get 12 nodes, as shown in the figure below:

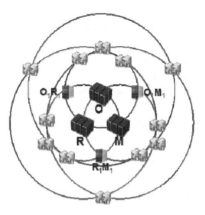

Fig. 5. The second call level schematic

However, in the second call, there will be a repetition of the name server nodes, at the same time, the second call appears far more than the number of server nodes level node number of the call. Therefore, in order to give a unique server every call node ID, and reduce the secondary call actual call node number, need to be seen, the consolidation of the intersections of secondary call when calling node 2r + 1r of server nodes with the primary server secondary invocation pattern of server nodes. Integration formula is as follows:

$$
\begin{pmatrix} O_2 \\ R_2 \\ M_2 \end{pmatrix} \begin{pmatrix} O_1O_2 & R_1R_2 & M_1M_2 \end{pmatrix} = \begin{pmatrix} O_2O_1O_2 & O_2R_1R_2 & O_2M_1M_2 \\ R_2O_1O_2 & R_2R_1R_2 & R_2M_1M_2 \\ M_2O_1O_2 & M_2R_1R_2 & M_2M_1M_2 \end{pmatrix}
\tag{1}
$$

Of 9 nodes, exist $O_2O_1O_2$ this only be invoked by a single master server nodes, is not in conformity with the ripples mechanism model, so get rid of this kind of node, end up with six nodes.Integration of the diagram as shown in the figure below:

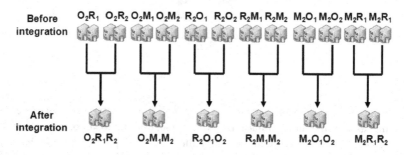

Fig. 6. Integrated schematic node

After transformation, the original 6, 12 point to reduce the secondary call start the number of servers, at the same time, reduce the waste of resources security can enough call other overload node auxiliary for request processing.

4.3   Multistage Call Level

When secondary calls are insufficient to meet the actual demand, ripples mechanism will continue to launch level 3 or higher level calls, the node call process and secondary call principle is the same. Verified, second class above invoked, the final call on the number of nodes per level is 6.

In the process of ripples mechanism called the lower server, the server is load that call at the server overload. The mechanism can be on the basis of the existing cloud storage server cluster resources, make full use of the cluster and has the following features:

• independent distribution server ID address;
• multi-level capacity manager;
• dynamically reallocated per demand based on sudden tasks;
• keep load balance.

4.4   Program Code

Input: a matrix of Server nodes ($V(G)$), number of current access (*access-current*), number of max access (*access-max*).
Output: reallocation.

```
begin
var LY-Cluster mechanism level(Y), average access(S),
number of new add node(N)
  FG ←|V(G)| × |V(G)|
  O ,R ,M ← FG(a, b)
  access-current←Threshold
  for each nodeList.size()∈ FG do
  if server.access-max < access-current then
    if Y = 1 then
      (O₁, R₁, M₁) ← FG₁(a₁, b₁) ←|V(G₁)| × |V(G₁)|
      if a₁!=b₁ and ( a₁,b₁ no repeat) then
        O₁ ← S₁=access-current₁/(N₁+1)
        R₁ ← S₂=access-current₂/(N₂+1)
        M₁ ← S₃=access-current₃/(N₃+1)
    else if Y !=1 then
      (Oᵧ, Rᵧ, Mᵧ) ← FG1(a₁, b₁, c₁) ← |V(G₁)| × |V(G₁)|
      if a₁!=b₁ and a₁!=c₁ and b₁!=c₁ and ( a₁,b₁,c₁ no
      repeat) then
          Oᵧ ← S₁=access-current₁/(N₁+1)
          Rᵧ ← S₂=access-current₂/(N₂+1)
          Mᵧ ← S₃=access-current₃/(N₃+1)
      end if
    end if
  end if
  return FG₁
end
```

5   The Experimental Scheme and Evaluation

5.1   Experimental Methods and Result

This study using data sets DAIM research group from Beijing university, the experiment by using it on HP DL380 server virtual building contains 30 gathers with the same capacity node group of environment, and simulation test in which the ripple-type load balancing algorithm for cloud storage services.

In order to ensure that the load when the trigger ripples mechanism accuracy, the first to set up the server in the cluster nodes to conduct stress tests. Set up by the node to make use of TCPCopy tools, concurrency is respectively 200, 800, 1600 and 2000 the number of the four processing speed and packet loss rate test. Test results are as follows:

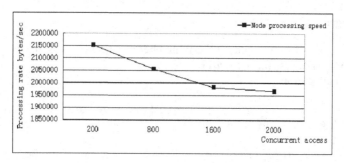

Fig. 7. Effect of concurrency on the processing speed

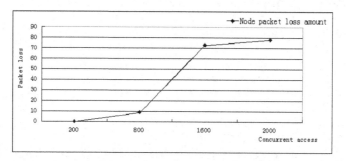

Fig. 8. Effect of concurrency on the loss ratio

Through the actual test results can be obtained, as the growth of the concurrency value, the server's processing speed is lower, the growing number of packet loss. When the server number of concurrent requests reached 2000 times per second, the server is down.Finally, through testing, each server node node pressure test value is 2000 times per second. Load balancing degree is refers to the total deviation of server cluster all backend server load and the ratio of the total load. This experiment uses the server of the absolute value of the difference between the actual load and the load average on the backend server load total deviation said.Load balancing calculation formula is as below:

$$B_L = \frac{\sum |L_i - \overline{L}|}{\sum L_i} \tag{2}$$

Among them, the said B_L load balancing degree, L_i for the ith a node in the cluster load, $\overline{L}$ for cluster load average. Thus, the smaller the load balance degree of equilibrium the better the results. Experiments by extracting the data set of valid data, in different traffic conditions, the cloud cluster load balancing test. Experiment to the partial copy content based scheduling algorithm of the principle of PRLARD (Partially Replicated the Locality - Aware Request Distribution) and this ripple-type load balancing algorithm for cloud storage services, comparing the load balancing server cluster algorithm contrast in order to guarantee the validity of the assumption that at the start of the test, all processing nodes are idle, and assigned to the lower load in a call node F 1000 times per second. The resulting load balancing degree for such as shown below:

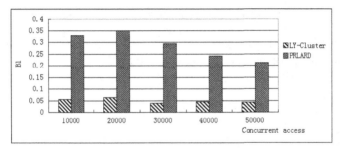

Fig. 9. Contrast of load balancing

In the above experiments, the two different algorithms of load balancing degree were compared. Server, in turn, launched a to 5 calls, to enhance the accuracy of the experiment, to avoid the accident of the experimental results, the experiment on the same level for concurrent call traffic within the scope of the different test values, the results shown in the figure below:

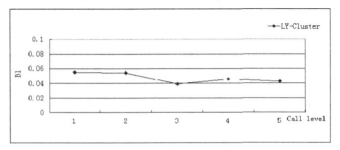

Fig. 10. Call test with at the same stage

Experimental results show that, under the same level pattern call different concurrent traffic corresponding load balance degree is almost the same, so as to show the results in Fig.9. Next to the server through the experiment under the different call level of load balancing cluster degree of contrast test, verify whether LY-Cluster algorithm can provide unlimited increase for the cloud platform server cluster load balancing. Test results as shown in the figure below:

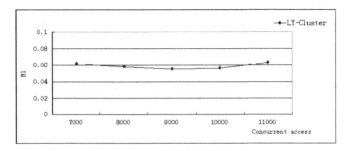

Fig. 11. Comparison of the different call-stage load balancing

Meanwhile, on the premise of concurrent access to the same amount, the experiment tests and compares the system response time and request processing speed of the cloud platforms during the use of two different load balancing algorithm. The results of simulation comparison chart is shown in Fig.12.

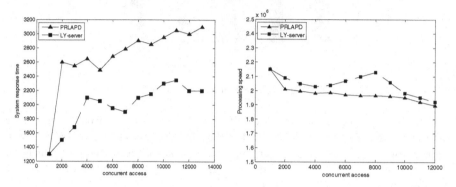

Fig. 12. Comparison of system performance testing

Fig.12 shows that compared with PRLARD algorithm, LY-Cluster algorithm has shorter system response time and higher data throughput rate. LY-Cluster algorithm can effectively guarantee the system in a state of smooth processing.

5.2   Discussion

Analyzing from the two load-balancing algorithms, the comparison chart of load-balancing degree draws the conclusion as below:

Firstly, LY-Cluster algorithm has better load balancing performance than PRLARD algorithm. LY-Cluster algorithm will evenly allocate the overloaded traffic of a certain node to its corresponding node cluster. When the normal node is used collectively by the primary server, the access amount of its corresponding node is comparatively even, reducing the deviation of average load in cluster. Thus, comparing with the PRLARD algorithm which uses the partial replication to replace role, the LY-Cluster algorithm has better load balancing degree. Secondly, the slight difference of the cluster load balancing degree during the use of LY-Cluster algorithm in different level, also can confirm that LY-Cluster algorithm can provide ever-increasingly reliable load balance.Finally, by the comparing analysis of the two parameters, which includes the system response time and request processing speed during the use of two different load balancing algorithm, LY-Cluster algorithm can significantly improve the processing performance of the system, and effectively enhance the current weak situation when the cloud platforms deal with large scale access.

6   Summary

Based on the systematic analysis of existing load balancing algorithm, our proposed LY-Cluster algorithm could automatically allocate services' IDs, multi-level capacity manager, and dynamically reallocated per demand based on sudden tasks over cloud storage services. Theoretical analysis and experimental results show that the algorithm has good capability of load balancing, which can provide ever-increasingly reliable load balance, and guarantee the cloud platform system effectively when dealing with large-scale accesses. In the future, we will go into details of private cloud system and research how to enhance the overall performance of the private cloud cluster.

Acknowledgements. This work was supported by the National Natural Science Foundation of China (No.51277023) and by the Jilin provincial department of science and technology (No.20120363).

References

1. Armbrust, M., Fox, A., Griffith, R., Joseph, A.D., Katz, R., Konwinski, A., Zaharia, M.: A view of cloud computing. Communications of the ACM. 53(4), 50-58 (2010)
2. Espadas, J., Molina, A., Jiménez, G., Molina, M., Ramírez, R., Concha, D.: A tenant-based resource allocation model for scaling Software-as-a-Service applications over cloud computing infrastructures. Future Generation Computer Systems. 29(1), 273-286 (2013)
3. Subramani, V., Kettimuthu, R., Srinivasan, S., Johnston, J., Sadayappan, P.: Selective buddy allocation for scheduling parallel jobs on clusters. In: 4th IEEE International Conference on Cluster Computing, pp. 107-116. IEEE Press, Illinois (2002)
4. Maheswaran, M., Ali, S., Siegel, H. J., Hensgen, D., Freund, R. F.: Dynamic mapping of a class of independent tasks onto heterogeneous computing systems. Journal of Parallel and distributed Computing, 59(2), 107-131 (1999)
5. Overman, R. E., Prins, J. F., Miller, L. A., Minion, M. L.: Dynamic Load Balancing of the Adaptive Fast Multipole Method in Heterogeneous Systems. In: 2013 IEEE 27th International Conference on Parallel and Distributed Processing Symposium Workshops & PhD Forum (IPDPSW), pp. 1126-1135. IEEE Press, Cambridge (2013)
6. Chen, L., Villa, O., Krishnamoorthy, S., Gao, G. R.: Dynamic load balancing on single- and multi-GPU systems. In: 2010 IEEE International Symposium on Parallel & Distributed Processing (IPDPS), pp. 1-12. IEEE Press, Atlanta (2010)
7. Ghemawat, S., Gobioff, H., Leung, S. T.: The Google file system. In: ACM SIGOPS Operating Systems Review. Vol. 37, No. 5, pp. 29-43. ACM (2003)
8. Ekanayake, J., Fox, G.: High performance parallel computing with clouds and cloud technologies. In: Avresky, D.R., Diaz, M., Bode, A., Ciciani, B., Dekel, E.(eds.) Cloudcomp 2009. LNICS, vol. 34, pp. 20-38. Springer, Heidelberg (2010)
9. Gufler, B., Augsten, N., Reiser, A., Kemper, A.: Load balancing in mapreduce based on scalable cardinality estimates. In: 2012 IEEE 28th International Conference on Data Engineering (ICDE), pp. 522-533. IEEE Press, Washington (2012)
10. Fan, K., Zhang, D., Li, H., Yang, Y.: An Adaptive Feedback Load Balancing Algorithm in HDFS. In: 5th International Conference on Intelligent Networking and Collaborative Systems (INCoS), pp. 23-29. IEEE Press, Xi'an (2013)
11. Fang, Y., Wang, F., Ge, J.: A task scheduling algorithm based on load balancing in cloud computing. In: Wang, F.L., Gong, z., Luo, X., Lei, J.(eds.)WISM 2010. LNCS, vol. 6318, pp. 271-277. Springer, Heidelberg (2010)
12. Venkata Krishna, P.: Honey bee behavior inspired load balancing of tasks in cloud computing environments. Applied Soft Computing, 13(5), 2292-2303 (2013)
13. Hu, J., Gu, J., Sun, G., Zhao, T.: A scheduling strategy on load balancing of virtual machine resources in cloud computing environment. In: 2010 Third International Symposium on Parallel Architectures, Algorithms and Programming (PAAP), pp. 89-96. IEEE Press, Dalian (2010)

An Effective Algorithm for Interference Minimization in Wireless Sensor Networks

Xiaoke Zhang and Weidong Chen[*]

School of Computer Science, South China Normal University, Guangzhou 510631, China
chenwd2007@hotmail.com

Abstract. Interference Reduction is one of fundamental issues to be addressed in wireless communication when energy conservation is considered. Through the topology control method, this is done by adjusting the transmission radii of nodes while preserving the network connectivity. Under the receiver-centric model, the problem of minimizing the maximum interference of a wireless sensor network has been proved to be NP-hard. In this paper, an effective algorithm called Best Neighbor Algorithm is proposed to solve the problem. Simulation results show that the proposed algorithm outperforms an existing popular algorithm called Nearest Neighbor Algorithm with regard to solution quality for almost all instances randomly generated.

Keywords: Interference minimization, wireless sensor network, topology control, Algorithm.

1   Introduction

In recent years, wireless sensor networks have received extensive attentions. A wireless sensor network consists of mobile nodes with limited processing ability, each equipped with a wireless radio and a power source. Wireless networks have great potential in many application scenarios such as battlefield, monitoring and surveillance, medical treatment, and city construction [3, 11]. Because the workplace of wireless sensor networks is typically not convenient to control directly for human beings and the power source of nodes is very limited, energy has become one of key factors limiting the service life of wireless networks. One way to conserve energy is to reduce interference that happens when close-by nodes transmit signals concurrently.

How to reduce interference is one of critical issues in wireless communication [7]. The topology control method is an important approach to deal with this issue, which involves assigning a suitable transmission radius to each node to form a connected network while minimizing some objective function of the radii [5]. Different models have been proposed to address the issue of reducing interference in wireless communication. One is called *the sender-centric model*, where interference refers to the interference of edge [2, 13]. Interference of this type considers the sender but not the

---

[*] Corresponding author.

D.-S. Huang et al. (Eds.): ICIC 2014, LNCS 8588, pp. 572–581, 2014.
© Springer International Publishing Switzerland 2014

receiver, and it is argued that such sender-centric perspective hardly reflects the reality. Thus, *the receiver-centric model* was proposed [8], where the interference on a node v is the number of other nodes whose transmission ranges cover node v. In this paper, we take the receiver-centric model into account. For interference minimization problems, the objective mainly includes minimizing the *average interference* and the *maximum interference*. To minimize the maximum interference while maintaining connectivity is one of the most well-known open algorithmic problems in wireless sensor network optimization.

It is NP-hard to compute the minimum maximum interference of a wireless sensor network in the plane while keeping the network connectivity [4]. In the case of grid networks, the problem has been solved by using dynamic programming [10]. An $O(\Delta^{1/2})$ bound for the maximum interference has been attained by using ideas from computational geometry and ε-net theory [6], where Δ is the interference of a uniform-radius ad-hoc network. A polynomial time approximation algorithm has been proposed which finds a connected topology with interference at most $O((opt \ln n)^2)$, where *opt* is the interference of the minimum interference connected network for an input instance of n nodes [1]. To the best of our knowledge, almost all previously proposed topology control algorithms are based on the so-called nearest neighbor algorithm, where every node establishes a link to at least its nearest neighbor [8, 9]. Despite some significant efforts, known solutions are not very satisfying.

In this paper, we only consider the topology control method to minimize the maximum interference of a wireless sensor network in the plane under the receiver-centric model. For the problem of finding a connected topology with the minimum maximum interference in the plane, we propose an improved greedy algorithm based on the well-known nearest neighbor algorithm, and compare their performance by examples and simulation.

2   Problem Model

Throughout this paper, the terms graph and network are used interchangeably since networks are modeled as graphs. Furthermore, a graph refers to a simple undirected graph $G=\langle V, E \rangle$, where V is the set of nodes and E is the set of edges or links. Related notation and terminology of graph theory follow those in [12].

A wireless sensor network in the plane can be described as a node set V with coordinates $(x(i), y(i))$ and a transmission radius $r(i) \in [0,1]$ for every node $i \in V$. Thus, the transmission relation between all these nodes, incurred by the transmission radius assignment r, can be modeled as a graph $G_r=\langle V, E_r \rangle$ such that for any $u, v \in V$, an edge $(u,v) \in E_r$ exists if and only if their transmission radii, $r(u)$ and $r(v)$, are not shorter than their Euclidean distance $d(u,v)$, i.e., $r(u) \geq d(u,v)$ and $r(v) \geq d(u,v)$. For convenience, the resulting graph $G_r=\langle V, E_r \rangle$ is called *the induced graph* of the transmission radius assignment r. In order to describe the problem model under consideration we need the following concepts and notations.

- Given a transmission radius threshold $R(i) \in [0,1]$ for every node $i \in V$, a *transmission radius assignment* $r: V \rightarrow [0,1]$ is a function on V satisfying $r(i) \in [0,1]$ and $r(i) \leq R[i]$ for every node $i \in V$.

- *The interference of node v in $G_r = \langle V, E_r \rangle$, denoted by $I_r(v)$, is the number of other nodes whose transmission ranges cover node v. That is,*

$$I_r(v) = |\{u \in V \backslash \{v\} | r(u) \geq d(u,v)\}|$$

- *The interference of the induced graph $G_r = \langle V, E_r \rangle$, denoted by I_r, is the maximum interference among all nodes. That is,*

$$I_r = \boldsymbol{max}\{I_r(v) | v \in V\}$$

Considering the simple connection requirement, we have the following problem model.

The Minimum Interference Problem (MIP): *Given a node set V in the plane with coordinates (x(i), y(i)) and a transmission radius threshold $R(i) \in [0,1]$ for every node $i \in V$ satisfying that the induced graph $G_R = \langle V, E_R \rangle$ is connected, we are asked to find a transmission radius assignment r such that the induced graph $G_r = (V, E_r)$ is a connected graph with the minimum interference.*

Note that for a transmission radius assignment r, G_r is a sub-graph of G_R. Furthermore, if we re-assign the distance between node v and its farthest neighbor in G_r as the transmission radius of node v, then the resulting induced graph has the same interference as G_r. Therefore, we can select a transmission radius for node v only from the different distances node v can reach in G_R. By doing so, the radius of one node is the farthest distance it can reach in G_r.

MIP has been proven to be NP-hard, and it appears very difficult to find an optimal solution to an instance of **MIP** even if all nodes are located on a line (called *the highway model* [11]) , although the NP-hardness result is not yet known. The difficultness results from the fact that the combination structure of the problem is so complicated that many intuitions, such as small degrees and near neighbors, are not applicable to yield satisfying solutions of the problem.

3   Nearest Neighbor Algorithm

Since **MIP** is NP-hard, heuristics based on intuitions are often considered. In order to reduce interference, one natural approach is the nearest neighbor method that establishes an edge between two nearest nodes, which can be done by selecting the distance from every node to one of its nearest neighbors as its transmission radius. Technically, this means that these topologies contain the so-called *nearest neighbor forest* as a sub-graph [8]. The idea is very intuitive, and can effectively solve many instances of **MIP**. In fact, among existing heuristics for **MIP**, the nearest neighbor algorithm is one of the most popular algorithms.

An implementation of the nearest neighbor method is similar to that of Prim's algorithm for finding a minimum cost spanning tree in a weighted graph. The basic idea is as follows. Starting from an arbitrary node, saying node 1, and the transmission radius assignment $r=0$, the algorithm grows the connected part containing node 1 step by step. The algorithm begins by creating two sets of nodes: $P = \{1\}$ and $Q = V \backslash \{1\}$. It then grows the connected part by moving one node from Q to P at a time. At each step, it finds a node $p \in P$ and a node $q \in Q$ that have the smallest distance, and then moves node q from Q to P. Meantime, it grows the connected part by selecting their distance $d(p,q)$ as the transmission radius of node q and adjusting the transmission radius $r(p)$ of node p to the distance $d(p,q)$ if $r(p)$ is smaller than the distance. This step is repeated until Q becomes empty. This algorithm, called Nearest Neighbor Algorithm for convenience, is outlined in Alg.1.

Alg.1. Nearest Neighbor Algorithm
INPUT: A node set $V = \{1,2,\ldots,n\}$ in the plane with coordinates $(x(i),y(i))$ and a maximum transmission radius $R(i) \in [0,1]$ for every node $i \in V$ satisfying that graph $G_R = \langle V, E_R \rangle$ is connected.
OUTPUT: A transmission radius assignment r such that $G_r = \langle V, E_r \rangle$ is a connected graph with a small interference.
1. $P \leftarrow \{1\}$, $Q \leftarrow V \setminus \{1\}$
2. $r \leftarrow 0$
3. $min\_I \leftarrow \infty$
4. **WHILE** $Q \neq \varnothing$ **DO**
5.   **FOR** every $(i,j) \in P \times Q$ **DO**
6. $r' \leftarrow r$
7. $r'(i) \leftarrow \mathbf{max}\{r(i), d(i,j)\}$, $r'(j) \leftarrow d(i,j)$
8.   **IF** $(I_{r'} < min\_I)$ **THEN** $min\_I \leftarrow I_{r'}$, $p \leftarrow i$, $q \leftarrow j$
9.   **END FOR**
10. $r(p) \leftarrow \mathbf{max}\{r(p), d(p,q)\}$, $r(q) \leftarrow d(p,q)$
11. $P \leftarrow P \cup \{q\}$, $Q \leftarrow Q \setminus \{q\}$
12. **END WHILE**

It is obvious that Alg. 1 returns a transmission radius assignment r satisfying graph $G_r = \langle V, E_r \rangle$ is connected, and has the time complexity $O(n^2)$. Thus, we have the following theorem.

Theorem 1. *Alg.1 yields a transmission radius assignment r in $O(n^2)$ time such that the induced graph $G_r = \langle V, E_r \rangle$ is connected.*

With regard to the quality of the solution obtained, Alg. 1 has a good effect for general networks. However, just as many greedy algorithms for NP-hard problems, Alg.1 performs relatively poor in some extreme cases. The authors of [8] have proved that Alg.1 (in fact any algorithm based on the nearest neighbor method) can have $\Omega(n)$ times larger interference than the interference of the optimum connected topology for certain kinds of networks.

Here we give two examples to illustrate that Alg.1 yields relatively poor solutions claimed above when run on the corresponding instances [5].

Example 1. For an n-node instance of the one dimension exponential node chain network, all nodes are located on a line such that the distances between two consecutive nodes grows exponentially $(2^0, 2^1, \ldots, 2^{n-1})$. The induced graph generated by Alg.1 has interference $n-1$. For a 6-node instance, the induced graph is shown in Fig. 1(a), which has interference 5. The minimum interference of the instance is 3, which is shown in Fig. 1(b).

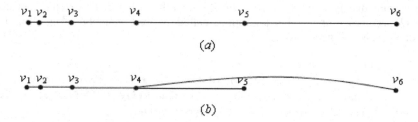

Fig. 1. (a) The induced graph generated by Alg.1 on a 6-node instance of the one dimension exponential node chain network. (b) The induced graph with interference 3 of the instance.

Example 2. An n-node instance of the two dimension exponential node chain network, shown in Fig.2, contains a horizontal exponential node chain (including h_i) and a diagonal exponential node chain (including v_i), and helper nodes t_i satisfying $d(h_i,t_i)>d(h_i,v_i)$. The induced graph generated by Alg.1 when run on the instance, depicted in Fig.3 (a), has interference $\Omega(n)$. For the instance, the induced graph with minimum interference (a constant with regard to n) is shown in Fig.3 (b).

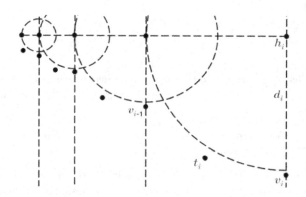

Fig. 2. An n-node instance of the two dimension exponential node chain network [8]

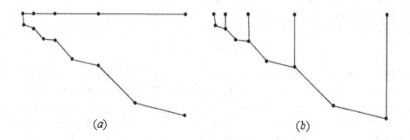

Fig. 3. (a) The induced graph by Alg.1 on the instance shown in Fig.2, which has interference $\Omega(n)$. (b) The induced graph with minimum interference of the instance.

4   An Improved Algorithm

From the above two examples, we observe that Alg. 1 has some obvious drawbacks. Each time one node q in Q is connected to one nearest node p in the connected part, we assign its transmission radius as $d(p,q)$, and must re-assign the transmission radius of node p if needed. Their transmission radii are no less than their distance $d(p,q)$. Intuitively, the move of node q to the connected part poses as small interference as possible to other nodes since $d(p,q)$ is the minimum among all pairs of nodes (i,j), where $i \in P$ and $j \in Q$. However, in certain extreme cases, such as in exponential node chain networks, every move or every several moves of nodes to the connected part will always increase interferences of some specific nodes, which will result in a large interference of these nodes, even in proportion to the number of nodes in the network.

Based on the above observation, we give the following strategies to improve Alg.1 while preserving its reasonable intuition.

- **S1**: At each step, we connect one node q in Q to one node p in the connected part such that assigning $r(p)$ and $r(q)$ incurs a minimum increase in the maximum interference of all nodes in that time. This strategy is clearly aimed at the goal function of **MIP**.
- **S2**: Each time we encounter a tie when using **S1**, we prefer to choose such a pair node (i,j) to break the tie that are two nearest nodes among all pairs $(i,j) \in P \times Q$. This strategy is the same as that used in Alg. 1.
- **S3**: Each time we encounter a tie when using **S1** and **S2**, we use a random choice to break the tie. Sometimes this strategy can avoid yielding relatively poor results.

The process of our algorithm is similar to that of Alg. 1 except for using successively strategies **S1**, **S2** and **S3** instead of only using **S1** in Alg. 1. This algorithm, called Best Neighbor Algorithm for convenience, is formally described in Alg.2. In Alg.2, min_I is the current minimum interference of the induced graph G_r by using strategy **S1** to the transmission radius assignment r, and min_D is the current minimum interference of the induced graph by using strategies **S1** and **S2** to the transmission radius assignment r. set_min_ID is the subset of node pairs in $P \times Q$ that have the same min_I and min_D. With regard to the Performance of Alg.2, we have the following Theorem 2.

Alg.2. Best Neighbor Algorithm
INPUT: A node set $V=\{1,2,...,n\}$ in the plane with coordinates $(x(i),y(i))$ and a maximum transmission radius $R(i) \in [0,1]$ for every node $i \in V$ satisfying that graph $G_R=\langle V, E_R\rangle$ is connected.
OUTPUT: A transmission radius assignment r such that $G_r=\langle V, E_r\rangle$ is a connected graph with a small interference.
1. $P \leftarrow \{1\}$, $Q \leftarrow V \setminus \{1\}$
2. $r \leftarrow 0$
3. $min\_I \leftarrow \infty$, $min\_D \leftarrow \infty$, $set\_min\_ID \leftarrow \emptyset$
4.   **WHILE** $Q \neq \emptyset$ **DO**
5.   **FOR** every $(i,j) \in P \times Q$ **DO**

```
6.      r'←r
7.      r'(i)←max{r(i),d(i,j)},  r'(j)←d(i,j)
8.      CASE(I_r<min_I):  min_I←I_r,, set_min_ID←{(i,j)}
9.      CASE(I_r=min_I∧d(i,j)<min_D):  min_D←d(i,j),set_min_ID←{(i,j)}
10.     CASE(I_r=min_I∧d(i,j)=min_D):  set_min_ID ←set_min_ID∪{(i,j)}
11.     END FOR
12.     Randomly choose (p,q)∈set_min_ID
13.     r(p)←max{r(p), d(p,q)},   r(q)← d(p,q)
14.     P←P∪{q},  Q←Q \{q}
15.     END WHILE
```

Theorem 2. *Alg. 2 yields a transmission radius assignment r in $O(n^2)$ time such that the induced graph $G_r=\langle V, E_r\rangle$ is connected.*

Proof. Firstly, at each step, *set_min_ID* is a nonempty set. Thus, Alg. 2 yields a transmission radius assignment r such that the induced graph $G_r=\langle V, E_r\rangle$ is connected. Secondly, Alg. 2 moves one node from Q to P in time $O(n)$. Therefore, Alg.2 has time complexity $O(n^2)$ since Alg.2 repeats the move operation $n-1$ times. The proof is complete.   ∎

With regard to the quality of the solutions Alg. 2 yields, we will estimate it by simulation in the next section. Here we give two examples to illustrate that it does overcome the drawback observed in Alg.1 while preserving good intuition.

Example 3. For the 6-node instance of the one-dimension exponential node chain network given in Example 1, we consider the execution of Alg. 2 when run on it. Alg.2 does five move operations, which are shown in Fig. 4. In the beginning, $P = \{v_1\}$, $Q =\{v_2,v_3,v_4,v_5,v_6\}$. After the 1st move operation, $P = \{v_1,v_2\}$, $Q = \{v_3,v_4,v_5,v_6\}$, $min\_I = 1$. After the 2nd move operation, $P = \{v_1,v_2,v_3\}$, $Q = \{v_4,v_5,v_6\}$, $min\_I= 2$. After the 3rd move operation, $P = \{v_1,v_2,v_3,v_4\}$, $Q = \{v_5,v_6\}$, $min\_I = 2$. After the 4th move operation, $P=\{v_1,v_2,v_3,v_4,v_5\}$, $Q = \{v_6\}$, $min\_I = 3$. After the 5th move operation, $P = \{v_1,v_2,v_3,v_4,v_5,v_6\}$, $Q =\varnothing$, $min\_I = 3$. The resulting induced graph attains the best interference 3. Alg.2 yields a constant interference when run on an n-node instance of the one dimension exponential node chain network, which is optimal in an asymptotic sense.

Example 4. We consider the execution of Alg. 2 when run on the instance shown in Fig. 2. We can display the result of Alg.2 after every move operation in a similar way as in Example 3, and the details are omitted. Fig. 5 gives the resulting induced graph by Alg.2, which has interference 3. Generally, Alg.2 yields a constant interference when run on an n-node instance of the two dimension exponential node chain network, which is optimal in an asymptotic sense.

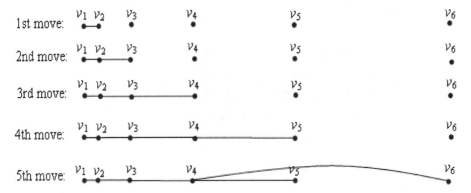

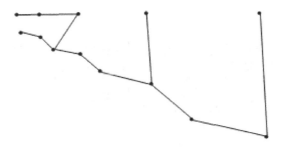

Fig. 4. A sample execution of Alg.2 on the 6-node instance shown in Fig. 1(a)

Fig. 5. The induced graph generated by Alg.2 on the instance shown in Fig. 2

5   Experiment

In this section, we compare Alg.1 and Alg.2 with regard to the quality of solutions these two algorithms respectively yield when run on the same instance.

Let n be the number of nodes in an instance, R be the threshold of the transmission radius of every node (here we always set R to be 1), and d be the average degree of nodes in the induced graph $G_R=\langle V, E_R\rangle$, i.e., the edge density of $G_R=\langle V, E_R\rangle$. Given n and d, we randomly generate n nodes in a $w \times w$ square as an instance for the pair (n, d), where $d=n\pi R^2/w^2$. For every instance, we compare the interferences of the induced graphs Alg.1 and Alg.2 yield respectively. For every pair (n, d), where n is 100 or 1000, and d is a small value in contrast to the node number n, we randomly generate 100 instances and used the average of interferences of these 100 instances as the final result of pair (n, d).

The experimental results are shown in Figure 6 and Figure 7, respectively giving a comparison of results by Alg.1 and Alg. 2 on instances with 100 and 1000 nodes. From these two figures, we easily see that the interferences calculated by Alg. 2 are smaller than those by Alg. 1 for all these instances generated randomly. This means Alg. 2 is more effective than Alg. 1 in our experiment.

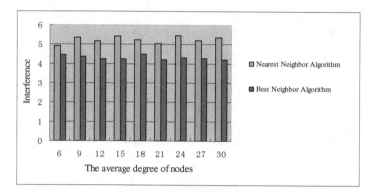

Fig. 6. A comparison between results by Alg.1 and Alg.2 on instances with 100 nodes

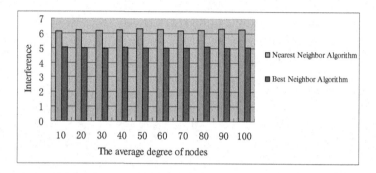

Fig. 7. A comparison between results by Alg.1 and Alg.2 on instances with 1000 nodes

6   Conclusion

In this paper effective algorithms are investigated for finding a minimum interference connected topology for a sensor wireless network in the plane under the receiver-centric model. We have proposed several strategies for improving the well-known greedy algorithm called Nearest Neighbor Algorithm based on analysis of its behavior in the worst cases, and then presented the improved version called Best Neighbor Algorithm. We have shown that with regard to the quality of solutions, Best Neighbor Algorithm outperforms the existing Nearest Neighbor Algorithm by examples and by simulation.

Acknowledgements. The research is supported in part by National Natural Science Foundation of China (No. 61370003), and by the Scientific Research Foundation for the Returned Overseas Chinese Scholars, State Education Ministry.

References

1. Aslanyan, H., Rolim, J.: Interference Minimization in Wireless Networks. In: IEEE/IFIP International Conference on Embedded and Ubiquitous Computing, pp. 444–449 (2010)
2. Benkert, M., Gudmundsson, J., Haverkort, H., Wolff, A.: Constructing Minimum-Interference Networks. Computational Geometry 40(3), 179–194 (2008)
3. Bilò, D., Proietti, G.: On the Complexity of Minimizing Interference in Ad-Hoc and Sensor Networks. In: Nikoletseas, S.E., Rolim, J.D.P. (eds.) ALGOSENSORS 2006. LNCS, vol. 4240, pp. 13–24. Springer, Heidelberg (2006)
4. Buchin, K.: Minimizing the Maximum Interference is Hard. Arxiv: 0802.2134v1 (2008)
5. Burkhart, M., von Rickenbach, P., Wattenhofer, R., Zollinger, A.: Does Topology Control Reduce Interference? In: 5th ACM Int. Symposium on Mobile Ad-Hoc Networking and Computing (MobiHoc), pp. 9–19 (2004)
6. Halldórsson, M.M., Tokuyama, T.: Minimizing Interference of a Wireless Ad-Hoc Network in a Plane. In: Nikoletseas, S.E., Rolim, J.D.P. (eds.) ALGOSENSORS 2006. LNCS, vol. 4240, pp. 71–82. Springer, Heidelberg (2006)
7. Kuhn, F., von Rickenbach, P., Wattenhofer, R., Welzl, E., Zollinger, A.: Interference in Cellular Networks: The Minimum Membership Set Cover Problem. In: Wang, L. (ed.) COCOON 2005. LNCS, vol. 3595, pp. 188–198. Springer, Heidelberg (2005)
8. von Rickenbach, P., Schmid, S., Wattenhofer, R., Zollinger, A.: A Robust Interference Model for Wireless Ad-Hoc Networks. In: 5th International Workshop on Algorithms for Wireless, Mobile, Ad Hoc and Sensor Networks, Wman (2005)
9. von Rickenbach, P., Wattenhofer, R., Zollinger, A.: Algorithmic Models of Interference in Wireless Ad Hoc and Sensor Networks. IEEE/ACM Transactions on Networking 17(1), 172–185 (2009)
10. Tan, H.: Minimizing Interference in Wireless Sensor Networks. Ph.D. Thesis, University of Hong Kong, pp. 51–64 (2011)
11. Tan, H., Lou, T., Lau, F.C.M., Wang, Y., Chen, S.: Minimizing Interference for the Highway Model in Wireless Ad-Hoc and Sensor Networks. In: Černá, I., Gyimóthy, T., Hromkovič, J., Jefferey, K., Královič, R., Vukolić, M., Wolf, S. (eds.) SOFSEM 2011. LNCS, vol. 6543, pp. 520–532. Springer, Heidelberg (2011)
12. West, D.B.: Introduction to Graph Theory. Prentice-Hall, Englewood Cliffs (2001)
13. Wu, K.-D., Liao, W.: On Constructing Low Interference Topology in Multi-Hop Wireless Networks. International Journal of Sensor Networks 2(5-6), 321–330 (2007)

Subjective Evaluation of Prototypes and Task Complexities for Mobile Phone Usability

Ibrahim Furkan Ince[1] and Ilker Yengin[2]

[1] Hanwul Multimedia Communication Co. Ltd. 1012-1015, Ace High Tech21,
1470, U-dong, Haeundae-gu, Busan, Republic of Korea
furkan@hanwul.com
[2] A*STAR, Institute of High Performance Computing,
1 Fusionopolis Way ,#16-16 Connexis North, 138632, Singapore
yengini@ihpc.a-star.edu.sg

Abstract. Different types of prototypes may have different effect on mobile phone interface design processes. Understanding these effects may help designers to evaluate design options in depth. Empirically derived research results may provide a realistic ways for evaluations. Confirming and extending our previous research study, this study investigates the effect of different mobile phone interface prototype types (paper, computer and fully operational device) on task complexities. Task complexity of mobile phone was measured by 5 scale self-evaluation instrument. Results showed that computer based prototype with complex task yielded the highest usability rate. This empirical based paper may help designers to evaluate what types of prototypes are usable for complex and non-complex tasks.

1 Introduction

User-centered design methods require designers to understand the users, analyze users' activities and asking users to use to participate into the early stages of design and development cycles [1]. Prototyping gives designers a practical way to incorporate user into design and development cycles [2]. In addition to necessity of providing user centered design, there are other challenges that designers should tackle. For instance, when designing new systems and devices, designers face lots of challenges such as hardware (e.g. small screens, battery life, input capabilities) and software limitations [3, 4 and 5]. Despite to the powerful mobile phone hardware and software availability, the design process is expensive, challenging and time consuming [6]. In order to have easier, cost and time effective design and development processes, prototypes are commonly used in iterative design tasks [7].

Research studies indicate that different prototyping techniques have different advantages and disadvantages [8-10] (explained in detail in "Background" section). Although these studies depict the picture relying on anecdotal methods, they don't provide any empirical evidence to understand the effect of different types of prototyping techniques. Relying on empirical studies based on collection of real world

D.-S. Huang et al. (Eds.): ICIC 2014, LNCS 8588, pp. 582–593, 2014.
© Springer International Publishing Switzerland 2014

users' data would provide well established evidences to fill the gap in our knowledge on the mobile phone prototyping techniques.

In order to understand the advantages and disadvantages of different prototyping techniques, we need more empirically derived understanding on the role of prototypes in design. Rather than just providing a list of suggestions based on theoretical and anecdotal methods, empirical studies provide the evidence based user interface design choices that are available to designers. Empirical studies help designers to understand concepts in details because empirical studies usually "focus in-depth on a particular aspect of design and attempt to situate that aspect in a larger design context" [11]. Many researchers in the human computer interaction field acknowledge the importance of empirical methods [12].

Mobile phone designers may benefit from studies that are documenting the roles and benefits of different prototypes and design choices available. Although there are guidelines available for designing prototypes for mobile phones, there is no (or little) study addressing the relationship and the effect between mobile phone interface prototype formats and task complexity.

In order to provide essential directions that address these issues, we had conducted a previous research study [13]. In the previous study, we had investigated the effects of different type of mobile phone prototypes on the usability of different level of tasks complexities. Our previous study's results showed that computer type prototype with complex task yielded highest usability rate while paper type prototype with simple task yielded lowest usability rates. The previous study suggested that computer type prototype had significantly higher usability rate than paper type prototype and there was a significant interaction between prototype and complexity.

Putting forward our previous study [13], we detailed our investigations of the issue in this current study. Our current study addresses the need of empirical based design options and answers the question of "How do we evaluate the effect of prototype of mobile phone and task complexity?" In simple terms, this study attempts to find out what types of prototypes are usable for complex and non-complex tasks.

The purpose of this empirical quantitative study is to evaluate the effect of different mobile phone interface prototypes (paper, computer, and a fully operational device) on the task complexity. The task complexity is defined in the form of usability for mobile phone users and obtained from subjective user evaluation by asking study participants to send text messages and save the schedules.

This study may help mobile phone designers to evaluate different format of mobile phone interface prototypes according to the relations with task complexity in order to provide empirical based design options. This study may also be beneficial to researchers and professionals in the human computer interaction field by demonstrating the methods to evaluate the right prototype techniques according to the defined tasks.

2   Background

This section presents the significance of using of prototypes. The first part introduces the importance of using prototypes in design and development in general sense. In the second part, the specific uses of prototypes for mobile phone interfaces are discussed. Also in the second part, a classification of different prototypes is presented.

2.1   Prototyping

In short definition, prototypes are the representation of interactions between users and a system [14]. According to Cambridge Online Dictionary, prototype is defined as "the first example of something, such as a machine or other industrial product, from which all later forms are developed". Dictionary.com provides an extended definition of the word as "The original or model on which something is based or formed" and/ord "Someone or something that serves to illustrate the typical qualities of a class; model; exemplar: She is the prototype of a student activist"

As the definitions suggests, prototypes are working models that gives us some ideas on the real object/system that is aimed to be built. Modelling techniques to construct prototypes helps us to visualize the ideas and evaluate system requirements and problems [15]. Thus, prototypes help the designer to see the overall picture and have a design space (or a playground) to play with the different parameters that may have possible effects on the model [16].

Especially in information technology related projects, playing with different parameters affecting the prototype helps designers to see the dynamic formation of the system, growth mechanisms and evaluations. Also, prototypes introduce the interactions between system players by highlighting the most important elements and requirements [17].

Before all else, system's users are the most important players in prototyping because they are the main source for detecting problems, advising alternative solutions and evaluation [17]. Therefore prototypes are especially useful for getting feedback from users [18], evaluating UI look and feel [19] and communicating with the users [20].

As suggested above paragraphs, prototyping enhance system quality and reduce development time of the products within more easier and cheaper ways than the traditional system development approaches [21]. For example, modeling the user interactions with a quick iterative approach in a prototype may take nearly just a day which is very quicker than the traditional approaches [22].

Different designers emphasize the various benefits of prototyping [23, 24, and 25]. There are lots of different prototyping formats where they can give similar results but each of them also has some side effects such as scalability, manageability, and controllability [26]. Using pencil and paper based sketching are the most common methods for prototyping in design process, especially in HCI field [27, 28]. In addition, there are high-fidelity functional prototypes that are nearly operates like the final system and emulates the UI of targeted final system [29]. Computer based applications are examples of [30, 31] high-fidelity prototypes.

2.2   Mobile Phone Interface Prototypes

In mobile phone software and hardware design and development activities, prototyping becomes a very important aspect [32]. The idea of using prototypes for mobile phone interfaces introduced by the early stage mobile phone manufacturing companies such as Nokia and Siemens [33,34]. Early research indicated that prototyping mobile phone interfaces may be effective both in the early design phases and product development cycles [35]. Early mobile phone developers also had the chance to follow the human centered design rules which were also identified by ISO [36]. Including different methods and tools for designing mobile phone interfaces, Kiljander [37] proposed a classification for mobile interface prototyping methods in 9 different categories which are "Using Scenarios", "State Transition Diagrams", "Paper Prototypes (Wizard of Oz)", "3D Paper Prototypes", "CAD Models", "Computer Simulations", "Hard Models (Physical Product Models)", "Virtual Reality" and "Hardware". Each category is explained in detail in the following list:

- **Using Scenarios:** Scenarios are used for helping to specify the functionalities of new products. Scenarios also may be in form of role playing [38]. Svanæs and Seland [39] argue that "role playing takes users and developers out of the chair and into the physical, social, and embodied reality of mobile computing."
- **State Transition Diagrams:** A state transition diagram is useful to show the sequences of screens. They can be in form of diagram, user interface map or storyboard [37].
- **Paper Prototypes (Wizard of Oz):** Paper prototypes are used in initial usability evaluations in product development stages that consist of control flow diagrams and sketches [37]. Usually, for mobile phone designs, using paper based Wizard of OZ approach could be very difficult to use [40, 41].
- **3D Paper Prototypes:** 3D papers are used to model the interface in 3D forms built in forms of paper, cardboard and foam; they are useful to model physical interaction. [42]. 3D prototypes help designers to evaluate the mobile phone interfaces and the software on the screen as a whole [43].
- **CAD Models:** CAD models can be used to create prototypes in later stages of product development using rapid prototyping machines [44].
- **Computer Simulations:** Computer simulations are used in cases where the user interfaces grow into a very complex interaction mechanism. Computer simulations can be on screen prototype operated with a mouse or touch screen [47].
- **Hard Models (Physical Product Models):** Hard models may be produced as the replicas with functional parts which are made from materials [47].
- **Virtual Reality:** They may simulate the functionality of interface software, behavior of the product and the users' interactions with the product in certain simulation virtual worlds [45].
- **Hardware:** Hardware prototypes are nearly to finish products which are good enough to use proposed hardware and software design. Usually, hardware prototypes are the speeding the process of getting feedback from the users [37].

In addition to the well-known traditional prototyping practices, researchers also make it possible to have new techniques for usability evaluation of mobile systems using different prototyping methods by creating the combinations of the methods very similar to the ones suggested above list [46]. Similarly, for different specific purposes, researchers also introduced other different prototyping applications (e.g. DENIM [47], SUEDE, [48], SILK [49], SketchWizard [50], CARD [51] etc.) to provide different ways of approaching human centered design and evaluation of mobile phones user interfaces.

Moreover, researchers also point to the necessity of using rapid prototyping techniques in mobile phone design to reduce the time required to evaluate the applications including several logical control flow re-directions between sensors, mobile device and servers [52]. In past, several rapid prototyping techniques for user interfaces were introduced for similar purposes [53 – 56].

3   Methodologies

This section presents the hypotheses, participants, and experiment design procedures (user study) to assess the effect of prototype of mobile phone and task complexity in our study.

3.1   Null Hypotheses

Before starting of our experimental design we had identified the null hypothesis for this study as follow:

- Prototype and task complexity in mobile phone has no interaction and relation between each other.
- All the prototypes have same amount of effect on task complexities in the total usability rates.
- Simple and complex tasks have the same amount of effect on the usability.

3.2   Participants

To make sure that participants have not performed the tasks included in the experiment, we used a purposeful sampling methodology. We selected participants with no previous experience with the tasks in our experiment. We tested a total of 20 participants, (8 male and 12 female). Most of them were under graduate or graduate students recruited through mailing lists from different universities. There were 6 participants with a background in sciences or engineering, 12 participants from the humanities or social sciences, and 2 participants with no academic background. All of them spoke Korean as their native language and English as a second language. They also owned a mobile phone and computer thus there was no difficulty for them to learn to operate the fully working devices and computer simulations. Except for 2 participants, all participants used their mobile on a daily basis. The average age of participants was 28.45 years, ranging from 22 to 35 (SD = 4.01) years.

3.3   Experimental Design (User Study)

In the experiments, three different kinds of mobile phones prototypes were used. First prototype was sketched on paper (Figure 1 - Left), second one was a simulation in the computer (Figure 1 - Middle), and the last one was a particular fully working device with full functional hardware and software (Figure 1 - Right).

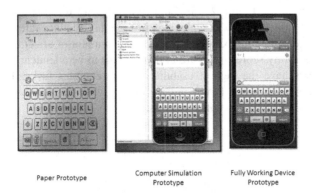

Paper Prototype   Computer Simulation   Fully Working Device
                                          Prototype   Prototype

Fig. 1. Prototypes tested in the experiment

Paper prototype sketches were drawn on paper by our user interface graphical artists. Also the workflow and the connections of the screen images were illustrated in storyboards and flow charts. Simulation was done using XCode iOS simulator [57]. XCode iOS Simulator was used for converting the sketches in paper prototypes to fully working software by linking the screens defined in sketches together. After a little coding process in the XCode iOS Simulator, our software was ready for using as a computer simulation prototype. Finally, the fully working device was an iPhone 4 with iOS version 6.1.5.

The task complexity is defined in the form of usability. Sending text messages and saving schedules were the two tasks accomplished by the participants of this study. Considering the users familiarity with sending text messages with similar mobile phones and the number of steps required (including the number of different screen to switch and the number of buttons to click) to finish the tasks, we assumed the sending text message as an easy task and the schedule saving as the difficult task. Simple and difficult tasks were performed by all the participants using three different prototypes: paper, computer simulations, fully working device.

The usability of the different prototypes was obtained by using a subjective user evaluation which was measured in 5 points scale. We collected only results for the explicit answers from a post-test questionnaire administered after all tasks were completed.

4   Experimental Results

The usability scores of different types of prototypes are optioned in 5 points scale. Results of usability scores are illustrated in the Figure 2 (higher scores means better usability). Also Table 1 depicts the descriptive statics for the usability scores of Figure 2.

As evident from the Figure 2, we could refute that all the prototypes types have same amount of effect on task complexities in the total usability rates. Similarly, we reject that simple and complex tasks have the same amount of effect on the usability. Based on the usability scores in Figure 2, there seems to be a tendency that computer based prototypes have the highest usability for the complex tasks. Thus we could interpret that computer based prototypes are more usable for mobile phone interface prototyping which includes complex tasks scenarios.

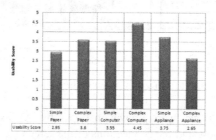

Fig. 2. Usability scores of different prototypes for different task complexities

Table 1. Descriptive Statistics

	Dependent Variable score:			
prototype	complexity	Mean	Std. Deviation	N
	simple	2.95	.945	20
paper	complex	3.60	.598	20
	Total	3.28	.847	40
	simple	3.55	.605	20
computer	complex	4.45	.686	20
	Total	4.00	.784	40
	simple	3.75	.444	20
appliance	complex	2.65	.489	20
	Total	3.20	.723	40
	simple	3.42	.766	60
Total	complex	3.57	.945	60
	Total	3.49	.860	120

Figure 2 suggest that prototype type (paper, computer, fully working device) and the task complexity (simple and complex) have effects on the usability. Also we performed a two way analysis of variance (ANOVA), to understand if there is an interaction between prototype type and task complexity (independent variables) on the mobile phone usability (dependent variable).

According to our analysis results (Table 2), there is a significant effect of prototype type on usability, F (0.05, 2, 114) = 18.564, p = 0.000 and a significant interaction effect between prototype type and task complexity, F (0.05, 2, 114) = 28.233, p = 0.000. This relationship also illustrated in Figure 3.

Table 2. Tests of Between-Subjects Effects

Dependent Variable score:					
Source	Sum of Squares	df	Mean Square	F	Sig.
Corrected Model	40.042[a]	5	8.008	19.040	.000
Intercept	1463.008	1	1463.008	3.478E3	.000
prototype	15.617	2	7.808	18.564	.000
complexity	.675	1	.675	1.605	.208
prototype * complexity	23.750	2	11.875	28.233	.000
Error	47.950	114	.421		
Total	1551.000	120			
Corrected Total	87.992	119			

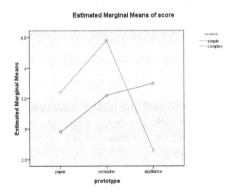

Fig. 3. Interaction effect between prototype and task complexity

According to our analysis, we should reject the hypothesis that suggest prototype and task complexity in mobile phone has no interaction and relation between each other. Moreover, we studied the effects of other factors, namely age, gender, and experience with advanced phone features. We did not find an interaction effect with interaction method for any of these factors.

We therefore conclude that, for the 20 participants in our study, the use of the computer-based prototypes with complex tasks was the most usable and fast interaction irrespective of their gender, age, or previous mobile phone experience.

5   Discussions

As results of analysis indicates, using computer based prototypes appear to be suitable to use for with complex tasks. Similar results also had been obtained in a previous study [23]. On the other hand, different than the previous study' results, it is interesting to note that there was no significant difference between fully working device to paper based prototype and computer based to fully working device based prototype in terms of effect on usability.

The only significant difference appears between paper based and computer based prototypes and computer based prototypes are yielding better results. Apparently, it is reasonable to suppose that using computer based prototypes for mobile phone interfaces would be more beneficial in terms of usability. Mobile phone producers and mobile phone interface designers can enjoy the benefits of using computer based prototypes in complex tasks. If designers choose the computer based prototypes for the complex task scenarios the participants using the prototype may perceive higher usability.

We could argue that this approach will allow participants to use cognitive resources to easily focus on task scenarios rather than spending effort on usability issues because of the prototype format. Interestingly, our results suggest that fully working device prototypes are less usable than the computer based prototypes. This may be due to participants' familiarity with the prototype format. For instance learning to operate within a fully working phone prototype interface may require significant amount of time than the computer prototype interface. This could be due to participants' familiarity to using computers but not the fully working phone prototype model. This effect should be investigated in detail in a future study.

6   Conclusions

Understanding the relationship and the effect between mobile phone interface prototype formats and task complexity helps designers to evaluate these options. Thus, it is important for designers to be aware of what types of prototypes are usable for complex and non-complex tasks.

In our empirical study, we evaluated the effect of different mobile phone interface prototypes (paper, computer, and a fully operational device) on the task complexity (simple and complex).

Experiment results yielded that computer prototype has significantly higher usability rate than paper type prototype and there is a significant interaction between prototype and complexity. This result is also similar to our previous study. Eventually, computer type prototype with complex task yielded highest rate while paper type prototype with simple task yielded lowest usability rates. Thus, designers may choose the computer based prototypes in complex tasks for better prototype usability.

References

1. Norman, D.A., Draper, S.: User-Centered System Design. Lawrence Erlbaum Assoc. (1986)
2. Torres, R.J.: User Interface Design and Development. Prentice Hall (2002)
3. de Sá, M., Carriço, L.: Designing for mobile devices: Requirements, low-fi prototyping and evaluation. In: Jacko, J.A. (ed.) HCI 2007. LNCS, vol. 4551, pp. 260–269. Springer, Heidelberg (2007)
4. Roduner, C., Langheinrich, M., Floerkemeier, C., Schwarzentrub, B.: Operating appliances with mobile phones – strengths and limits of a universal interaction device. In: LaMarca, A., Langheinrich, M., Truong, K.N. (eds.) Pervasive 2007. LNCS, vol. 4480, pp. 198–215. Springer, Heidelberg (2007)
5. Maragoudakis, M., Tselios, N.K., Fakotakis, N., Avouris, N.M.: Improving SMS usability using Bayesian networks. In: Vlahavas, I.P., Spyropoulos, C.D. (eds.) SETN 2002. LNCS (LNAI), vol. 2308, pp. 179–190. Springer, Heidelberg (2002)
6. Huebscher, M., Pryce, N., Dulay, N., Thompson, P.: Issues in developing ubicomp applications on Symbian phones. In: Proceedings of the FUMCA 2006: Future Mobile Computing Applications, pp. 51–56 (2006)
7. Nielsen, J.: Iterative user-interface design. Computer 26, 32–41 (2002)
8. Lim, Y.K., Stolterman, E., Tenenberg, J.: The anatomy of prototypes: Prototypes as filters, prototypes as manifestations of design ideas. ACM Transactions on Computer-Human Interaction (TOCHI) 15, 1–27 (2008)
9. Sefelin, R., Tscheligi, M., Giller, V.: Paper prototyping-what is it good for?: a comparison of paper-and computer-based low-fidelity prototyping. In: CHI 2003 Extended Abstracts on Human Factors in Computing Systems, pp. 778–779 (2003)
10. Arnowitz, J., Arent, M., Berger, N.: Effective Prototyping for Software Makers. he Morgan Kaufmann Series in Interactive Technologies. Morgan Kaufmann Publishers Inc., San Francisco (2006)
11. Mehalik, M., Schunn, C.: What constitutes good design? A review of empirical studies of design processes. International Journal of Engineering Education 22(3), 519 (2007)
12. Clarke, M.C.: Teaching the empirical approach to designing human-computer interaction via an experiential group project. ACM SIGCSE Bulletin 30(1), 198–201 (1998)
13. Ince, I.F., Salman, Y.B., Yildirim, M.E.: A user study: the effects of mobile phone prototypes and task complexities on usability. In: Proceedings of the 2nd International Conference on Interaction Sciences: Information Technology, Culture and Human, pp. 300–302. ACM (2009)
14. Carroll, J.M.: Scenario-based design: envisioning work and technology in system development. John Wiley & Sons, Inc., New York (1995)
15. Baskinger, M.: Pencils before pixels: a primer in hand-generated sketching. Interactions 15, 28–36 (2008)
16. Goel, V., Pirolli, P.: The structure of design problem spaces. Cognitive Science 16, 395–429 (1992)
17. Naumann, J.D., Jenkins, A.M.: Prototyping: the new paradigm for systems development. MIS Quarterly 6, 29–44 (1982)
18. Kelley, T., Littman, J., Hill, D.: The art of innovation. HarperCollinsBusiness (2001)
19. Houde, S., Hill, C.: What do prototypes prototype? In: Handbook of Human-Computer Interaction, vol. 2, pp. 367–381 (1997)
20. Warr, A., O'Neill, E.: Understanding design as a social creative process. In: Proceedings of the 5th Conference on Creativity & Cognition, pp. 118–127 (2005)

21. Kraushaar, J.M., Shirland, L.E.: A prototyping method for applications development by end users and information systems specialists. MIS Quarterly 9, 189–197 (1985)
22. Basili, V.R., Turner, A.J.: Iterative enhancement: A practical technique for software development. Foundations of Empirical Software Engineering: The Legacy of Victor R. Basili, p. 28 (2005)
23. Brown, T., Katz, B.: Change by design: how design thinking transforms organizations and inspires innovation. Harper Business (2009)
24. Laurel, B.: Design research: Methods and perspectives. MIT Press (2003)
25. Buchenau, M., Suri, J.F.: Experience prototyping. In: Proceedings of the 3rd Conference on Designing Interactive Systems: Processes, Practices, Methods, and Techniques, pp. 424–433 (2000)
26. Hennipman, E.J., Oppelaar, E.J.R.G., van der Veer, G.C., Bongers, B.: Rapid and rich prototyping: proof of concepts for experience. In: Proceedings of the 15th European Conference on Cognitive Ergonomics: The Ergonomics of Cool Interaction, pp. 1–6 (2008)
27. Laurila, J., Tuulos, V., Maclaverty, R.: Scripting environment for pervasive application exploration on mobile phones. In: Proceedings of the Pervasive 2006 (2006)
28. Buxton, B., Buxton, W.: Sketching user experiences: getting the design right and the right design. Morgan Kaufmann Pub. (2007)
29. Preece, J., Rogers, Y., Sharp, H.: Interaction design: beyond human-computer interaction (2002)
30. Holleis, P., Schmidt, A.: MAKEIT: Integrate user interaction times in the design process of mobile applications. In: Indulska, J., Patterson, D.J., Rodden, T., Ott, M. (eds.) PERVASIVE 2008. LNCS, vol. 5013, pp. 56–74. Springer, Heidelberg (2008)
31. Li, Y., Landay, J.A.: Activity-based prototyping of ubicomp applications for long-lived everyday human activities. In: Proceedings of the Twenty-sixth Annual SIGCHI Conference on Human Factors in Computing Systems, pp. 1303–1312 (2008)
32. Kjeldskov, J., Graham, C.: A Review of Mobile HCI Research Methods. In: Chittaro, L. (ed.) Mobile HCI 2003. LNCS, vol. 2795, pp. 317–335. Springer, Heidelberg (2003)
33. Lindholm, C., Keinonen, T.: Mobile Usability: How Nokia Changed the Face of the Cellular Phone. McGraw-Hill Inc. (2003)
34. Böcker, M., Suwita, A.: Evaluating the Siemens C10 mobile phone: going beyond "quick and dirty" usability testing. In: Proceedings of the HFT 1999 (1999)
35. Virzi, R.A., Sokolov, J.L., Karis, D.: Usability problem identification using both low-and high-fidelity prototypes. In: Proceedings of the SIGCHI Conference on Human Factors in Computing Systems, pp. 236–243. ACM (1996)
36. Maguire, M.: Methods to support human-centered design. International Journal of Human-computer Studies 55(4), 587–634 (2001)
37. Sasse, M.A., Johnson, C.: User interface prototyping methods in designing mobile handsets. In: Proceedings of the Interact 1999: 13th International Conference on Human-Computer Interaction, vol. 1, p. 118. IOS Press (1999)
38. Snyder, C.: Paper prototyping: The fast and easy way to design and refine user interfaces. Morgan Kaufmann (2003)
39. Svanaes, D., Seland, G.: Putting the users center stage: role playing and low-fi prototyping enable end users to design mobile systems. In: Proceedings of the SIGCHI Conference on Human Factors in Computing Systems, pp. 479–486 (2004)
40. De Sá, M., Carriço, L.: Low-fi prototyping for mobile devices. In: CHI 2006 Extended Abstracts on Human Factors in Computing Systems, pp. 694–699 (2006)

41. Abowd, G.D., Atkeson, C.G., Hong, J., Long, S., Kooper, R., Pinkerton, M.: Cyberguide: A mobile context-aware tour guide. Wireless Networks 3, 421–433 (1997)
42. Säde, S., Nieminen, M., Riihiaho, S.: Testing usability with 3D paper prototypes—Case Halton system. Applied Ergonomics 29(1), 67–73 (1998)
43. Mäkelä, A., Batterbee, K.: Applying usability methods to concept development of a future wireless communication device, case in Maypole. In: Proceedings of Human Factors in Telecommunications (1999)
44. Yan, X., Gu, P.E.N.G.: A review of rapid prototyping technologies and systems. Computer-Aided Design 28(4), 307–318 (1996)
45. Wang, G.G.: Definition and review of virtual prototyping. Journal of Computing and Information Science in Engineering (Transactions of the ASME) 2(3), 232–236 (2002)
46. Kjeldskov, J., Stage, J.: New techniques for usability evaluation of mobile systems. International Journal of Human-Computer Studies 60(5), 599–620 (2004)
47. Lin, J., et al.: Denim: Finding a Tighter Fit between Tools and Practice for Web Site Design. In: Proceedings of the CHI 2000. ACM (2000)
48. Klemmer, S.R.: SUEDE: A Wizard of Oz Prototyping Tool for Speech User Interfaces. In: Proceedings of the UIST 2000 (2000)
49. Landay, J.: SILK: Sketching Interfaces like Krazy. In: Proceedings of the CHI 1996, pp. 398–399 (1996)
50. David, R.: SketchWizard: Wizard of Oz Prototyping of Pen-Based User Interfaces. In: Proceedings of the UIST 2007, pp. 119–128 (2007)
51. Muller, M.J.: Layered participatory analysis: new developments in the CARD technique. In: Proceedings of the SIGCHI Conference on Human Factors in Computing Systems, pp. 90–97 (2001)
52. Aleksy, M.: An Approach to Rapid Prototyping of Mobile Applications. In: Proceedings of the AINA 2013: 27th IEEE International Conference on Advanced Information Networking and Applications (2013)
53. Long, S., Kooper, R., Abowd, G.D., Atkeson, C.G.: Rapid prototyping of mobile context-aware applications: the Cyberguide case study. In: Proceedings of the MobiCom 1996: 2nd Annual International Conference on Mobile Computing and Networking, pp. 97–107 (1996)
54. Placitelli, A.P., Gallo, L.: Toward a Framework for Rapid Prototyping of Touchless User Interfaces. In: Proceedings of the CISIS 2012: Sixth International Conference on Complex, Intelligent, and Software Intensive Systems, pp. 539–543 (2012)
55. Bannach, D., Amft, O., Lukowicz, P.: Rapid Prototyping of Activity Recognition Applications. Pervasive Computing 7(2), 22–31 (2008)
56. Weis, T., Knoll, M., Ulbrich, A., Mühl, G., Brändle, A.: Rapid Prototyping for Pervasive Applications. Pervasive Computing 6(2), 76–84 (2007)
57. Xcode. An Integrated Development Environment (IDE) for Developing Software for OS X and iOS, https://developer.apple.com/xcode/ (accessed January 20, 2014)

Kinect-Based Algorithms
for Hand Feature Points Localization

Yicheng Li and Zeyu Chen

E-Learning Lab
Shanghai Jiao Tong University
Shanghai, China
{liyicheng,zychen}@sjtu.edu.cn

Abstract. In recent years, the approaches of human computer interaction (HCI) are in rapid improvement. There is rich information in the hand which can be used to accomplish complex interactions in HCI. Taking advantage of the depth information from Kinect, this paper defines some feature points in hand and furthermore presents a set of Kinect-based algorithms to detect and locate these points to fulfill hand recognition. Our algorithms include three steps: wrist recognition, palm center detection and fingertips localization. The innovative wrist recognition method proposed in this paper is used to separate the hand and forearm region and is based on distance transformation algorithm and the geometric feature of wrist. As for fingertip localization, we propose the θ-curvature algorithm, which overcomes the shortcoming of the former K-curvature algorithm on the parameter selection. Our experiment results demonstrate that wrist, palm center and fingertips can be located quickly and accurately by the hand feature points localization algorithm.

Keywords: HCI, hand feature points, wrist recognition, fingertips localization, θ-curvature algorithm.

1   Introduction

Using hands to interact with a computer is probably the most natural way in the ever popular research field of HCI (Human Computer Interaction), since it allows people to communicate with computer directly with the knowledge they learned in their daily life [1]. There are many applications in this domain, for example, Ye et al [2] invent a system that allows people to input characters by writing in the air virtually; Tao et al [3] use popular gestures to scale, rotate and translate a virtual object with two fingers. All these complex interactions can be implemented on the basis of accurately detecting and localizing some feature points in hand such as fingertips. However, vision-based finger interaction is one of the greatest challenging tasks in this field and the method to implement it needs to be improved. In this paper, we define some feature points in hand and furthermore design a set of algorithms to locate the feature points of hand, which includes wrist recognition, palm center detection and fingertips localization.

D.-S. Huang et al. (Eds.): ICIC 2014, LNCS 8588, pp. 594–606, 2014.

2   Related Work

Hand segmentation is the first step to detect and locate the feature points in hand. Some researchers use skin-color model to extract the region of hand and face from background, but this method requires a strict environmental condition with proper illumination and a clear background with no other skin-colored objects. Since the advent of Kinect, researchers have put forward other approaches to hand segmentation, among which, the depth map generated by Kinect is often used to extract the region of hand. For example, reference [4] selects a certain range of the depth information to define the hand region. And Wang et al [5] use a square around the palm point provided by Kinect skeleton tracking system to extract the hand region.

It is often necessary to find the wrist to separate these two parts. The method proposed by [6] requires the user to wear a wrist band, but it goes against the principle of bare hand interaction. The way proposed by [7] to recognize the wrist is based on the fingertips detection while in many situations fingertips detection is a challenge task and the wrong localization of fingertips will lead to a wrong result of wrist recognition.

As for fingertips localization, there are three types of solutions which respectively based on contour, template and morphology. The method based on template [8] takes advantage of the geometrical features of fingertips. It predefines an image which is similar to fingertips as a template and searches those input hand images to find the template's counterparts as candidate regions of fingertips. The weakness of the method is the low calculation speed and it will generate some invalid points. The method based on morphology focuses on the slenderness of fingertips. Dung et al [9] perform the open operator on the hand image to extract the region of fingers and get the accurate positions of fingertips by geometrical calculation. One defect of this method is that it is hard to detect the fingers due to noise, including the size and position of the noise. The method based on contour is a popular and efficient way at present. Reference [10] calculates convex hull of hand contour, and then locates fingertips by convex hull. Reference [11] proposes a K-curvature algorithm. It calculates curvature of each pixel of hand contour and defines a point as a fingertip while its curvature is smaller than a predefined threshold value. However, the value of parameter K is hard to predefine because of the variety of the size of hand.

In this paper, we propose a set of Kinect-based algorithms for hand feature points localization. It overcomes the uncertainty of parameter selection in K-curvature algorithm. Experimental result shows that wrist, palm center and fingertips can be located quickly and accurately by our hand feature points localization algorithms.

3   Hand Segmentation

To segment the region of hand from the body is the first job of the whole localization process and it can be implemented conveniently and efficiently as follows.

Firstly, we use Kinect as an input device to capture a depth map. However, the image we get directly from Kinect is actually a much larger region with many invalid parts than the real hand region. In order to reduce computational complexity, we take advantage of skeleton tracking system of Kinect to get a smaller region (a primary hand region). The skeleton tracking system can provide 3D positions of a point on the hand and we call it the seed point. We capture a square around the seed point as a primary hand region. When the hand is close to the Kinect, the size of the square is big; otherwise the square is small. So the height and width of the square is a function of the depth value of the seed point. Taking the seed point as the center of the square, we can get the width of the square from the following equation:

$$Width(W(Z)) = Height(W(Z)) = 2*\min\{\max\{80-0.2*(z-640),60\},80\} \qquad (1)$$

In equation (1), z represents the depth of seed point. The square we get from the equation above is the primary hand region.

Then, we optimize the primary hand region according to the depth information of pixels from Kinect. In the depth map, every pixel is composed of 16 bits. The first 13 bits stand for the depth information which ranges from 0mm to 8192mm, and the last 3 bits stand for PlayerIndex, a value that shows whether a pixel is a part of human body or not. PlayerIndex only has two values: 0 and 1. Only when the value is 1, the pixel is indicated as a part of human body. Therefore, we can extract the human body from the background according to the information of PlayerIndex.

Combining the information of pixel depth and PlayerIndex, we use a flag to show if one pixel belongs to the ones that indicate the hand in the primary hand region. In equation (2), the PalmDepth is the depth of the seed point, α and β are two empirical parameters. According to our experiment results, $\alpha = 0.97$, $\beta = 1.03$.

$$hand\ flag = \begin{cases} 1, & if\ \alpha * PalmDepth < Depth < \beta * PalmDepth, and\ PlayerIndex = 1 \\ 0, & else \end{cases} \qquad (2)$$

We apply the hand segmentation algorithm above on the primary hand region, and consequently obtain a binary image in which the hand is revealed.

4   Wrist Recognition

We have got the hand region through the process of hand segmentation. However, in some cases, a large part of the forearm enters the valid depth range, so it is necessary to judge if any part of forearm is in the hand region and figure out how to separate the hand region and the forearm region. The detailed steps are shown below.

4.1   Distance Transformation

First of all, fingers need to be filtered out from the hand. Since the shape of fingers is long and thin compared to other parts of hand, we use distance transformation algorithm to achieve this goal. Distance transformation of image is defined as a new image, in which a pixel's value is set as the distance to its nearest zero pixel in the input image.

We use approximate template method to process distance transformation. The basic idea is to use a template to calculate the distance from a pixel to its nearest zero pixel in image. Fig. 1 is the approximate template in our algorithm.

3	2.5	2	2.5	3
2.5	1.5	1	1.5	2.5
2	1	0	1	2
2.5	1.5	1	1.5	2.5
3	2.5	2	2.5	3

Fig. 1. Approximate template

After applying the distance transformation to the input image, we get a gray scale image. We can find that the farther a pixel is from the boundary of hand region, the brighter it is. Thus, the finger and the edge of the forearm are darker than other parts. Therefore, we can filter these parts out by setting a threshold for each pixel value. Fig. 2 is the illustration of the process of this algorithm.

a. Original image   **b.** Image after distance   **c.** Image after threshold
                                transformation   filtering (inner hand region)

Fig. 2. The process of distance transformation and threshold filtering

4.2   Wrist Recognition

Now we can use the binary image after threshold filtering to locate the wrist part in the hand region. The shape of inner hand region is like a stick with a polygon at one side. Here, the stick is the remained forearm region after filtering and the polygon part is the palm region.We intend to find a straight line as the wrist line to separate the palm region and the forearm region. It is obvious that the pit point(s) of the contour of inner hand region should be point(s) on the wrist line no matter what shape the inner hand region is, which means we can at least find one point on the wrist line by looking for the defect point in the contour and subsequently locate the wrist line of hand. Following are steps to find the wrist line in the hand region.

1) Find the contour of inner hand region, and compute the convex hull of the contour. Here, the Graham scan algorithm is applied.

2) Find convexity defect points in the contour. Between two consecutive points of convex hull, the point in the contour which has the largest distance to the connecting line of these two points of convex hull is defined as a convexity defect point. Among the group of convexity defect points, we regard the point which has the

largest disctance from the correspoinding connecting line as one point in the wrist line. As shown in Fig.3(a), the orange point is a point in the wrist line we locate in the method above, and the orange line is the corresponding largest distance.

3) Get the gradient of the wrist line.The direction of wrist line should be almost vertical to the direction of the forearm region; but the requirement of the wrist line recognition is not very strict. According to these two presuppostions, we develop a projection-based algorithm to get the gradient of the wrist line. For every interval angle $\Delta\alpha$ ($0°$ to $180°$), the inner hand region is projected to two vertical directions (α and $\alpha+90°$) in the space, here, $\alpha = n\Delta\alpha\,(n=1,2,3...)$. In the two projected lines, we defined the longer one as height projection and the other as width projection. We get the ratio by height projection/width projection, and iterate the angles of projection to get the maximal value of ratio. If the ratio doesn't exceed a threshold, it can be predicated that there is no forearm region in the hand region; otherwise, the corresponding angle of the maximal ratio is defined as the angle of inclination of the wrist line. In Fig.3(b) the height and width of the rectangle around the inner hand region makes the ratio maximal. Normally, it's unnecessary to project the image in all directions because the result will not change too much when the projection angle interval is small than a certain value.

The interval angle $\Delta\alpha$ is set as $30°$, and the threshold is set as 1.2. Following are the procedures of the projection-based algorithm.

Algorithm 1. Projection-based algorithm

Input:

    Binary image of inner hand

    Projection angle interval $\Delta\alpha = 30°$

    A predefined threshold $\mu =1.2$

Output:

    If the forearm exists in the input image,output the angle of inclination of the wrist line.

```
1    ratio_max=0; output_angle= 0° ;
2    i=0; i<= 180/Δα ; i++
3        α =i*Δα , β =i * Δα +90° ;
4        Project the inner hand region in directions angle α and β , the
    lengths of projected lines are a and b;
5            height_projection=(a>b)?a:b;
6            width_projection=(a<=b)?a:b;
7            ratio_temp=height_projection/width_projection;
8            if (ratio_temp>ration_max)
9                ration_max=ratio_temp;
10               output_angle= α ;
11   if(ratio_max< μ )
12      the forearm doesn't exist in the inner hand region;
13   else
14      return output_angle;
```

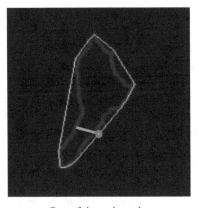
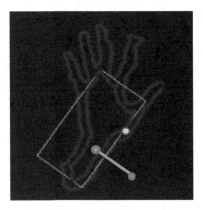

a. One of the wrist points   **b.** The wrist line

Fig. 3. Wrist recognition algorithm

4) With the wrist point we get in 2) and the gradient we get in 3), it is easy to draw a straight line in the space. And the two crossover points of the straight line and the contour of the hand region are the two endpoints of the wrist line. Connecting the two points, we can get the accurate wrist line,which is shown in Fig.3(b).

4.3   Estimating the Center of the Palm

The center of the palm could be considered as the center gravity of the optimized hand region. Formulas to calculate the center gravity are as follows.

$$MM_{00} = \sum_i \sum_j I(i, j) \tag{3}$$

$$MM_{10} = \sum_i \sum_j iI(i, j) \tag{4}$$

$$MM_{01} = \sum_i \sum_j jI(i, j) \tag{5}$$

$$i_c = \frac{MM_{10}}{MM_{00}}, \ j_c = \frac{MM_{01}}{MM_{00}} \tag{6}$$

In formulas above, $I(i,j)$ is the value of pixel whose coordinate is (i,j), and (i_c,j_c) is the coordinate of the center of the palm. The pink point in Fig.3(b) is the center of the palm calculated by the formulas mentioned above.

5   Fingertips Localization

Now we get the optimized hand region without the forearm part. With the palm center and the optimized hand contour, we will perform the fingertips localization algorithm to the hand region. The method based on the contour is a popular and effective way. The traditional K-curvature algorithm works as follows: for each point P_i in the contour, the algorithm gets its *kth* point in the left P_{i-k} and it *kth* point in the right P_{i+k}. For the two vectors $\overrightarrow{P_i P_{i-k}}$ and $\overrightarrow{P_i P_{i+k}}$, it calculates their intersection angle. If the cosine of the intersection angle exceeds the threshold, it is considered as a candidate fingertip.

K-curvature algorithm works quickly but there exists an obvious shortcoming that there isn't a universal standard to choose the value of parameter K. When the size of the hand region changes, the number of the points in the hand contour also changes and a constant K can't assure the correctness of fingertip localization. To overcome this shortcoming, we propose the θ-curvature algorithm.

The principle of θ-curvature algorithm can be elaborated as follows. For each finger contour, we pick two points which are farthest from the fingertip and are at different sides so that the contour between the two points is the corresponding finger contour. To be more vivid and explicit, as shown in Fig.5, each of the two points is linked to the center point of the palm with the same color. We manually draw lines in different colors to distinguish different fingertip regions. Two lines with the same color are considered as a pair, for example, the red pair is the part of little finger and the purple pair is the part of the thumb.

Fig. 4. Different part of each finger

The intersection angles of the five line pairs of vectors are $\theta_1, \theta_2, \theta_3, \theta_4$ and θ_5, we call them central angles. According to a large amount of observation, we find that values of these five central angles are very similar to each other ($\theta_1 \approx \theta_2 \approx \theta_3 \approx \theta_4 \approx \theta_5 \approx \theta$). Even if one or more fingers bend, the similarity between these five angles still stays the same. Since humans' hand shape is very similar, the central angle $\angle \theta$ keeps almost the same value when the size of hand varies. According to our measurement on different hand sizes, $\theta = 25°$ meets the requirement. Based on this conclusion, we propose the θ-curvature algorithm.

To better explain the θ-curvature algorithm, we convert the hand contours into a time-series curve which records the relative distance between each pixel on the contour and the palm center. As shown in Fig.6 and Fig.7, in our time-series figure, the vertical axis denotes the Euclidean distance between the contour pixels and the palm center, and the horizontal axis denotes the angle between each contour pixel and a certain pixel which we defined as the start point. The start and end points are two endpoints in the wrist line we get by the wrist recognition algorithm.

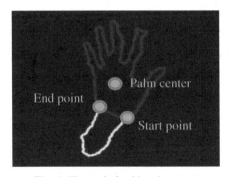

Fig. 6. The optimized hand contour

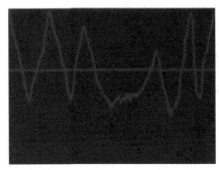

Fig. 7. The time-series curve

Here, we consider the hand contour as a continuous curve and it is an ideal model. Firstly, we sample the contour of hand between the start point and the end point to get a series of clockwise vertex points P_1, P_2..., P_n, making sure that $\angle P_{start}P_{center}P_1 > \frac{\theta}{2}$ and $\angle P_{end}P_{center}P_n > \frac{\theta}{2}$. As shown in Fig.8, for every P_i in the point sequences, we link it to the palm center point so that we get line segment $P_{center}P_i$. The length of $P_{center}P_i$ is recorded as L_i and the angle of inclination of $P_{center}P_i$ is β. Setting the palm

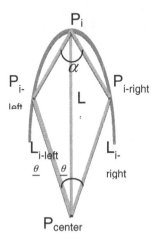

Fig. 8. θ-curvature algorithm

center point as the endpoint, $\beta - \frac{\theta}{2}$ and $\beta + \frac{\theta}{2}$ as two angles of inclination, we get two new lines at each side of the original line. The crossover points of the two lines and the hand contour are defined as $P_{i\text{-left}}$ and $P_{i\text{-left}}$, and each length of the two line segment is $L_{i\text{-left}}$ and $L_{i\text{-right}}$.

Since the angel of inclination and the length of each line can be directly known from Fig. 7, we don't need to figure out the coordinates of P_i, $P_{i\text{-left}}$ and $P_{i\text{-right}}$ to calculate α, which is the intersection angle of vector $\overrightarrow{P_i P_{i-left}}$ and $\overrightarrow{P_i P_{i-right}}$ .The value of α can been calculated by cosine theorem and sine theorem. Details are in equations (7) to (11).

$$P_i P_{i-left} = \sqrt{L_{i-left}^2 + L_i^2 - 2 * L_{i-left} * L_i * \cos\frac{\theta}{2}} \tag{7}$$

$$L_{i-left} * \sin\frac{\theta}{2} = P_i P_{i-left} * \sin\alpha 1 \tag{8}$$

$$P_i P_{i-right} = \sqrt{L_{i-right}^2 + L_i^2 - 2 * L_{i-right} * L_i * \cos\frac{\theta}{2}} \tag{9}$$

$$L_{i-right} * \sin\frac{\theta}{2} = P_i P_{i-right} * \sin\alpha 2 \tag{10}$$

$$\alpha = \alpha 1 + \alpha 2 \tag{11}$$

Here, $\alpha 1$ represents $\angle P_{i\text{-left}} P_i P_{center}$ and $\alpha 2$ represents $\angle P_{i\text{-right}} P_i P_{center}$.

If α is smaller than an angle threshold, then we regard P_i as a fingertip candidate. However, during this process, the valleys between fingers which are useless points will be picked out as well. To solve this problem, we set a condition that while L_i is the longest length among L_i, $L_{i\text{-right}}$, $L_{i\text{-left}}$, the point detected is the real fingertip. After the processing of θ-curvature algorithm, a series of candidate fingertips has been picked out. And by using clustering algorithm, we can get the final result of fingertips localization.

Nevertheless, the contour is actually represented by point sequences in system memory and there is a little difference in the algorithm implementation. Following is the detailed procedure of θ-curvature algorithm.

Algorithm 2. θ - curvature algorithm

Input:

The contour vertice $P_1, P_2, \ldots P_n$

Palm center point: P_{center}, $\beta_{threshold} = 25°$

Output:

$P_{fingertip}$ are the points that indicate the fingertips on hand contour

1   int start; int end; $P_{fingertip} = \{\}$

2   int i=0; i<n; i++

3     $\theta_i = \angle P_i P_{center} P_1$; $L_i = \left| \overrightarrow{P_i P_{center}} \right|$

4   If $\theta_i < \dfrac{\theta}{2}$ and $\theta_{i+1} \geq \dfrac{\theta}{2}$

5   Start=i+1

6   If $\angle P_n P_{center} P_1 - \theta_i \geq \dfrac{\theta}{2}$ and $\angle P_n P_{center} P_1 - \theta_{i+1} < \dfrac{\theta}{2}$

7   End= i;

8   Break;

9   j=start; j<=end; j++

10   m=j; m>=start; m--

11   if $\theta_j - \theta_m < \dfrac{\theta}{2}$ and $\theta_j - \theta_{m-1} \geq \dfrac{\theta}{2}$

12   left = m;

13   break;

14   n=j; n<end; n++

15   if $\theta_n - \theta_j < \dfrac{\theta}{2}$ and $\theta_{n+1} - \theta_j \geq \dfrac{\theta}{2}$

16   right = n;

17   break;

18   if $L_i \geq \max\{L_{left}, L_{right})$

19   calculate α with formulas (7) to (11)

20   if $\alpha < \beta_{threshold}$

21   P_i is selected as a fingertip candidate, and append P_i to $P_{fingertip}$

Finally, we apply the clustering algorithm to $P_{fingertips}$ and get the final result shown as Fig. 9.

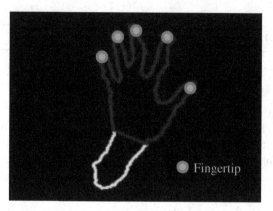

Fig. 9. The result of fingertips localization

6   Experiment Results

Running Time. There are 3 procedures during the whole process of algorithm: 1) Hand segmentation; 2) Wrist recognition; 3) Fingertips localization. For the 600 images, we figured out the average running time of each procedure. The result is shown in Fig.10.

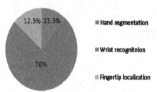

Fig. 10. The running time of every procedure

We infer that it is because the operation on every pixel of the image in distance transformation algorithm and gravity center calculation that the wrist recognition algorithm occupies most of the running time. And the total running time is about 30ms, which is of high efficiency and can support real-time interaction.

Accuracy Rate. We tested the accuracy of fingertips localization and wrist recognition algorithm. Here is the recognition accuracy of this experiment.

As shown in the table, the accuracy rate of wrist recognition algorithm is 91% and the average accuracy rate of θ-curvature algorithm is 91.33%. Since the θ-curvature algorithm is the improvement of K-curvature algorithm, we took K-curvature algorithm as a comparison, whose average accuracy is 88.8%. It can be seen from the comparison that the new algorithm has a distinct advantage in the performance.

Table 1. Recognition accuracy

Algorithm	The number of fingers	The number of images	Accuracy rate
θ-curvature algorithm	0	100	93%
	1	100	91%
	2	100	88%
	3	100	89%
	4	100	92%
	5	100	95%
K-curvature algorithm	0	100	88%
	1	100	91%
	2	100	87%
	3	100	85%
	4	100	90%
	5	100	92%
Wrist recognition algorithm	0,1,2,3,4,5	600	91%

7   Conclusion

In this paper, we proposed a set of effective Kinect-based algorithms to locate the feature points in the hand region. The method includes wrist recognition, palm center detection and fingertips localization. In the experiments, our algorithms have a good performance in terms of run-time and accuracy and are proved to be suitable for real-time application.

References

1. Wachs, J.P., Kosch, M., Stern, H., Edan, Y.: Vision-based hand gesture applications. Communications of the ACM, 60–71 (2011)
2. Zhichao, Y., Xin, Z., Lianwen, J., Ziyong, F., Shaojie, X.: Finger-Writing-in-the-Air System Using Kinect Sensor. In: Multimedia and Expo Workshops (ICMEW), pp. 1–4 (2010)
3. Hongyong, T., Youling, Y.: Finger Tracking and Gesture Interaction with Kinect. In: Computer and Information Technology (CIT), pp. 214–218 (2010)
4. Nikolaos, K., Antonis, A., Iason, O.: Efficient Model-based 3D Tracking of Hand Articulations using Kinect. In: Proceedings of the British Machine Vision Conference, pp. 1–11 (2011)
5. Baoliang, W., Zeyu, C., Jing, C.: Gesture Recognition by Using Kinect Skeleton Tracking System. In: Fifth International Conference on Intelligent Human-Machine System and Cybernetics, pp. 418–422 (2013)
6. Zhou, R., Junsong, Y., Zhengyou, Z.: Robust hand Gestue Recognition Based on Finger-Earth Mover's Distances with a Commodity Depth Camera. In: The 9th International Conference on Multimedia, pp. 1093–1096 (2011)

7. Zhengwei, Y., Zhigeng, P., Shuchang, X.: Wrist Recognition and the Center of the Palm Estimation Based on Depth Camera. In: International Conference on Virtual Reality and Visualization, pp. 100–105 (2013)
8. Sato, Y., Kobayashi, Y., Koike, H.: Fast Tracking of Hands and Fingertips in infrared Images for Augmented Desk Interface. In: Fourth IEEE International Conference on Automatic Face and Gesture Recognition, pp. 462–467 (2000)
9. Dung Duc, N., Thien Cong, P., Jae Wook, J.: Fingertip Detection with Morphology and Geometric Calculation. In: IEEE/RSJ International Conference on Intelligent Robots and System, pp. 1460–1465 (2009)
10. Shoubing, X., Guangda, S., Xiaolong, R., Qianqian, J., Fei, F.: Embedded implementation of real-time finger interaction system (in Chinese). In: Optics and Precision Engineering, pp. 1911–1918 (2011)
11. Zhigeng, P., Yang, L., Mingmin, Z., Chao, S., Kangde, G., Xing, T., Steven Zhiying, Z.: A Real-time Multi-cue Hand Tracking Algorithm Based on Computer Vision. In: Virtual Reality Conference (VR), pp. 219–222 (2010)

Designing Robot Eyes and Head and Their Motions for Gaze Communication

Tomomi Onuki[1], Kento Ida[1], Tomoka Ezure[1], Takafumi Ishinoda[1], Kaname Sano[1], Yoshinori Kobayashi[1,2], and Yoshinori Kuno[1]

[1] Department of Information and Computer Sciences, Saitama University, 255 Shimo-okubo, Sakura-ku, Saitama 338-8570, Japan
{t.onuki,kentida,ezure,t_ishinoda,
yosinori,kuno}@cv.ics.saitama-u.ac.jp
[2] Japan Science and Technology Agency (JST), PRESTO, 4-1-8 Honcho, Kawaguchi, Saitama 332-0012, Japan

Abstract. Human eyes not only serve the function of enabling us "to see" something, but also perform the vital role of allowing us "to show" our gaze for non-verbal communication. The gaze of service robots should therefore also perform this function of "showing" in order to facilitate communication with humans. We have already examined which shape of robot eyes is most suitable for gaze reading while giving the friendliest impression, through carrying out experiments where we altered the shape and iris size of robot eyes. However, we need to consider more factors for effective gaze communication. Eyes are facial parts on the head and move with it. Thus, we examine how the robot should move its head when it turns to look at something. Then, we investigate which shape of robot head is suitable for gaze communication. In addition, we consider how the robot move its eyes and head while not attending to any particular object. We also consider the coordination of head and eye motions and the effect of blinking while turning its head. We propose appropriate head and eye design and their motions and confirm their effectiveness through experiments using human participants.

Keywords: Robot eye, gaze reading, facial design.

1 Introduction

In the last decade, extensive research has been conducted into the use of robots in daily life that are able to engage in natural communication with people. In particular, service/communication robots are required to be able to perform not only verbal communication but also non-verbal communication. For human beings, one of the most important aspects of non-verbal communication is gaze, utilized, for example, to make eye contact and establish joint attention. We are developing a robot system that can attract a target person's attention, make eye contact, and then establish joint attention. We have revealed necessary robot actions depending on the positional relation between the human and the robot [7]. Our next target is to develop a robot system that

D.-S. Huang et al. (Eds.): ICIC 2014, LNCS 8588, pp. 607–618, 2014.

can effectively perform such actions for gaze communication. Before developing an actual robot, we need to find the design principles of robot's shape and motion. It is needless to say that the most important part related to gaze communication is the eye. However, since the eye is a facial part on the head, we also need to consider the head.

Kobayashi and Kohshima [5] found that among primates only human eyes have no pigment in the sclera; moreover, human eyes also have the horizontally longest shape with the largest exposed area of sclera. Various explanations were offered as to why other primates have sclera colored in a similar fashion to their irises or the outside of their eyes. But all the explanations were based on the consensus that primates may avoid clearly showing their gaze. In contrast, human eyes have sclera with clearly different colors from those of the irises and the outside of the eyes. This enables the human gaze to be readily comprehended by others. Based on these studies, we determined a shape of robot eyes which conveyed a friendly appearance and a readily-readable gaze [8].

As the next step, we mainly consider the head in this paper. First, we investigate how the robot should move its head when it turns to look at something. We observe human head turning motions and find that the head speed is not constant during the action. We show that such human-like head turning make robot actions felt more natural through experiments using human participants.

Then, we examine the face/head shape. Face direction can be considered as a cue for gaze direction. Funatsu et al. [4] reported that gaze direction can be read by others through face direction with respect to body orientation when at a distance. In human-human communication, gaze direction tends to be recognized through head direction rather than eye direction [6]. Delaunay et al. [2] proposed a retro-projected robotic face and conducted experiments assessing a user's ability to read gaze direction for a selection of different robotic face designs. Through these experiments, they concluded that gaze-reading by the user was performed most successfully with a 3D-shaped mask with a human appearance. In this paper, we extend their study in a more systematic way. We prepare three faces: a flat face, a spherical face, and a spherical face with a nose. We show is the spherical face with a nose is most suitable for gaze reading through experiments using human participants.

Next, we consider idling motion, that is, how the robot should move when it does not attend to any particular person or object. Our hypothesis is that if the robot's idling action is seen natural, the robot may be more effectively able to attract people's attention when it really looks at them. Chikaraishi et al. showed that the gaze idling motion reacting to objects around especially those emerging or moving is more natural idling motion than that doing nothing and not reacting to any visual information [1]. We compare three gaze idling motions: reacting to high visual saliency positions (saliency motion), turning the head from side to side at a uniform velocity (horizontal uniform motion), and doing nothing (static motion). Experimental results are promising for the salient idling motion to effectively attract attention of people and induce joint attention although the number of participants was small and we need further investigation.

We also need to consider the coordination of the eye movements and head motions. In the field of research on driver assistance systems, there have been a number of studies about this. Doshi et al. reported [3] that human attention shifts can be divided into two types. When surprised, humans shift their gaze by means of the head following a motion of the eyes. In other words, the head pursues the motion of the eyes during an unintentional gaze shift. On the other hand, the opposite is true during an intentional gaze shift, such as when looking at a side mirror; the eyes follow the motion of the head. In this paper, we only consider the differences between the head turning actions with or without eye movements about the impressions on human observers. Here we include blinking in eye movements. When we investigated human head turning action, we found that many people blink when turning their head. Thus, we assume that blinking while turning the head may enforce the naturalness in robot motion. We have verified this assumption through experiments by using human participants.

Lastly we develop a robot head based on our proposed design principles. This paper presents our design principles and experimental results to support the principles.

2   Eye Shape Design

This section briefly describes our robot eyes and summarizes the results of our research on static gaze. For details see [8].

2.1   Development of Robot Eyes

We have developed a robot head with eyes enabling the functions of both "seeing" and "showing" gaze. For the latter function we employ projectors to display various types of eye images generated by computer graphics. We attached cameras to the robot head to enable the robot "to see." Furthermore, to enable precise control of the eye gaze towards a certain object, we establish calibrations among CG coordinates, camera coordinates and real world coordinates. An overview of our prototype robot head is shown in Fig. 1 (left). As shown in Fig. 1 (right), each of the eyes consists of a projector, a mirror and a screen. The eye images, generated by CG, are projected onto the hemisphere screen via rear-projection.

Kobayashi et.al [5] considered two parameters of eyes: the lid distance and the iridal diameter. Thus, we have designed this robot head in such a way that these parameters can be adjusted. The hemisphere mask can be changed to alter the lid distance (outline shape of the eyes), while the iridal diameter (iris size) can be easily changed because the irises are generated by CG. For the experiments described in the next subsection, we prepared three types of outline eye shape based on the width/height ratio: "round" (1:1), "ellipse" (1:0.5), and "squint" (1:0.25). In the same manner, we changed the iris size based on the ratio of the eye width and iris diameter: "large" (1:0.75), "medium" (1:0.5), and "small" (1:0.25). Consequently, we had 9 types (the combination of 3 outline shapes and 3 iris sizes) of eye design, as shown in Fig. 2. Among these, Fig. 2-E is the most similar to the human eye.

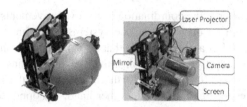

Fig. 1. (left) Overview of our proposed robot head;(right) the inside of the robot head consisting of laser projectors, mirrors and screens

Fig. 2. Candidate designs for robot eyes derived by varying lid distance and iridal diameter

2.2   Experimental Evaluation of Eye Shapes

We conducted two experiments. First, we performed an experiment to measure the perceived friendliness of robot eyes. We showed each of the nine pairs of robot eye design on an iPad to 105 participants (university students), and asked which seemed friendlier (Fig. 3). We analyzed the data by using Thurstone's method of paired comparison. According to the results, there are two factors involved in determining perceived friendliness. One is the height of the eyes, and the other is the ratio of the iris area to the white area of the eyes. The results show that the friendliest robot face is one with eyes featuring a round outline shape and a large iris (as in Fig. 2-A).

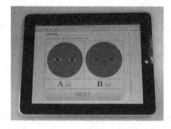

Fig. 3. Evaluating impression of robot eyes through the method of paired comparisons

We next performed an experiment to analyze the relationship between the accuracy of gaze reading and the shape of the robot eyes by using the robot head proposed in the previous subsection. We evaluated errors in gaze reading using the 9 types of design candidates for robot eyes shown in Fig. 2. We lined up a series of markers between the participant and the robot head, as shown in Fig. 4. We asked each participant to state at which marker the robot looked. We divided the 36 participants into 3 groups. Each group experienced one type of outline shape with three different sizes of iris projected in a random order. We calculated the average errors for each design candidate.

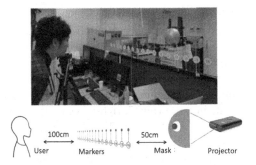

Fig. 4. Experimental scene from robot gaze reading

Fig. 5 shows the result. In general, errors were fewer for eyes with a larger iris with respect to the sclera. A robot's gaze is therefore most readable by human users when performed by robot eyes with a round outline shape and a large iris (Fig. 2-A). In this experiment, the robot eyes did not move; they just appeared suddenly on the screens. Thus, we conducted a further experiment where the robot eyes would dynamically move toward the target marker. We compared the average errors between the static gaze and the dynamic gaze. The average errors for the dynamic gazes were smaller than those of the static gazes in all cases. From these experiments, we can conclude that among the 9 design candidates for robot eyes, the design with a round outline shape and a large iris (Fig. 2-A) is most suitable for gaze reading and conveying an impression of friendliness, and that a dynamic gaze is more effective than a static one.

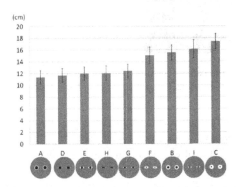

Fig. 5. Errors of gaze reading for each design candidate of robot eyes

3   Head Turning Design

3.1   Observing Human Attentive Action

To learn how the robot head should be turned, with appropriate eyes, to express its visual focus of attention for establishing joint attention, we observed human eye and head motion during gaze shifting. Fig. 6 depicts a scene from the observation experiment. A participant is on the right in Fig. 6. We lined up a series of successively-numbered markers in front of each participant. The distance between the participants and the markers was 50 cm. We asked the participants to look at the marker we indicated. The camera in front of the participants filmed their behavior looking at the marker. Fig. 7 (left) shows an example of a captured image. An accelerometer (Wii remote controller by Nintendo Co., Ltd.) on the participant's head measured the speed with which they turned their head. When the accelerometer was rotated horizontally, it calculated an angular velocity at the time instance (Fig. 7, right). We used 39 participants, consisting of male and female university students around 20 years of age.

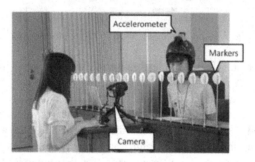

Fig. 6. Scene from experiment observing users turning their heads

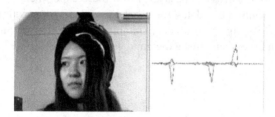

Fig. 7. (left) Image from the camera in front of a participant; (right) resulting graph of accelerometer's measurements (vertical axis: angular velocity, horizontal axis: time)

An example of a participant turning their head is shown in Fig. 8. The solid line shows the series of changes in the speed of head-turning in a large angle, while the dashed line shows the series of changes in the speed of head-turning in a small angle. As seen in Fig. 8, a large head turn and a small head turn take the same duration. Additionally, speed increases at first and then decreases, and a larger head turn involves a larger maximum speed. Most of the participants blinked their eyes when turning their heads. Almost all participants moved their heads before moving their eyes. Participants could likely predict the gaze destination because of the successive order of indicated markers.

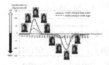

Fig. 8. Series of changes in speed of head-turning

3.2   Evaluating Impressions of Robot Head-Turning

Based on the findings mentioned in the subsection 3.1, we prepared a robot so that the robot turned its head like a human, and conducted an experiment to evaluate the impression it made. Fig. 9 shows the measured results of the robot's head-turning.

We asked participants to observe the robot's head-turning behavior, both with an accelerated speed like humans and with a constant speed. After observing the two types of robot behavior, participants answered a questionnaire about their impressions. Fig. 10 shows a scene from the experiment. Participants observed each motion and answered questions about their impression of each.

In the questionnaire we asked the following six questions:

1. Which motion do you like? (Likeability)
2. Which motion do you think gives the impression of the robot turning its head to see something? (Robot turns its head to see something)

Fig. 9. Graph showing robot head-turning speed

Fig. 10. Scene from experiment evaluating impressions of a robot's head-turning behavior

3. Which motion do you think gives the impression that the robot looks at you? (Robot looks at you)
4. Which motion do you think is more natural? (Natural)
5. Which motion do you think appears friendlier? (Friendly)
6. Which eyes do you think are more mechanical? (Mechanical)

The participants evaluated each motion on a seven-point Likert scale. Which motion each participant saw first was random. We used 24 participants consisting of male and female university students around 20 years of age.

The results are shown in Fig. 11. Regarding the questions "Likeability," "Robot turns its head to see something," "Robot looks at you," "Natural," and "Friendly," the accelerated speed version was frequently chosen by the participants. We analyzed the results using a sign test. Significant differences were recognized for all items except "Robot turns its head to see something" ($p<.05$). Especially, there were large differences for "Natural." And the constant speed version was felt "Mechanical." The results indicate that the accelerated speed version is preferable for service robots.

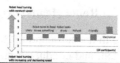

	Likely	Robot turns its head to see something	Robot looks at you	Natural	Friendly	Mechanical
Average	3.042	3.25	3.29	2.75	3.042	5.21
p-value	.027	.17	.041	.0071	.035	.027

Fig. 11. Results of evaluation comparing impressions of head-turning motion

4   Head Shape Design

Face/head direction is an important cue for gaze direction. Thus, we examined which shape of robot head is suitable for gaze communication. We prepared   three robot faces for comparison experiments. Fig. 12 shows these faces: a flat face, a hemisphere face, and a hemisphere face with a nose. A tablet PC was used to show the flat face's eyes. The rear-projection system described in Section 2 was used to display the eyes for the rest two faces.

Fig. 12. Developed robot faces. (a)Flat, (b) Hemisphere, and (c) Hemisphere with a nose

The same manner and settings in subsection 2.2 were used in the experiment.   Initially, the robot looks in front. Then the robot head rotates toward one of the markers at variable speed proposed in the previous section. The robot head returns back to look in the direction of the front marker at a uniform speed after 5 seconds. In this experiment, we move only the head. The eyes are fixed at the center.

A participant read robot gaze direction from each of three faces. We gathered 16 male and female participants around 20 years old.

The results are shown in Fig. 13. As shown in Fig. 13, the hemisphere face with a nose gives the smallest average error. We performed analysis of variance. Table 1 shows the results. There are significant differences in the results of Holm test (Table 2). These results indicate that the robot gaze direction is easier to be read in the order of the hemisphere face with a nose, the flat face, and the hemisphere face. Some participants even mentioned that they felt an arrow coming out from the nose of the hemisphere face with a nose.

Fig. 13. Gaze reading errors for three faces

Table 1. Analysis of variance for gaze reading accuracy

A = {Flat, Hemisphere, Hemisphere with Nose}

S.V	df	F	η^2	p
A	2	19.67	.466	.727E-06
s x A	45			

Table 2. Multiple comparisons by holm

Face	Number of Data	Mean (cm)
Flat	16	19.25
Hemisphere	16	26.06
Hemisphere + Nose	16	9.31
Comparison	P	
Flat < Hemisphere	.747E-02	
Flat > Hemisphere + Nose	.181E-03	
Hemisphere > Hemisphere + Nose	.243E-07	

5 Head and Eye Motion Design

5.1 Natural Idling Motion

We examined idling motion of the robot, that is, how the robot should move when it does not attend to any particular person or object. We hypothesized that if the robot's idling action is seen natural, the robot may be more effectively able to attract people's attention when it really looks at them. Thus we compared three gaze idling motions: reacting to high visual saliency positions (the saliency motion), turning the head from side to side at uniform velocity (the horizontal uniform motion), and doing nothing (the static motion).

Fig. 14 shows the experimental environment. Participants are asked to enter the room where the robot is placed and to observe the room and the robot so that they can answer questions about them later. They first stand 5 meters away from the robot and look around the room and the robot showing one of the three idling motions for 20seconds. Then they move to the side of the robot 1 meter away. The robot gazes at the target object indicated by the square in Fig. 14 when they approach. The experiment finishes 5 seconds later. We take the video for later analysis.

We prepare the following three idling motions. In the static motion, the robot continues gazing in front. In the horizontal uniform motion, the robot rotates its head from side to side at a constant velocity. In the saliency motion, the robot turns its head so that it may look at visually salient parts in the scene one by one. In Fig. 14 these parts that the robot looks at are indicated by the circles. In this motion, the robot shifts its face direction toward the participants when they are approaching the robot because moving objects show high visual saliency. Each participant experiences only one idling motion. The number of participants for each idling motion experiment is 10 (horizontal uniform motion) or 11(the others).

Fig. 14. Experimental environment

After the experiment we asked the participants to fill in a questionnaire about impressions of the robot. However, we cannot find any significant differences in the results. We then analyzed the participants' actions from the recorded video. The purpose of our research is to develop a robot that can effectively establish joint attention with humans. Thus we examined whether or not the robot can establish joint attention when the robot turns its gaze toward the object indicated by the square in Fig. 14. In the salience motion case, 6 participants out of 11 (54.5%) turned to look at the object and the joint attention was established. In the horizontal uniform motion case, 4 participants out of 10 (40%) did so, whereas only 2 out of 11 (18.2%) in the static motion case. The results are promising for the salient idling motion to effectively attract attention of people and induce joint attention although the number of participants was small and we need further investigation.

5.2 Coordination of Head and Eye Motions

We can move the eyes and the head to gaze at something. Thus we need to consider the coordination of these two motions. In this paper, we only consider the differences between the head turning actions with or without eye movements about the impressions on human observers. Here we include blinking in eye movements. When we investigated human head turning action described in section 3, we found that many people blink when turning their head. Thus, we assume that blinking while turning the head may enforce the naturalness in robot motion. We have verified this assumption through experiments by using human participants.

The setup of the experiment is shown in Fig. 15. A participant is seated in front of the robot. The robot explains to him/her the picture on the monitor. The robot looks in the direction of the monitor during explanation, but it turns its head toward the participant at some appropriate timings as we proposed in our guide robot study [9]. We prepare two motion patterns for this head turning. One is that the robot just turns its head by 60 degrees without any eye movements. The eyes are fixed in the center positions. In the other, the robot rotates both head and eyes each by 30 degrees simultaneously the robot also blinks during turning. Participants are just asked to listen to the robot's explanation. Each participant experience one of the two patterns. We use 51 participants, 24 for the head motion only and 27 for the head-eye combination motion.

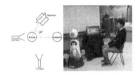

Fig. 15. The experiment setup

After the experimental session, participants were asked to evaluate the naturalness of the robot in 7 steps by the Likert-scale (the larger, the more natural). The average scores are 5 for the head turning with eye motion, and 3.96 for the head turning without eye motion. There are significant differences by t-test (p=.001). Although we cannot separate the effects of total eye motion and blinking from this experiment, we can conclude that head turning with eye motion improves the natural impression.

6   Combining Eye and Head Design

We have designed and made a robot head as shown in Fig. 16 based on our design principles. We have replaced the head of Robovie-R ver.3 by Vistone by this head. The inner structure of the head is the same as the one shown in Fig. 1. Although the original Robovie's face does not have a nose, the face center of the new robot head is protuberant like a nose. Furthermore, ear like parts are attached to both sides of the head to show the head direction more clearly.

Fig. 16. New robot head embedded in Robovie-R

7   Conclusion and Future Work

Gaze communication is important for service robots interacting with humans. We would like to design robots suitable for gaze communication. The major parts related to gaze communication are eyes and head. We need to consider their static shapes and their motions. We have already examined which shape of robot eyes is most suitable for gaze reading while giving the friendliest impression. In this paper, we have considered the head shape design and the head motion. We also consider the combination of head and eye motions. We have found that the head motion at changing speed of slow-fast-slow like humans do can give natural and friendly impressions. We have also revealed that we can estimate gaze direction more correctly for a face with a nose than faces without a nose. We have further examined how the robot should move its head while not attending to any particular object. We have confirmed that moving both head and eyes can give more natural impressions.

In this paper, we just compared the head motions with and without eye movements. We would like to examine more complicated coordination of head motion and eye motion.

Acknowledgment. This work was partly supported by JST CREST and JSPS KAKENHI Grant Number 23650094.

References

1. Chikaraishi, T., Nakamura, Y., Matumoto, Y., Ishiguro, H.: Gaze control for a natural idling motion in an android. In: The 13th Robotics Symposi., pp. 206–211 (2008) (in Japanese)
2. Delaunay, F., Greeff, J., Belpaeme, T.: A study of a retro-projected robotic face and its effectiveness for gaze reading by humans. In: Proc. HRI 2010, pp. 39–44 (2010)
3. Doshi, A., Trivedi, M.M.: Head and gaze dynamics in visual attention and context learning. In: CVPR Workshops 2009, pp. 77–84 (2009)
4. Funatsu, N., Takahashi, T., Deguchi, D., Ide, I., Murase, H.: Kao to karada no shisei joho o riyo shita jinbutsu douga-zo kara no shisen hoko suitei ni kansuru kento. In: MIRU 2012, IS3-59 (2012) (in Japanese)
5. Kobayashi, H., Kohshima, S.: Unique morphology of the human eye and its adaptive meaning: comparative studies on external morphology of the primate eye. Journal of Human Evolution 40, 419–435 (2001)
6. Kobayashi, H.: Gaze movement in establishing joint attention. Annual Report of the Research Institute for Science and Technology, Tokyo Denki University, 23, 219–222 (2004) (in Japanese)
7. Hoque, M.M., Onuki, T., Kobayashi, Y., Kuno, Y.: Effect of robot's gaze behaviors for attracting and controlling human attention. Advanced Robotics 27(11), 813–829 (2013)
8. Onuki, T., Ishinoda, T., Kobayashi, Y., Kuno, Y.: Designing robot eyes for gaze communication. In: FCV 2013, pp. 97–102 (2013)
9. Yamazaki, A., Yamazaki, K., Kuno, Y., Burdelski, M., Kawashima, M., Kuzuoka, H.: Precision timing in human-robot interaction: coordination of head movement and utterance. In: Proceedings of the SIGCHI Conference on Human Factors in Computing Systems, pp. 131–140. ACM (2008)

Robust Fingertip Tracking with Improved Kalman Filter

Chunyang Wang and Bo Yuan

Intelligent Computing Lab, Division of Informatics
Graduate School at Shenzhen, Tsinghua University
Shenzhen 518055, P.R. China
tsinglong2011@gmail.com, yuanb@sz.tsinghua.edu.cn

Abstract. This paper presents a novel approach to reliably tracking multiple fingertips simultaneously using a single optical camera. The proposed technique uses the skin color model to extract the hand region and identifies fingertips via curvature detection. It can remove different types of interfering points through the cross product of vectors and the distance transform. Finally, an improved Kalman filter is employed to predict the locations of fingertips in the current image frame and this information is exploited to associate fingertips with those in the previous image frame to build a complete trajectory. Experimental results show that this method can achieve robust continuous fingertip tracking in a real-time manner.

Keywords: multi-fingertip tracking, curvature detection, distance transform, Kalman filter, data association.

1   Introduction

In traditional human-computer interaction (HCI), users rely on devices such as key-boards, mice, remote controllers and touch screens as the control interface, causing various levels of inconvenience. With the popularization of smart TVs and smart phones, the desire for simple, natural and intuitive HCI techniques has been consistently increasing. Vision based HCI, which uses video to capture a user's information such as face, gait and gesture, has the inherent advantage of being natural and user friendly and has become a fast growing research area in recent years [1].

As a key component of vision based HCI, fingertip tracking features extensive application prospects. Previously, fingertip tracking has been realized using infrared camera [2], multiple cameras [3] and LED lights for marking fingers [4]. Although these methods did achieve fingertip tracking to some extent, they suffer from a number of issues: i) the use expensive cameras; ii) strict restrictions on the positions of fingers; iii) the need of auxiliary devices. By contrast, in this paper, we focus on real-time fingertip tracking using a single low cost optical camera and bare hands to improve the applicability and user experience.

For multi-target tracking, data association is the most commonly used method [5]. The typical methods of data association include nearest neighbor, probabilistic data association filter and joint probabilistic data association filter. A simple approach to

D.-S. Huang et al. (Eds.): ICIC 2014, LNCS 8588, pp. 619–629, 2014.

multi-fingertip tracking is by associating the detected fingertips in the current image frame with the corresponding nearest fingertips in the previous image frame [6]. However, it is prone to false association when fingertips move quickly. In order to solve the mutual interference and the interfering noise in multi-fingertip tracking and make full use of the motion information of fingertips, Particle filter and Kalman filter have been used to improve the effectiveness of fingertip tracking. For example, the K-means clustering algorithm can be combined with Particle filter to solve the problem of mutual interference [7]. Although this method can track multiple fingertips accurately, the cost of computation is significant and is not suitable for real-time applications. In the meantime, Kalman filter can be used to predict the locations of fingertips in the new image frame and associate fingertips between image frames using the predicted locations [2]. This method needs to measure the time interval between image frames and the velocity of fingertips and cannot properly handle the sudden change of locations caused by the acceleration of fingertips.

In this paper, we employ an improved Kalman filter to predict the locations of fingertips under the assumption of uniform acceleration instead of uniform speed. In addition, we propose a method using distance transform to remove interfering points. In the rest of this paper, Section 2 describes the procedure of extracting the hand region using the skin color model. Section 3 shows how to detect fingertips by curvature and remove interfering points using the cross product of vectors and distance transform. Section 4 gives the details of our approach to multi-fingertip tracking. Section 5 presents the main experimental results and analysis and this paper is concluded in Section 6.

2   Hand Extraction

The purpose of hand extraction is to separate hand region from background. To extract hand efficiently, we use the skin color model. In computer vision, the color space includes RGB, YCbCr, HSV, YIQ and so on. According to the distribution of human skin color [8] and previous research results [9–11], YCbCr is selected in our work. The conversion from RGB to YCbCr is as below:

$$\begin{bmatrix} Y \\ Cb \\ Cr \end{bmatrix} = \begin{bmatrix} 0 \\ 128 \\ 128 \end{bmatrix} + \begin{bmatrix} 0.2990 & 0.5870 & 0.1140 \\ -0.1686 & -0.3311 & 0.4997 \\ 0.4998 & -0.4185 & -0.0813 \end{bmatrix} \begin{bmatrix} R \\ G \\ B \end{bmatrix} \tag{1}$$

In YCbCr color space, Y represents illumination while Cb and Cr represent chrominance information. Each image is segmented into skin region and non-skin region based on the values of Cb and Cr. The threshold values of Cb and Cr are set as below according to previous studies:

$$\begin{cases} 77 \leq Cb \leq 127 \\ 133 \leq Cr \leq 173 \end{cases} \tag{2}$$

The original color image in a real-world environment is shown in Fig. 1(a) and the binary image as the result of skin color segmentation is shown in Fig. 1(b). It is easy to see that there are some noise and holes in the initial binary image. Next, morphological operation with a 5-by-5 structuring element is applied to remove small noise and the morphological closing operation is used to remove small holes. After morphological operation, we calculate the size of each connected region and the largest one is identified as the hand region, as shown in Fig. 1(c). Note that, in normal circumstances, the hand is much closer to the camera than the human face and it is reliable to use area information to remove the face region.

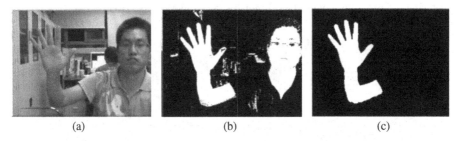

(a)	(b)	(c)

Fig. 1. Hand extraction: (a) Original image; (b) Skin color segmentation; (c) Binary hand image

3   Fingertip Detection

3.1   Curvature Detection

In order to detect fingertips from the contour of hand, the curvature-based algorithm is adopted [12–14]. The curvature of a contour point is represented by the cosine value of θ, which is the angle between $\overrightarrow{P_iP_{i-N}}$ and $\overrightarrow{P_iP_{i+N}}$:

$$\cos\theta = \frac{\overrightarrow{P_iP_{i-N}} \cdot \overrightarrow{P_iP_{i+N}}}{\left\|\overrightarrow{P_iP_{i-N}}\right\|\left\|\overrightarrow{P_iP_{i+N}}\right\|} \tag{3}$$

where P_i, P_{i-N} and P_{i+N} are the ith, the $(i-N)^{th}$ and the $(i+N)^{th}$ points in the contour, respectively. In practice, values of N between 5 and 25 work reasonably well.

Points with curvature values satisfying a predefined threshold value (close to 0) are selected as candidates for fingertips. An example of candidate fingertips after curvature detection is shown in Fig. 2(a). Note that the curvature values are also close to 0 for points located at places such as the valleys between fingers, the joint of arm and other peaks on the contour. These are regarded as interfering points and, in order to avoid false positive results, some measures should be taken to remove these points.

3.2   Fingertip Filtering

For points located at the valleys of the contour, the values of the cross product of vectors are different to those located at peaks (opposite sign). As a result, we can use this direction information to remove interfering points located at places such as the valleys between fingers and the joint of arm. The remaining candidate fingertips after being filtered by this method are shown in Fig. 2(b).

For interfering points located at the peaks of the contour, we can use distance transform to remove them [15].The result of distance transform on a binary image is a grayscale image as shown in Fig. 3(b), which will be further converted to a binary image. From Fig. 3(c), we can see that fingers are removed while the palm and arm are preserved by distance transform. For each candidate fingertip, we calculate its minimum distance to the contour and remove those candidates whose minimum distances are less than a threshold value. The candidate fingertips after distance transform are shown in Fig. 2(c). Finally, the locations of fingertips are obtained by clustering the remaining candidate fingertips, as shown in Fig. 2(d).

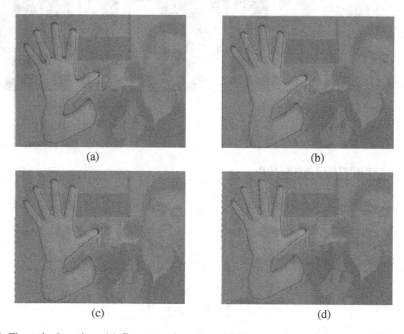

(a)   (b)

(c)   (d)

Fig. 2. Fingertip detection: (a) Curvature detection; (b) Cross product of vectors; (c) Distance transform; (d) Cluster analysis

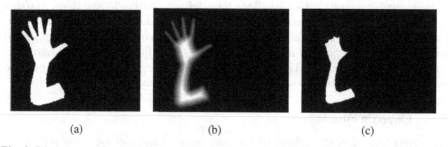

(a)   (b)   (c)

Fig. 3. Distance transform: (a) Original binary image; (b) Gray image after distance transform; (c) Binary image after distance transform

4   Fingertip Tracking

For the locations X_k of fingertips in the kth image frame, an improved Kalman filter is used to predict the locations $\hat{X}_{k+1}$ of fingertips in the $(k+1)^{th}$ image frame. Then we use the nearest neighbor rule to associate fingertips between frames by comparing the predicted locations $\hat{X}_{k+1}$ with the detected locations X_{k+1} in the $(k+1)^{th}$ image frame to achieve the robust tracking of multiple fingertips.

4.1   Fingertip Prediction

We take into account the location, velocity, and acceleration information of fingertips and use an improved Kalman filter to predict the locations of fingertips [2, 16–18].

The location, velocity and acceleration of a fingertip along a certain coordinate in the kth image frame are represented by x_k, v_k and a_k, respectively. The kinematic equations describing the motion of a fingertip are shown as follows:

$$x_k = x_{k-1} + v_{k-1}T + a_{k-1}T^2/2 \tag{4}$$

$$x_{k-1} = x_{k-2} + v_{k-2}T + a_{k-2}T^2/2 \tag{5}$$

$$x_{k-2} = x_{k-3} + v_{k-3}T + a_{k-3}T^2/2 \tag{6}$$

$$v_{k-1} = v_{k-2} + a_{k-2}T \tag{7}$$

$$v_{k-2} = v_{k-3} + a_{k-3}T \tag{8}$$

Since the time interval T between two consecutive image frames is very short, we assume that the acceleration of fingertips in three successive image frames is approximately constant. Therefore, the following equation is obtained by combining (4), (5), (6), (7) and (8):

$$x_k = 3x_{k-1} - 3x_{k-2} + x_{k-3} \tag{9}$$

Note that velocity, acceleration and time interval between image frames are all eliminated in (9). With this simplified representation, it is now possible to only rely on the locations of fingertips in the previous three image frames to predict the locations of fingertips in the current image frame.

We define the state vector x_k as the locations of a fingertip along one coordinate axis in three successive image frames, $x_k=[x_k, x_{k-1}, x_{k-2}]^T$, and the observation vector z_k as the locations of the fingertip in the kth image frame ($z_k=x_k$).

The state and observation equations of system are:

$$x_{k+1} = F_k x_k + \Gamma_k w_k \tag{10}$$

$$z_k = H_k x_k + v_k \tag{11}$$

In the above, F_k is the state transition matrix; Γ_k is the noise matrix of system; H_k is the observation matrix; w_k is system noise, which influences the state of system; v_k is observation noise, which is the error between real and detected locations of fingertip.

The definitions of F_k, Γ_k and H_k are as follows:

$$F_k = \begin{bmatrix} 3 & -3 & 1 \\ 1 & 0 & 0 \\ 0 & 1 & 0 \end{bmatrix} \tag{12}$$

$$\Gamma_k = \begin{bmatrix} 1 & 0 & 0 \end{bmatrix}^T \tag{13}$$

$$H_k = \begin{bmatrix} 1 & 0 & 0 \end{bmatrix} \tag{14}$$

Assume that w_k and v_k are the Gaussian white noise with zero mean and the covariance matrix of w_k and v_k are Q_k and R_k, respectively. In addition, w_k and v_k are independent of each other and also independent of the initial state of system.

The optimal predicted value and the covariance matrix of predicted error are:

$$\hat{X}_{k+1|k} = F_k \hat{X}_{k|k} \tag{15}$$

$$P_{k+1|k} = F_k P_{k|k} F_k^T + \Gamma_k Q_k \Gamma_k^T \tag{16}$$

In the above, $\hat{X}_{k+1|k}$ is the optimal predicted value in the $(k+1)^{th}$ image frame; $\hat{X}_{k|k}$ is the optimal estimated value in the kth image frame; $P_{k+1|k}$ is the covariance matrix of $\hat{X}_{k+1|k}$; $P_{k|k}$ is the covariance matrix of $\hat{X}_{k|k}$.

After obtaining the observation vector, the optimal estimated value and the covariance matrix of estimated error are calculated by (17), (18) and (19) where K_{k+1} is the Kalman gain in the $(k+1)^{th}$ image frame.

$$K_{k+1} = P_{k+1|k} H_{k+1}^T \left(H_{k+1} P_{k+1|k} H_{k+1}^T + R_{k+1} \right)^{-1} \tag{17}$$

$$\hat{X}_{k+1|k+1} = \hat{X}_{k+1|k} + K_{k+1} \left[z_{k+1} - H_{k+1} \hat{X}_{k+1|k} \right] \tag{18}$$

$$P_{k+1|k+1} = \left[I - K_{k+1} H_{k+1} \right] P_{k+1|k} \tag{19}$$

During fingertip tracking, we first substitute the optimal estimated value $\hat{X}_{k|k}$ in the kth image frame into (15) to get the optimal predicted value $\hat{X}_{k+1|k}$ in the $(k+1)^{th}$ image frame and obtain the predicted locations in the $(k+1)^{th}$ image frame. Second, we use data association to associate the trajectories of fingertips with the detected locations of fingertips in the $(k+1)^{th}$ image frame (see Section 4.2). Third, we substitute the detected locations of fingertips in the $(k+1)^{th}$ image frame into (18) to get the optimal estimated value in the $(k+1)^{th}$ image frame.

4.2   Fingertip Association

The nearest neighbor method is used for data association with Euclidean distance as the metric. The Euclidean distance between the predicted location and the detected location in the $(k+1)^{th}$ image frame is calculated by:

$$d = \sqrt{(x-x')^2 + (y-y')^2} \qquad (20)$$

The detected location of the jth fingertip in the $(k+1)^{th}$ image frame is (x_j, y_j) and the predicted locations of fingertips in the $(k+1)^{th}$ image frame are (x'_1, y'_1), $(x'_2, y'_2)...(x'_n, y'_n)$. The Euclidean distances between the detected location of the jth fingertip and all predicted locations of fingertips are calculated according to (20). If the nearest neighbor of the jth detected fingertip is the ith predicted fingertip and their distance is less than a threshold value, the ith fingertip in the kth image frame is matched to the jth fingertip in the $(k+1)^{th}$ image frame. Consequently, the jth fingertip in the $(k+1)^{th}$ image frame is associated with the ith fingertip in the kth image frame to form a trajectory. In this way, we can realize the tracking of multiple fingertips.

Note that new fingertips may appear and existing fingertips may disappear during the process of tracking. If a detected fingertip in the $(k+1)^{th}$ image frame is not close enough to any predicted fingertips, it is considered as a new fingertip. If the predicted location of a tracked fingertip in the $(k+1)^{th}$ image frame is not close to any detected fingertip, this fingertip is considered as being disappeared.

In summary, compared with the traditional Kalman filter, the improved Kalman filter eliminates the velocity and acceleration of fingertips as well as the time interval between image frames. In fact, it only uses the locations of fingertips in the previous three image frames to predict the locations of fingertips in the current image frame, which reduces the computational cost and avoids the prediction error caused by the change of time interval between image frames. The proposed method considers the influence of acceleration, which makes the predicted locations of fingertips more accurate and avoids the failure of tracking caused by the sudden change of movement.

5   Experiment and Analysis

Experiments were conducted using a desktop computer with Intel Core i5-2400 at 3.10 GHz CPU and a USB optical camera with 640×480 resolution in the normal lighting environment. The proposed methods were implemented using Visual C++ and OpenCV 2.4.3.

Efficiency. The tracking program achieved ~23 FPS with no finger extended and ~18 FPS when all fingers were extended. The average frame rate of the proposed method was ~20 FPS and the real-time constraint was satisfied reasonably well.

Accuracy. Table 1 shows the accuracy of fingertip detection and tracking with different fingertip numbers. Note that the accuracy of tracking was higher than the accuracy of detection. The reason is that even when there were some interfering points, they were likely to be far away from fingertips and might not be incorrectly associated with fingertips. The main reason of fingertip tracking failure is due to the false locations of fingertips caused by the environmental factors (e.g., illumination change), resulting in wrong fingertip association.

Table 1. Accuracy of Proposed Tracking Method

Number of fingertips	Number of frames	Number of correctly detected	Number of correctly tracked	Accuracy rate of detection	Accuracy rate of tracking
1	200	199	200	99.5%	100%
2	200	198	200	99.0%	100%
3	200	189	200	94.5%	100%
4	200	192	196	96.0%	98.0%
5	200	188	200	94.0%	100%
Total	1000	966	996	96.6%	99.6%

Number of correctly detected is the number of image frames that have only one detected fingertip for each finger and no interfering points;
Number of correctly tracked is the number of image frames that achieve continuous and accurate tracking for each fingertip.

In order to further demonstrate the effectiveness of the improved Kalman filter, a set of illustrative experiments was conducted using the mouse cursor instead of the fingertip. Fig. 4 shows the predication results under different conditions where 'I' represents the traditional Kalman filter and 'II' represents the improved Kalman filter. The red circle, green triangle and blue square are the detected location, the location predicted by the traditional Kalman filter and the location predicted by the improved Kalman filter, respectively.

Intuitively, the improved Kalman filter produced more accurate prediction results and its advantage became more distinct when there was velocity change or direction change. Table 2 shows the distances between the detected locations and the predicted locations using the traditional Kalman filter and the improved Kalman filter in different conditions.

Finally, Fig. 5 shows a typical example of the continuous tracking of five fingertips within around 100 frames in a practical environment. The five trajectories are marked in red and it is clear that although the motions of fingertips were highly nonlinear, the constructed trajectories were smooth and complete, confirming the effectiveness of the proposed tracking method.

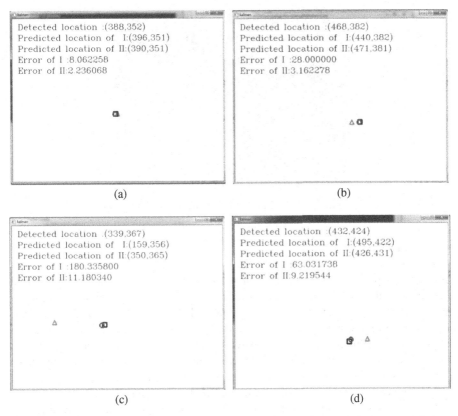

Fig. 4. Comparison of accuracy between the traditional Kalman filter and the improved Kalman filter under different conditions: (a) Low velocity; (b) High velocity; (c) Velocity change; (d) Direction change

Table 2. The Distances between the Detected Locations and the Predicted Locations

	Low velocity	High velocity	Velocity change	Direction change
The traditional Kalman filter	8.1	28.0	180.3	63.0
The improved Kalman filter	2.2	3.2	11.2	9.2

The unit of distance is pixel.

Fig. 5. An illustration of the continuous tracking of five fingertips

6   Conclusions

This paper proposed an effective fingertip tracking approach using a single entry-level USB camera and bare hands. It combines the location, velocity and acceleration information of fingertips and employs the improved Kalman filter to predict the locations of fingertips in the current image frame based on the locations of fingertips in the previous three image frames. Compared with the traditional Kalman filter, it does not need to calculate the velocity of fingertips and the time interval between image frames, which reduces the computational cost and avoids the prediction error caused by the change of time interval between frames. It also considers the influence of acceleration, which makes the predicted locations of fingertips more accurate and avoids the failure of fingertip tracking caused by the sudden change of movement effectively. Furthermore, this paper proposed a method for removing interfering points based on distance transform.

Experimental results show that the proposed method can achieve robust fingertip tracking in a real-time manner and properly handle situations with the appearance of new fingertips as well as the disappearance of existing fingertips. A potential direction for future work is to further improve the robustness of tracking against complex background using machine learning techniques and apply the proposed techniques to interesting real-world applications.

References

1. Wang, L., Hu, W., Tan, T.: Recent developments in human motion analysis. Pattern Recognition 36, 585–601 (2003)
2. Oka, K., Sato, Y., Koike, H.: Real-time fingertip tracking and gesture recognition. Computer Graphics and Applications 22, 64–71 (2002)
3. Xie, Q., Liang, G., Tang, C., Wu, X.: A fast and robust fingertips tracking algorithm for vision-based multi-touch interaction. In: 10th IEEE International Conference on Control and Automation, pp. 1346–1351 (2013)

4. Nakamura, T., Takahashi, S., Tanaka, J.: Double-crossing: A new interaction technique for hand gesture interfaces. In: Lee, S., Choo, H., Ha, S., Shin, I.C. (eds.) APCHI 2008. LNCS, vol. 5068, pp. 292–300. Springer, Heidelberg (2008)
5. Jaward, M., Mihaylova, L., Canagarajah, N., Bull, D.: A data association algorithm for multiple object tracking in video sequences. In: The IEE Seminar on Target Tracking: Algorithms and Applications, pp. 129–136 (2006)
6. Letessier, J., Bérard, F.: Visual tracking of bare fingers for interactive surfaces. In: 17th Annual ACM Symposium on User Interface Software and Technology, pp. 119–122 (2004)
7. Wang, X.Y., Zhang, X.W., Dai, G.Z.: An approach to tracking deformable hand gesture for real-time interaction. Journal of Software 18, 2423–2433 (2007)
8. Zarit, B.D., Super, B.J., Quek, F.K.: Comparison of five color models in skin pixel classification. In: International Workshop on Recognition, Analysis, and Tracking of Faces and Gestures in Real-Time Systems, pp. 58–63 (1999)
9. An, J.H., Hong, K.S.: Finger gesture-based mobile user interface using a rear-facing camera. In: 2011 IEEE International Conference on Consumer Electronics, pp. 303–304 (2011)
10. Gasparini, F., Schettini, R.: Skin segmentation using multiple thresholding. In: Internet Imaging VII, SPIE, vol. 6061, p. 60610F (2006)
11. Chai, D., Ngan, K.N.: Face segmentation using skin-color map in videophone applications. IEEE Trans. on Circuits and Systems for Video Technology 9, 551–564 (1999)
12. Lee, T., Hollerer, T., Handy, A.R.: Markerless inspection of augmented reality objects using fingertip tracking. In: 11th IEEE International Symposium on Wearable Computers, pp. 83–90 (2007)
13. Lee, B., Chun, J.: Manipulation of virtual objects in marker-less AR system by fingertip tracking and hand gesture recognition. In: 2nd International Conference on Interaction Sciences: Information Technology, Culture and Human, pp. 1110–1115 (2009)
14. Pan, Z., Li, Y., Zhang, M., Sun, C., Guo, K., Tang, X., Zhou, S.Z.: A real-time multi-cue hand tracking algorithm based on computer vision. In: 2010 IEEE Virtual Reality Conference, pp. 219–222 (2010)
15. Liao, Y., Zhou, Y., Zhou, H., Liang, Z.: Fingertips detection algorithm based on skin colour filtering and distance transformation. In: 12th International Conference on Quality Software, pp. 276–281 (2012)
16. Han, C.Z., Zhu, H., Duan, Z.S.: Multi-source information fusion, 2nd edn. Press of Tsinghua University, Beijing (2010)
17. Kalman, R.E.: A new approach to linear filtering and prediction problems. Journal of Basic Engineering 82, 35–45 (1960)
18. Schneider, N., Gavrila, D.M.: Pedestrian Path Prediction with Recursive Bayesian Filters: A Comparative Study. In: Weickert, J., Hein, M., Schiele, B. (eds.) GCPR 2013. LNCS, vol. 8142, pp. 174–183. Springer, Heidelberg (2013)

Monte Carlo-Discrete Wavelet Transform for Diagnosis of Inner/Outer Race Bearings Faults in Induction Motors

Amirhossein Ghods and Hong-Hee Lee

School of Electrical and Computer Engineering, University of Ulsan, South Korea
{amirhossein,hhlee}@mail.ulsan.ac.kr

Abstract. Detection and precise diagnosis of faults in induction motors has turned into a major field in the electric machinery world, due to the importance of these motors in industry and research. Faults are usually classified in two groups as mechanical and electrical faults. Bearing failures that consists more than half of mechanical faults can be recognized by applying high magnitude, short-time frequency bursts to supply sources. Methods like (short-time) Fourier, (continuous-discrete) Wavelet, and Park transforms are among the most common strategies for fault detection. The deficit that these methods usually face is usually related to the fact that either these methods cannot derive non-stationary behavior of fault; whereas most of bearing faults are non-stationary and low energy signals. In the proposed method, a new approach has been offered that consists of decomposing the frequency spectrum of output stator current signal into several levels. Considering this matter that faults have non-certain behavior, it seems necessary to model this non-deterministic behavior. Monte-Carlo model is a strong approach toward deterministic modeling. In this paper, the fault probabilistic behavior has been modeled using this approach.

Keywords: Discrete wavelet analysis, fault diagnosis, inner/outer race bearing faults, Monte-Carlo modeling, non-stationary faults.

1 Introduction

Bearing faults mostly result into vibrations along the inner- and outer-races that cause loss of energy in high maintenance costs [1]. Inner race faults have more severe effects and damages on the output signal of the induction motor rather than outer race ones [2], so inner race fault detection is a more time consuming process.

To address the issue of bearing fault diagnosis, it can be seen in [2], using Finite Element Analysis (FEA), bearing faults are detected with application of Fast Fourier and Wavelet Transforms. Authors of [3-4] tried to verify the fact that because most of bearing faults have non-stationary low-energy characteristic, it is almost impossible to use conventional methods to derive the behavior of the fault. By the term non-stationary, authors mean frequency characteristic varies during the time. Wavelet transform is introduced as a method to include both time and frequency characteristics together and therefore the deficit of Fourier transform is eliminated; as it has been shown in [5-6] that Continuous Wavelet Transform (CWT) and Discrete Wavelet

D.-S. Huang et al. (Eds.): ICIC 2014, LNCS 8588, pp. 630–636, 2014.
© Springer International Publishing Switzerland 2014

Transform (DWT) have application in analyzing the current signature of the defected induction motor. The major characteristic of CWT is keeping frequency resolution high in case of low time resolution, that makes it ideal for transient analysis.

The redundancy of fault signal in CWT increases the amount of computation time, while DWT provides enough information for analysis of signal. In DWT, the signal is only decomposed related to time axis and therefore, in case that the fault has a non-stationary behavior-such as bearing defects, the fault behavior cannot be derived precisely, because the major fault information are carried in the frequency domain.

This paper proposes a novel approach toward detection of bearing fault. A fault that happens in an induction motor usually has a non-stationary characteristic because of harmonics and noises [7]. In the proposed model, the uncertainty of fault behavior is simulated using Montel-Carlo model. Using this concept, bearing fault happening probability can be predicted and therefore possible losses and damages will be reduced.

This manuscript is structured as follows: in section 2 a SIMULINK model for an induction motor is developed; bearing faults are simulated on this model. In section 3, a new diagnosis method will be proposed that relies on frequency-domain analysis using DWT. In section 4, a probabilistic model for prediction bearing faults behavior using Monte Carlo method will be proposed. Finally, simulation results are brought in section 5 that explains the benefits of probabilistic modeling of faults, especially bearing faults.

2   Modeling of Bearing Faults in Induction Motor

For having knowledge of fault behavior, a model has been developed using MATLAB® SIMULINK. In this model, a PWM motor drive is used for controlling the speed of motor. This motor is a squirrel cage induction motor with nominal speed, nominal power, and voltage as 1430 rpm, 4 kW, and 400 kV respectively.

As in Fig. 2, general ball bearing configuration is shown; and therefore, inner and outer races and location of possible faults can be seen. There are two major frequency components that exist in stator currents; ball pass frequency of outer race f_o, ball pass frequency of inner race f_i[8]. These components are used for diagnosis in case a bearing fault happens

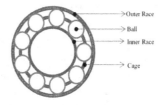

Fig. 1. Ball Bearing Configuration

Usually, the number of balls in a bearing system varies between 6 and 12 [11]; therefore, in this simulation 6 ball are considered. The characteristic frequencies given above may be approximated by (using 50 Hz as the major frequency):

$$f_0 = 0.4 \times n \times f_r = 0.4 \times 6 \times 50 = 120 \, Hz \tag{1}$$

$$f_i = 0.6 \times n \times f_r = 0.6 \times 6 \times 50 = 180 \, Hz \tag{2}$$

If a bearing fault occurs, there would be abnormality in these components; in other words, there are high magnitude components in these frequencies that represent there is a fault existing in the motor.

A common way of modeling bearing faults is applying sequence of high frequency bursts, which represent the impulse response of the signal transmission path, and can be repeated at a rate related to the fault that is interacting with whether inner race, outer race, or rolling elements [9]. Therefore, by applying these high frequency bursts, bearing fault may be realized.

3   DWT Frequency-Domain Analysis

The DWT provides sufficient information both for analysis and synthesis of the original signal, with a significant reduction in the computation time.

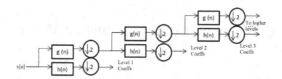

Fig. 2. DWT Decomposition Process

With attention to Fig. 3, the input signal is first passed through low-pass filters, so that high frequency information would be removed by using low-pass filter. Then signal is subsampled by 2; in other words, every other sample is discarded. By this action, the frequency resolution doubles, but the time-resolution remains unchanged. These two-steps are called one level of decomposition. In this paper, the output fault signal is a 128-sample long signal at 1 MHz. The first level of decomposition makes the signal into a 64-sample long one at 500 KHz.

It is important to calculate the required number of decomposition levels. As mentioned before, in each level that signal is passed through filters and then subsampled, the number of samples is halved and the resolution in frequency axis will be doubled. It is recommended [10] that the process should continue until there are only two samples left; in other words, the number of levels is n for a 2^n-sample long signal. The main purpose of this action is to focus as much as possible in frequency domain and be able to detect non-stationary faults that have extremely low energy characteristic, such as bearing faults. In this work, a 128 (2^7)-sample long fault signal is derived, so there exist 7 levels of decomposition in the DWT process. After each level of decomposition, the probabilistic model is applied and non-deterministic behavior of fault in each level of decomposition will be derived.

It is recommended to go as high as possible in the number of decomposition levels, in order to eliminate non-required high energy components and detect major fault information that have a low energy characteristic. In [11], it is shown that there are differences between types of bearing faults, in terms of signal energy. For faults that usually occur in inner race, the energy of signal is lower compared to the faults that happen in outer race or ball-bearings; This fact will be shown again in simulation results.

4   Monte Carlo Modeling of Faults in Induction Motors

Monte Carlo modeling is chosen as appropriate method because occurrence of fault mostly is random [12]; therefore, authors of this paper try to model the stochastic behavior of induction motor, so the effects of faults will be as close as possible to real case.

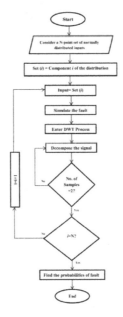

Fig. 3. Monte Carlo Probabilistic Process

As it can be seen in Fig. 4, the flow diagram shows the process of Monte Carlo modeling. First, a set of normally distributed points is defined, that each point represents the frequency of input voltage to the motor. In other words, in an N-points normally distributed set, each point is regarded as an independent scenario, in which it shows the frequency of voltage. By considering each point, the fault is simulated one time using the developed model that is shown in Fig. 1.

5   Simulation Studies

A Monte Carlo model has been used by applying DWT in frequency domain. It is assumed that the Gaussian distributed set of input voltage frequencies has a mean in 50 Hz and 10 Hz variance (∂). So, 99% of data is in range of -3∂ to +3 ∂, which is from 31 Hz to 78 Hz.

By considering the bearing fault, based on (1) and (2), two special spectra are supposed to be detected. In our model, these frequencies can be observed in 120 and 180 Hz. Therefore, the measure for accuracy of this method can be regarded as detectability of fault frequency components in 120 and 180 Hz.

In Fig. 5 the decomposed output signal (stator current) is shown in different DWT levels. In lower levels, the fault frequency components cannot be recognized (120 and 180 Hz), but as it follows to higher levels, these frequencies can been seen.

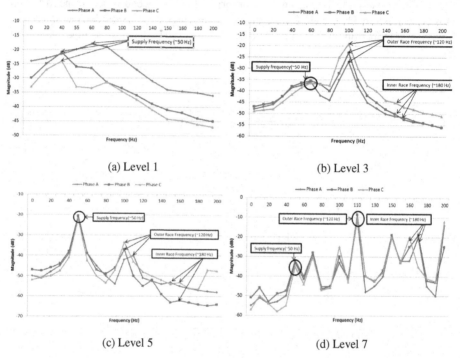

(a) Level 1   (b) Level 3

(c) Level 5   (d) Level 7

Fig. 4. 5th Repetition in Gaussian Distributed Set, Frequency Domain Analysis

Based on Fig. 6, if the input to a linear system follows a Normal distribution, the output will definitely follow a Normal distribution [13]. Therefore, if input voltage frequencies are Normally distributed, then fault frequencies in stator current will follow a Normal distribution.

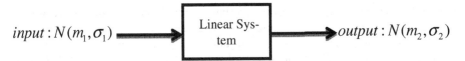

Fig. 5. Linear System Response to Gaussian Input

In Fig. 7, the probability distribution of fault is shown for decomposition in level 1, 3, 5, and 7. It can be clearly seen that in Fig. 7. a, the curve is definitely not looking like a Gaussian distribution around fault spectra, because the output signal includes of many noises and additional components.

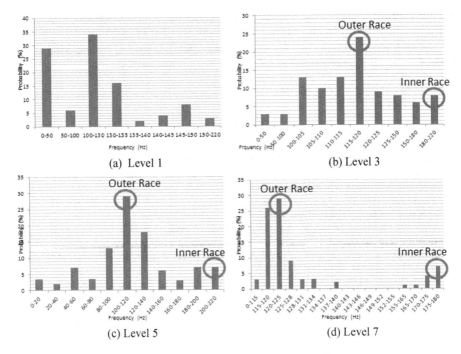

Fig. 6. Induction Motor Stator Current DWT-Processed Output Probability Distribution for Different Levels

In higher levels of decomposition like Fig. 7. b, c, and d, the probability distribution curve looks more to a Gaussian distribution around fault spectra and fault frequencies can be seen clearly. This shows that by considering the DWT process, additional components are eliminated and faults can be recognized exactly.

6   Conclusion

Because of the uncertainty that exists in the nature of fault, non-deterministic behavior of fault is simulated. For this purpose, a Monte Carlo approach has been proposed. A set of Normally distributed inputs are defined that will represent the supply voltage frequencies. In every iteration, the input voltage carries a special frequency, and then the output signal (stator current) is passed through a low-pass filter and sub-sampled for eliminating the high energy un-wanted components. As the level of decomposition goes higher, it can be seen that first of all, the outer race defect is observed in 120 Hz, and then inner race defect that has a lower energy characteristic is recognized in 180 Hz. This shows the majority of DWT process in frequency domain. In lower levels of decomposition, the probability distribution doesn't look like a Gaussian distribution; however, in higher levels, because of eliminating high energy additional components, this output would be more similar to a Gaussian distribution and fault frequencies are more visible.

Acknowledgment. This work was partly supported by the National Research Foundation of Korea Grant funded by the Korean Government (No.2013R1A2A2A01016398) and the Network-based Automation Research Centre (NARC) funded by the Ministry of Trade, Industry & Energy.

References

1. Onel, I.Y., Benbouzid, M.E.H.: Induction motors bearing failures detection and diagnosis: Park and Concordia transform approaches comparative study. In: IEEE Electric Machines and Drives Conference, Antalya, Turkey, pp. 1073–1078 (2007)
2. Li, W.: Detection of induction motor faults: a comparison of stator current, vibration and acoustic methods. J. Vibration and Control 12(2), 165–188 (2006)
3. Stack, J.R., Habetler, T.G., Harley, R.G.: Fault-signature modeling and detection of inner-race bearing faults. IEEE Trans. Ind. Appl. 42(1), 61–68 (2006)
4. Wang, Y., Kang, S., Jiang, Y., Yang, G., Song, L., Mikulovich, V.I.: Classification of fault location and the degree of performance degradation of a rolling bearing based on an improved hyper-sphere-structured multi-class support vector machine. Mechanical Sys. and Signal Process. 29, 404–414 (2012)
5. Prabhakar, S., Mohanty, A.R., Sekhar, A.S.: Application of discrete wavelet transform for detection of ball bearing race faults. Tribology Int. 35, 793–800 (2002)
6. Constantin, A.I., Fireteanu, V., Leconte, V.: Effects of the short-circuit faults in the stator winding of induction motors and fault detection through the magnetic field harmonics. In: Advanced Topics in Electrical Engineering Symposium, Bucharest, Bulgaria, pp. 1–6 (2013)
7. Zarei, J., Poshtan, J.: Bearing fault detection using wavelet packet transform of induction motor stator current. Tribology Int. 40, 763–769 (2007)
8. McFadden, P.D., Smith, J.D.: Vibration monitoring of rolling element bearings by the high frequency resonance technique. Tribology Int. 11, 3–10 (1984)
9. Polikar, R.: The wavelet tutorial (1999), http://users.rowan.edu/~polikar/WAVELETS/WTtutorial.html (April 29, 2014)
10. Kannan, K., Perumal, S.A., Arulmozhi, K.: Optimal decomposition level of discrete, stationary and dual tree complex wavelet transform for pixel based fusion of multi-focused images. Serbian J. Elect. Eng. 7, 81–93 (2010)
11. Ghods, A., Lee, H.H.: A frequency-based approach to detect bearing faults in induction motors using discrete wavelet transform. In: IEEE International Conference on Industrial Technology, Busan, Korea, pp. 1–5 (2014)
12. Galton, F.: Natural inheritance. Macmillan and Co., London (1889)
13. Grigoriu, M.: Transient response of linear systems to stationary Gaussian inputs. Probabilistic Eng. Mech. 7(3), 159–164 (1992)

Impact of Using Information Gain in Software Defect Prediction Models

Zeeshan Ali Rana[1,2], Mian M. Awais[1], and Shafay Shamail[1]

[1] Department of Computer Science, SBA School of Science and Engineering,
Lahore University of Management Sciences (LUMS), Lahore, Pakistan
{zeeshanr,awais,sshamail}@lums.edu.pk
[2] Faculty of Information Technology, University of Central Punjab, Lahore

Abstract. Presence or absence of defective modules in software is an indicator of quality of the software. Every company aspires to deliver good quality software with minimum number of defective modules. To achieve this goal, defect prediction models are used in different phases of software lifecycle. These models have to deal with a large number software metrics (as input parameters to the models). These metrics have correlation issues that affect a model's performance. Also, in some cases using all the metrics negatively impacts the models' performances. In order to reduce size of input space and resolve the possible issues of correlation in input data, models reported in literature use Principal Component Analysis (PCA) and Information Gain (IG) based dimension reduction. PCA reduces the dimensions but keeps the representation of all the input variables intact. Use of PCA is not suitable where representation of all the metrics is declining a model's performance. To handle such situations, this paper advocates use of Information Gain (IG) based technique to reduce size of input space by dropping the irrelevant metrics. Afterwards, only relevant metrics are used to develop a prediction model. This paper compares the PCA and IC based techniques to develop classification tree and fuzzy inferencing system based models. In order to study the impact of using IG, percentage improvement in Recall, Accuracy and Misclassification Rate have been calculated for the aforementioned models. The results show that use of IG improves the models' performances more often than PCA does.

1   Introduction

Defectiveness of software is widely considered as a parameter to judge quality of the software [5], [7], [8], [14], [25]. It is desired to produce a good quality software which has minimum post release defects. In order to reduce post release defects and catch as many defects as possible prior to the release, defect prediction is performed at various stages of software lifecycle [10]. Models used for the defect prediction usually use software metrics and predict the defectiveness of the software. The defect data of NASA projects is publicly available, in form of datasets [16]. These models use various data mining based techniques such as classification trees, neural networks and rule based systems, for example [4], [10], [12], [26]. These data mining techniques are capable of extracting useful and meaningful information from large data repositories. So the

D.-S. Huang et al. (Eds.): ICIC 2014, LNCS 8588, pp. 637–648, 2014.
© Springer International Publishing Switzerland 2014

models based on these techniques are used to extract the defectiveness information from the software measures collected during development of the software [10].

All data mining based defect prediction models go through the stages of problem statement, data collection, data preprocessing, model development, model evaluation and reporting of results [3]. Data preprocessing stage includes the resolution of data compatibility issues, correlation issues and dimensionality reduction (DR). After the required preprocessing, data mining models are developed using the reduced input space. Afterwards, each model is tested and certain test performance measures are recorded to evaluate the model performance before the reporting phase.

Principle Component Analysis (PCA) and Factor Analysis (FA) have been used for dimension reduction in defect prediction [6], [11], [13], [15]. Other comprehensive studies on dimension reduction techniques have also been done [1]. Information Gain (IG) [18] is another technique that has been investigated for dimension reduction [9]. Kao et al. have demonstrated the utility of IG on eclipse based defect data of two releases of Eclipse project [9]. Roobaert et al [24] have combined IG with feature selection matrix to reduce dimensions of input space in financial analytics domain, and developed a support vector machine based classification model. Furthermore, IG has been reported as a useful candidate for dimension reduction in different domains [2], [19], [20].

PCA works on the principle of transforming the larger input space to a smaller input space such that all the variables of the larger input space are represented in the reduced input space. We argue that in some cases using influence of all variables in the input space does not help. Rather using a subset of the variables for model development can give better results, for example [22]. Furthermore, there are scenarios where the data is not suitable for direct PCA. In such cases some additional preprocessing is needed to make data suitable for PCA. So for these scenarios, there should be an alternate approach as well.

In this paper we compare an Information Gain (IG) [18] based dimensionality reduction approach for the aforementioned scenarios. We drop the irrelevant metrics using an IG based algorithm and select relevant metrics for defect prediction models. We study the impact of using this approach on two defect prediction models i.e. Classification Trees (CT) [18] and a Fuzzy Inference System (FIS) based model [21]. Performance of both the models has improved after the IG based preprocessing. We develop the models using public datasets [16]. These datasets are other than the Eclipse project datasets used in earlier studies like [9]. We compare the results of using PCA and IG for both the models and notice that using IG produces better prediction results in terms of misclassification rate and recall.

Section 2 describes our methodology, the suggested dimensionality reduction approach and the evaluation parameters for the comparison. Section 3 presents the experimental results and section 4 provides an analysis of these results. Section 5 identifies the future directions and concludes the paper.

2   Methodology

This paper suggests an Information Gain (IG) [18] based approach to select relevant input measures from software defect data [16]. This paper develops classification trees and fuzzy inference system based models for identification of defect prone

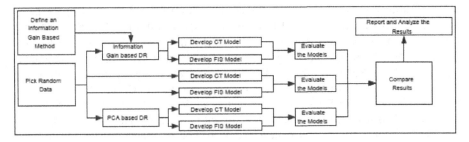

Fig. 1. Our Methodology

Table 1. Datasets List

Serial	Dataset	Parameters	Instances	Variant	%age of *ND* Modules	SVP[*]
ds1	kc1	21	2109	No	84.54%	2
ds2	kc1-classlevel	94	145	No	58.62%	8
ds3	kc1-classlevel-oo	8	145	Yes	58.62%	0
ds4	kc1-originally-classlevel	10	145	Yes	58.62%	1
ds5	kc2	21	522	No	79.5%	0
ds6	kc3	39	458	No	90.61%	0
ds7	pc1	21	1109	No	93.05%	0

*SVP = Same Valued Parameter

modules Fig. 1 shows that PCA and IG have been compared as dimension reduction approaches for these two models. Rest of the section discusses the steps mentioned in Fig. 1 after a brief discussion on the datasets used for this study.

2.1 Datasets

Datasets used in this study describe the defectiveness information regarding various software modules. From the 7 datasets, listed in table 1 and used in this study, 2 have been modified prior their use, and 5 have been used without any modifications. Number of parameters and instances in each of the dataset has been mentioned in table 1. The parameters are the independent variables and an additional classification parameter is used as output to indicate if the software class is defect prone (*D*) or not-defect prone (*ND*). Each instance in a dataset represents a software class (or module) and the parameters are software measures calculated for that module. A dataset is termed as a variant if the number of parameters of the dataset has been changed for this study. Here, ds3 and ds4 are variants of ds2. Furthermore, percentage of negative instances in each dataset is also mentioned in the table to show any imbalance in a dataset. The last column mentions the number of parameters that hold same value for all the instances. Such parameters are not useful in prediction, rather they can potentially have a negative effect on the prediction.

Two variants of ds2 (kc1-classlevel data) have been formed by dividing the parameters into three groups. *Group A* has 8 parameters, *Group B* has 10 parameters and *Group C* has 84 parameters, *Group A* $\subset$ *Group B* and *Group B* $\cap$ *Group C* = ϕ. Parameters in *Group A* are the most commonly used parameters for defect prediction identified in [17]. The ds3 (kc1-classlevel-oo) comprises of *Group A* parameters only and is marked as variant in table 1. Values of the parameters in *Group B* are originally measured at module level and these parameters are used to form ds4 (kc1-originally-classlevel). This dataset is also marked as variant in the table. The values of the parameters in *Group C* are originally measured at method level and are later transformed to module level before making the dataset available at PROMSIE website [16]. Parameters used in ds2 are a union of *Group B* and *Group C* parameters.

2.2   Information Gain (IG) Based Preprocessing Approach

The IG based approach, presented in algorithm 1, works on the principle of dropping the irrelevant input parameters (independent variables) and using the rest of the parameters for model development. Irrelevant parameters are those which do not individually contribute much in prediction.

As seen from algorithm 1, three input values are required for this approach: two of these are non-scalar and are underlined, third is a scalar value. First of the two non-scalar inputs, *AllParameters*, is a $n{\times}m$ matrix with n software modules and m software measures (or independent variables) used to predict defectiveness of the software. Second input, *Target*, is an $n \times 1$ matrix representing the defectiveness of n software modules. The scalar α s.t. $0 \leq \alpha \leq 1$, represents a threshold to decide whether a certain parameter needs to be dropped or not. The output, <u>*RelevantP arameters*</u>, is an $n \times k$ matrix with $k \leq m$ and <u>*RelevantP arameters*</u> $\subseteq$ <u>*AllParameters*</u>.

Calculating Entropy: Entropy of a set S are defined as follows [18]:

$$Entropy(S) = \sum_{i=1}^{k} - p_i \log_2 p_i$$

$$(1)$$

where k is total number of classes and p_i is proportion of examples that belong to class i. In this paper we are focusing on a binary class problem (i.e. *D* or *ND* modules), hence $k=2$ in our case.

Calculating Information Gain: Information gain of an attribute A in a set S is defined as follows [18]:

$$IG(S,A) = Entropy(S) - \sum_{i=1}^{v} \frac{|S_i|}{|S|} Entropy(S_i)$$

$$(2)$$

where v is total number of distinct values in attribute A (i.e. domain of A) and S_i is set of examples that hold a certain value from domain of A.

Algorithm 1 IG-based algorithm to select relevant measures for defect prediction models

---

Require: $AllParameters, Target, \alpha$
Ensure: $RelevantParameters$ is $returned$ $s.t.$ $RelevantParameters$ $\subseteq$
    $AllParameters$
    {Steps:}
    $Entropy \leftarrow calculateEntropy(Target)$
    $RelevantParameters \leftarrow AllParameters$
    $paramCount \leftarrow countParams(AllParameters)$
    for $i = 1$ to $paramCount$ **do**
      $IG[i] \leftarrow calculateInformationGain(AllParameters[i], Entropy)$
      if $IG[i] < \alpha$ **then**
        Drop this parameter from $RelevantParameters$
      end if
    end for
    return $RelevantParameters$

2.3   Randomizing and Preprocessing Data

Before data preprocessing and dimension reduction, we perform randomization of the data so that random data is used for training and testing. The same random set of data is fed into both the prediction models to compare PCA and IG.

2.4   Development and Evaluation of the Defect Prediction Models

Two defect prediction models have been developed for all the datasets. One is a binary CT whereas the second is an FIS based model. Each model has been developed with and without dimension reduction. Three models have been developed for each dataset, i.e. one with each of the dimension reduction approach and one without any dimension reduction. After development of a model based on training data, the model is evaluated using 10-fold validation.

2.5   Comparison Framework

Performance of the proposed approach must be compared with PCA to see the potential benefit in using the new approach. We compare these two preprocessing approaches to see impact of IG on the models' performances. We perform this comparison by developing prediction models without any dimension reduction and then calculating the percentage change in the model's performance when PCA and IG based dimension reduction is performed as shown in Fig. 1.

The comparison mentioned above has been done in terms of the evaluation parameters listed in Table 2. *TNRate* and *TPRate* (or *Recall*) are obtained while computing a confusion matrix. *Acc* and *MC* are important and widely used model performance measures when confusion matrix is available. It is worthwhile to mention that performance in terms of *Recall* is significant as it helps in focusing on the problematic areas in the software and thus better resource planning can be done.

And our goal is to detect the defect prone modules, so we consider *Recall* as our primary measure of comparison and then we compare the results on the basis of *MC*. In order to gauge the performance of each approach, percentage change (increase/decrease) in *Recall*, *Acc* and *MC* have been calculated using the following expression:

Table 2. Evaluation Parameters Used for Comparison

Evaluation Parameter	Abbreviation
True Negative Rate	*TN*
True Positive Rate (Recall)	*TP (Recall)*
Accuracy	*Acc*
Misclassification Rate	*MC*
Percentage Change in Recall	*RC Change (%)*
Percentage Change in Acc	*Acc Change (%)*
Percentage Change in MC	*MC Change (%)*

$$\%age\ change\ =\ \frac{(DR\ Reading - NoDR\ Reading)}{NoDR\ Reading} \times 100 \qquad (3)$$

where *NoDR Reading* is value of evaluation parameter when model is developed without any dimension reduction and *DR Reading* is value of evaluation parameter when models are developed after dimension reduction. Positive values of *RCChange(%)* and *Acc Change(%)* indicate improvement in models' performances whereas a positive value of *MC Change(%)* indicate that a model's performance has declined.

In addition to these parameters we have also calculated the average *RC Change(%)*, *Acc Change(%)* and *MC Change(%)*. These averages will help to determine the net effect of using IG instead of PCA for the 7 datasets used in this study. For highly imbalanced datasets, we have also considered F measures [23] with equal importance to *Recall* and *Precision*.

3   Results

This section presents the best results of each model in terms of misclassification rate. All the CT and FIS models have been developed using Matlab. Dimension reduction is performed on random data. After randomizing data, PCA based dimension reduction is performed by keeping maximum fraction of variance for removed rows = 0.02 and IG based dimension reduction is carried out with $\alpha = 0.005$. Once the CT based models are developed using the training data their evaluation and testing is performed. Afterwards the FIS models have been generated with the same fraction of variance and α, and evaluated using the same performance measures. Results obtained after testing of the CT and FIS models have been reported in Table 3. Results show that use of information gain before development of CT and FIS based defect prediction models for 7 datasets has been encouraging. In the following two sections we report the models' performances with PCA and IG respectively.

Table 3. Test Performance of CT and FIS

Data-set	P repro-cess	CT				FIS			
		T N	Recall	Acc.	MC	T N	Re-call	Acc.	MC
ds1	None	0.885	0.461	0.822	0.178	0.99	0.139	0.842	0.158
	PCA	0.893	0.308	0.806	0.193	0.998	0	0.825	0.175
	IG	0.885	0.462	0.822	0.178	0.985	0.131	0.836	0.164
ds2	None	0.851	0.571	0.729	0.271	0.074	0.952	0.458	0.542
	PCA	1	0	0.563	0.438	1	0.191	0.646	0.354
	IG	0.704	0.619	0.667	0.333	0.333	0.905	0.583	0.417
ds3	None	0.625	0.625	0.625	0.375	0.781	0.688	0.75	0.25
	PCA	1	0.0625	0.688	0.313	0.938	0.125	0.667	0.333
	IG	0.625	0.625	0.625	0.375	0.781	0.688	0.75	0.25
ds4	None	0.8	0.609	0.708	0.292	0.84	0.652	0.75	0.25
	PCA	1	0.044	0.542	0.46	0.96	0.261	0.625	0.375
	IG	0.88	0.304	0.605	0.40	0.84	0.652	0.75	0.25
ds5	None	0.830	0.333	0.719	0.282	1	0.026	0.782	0.218
	PCA	0.882	0.410	0.776	0.224	1	0.026	0.782	0.218
	IG	0.83	0.333	0.719	0.282	1	0.026	0.782	0.218
ds6	None	0.936	0.167	0.875	0.125	0.986	0.083	0.915	0.085
	PCA	0.921	0.167	0.862	0.14	0.964	0.167	0.901	0.099
	IG	0.943	0.167	0.882	0.118	0.986	0.083	0.915	0.086
ds7	None	0.95	0.458	0.916	0.084	1	0	0.935	0.065
	PCA	0.933	0.25	0.889	0.1111	0.945	0.245	0.894	0.105
	IG	0.95	0.458	0.916	0.084	0.983	0.1667	0.93	0.071

3.1   Impact of PCA

Use of PCA has improved *Recall* of CT model for 2 datasets (ds5 and ds7), has decreased it for 4 out of 7 datasets (ds1, ds2, ds3 and ds4) and has resulted in no change for one dataset (ds6). Maximum positive *RC Change* (%) has been 59.99% for ds7 and maximum *RC Change(%)* in negative direction has been for ds2 when the CT based model failed to identify even a single *D* module, resulting in *Recall* = 0. Average *RC Change* (%) has been recorded as -48.37 %. Use of PCA has improved *Accuracy* of CT model for 3 datasets (ds3, ds5 and ds7). Maximum increase in *Acc Change* (%) has been 10%. The *Accuracy* has decreased for rest of the datasets and accuracy shortfall of up to 23.52% has been recorded. Average *Acc Change* (%) of -4.96% has been observed.*MC Change* (%) has been positive for 3 datasets and negative for 4 datasets. The maximum decline in *MC* has been for ds2 where *MC* has gone up by 61.56%. The maximum increase in *MC Change(%)* has been for ds7 when *MC* improved by 23.08%. Average value of *MC Change(%)* has been 19.03%.

Performance of FIS based prediction model, in terms of *Recall*, has declined 4 out of 7 times and improved twice when PCA based dimension reduction was performed. The performance remained unchanged in case of ds5. Maximum negative *RC Change* (%) has been recorded for ds1 where the FIS model could not identify the *D* modules. Maximum positive *RC Change* (%) value has been recorded for ds6 where the identification of *D* modules became 2 times better due to PCA. Average *RC*

Table 4. Winners in terms of *Recall* and *Acc*

Dataset	Recall		Accuracy		Overall
	CT	FIS	CT	FIS	
ds1	IG	IG	IG	IG	IG
ds2	IG	IG	IG	PCA	IG
ds3	IG	IG	PCA	IG	IG
ds4	IG	IG	IG	IG	IG
ds5	PCA	None	PCA	None	PCA
ds6	None	PCA	IG	IG	IG
ds7	None	PCA	IG	IG	IG

Change (%) value has been negative i.e. -36.94%. In case of ds2, *Accuracy* of the FIS model increased by 40.91% due to PCA. But PCA negatively affected the FIS *Accuracy* at 4 occasions (ds1, ds3, ds4 and ds6). Maximum decrease in *Acc Change*(%) was 16.67% in case of ds4. Average value of *Acc Change*(%) has been 0.76%. Use of PCA, could not help the FIS model to improve *MC Rate* at more than one occasions. The best, worst and average values of *MC Change* (%) have been -34.61% (ds2), 50% (ds4) and 19.59% respectively.

3.2   Impact of IG

Use of IG had no effect on *Recall* of the CT model for 5 datasets (ds1, ds3, ds5, ds6 and ds7). Increase in *RC Change*(%) has been 8.33% for ds2. *RC Change*(%) decreased by 50% in case of ds4. Average *RC Change*(%) has been -5.95%. *Acc Change*(%) remained unchanged for 4 datasets (ds1, ds3, ds5 and ds7) and has positive values for ds6. Use of IG has negatively affected the performance of CT model twice (ds2 and ds4). Maximum accuracy shortfall of 14.7% has been observed for ds4. Maximum increase in *Acc Change*(%) has been 0.75% and average *Acc Change*(%) has been -3.21%. *MC* of CT has deteriorated at 2 occasions (ds2 and ds4) and improved in case of ds6. Maximum increase and decrease in *MC Change*(%) have been 35.68% and 5.28% respectively. Average value of *MC Change*(%) has been 7.74%.

Use of IG has declined *Recall* of FIS model only twice (ds1 and ds2). For the rest of the datasets, the *Recall* either remained unchanged or has improved. Average *RC Change*(%) of FIS has been -1.81%. Maximum decrease in *RC Change*(%) is recorded to be 5.89% in case of ds1. *Accuracy* of the FIS remained unchanged for most of the datasets and decreased for ds1 and ds2. The maximum shortfall in *Accuracy* has been very low i.e. 0.68%. On the other hand the maximum increase in *Acc Change* (%) has been 27.27% for ds2. Average *Acc Change*(%) has been 3.71%. Results of using IG with FIS are similar in terms of *MC*. Maximum increase in *MC Change(%)* has been recorded as 8.46 and the maximum improvement in *MC* has been 23.08%. Average *MC Change(%)* has been -1.57%.

4   Analysis and Discussion

For most of the datasets performance shortfall incurred in case of Information Gain (IG) has been smaller as compared to performance shortfall observed with Principal Component Analysis (PCA). Average values of percentage changes in the three evaluation parameters indicate that on average IG has been better choice for these datasets.

We have compared the two dimension reduction approaches dataset wise and presented the results in Table 4. IG has emerged as better approach for majority of these datasets. For ds1, use of IG has improved performance of both the models and therefore IG is listed as overall winner. This is to be mentioned here that the largest dataset ds1 is dominated by 84.5% negative examples i.e. *ND* modules. Therefore none of the models could perform very well in identifying the *D* modules. Moreover this dataset had a few attributes with same values. FIS model cannot cope with such attributes but PCA keeps representation of these input attributes in the resultant input space. Hence using PCA resulted in *Recall* = 0 for FIS model. In case of ds2, IG wins once again despite the fact that PCA has helped FIS improve *Acc* and *MC*. ds2 is a small dataset with only 145 instances but has the largest number of input parameters i.e. 94. There are 8 parameters in ds2 which had same values for all the instances. From rest of the parameters, there are many parameters that are not good predictors of *D* modules for this dataset [22]. But PCA still keeps their contribution in the smaller input space hence deteriorating model performances. That is why we see that use of PCA has resulted in *Recall* = 0 for CT model. IG has dropped the irrelevant parameters and has resulted in better *Recall* for both the models. IG has outperformed PCA in case of ds3 and ds4 as well. One reason for IG to win here is that both the datasets are variants of ds2 and have small size. Another reason is that the input space here is very small too, and the input parameters selected here are on the basis of common use of these parameters in defect prediction literature. Hence using all these parameters without any dimension reduction gives reasonable results. Use of both the dimension reduction techniques degrades the models' performances most of the times. But IG is reported as a winner because the performance shortfall in case of IG is smaller than the performance shortfall incurred in case of PCA. PCA wins in case of ds5 which is the third largest dataset with 21 attributes. For this dataset, the suggested IG based method failed to drop any irrelevant input parameters and therefore performed just as if no dimension reduction has taken place. On the other hand, PCA has handled the situation because now all the variables are relevant ones, there is correlation in the input parameters and PCA is best known for such scenarios. In case of ds6, performances of PCA and IG are very close. Use of PCA has significantly improved the *Recall* of FIS but at the same time reduced the *Accuracy*. This dataset is dominated by negative examples with approximately 90% *ND* instances. The models trained on such an imbalanced dataset did not detect *D* modules very well at the time of testing. IG could not improve the *Recall* of both the models but slightly improved *Accuracy* and *MC* of CT model and did not incur any performance shortfall in case of FIS model. Furthermore, for ds6, F value of IG has been 0.18 and 0.13 for CT and FIS models respectively in comparison to F values of PCA which are 0.16 and 0.21 for CT and FIS respectively. The second largest dataset ds7 is also an imbalanced dataset with approximately 93% *ND* modules. For this reason FIS model without any preprocessing could not identify a single *D* module but still has the best accuracy. PCA increased the FIS ability to detect *D* modules in ds7 but has lesser accuracy. Performance wise PCA is not very far behind IG in case of ds6 and ds7. In case of ds7 PCA has F values 0.23 and 0.25 for CT and FIS and IG has 0.41 and 0.23 respectively.

Higher values of *TN* and relatively lower values of *Recall* and *Acc* for most of the datasets show that the CT model could not detect the *D* modules very well when PCA based dimension reduction was performed. On the other hand use of IG has resulted in decreased *TN* but reasonable accuracy for certain datasets. Meaning thereby the IG based dimension reduction helped CT model to correctly identify the *D* modules.

Impact of using IG based preprocessing for CT and FIS models has been very positive. It has improved the CT based model in learning *D* modules and helped the FIS model decrease its *FP* and learn *ND* modules at the same time. Correct detection of *D* modules is very important since it helps focus on problem prone areas. Reduction of *FP* and correct detection of *ND* saves the efforts from being wasted due to false alarms.

On the basis of the above comparison we can suggest that if there is a small dataset with lesser issues of correlation, then dimension reduction should be performed using IG technique. If it is desired to identify *D* modules more accurately then IG should be preferred on PCA. In case of large sized imbalanced datasets dominated by negative examples, performance of both the techniques is mixed, but the highest recorded F value (0.41) has been obtained through IG.

5   Conclusions and Future Work

When predicting software defect Principal Component Analysis (PCA) is used to reduce dimension of input data. PCA reduces dimensions of the input data, and at the same time, it keeps the representation of all the input variables intact. In some cases the representation of all the input variables is not very helpful rather it can negatively affect the prediction. Therefore, this paper suggests a dimensionality reduction approach that works on the principle of dropping the input variables which seem to be irrelevant for prediction. This approach calculates information gain (IG) of each of the input variable and drops the variables which have IG values below a threshold value α. The proposed technique has been used to develop classification tree (CT) and fuzzy inference system (FIS) based models for 7 datasets. The models have been developed with $\alpha = 0.005$ and the results have been compared with the same models developed using PCA. The results show that the IG based approach has enabled the models to perform better in terms of *Recall* and misclassification rate (*MC*) than the models developed after PCA. Moreover, the IG based approach has improved the detection of *D* modules for CT. The proposed approach resulted in lesser performance shortfall as compared to PCA in case of smaller datasets with a large number of independent variables. IG based approach also showed better results in case of large imbalanced datasets.

Results presented here cannot be generalized. We plan to verify these results by using more datasets and conducting more experiments with different values of α and fraction of variance. We also plan to study characteristics of the parameters dropped by the suggested approach to find the exact reason for better performance of the IG based approach.

Acknowledgment. The authors would like to thank Higher Education Commission (HEC) of Pakistan and Department of Computer Science at LUMS for funding this research, and Dean Faculty of IT, University of Central Punjab (UCP) for providing support to ensure timely registration.

References

1. Altidor, W., Khoshgoftaar, T.M., Van Hulse, J.: An empirical study on wrapper-based feature ranking. In: 21st International Conference on Tools with Artificial Intelligence, ICTAI 2009, pp. 75–82 (November 2009)
2. Azhagusundari, B., Thanamani, A.S.: Feature selection based on information gain. International Journal of Innovative Technology and Exploring Engineering (IJITEE) 2(2) (January 2013)
3. BPB Editorial Board. Data Mining: Data Mining: Typical Data Mining Process for Predictive Modeling, 1st edn. BPB Publications, Connaught Place (2004)
4. Bouktif, S., Azar, D., Precup, D., Sahraoui, H., Kegl, B.: Improving rule set based software quality prediction: A genetic algorithm-based approach. Journal of Object Technology 3(4), 227–241 (2004)
5. Briand, L.C., Wst, J., Daly, J.W., Victor Porter, D.: Exploring the relationship between design measures and software quality in object-oriented systems. Journal of Systems and Software 51(3), 245–273 (2000)
6. Challagulla, V.U.B., Bastani, F.B., Paul, R.A.: Empirical assessment of machine learning based sofwtare defect prediction techniques. In: Proceedings of 10th Workshop on Object-Oriented Real-Time Dependable Systems (WORDS 2005), Washington, DC, USA, pp. 263–270. IEEE Computer Society (2005)
7. Fenton, N.E., Neil, M.: A critique of software defect prediction models. IEEE Transactions on Software Engineering 25(5), 675–687 (1999)
8. Ganesan, K., Khosgoftaar, T.M., Allen, E.B.: Case-based software quality prediction. International Journal of Software Engineering and Knowledge Engineering 10(2), 139–152 (2000)
9. Gao, K., Khoshgoftaar, T.M.: Software defect prediction for high-dimensional and class-imbalanced data. In: SEKE, pp. 89–94, Knowledge Systems Institute Graduate School (2011)
10. Jiang, Y., Cukic, B., Menzies, T., Bartlow, N.: Comparing design and code metrics for software quality prediction. In: Proceedings of PROMISE 2008. ACM (May 2008)
11. Khosgoftaar, T.M., Munson, J.C.: Predicting software development errors using software complexity metrics. IEEE Journal on Selected Areas In Communications 8(2), 253–261 (1990)

12. Khoshgoftaar, T.M., Allen, E.B.: Predicting fault-prone software modules in embedded systems with classification trees. In: Proceedings of the 4th IEEE International Symposium on High-Assurance Systems Engineering. IEEE Computer Society (1999)
13. Khoshgoftaar, T.M., Allen, E.B., Kalaichelvan, K.S., Goel, N.: Early quality prediction: A case study in telecommunications. IEEE Software Early Quality Prediction: A Case Study in Telecommunications 13(1), 65–71 (1996)
14. Khoshgoftaar, T.M., Cukic, B., Seliya, N.: Predicting fault-prone modules in embedded systems using analogy-based classification models. International Journal of Software Engineering and Knowledge Engineering 12, 201–221 (2002)
15. Khoshgoftaar, T.M., Seliya, N.: Fault prediction modeling for software quality estimation: Comparing commonly used techniques. Empirical Software Engineering 8(3), 255–283 (2003)
16. Menzies, T., Caglayan, B., He, Z., Kocaguneli, E., Krall, J., Peters, F., Turhan, B.: The promise repository of empirical software engineering data (June 2012)
17. Menzies, T., Di Stefano, J.S., Chapman, M.: Learning early lifecycle ivy quality indicators. In: Proceedings of IEEE Metrics 2003. IEEE (2003)
18. Mitchell, T.M.: Machine Learning. McGraw-Hill, New York (1997)
19. Pacharaney, U.S., Salankar, P.S., Mandalapu, S.: Dimensionality reduction for fast and accurate video search and retrieval in a large scale database. In: 2013 Nirma University International Conference on Engineering (NUiCONE), pp. 1–9 (November 2013)
20. Palghamol, T.N., Metkar, S.P.: Constant dimensionality reduction for large databases using localized pca with an application to face recognition. In: 2013 IEEE Second International Conference on Image Information Processing (ICIIP), pp. 560–565 (December 2013)
21. Rana, Z.A., Awais, M.M., Shamail, S.: An FIS for early detection of defect prone modules. In: Huang, D.-S., Jo, K.-H., Lee, H.-H., Kang, H.-J., Bevilacqua, V. (eds.) ICIC 2009. LNCS, vol. 5755, pp. 144–153. Springer, Heidelberg (2009)
22. Rana, Z.A., Shamail, S., Awais, M.M.: Ineffectiveness of use of software science metrics as predictors of defects in object oriented software. In: WCSE 2009: Proceedings of the 2009 WRI World Congress on Software Engineering, May 19-21, pp. 3–7. IEEE Computer Society, Washington, DC (2009)
23. Van Rijsbergen, C.J.: Information Retrieval, 2nd edn. Butterworth-Heinemann, Newton (1979)
24. Roobaert, D., Karakoulas, G., Chawla, N.V.: Information Gain, Correlation and Support Vector Machines. In: Guyon, I., Nikravesh, M., Gunn, S., Zadeh, L.A. (eds.) Feature Extraction. STUDFUZZ, vol. 207, pp. 463–470. Springer, Heidelberg (2006)
25. Seliya, N., Khoshgoftaar, T.M.: Software quality estimation with limited fault data: A semi-supervised learning perspective. Software Quality Journal 15, 327–344 (2007)
26. Wang, Q., Zhu, J., Yu, B.: Extract rules from software quality prediction model based on neural network. In: Proceedings of the 11th International Conference on Evaluation and Assessment in Software Engineering, EASE (April 2007)

CosSimReg: An Effective Transfer Learning Method in Social Recommender System

Hailong Wen[1], Cong Liu[1], Guiguang Ding[2], and Qiang Liu[2]

[1] Department of Computer Science and Technology, Tsinghua University, Beijing, China
[2] School of Software, Tsinghua University, Beijing, China
{wenhl11,cong-liu11}@mails.tsinghua.edu.cn,
{dinggg,liuqiang}@tsinghua.edu.cn

Abstract. Traditional recommender systems perform poorly when training data is sparse. During past few years, researchers have proposed several social-based methods to alleviate this sparsity problem. The basic assumption of these social recommender systems is that friends should have similar interests. However, this assumption does not always hold due to the heterogeneity between recommendation domain and social domain. Thus, knowledge transferred from social network often contains noises. To solve this problem, in this paper, we analyze and identify what knowledge is useful during transfer learning process, and develop a method, called Cosine Similarity Regularization (CosSimReg), to transfer only useful information from social domain. CosSimReg is able to minimize the negative effects of noisy data in social network, thus improving the performance. Experiments on two real life datasets demonstrate that CosSimReg performs better than the state-of-the-art approaches.

Keywords: Recommender Systems, Cosine Similarity, Collaborative Filtering, Social Network, Matrix Factorization.

1   Introduction

In the last decades, recommender systems have become very popular and are widely deployed in major large scale commercial websites like Amazon, Ebay and Taobao. Recommender systems of these websites analyze behaviors of individual user and autonomously identify attractive items with high potential that he or she may want to purchase. This active process increases interactions between users and products and reduces users' efforts on product searching, thus promoting overall sales and enhancing user experience. However, these recommender systems often suffer from sparsity problem: the amount of products is too large that most users have only rated a portion database, which makes it difficult to give accurate recommendation.

Along with the flourish of Web 2.0 technique, online social network websites like Facebook, Twitter, Renren and Weibo have boomed and attracted millions of users. By incorporating social network information, recommender system could achieve higher accuracy. These systems are called social recommender systems. The basic assumption behind social recommender systems is that users linked with each other in

D.-S. Huang et al. (Eds.): ICIC 2014, LNCS 8588, pp. 649–660, 2014.
© Springer International Publishing Switzerland 2014

social networks tend to share certain common tastes or have similar interests. Users in recommender system with no or limited history behaviors could benefit from his or her friends/followers in social network. Previous studies have studied how to utilize social link information in the recommender system [1-11]. However, their works have 3 major shortcomings:

1. Their methods ignored the heterogeneity between social network and recommendation. The internal factors of people's social activities are different from those of buying activities, which means that tastes of friends are not always same or similar. These cases create noises in the knowledge transferring process. Liu et al. reports that social network contributes nothing to the final result in particular dataset [12], which shows that noisy social data could not help to improve the performance.
2. Their methods rely on fine-tuning parameters to balance between social domain and recommendation domain.
3. Their methods usually have high complexity and could not be easily implemented and scaled. Most of the models mentioned above involve time-consuming network computation, which would be a challenge when the amount of users/items goes up. This paper aims to solve the above problems.

This paper aims to research and find a new approach to solve the above problems. The contributions of our work are:

1. We investigate the matrix factorization model used by most previous studies, and interpret it in an information retrieval way and identify what information are useful and should be transferred. This work addresses the problem of "what to transfer" in transfer learning.
2. We developed a method called Cosine Similarity Regularization (CosSimReg) based on the previous investigation. This work addresses the problem of "how to transfer" in transfer learning.
3. The proposed CosSimReg performs better than the state-of-the-art approaches. Besides, CosSimReg is easy to implement, and the parameter of CosSimReg is less sensitive and needs less tuning. This work addresses the scalability and robustness in social recommender system.

The rest of this paper is organized as follows. Section 2 gives an overview of recommender systems based on matrix factorization and how they incorporate with social network. Section 3 elaborates the motivation and algorithm of CosSimReg. Section 5 shows the experimental results of CosSimReg, which is followed by conclusions in section 6.

2   Related Work

2.1   Low-Rank Matrix Factorization

Low-rank matrix factorization is an effective and efficient approach to do collaborative filtering [13-16]. In this model, all users' past behaviors could be represented as a rating matrix $R \in \mathbb{R}^{n \times m}$, where n is the number of users, and m is the number of items. Each entry $R_{i,j}$ in R denotes the interaction of user i with item j. This interaction could be explicit feedbacks, e.g. ratings, or implicit feedback, e.g. number of clicks. Matrix factorization assumes that only a small number of latent factors affect users' preference. Let $U \in \mathbb{R}^{n \times k}$ denotes the user preference matrix, with each row of U_i representing the interests or tastes of user i on each factor. Let $V \in \mathbb{R}^{m \times k}$ denotes the item feature matrix, with each row V_j describing the feature and compositions. The rating of an item j given by user i is determined by $\tilde{R}_{i,j} = U_i V_j^T$, which could be explained by the sum of impact of item's feature vector applies to the user preference vector. The objective function of matrix factorization is to minimize

$$L = \frac{1}{2}\sum_{i,j} I_{i,j}\left(R_{i,j} - U_i V_j^T\right)^2 + \frac{\lambda_u}{2}\|U\|_F^2 + \frac{\lambda_v}{2}\|V\|_F^2, \qquad (1)$$

where $\left(R_{i,j} - U_i V_j^T\right)^2$ stands for the reconstruction error, λ_u and λ_v are parameters to constrain the model complexity, and $\|\cdot\|_F$ is the Frobenius norm. This function does not have a global optimal, but a local optimal could be acquired by performing alternative gradient descent on U and V. Some researchers propose to enforce non-negative constraint on (1), which has achieved good results in computer vision fields, called the Non-negative Matrix Factorization (NMF) [20]. But, when applied in recommendation systems, NMF does not perform well most of the time [11]. Thus, current recommendation system mainly uses matrix factorization technique without non-negative constraints.

2.2   Social Recommendation

Recent studies on social recommendation systems are mainly based on matrix factorization [1-11, 17-19]. The underlying assumption is quite intuitive: users directly linked in the social network may share certain common tastes or have similar interests. Thus, network-based similarity properties could be used to transfer knowledge between social and recommendation domains.

Ma et al. proposed the SoRec model [5], which employs matrix factorization on the social network matrix C. They use user feature matrix $Z \in \mathbb{R}^{n \times k}$ to represent users' social feature and use $C_{i,f} = U_i Z_f^T$ to recover the interaction between user i and user f. The objective function is in a collective matrix factorization form:

$$L = \frac{1}{2}\sum_{i,j} I_{i,j}\left(R_{i,j} - U_i V_j^T\right)^2 + \frac{\lambda_u}{2}\|U\|_F^2 + \frac{\lambda_v}{2}\|V\|_F^2 + \frac{\beta}{2}\sum_{i,j} I_{i,j}^C\left(C_{i,j} - U_i Z_j^T\right)^2, \qquad (2)$$

By sharing latent preference vector U, SoRec transfers feature representation of U from one domain to the other domain. The problem of this model is that latent factors of social activities are not all the same with those of buying activities, thus the assumption of shared U is too strict.

Not long after that, Ma et al. developed another model called Social Trust Ensemble (STE) [3]. STE views $R_{i,j}$ as a combination of ratings given by the user and his friends. The predicted rating is estimated by a weighted voting strategy:

$$L = \frac{1}{2}\sum_{i,j} I_{i,j}\left(R_{i,j} - \left(\alpha U_i V_j^T + (1-\alpha)\sum_{k \in N_i} T_{i,k} U_k V_j^T\right)\right)^2 + \frac{\lambda_u}{2}\|U\|_F^2 + \frac{\lambda_v}{2}\|V\|_F^2, \quad (3)$$

where α is a parameter controlling to what extent a user insists on his or her ratings, N_i is the neighborhood of user i, and $T_{i,k}$ is the similarity between user i and k.

Gu et al. proposed the graph regularized nonnegative matrix factorization ([19]). They use Laplacian matrix of U and V to constrain user preference vectors and item feature vectors:

$$L = \frac{1}{2}\sum_{i,j} I_{i,j}\left(R_{i,j} - U_i V_j^T\right)^2 + \frac{\lambda_u}{2}\|U\|_F^2 + \frac{\lambda_v}{2}\|V\|_F^2 + \frac{\beta_u}{2} tr\left(U^T L_U U\right) + \frac{\beta_v}{2} tr\left(V^T L_V V\right), \quad (4)$$

where $L = D - W$ is the Laplacian matrix with W representing the adjacent matrix and D is an diagonal matrix with $D_{ii} = \sum_j W_{i,j}$, the symbol $tr(\cdot)$ denotes trace of a matrix, and β_u, β_v is the weight of the regularization. They also prove that this Laplacian regularization equals to $\sum_{i,j} W_{i,j}\left(U_i - U_f\right)^2$, i.e. minimizing the sum of weighted difference between different users. They mentioned that user similarities extracted from social network could fill in W. However, this method could not be easily scaled up due to the high complexity of Laplacian matrix computation.

Jamali et al. pointed out that STE fails to handle trust propagation, and then proposed a similar model called Social MF ([1]). Their model assumes that the latent feature vector of user i is dependent on the latent feature vector of his entire direct neighborhood,

$$L = \frac{1}{2}\sum_{i,j} I_{i,j}\left(R_{i,j} - U_i V_j^T\right)^2 + \frac{\lambda_u}{2}\|U\|_F^2 + \frac{\lambda_v}{2}\|V\|_F^2 + \frac{\beta}{2}\sum_{i=1}^n \sum_{f \in N_i^+}\left\|U_i - T_{i,f}U_f\right\|_{Fro}^2, \quad (5)$$

where $T_{i,k}$ stands for the similarity between user i and k, with $T_{i,k} = 1/|F_i|$, $|F_i|$ denoting the amount of his friends.

Ma et al. then improved STE to Social Regularization [11]. The assumption is slightly different from Social MF that the latent feature vector of user i is similar to that of his friends f to the extent of $Sim(i,f)$. The objective function is given by

$$L = \frac{1}{2}\sum_{i,j} I_{i,j} \left(R_{i,j} - U_i V_j^T \right)^2 + \frac{\lambda_u}{2}\|U\|_F^2 + \frac{\lambda_v}{2}\|V\|_F^2 + \frac{\beta}{2}\sum_{i=1}^n \sum_{f \in N_i^+} Sim(i,f)\|U_i - U_f\|_{Fro}^2 \quad (6)$$

Pan et al. adopted Transfer Learning method to predict missing values [2] using different auxiliary datasets, including social network data. This method uses matrix tri-factorization instead of bi-factorization. Domain specific knowledge is kept in matrix B and $\hat{B}$, and common knowledge is transferred through U and V.

$$L = \frac{1}{2}\left\| I \odot \left(R - UBV^T \right) \right\|_F^2 + \frac{\beta}{2}\left\| \hat{I} \odot \left(\hat{R} - U\hat{B}V^T \right) \right\|_F^2 + \frac{\lambda_u}{2}\|U\|_F^2 + \frac{\lambda_v}{2}\|V\|_F^2 + \frac{\lambda_B}{2}\|B\|_F^2 \quad (7)$$

3 Social Recommendation with Cosine Similarity Regularization

In this section, we first give an alternative interpretation of matrix factorization in the perspective of information retrieval. This interpretation allows us to view interactions between users and items as the product of cosine similarity (taste direction) and vector norm (strength), which further motivates us to use cosine similarity to help solve **Problem 1**. Then, we propose to weight these similarities by the correlation of two users to solve **Problem 2**. The proposed method is called Cosine Similarity Regularization (CosSimReg).

To conclude, all the previous mentioned social recommendation approaches manage to build bridges to transfer information by mainly two ways: 1) Use constraints on latent vector: imposing constraint on difference between U_i and his friends' U_f. 2) Use common latent vector space: imposing common latent vector U between two domains. These two ways are too strict to be restricted in real life, and will lead to two problems:

— **Problem 1. Individual Differences** Although users linked in social network could share common tastes, their strengths of these tastes may not be the same. For example, user i is a car fan; user j likes cars a little and he follows user i in Twitter just because he may get newest information about cars from user i. In this case, user j "like cars" does not mean user j "like cars very much". Constraints like $U_i - U_f$ fails to describe this situation.

— **Problem 2. Social Relationship** Users in online social network such as Facebook and Renren establish their relationship basically because they are friends, classmates, families or even business partners. Latent space of this kind of social network contains latent factors relate to social groups or circles; while latent factors of buying activities are all relate to item features. This is why imposing common latent vector between social and recommendation domain performs poorly.

3.1   Addressing Individual Differences Problem

In matrix factorization, $R_{i,j}$ is represented as the inner product of U_i and V_j, which could be rewritten as:

$$R_{i,j} = U_i V_j^T = \|U_i\| \|V_j\| \cos(U_i, V_j),$$

(8)

where $\|\cdot\|$ denotes the L2-norm of a vector, and $\cos(\cdot, \cdot)$ denotes the cosine similarity of two vectors. Equation (8) breaks the low-rank preference approximation in two parts, the vector norms and cosine similarity. Vector norms are macroscopic magnitudes with information of the strength of user i's preference vector and item j's feature vector. If $\|U_i\|$ is low, user i may generally has low preference on most of the items, and if $\|V_j\|$ is low, item j may generally has low attraction to most of the users. Cosine similarity is a microscopic measure of the pairwise inter-action between U_i and V_j. If $\cos(U_i, V_j)$ is low, the feature of item j may not appeal to user i's preference.

Just as **Problem 1** states, two friends may have similar interests' direction, but the strength of their interests may not be the same. We should not impose constraints on strengths information in transferring process when measuring similarities between users. This motivates us to take away vector norm and use only cosine similarity to denote the similarity between two user latent vector:

$$Sim_{rec}(U_i, U_f) = \cos(U_i, U_f) = \frac{U_i U_f^T}{\|U_i\| \|U_f\|}$$

(9)

3.2   Addressing Social Relationship Problem

The basic assumption of social recommendation is that if two users are linked in the social domain, they are likely to be similar in the recommendation domain. This likelihood, or called the correlation between recommender system and social network, is modeled as $Cor(i, f)$. If two users become friend because of they share common interests, their correlation should be high; otherwise, e.g., these two user are classmates off-line, their correlation should be evaluated according to different situations: if their interests do not intersect, the correlation should be low; if they happen to like similar things, this social relationship can still contribute to the recommender system, thus correlation still should be high.

There is no direct way to measure $Cor(i, f)$. But it could be inferred from the user rating history by their Pearson-Correlation-Coefficient:

$$Cor(i,f) = \frac{\sum\limits_{j \in I(i) \cap I(f)} (R_{i,j} - \bar{R}_i) \cdot (R_{f,j} - \bar{R}_f)}{\sqrt{\sum\limits_{j \in I(i) \cap I(f)} (R_{i,j} - \bar{R}_i)^2} \cdot \sqrt{\sum\limits_{j \in I(i) \cap I(f)} (R_{f,j} - \bar{R}_f)^2}}, \tag{10}$$

which means that if user i and j have bought lots of common things and rated them in similar way, their relationship are mainly determined by item feature factors instead of social factors.

Bringing $Cor(i,f)$ helps to minimize the negative effects of pure social relationship and reduces irrelevant information during knowledge transfer, thus alleviating **Problem 2**.

3.3   Cosine Similarity Regularization

To transfer similarity information from social domain to recommendation domain, we need to minimize the weighted difference of latent feature similarities between social domain and recommendation domain, which yields the objective function of Cosine Similarity Regularization:

$$L = \frac{1}{2}\sum_{i,j} I_{i,j} \left(R_{i,j} - U_i V_j^T\right)^2 + \frac{\lambda_u}{2}\|U\|_F^2 + \frac{\lambda_v}{2}\|V\|_F^2$$
$$+ \frac{\beta}{2}\sum_{i=1}^n \sum_{f \in N_i^+} Cor(i,f) Sim(i,f)\left(1 - \cos\left(U_i, U_f\right)\right)^2, \tag{11}$$

where λ_u and λ_v has the meaning with (1), and β is the parameter to balance between recommendation domain and social domain.

A local minimum of (11) could be found by performing gradient descent in feature vectors U_i and V_j. The derivation of U_i and V_j of (11) are given by the following equations:

$$\frac{\partial L}{\partial U_i} = -\sum_{j=1}^m I_{i,j}\left(R_{i,j} - U_i V_j^T\right)^2 V_j + \lambda_u U_i$$
$$- \beta \sum_{f \in N_i^+} Cor(i,f) Sim(i,f)\left(1 - \frac{U_i U_f^T}{\|U_i\|\|U_f\|}\right)\left(U_f - \frac{U_i U_f^T}{\|U_i\|\|U_f\|}U_i\right) \tag{12}$$
$$- \beta \sum_{g \in N_i^-} Cor(i,g) Sim(i,f)\left(1 - \frac{U_i U_g^T}{\|U_i\|\|U_g\|}\right)\left(U_g - \frac{U_i U_g^T}{\|U_i\|\|U_g\|}U_i\right)$$

$$\frac{\partial L}{\partial V_j} = -\sum_{i=1}^n I_{i,j}\left(R_{i,j} - U_i V_j^T\right)^2 U_i + \lambda_v V_j \tag{13}$$

Table 1. Dataset statistics

	Sampling	# of ratings	sparsity	# of social data
Epinions	80%	531,859	9.5e-5	487,183
Douban	20%	3,366,168	4.5e-4	1,692,952

The algorithm uses random initialized value to fill in U and V. Then, it first fix V and update U using (12), and then fix U and update V using (13). After one iteration is done, (11) is calculated. If the improvement of (11) is within a given threshold, the algorithm could be considered as converged, and we can predict rating of user i on item j using $\tilde{R}_{i,j} = U_i V_j^T$.

4   Experimental Analysis

4.1   Datasets and Metrics

We use two public real life dataset to testify the performance of our proposed method. The statistics of datasets could be found in Table 1.

Epinions[1] [9] Epinions.com is a well-known web site for customers to read and post reviews on a large collection of items. Each review relates to a rating ranging from 1 to 5. Users of Epinions could choose to trust or distrust someone's review (for simplicity, we only apply trust information). There are 40,163 users, 139,738 items, 664,824 ratings and 487,183 trust data in this dataset.

Douban[2] [11] Douban.com is a Chinese website for users to share their interests on films, books, music and events. Each item is scored from 1 to 5. Users in Douban could add others as their friend as what people do in Facebook, i.e., the relationship is symmetric. It consists of 129,490 users, 58541 items, 16,830,839 ratings and 1,692,952 friend statements.

We adopt two widely used evaluation metrics for our experiment, the Mean Absolute Error (MAE) and the Root Mean Square Error (RMSE):

$$MAE = \frac{\sum_{i,j} \left| R_{i,j} - \tilde{R}_{i,j} \right|}{\left| T_E \right|} \qquad RMSE = \sqrt{\frac{\sum_{i,j} \left(R_{i,j} - \tilde{R}_{i,j} \right)^2}{\left| T_E \right|}},$$

where $R_{i,j}$ denotes the actual rating given by user i to item j and $\tilde{R}_{i,j}$ denotes the predicted rating. $\left| T_E \right|$ is the amount of ratings in test data set. By definition, smaller MAE and RMSE are better.

---

[1] http://www.trustlet.org/wiki/Epinions datasets
[2] https://www.cse.cuhk.edu.hk/irwin.king/pub/data/douban

Table 2. Performance of different algorithms

Dataset	Metric	BaseMF	SocialMF	SR	CosSimRec
Epinions 80%	MAE	0.9145	0.8744	0.8791	0.8560
	Improve	-	4.37%	3.88%	6.40%
	RMSE	1.185	1.126	1.133	1.107
	Improve	-	4.98%	4.43%	6.54%
Douban 20%	MAE	0.6155	0.6120	0.6012	0.5943
	Improve	-	0.57%	2.36%	3.49%
	RMSE	0.7854	0.7821	0.7596	0.7575
	Improve	-	0.45%	3.31%	3.56%

4.2 Baseline and Parameters

BaseMF [15], SocialMF [1] and Social Regularization [11] are chosen as the experiment baseline. The objective functions of these methods are listed in (2), (5) and (6) respectively. Since different models have different parameters setting, we only show the best result of these models. Different balance parameter $\beta = 0.1, 0.2, 0.5, 1, 2, 5, 10, 20, 50, 100, 200, 500$ and 1000 are tried for SocialMF, Social Regularization and Cosine Similarity Regularization. The learning rate of the each algorithm is set to 0.01 on every dataset. The number of latent factors is fixed to 10, and the regularization parameter λ_u, λ_v are set to 0.001 empirically. Each algorithm is performed 10 times, and we only report the average MAE and RMSE.

4.3 Performance

The performances of different algorithms are shown in Table 2. The standard deviations of the results are all around 0.004.

From Table 2, we observe that all social recommendation methods have better MAE and RMSE than the original BaseMF, which means that social network information do help the recommendation system to alleviate sparsity problem. Since Epinions is sparser than Douban, the improvement of social-based methods in Epinions is also higher than that in Douban.

We should notice that the performances of SocialMF and Social Regularization are different in these two datasets. In Epinions, SocialMF performs better, while in Douban, Social Regularization performs better. This is because items in Epinions are all objective reviews while items in Douban are subjective movies, books and music. In Douban, SocialMF uses same similarity for all friends for a user, which ignores the individual differences between different friends.

The CosSimReg outperforms SocialMF and Social Regularization in both datasets, and proves to be an effective way to transfer social network knowledge to recommender system.

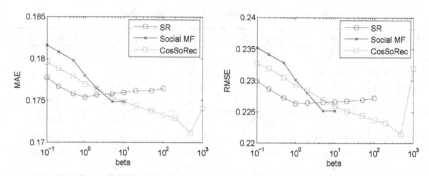

Fig. 1. Performances of different methods with different β settings on Epinions

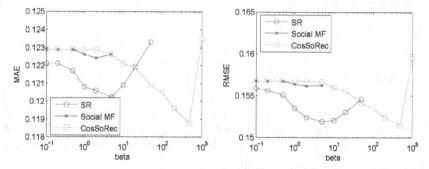

Fig. 2. Performances of different methods with different β settings on Douban

4.4   Parameter Sensitivity

We investigate the parameter sensitivity of parameter β by varying its values in different methods. From Figure 1 and Figure 2 we can see that social-based recommender systems needs fine β tuning to reach the best result. β should not be too high, otherwise the noises of social network would override the useful information. If the β gets too large, the gradient descent would take too big step and could not converge at all, which forms the sharp rise of every curve. The SocialMF does not converge when β is larger than 10, and the value for Social Regularization is about 100, and for CosSimReg, the value is 1000.

We can also find that different algorithms have different optimal parameter choices. The CosSimReg constrains the cosine similarity, which is very weak. Thus, the performance of CosSimReg improves slower than other methods as β changes. Social Regularization imposes stronger constraints, and it is more sensitive to parameter changes. SocialMF uses same weight for different friends, and individual differences are more easily to be magnified as parameter increases, so SocialMF is extremely sensitive to parameter changes. A slower change with parameter β

means that the algorithm is more stable, and the parameter acquired by cross-validation could be applied to other scenarios without modification or tuning. This is very important for industrial applications. In the experiment, the best β for SocialMF are 5 for Epinions and 2 for Douban; the best β for Social Regularization are 1 for Epinions and 5 for Douban; while the best β for CosSimReg are all 500. We could empirically set $\beta = 500$ for future use of CosSimReg since it is stable.

5 Conclusion

In this paper, we analyzed the shortcomings of previous social recommender systems, and proposed a novel approach, Cosine Similarity Regularization to alleviate those problems. Previous works often ignored the individual differences brought by heterogeneity between recommendation domain and social domain. We gave an alternative interpretation for matrix factorization and pointed out that individual differences just affect the vector magnitudes, and proposed to use cosine function to effectively constrain user similarity. Previous works often assumed that all social network data is useful, while we use correlation to denote the contribution that each social data could make. The proposed method only added a regularization term to original matrix factorization, and is easily to be implemented. Experimental results on Epinions and Douban proves the effectiveness, robustness and scalability of our proposed method.

Acknowledgement. This research was supported by the National Basic Research Project of China (Grant No. 2011CB70700), the National Natural Science Foundation of China (Grant No. 61271394). And the author would like to thank the reviewers for their valuable comments.

References

1. Jamali, M., Ester, M.: A Transitivity Aware Matrix Factorization Model for Recommendation in Social Networks. In: Proceedings of the Twenty-Second International Joint Conference on Artificial Intelligence, vol. 3, pp. 2644–2649. AAAI Press (2011)
2. Pan, W., Liu, N.N., Xiang, E.W., Yang, Q.: Transfer Learning to Predict Missing Ratings Via Heterogeneous User Feedbacks. In: Proceedings Of the Twenty-Second International Joint Conference on Artificial Intelligence, vol. 3, pp. 2318–2323. AAAI Press, Menlo Park (2011)
3. Ma, H., King, I., Lyu, M.R.: Learning to Recommend With Social Trust Ensemble. In: Proceedings of the 32nd International ACM SIGIR Conference on Research and Development in Information Retrieval, pp. 203–210. ACM (2009)
4. Ma, H., King, I., Lyu, M.R.: Learning to Recommend With Explicit and Implicit Social Relations. ACM Transactions on Intelligent Systems and Technology (TIST) 2(3), 29 (2011)
5. Ma, H., Yang, H., Lyu, M.R., King, I.: Sorec: Social Recommendation Using Probabilistic Matrix Factorization. In: Proceedings of the 17th ACM Conference on Information and Knowledge Management, pp. 931–940. ACM (2008)

6. Ma, H., Lyu, M.R., King, I.: Learning to Recommend With Trust and Distrust Relationships. In: Proceedings of the Third ACM Conference on Recommender Systems, pp. 189–196. ACM (2009)
7. Konstas, I., Stathopoulos, V., Jose, J.M.: On Social Networks and Collaborative Recommendation. In: Proceedings of the 32nd International ACM SIGIR Conference on Research and Development in Information Retrieval, pp. 195–202. ACM (2009)
8. Shen, Y., Jin, R.: Learning Personal+ Social Latent Factor Model for Social Recommendation. In: Proceedings of the 18th ACM SIGKDD International Conference on Knowledge Discovery and Data Mining, pp. 1303–1311. ACM (2012)
9. Massa, P., Avesani, P.: Trust-aware bootstrapping of recommender systems. In: ECAI Workshop on Recommender Systems, pp. 29–33. Citeseer (2006)
10. Jamali, M., Ester, M.: Trustwalker: a random walk model for combining trust-based and item-based recommendation. In: Proceedings of the 15th ACM SIGKDD International Conference on Knowledge Discovery and Data Mining, pp. 397–406. ACM (2009)
11. Ma, H., Zhou, D., Liu, C., Lyu, M.R., King, I.: Recommender systems with social regularization. In: Proceedings of the Fourth ACM International Conference on Web Search and Data Mining, pp. 287–296. ACM (2011)
12. Liu, N.N., Cao, B., Zhao, M., Yang, Q.: Adapting neighborhood and matrix factorization models for context aware recommendation. In: Proceedings of the Workshop on Context-Aware Movie Recommendation, pp. 7–13. ACM (2010)
13. Takacs, G., Pilaszy, I., Nemeth, B., Tikk, D.: Investigation of various matrix factorization methods for large recommender systems. In: Data Mining Workshops, IEEE International Conference on ICDMW 2008, pp. 553–562. IEEE (2008)
14. Takacs, G., Pilaszy, I., Nemeth, B., Tikk, D.: Matrix factorization and neighbor based algorithms for the netflix prize problem. In: Proceedings of the 2008 ACM Conference on Recommender Systems, pp. 267–274. ACM (2008)
15. Koren, Y., Bell, R., Volinsky, C.: Matrix factorization techniques for recommender systems. Computer 42(8), 30–37 (2009)
16. Srebro, N., Jaakkola, T., et al.: Weighted low-rank approximations. In: ICML, vol. 3, pp. 720–727 (2003)
17. Shi, J., Long, M., Liu, Q., Ding, G., Wang, J.: Twin bridge transfer learning for sparse collaborative filtering. In: Pei, J., Tseng, V.S., Cao, L., Motoda, H., Xu, G. (eds.) PAKDD 2013, Part I. LNCS, vol. 7818, pp. 496–507. Springer, Heidelberg (2013)
18. Gu, Q., Zhou, J.: Neighborhood preserving nonnegative matrix factorization. In: BMVC, pp. 1–10 (2009)
19. Gu, Q., Zhou, J., Ding, C.H.: Collaborative filtering: Weighted nonnegative matrix factorization incorporating user and item graphs. In: SDM, pp. 199–210. SIAM (2010)
20. Lee, D.D., Seung, H.S.: Algorithms for non-negative matrix factorization. In: Advances in Neural Information Processing Systems, pp. 556–562 (2000)

Chaotic Features Identification
and Analysis in Liujiang River Runoff

Hong Ding[1,2], Wenyong Dong[3,*], and Demin Wu[3]

[1] School of Information Engineering, Wuhan University of Technology,
Wuhan, 430070, Hubei, P.R. China
[2] Department of Physics and Information Science, Liuzhou Teachers College, Liuzhou, Guangxi,
545004, P.R. China
[3] Computer school, Wuhan University, Wuhan, 430072, Hubei, P.R. China
{dhong20123,woodmean}@163.com, dwy@wu.edu.cn

Abstract. Because many factors, such as hydrology, riverbed, topography, impose hugely on the evolution of tendency in Liujiang river runoff, so its intrinsic dynamic behavior implied in runoff time series emerges characteristics of dissipative nonlinear systems. This paper firstly presents chaos theory to identify and analyze the time series of Liujiang river runoff. To identify characteristics in different riverbed, season and topography, runoff time series from typical sites and seasons were analyzed in phase space, auto correlation algorithm was employed to analyze the time delay, and the correlation dimension of runoff time series was calculated by Grassberger and Procaccia algorithm. To reduce complexity, small data sets algorithm was adopted to calculate the maximum Lyapunov exponents after the phase space reconstructing. Some crucial conclusions are drawn from chaos theory, and the relationships between evolved tendency of runoff and factors, such as topography and seasons, are deeply analysis.

Keywords: Time delay, correlation dimension, the maximum Lyapunov exponents, Liujiang River runoff.

1   Introduction

Convincing evidence for deterministic chaos has come from a variety of recent experiments on dissipative nonlinear systems; therefore, how to detect and quantify chaos data sets has become an important issue. The rainfall-runoff is one of the most complicated and incomprehensible hydrologic phenomena due to the variability in time and space, space complicated watershed characteristics and multiple precipitation patterns. Recently, some hydrologic phenomena were embedded with chaotic features by many researchers. For example, the stream flow series of Danjiangkou, Cuntan were reconstructed into phase space and forecasted as chaotic series [1], Millán used chaotic theory to describe different time series of daily precipitations [2], Ding *et al.* [3][4]

---

[*] Coresepoding author.

D.-S. Huang et al. (Eds.): ICIC 2014, LNCS 8588, pp. 661–667, 2014.

explored the chaotic properties of daily flow variation in the Yangtze River, and calculated its time delay and embedding dimensions. Wen *et al.* [10~12] presented the global exponential synchronization of a class of memristor-based recurrent neural networks with time-varying delays based on the fuzzy theory and Lyapunov method. Therefore, chaos has provided a new way to study complicated and changeable non-linear hydrology time series.

Phase space reconstruction (PSR) is a first essential step in the correlation dimension method, chaos identification and prediction method. Even though the sensitivity of initial conditions limits the predictability in chaotic systems, PSR can reduce the prediction uncertainty to some extent. Dhanya focused on an ensemble prediction approach by reconstructing the phase space using different combinations of chaotic parameters, i.e., embedding dimension and delay time to quantify the uncertainty in initial conditions [5]. According to phase space reconstruction theory carrying on reconstruction of phase space to monthly surface flow course, Yu *et al.* had discussed the non-linear prediction model of time series of Chaos of the support vector machine, which was an application in the monthly surface flow [6]. Kim *et al.* used the C-C method to estimate both the delay time and delay time window to make it clear whether hydrologic time series depends on its linear or nonlinear characteristics, which are very important for modeling and forecasting of underlying system [7].

In this paper, the chaotic features of Liujiang River runoff collected from three different areas and three different periods were analyzed. Auto correlation algorithm was used to analyze the time delay of the runoff time series, the correlation dimension of runoff time series was calculated by Grassberger and Procaccia algorithm and Small data sets algorithm was adopted to calculate the maximum Lyapunov exponents after the phase space reconstructing. Results show that the different chaotic characteristics of the runoff time series emerge with different areas and periods.

This paper is organized as follows. The theories and methods to identify the runoff time series chaotic features are described in Section 2. The results of the chaotic feature analysis, including time delay, embedding dimension, the maximum Lyapunov exponents and phase space reconstructing, for Liujiang River runoff, were deeply studied in section 3. Finally, some crucial conclusions and future issues are summarized in Section 4.

2   Identification Theory of Runoff Chaotic Feature

2.1   Phase Space Reconstruction

Phase space reconstruction enabled the visualization of trajectories embedded on a chaotic attractor. Contrary to the case of numerical simulations, the phase space of experimental data was unknown in a true physical experiment, e.g. the fluid dynamics. Therefore, the reconstructed trajectory, X, can be expressed by a matrix, in which each row is a phase-space vector [8]. That is,

$$\mathbf{X} = \left[X_1, X_2, \cdots, X_i, \cdots, X_M \right]^T \tag{1}$$

where X_i is the state of the system at discrete time i. For an N-point time series $\{x_1, x_2, \ldots, x_N\}$, and each X_i is given by

$$\mathbf{X}_i = \left[x_i, x_{i+\tau}, \cdots, x_{i+(m-1)\tau} \right] \tag{2}$$

where τ is the lag or delay time, and m is the embedding dimension. Thus, X is an $M \times m$ matrix, and the relationship among all the constants, such as, m, M, τ, and N, can be written as

$$M = N - (m-1)\tau \tag{3}$$

2.2   Measure Delay Time

Theoretically any value of τ is acceptable for the choice of the delay time. However, the appearance of the reconstructed attractor depends strongly on the choice of embedding lag. A suitable value for τ must bear the function to sufficiently separate the data in the time series as to have a smooth reconstruction of the attractor, therefore larger values were always favored. On the other hand, the lag must be limited at a certain level, to avoid deluding the adjacent relation between points in the reconstructed trajectory. In this sense, a useful measure to establish τ is to take the value for which the Mutual Information (MI) has its first minimum.

Auto correlation algorithm is widely used to ascertain delay time, as one of the most important parameters, in reconstructing phase space from nonlinear time series [5], which is identified as

$$C(\tau) = \frac{\frac{1}{N} \sum_{i=1}^{N} [x(i+\tau) - \overline{x}][x(i) - \overline{x}]}{\frac{1}{N} [x(i) - \overline{x}]^2} \tag{4}$$

where $\overline{x} = \frac{1}{N} \sum_{i=1}^{N} x(i)$.

2.3   Measure Embedding Dimension

The correlation dimension of runoff time series recorded at Liuzhou station was calculated by Grassberger and Procaccia algorithm (G-P algorithm). In the phase space, a couple of phase points were constructed as follows:

$$x(t_i) = (x(t_i), x(t_i + \tau), \cdots, x(t_i + (m-1)\tau)) \tag{5}$$

$$x(t_j) = (x(t_j), x(t_j + \tau), \cdots, x(t_j + (m-1)\tau)) \tag{6}$$

where τ is an arbitrary constant called delay time constant. x provides the coordinate of a point in an n–dimensional space, where m is the embedding dimension. One of the reference points from the data set was inspected, and taken into further calculation of all the distances $r_{ij}(m) = \| x(t_i) - x(t_j) \|$ from the N-m-1 remaining points. The total amount of data points within a given distance r in the phase space was then counted. Repeating the procedure for all values of i, one reached the expression [9]

$$C(r,m) = \frac{1}{N(N-1)} \sum_{i,j,i \neq j} H(r - \| x(t_i) - x(t_j) \|) \tag{7}$$

where $C(r,m)$ is given the name the correlation integral of the attractor, H is the Heaviside step function, expressed as

$$H(x) = \begin{cases} 1, x > 0 \\ 0, x \leq 0 \end{cases} \tag{8}$$

An attractor is a compact set in the phase space toward where the systems converge. It has a zero volume, but the volume of its basin of attraction is nonzero or even infinite. It is possible to show that if the size of r/δ falls between the size of the attractor and the smallest spacing between points, then

$$C(r,m) \propto (r/\delta)^D \tag{9}$$

Where δ is a microscale which used to explore the attractor structure, and

$$D = \lim_{r \to 0} \frac{\log C(r,m)}{\log r} \tag{10}$$

The dimension D of the attractor is the correlation dimension. This procedure must be repeated for increasing values of m. If the dependence of D versus m reaches the saturation, m go beyond some relatively small value, the system represented by the time series should possess an attractor. The saturation value of D is defined as the GP correlation dimension of the attractor represented by the time series. Another expression was drawn by Shizhong Hong and Shiming Hong, as

$$N_{min} > \sqrt{2 \times (27.5)^D} \tag{11}$$

When D is equal 2, $N_{min} > 39$, while $D = 3$, $N_{min} > 204$.

2.4   The Maximum Lyapunov Exponent (MLE)

In order to study in more depth specific features to identify deterministic features, the Lyapunov exponents measure, specifically the maximum Lyapunov exponent (MLE) (as it can be computed with high accuracy) was explored to further looking into more specific traces, as for identification possible deterministic features. The numerical value of the MLE defined the possible logarithmic separation of trajectories in the phase space of the system. For time series produced by dynamical systems, the presence of a positive MLE indicated the emerging of chaos. To a one-dimensional dynamical system $x_{n+1} = F(x_n)$, Lyapunov exponents was simplified as follows:

$$\lambda = \lim_{n \to \infty} \frac{1}{n} \sum_{i=0}^{n-1} \ln \left| \frac{dF(x)}{dx} \right|_{x=x_i} \tag{12}$$

Lyapunov exponents were taken to quantify the exponential divergence of initially close state-space trajectories and estimate the degree of chaos in a system.

The MLE was obtained by Wolf algorithm, Jacobian algorithm and small data sets, etc. Small data sets algorithm was adopted by this paper.

3   Experimental Results and Discussion

3.1   Description of Study Data

Data used in this paper are water level time series of the whole 2007 year, gathered from Liuzhou station, Rongshui station and Xiaochangan station in Liujiang basin, as shown in Fig 1. To each time series, there were three parts to be studied, namely the whole year, the dry season and the wed season. To the water level of Liuzhou station, the period from record point 1 to record 1396 was study as a dry season and the period from record point 1397 to record 2607 was study as a wed season. To the water level of Rongshui station, the period from record point 1 to record 1297 was study as a dry season and the period from record point 1298 to record 2367 was study as a wed season. While to the water level of Xiaochangan station, the period from record point 1 to record 513 was study as a dry season and the period from record point 514 to record 1134 was study as a wed season.

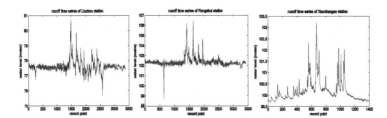

Fig. 1. Water level time series of the three stations in 2007 year

3.2   Results Analysis

The chaotic characteristic analysis results including time delay, correlation dimension and the maximum Lyapunov exponents were shown in following figures 2 and table 1. To each station, the time delay and the correlation dimension of the whole year, the wed season and the dry season respectively, were shown different waves which indicated different characteristic of them. The chaotic characteristic of wed season influenced the whole year's chaotic characteristic. From the Fig. 2 and table 1, the maximum Lyapunov exponents of the whole year and the wed season of the Liuzhou station were positive, but the correlation dimension of the dry season is none. This indicated the runoff time series of the whole year and the wed season of the Liuzhou station were chaotic, while the series of the dry season was stable. While it indicated three chaotic time series by the results that maximum Lyapunov exponents of all series of the Rongshui station were all positive, because of its more complex change produced by other branch nearby running into the river.

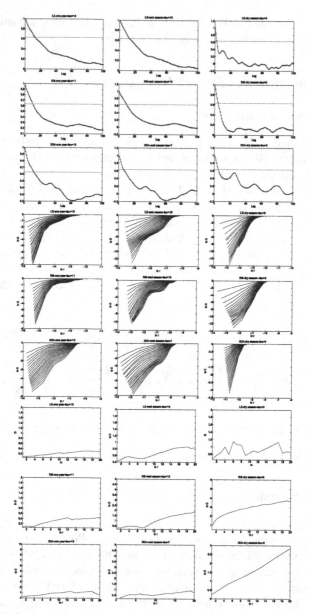

Fig. 2. The chaotic characteristic analysis results for the three parts of the three stations

Table 1. The MLEs of Liujiang River time series (Bits/s)

	one year	wed season	dry season
Liuzhou	0.0551	0.0785	None
Rongshui	0.0690	0.0812	0.0385
Xiaochangan	0.0593	0.0619	None

4   Conclusion

The conclusions were summarized as follows: Liujiang runoff time series has the chaotic characteristic of different degrees in different stations, which reason included climate, geographical environment and manual intervention, etc. Further reason should be studied in the future.

Acknowledgement. The authors would like to extend sincere gratitude to the editor and anonymous reviewers' comments for suggestions of improving this paper. This paper was supported by the National Natural Science Foundation of China under Grant No. 61170305, the National Natural Science Foundation of China under Grant No.11161029, the National Natural Science Foundation of China under Grant No. 60873114, the Foundation of Guangxi ministry of Education under Grant No. 2012041x501.

References

1. Wu, C.L., Chau, K.W.: Data-driven models for monthly stream flow time series prediction. Engineering Applications of Artificial Intelligence 23, 1350–1367 (2010)
2. Millán, H., Rodríguez, J., Ghanbarian-Alavijeh, B., Biondi, R., Llerena, G.: Temporal complexity of daily precipitation records from different atmospheric environments: Chaotic and Lévy stable parameters. Atmospheric Res. 101, 879–892 (2011)
3. Ding, J., Wang, W.S., Zhao, Y.L.: Characteristics of daily flow variation in the Yangtze River, optimum determination of delay time for reconstruction of a phase space. Advances in Water Science 14(4), 407–411 (2003)
4. Ding, J., Wang, W.S., Zhao, Y.L.: Characteristics of daily flow variation in the Yangtze River, optimum determination of embedding dimension for reconstruction of a phase space. Advances in Water Science 14(4), 412–416 (2003b)
5. Sivakumar, B., Berndtsson, R., Persson, M.: Monthly runoff prediction using phase space reconstruction. . Hydrological Sciences Journal 46(3), 377–387 (2001)
6. Yu, G.R., Yang, J.R., Xia, Z.Q.: The prediction model of chaotic series based on support vector machine and its application to Runoff. Advanced Materials Research 255, 3594–3599 (2011)
7. Kim, H.S., Lee, K.H., Kyoung, M.S., Sivakumar, B., Lee, E.T.: Measuring nonlinear dependence in hydrologic time series. Stochastic Environmental Research and Risk Assessment 23(7), 907–916 (1993)
8. Rosenstein, M.T., Collins, J.J., De Luca, C.J.: A practical method for calculating largest Lyapunov exponents from small data sets. Physica D: Nonlinear Phenomena 65(1), 117–134 (2009)
9. Almog, Y., Oz, O., Akselrod, S.: Correlation dimension estimation: Can this nonlinear description contribute to the characterization of blood pressure control in rats? IEEE Transactions on Biomedical Engineering 46(5), 535–547 (1999)
10. Wen, S., Zeng, Z., Huang, T.: Adaptive synchronization of memristor-based Chua's circuits. Physics Letters A 376(44), 2775–2780 (2012)
11. Wen, S., Bao, G., Zeng, Z., Chen, Y., Huang, T.: Global exponential synchronization of memristor-based recurrent neural networks with time-varying delays. Neural Networks 48, 195–203 (2013)

Log-Cumulant Parameter Estimator
of Log-Normal Distribution

Zengguo Sun and Ji-Xiang Du

College of Computer Science and Technology, Huaqiao University,
Xiamen, China
{sunzg,jxdu}@hqu.edu.cn

Abstract. Based on the second-kind statistics, the log-cumulant estimator is proposed for parameter estimation of the log-normal distribution. The estimates of the shape parameter and scale parameter are independent for the log-cumulant estimator. Parameter estimation results from Monte Carlo simulation demonstrate that, the log-cumulant estimator is insensitive to samples and it leads to better performance when compared to the moment estimator.

Keywords: Log-normal distribution, parameter estimation, second-kind statistics, log-cumulant estimator.

1   Introduction

The log-normal distribution is commonly used to model the high resolution sea clutter and the regions of strong spatial variation such as urban areas [1-4]. With two parameters (shape parameter and scale parameter), the log-normal distribution can fit the experimental data better than the one-parameter distributions such as Rayleigh. For example, the Rayleigh distribution usually applies when the radar resolution cell is large so that it contains many scatterers, with no one scatterer dominant. However, when the resolution cell size and the grazing angle are small, the log-normal distribution can describe the heavy tails of the clutter more precisely compared to the Rayleigh [2]. In order to use the log-normal distribution in practical applications, its parameters should be estimated accurately. The moment method can be used to estimate the parameters of this distribution [1, 5], but it is sensitive to samples. For the severe impulsive samples, which correspond to the large values of the shape parameter, the performance of the moment estimator is degraded seriously. In this paper, the log-cumulant estimator is proposed for the log-normal distribution based on second-kind statistics, which relies on the Mellin transform [6-8]. We compare the performances of the log-cumulant estimator and the moment estimator, and we have observed that the log-cumulant estimator leads to high estimation accuracy no matter what values are chosen for the shape parameter, which is validated by parameter estimation results from Monte Carlo simulation. Consequently, we recommend the log-cumulant estimator compared to the moment estimator.

D.-S. Huang et al. (Eds.): ICIC 2014, LNCS 8588, pp. 668–674, 2014.
© Springer International Publishing Switzerland 2014

2   Log-Normal Distribution

The log-normal distribution has the following probability density function (pdf) [1, 2]

$$f(x) = \frac{1}{x\sqrt{2\pi V}} \exp\left[-\frac{(\log x - \beta)^2}{2V} \right], \quad x \geq 0, \tag{1}$$

where V ($V > 0$) is the shape parameter and β ($-\infty < \beta < +\infty$) is the scale parameter. Denoting X as a log-normal random variable with parameters V and β, its natural logarithm Y ($Y = \log X$) can be represented by the following pdf

$$f_Y(x) = \frac{1}{\sqrt{2\pi V}} \exp\left[-\frac{(x - \beta)^2}{2V} \right]. \tag{2}$$

Obviously, Y follows the classical Gaussian distribution with mean β and variance V. For various values of V, the pdf of log-normal distribution is plotted in Fig. 1. Obviously, the value of V controls the shape of the pdf. The larger the value of V is, the severer skewness the distribution has.

Since the natural logarithm of the log-normal distribution is the Gaussian distribution, the log-normal distribution can be simulated by

$$X = \exp\left(\sqrt{V}Z + \beta\right), \tag{3}$$

where X is the log-normal random variable with parameters V and β, and Z is the standardized Gaussian random variable with mean 0 and variance 1. With the help of (3), the log-normally distributed samples can be simulated, which are shown in Fig. 2 for various values of V. It is apparent that the log-normal samples with

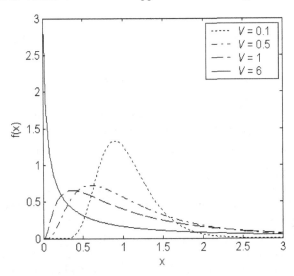

Fig. 1. Pdfs of log-normal distribution ($\beta = 0$)

$V = 6$ show much severer impulsiveness than the ones with $V = 0.1$. In general, the larger the value of V is, the more impulsive the log-normal samples are. Since the log-normally distributed samples can be simulated readily, we can use the Monte Carlo simulation to compare the performance of various parameter estimators.

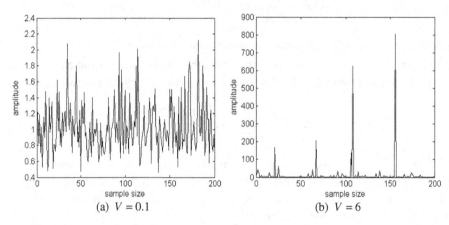

(a) $V = 0.1$ (b) $V = 6$

Fig. 2. Log-normally distributed samples ($\beta = 0$ and the number of samples is 200)

3   Moment Estimator

The n th order moment of log-normal distribution can be written as

$$E\left(X^{n}\right) = \exp\left(n\beta + \frac{n^{2}V}{2}\right), \quad n = 1, 2, \ldots, \tag{4}$$

where X is the log-normal random variable with parameters V and β. Hence, the moment estimator for the log-normal distribution is straightforward as follows [1]:

$$\frac{E\left(X^{2}\right)}{E^{2}\left(X\right)} = \exp(V), \quad E\left(X\right) = \exp\left(\beta + \frac{V}{2}\right). \tag{5}$$

By replacing the actual moments with the sample moments, the shape parameter V and the scale parameter β can be estimated from (5). It should be noted that the estimate of β relies on the estimated values of V, so the error of estimate of V may degrade the estimation accuracy of β.

The moment estimator was tested for various true values of parameter V according to Monte Carlo simulation. The log-normally distributed samples were simulated independently by using (3), and the number of samples is 10000. For each true parameter V, the Monte Carlo simulation experiment was repeated 100 times independently, and then the average and standard deviation of the estimates were computed. The results are shown in Table 1 with standard deviations in parentheses. Obviously, the performance of the moment estimator relies on the true values of V.

For the smaller values of V, the moment estimator can lead to high estimation accuracy (e.g., $V = 0.1$). However, if the larger values are chosen for the V (e.g., $V = 6$), the moment estimator results in poor performance. In other words, the moment estimator is sensitive to samples. If the samples show severe impulsiveness (e.g., $V = 6$), the moment estimator cannot achieve high estimation accuracy.

Table 1. Monte Carlo simulation of moment estimator (true $\beta = 0$)

True Value	$V = 0.1$	$V = 0.5$	$V = 1$	$V = 2$	$V = 5$	$V = 6$
$\hat{V}$	0.0998	0.4999	0.9971	1.9385	4.1894	4.6268
	(0.0017)	(0.0138)	(0.0467)	(0.2015)	(0.5715)	(0.6755)
$\hat{\beta}$	0.0003	0.0010	0.0022	0.0303	0.3994	0.6625
	(0.0032)	(0.0089)	(0.0218)	(0.0916)	(0.2332)	(0.2449)

4   Log-Cumulant Estimator

The log-cumulant estimator is based on second-kind statistics, which relies on the Mellin transform [6-8]. Denoting g as a function defined over $[0, +\infty)$, its Mellin transform is defined as

$$\mathrm{M}\big[g(x)\big](s) = \int_0^{+\infty} x^{s-1} g(x)\, dx, \tag{6}$$

where s is the complex variable of the transform. Specifically, for a pdf f defined in $[0, +\infty)$, the second-kind first characteristic function is defined by

$$\Phi(s) = \int_0^{+\infty} x^{s-1} f(x)\, dx. \tag{7}$$

Then, the second-kind second characteristic function is defined by

$$\Psi(s) = \log\big(\Phi(s)\big). \tag{8}$$

Finally, the r th order second-kind cumulant (log-cumulant) is defined by

$$\tilde{k}_r = \left.\frac{d^r \Psi(s)}{ds^r}\right|_{s=1}. \tag{9}$$

The first two log-cumulants $\tilde{k}_1$ and $\tilde{k}_2$ can be estimated empirically from N samples y_i as follows:

$$\hat{\tilde{k}}_1 = \frac{1}{N}\sum_{i=1}^{N}\big[\log(y_i)\big], \qquad \hat{\tilde{k}}_2 = \frac{1}{N}\sum_{i=1}^{N}\left[\big(\log(y_i) - \hat{\tilde{k}}_1\big)^2\right]. \tag{10}$$

By substituting (1) into (7) and after some manipulation, the second-kind first characteristic function of log-normal distribution is given by

$$\Phi(s) = \exp\left[\beta(s-1) + \frac{(s-1)^2 V}{2}\right]. \tag{11}$$

Then, substituting (11) into (8) and subsequently into (9), the log-cumulant estimator (i.e., the first and second orders log-cumulants) for the log-normal distribution is finally obtained as follows:

$$\tilde{k}_1 = \beta, \quad \tilde{k}_2 = V. \tag{12}$$

By replacing the actual log-cumulants with the sample log-cumulants (equation (10)), parameters β and V can be easily estimated from (12). Compared to the moment estimator (equation (5)), the estimate of the scale parameter β in the log-cumulant estimator is independent with that of the shape parameter V. In addition, the estimated parameters of log-cumulant estimator are only related with the sample log-cumulants, which are calculated from the log-transformed samples. Therefore, the impulsiveness of the log-normal distributed samples especially for the case of larger shape parameter is degraded, and the estimation performance of the log-cumulant estimator can be improved in comparison with the moment estimator.

The log-cumulant estimator was tested on Monte Carlo simulations for various true values of parameter V. For each true parameter V, the Monte Carlo simulation experiment was repeated 100 times independently, and the number of samples was 10000 for each time. Table 2 illustrates the average and standard deviation values (in parentheses) of Monte Carlo simulation results based on the log-cumulant estimator, and Fig. 3 shows the performance comparison of the log-cumulant estimator and the moment estimator as a function of true V. Obviously, the log-cumulant estimator leads to high estimation accuracy no matter what values are chosen for the true V. Therefore, the log-cumulant estimator is robust and not sensitive to samples. Even for the samples with severe impulsiveness (e.g., $V = 6$), the log-cumulant estimator can achieve high estimation accuracy. For the moment estimator, on the other hand, the estimated parameters may be close to that of the log-cumulant estimator for the smaller values of V (e.g., $V = 0.1$). However, as shown in Table 1, the performance of the moment estimator is deteriorated for the larger values of V (e.g., $V = 6$). In a word, the log-cumulant estimator is superior to the moment estimator, which is validated by the Monte Carlo simulations.

Table 2. Monte Carlo simulation of log-cumulant estimator (true $\beta = 0$)

True Value	$V = 0.1$	$V = 0.5$	$V = 1$	$V = 2$	$V = 5$	$V = 6$
$\hat{V}$	0.1000	0.5000	1.0006	2.0001	5.0001	6.0006
	(0.0015)	(0.0061)	(0.0143)	(0.0244)	(0.0793)	(0.0793)
$\hat{\beta}$	0.0002	0.0003	0.0010	0.0007	0.0010	0.0013
	(0.0028)	(0.0068)	(0.0089)	(0.0136)	(0.0221)	(0.0238)

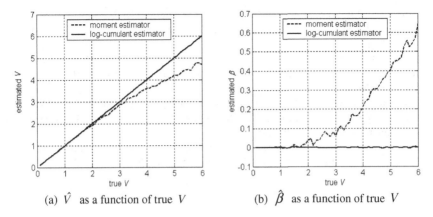

(a) $\hat{V}$ as a function of true V (b) $\hat{\beta}$ as a function of true V

Fig. 3. Monte Carlo simulation comparison of log-cumulant estimator and moment estimator (true $\beta = 0$)

5   Conclusion

The log-cumulant estimator based on the second-kind statistics is proposed to estimate the parameters of log-normal distribution in this paper. Compared to the moment estimator, the estimates of the shape parameter and scale parameter in the log-cumulant estimator are independent with each other, and they only rely on the sample log-cumulants calculated from the log-transformed samples instead of the impulsive samples themselves. Therefore, the log-cumulant estimator leads to high estimation accuracy no matter what values are chosen for the shape parameter, which is demonstrated by parameter estimation results from Monte Carlo simulations. The log-cumulant estimator is insensitive to samples, and it can achieve good performance even for the severely impulsive samples. Therefore, the log-cumulant estimator is recommended compared to the moment estimator.

Acknowledgments. This work was supported by the National Natural Science Foundation of China (Grant No. 61102163) and the Natural Science Foundation of Fujian Province, China (Grant No. 2012J01271).

References

1. Oliver, C., Quegan, S.: Understanding Synthetic Aperture Radar Images. Artech House, Boston (1998)
2. Skolnik, M.I.: Introduction to Radar Systems, 3rd edn. McGraw-Hill, New York (2001)
3. Ulaby, F.T., Dobson, M.C.: Handbook of Radar Scattering Statistics for Terrain. Artech House, Norwood (1989)
4. Gao, G.: Statistical Modeling of SAR Images: A Survey. Sensors 10, 775–795 (2010)

5. Billingsley, J.B., Farina, A., Gini, F., Greco, M.V., Verrazzani, L.: Statistical Analyses of Measured Radar Ground Clutter Data. IEEE Trans. Aerospace and Electronic Systems 35, 579–593 (1999)

6. Nicolas, J.M.: Alpha-Stable Positive Distributions: A New Approach Based on Second Kind Statistics. In: European Signal Processing Conference, EURASIP, Toulouse, pp. 197–200 (2002)

7. Tison, C., Nicolas, J.M., Tupin, F., Maitre, H.: A New Statistical Model for Markovian Classification of Urban Areas in High-Resolution SAR Images. IEEE Trans. Geoscience and Remote Sensing 42, 2046–2057 (2004)

8. Achim, A., Kuruoglu, E.E., Zerubia, J.: SAR Image Filtering Based on the Heavy-Tailed Rayleigh Model. IEEE Trans. Image Processing 15, 2686–2693 (2006)

Noise Removal Using Fourth Order PDEs Based on Nonlocal Derivative

Chang-xiong Zhou[1], Shufen Lui[1], Ting-qin Yan[1,2], and Wenlin Tao[1]

[1] Department of Electronic and Informational Engineering,
Suzhou Vocational University, Suzhou, Jiangsu, China, 215104
handzhou@sina.com

[2] Suzhou High-tech Key Laboratory of Cloud Computing & Intelligent Information Processing,
Suzhou, Jiangsu, China, 215104

Abstract. Since fourth order partial differential equations (PDEs) are able to avoid the blocky effects widely seen in images processed by second order PDEs, and nonlocal derivative (NLD) can effectively capture fundamental features such as edges, the improve PDE that is able to remove noise while preserving edge features will be built and its numerical scheme will be given. In the proposed model, the performances of removing noise and preserving edge are measured according to PSNR values and keeping edges index (KEI) respectively, and the experiments show that our method has a better performance than others.

Keywords: image de-noising, partial differential equations, diffusion coefficient, nonlocal derivative.

1 Introduction

Second order PDEs have been studied as a useful tool for image noise removal. They include anisotropic diffusion equations as formulated in [1]–[3] and derived from total variation minimization in [4, 5]. Although these techniques have been demonstrated to be able to remove noise, they tend to cause the processed image to look "blocky" as can be seen from the images in [1], [4]. This effect is visually unpleasant and is likely to cause a computer vision system to falsely recognize. Fourth order PDEs were derived as a process that seeks to minimize a functional proportional to the absolute value of the Laplacian of the image intensity function in [5]. When removing noise, the time evolution of these PDEs will converge to piecewise planar images which look more natural than that of second order PDEs. As well known, these PDEs are defined as the case where the diffusivity is a scalar function varying with the location in the image [1, 5]. If diffusion coefficient (DC) is effective in detecting edge features and insensitive to noise [6], the improved models with DC will remove noise well and at the same time can protect the edge features and details of images.

The remainder of this paper is structured as follows. In Section 2, we introduce second order PDEs and fourth order PDEs. In section 3, a diffusion coefficient with

D.-S. Huang et al. (Eds.): ICIC 2014, LNCS 8588, pp. 675–683, 2014.
© Springer International Publishing Switzerland 2014

nonlocal derivative (NLD) will be built and the improved fourth order PDE is proposed. In section 4, we analyze and discuss our method by comparative results on noisy images. Finally, conclusions are drawn in Section 5.

2   Second-Order PDEs and Fourth-Order PDEs

The theory of image de-noising based on PDEs is often linked with heat conduction equation, and this equation is associated with the following energy functional in [1]

$$E_1(u) = \int_\Omega f(|\nabla u|) d\Omega .$$

(1)

Where ∇ is the gradient operator, u is gray value of image and Ω is the image support, and f is an increasing function associated with the diffusion coefficient c as

$$c(s) = \frac{f'(s)}{s} .$$

(2)

Anisotropic diffusion is then shown to be an energy-dissipating process that seeks the minimum of the energy functional. This functional minimization is a special form of the general variation problem and the equivalent Euler equation is

$$div(c(|\nabla u|)\nabla u) = 0 .$$

(3)

Let t denote the time, the anisotropic diffusion as formulated in [1] may be presented as

$$\frac{\partial u}{\partial t} = div(c(|\nabla u|)\nabla u) .$$

(4)

Where div is the divergence operator, Eq. (4) is called P-M model which is second order PDEs. Since anisotropic diffusion is designed such that smooth areas are diffused faster than less smooth ones, blocky effects will appear in the early stage of diffusion, even though all the blocks will finally merge to form a level image. When there is backward diffusion, however, any step image is a global minimum of the energy functional, so blocks will appear in the early stage of the diffusion and will remain as such. Let us first consider the following functional defined in the space of continuous images over a support of Ω

$$E_2(u) = \int_\Omega f(|\nabla^2 u|) d\Omega .$$

(5)

Where ∇^2 denotes Laplacian operator. We require that the function $f(\cdot) \geq 0$ and is an increasing function

$$f'(\cdot) > 0 .$$

(6)

So that the functional is an increasing function with respect to the smoothness of the image as measured by $\left|\nabla^2\right|$. Therefore, the minimization of the functional is equivalent to smoothing the image. The equivalent Euler equation is

$$\nabla^2[c(|\nabla^2 u|\nabla^2 u)] = 0 \quad . \tag{7}$$

The Euler equation may be solved through the following gradient descent procedure:

$$\frac{\partial u}{\partial t} = -\nabla^2[c(|\nabla^2 u|)\nabla^2 u] \,. \tag{8}$$

Eq. (8) is called YK mode. The solution is arrived when $t \to \infty$, but the time evolution may be stopped earlier to achieve an optimal tradeoff between noise removal and edge preservation in [5].

3   Our Proposed Method

3.1   Nonlocal Derivative and Diffusion Coefficient

By nonlocal derivative (NLD) the grayscale difference between two pixels is measured by two regions (patches) centered at the pixels, the new feature detector which measures image intensity contrasts between neighboring patches in a more sophisticated manner and can effectively capture fundamental edge features [6].

We describe the concept of the nonlocal difference in a one-dimensional (1-D) image. Extension to the two-dimensional (2-D) case is straightforward and will be discussed later. Let $I : \Omega \to R^1$ be a 1-D scalar image defined on the discrete domain Ω and $x \in \Omega$ is the pixel position, $x = [x_1, x_2,..., x_i,..., x_N]$. For each pixel x, we define a neighborhood region N_x , which comprises W pixels centered on x. We further define a patch P_x , which is a vector comprising gray-level value of all pixels within the neighborhood region N_x ,

$$P_{x_i} = [x_{i-(W-1)/2}, x_{i-(W-1)/2+1},...x_i,...x_{i+(W-1)/2}] \quad , \tag{9}$$

here W is assumed to be an odd number for symmetry consideration. The nonlocal difference between two signal samples $I(x_i)$ and $I(x_j)$ can be measured as a Gaussian weighted Euclidean difference,

$$d_{NL}(x_i, x_j) = \left\|P_{x_i} - P_{x_j}\right\|_{2,\sigma} \tag{10}$$

of two vectors P and in a W-dimensional space, where σ is the standard deviation (Std) of a normalized Gaussian kernel.

$$\left| \nabla_{NL} x_i \right| = \left\| P_{x_{i-(W+1)/2}} - P_{x_{i+(W+1)/2}} \right\|_{2,\sigma} =$$

$$\left(\sum_{k=-(W-1)/2}^{k=(W-1)/2} G_\sigma(k) \left(x_{i-(W+1)/2+k} - x_{i+(W+1)/2+k} \right)^2 \right)^{1/2} . \tag{11}$$

From Eq. (11), we can define a nonlocal derivative at pixel x_i when x_j approaches x_i. Where G_σ is a Gaussian kernel with Std. σ. For two-dimensional image $u(i, j)$, we may define that the nonlocal derivative $\left| \nabla_{NL} u(i, j) \right|$ at pixel (i, j) is the corresponding maximum value in horizontal, vertical and both diagonal directions,

$$\left| \nabla_{NL} u(i, j) \right| = \max(\left| \nabla_{NL} u(i, j) \right|_{\text{horizontal , vertical and both diagonal directions}}). \tag{12}$$

The nonlocal derivative involves the difference between two adjacent patches. For the same reason, Eq. (11) and Eq. (12) are more reliable than the pixel-level gradient operator involving two pixels to measure edges under noise contamination. In Eq. (4), if its diffusion coefficient is replaced by the nonlocal derivative, then it becomes Eq. (13) as follow:

$$c(\cdot) = 1/((\left| \nabla u_{NL} \right| / k)^2 + 1) . \tag{13}$$

So the improved second PDEs called NLM-PM in [7] may be presented as

$$\frac{\partial u}{\partial t} = div \, (c(\left| \nabla u_{NL} \right|) \nabla u) . \tag{14}$$

3.2   A New PDE Based on Nonlocal Derivative

To better preserve the geometric properties of image features and remove noise, the orientations of features should be taken into account when we apply the PDEs. using (13), we may represent Eq. (8) as

$$\frac{\partial u}{\partial t} = -\nabla^2 [c(\left| \nabla u_{NL} \right|) \nabla^2 u] . \tag{15}$$

The Eq. (15) is called the NLD-YK model. The differential equation (15) may be solved numerically using an iterative approach. Assuming a time step size of τ and a space grid size of h, we quantize the time $t = n\tau$ and space coordinates $y = ih, x = jh$. Let h be one, we have

$$\nabla u = (u_x, u_y)^T , \tag{16}$$

$$u_x = 0.5(u_{i,j+1} - u_{i,j-1}) , \tag{17}$$

$$u_y = 0.5(u_{i+1,j} - u_{i-1,j}) \ . \tag{18}$$

We can calculate the Laplacian of the image intensity function as

$$\nabla^2 u = u_{i+1,j} + u_{i-1,j} + u_{i,j+1} + u_{i,j-1} - 4u_{i,j} \ . \tag{19}$$

Let g denote as follow:

$$g = c(|\nabla u_{NL}|)\nabla^2 u \ . \tag{20}$$

We will get the Laplacian of g as following:

$$\nabla^2 g = g_{i+1,j} + g_{i-1,j} + g_{i,j+1} + g_{i,j-1} - 4g_{i,j} \tag{21}$$

The Eq. (15) may be presented as

$$u^{n+1} = u^n - \tau \left(g_{i+1,j} + g_{i-1,j} + g_{i,j+1} + g_{i,j-1} - 4g_{i,j}\right) \ . \tag{22}$$

This is the numerical scheme of NLD-YK in Eq. (15). We will evaluate the performance of different methods, the PM, YK, NLM-PM, and the proposed NLD-YK models, using the peak signal-to-noise ratio (PSNR in db) and keeping edges index (KEI) respectively. PSNR may be defined as

$$PSNR = 10 \log_{10}\left(\frac{255^2}{MSE}\right), \tag{23}$$

with $MSE = |\Omega|^{-1}\sum_{x \in \Omega}(I(x) - u(x))^2$ where I is the noise-free original image, and KEI will be defined as

$$KEI = \frac{\sum\left(|u_{i+1,j} - u_{i,j}| + |u_{i,j+1} - u_{i,j}|\right)}{\sum\left(|I_{i+1,j} - I_{i,j}| + |I_{i,j+1} - I_{i,j}|\right)} \ . \tag{24}$$

The de-noising goal is to maximum the PSNR and the effect of preserving image features is to make KEI close one.

4   Experiments and Comparisons

This section aims at demonstrating the de-noising performance of the proposed NLD-YK. We show the experiment results with four methods of the PM (Eq. (4)), NLM-PM (Eq. (14)), and YK (Eq. (8)) and improved NLD-YK (Eq. (15)) in various noise levels. All of the experiments were performed on images of LENA and Camera.

We study the PSNR performance of the four methods using images corrupted with Gaussian noises (mean = 0 and variance = 10 to 25), respectively.

Table 1 shows the PSNR values of four methods for the images (LENA and CAMERA) corrupted with Gaussian noises. Data in the table 1 shows that the PSNR values of the image denoised with NLM-PM and NLD-YK are higher than that with PM and YK in different noise levels, since the two previous methods involve the NLD between two adjacent patches and are more reliable than two methods behind. Experimental results also show that the NLD-YK is the best one in four de-noising methods, since it is an integrated algorithm which has advantages of fourth-order PDEs and the NLD for image de-noising.

Table 1. Comparison of PSNR of images (Lena/ Camera) corrupted with Gaussian noises by four methods

Noise variance	Noisy image	PM	NLM-PM	YK	NLD-YK
10	39.11/39.23	40.51/40.60	40.52/40.38	40.11/39.83	41.24/41.22
15	37.37/37.46	39.40/39.14	39.51/39.15	39.28/38.97	40.19/39.96
20	36.15/36.27	38.50/38.23	38.79/38.36	38.76/38.24	39.43/39.05
25	35.25/35.36	37.78/37.36	38.30/37.71	38.32/37.69	38.84/38.27

Table 2 shows the KEI values of four methods for the images (LENA and CAMERA) corrupted with Gaussian noises. Data in the table 2 shows that the KEI values of the image denoised with NLM-PM and NLD-YK are more close to one than that with PM and YK in different noise density levels, and shows that NLM-PM and NLD-YK methods performance is inferior to that of the PM and YK method. Furthermore, the KEI values of the proposed NLD-YK are most approximate to one and its performance is the best one in four PDEs.

Table 2. Comparison of KEI of images (Lena/ Camera) corrupted with Gaussian noises by four methods

Noise variance	Noisy image	PM	NLM-PM	YK	NLD-YK
10	1.842/1.698	1.370/1.342	1.202/1.181	1.243/1.222	0.991/0.992
15	2.426/2.146	1.521/1.552	1.174/1.231	1.250/1.225	0.971/1.003
20	3.022/2.584	1.732/1.720	1.188/1.189	1.259/1.221	0.977/1.001
25	3.587/3.008	1.966/1.919	1.209/1.177	1.271/1.224	1.002/1.001

Experiment results on Lena are shown as Fig.1.

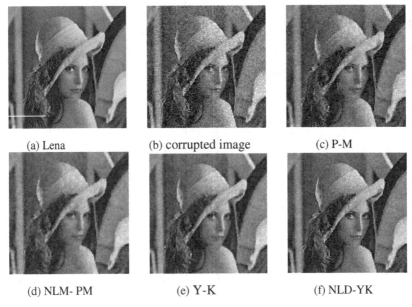

(a) Lena   (b) corrupted image   (c) P-M

(d) NLM- PM   (e) Y-K   (f) NLD-YK

Fig. 1. Noisy image of Lena and results of denoising by four methods

From visual sense effect of human eyes, Fig.1 shows that the NLD-YK has a good de-noising performance for Lena compared with PM and YK and NLM-PM models. The proposed approach not only removes the noise well but also reserves edges and details information. From Fig. 1, the NLD-YK is the best one in four de-noising methods, since Fig.1 (f) has the less noise and the clearest details.

Fig.2 is an intensity distribution from 1 to 80 columns at the 230th rank showing as horizontal white band in Fig.1 (a). From Fig.2, the NLM-PM and NLD-YK have good de-noising performance on a flat area compared with PM and YK model because of DC using the NLD. Meanwhile, YK and NLD-YK with fourth order PDEs have many details than that of PM and NLM-PM model. So the NLD-YK has the best performance of removing noise and keeping details within four methods.

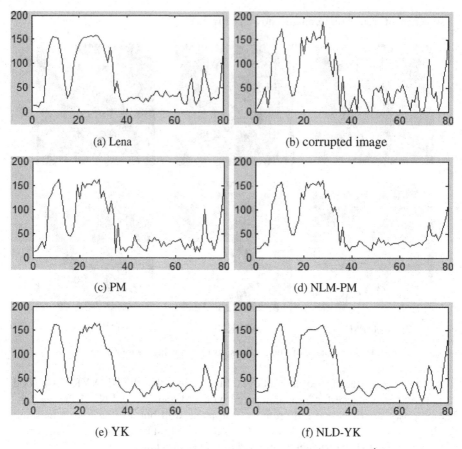

Fig. 2. Lena intensity distribution from 1 to 80 columns at the 230th rank

5   Conclusions

In this paper, we introduce nonlocal derivative and construct the fourth order PDEs with NLD, so-called NLD-YK, which can reject diffusion at edges and permit smoothing in other places, so remove noise of images while preserving edge features, as well as give the numerical scheme of the improved method. Compare to the PM, YK and NLM-PM models, simulation experiments show that the NLD-YK is the best one in performance of PSNR and KEI.

Acknowledgements. This research was sponsored by the grants of Natural Science Foundation of China (No.61373098), Innovative Team Foundation of Suzhou Vocational University (No.3100125), and the Opening Project of Suzhou High-tech Key Laboratory of Cloud Computing & Intelligent Information Processing (No. SXZ201304).

References

1. Perona, P., Malik, J.: Scale-space and edge detection using anisotropic diffusion. IEEE Trans. Pattern Anal. Machine Intell. (1990)
2. Gilboa, G., Sochen, N., Zeevi, Y.Y.: Image enhancement and denoising by complex diffusion process. IEEE Trans. Patt. Anal. Mach. Intell. (2004)
3. Cao, Z., Zhang, X.: PDE-based Non-linear anisotropic diffusion techniques for medical image denoising. In: 2012 Spring World Congress on Engineering and Technology, Xi'an, China, May 27-30 (2012)
4. Rudin, L.I., Osher, S., Fatemi, E.: Nonlinear total variation based noise removal algorithms. Physica D 60 (1992)
5. You, Y.-L., Kaveh, M.: Fourth-Order Partial Differential Equations for Noise Removal. IEEE Transactions on Image Processing 9(10) (October 2000)
6. Qiu, Z., Yang, L., Lu, W.: A new feature-preserving nonlinear anisotropic diffusion for denoising images containing blobs and ridges. Pattern Recognition Letters (2012)
7. Deng, S.-W., Jin, L.-Y., Xu, X.-Q., Ma, W.-B.: Image denoising study on Nonlocal-Means Filter. In: 2012 International Conference on Image Analysis and Signal Processing (IASP), November 9-11, pp. 1–4 (2012)

The Research of the Transient Feature Extraction by Resonance-Based Method Using Double-TQWT

Weiwei Xiang[1], Gaigai Cai[1,*], Wei Fan[1], Weiguo Huang[1],
Li Shang[2], and Zhongkui Zhu[1]

[1] School of Urban Rail Transportation, Soochow University, Suzhou 215137, China
[2] Department of Electronic Information Engineering, Suzhou Vocational University,
Suzhou 215104, China
ggcai@suda.edu.cn

Abstract. Signal processing aims to extract useful features from signals. However, the useful features are usually so weak, and corrupted by strong background noise, so it is difficult to extract by traditional linear methods. In this paper, a resonance-based method using double tunable Q-factor wavelet transform (TQWT) is applied for transient feature extraction. With the double-TQWT, the non-stationary signal is represented as the mixture of high resonance components and low resonance components based on the different resonance. The transient feature has a low Q-factor and belongs to low resonance components. Results of applications in transient feature extraction for simulation signal and bearing fault signal show the new method outperforms the average filtering method and the wavelet threshold algorithm, which further confirms the validity and superiority of this method for transient feature extraction.

Keywords: transient feature extraction, double-TQWT, resonance.

1 Introduction

The task of extracting transient features from signals is of great interest in many applications. For instance, a prominent practical problem is represented by the extraction of the rotating machines fault transient features from the strong background noise in severe working conditions. Successful methods for signal processing play a vital role in many applications of signal processing and pattern recognition. Frequency-based analysis and filtering are fundamental tools in signal processing [1], such as Fourier transform, short-time Fourier transform, wavelet transform (WT), empirical mode decomposition (EMD), and spectral kurtosis(SK). However, the traditional methods all suffer from some disadvantages. Here are three main reasons: (1) In addition to being non-stationary, numerous measured signals are a mixture of oscillatory and non-oscillatory behaviors; (2) Interference elements, such as noise, exist in the whole frequency range. Conventional frequency-based methods

---

* Corresponding author.

D.-S. Huang et al. (Eds.): ICIC 2014, LNCS 8588, pp. 684–692, 2014.
© Springer International Publishing Switzerland 2014

cannot effectively remove interference elements which in the same frequency band with the target features; (3) Target transient features are usually very weak and embedded in strong background noise.

WT provides plentiful non-stationary information of signals, and has become one of the most outstanding techniques in signal processing. During the past decade, great progress has been made in the theory and many publications have been seen in the field of signal processing [2,3]. WT is widely known as a type of inner product transform that analyzes non-stationary features in the signals through a predetermined wavelet bases. Generally, when processing the oscillatory signals, the WT should have a relatively high Q-factor. On the other hand, the WT should have a low Q-factor when the signal measured has little oscillatory behavior. However, traditional WT is commonly known as constant-Q transform which the Q-factor cannot adjust flexibly [4,5].

In 2011, Selesnick proposed a new wavelet transform named tunable Q-factor wavelet transform, which is based on the oscillatory behaviors rather than the frequency [4]. The TQWT is developed for discrete-time signals designed so that the Q-factor is easily and continuous adjustable [5,6,7]. When the target features have the similar center frequency to interference elements, they cannot be distinguished by frequency-based methods. However, the TQWT can solve the problem based on the different oscillatory behaviors of the signals rather than on the frequency band or scale.

The actual measured signals arising from physiological and physical processes usually exhibit a mixture of sustained oscillations and non-oscillatory transient behaviors. For example, speech, biomedical (EEG), and rotating machinery fault signals all possess both sustained oscillatory behavior and non-oscillatory transient behaviors. EEG signals contain rhythmic oscillations but they also contain transients due to measurement artifacts and non-rhythmic brain activity. Bearing fault signals are composed of harmonic one, transient impulse features and noise [1] [7]. In order to extract the target features from the complex non-stationary signals which composed of sustained oscillations and non-oscillatory transient behaviors, this paper studied a novel resonance-based method for transient feature extraction using double-TQWT. Firstly, two Q-factor wavelet bases are built, one characterized by a high Q-factor and the other characterized by a low Q-factor, according to the oscillation characteristics of the measured signal. Then the signal can be represented by the two Q-factor wavelet bases, and the corresponding coefficients of TQWT are obtained. Finally, morphological component analysis (MCA) is employed to separate the different oscillatory components and extract the transient features[1][6]. The effectiveness of the proposed method has been demonstrated by both simulation and experimental analyses. What's more, in order to demonstrate the superiority of the proposed method, the result is compared with the analysis results of average filtering method and wavelet threshold method.

2   Double Tunable Q-Factor Wavelet Transform

2.1   Summary of TQWT

Roughly, the Q-factor is a measure of the number of oscillations the wavelet exhibits. For Q, a value of 1.0 or greater can be continuously specified. Redundancy γ must be greater than 1.0, and a value of 3 or greater is recommended. A given (Q, γ) pair, the decomposition levels J can't be greater than the maximum number J_{max} [4]. The TQWT is implemented using a reversible over-sampled filter bank with real-valued sampling factors. The analysis filter banks are shown in Fig.1.

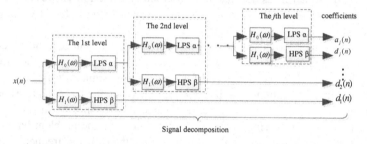

Fig. 1. Analysis filter banks: LPS and HPS represent low-pass、high-pass scaling

The variables α and β are the low-pass scaling parameter and high-pass scaling parameter respectively. The relationship between the scaling parameters and the main parameters of the TQWT are :

$$\beta = \frac{2}{(Q+1)}, \qquad \alpha = 1 - \frac{\beta}{\gamma} \tag{1}$$

The functions $H_0(\omega)$ and $H_1(\omega)$ are the frequency responses of the low-pass filter and high-pass filter. Considering the perfect reconstruction $|H_0(\omega)|^2 + |H_1(\omega)|^2 = 1$, the frequency responses $H_0(\omega)$ and $H_1(\omega)$ are generally defined as:

$$H_0(\omega) = \begin{cases} 1 & |\omega| \le (1-\beta)\pi \\ \theta(\dfrac{\omega+(\beta-1)\pi}{\alpha+\beta-1}) & (1-\beta)\pi \le |\omega| < \alpha\pi \\ 0 & \alpha\pi \le |\omega| \le \pi \end{cases} \tag{2}$$

$$H_1(\omega) = \begin{cases} 0 & |\omega| \le (1-\beta)\pi \\ \theta(\dfrac{\alpha\pi-\omega}{\alpha+\beta-1}) & (1-\beta)\pi \le |\omega| < \alpha\pi \\ 1 & \alpha\pi \le |\omega| \le \pi \end{cases} \tag{3}$$

where $\theta(\omega) = 0.5(1+\cos\omega)\sqrt{2-\cos\omega}, |\omega| \leq \pi$, and $\theta(\omega)$ originates from Daubechies filter with two vanishing moments.

2.2   Summary of Double-TQWT

TQWT is also a type of inner product transform that analyzes non-stationary features in the signals through a predetermined wavelet bases. Appropriate Q-factor wavelet bases are necessary for the feature extraction. When processing the oscillatory signals, the TQWT should have a relatively high Q-factor. On the other hand, the TQWT should have a low Q-factor when the signal has little oscillatory behavior. Thereby, One TQWT is helpless when processing signals such as the rotating machines fault vibration signals which exhibit a mixture of sustained oscillations and non-oscillatory transient behaviors. However, the signal can be represented as a mixture of high and low resonance components based on the different resonance by using double-TQWT. And this paper studied a novel resonance-based method for transient feature extraction using double-TQWT.

As is shown in Fig.2, signal (a) has a low Q-factor with a center frequency of 0.03 Hz and signal (b) has a high Q-factor also with a center frequency of 0.03 Hz. Conventional frequency-based methods separate a signal into low, mid and high frequency components by using different pass filters. However, a signal, composed of signal (a) and (b) which have the similar center frequencies, cannot be solved by the traditional frequency-based methods. However, signals (a) and (b) have the different oscillatory behaviors, so resonance-based method can separate them based on the different oscillatory behaviors.

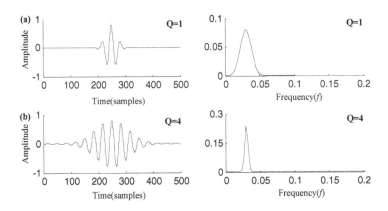

Fig. 2. The time domain waveform and its spectrum of different signals characterized by its Q-factor: (a) the time domain waveform of a Q=1 signal with a center frequency of 0.03 Hz; (b) the time domain waveform of a Q=4 signal with a center frequency of 0.03 Hz

The first step of the resonance-based signal method using double-TQWT is to choose two Q-factor wavelet bases according to the resonance of the measured signal, one characterized by a high Q-factor and the other characterized by a low Q-factor.

Then the different oscillatory components of the signal can be represented by the two Q-factor wavelet bases, and the corresponding coefficients of TQWT are obtained.

Then MCA is employed to separate the different resonance components. The main idea of MCA is to separate a signal or image into its building blocks considering that there is morphological diversity among a signal or an image's components.

A signal y is composed of a high-oscillatory component x_1 and a low-oscillatory component x_2 and noise. The goal to extract component x_1 or component x_2 can be replaced by solving an unconstrained optimization problem:

$$J(w_H, w_L) = 0.5 \left\| y - \Phi_H^* w_H - \Phi_L^* w_L \right\|_2 + \lambda_H \left\| w_H \right\|_1 + \lambda_L \left\| w_L \right\|_1 \tag{4}$$

where Φ_H^* and Φ_L^* are high Q-factor wavelet bases and low Q-factor wavelet bases, λ_H and λ_L are Lagrange multipliers which is associated with the signal and noise, the $\hat{w}_H$ and $\hat{w}_L$ are the coefficients vectors.

Split augmented Lagrangian shrinkage algorithm (SALSA) [10] has been proposed to minimize this kind of objective function. So SALSA is applied to solve the optimization problem. Then MCA can extract the high-oscillatory component or low-oscillatory component of the signal y

$$\hat{x}_H = TQWT_H^{-1}(\hat{w}_H), \quad \hat{x}_L = TQWT_L^{-1}(\hat{w}_L) \tag{5}$$

3   Simulation Study

In this section, we use a simulation signal with white Gaussian noise to demonstrate the effectiveness of the proposed resonance-based method. The simulated periodic impulse is generated as Eq.(6). And the simulation signal is sampled at 32768 samples s^{-1}. The sampling time is $1 \, s$.

$$x(t) = \exp(-445 \times t) \times \sin(2\pi \times 490 \times t) \tag{6}$$

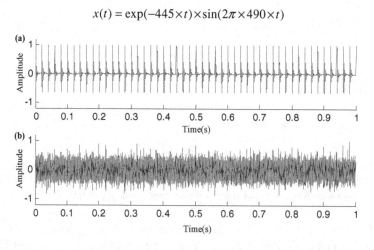

Fig. 3. Simulation signal: (a) simulation signal without noise; (b) simulation signal with noise

The simulated periodic impulse signal is plotted in Fig.3(a). The white Gaussian noise is directly added to the simulated periodic impulse signal, and the signal-to-noise ratio (SNR) of the simulated signal is about -6.5611. As is shown in Fig.3(b), the periodic transient features are submerged in the noise.

The result of applying the proposed resonance-based method is illustrated in Fig.4(c). For comparison, average filtering method, wavelet threshold methods are applied to the same simulation signal and the transient features are obtained in Fig.4(a) and Fig.4(b). The resonance-based method using double-TQWT has better performance in transient feature extraction.

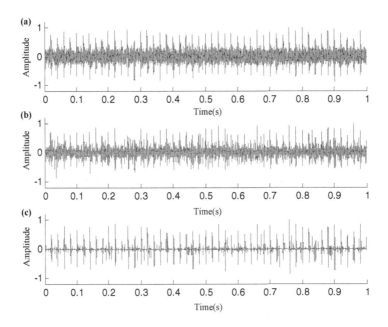

Fig. 4. Simulation signal analysis: (a) average filtering method; (b) wavelet threshold method; and (c) resonance-based method

4   Application in Transient Fault Feature Extraction

In this section, the proposed method is applied to fault feature extraction for bearing fault signal. The bearing type is N203 cylindrical rolling bearing. The outer fault characteristic frequencies of N203 cylindrical rolling bearing is 117.8 Hz .The sampling frequency is 32 KHz and sampling time is $1\,s$. Motor rotating speed was set as 1750 r/min.

As we can see from the Fig.5, transient features are submerged in strong noise, and we cannot detect the location of faults by the original signal in time domain. Average filtering method, wavelet threshold method and the proposed resonance-based method are applied to the original vibration signal respectively leading to the results shown in the Fig.6. And each Hilbert demodulation spectrum is presented in Fig.7.

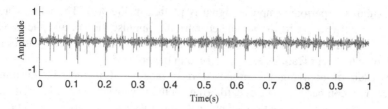

Fig. 5. Time-domain wavelet of the original signal of outer fault

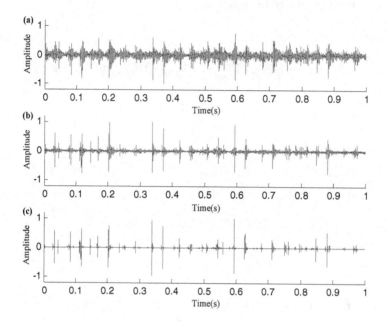

Fig. 6. Transient feature extraction of outer fault: (a) average filtering method; (b) wavelet threshold method; and (c) resonance-based method

The characteristic frequency of outer fault 117.5 Hz can be observed from the Hilbert demodulation spectrum in Fig.7. So we can conclude that the defects occurred in the outer. What's more, we can conclude that the new proposed resonance-based method outperforms average filtering method and wavelet threshold method in outer fault feature extraction.

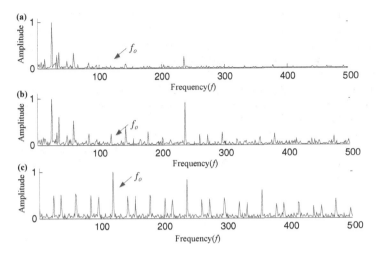

Fig. 7. Spectrum of the outer fault transient feature with Hilbert analysis: (a) average filtering method; (b) wavelet threshold method; and (c) resonance-based method

5   Conclusions

The main contribution of this paper is to explore a new transient feature extraction method based not on the frequency or scale like conventional frequency-based methods but on the oscillatory information of signals. The Q-factor of the TQWT is continuously tunable, offering the greater chance to select the most similar wavelet bases and wavelet parameters with the oscillatory behavior of the measured signal.

The effectiveness and superiority of the resonance-based method using double tunable Q-factor wavelet transform for transient feature extraction are verified and supported by the results of simulation study and the practical application in fault feature extraction for rolling element bearings.

Acknowledgements. The authors gratefully acknowledge the support from the National Science Foundation China (No. 51375322 and No. 61373098).

References

1. Selesnick, I.W.: Resonance-based Signal Decomposition: A New Sparsity-enabled Signal Analysis Method. Signal Processing 91, 2793–2809 (2011)
2. Wang, S., Huang, W., Zhu, Z.: Transient Modeling and Parameter Identification based on Wavelet and Correlation Filtering for Rotating Machine Fault Diagnosis. Mechanical Systems and Signal Processing 25, 1299–1320 (2011)
3. Zhu, Z., Yan, R., et al.: Detection of Signal Transients based on Wavelet and Statistics for Machine Fault Diagnosis. Mechanical Systems and Signal Processing 23, 1076–1097 (2009)

4. Selesnick, I.W.: Wavelet Transform with Tunable Q-factor. IEEE Transaction Signal Processing 59, 3560–3575 (2011)
5. He, W., Zi, Y., Chen, B., et al.: Tunable Q-factor Wavelet Transform Denoising with Neighboring Coefficients and its Application to Rotating Machinery Fault Diagnosis. Science China Technological Sciences 56, 1956–1965 (2013)
6. Cai, G., Chen, X., He, Z.: Sparsity-enabled Signal Decomposition using Tunable Q-factor Wavelet Transform for Fault Feature Extraction of Gearbox. Mechanical Systems and Signal Processing 41, 34–53 (2013)
7. Chen, X., Yu, D., Luo, J.: Envelope demodulation method based on resonance-based sparse signal decomposition and its application in roller bearing fault diagnosis. Journal of Vibration Engineering 6, 628–636 (2012)

Fast Mode and Depth Decision Algorithm
for Intra Prediction of Quality SHVC

Dayong Wang[1], Chun Yuan[1,*], Yu Sun[2], Jian Zhang[3], and Hanning Zhou[4]

[1] Graduate School at Shenzhen, Tsinghua University, Shenzhen, China
[2] Department of Computer Science, The University of Central Arkansas, Conway, USA
[3] School of Software, University of Technology, Sydney, Australian
[4] Zhigu Technology, Beijing, China
yuanc@sz.tsinghua.edu.cn

Abstract. Scalable High-Efficiency Video Coding (SHVC) is an extension of High Efficiency Video Coding (HEVC). Since the coding procedure for HEVC is very complex, the coding procedure for SHVC is even more complex, it is very important to improve its coding speed. In this paper, we have proposed a fast mode and depth decision algorithm for Intra prediction of Quality SHVC. Initially, only partial modes are checked to determine the local minimum points (LMPs) based on the relationships between the modes and their corresponding Hadamard Costs (HC); and then only partial depths are checked by skipping depths with low possibilities indicated based on their inter-layer correlations and textural features. The experimental results showed that the proposed algorithm could improve coding speed by 61.31% on average with negligible coding efficiency losses.

Keywords: SHVC, mode decision, depth decision, Intra prediction.

1 Introduction

As high-resolution video applications and services are becoming increasingly more popular, the Joint Collaborative Team on Video Coding (JCT-VC) have developed the next generation video coding standards, called High Efficiency Video Coding (HEVC). HEVC has evolved from the previous standards by adding more advanced features and higher-efficiency coding tools to improve compression ratio without affecting the image quality. HEVC is already capable of reducing bit rates by 50% in comparison to H.264/AVC, while still maintaining the same video quality level [1]. Obviously, HEVC will be widely used in the future. Currently, the number of video applications continues to increase. This wide range of applications requires various resolution levels, and imposes different types of resource limitations [2]. Obviously, HEVC are not capable of effectively meeting all these needs. In order to answer to various needs, the Joint Collaborative Team on Video Coding (JCTVC) is developing a new coding standard called SHVC as a scalable extension of HEVC. Although

---

* Corresponding author.

D.-S. Huang et al. (Eds.): ICIC 2014, LNCS 8588, pp. 693–699, 2014.
© Springer International Publishing Switzerland 2014

HEVC has very high coding efficiency, the coding complexity is also very high. The computational complexity of HEVC is about two to four times that of H.264/AVC [3]. Similarly, SHVC also has higher coding complexity. Therefore, it's very important to improve the coding speed to promote its adoptions. Since SHVC is a very new video coding standard, there has been very little related research work done, to date. Currently, there exist a number of fast algorithms for HEVC [4-6]. However, the difference between the coding procedures for HEVC and SHVC is very large. Therefore, it is very important to improve the coding speed of SHVC based on its specific features. In this paper, we focus on the Intra algorithm for EL of Quality SHVC. In order to improve the coding speed, the low modes and depths with low possibilities were excluded.

2   Mode Decision

The Intra encoding process for SHVC is the same as that for HEVC. In the rough mode decision (RMD) process, the first best N candidates are selected from all 35 modes. Subsequently the most probable modes (MPMs) which are obtained from the left side and upper prediction units (PUs) are adopted as candidate modes. Those candidates would then enter the rate distortion optimization (RDO) process for selection of the optimal mode.

Since there are 35 modes in every CU, there is a very high coding complexity. Modes with low possibilities can be excluded to improve the coding speed and maintain the coding efficiency. By first studying the relationships between the modes and their corresponding HC values, the candidate modes can then be predicted based on their relationships. Figure 1 shows the relationships between the modes and their corresponding HC values with FlowerVase sequence in a 416×240 format.

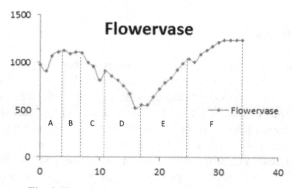

Fig. 1. The relationship between the modes and HC

In Figure 1, the horizontal ordinate represents modes, and the vertical ordinate represents the corresponding HC values. From Fig. 1, in each of the six regions A, B, C, D, E and F, there exists a LMP. Obviously, the smallest point should be found within these LMPs, and the best mode is very likely to be the vertical ordinate of the smallest point, so the search should only locate these LMPs. In every region, it was found that the left and right points were all monotonically decreasing when approaching its LMP. Experiments showed that there existed several regions in all 35 points of every CU in all the sequences, and in every region the left and right points were all monotonically decreasing while approaching its LMP. Therefore, the LMP within a region could be obtained by following the direction of descent of the HC values. If the HC value of a point was smaller than that of its left and right points, the point was the LMP of its region. In order to improve the coding speed and maintain the coding efficiency, the key was how to select the initial points in the search for LMPs. As is already known, relative CUs which include neighboring CUs and the co-located CU in base layer(BL) have similar modes with current CU, thus the modes of three relative CUs can be selected as the initial points, as shown in Figure 2.

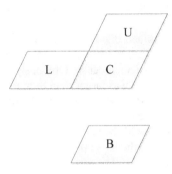

Fig. 2. Relative CUs

In Fig .2, C is the current MB, L and U are its neighboring MBs; B is the co-located MB of current MB in BL. Since there are only three initial points, and these points usually coincide with each other, the number of initial points seems to be too few. In order to maintain the coding efficiency, the even points whose horizontal ordinates were 2, 10, 18, 26, and 34 were also selected to be checked, and the point with the smallest HC was further selected as the initial point.

Initial points were obtained through the process described above. Duplicate points might exist, therefore in order to avoid a repeated search, redundant points were discarded. LMPs were obtained by following the descent direction of the HC values from the initial points. In addition, modes 0 and 1 had not yet been considered, thus 0 and 1 were also selected as LMPs, and checked to obtain their HC values. All LMPs were ordered from small to large, according to their HC values.

All LMPs were thus obtained. In order to test the effectiveness of this approach, the percentage of the best point of each set of LMPs was calculated. Extensive simulations were conducted using eight video sequences, with different resolutions. The test conditions were as follows: the Quantization Parameters (QPs) were 26 and 30 in the BL, and 22 and 26 in the EL, the Max CU Size was 64, the MaxPartDepth was 4, and the number of coded frames was 50. Table 1 shows the percentage of the best points from the LMPs, as shown below.

Table 1. The percentage of the best points from the LMPs

Resolutions	Sequences	Percentage (%)
416×240	BlowingBubbles	99
416×240	RaceHorses	99
832×480	BasketballDrillText	99
832×480	PartyScene	99
1280×720	Parkruner	100
1280×720	Town	99
1920×1088	Sunflower	100
1920×1088	Tractor	99
Average		99

From Table 1, it was determined that the percentage was 99%, indicating that the proposed approach was very effective. After the RMD process, the best point would be selected from LMPs in the RDO process. Obviously, the number of LMPs was between 2 and 5. When the PU size was larger than 8×8, the 3+MPM modes would be checked. If there were only 2 LMPs, we could save 1+MPM points to check. When the PU size was smaller than or equal to 8×8, 8+MPM modes should be checked, so we could save from 3+MPM to 6+MPM points to check. Obviously, the coding speed would be further improved during the RDO process.

3   Depth Decision

Scalable HEVC Test Model (SHM) usually allows the maximum CU size to equal 64, and the depth level range would be between 0 and 3. Since the frame resolutions in the BL and EL are the same, Inter-layer correlations between the BL and EL are very strong. We can predict the candidate depths of the current CU in EL based on that of the co-located CU in BL, excluding low possible depths to improve the coding speed. By exploiting the exhaustive depth decision in SHM3.0 which is from MPEG Website under the test conditions mentioned in Section 2, the distribution of the depths between a CU in EL and the co-located CU in BL is shown in Table 2.

Table 2. Distribution of depth levels between the EL's and BL's

PE / Sequences	B0				B1				B2				B3			
	E0	E1	E2	E3	E0	E1	E2	E3	E0	E1	E2	E3	E0	E1	E2	E3
Blowing	-	-	-	-	1	80	17	2	1	22	65	12	0	12	52	36
RaceHorses	-	-	-	-	1	70	23	6	1	23	59	18	1	17	46	36
BaskeballDrill	-	-	-	-	9	61	16	14	8	40	38	14	5	31	29	35
PartyScene	-	-	-	-	7	85	6	2	3	22	50	25	1	8	46	45
Parkruner	64	4	23	9	0	23	56	21	2	25	50	23	1	21	57	21
Town	79	17	3	1	44	39	10	7	3	23	41	33	1	20	33	46
Sunflower	68	24	6	2	61	29	9	1	47	37	13	3	23	34	15	29
Tractor	23	41	26	9	20	44	27	9	17	48	25	10	11	40	26	23
Average	59	22	15	5	18	54	21	8	10	30	43	17	5	23	38	34

In Table 2, B0, B1, B2 and B3 are the depths of 0, 1, 2 and 3 in BL, and E0, E1, E2 and E3 are the depths 0, 1, 2 and 3 in EL. PE represents a percentage of a depth in EL, and PB represents a percentage of a depth in   BL. Since PB with 0 is very small in the first four sequences, the corresponding PE has no statistical significance and was denoted as "-". For convenience, BH and EH represent a depth in BL and EL, respectively. From Table 2, we deduced the following observations for the depth distribution:

(1) If BH is smaller than 2, the percentage of the co-located CUs in EL that would select EH with 3 is very low, so it can be skipped.

(2) If BH is larger than 1, the percentage of the co-located CUs in EL that would select EH with 0 is very low, so it can be skipped.

(3) There also exist some PE with small percentages, such as EH with 2 under BH with 0; EH with 3 under BH with 2, and EH with 1 under BH with 3. Experiments show that by directly skipping these depths, the coding efficiency would obviously degrade. We can make further predictions based on the textural features, as measured by textural variances. A CU was first divided into four even sub-CUs, and then the variance of the CU was compared with the average variance of its four sub-CUs to determine whether to check or skip these depths. The procedure is as follows:

The variance of a CU is calculated as follows:

$$\sigma^2 = \frac{\sum_{i=0}^{n}(x_i - \overline{x})}{n-1} \tag{1}$$

Where σ^2 is the variance, x_i is the ith pixel value, $\overline{x}$ is the average of all pixel values, and n is the number of all pixels. Let σ_W^2 represent the variance of a whole CU, σ_1^2, σ_2^2, σ_3^2 and σ_4^2 represent the variance of its four sub-CUs, respectively. Average variances of the four sub-CUs denoted as σ_S^2 is:

$$\sigma_S^2 = \left(\sigma_1^2 + \sigma_2^2 + \sigma_3^2 + \sigma_4^2\right)\big/4 \tag{2}$$

We can check or skip these three depths based on the relationships between σ_W^2 and σ_S^2 as follows:

(a) For EH with 1 under BH with 3, if $\sigma_W \geq \sigma_S$, the texture is indicated to be very complex and this depth should be checked to maintain coding efficiency; otherwise, this depth should be skipped to improve the coding speed.

(b) For EH with 2 under BH with 0, when the parent depth which was EH with 1 under BH with 0 was checked, if $\sigma_W \geq \sigma_S$, then obviously it was better to split further, so EH with 2 under BH with 0 should be checked. Otherwise, the depth should be skipped. Similarly, for EH with 3 under BH with 2, when we check its parent depth, if $\sigma_W \geq \sigma_S$, the depth should be checked; otherwise, the depth should be skipped.

4   Experiment Results

In order to evaluate the performance of the proposed fast Intra prediction algorithm for Quality SHVC, we implemented on the recent SHVC reference software (SHM 3.0) with Main Profile. The performance of the proposed algorithm was as shown in Table 3. Experiments were carried out for all I-Frame sequences. In order to verify the effective of the proposed algorithm, we select five different resolutions sequences to test and each resolution includes two sequences. The Coding Treeblock has a fixed size of 64×64 pixels (for Luma), and a maximum depth level of 4, resulting in a minimum CU size of 8×8 pixels. The proposed algorithm was evaluated with QPs 24, 28, 32 and 36 in EL, and 28, 32, 36 and 40 in BL. Coding efficiency was measured with PSNR and bit rates, and computational complexity was measured by consumed coding time in EL. BDPSNR (dB) and BDBR (%) were used to represent the average PSNR and bitrate differences [7], and "TS (%)" was used to represent the percentage of coding time savings of EL. To the best of our knowledge, this is the first work on improving coding speed for intra quality SHVC, so we can only conduct performance comparisons with the reference software.

Table 3. Average performance results

Resolutions	Sequences	BDPSNR	BDBR	TS
2560x1600	PeopleOnStreet	-0.011	0.21	67.75
2560x1600	Traffic	-0.007	0.13	65.94
1920x1080	BasketballDrive	-0.004	0.14	59.32
1920x1080	Kimono	-0.015	0.34	52.91
832x480	Mobisode	-0.017	0.75	52.62
832x480	RaceHorses	-0.073	0.80	68.21
416x240	BasketballPass	-0.010	0.14	65.66
416x240	Keiba	-0.008	0.12	69.94
1280x720	Johnny	-0.020	0.50	53.63
1280x720	KristenAndSara	-0.010	0.17	57.12
Average		-0.018	0.33	61.31

Table 3 shows performances of the proposed fast Intra prediction algorithm. The proposed algorithm could greatly reduce coding time for all sequences. The proposed algorithm could reduce the coding time by an average of 61 %, with a maximum of 67.75 %, and with a minimum of 52.62 %. The average coding efficiency loss in terms of PSNR was about 0.0018 dB with a minimum of 0.007 dB and a maximum of 0.073 dB, while the average Bitrate was 0.33% with a minimum of 0.12% and a maximum of 0.80%. Therefore, the proposed algorithm could improve coding speed significantly with negligible coding efficiency losses.

5   Conclusion

In this paper, we proposed a fast mode and depth decision algorithm for Intra prediction to reduce the computational complexity of Quality SHVC. Based on the relationships between the modes and their corresponding HCs, only partial modes were checked; and based on inter-layer correlations and textural features, low possible depths were skipped. Experiments showed that the proposed algorithm could improve the coding speed significantly with negligible coding efficiency losses.

Acknowledgement. The work is supported by National Significant Science and Technology Projects of China under Grant No. 2013ZX01039001-002-003, National Natural Science Foundation of China Project under Grant No.61170253, Promotion Project of Shenzhen Key Laboratory on Information Science and Technology 2012, and China Postdoctoral Science Foundation (2013M 540953).

References

1. Corrêa, G., Assunção, P., Agostini, L., da Silva Cruz, L.A.: Performance and Computational Complexity Assessment of High-Efficiency Video Encoders. IEEE Transactions on Circuits and Systems for Video Technology 22(12), 1899–1909 (2012)
2. Shi, Z.B., Sun, X.Y., Xu, J.Z.: CGS Quality Scalability for HEVC. In: IEEE International Workshop on Multimedia Signal Processing, pp. 1–6 (2011)
3. Yan, S.Q., Hong, L., He, W.F., Wang, Q.: Group-Based Fast Mode Decision Algorithm for Intra Prediction in HEVC. In: IEEE International Conference on Signal Image Technology and Internet Based Systems, vol. (8), pp. 225–229 (2012)
4. Shen, L.Q., Zhang, Z.Y., An, P.: Fast CU Size Decision and Mode Decision Algorithm for HEVC Intra Coding. IEEE Transactions on Consumer Electronics 59(1), 207–213 (2013)
5. Shen, L.Q., Liu, Z., Zhang, X.P., Zhao, W.Q., Zhang, Z.Y.: An Effective CU Size Decision Method for HEVC Encoders. IEEE Transactions on Multimedia 15(2), 465–470 (2013)
6. Cho, S., Kim, M.: Fast CU Splitting and Pruning for Suboptimal CU Partitioning in HEVC Intra Coding. IEEE Transactions on Circuits and Systems for Video Technology 23(9), 1555–1564 (2013)
7. Bjontegaard, G.: Calculation of average PSNR difference between RD-curves. In: 13th VCEG-M33 Meeting, Austin, TX, April 2-4 (2001)

A New Local Binary Pattern in Texture Classification

Haibin Wei[1], Hao-Dong Zhu[2], Yong Gan[2], and Li Shang[3]

[1] College of Electronics and Information Engineering
Tongji University Shanghai, China
[2] School of Computer and Communication Engineering ZhengZhou University of Light Industry
Henan Zhengzhou, 450002, China
[3] Department of Communication Technology, College of Electronic Information Engineering,
Suzhou Vocational University, Suzhou 215104, Jiangsu, China

Abstract. The e LBPs extract local structure information by establishing a relationship between the central pixel and its adjacent pixels. However, it is very sensitive to the change of the central pixel .In this paper,we choose a circle with the radius of 1 instead of a single center. The method is proposed for texture classification by comparing the information between the neighbors and the new center pixel. In order to decrease the feature size and increase the classification accuracy, both LBC-like feature and CLBP feature were used in the proposed method .Experiments are carried out on Outex and UIUC databases. The experimental results demonstrate that this method perform effectively.

Keywords: CLBP, LBP, LBC, texture classification.

1 Introduction

The local binary pattern (LBP) descriptor, firstly proposed by Ojala et al [1], is a popular texture feature. It probes texture images by first comparing the greyscale values between a given pixel and its neighboring pixels. Since Ojala's work, Tan and Triggs [2] proposed Local Ternary Pattern (LTP) by extending original LBP to 3-valued codes. Recently, Guo et al [3] proposed the completed LBP (CLBP) by combining the conventional LBP with the measures of local intensity difference and central gray level. Zhao and Huang [4] proposed to use local binary count (LBC) to extract the local neighborhood distribution.

While traditional methods only encode the greyscale values between a given pixel and its neighboring pixels. These methods are sensitive to noise and the results were unsatisfactory. Motivated by the work of Guo et al. [3], Zhao and Huang [4], we proposed in this letter a new pattern to texture information by shaping a new center.This method captured more detailed information by exploring gray-scale properties between neighborhoods and more pixels. Experimental results illustrate that the supplement information can effectively increase the classification accuracy.

D.-S. Huang et al. (Eds.): ICIC 2014, LNCS 8588, pp. 700–705, 2014.

2   Brief Review of the LBP, LBC and the CLBP

2.1   Local Binary Pattern(LBP)

LBP has shown excellent performance in many comparative studies[5]. We can describe the algorithm of LBP in three main steps. 1) The values of neighbor pixels are turned to binary values by comparing them with the central pixel. 2) We encode the binary numbers and then the code is transformed into decimal number. 3) A histogram will be built to represent the texture image. The LBP coding strategy can be described as follows:

$$LBP_{P,R} = \sum_{p=0}^{P-1} s(g_p - g_c)2^P, \qquad s(x) = \begin{cases} 1, x \geq 0 \\ 0, x < 0 \end{cases} \tag{1}$$

where g_c represents the gray value of the center pixel and $g_p(p=0,\ldots,P-1)$ denotes the gray value of the neighbor pixel on a circle of radius R, and P is the total number of the neighbors.

2.2   Completed LBP (CLBP)

Guo et al [3] proposed CLBP descriptor to further improve the discrimination capability of LBP descriptor by decomposing the image local differences into two complementary components which named CLBP_S and CLBP_M where the CLBP_S is equivalent to the conventional LBP and the CLBP_M can be defined as follows:

$$CLBP\_M_{P,R} = \sum_{p=0}^{P-1} s(m_p - c)2^P, \qquad m_p = |g_p - g_c| \tag{2}$$

where g_p, g_c and s(x) are defined as in Eqn. (1), the threshold c is set as the mean value of m_p. Guo observed that the center pixel also has discriminative information. Thus, they defined an operator named CLBP_C to extract the local central information :

By combining the three operators of the CLBP_S, the CLBP_M and the CLBP_C, significant improvement is made for rotation invariant texture classification.

2.3   Local Bianary Count (LBC)

Zhao and Huang[4] proposed to use a local counting method to encode the local distribution after quantizing local neighbors into two values(0 or 1). In LBC method, they only count the number of value 1's in the binary neighbor sets.

The LBC reveals a cursory encoding method to characterize the local neighborhood distribution and the LBC-like encoding is easy to expand. LBC obtains the same performance as LBP in rotation invariant cases.

3   New Local Binary Pattern

Looking at Fig.1, if we hit the bull's eye, then we got out, rather than have to shoot to the center of the target.

Fig. 1. .Target   **Fig. 2.** (P=24,R=3)   **Fig.2.**(P=32,R=4)

Motivated by this, we shape a new center, a circle with radius of 1, As illustrated in Fig.2,We shape the new center with 9 pixels or 5 pixels, in other words, it is just 8 or 4 neighbors with a central pixel when the radius equals to 1.

Analyzing LBP from the view point of the local structures, we proposed a new local binary pattern in order to establish a relationship between the neighborhood pixels and the new center. Given a local structure (N=24, R=3), this means that the nearest neighborhood pixels also act as center pixel in the local region. Thus this method explores gray-scale properties between neighborhoods and the central pixel as well as gray-scale properties between neighborhoods and the nearest neighbors.

CLBP combined the three main componentsto to create a joint-histogram. In our method, The new center will generate the histogram as CLBP. For N = 24, the feature size of CLBP (S_MC) is (26+26*2)=78. Therefore, the proposed model will increase the feature size to 9*(26+26*2) =702 or 5*(26+26*2) =390.

As discussed in [4], the micro-structure was not absolutely invariant to rotation under the illumination changes. Therefore, [4] introduced the local binary count (LBC) by abandoning the micro-structure and obtained the same performance as LBP .

Motivated by CLBP and LBC, we applied this idea to our new local binary pattern . To decrease feature size, we can choose m pixels encoded in CLBP, and the rest of 9 or 5 central pixels are encoded in LBC. In this text we choose the original center encoded in CLBP and the 4 or 8 nearest neighbors in LBC. We create this complementary component named CLBP_L by summing the 'LBC' patterns. A simple example is conducted to demonstrate the proposed model In Fig.3

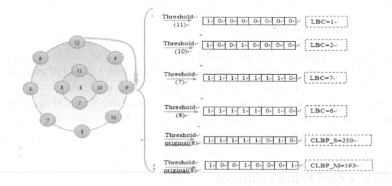

Fig. 3. (P=8,R=2); new center{8,7,10,11,8}

So CLBP_L=$\sum_{p=1}^{n} LBC(g_p)$ =16.

The final feature was a joint-histogram among CLBP_S, CLBP_M, CLBP_C, CLBP_L; such as CLBP_S/M/C_L or CLBP_S_M/C_L;

4   Experimental Results

To evaluate the effectiveness of the proposed method, we carried out a series of experiments on serval representative databases: the UIUC database [6] and Outex database [7].

4.1   The Methods in Comparison

In the experiments, we choose the original center encoded in CLBP and the 4 or 8 nearest neighbors in LBC, the NN classifier were used for all methods here.

4.2   Experimental Results on Outex Database

Table I lists the experimental results by different schemes. Under TC12, "T" represents the test setup of illuminant "t184" and "H" represents "horizon". (O=1,L=4/8) represents that the original center was encoded in CLBP and the 4 /8 nearest neighbors of the new center were encoded in LBC; We could make the following findings. First, CLBP_S_L achieves much better result than CLBP_S in most cases. It is in accordance with our analysis that the new center is more informative than the original center. Second, the comparing between the nearest neighbors and the neighborhood pixels contains additional discriminant information as CLBP_S_L could get much better results than CLBP_S, and CLBP_S/M/C_L gets better results than CLBP_S/M/C.

Table 1. The classification rates (%) on Outex database

	R=2,P=16				R=3,P=24			
	TC10	TC12/T	TC12/H	AVG	TC10	TC12/T	TC12/H	AVG
CLBP_S	89.40	82.26	75.20	82.28	95.07	85.04	80.78	86.96
CLBP_S_L(O=1,L=4)	92.86	86.32	82.57	87.25	96.77	89.24	85.90	90.64
CLBP_S_L(O=1,L=8)	93.20	86.92	82.36	87.45	96.38	88.80	86.78	90.65
CLBP_SM	97.89	90.55	91.11	93.18	99.32	93.58	93.35	95.41
CLBP_SM_L(O=1,L=4)	98.07	92.15	92.06	94.09	99.43	95.25	94.79	96.49
CLBP_SM_L(O=1,L=8)	98.15	92.04	92.08	94.09	99.30	95.05	94.70	96.35
CLBC_SMC	98.83	93.59	94.26	95.56	98.96	95.37	94.72	96.35
CLBP_SMC	98.72	93.54	93.91	95.39	98.93	95.32	94.53	96.26
CLBP_SMC_L(O=1,L=4)	98.80	94.26	94.28	95.78	99.35	96.44	95.58	97.12
CLBP_SMC_L(O=1,L=8)	98.80	94.51	94.10	95.80	99.30	96.30	95.46	97.02

4.3   Experimental Results on UIUC Database

The UIUC texture database contains materials imaged under significant viewpoint variations. In our experiment, *20* training images are randomly chosen from each class while the remaining 20 images are used as the test set. The average accuracy over 100 randomly splits are listed in Table 2. Similar findings to those in part B of section 4 can be found in Table 2. The classification rates were improved significantly with another component to supply LBP.

Table 2. The classification rates (%) on UIUC database

	R=1,P=8	R=2,P=16	R=3,P=24
CLBP_S	54.79	61.04	64.11
CLBP_S_L(O=1,L=4)	60.74	68.05	70.34
CLBP_S_L(O=1,L=8)	60.7	67.99	70.4
CLBP_SM	81.8	87.87	89.18
CLBP_SM_L(O=1,L=4)	**82.43**	**88.34**	**90.40**
CLBP_SM_L(O=1,L=8)	82.62	88.44	90.60
CLBC_SMC	87.64	91.04	91.2
CLBP_SMC	87.61	91.03	91.19
CLBP_SMC_L(O=1,L=4)	**88**	**91.75**	**92.31**
CLBP_SMC_L(O=1,L=8)	88.2	91.85	92.47

5   Conclusions

In this paper, we try to describe the local gray-level distribution by proposing a new center; A LBC-like feature is used in the proposed method. The experiment on two real-captured texture databases illustrated that this new center could effectively improve the classification rates.

Acknowledgment. The authors would like to sincerely thank T. Ojala and Z. Guo for sharing the source codes .This work is supported by the Science and Technology Innovation Outstanding Talent Plan Project of Henan Province of China under Professor De-Shuang Huang (No.134200510025), the National Natural Science Foundation of China (No. 61373105, 61373098 & 61133010).

References

1. Ojala, T., Pietikainen, M., Maenpaa, T.: Multiresolution gray-scale and rotation invariant texture classification with local binary patterns. IEEE Transactions on Pattern Analysis and Machine Intelligence 24(7), 971–987 (2002)
2. Tan, X.Y., Triggs, B.: Enhanced local texture feature sets for face recognition under difficult lighting conditions. IEEE Transactions on Image Processing 19(6), 1635–1650 (2010)
3. Guo, Z.H., Zhang, L., Zhang, D.: A completed modeling of local binary pattern operator for texture classification. IEEE Transactions on Image Processing 19(6), 1657–1663 (2010)
4. Zhao, Y., Huang, D.S., Jia, W.: Completed Local Binary Count for rotation invariant texture classification. IEEE transaction on Image Processing 21, 4492–4497 (2012)
5. Zhao, G., Pietikäinen, M.: Dynamic texture recognition using local binary patterns with an application to facial expressions. IEEE Trans. Pattern Anal. Mach. Intell. 29(6), 915–928 (2007)
6. http://www.cfar.umd.edu/~fer/High-resolution-data-base/hr_da tabase.htm
7. Ojala, T., Maenpaa, T., Pietikainen, M., Viertola, J., Kyllönen, J., Huovinen, S.: Outex - new framework for empirical evaluation of texture analysis algorithms. In: Proc. 16th Int. Conf. Pattern Recognit., vol. 1, pp. 701–706 (2002)

Improved Emotion Recognition
with Novel Task-Oriented Wavelet Packet Features

Yongming Huang[1,2], Guobao Zhang[1,2], Yue Li[1,2], and Ao Wu[1,2]

[1] School of Automation, Southeast University, Nanjing 210096, China
[2] Key Laboratory of Measurement
and Control of Complex Systems of Engineering, Ministry of Education

Abstract. In this paper, a wavelet packet based adaptive filter-bank construction method is proposed for speech signal processing. On this basis, a set of acoustic features are proposed for speech emotion recognition, namely Wavelet Packet Cepstral Coefficients (WPCC). The former extends the conventional Mel-Frequency Cepstral Coefficients (MFCC) by adapting the filter-bank structure according to the decision task; while the later aims at selecting the most crucial frequency bands where the most discriminative emotion information is located. Speech emotion recognition system is constructed with the two proposed feature sets and Gaussian mixture model as classifier. Experimental results on Berlin emotional speech database show that the proposed features improve emotion recognition performance over the conventional MFCC feature. The proposed feature extraction scheme has encouraging prospects since it can be extended to 2D image processing with 2D wavelet packets and hence extended to audio-visual bimodal emotion recognition application.

1 Instruction

Human-computer interface (HCI) of the new generation is expected to change from "machine centered" to "human centered" so that a more natural human-computer interaction can be achieved. In recent years, with the study on affective computing going deep, automatic emotion recognition is drawing wide attentions from research fields including psychology, linguistics, neuroscience and computer science as an Inter disciplinary subject [1].

In recent years, wavelet packets (WP) has emerged as an important signal analysis and representation scheme impacting compression, detection and classification[2, 3]. With the ability of generating a multi-resolution type of signal analysis [4], WP is drawing intense interest in speech processing as an powerful signal processing technique alternative to conventional Fourier transform. In this paper, the problem of constructing proper tree-structured WP basis that adapt the frequency partition solution to the emotion classification task is explored. A fast algorithm for tree pruning presented by Scott [5] is applied with different discrimination measure functionals adopted as tree pruning criteria. On this basis, novel WP-based acoustic features are proposed for speech emotion classification.

D.-S. Huang et al. (Eds.): ICIC 2014, LNCS 8588, pp. 706–714, 2014.

2   Wavelet Packets and Class-Separability Based Tree Pruning

Introduced by Coifman, Meyer, and Wickerhauser [6], the wavelet packets are known to offer flexible frequency segmentation and consequently the capability of complex signal analysis like speech processing. In this section, a brief introduction of the tree-structured wavelet packet is provided. For excellent expositions, readers are referred to [2, 6, 7]. And then a WP tree pruning scheme is proposed with class-separability based criteria to construct filter-bank structure suitable for the classification task.

2.1   Admissible Wavelet Packet Tree

Let $\mathbf{V}_J \in L^2(R)$ denote a finite resolution space for signals that are approximated at scale 2^{J_0}. A theorem given in [7] shows that $\mathbf{V}_{J_0}$ can be divided into two orthogonal sub-spaces using a pair of conjugate mirror filters $(h[n], g[n])$. By iterating the application of conjugate mirror filter pair, space V_{J_0} is recursively decomposed and the decomposition can be represented in a binary tree structure. Any node of the binary tree is labeled by (j, p), where $j \geq 0$ represents the depth of the node in the tree, and $p \in \{0, 1, ..., 2^j - 1\}$ is the number of nodes that are on its left at the same depth j. The sub-space associated to node (j, p) is denoted as W_j^p, and for the root node $(0 \cdot 0)$, we have $W_0^0 = V_{J_0}$.

A binary tree where each node has either zero or two children is called an admissible binary tree [7]. Let $\{(j_i, p_i)\}_{1 \leq i \leq I}$ denote the leaf nodes of an admissible binary tree, it has been proved that the corresponding sub-spaces $\{W_{j_i}^{p_i}\}_{1 \leq i \leq I}$ are mutually orthogonal and add up to $V_{J_0} = W_0^0$ [7]:

$$V_{J_0} = W_0^0 = \oplus_{i=1}^I W_{j_i}^{p_i} \tag{1}$$

2.2   Wavelet Packet Filter-Banks

Conjugate mirror filters $(h[n], g[n])$ play an important role in the WP decomposition. They represent the low-pass and high-pass filters respectively, and divide the frequency support of the wavelet packets into two [7]. By iterating the application of conjugate mirror filters in a particular way, a frequency axis partition manner is achieved in the form of corresponding admissible binary tree structure. Different admissible tree structures can be interpreted as different frequency partition solutions.

As presented in [3], from a filter-bank point of view, the process of projecting an original signal into the WP sub-spaces $\{W_{j_i}^{p_i}\}_{1 \leq i \leq I}$ is equivalent with an I-channel filter-bank followed by a set of aggregated down-samplers. This is also known as the wavelet packet's multi-rate filter-bank property [8].

For speech emotion recognition, our goal is to adapt the frequency partition manner according to the classification task, so that we can locate the emotion information in several specific frequency bands and acquire emotion-related acoustic features by multi-channel filtering with the filter-bank structure obtained. The issue of determining an optimal frequency partition solution is interpreted as the WP tree pruning problem and will be investigated below.

2.3   Frequency Ordering

Because of the side lobes of the conjugate mirror filter pair that are brought into interval $\left[-2^{-J_0}\pi, 2^{-J_0}\pi\right]$ by the dilation, the filter pair h and g do not always play the role of low-pass and high-pass filters, respectively [3, 7]. Actually, it has been proved in [7] that the frequency support of the basis of W_j^p is

$$I_j^q = [-(q+1)\pi 2^{-j}, -q\pi 2^{-j}] \cup [q\pi 2^{-j}, (q+1)\pi 2^{-j}] \tag{2}$$

with the permutation $q = G[p]$.

Frequency ordering is implemented by changing the position of node (j, p) to (j, p), and the resulting WP binary tree is frequency ordered with the left node corresponding to a lower-frequency wavelet packet while the right child to a higherfrequency one [7]. The permutation $q = G[p]$ is implemented on binary strings, if p_i is the ith binary digit of the integer p and q^i the ith digit of $q = G[p]$, then

$$q_{(i)} = (\sum_{c=i}^{+\infty} p_{(c)}) \bmod 2 \tag{3}$$

2.4   WP Tree Pruning Problem

For classification problems, adapting the WP filter-bank structure to the decision task is of particular significance for improving classification performance. For our speech emotion recognition task, choosing an optimal tree topology within the collection of tree-structured WP bases that best captures discriminative information fromoriginal signal space will help to give an insight into dominant frequency bands where the most significant emotion information is located and provide better comprehension of humans emotion perception mechanism.

Tree pruning algorithm is first and most widely used for classification and regresion trees (CART) [9], and then studied for more general binary tree pruning problem by Saito et al. [10], and more recently Scott [5]. In [10], a bottom-up search algorithm is proposed for the tree pruning problem, with additive discriminant measures used as cost functional. Before the algorithm is given, some notations are introduced first, following [3] and [11].

Referring to [5, 10], a bottom-up tree pruning algorithm with additive pruning criterion is presented in Algorithm 1. By applying the algorithm, a sequence of $\left\{\mathscr{T}^k\right\}_{k=1}^{|\mathscr{T}|}$ with different number of leaf nodes is constructed.

Algorithm 1 Fast tree pruning algorithm with additive discrimination measure

Initialization: Initial binary tree $\mathscr{T} = \mathscr{T}_{\text{full}}$
Iteration:
1: **for** $j = J$ to 0 **do**
2:    **for** $p = 2^j - 1$ to 0 **do**
3:       Set $v = (j, p)$
4:       Set $\mathscr{T}_v^1 = \{v\}$
5:       **if** v is not a terminal node **then**
6:          **for** $k = 2$ to $|\mathscr{T}_v|$ **do**
7:             Set MaxDiscrimination $= 0$
8:             **for** $s = \max(1, k - |\mathscr{T}_{r(v)}|)$ to $\min(|\mathscr{T}_{l(v)}|, k-1)$ **do**
9:                Set $t = k - s$
10:               Set Discrimination $= D(\mathscr{T}_{l(v)}^s) + D(\mathscr{T}_{r(v)}^t)$
11:               **if** Discrimination $>$ MaxDiscrimination **then**
12:                  Set MaxDiscrimination $=$ Discrimination
13:                  Set $\mathscr{T}_v^k = [v, \mathscr{T}_{l(v)}^s, \mathscr{T}_{r(v)}^t]$
14:               **end if**
15:            **end for**
16:         **end for**
17:      **end if**
18:   **end for**
19: **end for**
Output: The target trees $\mathscr{T}^k = \mathscr{T}_{\text{root}}^k, k = 1, 2, \ldots, |\mathscr{T}|$

---

3   Wavelet Packets and Class-Separability Based Tree Pruning

In this section, we describe the details of the proposed speech emotion recognition system with emphasis on the WP filter-bank based acoustic feature extraction.

3.1   Overview of the System

Framework of the proposed speech emotion recognition system is shown in Fig.1. The following subsections give detailed description of each part of the proposed system.

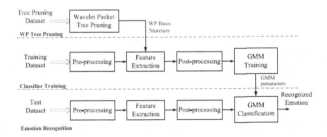

Fig. 1. Block diagram of the proposed system

3.2   Dataset Division

In this work, the whole emotional speech database is divided into three parts. The tree-pruning dataset, training dataset and test dataset are used for WP tree pruning, classifier training and emotion recognition respectively, and are not overlapped with each other.

The size of tree-pruning dataset may influence performance of the WP tree pruning procedure. The more adequate the dataset is, the more accurate class discriminative WP filter-bank structure will be achieved. However, the size of tree-pruning dataset is limited by the size of whole database. Therefore a proper dataset partition manner should be adopted to appropriately evaluate performance of the proposed system.

3.3   Wavelet Packet Tree Pruning

In this step, an optimal filter-bank structure is obtained using the tree pruning algorithm. The pruned WP tree structure represents a frequency band division method that is adapted to the speech emotion classification task. The optimalWP filter-bank is then applied on the training and test samples to calculate novel acoustic features. As mentioned above, the de-correlation ability of wavelet packet depends on the type and order of the conjugate mirror filters. In this paper, Daubechies filters of different orders are chosen and the Daubechies wavelets are calculated, which are orthogonal wavelets with compact support [2], and easy to put into practice.

With the fast tree pruning algorithm, a sequence of binary tree topologies with different number of leaf nodes is obtained, and correspondingly the set of filter-bank structures with different number of sub-bands. In this study, different filter-bank structures with various sub-band numbers are compared to find an optimal filterbank structure. The optimal filter-bank structure obtained fromWP tree pruning procedure is then used in the feature extraction step to capture emotion-discriminative information from original speech signal.

3.4   Pre- and Post-processing

Before we focus on the feature extraction scheme, some pre-processing and postprocessing techniques required for speech processing and feature optimization are given first.

3.4.1   Pre-processing
Before feature extraction, conventional speech signal processing operations including pre-emphasis, frame blocking and windowing are performed first.

Pre-emphasis.
The speech signal is first put through a pre-emphasis processor to spectrally flatten the signal and to make it less susceptible to finite precision effects later in the signal processing [12]. The most widely used pre-emphasis network is the first-order FIR filter with transfer function:

$$H(z) = 1 - \alpha z^{-1} \quad 0.9 \leq \alpha \leq 1.0 \tag{6}$$

In this paper $\alpha = 0.9375$ is used.

Frame blocking and windowing

The pre-emphasized speech signal is then blocked into frames of K samples with an overlap of K' samples between adjacent frames. Here we use $K = 256$ and $K' = K/2$. And each individual frame is multiplied by a Hamming window to reduce ripples in the spectrum [13].

3.4.2   Post-processing

In this paper, speaker normalization (SN) is investigated as post-processing operation applied on the features extracted, aiming at excluding the speaker identity information in speech emotion analysis. SN is realized through mean-subtracting and variance normalization of each feature within the scope of individual speaker, as performed in [14]. Since SN requires the system to be aware of the speaker information in advance, it is workable only in applications where speaker recognition is performed at first. Nevertheless, SN is still investigated here as an upper benchmark for ideal situations [14].

3.5   Feature Extraction

A type of wavelet packet based features are proposed for the speech emotion recognition task, namely the Wavelet Packet Cepstral Coefficients (WPCC).

The first feature set WPCC is implemented in a similar way to the conventional MFCC features. First, the windowed speech frame is passed through a filter-bank obtained from WP tree pruning step, and then the sub-band log-energy coefficients are calculated and frequency ordered, followed by the Discrete Cosine Transform (DCT) to de-correlate the log filter-bank energy. Finally, the first N_C DCT coefficients together with the log-energy of the frame are adopted, with the first and second order derivatives appended to the feature vector. The implementation of WPCC is illustrated in Fig.2.

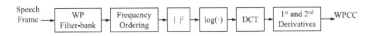

Fig. 2. WPCC feature extraction

3.6   Classifier

Gaussian mixture model (GMM) is adopted for speech emotion classification in this paper. For each emotion class a GMM is trained with the training dataset. The expectation-maximization (EM) algorithm used here for Gaussian mixture estimation was proposed by Figueiredo and Jain [15], which adjusts the number of components during estimation by annihilating components that are not supported by the data. Since the Figueiredo-Jain (FJ) algorithm determines the number of components by itself, the number of Gaussian components is not fixed and can vary among different classes. The implementation of the GMM classifier is provided by a publicly available Matlab toolbox named GMMBayes Matlab Toolbox [16].

4   Wavelet Packets and Class-Separability Based Tree Pruning

4.1   Emotional Speech Database and Experimental Setup

The proposed feature sets were evaluated on the Berlin emotional speech database[17], The data is available at a sampling rate of 16kHz with 16 bit resolution in wav format. In this paper, six emotions (no disgust) with a sum of 489 utterances were used for the classification task.

20% of the database was randomly selected to form the tree-pruning dataset, and we applied 5-fold cross validation on the remaining 80% utterances to assess the classification performance, i.e. the experiment is speaker-dependent. The maximum WP decomposition level was set to $J = 5$, and the Daubechies wavelets of order 10, 20, 30 and 40 were chosen to generate wavelet packets.

4.2   Experiments with the WPCC Feature Set

The effect of taking speaker normalization step after feature extraction was also investigated, where features were first mean-subtracted and variance-normalized for each speaker individually before data partitioning for cross validation. Daubechies filter of order 40 was used to generate wavelet packets. Experimentswere conducted with the number of Cepstral coefficients fixed to 12 (i.e. feature dimension is 39) and the number of sub-bands varied from 15 to 25. Classification result is shown in Fig.3. From Fig.3, it is noticeable that by performing speaker normalization, the classification accuracy has been improved considerably.

It is also worth noting that theWP filter-bank structureswith sub-band number 17 and 21 generally perform better among different feature configurations. This result is consistent for different tree pruning criteria and filter orders. The corresponding frequency partition solution gives us an insight into the emotion-related information distribution along frequency axis and provides us with better knowledge of humans emotion perception mechanism through acoustic signal.

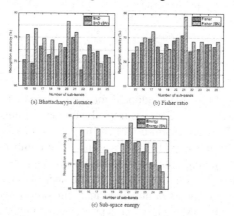

(a) Bhattacharyya distance   (b) Fisher ratio

(c) Sub-space energy

Fig. 3. Six-class emotion classification accuracies of WPCC with different tree pruning criteria as a function of sub-band number

A highest classification accuracy of 79.34% is achieved for the proposedWPCC feature set, with Daubechies filter of order 40, Fisher ratio as tree pruning criterion, 21-channel filter-bank 12 Cepstral coefficients and speaker normalization applied.

For the WPCC feature with the best classification performance, the corresponding pruned wavelet packet tree structure and frequency partition solution of [0,8kHz] is illustrated in Fig.4.

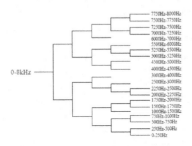

Fig. 4. Optimal WP tree structure for WPCC feature

5   Conclusion and Future Work

In this paper we explored the wavelet packet based acoustic feature extraction approach for speech emotion recognition.Wavelet packet tree pruning algorithm was applied with three types of discrimination based pruning criteria to adapt the WP filter-bank structure to the decision task.

A type of short-time acoustic features were proposed based on the obtained WP filter-bank structure, namely Wavelet Packet Cepstral Coefficients (WPCC). The WPCC feature is calculated following the conventional Mel-frequency Cepstral analysis paradigm. Finally, a speech emotion recognition system was built and experiments were carried out on the Berlin emotional speech database to evaluate the proposed feature sets. Extensive experiments were conducted with a set of filterbank structures and different feature dimensions adopted and compared. The benefit of performing speaker normalization was also investigated. Experimental results demonstrate the superiority of the proposed feature set over conventional MFCC feature.

Future work also includes investigating effectiveWP tree pruning scheme which, for example, not only takes the discriminating ability as tree pruning criterion, but also has the size ofWP tree controlled in a reasonable range. Apart from this, seeking for robust feature representation is also considered as part of the ongoing research, as well as efficient classification techniques for automatic speech emotion recognition.

Acknowledgements. This work was supported by open Fund of the Key Laboratory of Measurement and Control of CSE (School of Automation, Southeast University), Ministry of Education(no. MCCSE2013B03)

References

1. Marlow, C., Naaman, M., Boyd, D., Davis, M.: Position paper, tagging, taxonomy, flickr, article, toread. In: Collaborative Web Tagging Workshop at WWW 2006, Edinburgh, Scotland (2006)
2. Mika, P.: Ontologies are us: A unified model of social networks and semantics. In: Gil, Y., Motta, E., Benjamins, V.R., Musen, M.A. (eds.) ISWC 2005. LNCS, vol. 3729, pp. 522–536. Springer, Heidelberg (2005)
3. Wu, X., Zhang, L., Yu, Y.: Exploring social annotations for the semantic web. In: WWW 2006, pp. 417–426 (2006)
4. Resnick, P., Varian, H.R.: Recommender Systems. Communications of the ACM 40(3), 56–58 (1997)
5. Nakamoto, R., Nakajima, S., Miyazaki, J., Uemura, S.: Tag-based contextual collaborative filtering. IAENG International Journal of Computer Science 34(2), 214–219 (2007)
6. Fellbaum, C.: WordNet: An Electronic Lexical Database. MIT Press (1998)
7. Banerjee, S., Pedersen, T.: An adapted lesk algorithm for word sense disambiguation using wordNet. In: Gelbukh, A. (ed.) CICLing 2002. LNCS, vol. 2276, pp. 136–145. Springer, Heidelberg (2002)
8. Locoro, A.: Tagging ontologies with fuzzy wordNet domains. In: Petrosino, A. (ed.) WILF 2011. LNCS (LNAI), vol. 6857, pp. 107–114. Springer, Heidelberg (2011)
9. Dang, H., Lin, J., Kelly, D.: Overview of the TREC 2006 question answering track. In: Proc. of TREC 2006 (2006)
10. Song, Y., Zhuang, Z., Li, H., Zhao, Q., Li, J., Lee, W.-C., Giles, C.L.: Real-time automatic tag recommendation. In: SIGIR 2008, pp. 515–522 (2008)
11. Zeng, Z.H., Pantic, M., Roisman, G.I., Huang, T.S.: A survey of affect recognition methods: Audio, visual, and spontaneous expressions (2009)
12. Daubechies, I.: Ten Lectures on Wavelets. Society for Industrial and Applied Mathematics, Philadelphia (1992)
13. Martin, V., Jelena, K.: Wavelet and Subband Coding. Prentice Hall PTR, Englewood Cliffs (1995)
14. Conde, J.M., Vallet, D., Castells, P.: Inferring user intent in web search by exploiting social annotations. In: Proc. of International ACM SIGIR Conference on Research and Development in Information Retrieval, pp. 827–828 (2010)
15. Goldberg, D., Nichols, D., Oki, B., Terry, D.: Using collaborative filtering to weave an information tapestry. Communications of the ACM (1992)
16. Vallet, D., Cantador, I., Jose, J.M.: Personalizing web search with folksonomy-based user and document profiles. In: Gurrin, C., He, Y., Kazai, G., Kruschwitz, U., Little, S., Roelleke, T., Rüger, S., van Rijsbergen, K. (eds.) ECIR 2010. LNCS, vol. 5993, pp. 420–431. Springer, Heidelberg (2010)
17. Xu, S., Bao, S., Fei, B., Su, Z., Yu, Y.: Exploring folksonomy for personalized search. In: Proc. of International ACM SIGIR Conference on Research and Development in Information Retrieval, pp. 155–162 (2008)

Pedestrian Detection Based on HOG and LBP

Wen-Juan Pei[1], Yu-Lan Zhang[2], Yan Zhang[1], and Chun-Hou Zheng[1,*]

[1] College of Electrical Engineering and Automation, Anhui University, Hefei, China
zhengch99@126.com
[2] College of Jia Sixie Agriculture, Weifang University of Science and Technoloty,
Shouguang, China

Abstract. In this paper, we present a feature extraction approach for pedestrian detection by extracting the sparse representation of histograms of oriented gradients (HOG) feature and local binary pattern (LBP) feature using K-SVD. Moreover, we use PCA to reduce the dimension of HOG and LBP. We combine the low dimension principal features with the sparse representations of HOG feature directly for fast pedestrian detection from images. In addition, we compare the performance of sparse representations and PCA based features. Experimental results on INRIA databases show that the proposed approach provides a better detection result and spends less time.

Keywords: Pedestrian Detection, Local Binary Patterns, Histogram of Oriented, Sparse Representation, K-SVD.

1 Introduction

Pedestrian detection is an important branch of pattern recognition which can be applied in monitoring, intelligent machines, aircraft, image retrieval, smart cameras [7] etc. The locally normalized HOG (histograms of oriented gradients) [1] descriptor is one of the most popular feature extraction methods. In this literature, the HOG feature vectors are extracted in a grid of overlapping spatial blocks and then collected over the detection windows, finally fed into a linear Support Vector Machine (SVM) for pedestrian detection. The Local Binary Pattern (LBP) texture descriptors [10] has been widely used as texture classification and face recognition for its invariance to gray level changes and good anti-noise performance.

In recent years, plenty of improved LBP features have been proposed [4, 8]. E.g., Wang et al. [6] propose a texture descriptor (HOG-LBP features) which combines edge information (HOG) and the texture information (cell-structured LBP) for human detection. Local Structured HOG (LSHOG), Local Structured LBP (LSLBP) and the boosted Local Structures HOG-LBP [3] are proposed by Zhang et al. for object localization.

---

* Corresponding author.

D.-S. Huang et al. (Eds.): ICIC 2014, LNCS 8588, pp. 715–720, 2014.

Sparse representation method has been extensively used in object recognition and detection. The dictionary updating method of K-SVD [12] algorithm can be used to better represent the data. Histograms of Spare Codes (HSC) [11] extracts the local histograms by aggregating sparse codes based on K-SVD of each pixel.In this paper, we extract the sparse feature of HOG feature and LBP feature using dictionary learned from K-SVD. Then we use the lib-SVM (Support Vector Machine) [2] to classify the features respectively. The combined HOG-LBP [6] feature is used as comparison. Then we combine K-SVD-HOG and PCA-HOG which reduce the dimension of the feature set and time consumption yet without sacrificing detection performance significantly. Experiment on PCA-HOG & PCA-LBP shows the similarity performance.

2   Our Implementation

PCA is a widely used method for dimension reduction. In [5], the authors use PCA to reduce the HOG feature vector dimensions while keeping the recognition rate essentially the same and gain a practical filter tracking. Inspired by them, we firstly survey the effects of combining the PCA with the LBP and HOG features respectively, thus obtaining the low dimensional feature of PCA-HOG and PCA-LBP. The LBP and the HOG features will be used as the comparisons group respectively. Secondly, we apply the efficient K-SVD algorithm [12] to produce the sparse representation of the HOG and LBP features set separately by learning the dictionary. We denoted them as K-SVD-HOG and K-SVD-LBP. Our experiment processing is shown in figure 1.

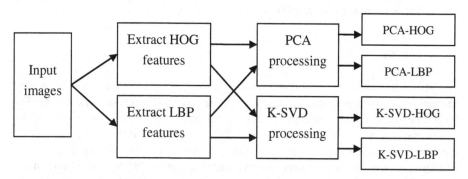

Fig. 1. The overview of our experiment(without HOG and LBP contrasting procedure)

2.1   Dimension Reduction by PCA

In our implementation, we follow the procedure in [1] and [6] to extract the traditional HOG and the cell-structured LBP from the dataset at first. We use the notion $A = [a_1, a_2, ..., a_M] \in R^{N \times M}$ to describe the HOG feature or the LBP feature of the training samples, and $B = [b_1, b_2, ..., b_L] \in R^{N \times K}$ for the testing samples (where M is the number of training samples, K is the number of testing samples, N is the feature dimension). PCA is the simplest of the true eigenvector-based multivariate analyses. PCA dimension reduction methods can be concisely described as:

$$A = WH \tag{1}$$

where $W \in R^{N \times J}$. $H \in R^{J \times M}$ is the reduced feature matrix.

2.2 Sparse Representation Using K-SVD

Due to the purpose of K-SVD is to find the common direction of energy distribution of the dataset, the feature set of spare representation mapped by the over-completed matrix can be used for classification and compression, etc. In this study, firstly, we establish the over-completed dictionary from a portion of training feature set, and then produce projection coefficients for training and testing set in the sparse space spanned by the dictionary. After we get the original (HOG and LBP) feature $A = [a_1, a_2, ..., a_M] \in R^{N \times M}$ of the training set ($B = [b_1, b_2, ..., b_K] \in R^{N \times K}$ denoting the testing set feature). We randomly select M^+ positive feature set to train the dictionary. Given the feature set (HOG or LBP) $A^+ = [a_1, a_2, ..., a_{M^+}] \in R^{N \times M^+}$ of the training set, thus the number of dictionary elements to train is M^+, the iterative implementation K-SVD try to find the best dictionary $D \in R^{N \times M^+}$ by solving :

$$\min_{D,S} \{ \|A^+ - DS\|_F^2 \} \quad \text{s.t.} \quad \forall_i, \|s_i\|_0 \le L_0. \tag{2}$$

Then the sparse representations of original feature set can be simply presented as:

$$\hat{A} = A^T D = [\hat{a}_1, \hat{a}_2, ..., \hat{a}_M] \in R^{M^+ \times M} \tag{3}$$

$$\hat{B} = B^T D = [\hat{b}_1, \hat{b}_2, ..., \hat{b}_K] \in R^{M^+ \times K} \tag{4}$$

where $\hat{a}_i$ and $\hat{b}_i$ are the sparse coding coefficients.

2.3 Combination of the Feature Set

As is well known, PCA pursues keeping the maximize data information after dimension reduction. By calculateing the variance of the data on the projection direction, it can measure the importance of the direction. But the classification distinction between the data and the projection data is often disagreement. On the contrary, the data points may be mixed together and indistinguishable. K-SVD uses an over-complete dictionary to describe the data by sparse linear combinations and could be applied for compression and feature extraction, etc.

By combining PCA and K-SVD, we can make full use of the advantages of the two methods. Thus, we proposed the K-SVD-HOG & PCA-HOG feature, which is the join combination of K-SVD-HOG and PCA-HOG, to keep the advantage of texture feature and the gradients feature. We do experiments on the K-SVD-HOG & K-SVD-LBP and PCA-HOG & PCA-LBP features, which are combined directly.

3   Experiment Results

3.1   The Experimental Dataset

In this paper, we perform the experiments on the INRIA database. We collected 10 negative feature vectors randomly from each person-free images with a fixed 64×128 window. Additional retraining process by searching exhaustively for hard examples is time-consuming and slightly improving the performance. Classification performances are evaluated on 1126 positive and 4530 negative patches.

In our experiment, the widely used lib-SVM (support vector machine) [2] is trained as the classifier. To quantify the classification performance of the detection algorithm, we choose the Detection Error Tradeoff (DET) curves (a log-log scale, miss rate versus FPPW (False Positives Per Window)) [1] to evaluate the method.

3.2   The Combined HOG Features of Different Projections

The results of K-SVD-HOG are shown in figure 2. Firstly, we calculate the dictionary with different number (500, 800, 1000, 1200) of random positive training features through 40 iterations while keep the total error within the scope of 0.065. The performance of K-SVD-HOG with different dimension are not obvious distinguished.

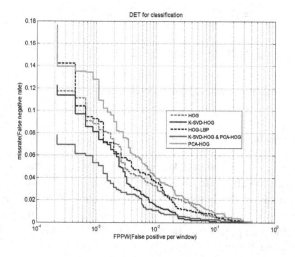

Fig. 2. The DET curves of K-SVD-HOG and PCA-HOG.

In this experiment, the dictionary size is 800. To compare with K-SVD-HOG, the dimension of PCA-HOG is selected as 800. Figure 2 shows that K-SVD-HOG and PCA-HOG can not improve the performance compared with the original HOG feature. We then investigate whether the simple combined K-SVD-HOG & PCA-HOG can performance better. It is inspiring to see that error detection rate of the combined feature greatly decrease and the performance is even better than HOG-LBP feature.

3.3   The Combined Features of Texture and Gradients

In this section, we research the ability of combined feature with the same projection of the texture feature and the gradients feature. We compare the performances of the combined K-SVD-HOG & K-SVD-LBP feature and PCA-HOG & PCA-LBP feature. The results are shown in figure 3. From this experiment it can be found that the performance of PCA-HOG & PCA-LBP feature is best.

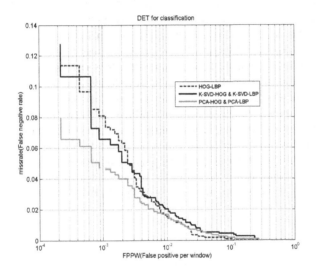

Fig. 3. Miss rate under varying combined features

4   Conclusions

In this paper, we demonstrated the improvement to the well known features, i.e., HOG and LBP, by utilizing K-SVD and PCA to extracted new features and combining them for pedestrian detection. We test our methods on INRIA, the experimental results show that our combined algorithms (KSVD-HOG & PCA-HOG and PCA-HOG & PCA-LBP) are effective and outperform the combined original HOG-LBP feature. In summary, by effective dictionary learning, PCA based feature extraction and effective feature combination, we may construct new features with richer structures than the original combined HOG-LBP.

720 W.-J. Pei et al.

Acknowledgments. This work was supported by the National Science Foundation of China under Grant No. 61272339, and the Key Project of Anhui Educational Committee, under Grant No. KJ2012A005.

References

1. Dalal, N., Triggers, B.: Histograms of Oriented Gradients for Human Detection. In: IEEE Conf. CVPR, vol. 1, pp. 886–893 (2005)
2. Chang, C.C., Lin, C.J.: LIBSVM: A Library for Support Vector Machines. ACM Trans. Intelligent Systems and Technology 27, 1–27 (2011)
3. Zhang, J., Huang, K., Yu, Y., Tan, T.: Boosted Local Structured HOG-LBP for Object Localization. In: IEEE Conf. PAMI, pp. 1393–1400 (2011)
4. Mu, Y., Yan, S., Liu, Y., Huang, T., Zhou, B.: Discriminative Local Binary Patterns for Human Detection in Personal Album. In: IEEE Conf. CVPR, pp. 1–8 (2008)
5. Lu, W.L., Little, J.J.: Simultaneous Tracking and Action Recognition using the PCA-HOG Descriptor. In: The 3rd Canadian Conf. Computer and Robot Vision, p. 6 (2006)
6. Wang, X., Han, T.X., Yan, S.: An HOG-LBP Human Detector with Partial Occlusion Handling. In: IEEE Conf. Computer Vision, pp. 32–39 (2009)
7. Dollár, P., Wojek, C., Schiele, B., Perona, P.: Pedestrian Detection: An Evaluation of the State of the Art. IEEE Trans. PAMI 34, 743–761 (2011)
8. Ahonen, T., Hadid, A., Pietikäinen, M.: Face Recognition with Local Binary Patterns: Application to Face Recognition. IEEE Trans. PAMI 28, 2037–2041 (2006)
9. Zhao, G., Pietikäinen, M.: Dynamic Texture Recognition using Local Binary Patterns with an Application to Facial Expressions. IEEE Trans. PAMI 27, 915–928 (2007)
10. Ojala, T., Pietikainen, M., Maenpaa, T.: Multiresolution Gray-Scale and Rotation Invariant Texture Classification with Local Binary Patterns. IEEE Trans. CVPR 24, 971–987 (2002)
11. Ren, X., Ramanan, D.: Histograms of Sparse Codes for Object Detection. In: IEEE Conf. CVPR (2013)
12. Aharon, M., Elad, M., Bruckstein, A.: K-SVD: An Algorithm for Designing Overcomplete Dictionaries for Sparse Representation. IEEE Trans. Signal Processing 54, 4311–4322 (2006)

A Novel Salient Line Detection Approach Based on Ant Colony Optimization Algorithm

Xutang Zhang[1], Yan Liu[2,*], Yinling Ma[1], Hong Liu[3], and Ting Zhuang[1]

[1]School of Mechatronics Engineering, Harbin Institute of Technology, China
zxt@hit.edu.cn
[2] College of Science, Harbin Institute of Technology, China
snacly@163.com
[3] Department of Foundation, Harbin Finance University, China
jrxyliuhong@163.com

Abstract. In this paper, we propose a novel salient line detection algorithm for 3-dimensional (3D) mesh models based on ant colony optimization algorithm. In proposed method, the heuristic function in ACO transition function is improved based on tensor feature and the relationship between mean curvature information via vertex geometric feature. In addition, the proposed method establishes a similarity measurement criterion of feature points by combining the normal tensor voting with discrete mean curvature. Moreover, a direction control function is utilized to conduct the detection process to a expect direction, which can achieve an interaction between operator and computer. The experimental results show that the proposed method can extract the salient line effectively and precisely.

Keywords: salient line detection, 3-dimensional mesh models, ant colony optimization.

1 Introduction

With the precision rapid growth of 3D data collection equipment and geometric models with high precision are despised to pursuit in graph applications, geometric 3D models are more and more used in computer graphics and virtual reality. Due to its simplicity and efficiency, triangular mesh has become one of the most effective and popular used approaches to describe 3D model. In the process of 3D model analysis, the features such as ridges, ravines, corners and boundaries etc, can provide important shape information and frequently require special treatment to achieve high accuracy and reliability, which have became an important part in model processing and understanding.

Feature detection takes an important position in 3D model processing and application, it can serve as a pre-processing technique which seeks and extracts features for various applications such as feature –preserving mesh simplification, 3D morphing

D.-S. Huang et al. (Eds.): ICIC 2014, LNCS 8588, pp. 721–728, 2014.

ensuring rigid correspondences between features of meshes, mesh enrichment striving to smooth out roughness of meshes [1]. Several feature detect algorithms have been proposed to detect different features such as point, line, surface, etc [2-5]. For the core of feature detection, salient line detection is an essential and important step in the research and application of mesh model, which can be used to partition measured data. Recently, several research efforts have shown growingly popularity of salient line detection methods based on different frameworks. However, it is still a challenge to extract the salient line with complicate structure properly. On one hand, existing salient line detection methods usually model the boundary features by considering the triangular mesh vertex vector information. Vertex vector information is helpful to detect the boundary of 3D mesh model with relative regular structure. For some 3D mesh models with complicated structures (e.g. 3D tooth mesh model), considering the vertex vector information only may get a poor result. One the other hand, most of previous works mainly focused on computing the triangular mesh vertex vector via curvature analysis. Since triangular mesh vertex vector includes a variety of information, computing the curvature information only is inefficient to reflect the triangular mesh vertex vector precisely.

In this paper, a novel feature detection line approach based on ant colony optimization (ACO) algorithm is proposed for detecting features on 3D triangular meshes [6]. Inspired by the idea of optimal path in ACO, the heuristic function of proposed method is improved based on the tensor feature and the relationship between mean curvature information via vertex geometric features [7-10]. In addition, the proposed method establishes the similarity measurement criterion of feature points by combining the normal tensor voting with discrete mean curvature. Moreover, a direction control function is utilized to conduct the detection process to a expect direction, which can achieve an interaction between operator and computer. The experimental results show that the proposed method can detect the salient line effectively and precisely.

The rest of this paper is organized as follows. Section 2 describes the related works in the area. Section 3 we briefly introduce the salient line detection method based on ACO algorithm. The results are given in Section 4 and Section 5 concludes the paper.

2   Related Work

The ant colony algorithm is a population-based search technique for the finding the optimal paths based on the behavior of ants searching for food [6]. At the beginning, each ant wanders randomly. From Fig. 1(a) it is seen that, when an ant finds a source of food, it walks back to the colony leaving "markers" (pheromones) that show the path has food. When other ants come across the markers, they are likely to follow the path with a certain probability. If they do, they then populate the path with their own markers as they bring the food back. As more ants find the path, as shown in Fig. 1(b), it gets stronger until there are a couple streams of ants traveling to various food sources near the colony.

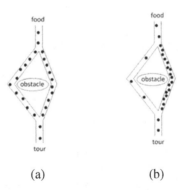

Fig. 1. The work flow of ACO

In ACO algorithm, each ant is a simple computational agent. At each time step, it iteratively constructs a solution for the problem at hand. The intermediate solutions are referred to as solution states. At each iteration of the algorithm, each ant moves from a state x to state y, corresponding to a more complete intermediate solution. Thus, each ant computes a set of feasible expansions to its current state in each iteration, and moves to one of these in probability. For ant k, the probability p_{xy}^k of moving from state x to state y depends on the combination of two values. The attractiveness η_{xy} of the move, as computed by some heuristic indicating the prior desirability of that move and the trail level τ_{xy} of the move, indicating how proficient it has been in the past to make that particular move. The trail level represents a posteriori indication of the desirability of that move. Trails are updated usually when all ants have completed their solution, increasing or decreasing the level of trails corresponding to moves that were part of "good" or "bad" solutions, respectively.

3   The Proposed Method

The 3D triangular mesh model can be defined as a triples $M = (V, T, E)$, where $V = \{v_0, v_1, \cdots, v_n\}$ denotes the set of vertices, $T = \{t_1, t_2, \cdots, t_m\}$ is a set of triangles. Each vertex $v \in V$ is respresented using Cartesian coordinates, denoted by $v = (v_i, v_j, v_k)$. Then the triangular face $t \in T$ can be defined by the vertices $t = \{(v_i, v_j, v_k) \mid v_i, v_j, v_k \in V, v_i \neq v_j \neq v_k\}$. $E = \{(v_i, v_j) \mid v_i, v_j \in V, v_i \neq v_j\}$ denotes the set of edges. Let $w \in V$ be the 1-ring neighbor point of vertex $v \in V$, and $t \in T$ be the 1-ring neighbor trangle.

In ACO algorithm, the heuristic function is computed by the reciprocal of distance between point i and point j : $\eta_{i,j} = 1 / d_{i,j}$. Since the function of ACO is to find the optimal path between two points, when the key points are distributed on a non-shortest path, the result may be not optimal. For solving the problem, we integrated the tensor feature into the ACO heuristic function, which can consider the similarity between two points more precisely:

$$\eta(i,j) = w_e FE_{i,j} + w_c FC_{i,j} \tag{1}$$

where $FE_{i,j}$ denotes the relationship of feature values, $FC_{i,j}$ denotes the average curvature values between point i and point j.

$$\mu(i,j) = w_{f_1} FD_1 + w_{f_2} FD_2 \tag{2}$$

In addition, to ensure that the ants can reach the expected target point, in this paper, a direction control function is proposed by considering the relationship between the current point i, the next point j and target expected target point tp:

$$\mu(i,j) = w_{f_1} FD_1 + w_{f_2} FD_2 \tag{3}$$

where FD_1 is the angle between vector (i,j) and vector (i,tp). $w_e, w_c, w_{f_1}, w_{f_2}$ is weight of each impact factor, respectively.

After visiting the vertex i, the next vertex j to be visited by the ant k can be selected from the 1-ring neighbor point sets $N_k(i)$ according to the state transition probability $p_{i,j}^k$:

$$p_{i,j}^k = \begin{cases} \dfrac{(\tau_{i,j})^\alpha (\eta_{i,j})^\beta (\mu_{i,j})^\gamma}{\sum\limits_{g \in N_k(i)} (\tau_{i,g})^\alpha (\eta_{i,g})^\beta (\mu_{i,g})^\gamma} , & j \in N_k(i) \\ 0 & \text{el se} \end{cases} \tag{4}$$

where $\tau_{i,j}$ is the number of pheromone deposited for transition, η_{ij} is the desirability of state transition from i to j, respectively. $\alpha \geq 0$ is the pheromone heuristic factor to control the influence of $\tau_{i,j}$, $\beta \geq 1$ is the similarity heuristic factor to control the influence of $\eta_{i,j}$ and γ is the direction heuristic factor to control the influence of direction.

When all the ants have completed a solution, the trails are updated by:

$$\tau_{i,j} \leftarrow (1-\rho)\tau_{i,j} + \sum_k \Delta\tau_{i,j}^k \tag{5}$$

In Eq. (5), ρ is the pheromone evaporation coefficient and $\Delta\tau_{xy}^k$ is the amount of pheromone deposited by k th ant, typically given for a Travel Saleman problem (TSP):

$$\Delta \tau_{xy}^{k} = \begin{cases} \lambda / L_{k} & \text{if ant } k \text{ uses curve } xy \text{ in its tour} \\ 0 & \text{otherwise} \end{cases} \tag{6}$$

where L_{k} is the cost of the k th ant's tour and λ is a constant parameter.

The algorithm flow of the proposed method is shown in table 1.

Table 1. The algorithm flow of the proposed method

```
Salient line detection
begin
    Initialize the parameters: α、β、γ、ρ ,Tao(n, n)
    Compute the feature value of each vertex: λ(n,3)
    Compute the curvature value of each vertex: K_H(n)
    While( iteration numbers NC<= NCmax)
    {
    Tabu {m,1}, Taboo table of m ants
    %Ant path search
        For (ant k=1 to m)
        {
        Tabu{k,1}=BeginPoint
            While (Tabu{k,1}(end)~=EndPoint)
            {
            Visited=Tabu{k,1}
            Can_visit=Neigh(Visited(end))-Visited
            For (j=1 to h )
                {
                        Compute the similarity function, direction function
                        Compute the state transition probabilities: P(j)
                }
                    To_visit=(P>rand)
                    Tabu{k,1}=[Tabu{k,1}, To_visit]
                }
        }
        % Pheromone update
        For(ant k=1 to m)
        {
            Route=Tabu{k,1}
            Compute Lk(Route)
            Compute Pheromone Incremental: Delta_Tao(Route)
        }
        Tao=(1-Rou)*Tao+Delta_Tao
    end
end
```

4   Results

In the experiment, a fandisk 3D triangular mesh model (including 6475 vertices, 12946 triangular faces) is firstly utilized to evaluate the performance of the proposed method. For the parameters of ACO, the heuristic factors are set to $\alpha=1$, $\beta=2$, $\gamma=0.5$, $\rho=0.5$. The number of ants k is set to 20 and the iteration number NC max is set to 100. The parameters of similarity function and direction control function are set to $\omega_e=1$, $\omega_c=0.8$, $\omega_{f_1}=0.5$, $\omega_{f_2}=0.8$. In order to validate the effectiveness of the proposed method comprehensively, we divided the selection of start point and end point into three categories: two points on a smooth curve, two points on a coplanar line and two points on a non-coplanar line. The detection results of three types of salient lines are shown in Fig. 2.

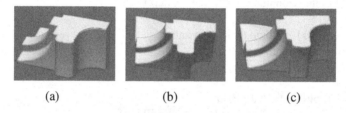

(a)   (b)   (c)

Fig. 2. The detection results of three types of salient lines on a 3D triangular mesh model: (a) two points on a smooth curve (b) two points on a coplanar line (c) two points on a non-coplanar line

From Fig. 2 it is seen that the proposed method can detect the salient lines between different types of two points well. However, for some cases, the curves on some model areas are closed, which is difficult to specify the start point and end point. By employing the proposed method, the experimental results are shown in Fig. 3. From the detection result it is seen that the proposed method is also useful in handling the cases with closed curves.

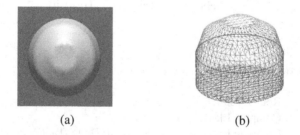

(a)   (b)

Fig. 3. The detection results of a salient line with closed curves

To further evaluate the validity of the proposed approach, a series of samples with poor quality meshes and blunt margins are utilized in the experiment. The detection results of the proposed method are shown in Fig. 4.

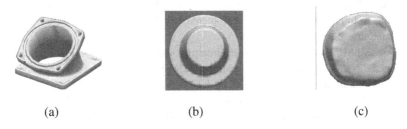

(a)   (b)   (c)

Fig. 4. The detection results of samples with poor quality meshes and blunt margins

The above experimental results show that the proposed method can handle the samples with poor quality meshes and blunt margins well. The proposed method produces a good result for the following reasons: (1) Based on iteratively search approach of ACO framework, the proposed method can adjust the weights of average curvature and angle direction adaptively and avoid the error produced by the one-off calculation. (2) The proposed method integrated both global and local features into the heuristic function, which can plan the optimal path effectively and accurately. (3) The proposed method calculate the 1-ring neighbors of vertex by average curvature and voronoi method, which can consider the impact of irregular triangular mesh model. Therefore, the proposed method can handle the cases with complicate structures.

5   Conclusion

In this paper, a novel salient line detection approach based on ACO algorithm is proposed. The proposed approach re-estimates heuristic function in ACO state transition probabilities the by integrating the tensor feature and mean curvature feature. In addition, the proposed method proposed a direction control function in basis of heuristic function, which can ensure the ants to reach the expected target point. In the experiment, a series of 3D mesh models are applied to validate the performance of proposed approach. Experimental results show that the proposed method can detect the salient line accurately and effectively, especially when the structures of 3D mesh models are complicate.

References

1. Golder, S.A., Huberman, B.A.: Usage patterns of collaborative tagging systems. J. Inf. Sci. 32(2), 198–208 (2006)
2. Quintarelli, E.: Folksonomies: power to the people. In: ISKO Italy-UniMIB Meeting (2005)
3. Bao, S., Xue, G., Wu, X., Yu, Y., Fei, B., Su, Z.: Optimizing web search using social annotations. In: WWW 2007, pp. 501–510 (2007)
4. Zhou, D., Bian, J., Zheng, S., Zha, H., Giles, C.L.: Exploring social annotations for information retrieval. In: WWW 2008, pp. 715–724 (2008)

5. Xu, S., Bao, S., Fei, B., Su, Z., Yu, Y.: Exploring folksonomy for personalized search. In: Proc. of International ACM SIGIR Conference on Research and Development in Information Retrieval, pp. 155–162 (2008)

6. Conde, J.M., Vallet, D., Castells, P.: Inferring user intent in web search by exploiting social annotations. In: Proc. of International ACM SIGIR Conference on Research and Development in Information Retrieval, pp. 827–828 (2010)

7. Vallet, D., Cantador, I., Jose, J.M.: Personalizing web search with folksonomy-based user and document profiles. In: Gurrin, C., He, Y., Kazai, G., Kruschwitz, U., Little, S., Roelleke, T., Rüger, S., van Rijsbergen, K. (eds.) ECIR 2010. LNCS, vol. 5993, pp. 420–431. Springer, Heidelberg (2010)

8. Guo, J., Cheng, X., Xu, G., Shen, H.: A structured approach to query recommendation with social annotation data. In: Proc. of the ACM Conference on Information and Knowledge Management, pp. 619–628 (2010)

9. Heymann, P., Ramage, D., Garcia-Molina, H.: Social tag prediction. In: SIGIR 2008: Proceedings of the 31st Annual International ACM SIGIR Conference on Research and Development in Information Retrieval, pp. 531–538 (2008)

10. Lu, Y.-T., Yu, S.-I., Chang, T.-C., Hsu, J.Y.: A content-based method to enhance tag recommendation. In: Proc. of IJCAI 2009, pp. 2064–2069 (2009)

Error Tracing and Analysis
of Vision Measurement System

Yu Song[1], Fei Wang[1], Sheng Gao[2], Haiwei Yang[1], and Yongjian He[3]

[1] Xi'an Jiaotong University, Xi'an, Shaanxi Province, China
[2] Beijing Institute of Spacecraft System Engineering, Beijing, China
[3] Xi'an Communication Institute, Xi'an, China
{songyuxjtu,yanghw.2005}@stu.xjtu.edu.cn, wfx@mail.xjtu.edu.cn,
castgaosheng@163.com, heshifu@gmail.com

Abstract. Due to the important role that accurate pose estimation plays in the field of vision measurement, a careful tracing and analysis of errors introduced in all aspects is necessary. In this paper, we launch an in-depth and meticulous research to seek for and analyze factors causing pose errors. These factors contain camera intrinsic parameters, 3D world coordinates of feature points, corresponding 2D image coordinates and different pose estimation algorithms. They are usually ignored in practical applications, leading to inaccurate pose. We develop mathematical tools to construct error models to estimate the actual error occurring in the implementation of a pose estimation algorithm. Our central idea is to make only one variable different at a time by controlling various variables. Then the effect of that single factor can be determined. Afterwards our approach is applied to ample synthetic experiments. The relationship between the influencing factors and pose estimation error is established and provides reference and basis to control errors and achieve more accurate pose.

Keywords: Vision Measurement, Error Tracing, Error Analysis, Pose Estimation.

1   Introduction

Computer vision is an emerging discipline. With the rapid development of computer hardware, software, image acquisition and processing techniques, the theories and techniques of computer vision have been widely used in medical image processing, robotics, character recognition, industrial inspection, military reconnaissance, geographical surveying and mapping, on-site measurement and so on. The application of computer vision in a variety of measurement is a quantitative analysis system with special accuracy requirements.

Behrooz Kamgar-Parsi [1] developed the mathematical tools to compute the average or expected error due to quantization. They derived the analytic expression for the probability density of error distribution of a function with an arbitrarily large number of independently quantized variables. Sunil Kopparapu [2] derived analytically the behavior of the camera calibration matrix under noisy conditions and further showed

D.-S. Huang et al. (Eds.): ICIC 2014, LNCS 8588, pp. 729–740, 2014.
© Springer International Publishing Switzerland 2014

that the elements of the camera calibration matrix have a Gaussian distribution if the noise introduced into the measurement system is Gaussian. Jeffrey Rodriguez [3] derived the probability density function of the range estimation error and the expected value of the range error magnitude in terms of the various design parameters. Hao Yingming and Zhu Feng [4], studied the relationship between the error of P3P pose estimation methods and the input parameters from the view of engineering application under the condition that three control points forming an isosceles triangle.

We launch an in-depth and meticulous research to seek for factors causing errors in vision measurement process. These factors are usually ignored in practical applications. We seek out all the possible influencing factors in the whole process, containing intrinsic parameters in camera calibration, 3D world coordinates of feature points, corresponding 2D image coordinates and different algorithms of pose estimation. The accuracy of camera calibration result, namely the intrinsic parameters, affects the final measurement result directly. In this paper we use Matlab Camera calibration toolbox [7] based on Zhang Zhengyou's calibration technique [6] to calibrate a camera. 3D world coordinates and 2D image coordinates of feature points are also directly involved in pose estimation. In addition, the accuracy of 3D points mainly relies on the design rationality of the visual landmark and the precision of the surveying instruments. And the accuracy of 2D points is connected with the imaging quality of optical imaging system. Different algorithms of pose estimation have respective advantages and weaknesses. We choose six widely used algorithms, namely linear algorithms including EPnP [8, 9], Fiore [10], Ansar [11], DLT (direct linear transform) [5, 16], and nonlinear algorithms including Lu [12] and SoftPOSIT [13]. We will discuss them detailedly in part 3.4.

In this paper, firstly we explain the camera optical imaging model systematically, then introduce all possible influencing factors and establish error models, finally conduct ample experiments to validate the theoretical analysis. Various variables are under control and only one variable is different at a time, so that the effect of that single factor can be determined. Afterwards our approach is applied to ample synthetic experiments and satisfying results are obtained. The relationship between the influencing factors and pose estimation error is established through theoretical derivation and experimental verification effectively. As a consequence, this paper provides reference and basis for error controlling during vision measuring process subsequently, and more precise pose can be achieved.

2   Camera Model

The pinhole camera model is used in this paper. In this model, a scene view is formed by projecting 3D points onto the image plane using a perspective projection transformation. The model describes the mathematical relationship between the coordinates of a 3D point and its projection onto the image plane, which is given by

$$s\mathbf{p} = \mathbf{A} \cdot (\mathbf{R} \cdot \mathbf{P}_w + \mathbf{t}), \text{ with } \mathbf{A} = \begin{bmatrix} f_x & 0 & u_0 \\ 0 & f_y & v_0 \\ 0 & 0 & 1 \end{bmatrix}, \tag{1}$$

where s is the scale factor, $\mathbf{R}$ and $\mathbf{t}$ are extrinsic parameters, i.e. pose, representing rotation matrix and translation vector respectively, $\mathbf{P}_w = (X_w, Y_w, Z_w)^T$ are coordinates of a 3D point in world coordinate system, $\mathbf{p} = (u, v)^T$ are coordinates of the projection point in pixels, and $\mathbf{A}$ is the matrix of intrinsic parameters, with (u_0, v_0) the principal point and (f_x, f_y) the equivalent focal lengths expressed in pixels.

(u, v) mentioned above are coordinates of the projection point in pixels, and (x, y) are its coordinates in millimeters. The relationship between them is given by

$$\begin{bmatrix} u \\ v \\ 1 \end{bmatrix} = \begin{bmatrix} 1/dx & 0 & u_0 \\ 0 & 1/dy & v_0 \\ 0 & 0 & 1 \end{bmatrix} \begin{bmatrix} x \\ y \\ 1 \end{bmatrix}, \tag{2}$$

where the physical sizes of every pixel in X and Y axis direction are dx and dy. $\mathbf{P}_w$ are defined above, and $\mathbf{P}_c = (X_c, Y_c, Z_c)^T$ are the coordinates of a 3D point in camera coordinate system. The relationship between these two coordinate systems is given by

$$\mathbf{P}_c = \mathbf{R} \cdot \mathbf{P}_w + \mathbf{t}. \tag{3}$$

The projection equation of the pinhole camera model is given by

$$s \begin{bmatrix} x \\ y \\ 1 \end{bmatrix} = \begin{bmatrix} f & 0 & 0 \\ 0 & f & 0 \\ 0 & 0 & 1 \end{bmatrix} \cdot \mathbf{P}_c, \tag{4}$$

where the scale factor s is equal to Z_c. From (2), (3), (4), the projection from 3D world points to 2D image points is established as shown in (1).

3   Error Trace and Analysis

3.1   Intrinsic Parameters

Camera acts as the original image acquisition device and should be calibrated in advance in vision measurement. Camera calibration is the first and a significant step, whose accuracy influences the accuracy of the final measurement result directly.

According to (1) and (3), we obtain $s\mathbf{p} = \mathbf{A} \cdot \mathbf{P}_c$, which indicates that the intrinsic parameters affect the accuracy of the 3D points in camera coordinate system, when the 2D image points $\mathbf{p}$ are acquired precisely. Then we have

$$X_c = \frac{u - u_0}{f_x} Z_c \tag{5}$$

$$Y_c = \frac{v - v_0}{f_y} Z_c \tag{6}$$

The Taylor expansions of (5) at true values of f_x and u_0 respectively are

$$X_c = \overline{X}_c + \frac{(-1)(u - u_0)Z_c}{\overline{f}_x^2}(f_x - \overline{f}_x) + \frac{(-1)^2(u - u_0)Z_c}{\overline{f}_x^3}(f_x - \overline{f}_x)^2 + \cdots$$
$$+ \frac{(-1)^n(u - u_0)Z_c}{\overline{f}_x^{n+1}}(f_x - \overline{f}_x)^n + R_n(f_x) \tag{7}$$

$$X_c = \overline{X}_c + \frac{Z_c}{\overline{f}_x}(u_0 - \overline{u}_0), \tag{8}$$

where variables with overline are true values of the corresponding ones, and R_n is the remainder of the Taylor formula. The calibration result f_x is far greater than the error $(f_x - \overline{f}_x)$. Therefore we can ignore those high order terms in the Taylor expansion (7). It can be obtained that, the error of $\mathbf{P_c}$ has inverse relationship with the error of the equivalent focal lengths approximately, when the principal point coordinates are constant values. From (8), the error of $\mathbf{P_c}$ is linear with respect to the error of principal point strictly, when the equivalent focal lengths are constant.

It also applies to (6) to be expanded at f_y and v_0 respectively.

3.2   3D World Coordinates

3D world coordinates of feature points are directly involved in pose estimation. The accuracy of them mainly relies on the design rationality of the visual landmark and the precision of the surveying instruments.

Visual Landmark. Visual landmarks are used to calculate the instantaneous pose of cooperative target. Design methods of visual landmarks can be summarized as follows.

1. Appearance: Circular points are used generally, whose diameters should not be too small under the precondition of clear imaging.
2. Color: Colors with high contrast such as black and white are recommended as colors of the mark points and background.
3. Material: It is recommended that light absorbing material is applied to black patterns to decrease the reflectance of visible light spectrum as far as possible so that the mark points can be clearly observed in strong light.

4. Number of points: It is determined by pose algorithm but should not too many in order to reduce the risk of misrecognition.
5. Distribution: The designer had better not arrange all mark points on the same plane to avoid ambiguity of pose estimation.

Surveying Instrument. It is impossible to obtain true value in reality, so the average value of several measurements is treated as true value. We use two Leica TM5100 theodolites with high precision (precision: 0.5 ") to measure three-dimensional coordinates of mark points in world coordinate system through optical triangulation method. The measurement error is given by

$$\Delta = \frac{2\pi D\lambda}{60 \times 60 \times 360},\qquad(9)$$

where D is the distance between the surveying instrument and the landmark to be measured, and λ is the precision of the surveying instrument (unit: second). In fact the disturbance of operator's factors on aiming at the centers of mark points and vibration of instrument should be taken into consideration.

3.3   2D Image Coordinates

The accuracy of 2D image coordinates mainly relies on image quality. The boundary a circular becomes unsmooth due to image discretization as shown in Fig 1. Imaging under long distance will cause that circles in the image occupy very small areas. The smaller the area is, the less it looks like a circle. It is extremely easy to cause inaccurate centers detected by centroid method under uneven illumination. In order to evaluate the imaging effect of circle point, circular degree is introduced to describe the similarity between a circle's image and a perfect circle. Circular degree is defined by

$$\psi = \frac{4\pi A}{l^2},\qquad(10)$$

where A and l are the area and perimeter of a circle respectively. The circular degree of a perfect circle is 1 while that of other planar closed figure is less than 1.

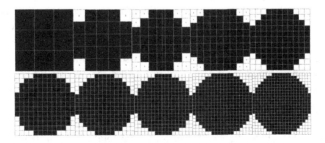

Fig. 1. Circle images with different diameters

Table 1. Circular degree

	Diameter of Circle (pix)	Circular Degree
1	3	0.7854
2	5	0.8465
3	7	0.8554
4	9	0.8843
5	11	0.8918
6	13	0.9079
7	15	0.9188
8	17	0.9088
9	19	0.9318
10	21	0.9342

According to table 1, the circular degree tends to be stable when the diameter of a circle increases to a certain extent. The diameter of a circle projected on the image had better not be less than 13 pixels in order to acquire an approximate perfect circle.

In real scene, as the illumination direction cannot be guaranteed perpendicular to the landmark, the image gray value of point close to light source is larger than that far from light source. It cause ambiguity at the border during image binaryzation, namely point centers shift along the direction of light. There is no effective solution to compensate this kind of error up to now. Thus we should carry out vision measurement under uniform illumination as far as possible.

3.4   Algorithms of Pose Estimation

In this paper we compare six widely used algorithms, namely linear algorithms including EPnP, Fiore, Ansar, DLT (direct linear transform), and nonlinear algorithms including Lu and SoftPOSIT.

EPnP [8, 9] is a linear method with high accuracy and low computation complexity. It is applicable for both planar and non-planar configurations whose number of feature points is no less than 4. It expresses 3D points as a weighted sum of four virtual control points. Estimating the coordinates of control points in camera reference can be done by expressing them as weighted sum of eigenvectors and picking right weights through quadratic equations.

Fiore et al [10] proposed that orthogonal decompositions were used to first isolate the unknown depths of feature points in the camera reference frame, allowing the problem to be reduced to an absolute orientation with scale problem, which is solved using the SVD. Fiore's algorithm needs 6 points at least.

Ansar et al [11] proposed an algorithm for pose estimation from n points or n lines. The solution is developed from a general procedure for linearizing quadratic systems. It fails in those cases where there are multiple discrete solutions caused by world objects lying in a critical configuration.

For a larger number of points, a well-known approximate method is the direct linear transform (DLT) [5, 16]. It solves for camera rotation and translation by ignoring orthonormality constraints on the rotation matrix, leading to a set of linear equations.

Lu et al [12] reformulated the pose estimation problem as minimizing an object space collinearity error in an iterative approach. This method relies on a good initial guess to converge to the correct solution. There is no guarantee that the algorithm will eventually converge or that it will converge to the correct solution.

The SoftPOSIT [13] algorithm combines the iterative softassign algorithm [14] for computing correspondences and the iterative POSIT algorithm [15] for computing object pose under a full perspective camera model. It does not have to hypothesize small sets of matches and then verify the remaining image points. Instead, all possible matches are treated identically throughout the search for an optimal pose.

4   Experiments

Following we use the controlling variable method to analyze the relationship between the pose error and the influencing factors.

In the simulation experiment, firstly we produce synthetic 3D to 2D correspondences in a 640×480 image acquired using a virtual calibrated camera with an equivalent focal length of 800 and a principal point at (320, 240). Then we generated the 3D word points which are uniformly distributed in a cube. Given the true rotation and translation, we obtain 2D image points by a perspective projection. Next we calculate pose by intrinsic parameters, 3D points and 2D points using different algorithms.

The reprojection error is defined by

$$E_{Rep} = \frac{1}{n}\sum_{i=1}^{n} \| \mathbf{A}(\mathbf{R}' \cdot \mathbf{P}_w^i + \mathbf{t}'), \mathbf{p}_i \|^2 , \tag{11}$$

where $\mathbf{R}'$, $\mathbf{t}'$ are calculated rotation matrix and translation vector, $\mathbf{p}_i$ are true values of image points. And the error of the rotation matrix $\mathbf{R}$ is represented by the corresponding angle error $\mathbf{q}$, which is defined by

$$E_q = \| \mathbf{q}' - \mathbf{q} \|, \tag{12}$$

and the error of the translation vector $\mathbf{t}$ is defined by

$$E_t = \| \mathbf{t}' - \mathbf{t} \|, \tag{13}$$

where $\mathbf{t}'$, $\mathbf{q}'$ are calculated translation vector and Euler angle respectively, and $\mathbf{t}$, $\mathbf{q}$ are corresponding true values.

4.1   Intrinsic Parameters

Random deviations distributed in the interval [−10, 10] with a step of 0.2 are generated. Then they are added to the equivalent focal lengths (f_x, f_y), and the principal point (u_0, v_0). We calculate the camera pose using the intrinsic parameters attached with deviations and compute errors of $\mathbf{R}$ (represented by Euler angle $\mathbf{q}$) and $\mathbf{t}$.

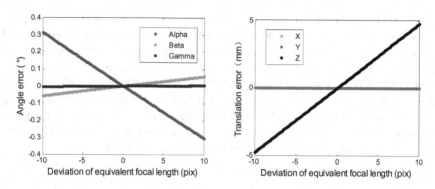

Fig. 2. Pose errors caused by deviations of the equivalent focal length

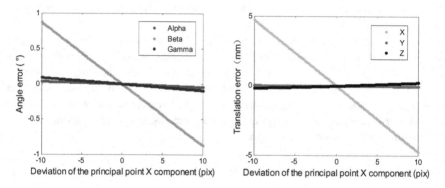

Fig. 3. Pose errors caused by deviations of the principal point X component u_0

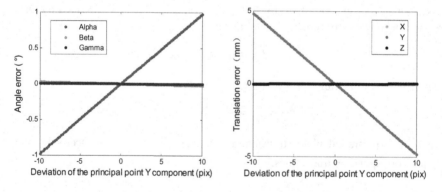

Fig. 4. Pose errors caused by deviations of the principal point Y component v_0

As shown in figures above, errors of all pose components are linear with the deviations. Moreover the focal length has a great effect on α and z, the principal point in X axis effects β and x, and the principal point in Y axis effects α and y.

4.2 World Coordinates

The influence of 3D world coordinates is displayed by the following simulation. Firstly we calculate the camera pose using the original size of the landmark, then add deviations to the true data, then recalculate pose, finally derive the errors of **R** (represented by Euler angle **q**) and **t**. The deviations increase from 0 to 0.5 with a step of 0.01 in millimeters.

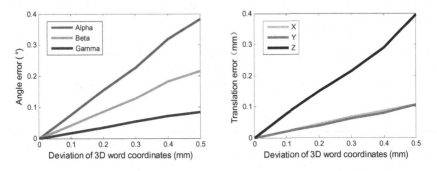

Fig. 5. Pose errors caused by 3D world coordinates

As shown in Fig 5, pose errors increase with the growth of deviations, satisfying an approximate linear relationship. The accuracy of the surveying instrument has the same effect as the accuracy of visual landmarks, because both of them introduce deviations into the 3D world coordinates of feature points.

4.3 Image Coordinates

All image feature points are added with random deviations in the same random direction in the range of 360 degrees. The sizes of deviations are distributed in the interval [0, 5].

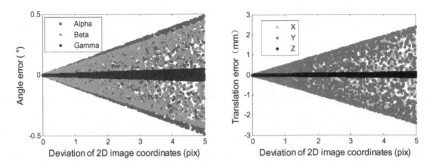

Fig. 6. Pose errors caused by 2D image coordinates

Fig 6 displays the enveloping lines of pose errors caused by image points. If we attempt to control the error less than a value, we have to guarantee the accuracy of image point coordinates larger than the corresponding value.

4.4   Algorithms

We compare the reprojection error, angle error, time cost of the six algorithms and the influence of the number of feature points and different levels of noise on pose calculation result.

All the plots discussed in this section were created by running 300 independent MATLAB simulations. To estimate running times, we ran the code 300 times for each algorithm and recorded the average run time in second. We also added Gaussian noise to the corresponding 2D image points to test robustness of each algorithm.

Table 2. Simulation Results

		R			t	E_{Rep}	E_q	Time
	True Value	0.8107	0.4694	0.3499	-12			
		0.5852	0.6336	0.5060	-7	-	-	-
		0.0158	0.6149	0.7884	5			
Linear	EPnP	0.8270	0.4204	0.3732	11.6887			
		0.5607	0.6643	0.4943	6.9165	7.801	0.0634	0.0097
		0.0401	0.6181	0.7851	4.5858			
	Fiore	0.8283	0.3668	0.4235	11.8018			
		0.5508	0.6713	0.4959	7.0564	5.489	0.0111	0.0187
		0.1025	0.6440	0.7581	5.4400			
	Ansar	0.8217	0.4383	0.3642	11.6899			
		0.5699	0.6323	0.5248	6.9239	8.906	0.0392	0.0287
		0.0002	0.6388	0.7693	5.0052			
	DLT	0.7649	0.3102	0.5646	10.8884			
		0.5029	0.7845	0.3628	8.3953	59.509	0.3656	0.0028
		0.3304	0.5614	0.7560	5.4983			
Nonlinear	Lu	0.8214	0.4516	0.3483	11.6404			
		0.5703	0.6418	0.5127	6.8669	5.632	0.0208	0.0236
		0.0080	0.6197	0.7848	4.3516			
	Soft POSIT	0.8257	0.4462	0.3451	11.4101			
		0.5640	0.6418	0.5196	6.6489	12.852	0.0286	0.0150
		0.0104	0.6236	0.7816	4.2929			

As shown in Fig 7, the pose errors decrease with the growth of the number of feature points used to estimate pose. While errors increase with the growth of Gaussian noise added to image points in Fig 8. In addition, Lu's method relies on a good initial guess to converge to the correct solution so that it is not stable. According to all simulation results, EPnP has the best performance compared with other algorithms in aspects of accuracy and computation complexity.

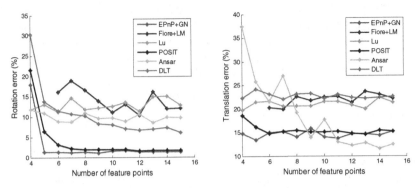

Fig. 7. Influence of number of feature points on pose error

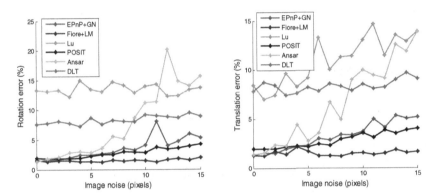

Fig. 8. Influence of noise level on pose error

5   Conclusion

In this paper we launch an in-depth and meticulous research to seek for and analyze factors causing errors in vision measurement process. These factors are usually ignored in practical applications and contain the accuracy of intrinsic parameters in camera calibration, 3D world coordinates of feature points, corresponding 2D image coordinates, and different algorithms of pose estimation.

Theoretical analysis is carried out by controlling various variables and only one variable is different at a time to determine the effect of that single factor. This paper constructs error models for every parameter. Afterwards our approach is applied to a large number of synthetic experiments, and satisfying results are obtained. Theoretical analysis and experiments validate the performance of our approach. The relationship between the influencing factors and pose estimation error is established through theoretical derivation and experimental verification effectively. As a consequence, this paper provides reference and basis for error controlling during vision measuring process subsequently to achieve more precise pose.

Acknowledgements. The authors would like to thank Vincent Lepetit for providing the code of the EPnP algorithm as well as Fei Wang and Haiwei Yang for their encouragement, discussion and assistance. This work was supported by National Natural Science Foundation of China (61273366) and National High Technology Research and Development Program of China (2013AA014601).

References

1. Kamgar-Parsi, B.: Evaluation of quantization error in computer vision. IEEE Transactions on Pattern Analysis and Machine Intelligence (1989)
2. Kopparapu, S.K., Corke, P.: The effect of measurement noise on intrinsic camera calibration parameters. In: Proceedings of IEEE International Conference on Robotics and Automation (1999)
3. Rodriguez, J.J., Aggarwal, J.K.: Quantization error in stereo imaging. In: Proceedings of the Computer Society Conference on Computer Vision and Pattern Recognition, CVPR 1988. IEEE (1988)
4. Hao, Y.M., Zhu, F.: Error analysis of P3P pose estimation. Computer Engineering and Applications (2008)
5. Ma, S.D., Zhang, Z.Y.: Computer vision – computation theory and algorithm foundation. Science Press, Beijing (1998)
6. Zhang, Z.Y.: A flexible new technique for camera calibration. IEEE Transactions on Pattern Analysis and Machine Intelligence 22(11), 1330–1334 (2000)
7. Camera calibration toolbox for matlab [CP/OL], http://www.vision.caltech.edu
8. Lepetit, V., Noguer, M.F., Fua, P.: EPnP: An Accurate O(n) Solution to the PnP Problem. International Journal of Computer Vision 2000, 155–166 (2000)
9. Noguer, M.F., Lepetit, V., Fua, P.: Accurate Non-Iterative O(n) Solution to the PnP Problem. In: Proceedings of International Conference on Computer Vision (ICCV 2007), pp. 1–8 (2007)
10. Fiore, P.D.: Efficient Linear Solution of Exterior Orientation. IEEE Transactions on Pattern Analysis and Machine Intelligence 23(2), 140–148 (2001)
11. Ansar, A., Danniilidis, K.: Linear Pose Estimation from Points or Lines. IEEE Transactions on Pattern Analysis and Machine Intelligence 25(5), 578–589 (2003)
12. Lu, C., Hager, G.D.: Fast and Globally Convergent Pose Estimation from Video Images. IEEE Transactions on Pattern Analysis and Machine Intelligence 22(6), 610–622
13. DeMenthon, D., David, P., Samet, H.: SoftPOSIT: An algorithm for registration of 3D models to noisy perspective images combining softassign and POSIT. University of Maryland, College Park, MD, Report CS-TR-969, CS-TR 4257 (2001)
14. Gold, S., Rangarajan, A.: A graduated assignment algorithm for graph matching. IEEE Transactions on Pattern Analysis and Machine Intelligence 18(4), 377–388 (1996)
15. DeMenthon, D., Davis, L.S.: Model-based object pose in 25 lines of code. International Journal of Computer Vision (1995)
16. Hartley, R., Zisserman, A.: Multiple view geometry in computer vision, vol. 15(1/2), pp. 123–141. Cambridge University Press, Cambridge (2004)

Online Model-Based Twin Parametric-Margin Support Vector Machine[*]

Xinjun Peng[1,2,**], Lingyan Kong[1], Dongjing Chen[1], and Dong Xu[1,2]

[1] Department of Mathematics, Shanghai Normal University, 200234, P.R. China
[2] Scientific Computing Key Laboratory of Shanghai Universities, 200234, P.R. China
xjpeng@shnu.edu.cn

Abstract. In this paper, an online model-based TPMSVM (OTPMSVM) is proposed in a reproducing Kernel Hilbert space (RKHS). By exploiting the techniques of the Lagrange dual problem like SVM, the solution of this OTPMSVM can be obtained iteratively and the solution of each iteration can be much efficiently determined. It is illustrated that, based on the experimental results, the accuracy of OTPMSVM is comparable with SVM and other state-of-the-art algorithms. On the other hand, the computational cost of this OTPMSVM algorithm is comparable with or slightly lower than existing online methods, but much lower than that of SVM-based algorithms.

Keywords: Twin parametric-margin support vector machine, parametric margin, reproducing Kernel Hilbert Space, kernel, online learning.

1   Introduction

Support vector machine (SVM) is an excellent kernel-based tool for classification and regression [1-3]. This learning strategy introduced is a principled and powerful method in machine-learning field. Within a few years after its introduction SVM has already outperformed most other tools in a wide variety of applications [4,5]. The most important challenge in SVM is the large computational complexity of quadratic programming problem (QPP). The long training time not only causes SVM to take a long time to train on a large data, but also prevents it from locating the optimal parameter set from a very fine grid of parameters over a large span. To reduce the learning complexity of SVM, various algorithms have been reported, such as the decomposition [6], sequential minimal optimization (SMO) [7], and SVMlight [8].

Here, we focus on a class of SVM classifiers named twin SVM (TwSVM) classifiers. The aim of original TwSVM [9] is to find a pair of nonparallel hyperplanes by solving two smaller sized QPPs compared with SVM, which makes the learning speed of TwSVM be approximately 4 times faster than SVM. However, the dual QPPs of

---

[*] This work is supported by the program of Shanghai Normal University (DZL121), the National Natural Science Foundation of Shanghai (12ZR1447100), and the National Natural Science Foundation of China (61202156).

[**] Corresponding author.

D.-S. Huang et al. (Eds.): ICIC 2014, LNCS 8588, pp. 741–752, 2014.

TwSVM require inversion of matrix of size $(m+1)$ or $(l+1)$ twice, where m is the data dimension and l is the data number, which not only enlarges the learning cost, but also loses the sparsity in the nonparallel and decisional hyperplanes. Also, it makes to difficultly extend some efficient SVM algorithms on TwSVM, which leads it to be not suitable for large scale problems. On the other hand, the nonparallel hyperplanes of TwSVM, each hyperplane is as close as possible to the corresponding class and is at a distance of at least one from the opposite class, cannot effectively depict the data scatters of two classes. For instance, it has a trouble in using two hyperplanes to depict the samples generated by Gauss distributions with different covariances.

In the spirit of TwSVM classifier, we proposed some improved and extended TwSVM models for data regression and classification. For example, we proposed the twin support vector regression (TSVR) model [10] and the twin parametric insensitive support vector regression (TPISVR) model [11] for data regression. For classification, we proposed a twin-hypersphere SVM (THSVM) [12]. In this THSVM, the pair of hyperspheres are substituted for the nonparallel hyperplanes to overcome the shortcomings of TwSVM. Another improvement on TwSVM is twin parametric-margin SVM (TPMSVM) [13]. Unlike the parallel-margin hyperplane in SVM, it determines the nonparallel positive and negative parametric-margin hyperplanes through two smaller sized QPPs. A merit compared with TwSVM is it avoids the inversions of matrices in the dual QPPs. Computational results indicate that this TPMSVM is not only fast, but also shows comparable generalization. However, similar to TwSVM, this TPMSVM also loses the sparsity, which makes it cannot be applied into large-scale problems.

In this paper, we proposed an online model-based TPMSVM (OTPMSVM) for data classification. In OTPMSVM, similar to TPMSVM, the optimal solution of the optimization problem can be achieved by exploiting the techniques of Lagrange dual problem. However, a major difference existing in the learning process is that the data points are processed one by one, whereas the Lagrange dual problem is solved just by maximizing a quadratic equation in a given interval. As such, our learning algorithm can be easily realized by a simple update process with a very low computational cost. Computational comparisons of the OTPMSVM with some other algorithms in terms of classification accuracy and learning time have been made on several UCI datasets, indicating our OTPMSVM derives significant generalization and very fast learning speed.

The rest of this paper is organized as follows: Section 2 briefly dwells on the basic properties of RKHS and TPMSVM. Section 3 introduces this online model-based twin parametric-margin support vector machine. Section 4 lists the experimental results of the OTPMSVM on some benchmark datasets. Section 5 concludes this paper.

2   Background

In this section, we simply dwells on the background knowledge of this online algorithm. First, we review the basic properties of RKHS. Then, we briefly dwells on the TPMSVM classifier.

2.1 Basic Properties of RKHS

Let χ be a set and $\mathcal{H}$ be a Hilbert space of functions on χ. We say that $\mathcal{H}$ is a RKHS if for every x in χ. there exists a unique kernel $k(x,.)$ of $\mathcal{H}$ with the property that: $f(x) = \langle f, k(x,.) \rangle_{\mathcal{H}}$ for all $f(x) \in \mathcal{H}$, where $\langle .,. \rangle_{\mathcal{H}}$ denotes dot product in $\mathcal{H}$. Note that an RKHS $\mathcal{H}$ has the following properties[14,15]:

(1) $k(x,y) = \langle k(x,.), k(.,y) \rangle_{\mathcal{H}}$ where $x, y \in \chi$;

(2) $k(x,y)$ has the reproducing property that $\langle f, k(x,.) \rangle_{\mathcal{H}} = f(x)$ where $x \in \chi$ and $f \in \mathcal{H}$;

(3) $\mathcal{H}$ is the closure of the span of all $k(x,.)$ with $x \in \chi$. That is, all $f \in \mathcal{H}$ are linear combinations of kernel functions;

(4) The inner product induces a norm of f in $\mathcal{H}$, i.e., $\|f\|_{\mathcal{H}}^2 = \langle f, f \rangle_{\mathcal{H}}$;

(5) Let $\varphi(.) = \{\varphi_i\}_{i=1}^{\infty}$ be an orthonormal sequence such that the closure of its span is equal to $\mathcal{H}$, then $k(x,y) = \sum_{i=1}^{\infty} \varphi_i(x) \varphi_i(y) = \varphi(x)^T \varphi(y)$, where $\varphi(x)$ denotes a map mapping x from the input space into a feature space;

(6) For $f \in \mathcal{H}$, we have $d \langle f(.), k(x,.) \rangle_{\mathcal{H}} / df = k(x,.)$ and $d \langle f, f \rangle_{\mathcal{H}} / df = 2f$.

2.2 Twin Parametric-Margin Support Vector Machine

Let the samples to be classified be denoted by a set of l vectors x_i, $i = 1, 2, \ldots, l$ in the m-dimensional real space $\chi \subset \mathcal{R}^m$, and $y_i \in \mathcal{Y} = \{-1, 1\}$ denotes the class to which the ith sample belongs. Without loss of generality, we use $I_{\pm}$ to represent the index sets of the positive/negative classes of points, and use I to represent the index set of all points, i.e., $I = \{1, 2, \ldots, l\} = I_+ \cup I_-$. We also denote $l_{\pm} = |I_{\pm}|$ as the point numbers of two classes, i.e., $l = l_+ + l_-$.

To merge into the linear and nonlinear cases, in the sequel we only consider the nonlinear case. That is, we map x to $\varphi(x) \in \mathcal{H}$ for all points in χ. Note that this can be used into the linear case if we set $\varphi(x) = x$.

TPMSVM finds two hyperplanes in the feature space $\mathcal{H}$:

$$f_{\pm}(x) = (w_{\pm}, \varphi(x))_H + b_{\pm} = 0, \tag{1}$$

where $w_{\pm} \in \mathcal{H}$ and $b_{\pm} \in \mathcal{R}$, and $\varphi(.)$ is a map determined by a chosen Mercer kernel $k(x,y)$. We call $f_{\pm}(x)$ the positive/negative parametric-margin hyperplanes, respectively. To obtain the two hyperplanes, TPMSVM considers the following pair of problems:

$$\min \frac{1}{2}\|f_+\|_{\mathcal{H}}^2 + v_1 \sum_{j \in I_-} f_+(x_j) + c_1 \sum_{i \in I_+} \xi_i$$
$$s.t.\ f_+(x_i) \geq 0 - \xi_i, \xi_i \geq 0, i \in I_+,$$
(2)

$$\min \frac{1}{2}\|f_-\|_{\mathcal{H}}^2 - v_2 \sum_{i \in I_+} f_-(x_i) + c_2 \sum_{j \in I_-} \eta_j$$
$$s.t.\ f_-(x_j) \leq 0 + \eta_j, \eta_j \geq 0, j \in I_-,$$
(3)

where $c_1, c_2, v_1, v_2 > 0$ are user-given regularization parameters which determine the penalty weights, ξ_i, η_j are slack variables, and $\|f_\pm\|_{\mathcal{H}}^2 / 2$ control the functional complexity. In an RKHS $\mathcal{H}$, $f_\pm$ are the elements or vectors denoted as $f_\pm \in \mathcal{H}$. Strictly speaking, the dimension of vectors $f_\pm$ in RKHS can be infinity. But TPMSVM searches for the optimal solutions of (2) and (3) in a subspace spanned by the basis functions constructed by the training data. In this case, we have $\|f_\pm\|_{\mathcal{H}}^2 = \langle w_\pm, w_\pm \rangle_{\mathcal{H}}$.

By introducing the Lagrange functions for (2) and (3), we obtain the duals as:

$$\min \frac{1}{2} \sum_{i,j \in I_+} \alpha_i \alpha_j k(x_i, x_j) - v_1 \sum_{i \in I_+, j \in I_-} \alpha_i k(x_i, x_j)$$
$$s.t.\ \sum_{i \in I_+} \alpha_i = l_- v_1, 0 \leq \alpha_i \leq c_1, i \in I_+,$$
(4)

$$\min \frac{1}{2} \sum_{i,j \in I_-} \alpha_i \alpha_j k(x_i, x_j) - v_2 \sum_{i \in I_-, j \in I_+} \alpha_i k(x_i, x_j)$$
$$s.t.\ \sum_{i \in I_-} \alpha_i = l_+ v_2, 0 \leq \alpha_i \leq c_2, i \in I_-,$$
(5)

After optimizing (4) and (5), we obtain the weight vectors according to the Karush-Kuhn-Tucker (KKT) necessary and sufficient optimality conditions, which are

$$w_+ = \sum_{i \in I_+} \alpha_i \varphi(x_i) - v_1 \sum_{j \in I_-} \varphi(x_j), w_- = v_2 \sum_{i \in I_+} \varphi(x_i) - \sum_{j \in I_-} \alpha_j \varphi(x_j).$$
(6)

Then, the final decision function is

$$f(x) = \text{sign}\left((\hat{w}_+ + \hat{w}_-)^T x + (\hat{b}_+ + \hat{b}_-)\right),$$
(7)

where $\hat{w}_{\pm} = w_{\pm} / \|w_{\pm}\|_{\mathcal{H}}$ and $\hat{b}_{\pm} = b_{\pm} / \|w_{\pm}\|_{\mathcal{H}}$. The regions between the parametric-margin hyperplanes $f_{\pm}(x) = 0$ and $f(x) = 0$ are the positive/negative parametric margins.

3   Online TPMSVM

In this section we propose and derive a new online TPMSVM (OTPMSVM) model in an RKHS $\mathcal{H}$ based on the Lagrange dual problem theory.

3.1   Derivation of OTPMSVM

Assume that the current learning data for classification are (x_1, y_1), $\dots, (x_s, y_s), \dots, (x_t, y_t)$, where $x_s \in \chi, y_s \in \mathcal{Y}$, and the current hyperplanes are $f_{\pm,t}(x)$. In online learning, we wish to update the hyperplanes $f_{\pm,t+1}$ based on $f_{\pm,t}(x)$ when the next data point (x_{t+1}, y_{t+1}) becomes available. The learning algorithm starts from one data point (x_1, y_1). In the beginning, $f_{\pm,0}$ are set as $f_{\pm,t+1} = 0$. Without loss of generality, we only consider the update of the positive parametric-margin hyperplane and abbreviate it as $f_t(x)$ during the update process because of the similarity of two formulations. Note that f_t and f_{t+1} can be written as follows without b in $\mathcal{H}$:

$$f_t(.) = \sum_{s=1}^{t} \alpha_s^* k(x_s,.)$$
$$f_{t+1}(.) = \sum_{s=1}^{t} \tilde{\alpha}_s^* k(x_s,.) + \tilde{\alpha}_{t+1}^* k(x_{t+1},.). \tag{8}$$

Note that $k(.,.)$ is a chosen kernel such that $k(u,v) = 1$ when $u = v$, such as the Gauss kernel. If not, we can normalize $k(u,v) := k(u,v) / \sqrt{k(u,u) \cdot k(v,v)}$. Then, the objective is to update $\{\tilde{\alpha}_s^*\}_{s=1}^{t+1}$ from $\{\alpha_s^*\}_{s=1}^{t}$ based on a proper algorithm to be derived. For simplicity, we let f denote f_{t+1} that needs to be determined. Similar to the key idea of TPMSVM in compromising the empirical risk and the complexity of f, our proposed model updating f from f_t can be formulated as follows:

$$\min W = \frac{1}{2} \|f - f_t\|_{\mathcal{H}}^2 + \frac{r}{2} \|f\|_{\mathcal{H}}^2 + R_{emp}(f, x_{t+1}, y_{t+1}), \tag{9}$$

where $(1/2)\|f - f_t\|^2_{\mathscr{H}}$ measures the distance of a predicted f from previous f_t, the term $\|f\|^2_{\mathscr{H}}$ controls the complexity of the predicted f, and $R_{emp}(f, x_{t+1}, y_{t+1})$ is the possible empirical risk for the newly arrived data (x_{t+1}, y_{t+1}) after f is determined. Here, r is parameter reflecting the weight compromising complexity and the possible empirical risk. Actually r also plays the role of a forgetting factor. In the next part, it will be interpreted clearly after deriving the formulations of $\{\tilde{\alpha}_s^*\}_{s=1}^{t+1}$. Due to the forgetting factor, controlling the number of support vectors becomes possible, as discussed in [17]. Also, such a forgetting factor may make it possible that OTPMSVM has the capability of dealing with non-stationary or time varying problems. In the following, we consider two cases to deal with the problem (9).

Case I: (x_{t+1}, y_{t+1}) is a point in the negative class, i.e., $y_{t+1} = -1$.

For this case, based on the idea in TPMSVM, the optimization problem (9) can be re-written as the following unconstrained problem:

$$\min W = \frac{1}{2}\|f - f_t\|^2_{\mathscr{H}} + \frac{r}{2}\|f\|^2_{\mathscr{H}} + vf(x_{t+1}), \tag{10}$$

where v is a positive regularization parameter. Note that $f(.) = w^T \varphi(.)$, $f_t(.) = w_t^T \varphi(.)$ and $f_{t+1}(.) = w_{t+1}^T \varphi(.)$ in RKHS. This implies that there is a one-to-one mapping between a vector w and $f(.)$. More specifically, we actually project $f_{t+1}(x)$ and $f_t(x)$ onto the space spanned by the basis functions $\{k(x_s, x)\}_{s=1}^{t+1}$ and $\{k(x_s, x)\}_{s=1}^{t}$, respectively, in (8). So $f(.)$ can be treated as a vector in a finite dimensional space. In this sense, $f_t(x)$ is a previous known vector and $f_{t+1}(x)$ is a vector to be determined at $t+1$. And $f(x_{t+1})$ is the inner product of two vectors in $\mathscr{H}$: $\langle f, k(x_{t+1}, .)\rangle_{\mathscr{H}} = f(x_{t+1})$. Thus, the optimization problem (10) is actually an unconstrained convex optimization with respect to an unknown finite dimensional vector.

Therefore, we can obtain the optimal solution by setting $\nabla_f W = 0$. Recall from [15] and [15] that

$$\frac{df(x_{t+1})}{df} = \frac{d\langle f, k(x_{t+1}, .)\rangle}{df} = k(x_{t+1}, .). \tag{11}$$

Then, we have

$$\nabla_f W = f - f_t + rf + vk(x_{t+1}, .) = 0 \Rightarrow$$
$$f = \frac{1}{1+r}f_t - \frac{v}{1+r}k(x_{t+1}, .) = \sum_{s=1}^{t}\frac{\alpha_s^*}{1+r}k(x_{t+1}, .) - \frac{v}{1+r}k(x_{t+1}, .) \tag{12}$$

Note that (12) gives the update rule for $\tilde{\alpha}_s^*$ from 1 to $t+1$:

$$\tilde{\alpha}_s^* = \frac{1}{1+r}a_s^*, s=1,\dots,t, \tilde{a}_{t+1}^* = -\frac{v}{1+r}, \tag{13}$$

where r and $a_s^*, s=1,\dots,t$ are known. Thus, the parameter r can be considered as a forgetting factor as seen in (13).

Thus, for this case, i.e., $y_{t+1}=-1$, we can easily update $\left\{\tilde{\alpha}_{s+1}^*\right\}_{s=1}^{t+1}$ from $\left\{\tilde{\alpha}_s^*\right\}_{s=1}^t$ by (13) without any optimization technique.

Case II: (x_{t+1}, y_{t+1}) is a point in the positive class, i.e., $y_{t+1}=1$.

If the newly arrived point is in the positive class, according to the idea in TPMSVM, we write the optimization problem (9) as following:

$$\min W = \frac{1}{2}\|f-f_t\|_H^2 + \frac{r}{2}\|f\|_H^2 + C\xi_{t+1}, \tag{14}$$
$$s.t. \ f(x_{t+1}) \geq 0 - \xi_{t+1}, \xi_{t+1} \geq 0,$$

where C is a regularization factor defined by users. Similar to the above case, (14) is a convex optimization problem. Then we can introduce the Lagrange multipliers to optimize it. Specifically, we have

$$L = \frac{1}{2}\|f-f_t\|_H^2 + \frac{r}{2}\|f\|_H^2 + C\xi_{t+1} - \alpha_{t+1}\left(f(x_{t+1})+\xi_{t+1}\right) - \beta_{t+1}\xi_{t+1,} \tag{15}$$

where α_{t+1} and β_{t+1} are nonnegative Lagrange multipliers corresponding to the constraints. Similarly, L attains its minimum with respect to f and ξ_{t+1} if and only if the following conditions are satisfied:

$$\nabla_f L = 0, \quad \nabla_{\xi_{t+1}} L = 0, \tag{16}$$

Using (11), we have

$$\nabla_f L = f - f_t + rf - \alpha_{t+1}k(x_{t+1,.}) = 0 \Rightarrow$$
$$f = \frac{1}{1+r}f_t + \frac{\alpha_{t+1}}{1+r}k(x_{t+1,.}) = \sum_{s=1}^t \frac{\alpha_s^*}{1+r}k(x_s,.) + \frac{\alpha_{t+1}}{1+r}k(x_{t+1},.). \tag{17}$$

Also, we have

$$\nabla_{\xi_{t+1}} L = 0 \Rightarrow C - \alpha_{t+1} - \beta_{t+1} = 0 \Rightarrow 0 \leq \alpha_{t+1} \leq C. \tag{18}$$

Then, (17) gives the update rule for $\tilde{\alpha}_s^*$ from 1 to $t+1$:

$$\tilde{\alpha}_s^* = \frac{\alpha_s^*}{1+r}, s=1,\dots,t, \ \tilde{a}_{t+1}^* = \frac{a_{t+1}}{1+r} \tag{19}$$

Again, the parameter r can be considered as a forgetting factor as seen in (18). From (17), we have

$$f\left(x_{t+1}\right)=\frac{1}{1+r}f_t\left(x_{t+1}\right)+\frac{\alpha_{t+1}}{1+r}k\left(x_{t+1},x_{t+1}\right)=\frac{1}{1+r}f_t\left(x_{t+1}\right)+\frac{\alpha_{t+1}}{1+r}, \tag{20}$$

since $k\left(x,x\right)\equiv1$ for any x. Substituting (17) and (20) into (15) and using the properties of RKHS, we have

$$\frac{1}{2}\left\|f-f_t\right\|_{\mathcal{H}}^2=\frac{1}{2}\left\|\frac{\alpha_{t+1}}{1+r}k\left(x_{t+1},\cdot\right)-\frac{1}{1+r}f_t\right\|_{\mathcal{H}}^2$$
$$=\frac{\alpha_{t+1}^2}{2\left(1+r\right)^2}-\frac{r}{\left(1+r\right)^2}\alpha_{t+1}f_t\left(x_{t+1}\right)+\frac{r^2}{2\left(1+r\right)^2}\left\|f_t\right\|_{\mathcal{H}}^2 \tag{21}$$

$$\frac{r}{2}\left\|f\right\|_{\mathcal{H}}^2=\frac{r}{2}\left\|\frac{1}{1+r}f_t+\frac{\alpha_{t+1}}{1+r}k\left(x_{t+1},\cdot\right)\right\|_{\mathcal{H}}^2$$
$$=\frac{r}{2\left(1+r\right)^2}\alpha_{t+1}^2+\frac{r}{\left(1+r\right)^2}\alpha_{t+1}f_t\left(x_{t+1}\right)+\frac{r}{2\left(1+r\right)^2}\left\|f_t\right\|_{\mathcal{H}}^2 \tag{22}$$

Then, we can obtain the dual optimization problem for (14) by combining with (18), which is:

$$\min\frac{1}{2\left(1+r\right)}\alpha_{t+1}^2+\frac{\alpha_{t+1}}{\left(1+r\right)}f_t\left(x_{t+1}\right)-c_0$$
$$s.t.\,0\le\alpha_{t+1}\le C, \tag{23}$$

where $c_0=\dfrac{r}{2\left(1+r\right)}\left\|f_t\right\|_{\mathcal{H}}^2$. Note that (23) is a single-variable quadratic programming problem, we can first find the stationary point $\bar{\alpha}_{t+1}$ satisfied with $dL\,/\,d\alpha_{t+1}\,|_{\alpha_{t+1}=\bar{\alpha}_{t+1}}=0$, which leads to $\bar{\alpha}_{t+1}=-f_t\left(x_{t+1}\right)$. Then the optimal solution of α_{t+1} in (23), still denoted by α_{t+1} for simplicity, is

$$\alpha_{t+1}=\begin{cases}0, & \text{if } -\bar{\alpha}_{t+1}<0,\\ C, & \text{if } \bar{\alpha}_{t+1}>C,\\ \bar{\alpha}_{t+1} & \text{otherwise.}\end{cases} \tag{24}$$

Combining (8) and (17), we have

$$f\left(x\right)=\sum_{s=1}^t\tilde{\alpha}_s^*k\left(x_s,x\right)+\tilde{\alpha}_{t+1}^*k\left(x_{t+1},x\right). \tag{25}$$

where

$$\tilde{\alpha}_s^* = \frac{\alpha_s^*}{1+r}, s=1,\ldots,t, \ \tilde{\alpha}_{t+1}^* = \begin{cases} 0, & \text{if } -f_t\left(x_{t+1}\right) \leq 0, \\ \dfrac{C}{1+r}, & \text{if } -f_t\left(x_{t+1}\right) \geq C, \\ -\dfrac{f_t\left(x_{t+1}\right)}{1+r}, & \text{otherwise.} \end{cases} \tag{26}$$

3.2   Algorithm and Discussion

In this section, we present the implementation procedure of the algorithm and then give some discussions.

Let t be the tth step in learning the training data $\left(x_1, y_1\right),\ldots,\left(x_l, y_l\right)$. The steps for OTPMSVM are illustrated as follows.

Online TPMSVM Algorithm

Step1: Set $f_{\pm,0}(x)=0$, collect $\left(x_1, y_1\right)$ and compute $\alpha_{\pm,1}^*$ by (13) or (26), and set $t=1$;

Step $t+1$: Update $f_{\pm,t+1}(x)=\sum_{s=1}^{t+1}\tilde{\alpha}_{\pm,s}^* k\left(x_s, x\right)$ from $f_{\pm,t}(x)=\sum_{s=1}^{t}\tilde{\alpha}_{\pm,s}^* k\left(x_s, x\right)$ for $t=1,\ldots,l-1:$, where the coefficients $\tilde{\alpha}_{\pm,s}^*, s=1,\ldots,t+1$, are computed by (13) or (26);

Step l: Output the parametric-margin hyperplanes $f_{\pm}(x) = f_{\pm,l}(x)$ and end.

In the following, some remarks will help readers to further understand this OTPMSVM algorithm.

Remark 1: It is easy to see that the parameter r plays a role of forgetting factor from (13) and (26). This implies that this OTPMSVM has the ability of dealing with time varying (non-stationary) problems, since the choice of r allows us to give higher weight to more recent data. To this end, variation rate r needs to be known or estimated before applying this OTPMSVM. In addition, the parameter r plays a role in the control of support vectors since we have $\left|\tilde{\alpha}_s^*\right| = \left|\alpha_s^* / (1+r)\right| < \left|\alpha_s^*\right|$.

Remark 2: From (25) or (26), for the newly arrived point $\left(x_{t+1}, y_{t+1}\right), y_{t+1}=1$, if the current positive parametric-margin hyperplane correctly classify this point, i.e., $f_{t+1}\left(x_{t+1}\right) \geq 0$, then this algorithm need not update the plane, i.e., $\tilde{\alpha}_s^*=0$. Factually, the current hyperplane will keep unchanged since the forgetting factor r is useless for this case.

Remark 3: This algorithm can be used for off-line learning. To improve the performance, we can repeatedly train the input points as the learning of perceptron. Fortunately, the computational cost of one learning loop is only $O(l)$.

Remark 4: This algorithm is suitable for linear case by the following strategy: Let's denote $f_{\pm,t}(x) = w_{\pm,t}^T \hat{x}$ as the positive/negative parametric-margin hyperplanes given the training points $(x_1, y_1), \ldots, (x_t, y_t)$, where $\hat{x} = (x;1)/\|(x;1)\|$. Then, given the arrived point $(x_{t+1}; y_{t+1})$, we update $w_{+,t+1}$ as

$$
w_{+,t+1} = \begin{cases} \dfrac{w_{+,t} + \max\left(\min\left(-w_{+,t}^T \hat{x}_{t+1}, C\right), 0\right) x_{t+1}}{1+r}, & \text{if } y_{t+1} = +1, \\ \dfrac{w_{+,t} - v x_{t+1}}{1+r}, & \text{otherwise} \end{cases}
\tag{27}
$$

Clearly, this linear case only needs a very small cache space and has a very small computational cost at each iteration. Then, it is very suitable for large-scale linear problems.

4   Experiments

In this section, we investigate the performance of this proposed OTPMSVM on several publicly available 13 benchmark datasets [16] in that order: Titanic (T), Heart (H), Diabetes (D), Breast Cancer (BC), Thyroid (Th), Flare (F), Splice (S), Image (I), German (G), Banana (B), Twonorm (Tw), Ringnorm (R), and Waveform (W). In particular, we use in each problem the train-test splits given in that reference. Table 1 contains the data dimensions and the training and test data sizes of these data sets. In the experiments, all methods are implemented in MATLAB on Windows running on a PC.

Table 1 reports the learning results of these algorithms with Gaussian kernels, including the average prediction accuracies and the learning time of one-run. From Table 1 it can be seen that the OLK_C[17] and OTPMSVM obtain the almost test accuracies as the corresponding algorithms, i.e., SVM and TPMSVM, on most of these datasets, although they are slightly worse than the latter. To further explain the performance of the proposed method, we introduce the p-values calculated from the paired t-test on these methods. The experimental results indicate that the average testing accuracies are not significant in 5% confidence level for those methods. This shows that the proposed method has the comparable performance with the other methods. As for the training time of these methods, the results in Table 1 indicate that our method and OLK_C need very less time than TPMSVM and the other methods. That is, they are much faster than these methods. In fact, this is because our method and OLK_C only have the computational cost $O(l)$ while the other methods need to optimize the QPPs. In summary, the proposed OTPMSVM obtains the comparable generalization with a much faster training speed.

Table 1. Results of SVM, OLK$_C$, TwSVM, TPMSVM, and OTPMSVM

Data size	SVM Acc. -	OLK$_C$ Acc. Time/p-value	TwSVM Acc. Time/p-value	TPMSVM Acc. Time/p-value	OTPMSVM Acc. Time
T (150+2051)*3	77.3±0.7 -	76.95±1.26 0.005/0.864	77.78±0.60 0.127/0.726	77.62±0.86 0.102/0.802	77.14±1.28 0.004
H (170+100)*13	84.1±3.2 -	83.70±4.12 0.006/0.855	83.75±4.36 0.092/0.867	84.01±3.20 0.075/0.924	83.96±3.50 0.005
D (468+300)*8	76.4±1.8 -	76.15±2.57 0.012/0.886	77.10±2.15 1.427/0.470	76.36±1.86 0.584/0.893	76.29±2.20 0.013
BC (200+77)*9	72.4±4.9 -	70.36±5.24 0.005/0.247	70.13±4.37 0.852/0.470	72.31±4.15 1.122/0.581	71.75±4.95 0.006
Th (140+75)*5	95.5±2.0 -	94.85±2.50 0.004/0.814	95.61±2.12 0.132/0.870	95.72±2.25 0.105/0.805	95.20±2.62 0.005
F (666+400)*9	67.1±1.6 -	65.24±4.26 0.017/0.090	66.72±2.52 3.872/0.875	67.81±2.56 4.459/0.312	66.97±3.55 0.019
S (1000+2175)*60	89.2±0.6 -	88.16±1.65 0.029/0.775	88.83±0.97 4.071/0.769	89.86±0.80 3.574/0.219	88.49±1.27 0.027
I (1300+1010)*18	96.2±0.6 -	05.75±1.17 0.046/0.138	97.02±0.79 13.154/0.905	96.56±0.86 17.486/0.896	96.80±0.87 0.044
G (700+300)*2	76.2±2.1 -	75.36±3.46 0.021/0.632	76.56±2.53 2.267/0.425	76.85±3.15 2.896/0.218	75.84±3.75 0.020
B (400+4900)*2	89.3±0.5	88.45±0.93 0.010/0.835	88.27±0.87 0.420/0.725	88.84±0.60 0.364/0.764	88.59±0.80 0.011
Tw (400+7000)*20	97.1±0.2 -	96.45±0.76 0.013/0.615	97.17±0.23 0.502/0.814	97.50±0.20 0.373/0.704	96.94±0.47 0.012
R (400+7000)*20	98.3±0.1 -	97.80±0.65 0.011/0.760	97.30±0.35 0.455/0.622	98.20±0.32 0.261/0.871	98.16±0.62 0.010
W (400+4600)*21	90.2±0.4 -	88.75±0.72 0.012/0.050	88.74±2.48 0.697/0.047	90.60±0.42 0.541/0.802	90.28±0.56 0.013

5   Conclusions

In this paper, we have considered an online model-based twin parametric-margin support vector machine with kernels in an RKHS space. This OTPMSVM can obtain the optimal solution by exploiting the techniques of Lagrange dual problem. In addition, this learning algorithm can be easily realized by a simple update process with a very low computational cost. Computational comparisons of the OTPMSVM with some other algorithms in terms of classification accuracy and learning time have been made on several UCI datasets, indicating our OTPMSVM derives significant generalization and very fast learning speed. The idea in this method can be easily extended to some other TwSVM classifiers and regressors.

References

1. Boser, B., Guyon, L., Vapnik, V.N.: A training algorithm for optimal margin classifiers. In: Proc. 5th Ann. Work. Comp. Lear. Theo., pp. 144–152. ACM Press, Pittsburgh (1992)
2. Vapnik, V.N.: The natural of statistical learning theory. Springer, New York (1995)
3. Vapnik, V.N.: Statistical learning theory. Wiley, New York (1998)
4. Osuna, E., Freund, R., Girosi, F.: Training support vector machines: an application to face detection. In: Proc. IEEE Comp. Visi. Patt. Reco., San Juan, Puerto Rico, pp. 130–136 (1997)
5. Joachims, T., Ndellec, C., Rouveriol, C.: Text categorization with support vector machines: Learning with many relevant features. In: Nédellec, C., Rouveirol, C. (eds.) ECML 1998. LNCS, vol. 1398, pp. 137–142. Springer, Heidelberg (1998)
6. Osuna, E., Freund, R., Girosi, F.: An improved training algorithm for support vector machines. In: Proc. IEEE Work. Neur. Netw. Sign. Proc., Amelia Island, FL, USA, pp. 276–285 (1997)
7. Platt, J.: Fast training of support vector machines using sequential minimal optimization. In: Adv. Kern. Meth.-Support Vector Learning, pp. 185–208. MIT Press, Cambridge (1999)
8. Joachims, T.: Making large-scale SVM learning practical. In: Adv. Kern. Meth.-Support Vector Machine, pp. 169–184. MIT Press, Cambridge (1999)
9. Jayadeva, R., Khemchandani, S., Chandra: Twin support vector machines for pattern classification. IEEE Trans. Patte. Anal. Mach. Intel. 29, 905–910 (2007)
10. Peng, X.: TSVR: an efficient twin support vector machine for regression. Neur. Netw. 23, 365–372 (2010)
11. Peng, X.: Efficient twin parametric insensitive support vector regression model. Neurocomputing 79, 26–38 (2012)
12. Peng, X., Xu, D.: A twin-hypersphere support vector machine classifier and the fast learning algorithm. Infor. Scie. 221, 12–27 (2013)
13. Peng, X.: TPMSVM: A novel twin parametric-margin support vector machine for pattern recognition. Patt. Reco. 44, 2678–2692 (2011)
14. Scholkopf, B., Smola, A.: Learning with Kernels: Support Vector Machines, Regularization, Optimization and Beyond. MIT Press, Cambridge (2002)
15. Kivinen, J., Smola, A., Williamson, R.: Online learning with Kernels. IEEE Trans. Signal Proc. 100, 2165–2176 (2004)
16. Ratsch, G.: Benchmark repository (2000), datasets available at http://ida.first.fhg.de/projects/bench/benchmarks.htm
17. Li, G., Wen, C., Li, Z.: Model-based online learning with kernels. IEEE Trans. Neur. Netw. Lear. Syst. 24(3), 356–369 (2013)

Action Segmentation and Recognition Based on Depth HOG and Probability Distribution Difference

Rui Yang[1,2] and Ruoyu Yang[1,2]

[1] State Key Laboratory for Novel Software Technology, Nanjing University, China
[2] Department of Computer Science and Technology, Nanjing University, China
ryang@smail.nju.edu.cn, yangry@nju.edu.cn

Abstract. The paper presents a method to automatically separate consecutive human actions into subsegments and recognize them. The 3D positions of the joints tracked by depth camera like Kinect sensors and the depth motion maps (DMMs) are used in the method. Both of the two types of data contain useful information to help us extract features for each action video. However, they are also full of noise. So we combine the pairwise relative positions of the 3D joints (Skeleton Joints) and Histograms of Oriented Gradients (HOG) calculated from the DMM together to improve the feature representation. A SVM-based classification ensemble is built to achieve the recognition result. We also build a Probability-Distribution-Difference (PDD) based dynamic boundary detection framework to segment consecutive actions before applying recognition. The segmentation framework is online and reliable. The experimental results applied to the Microsoft Research Action3D dataset outperform the state of the art methods.

1 Introduction

As a key step in an overall human action understanding system, action recognition is applied for many common applications including human computer interaction, video surveillance, game control system and etc [1, 2]. Based on the traditional video sequences captured by RGB cameras, spatio-temporal features are widely used for the recognition task [3]. With the recent development of high-speed depth cameras, we can capture depth information of body's 3D position and motion in real-time [4]. Compared with the traditional 2D information, the depth information has obvious advantages because it can distinguish the human actions from more than one view and the z-index displacement information which is lost in 2D frames is valued here.

On the other hand, action segmentation appears to be a problem no less difficult than action recognition itself [5]. Most of the state of the art human action recognition methods [1, 6-9] based on depth video sequences focus on using video clips which are segmented manually. The clips typically contain only one complete action. We can get quite satisfactory accuracy by training and testing on these clips, but the effect is hard to maintain on the real-time video stream and synthetic consecutive action clips. From the perspective of efficiency, some of these approaches' computational cost will rise exponentially when dealing with sequential video streams.

D.-S. Huang et al. (Eds.): ICIC 2014, LNCS 8588, pp. 753–763, 2014.
© Springer International Publishing Switzerland 2014

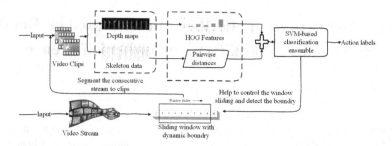

Fig. 1. An overall architecture of our approach shows a complete process including segmentation of video stream and action recognition based on clips

In this paper, we presented a complete system for segmentation and recognition of 3D human actions and an overall architecture of the system is shown in Figure 1. We firstly proposed a novel human action recognition method utilizing both of skeleton joints' 3D position data and depth maps captured by Kinect depth camera. We extract features from skeleton data and depth maps in different ways and utilize them to train the classifiers separately. For skeleton data, we track the pairwise relative positions of the 3D joints and utilize the mutual distances of them as the representation of the actions. We also build Histograms of Oriented Gradients (HOG) descriptors [10] from Depth Motion Maps (DMM) to train the action recognition classifier. Utilizing HOG to describe depth motion information was first proposed by Yang et al [7] and we put it into our system and made some minor changes to match with the former classifier based on Skeleton data better. With the steps of feature extracting work done, a SVM-based classification ensemble is built to achieve the recognition result.

Secondly we propose a Probability-Distribution-Difference (PDD) based dynamic boundary detection framework to separate consecutive actions. Different from traditional boundary detection methods [11, 12], we utilize the idea of the sliding window to detect the boundary instead of searching strict predefined start and end points of action in the video stream. Starting from the first frame of video, a fixed-size window is used to slide to the end with a fixed stride length. Whenever the window is moving, we are able to get a new clip. Then we apply the classification ensemble obtained in first step to recognize these clips. The recognition results are not accurate enough because the fixed window size may not be compatible with the length of real action or a single window may consist of parts of two different actions. However, we note that the probability distribution output by the classifier tempestuously changes when the window slides from one action to another action. So seeking out the significant difference between adjacent probability distributions can help us to get the window which contains the boundary and finally locate it. The boundaries we find are real "dynamic" for the reason that it has no definite shape characteristics compared with [11]. Even when there is no obvious gap between two actions, we can also find a boundary to separate them.

2   Related Work

Li et al [1] model the dynamics of the action by building an action graph and describe the salient postures by a bag-of-points (BOPs). It's an effective method which is similar to some traditional 2D silhouette-based action recognition methods. The method does not perform well in the cross subject test due to some significant variations in different subjects from MSR Action3D dataset.

Yang et al [7] are motivated by the success of Histograms of Oriented Gradients (HOG) in human detection. They extract Multi-perspective HOG descriptors from DMM as representations of human actions. They also illustrate how many frames are sufficient to build DMM-HOG representation and give satisfactory experimental results on MSR Action3D dataset. Before that, they have proposed an EigenJoints-based action recognition system by using a NBNN classifier [6] with the same goal.

In order to deal with the problems of noise and occlusion in depth maps, Wang et al extracts semi-local features called random occupancy pattern (ROP) features [8]. They propose a weighted sampling algorithm to reduce the computational cost and claim that their method performs better in accuracy and computationally efficiency than SVM trained by raw data. After that they further propose Local Occupancy Patterns (LOP) features [9] which are similar to ROP in some case and improve their results to some extent.

3   Recognition Based on Clips

This section describes our feature extraction method for video clips consists of Skeleton data and depth maps sequences in detail. Then it shows how we utilize the features to train a SVM-based classification ensemble.

3.1   Concise Feature for Skeleton Joints

The human skeleton tracked by Kinect system is the direct input for our first features extraction method. 20 body joints such as spine, hands and feet form a complete skeleton. The raw data is a sequence of this joints' positions expressed as x, y, z coordinates. For a given skeleton, we denote each joint as $p_i = (x_i, y_i, z_i)$. By these tuples, we utilizing Euclidean similarity to calculate the distance between any two joints:

$$Distance_{ij} = EuclideanSimilarity(p_i, p_j).$$

We can get N skeletons from a labeled video clip which contains N frames. Then for clip k, the feature is defined as:

$$f(k) = \{\overline{Distance_{ij}}, \sigma_{ij} \mid i \neq j\},$$

where the $\overline{Distance_{ij}}$ is the mean of all the $Distance_{ij}$ in these skeletons from different frames and the σ_{ij} is the variance.

However, the human's body is different from each other. Directly using the raw distances may bring some errors due to the natural differences produced by the body itself. So a simple normalization is applied to the data. We firstly choose a stable bone set: $\{(H,CS),(CS,S),(LH,LK),(RH,RK)\}$, H,CS,S,LH,LK,RH,RK represent head, center shoulder, spine, left hip, left knee, right hip and right knee. Each stable bone is defined by its two end-joints. We called them stable bones because unlike hand and wrist, these bones are usually not occluded by the human body. The length of the stable bones is rarely affected by the tracking error made by Kinect sensors and to some extent it can describe characteristics of a person's body shape. Finally, we compute total length for each skeleton's stable bone set in all training clips, then use the mean as the scaling factor λ. The final feature is denoted as:

$$normalized\_f(k) = f(k) * \frac{\lambda}{StableLength(k)},$$

$$StableLength(k) = \sum_{j=1}^{N} (\sum_{b_i \in S_j} length(b_i)/ \mid S_j \mid),$$

where N is the number of frames in clips k, S_j is the stable bones set of skeleton in jth frame and $|S_j|$ is the number of bones in S_j.

3.2   DMM-HOG

Depth Motion Maps. To put it simply, DMMs are used to summarize the difference between each two consecutive depth maps in a clip. Each 3D depth frame generates three 2D binary maps including front views map map_f, side views map map_s, and top views map map_t. Then DMM is denoted as:

$$DMM_v = \sum_{i=1}^{N-1} \mid map_v^i - map_v^{i+1} \mid,$$

where $v \in \{front, side, top\}$ and N is the number of frames in a given clip. Figure 2 shows an example of three DMM maps generated by the depth maps sequence which performs action "Hammer".

HOG Descriptor. HOG descriptor is used to characterize the DMMs. We resize each DMM to 240*160 pixels and set the block size to 40*40. Block stride and cell size are set to 20*20. The number gradient orientation bins is 8. The specific implementation is based on *HOGDescriptor()* from OpenCV API [13]. Finally we get HOG vector with the dimension of 2464 for each DMM and concatenate the three-views DMM-HOGs $\{DMM_{front}HOG, DMM_{top}HOG, DMM_{side}HOG\}$ as the feature for the given clip. The HOG vector is much smaller than [7].

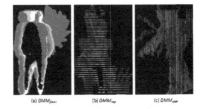

(a) DMM_{front} (b) DMM_{top} (c) DMM_{side}

Fig. 2. Three DMM calculated from the depth maps sequence which performs action "Hammer". They summarize the motion of the sequence from three orthogonal views.

3.3   SVM-Based Classification Ensemble

In order to improve predictive performance, building classification ensembles are a frequently used method. We separately use the features mentioned in the above two subsections to train two SVM classifiers. Different training process makes the two classifiers quite different from each other. In fact, both of them perform quite well and the errors they produce are quite diverse so the result of the combination is well-founded valuable [14].

Because we have only two classifiers, a large number of training iteration may cause low performance-cost ratio. A linear weighted combination is good enough to achieve a satisfactory accuracy. We do a sampling with replacement on the training set and use the samples to do a pre-testing based on the two classifiers. The pre-testing accuracy given by each classifier is set as its weight and then the combination is done.

4   Segmentation Based on Stream

As discussed in [5], boundary detection methods are attractive because their segmentation result is not dependent on the action classes so that it's generic for all videos. Sliding window methods make less assumptions, i.e. no special boundary criteria is needed. However, they usually cause more computational burden than boundary detection methods on account of the dynamic window size. In this section, we explain our segmentation method, Probability-Distribution-Difference (PDD) based dynamic boundary detection framework, in detail. Our method is able to combine the advantages of boundary detection and sliding windows.

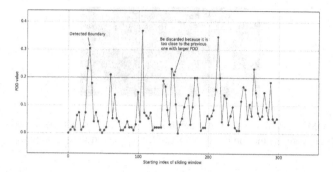

Fig. 3. A line chart of PDD values corresponding to segmentation results in Figure 4. The red line is the threshold line and the boundary detection strategy is illustrated in the Algorithm 1.

4.1   PDD Definition

We denote a sliding window with fixed size t as $W_t(i)$ where i indicates the location of the window's start point in the whole video stream. In our experiment t is set to 35 which near to the minimum length of an action in MSR action3D dataset. The optimal set of t is dependent on the specific dataset. We believe that the minimum length of action (or the approximation) is a generally good chose because one sliding window will not contain two or more boundaries in this case. In order to reduce the computational cost, the window will slide according to the given sliding stride length l instead of moving frame by frame. When the window slides, we regard frames in each window as different clips and apply the classification ensemble obtained in the above section to recognize these clips. Then we get output probability distribution PD_W for a certain window W. The PDD is finally denoted as

$$PDD_i = PD_{W_t(i+l)} - PD_{W_t(i)},$$

$$PD_A - PD_B = \sum_{i=0}^{K} | P_A(c_i) - P_B(c_i) |,$$

where K is the number of action types, c_i indicates an action type and $P(c)$ is the probability for type c. In other words, PD_W is a K-dimension vector representing probability distribution of actions, and PDD is to compute the L1 distance between PDs of two consecutive clips.

4.2   Boundary Detection Strategy

Now as we have mentioned earlier, the probability distribution for actions tremendous changes (embodied in large PDD value) when the window slides from one action to another action. So the boundary we want to detect is very likely to appear in the window with high PDD value. Finally we employ a framework to exactly locate the boundaries as Algorithm 1.

Algorithm 1. PDD based boundary detection framework

Input:
The sliding window's size, t;
The total video's length, T;
The sliding stride length, l;
The threshold to determine whether the window contains the boundary, TH;
The trusted minimum gap length between two boundaries, MINGAP;

Output:
An array contains all the boundaries algorithm detected, boundaryArray;
1: Initialize starting index i = 0;
2: Initialize boundaryArray = null;
3: **while** $i + t < T$ **do**
4:      **if** $PDD(Wt(i)) > TH$ and $i\ell o$ **then**
5:          /*Finding the index of the previous boundary and the index equals -1 if no
              such boundary.*/
6:          $ipre = boundaryArray.lastElement$;
7:          **if** $i - ipre <$MINGAP and $ipre \neq -1$ **then**
8:              /*Two boundaries are too close to each other so we choose the one with
                  larger PDD.*/
9:              **if** $PDD(Wt(i)) > PDD(Wt(ipre))$ **then**
10:                 $boundaryArray.remove(ipre)$;
11:                 $boundaryArray.add(i)$;
12:             **end if**
13:         **else**
14:             $boundaryArray.add(i)$;
15:         **end if**
16:     **end if**
17:     $i = i + l$;
18: **end while**
19: **return** $boundaryArray$;

It is noted that the *boundaryArray* stores the detected windows' starting indexes instead of boundary indexes. In our experiment, we eventually recognize the center point of the detected window as the boundary. So starting index i plus $t/2$ is the predicted boundary. Our framework takes a single scan on the whole video stream. Different from traditional boundary detection strategy, the boundary has been detected here is only affected by the adjacent boundaries regardless of the overall stream. So our framework is an online video segmentation method and its high reliability is testified by the experiments. A line chart example of PDD values is shown in Figure 3.

5   Experimental Results and Discussion

In this section, we show our experimental results produced by applying our method to the public domain MSR Action3D dataset [1] and compare them with the existing methods.

There are 20 action types in the MSR Action3D dataset. Each action type contains 10 subjects and they performed the same action 2-3 times for each subject. 567 depth map sequences are provided in total and 10 of them are abandoned because the skeletons are either missing or too erroneous. The resolution is 320*240. Each depth map sequence we used is regarded as a clip.

5.1   Recognition Accuracy

As same as [1], we split the 20 action types into three subsets as listed in Table 1 and do three different tests for each subset. In test one, 1/3 of the samples were used as training samples and the rest as testing samples; in test two, 2/3 of the samples were used as training samples and the rest as testing samples; in Cross Subject Test, half subjects are used for training and the rest ones used for testing.

Table 1. The three subsets of actions

Action Set 1 (AS1)	Action Set 2 (AS2)	Action Set 3 (AS3)
Horizontal arm wave	High arm wave	High throw
Hammer	Hand catch	Forward kick
Forward punch	Draw x	Side kick
High throw	Draw tick	Jogging
Hand clap	Draw circle	Tennis swing
Bend	Two hand wave	Tennis serve
Tennis serve	Forward kick	Golf swing
Pickup & throw	Side boxing	Pickup & throw

The performances of our method and several the state of the art methods are shown in Table 1. Intuitively, we see that our method achieve higher recognition rates than others except on AS1 Test One, AS2 Test One and AS3 Cross Subjection Test. Three failing cases' recognition rates are slightly lower. For the most challenging Cross Subject Test, the performance of our method is very stable (all over 90%) and is better than other methods on the average. The results of Yang et al [7] are close to ours, but it should be noted that we have reduced the dimension of HOG descriptor compares with [7] and the concise feature for skeleton joints is calculated very fast. Therefore, we get a substantial increase in the overall computational efficiency. As [1] said, the AS3 was made to group complex actions together. In fact, our method does not perform well enough on AS3 Cross Subject Test (slightly lower than [7] and [6]). There are at least two reasons for that, one of which is that mean and std of the distances extract from the skeleton data may not be suitable for summarizing complex movements. For example, "Pickup & throw" and "Hammer" are quite similar from the point of view of the average positions of skeleton joints. Another reason is that the noise problem. Although 10 sequences are not used in the experiment because the skeletons are either missing or too erroneous, we still find a lot of noise especially in subjects of "Pickup & throw" and "Golf swing". Some of the skeleton are completely wrong or even missed in a continuous period of time. In this case, the feature for skeleton data is difficult to play a role in the learning process.

Table 2. Recognition accuracies (%) comparison based on MSR Action3d dataset

	Li et al [1]	Yang et al [6]	Yang et al [7]	ours
AS1One	89.5	94.7	97.3	91.0
AS2One	89.0	95.4	92.2	90.9
AS3One	96.3	97.3	98.0	98.6
AS1Two	93.4	97.3	98.7	98.7
AS2Two	92.9	98.7	94.7	100.0
AS3Two	96.3	97.3	98.7	100.0
AS1CrSub	72.9	74.5	96.2	96.2
AS2CrSub	71.9	76.1	84.1	90.8
AS3CrSub	79.2	96.4	94.6	92.9

5.2   Temporal Segmentation Results

Since each clip provided by MSR Action3d dataset only contains a single action, we concatenate these ones to simulate real video streams and apply our segmentation method to them. Our strategy of concatenating is concise, open and repeatable without any manual optimization. In Section 5.1, we split the 20 action types into three subsets which consist of eight action types, AS1, AS2 and AS3. Now we iteratively choose eight clips with different actions from the same subset and concatenate them together just according to the order in table 1. Each eight clips concatenated together should belong to the same subject and performance turn. For example, the first video stream we simulated is composed of the clips set $\{a_i\_s_2\_e_1 \mid a_i \in AS1\}$ as shown in Figure 4. The $s_2\_e_1$ indicates that all the clips in the set are the first performance belonging to the subject two.

As like as in the previous subsection, we use half of the subjects to train the classification ensemble. The other half are used to do the simulation. So the segmentation test is a severe cross-subject test as before. In the experiment, we employ a sliding window with the length of 35 to take a single scan on each simulated stream. The stride length is 3. Whenever the window is detected, its start index plus 17 (half length of the window) is a predicted boundary location.

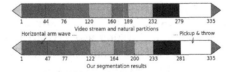

Fig. 4. A simulated video stream composed of the clips set $\{a_i\_s_2\_e_1 \mid a_i \in AS1\}$ and the segmentation results obtained by applying our PDD based framework to the stream. The corresponding line chart of PDD values is shown in Figure 3.

A test case is shown in Figure 4. We can see that the boundaries we detected are very close to the real ones (most of the offsets are less than 4). What is very interesting to see is that almost all of the detected boundaries have slight lags compared with the real ones. It is probably normal because there is a significant characteristic of clips in MSR action3D dataset. Each motion sequence has an obvious preparation time, but ends immediately when the action is done. So the "real natural" boundary is more likely to occur at the preparation time after the end of the previous action. However, annotating all the "real natural" boundaries in the testing set manually is far more complicated than using the original start and end points of clips as boundaries. It may also vary from person to person and is hard to absolutely avoid artificial errors. Since the slight offsets will not cause a big impact on the segmentation results, we do not change our simulation strategy.

Finally, we get 30 simulated video streams in total following our concatenation strategy. In our experiment, the detected boundary is determined to be wrong if the offset from itself to the corresponding natural boundary is more than 10, i.e., the detected boundary with an index of 200 in Figure 4 is a failure case. Figure 5 shows error rates of segmentation test for various tolerances of the offset. The result is satisfactory when offset tolerance gets to 10. Our segmentation method leads to a 9.07% of overall action detection error rate.

Nevertheless, we have to admit that the robustness of our segmentation is highly dependent on the classification ensemble built in the first step. If the probability distribution outputted by the classifier is unlikelihood, the boundary detection result will be meaningless.

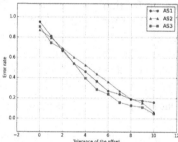

Fig. 5. Error rates of segmentation test for various tolerances of the offset

6   Conclusion

We presented a complete system for segmentation and recognition of 3D human actions based on 3D joints and DMM. We provide two different feature extraction methods, pairwise distances with normalization and DMM-HOG descriptor, for two kinds of data source captured by the latest depth camera. A SVM-based classification ensemble utilizing the combination of these features is proposed. The recognition process is effective and the recognition accuracy is impressive. For video segmentation, we developed PDD based dynamic boundary detection framework. The boundaries we find have no definite shape characteristics. If there is no interval between two different actions, we can also detect the boundary and separate them. Our framework is effective, online and reliable and leads to a detection error rate less than 10%. Each stage of the whole system is easy to implement. No expensive data processing and preferences setting are needed here.

Acknowledgment. This work was supported in part by the National Natural Science Foundation of China under Grant Nos. 61321491 and 61272218.

References

1. Li, W., Zhang, Z., Liu, Z.: Action recognition based on a bag of 3d points. In: 2010 IEEE Computer Society Conference on Computer Vision and Pattern Recognition Workshops (CVPRW), pp. 9–14. IEEE (2010)
2. Shotton, J., Sharp, T., Kipman, A., Fitzgibbon, A., Finocchio, M., Blake, A., Cook, M., Moore, R.: Real-time human pose recognition in parts from single depth images. Communications of the ACM 56, 116–124 (2013)
3. Li, W., Zhang, Z., Liu, Z.: Expandable data-driven graphical modeling of human actions based on salient postures. IEEE Transactions on Circuits and Systems for Video Technology 18, 1499–1510 (2008)
4. Zhang, S.: Recent progresses on real-time 3d shape measurement using digital fringe projection techniques. Optics and Lasers in Engineering 48, 149–158 (2010)
5. Weinland, D., Ronfard, R., Boyer, E.: A survey of vision-based methods for action representation, segmentation and recognition. Computer Vision and Image Understanding 115, 224–241 (2011)
6. Yang, X., Tian, Y.: Eigenjoints-based action recognition using naive-bayesnearest-neighbor. In: IEEE Computer Society Conference on Computer Vision and Pattern Recognition Workshops (CVPRW), pp. 14–19. IEEE (2012)
7. Yang, X., Zhang, C., Tian, Y.: Recognizing actions using depth motion maps-based histograms of oriented gradients. In: Proceedings of the 20th ACM International Conference on Multimedia, pp. 1057–1060. ACM (2012)
8. Wang, J., Liu, Z., Chorowski, J., Chen, Z., Wu, Y.: Robust 3D action recognition with random occupancy patterns. In: Fitzgibbon, A., Lazebnik, S., Perona, P., Sato, Y., Schmid, C. (eds.) ECCV 2012, Part II. LNCS, vol. 7573, pp. 872–885. Springer, Heidelberg (2012)
9. Wang, J., Liu, Z., Wu, Y., Yuan, J.: Mining actionlet ensemble for action recognition with depth cameras. In: 2012 IEEE Conference on Computer Vision and Pattern Recognition (CVPR), pp. 1290–1297. IEEE (2012)
10. Dalal, N., Triggs, B.: Histograms of oriented gradients for human detection. In: IEEE Computer Society Conference on Computer Vision and Pattern Recognition, CVPR 2005, vol. 1, pp. 886–893. IEEE (2005)
11. Wu, D., Zhu, F., Shao, L.: One shot learning gesture recognition from rgbd images. In: 2012 IEEE Computer Society Conference on Computer Vision and Pattern Recognition Workshops (CVPRW), pp. 7–12. IEEE (2012)
12. Rubin, J.M., Richards, W.A.: Boundaries of visual motion (1985)
13. Bradski, G., Kaehler, A.: Learning OpenCV: Computer vision with the OpenCV library. O'Reilly (2008)
14. Bauer, E., Kohavi, R.: An empirical comparison of voting classification algorithms: Bagging, boosting, and variants. Machine Learning 36, 105–139 (1999)

Weighted Deformable Part Model
for Robust Human Detection

Tianshuo Li[1], Yanwei Pang[1], Jing Pan[1,2], and Changshu Liu[1]

[1] School of Electronic Information Engineering, Tianjin University, Tianjin 300072, China
[2] School of Electronic Engineering,
Tianjin University of Education and Technology, Tianjin 300222, China

Abstract. Due to human pose articulation, variation in human shapes and appearances, especially occlusion between human and objects, one challenging problem in human detection is detect partially or completely occluded humans. In this paper, we propose a novel human detection approach capable of handling partial occlusion by combining the deformable part model and mean-shift algorithm. Since the training of root and part filters in deformable part model is separated, we construct the occlusion confidence map through root filter. Then mean-shift algorithm is used to segment the confidence map. The segmented portion of the window with a majority of negative response is referred to the partial occlusion. By constructing the relationship between deformable part model and the segmented sliding window, the negative influence of occlusion part is significantly weakened. This method dramatically decreases the effect of HOG feature of any occluded part during the decision-making process. Experimental results on the well-known INRIA data set demonstrate the effectiveness of the proposed method to solve the occlusion problem.

Keywords: Partial occlusion, human detection, deformable part model.

1 Introduction

Object detection is one of the fundamental challenges in computer vision. Reliable object detection is essential to image understanding and video analysis [7]. Among the object detection, Human detection plays a more important role and attracts more attention in applications of video surveillance, event analysis, smart room and wearable computing [14].

Human detection is challenging due to human pose articulation, variation in human shapes and appearances, especially partial occlusion [8]. Partial occlusion still poses a major challenge to state-of-art of the detector, as becomes apparent when analyzing the results of current benchmark datasets. The position of occlusion is arbitrary, sliding window method can not accurately locate the position of occlusion, resulting in degraded classification results.

Deformable Part Model (DPM) [9] is an efficient object detector that achieves state-of-the-art human detection results. DPM scans the image in a sliding window manner. The score of a detection window is the score of the root filter on the window

D.-S. Huang et al. (Eds.): ICIC 2014, LNCS 8588, pp. 764–775, 2014.

plus the sum over parts minus the deformation cost. The score of the root and part filters is calculated by the dot product between a set of weights and histogram of gradient features within a window. The features for the part filters are computed at twice the spatial resolution of the root filter. The model is defined at a fixed scale, so the object has been detected by searching over an image pyramid.

Although DPM method achieves a better performance in human detection, if a portion of the scanning window is occluded, the features inside the scanning window are densely selected, then the features corresponding to the occluded area are inherently noisy. This disturbance deteriorates the classification result of the score of DPM model and then seriously reduce the performance of the classifier. In order to avoid the negative influence of the occluded part score, we propose two ways to solve the problem of occlusion.

In DPM, the weights of the root filter and part filters are the same. However, the block energy of part filter for classification is different from each other. So the weights of the root filter and part filters should be different accordingly. According to the score map, if the region of a part filter is occluded, the score of the part model will be lower than unoccluded. The weights of unoccluded part filters should be increased and occluded parts should be decreased. In this way, the score of the unoccluded parts plays a leading role, so that the negative affect on classification can be reduced significantly.

Fig. 1. Some example proposed by DPM and our method. The first column shows the results of DPM, while the second are our approach.

According to the HOG-LBP [1] method, the Histogram of Oriented Gradients (HOG) vector in scanning window is constructed by 75 gradient histograms extracted from 15*5 blocks. By noticing the linearity of the scalar product, the linear SVM score of each scanning window is actually an inner product between the HOG vector and the weight vector of SVM. For each sliding window with ambiguous classification score, the possible occlusion regions can be segmented out by executing image segmentation algorithm on the binary occlusion likelihood image [1]. The mean-shift algorithm is applied to segment the binary image for each sliding window. The segmented regions with a negative overall response are defined as an occluded region. If the occluded part and unoccluded part in the sliding windows can be determined, the weight of occluded part filter can be reduced manually. Fig. 1. shows the result of DPM and our approach.

In this paper, we generalize and extend the above method to DPM. Specifically, the contribution of the paper is two-fold: (1) We propose a method of improving the score of root and part filters by strengthening the weights of key part and reducing the occluded one. (2) Through occlusion inference on sliding window classification results, we propose an approach to adjust the weight of part filters to reduce the negative influence of the occluded area.

2   Related Work

Wu et al. [16] used Bayesian combination to combine the part detector to get a robust detection in the situation of partial occlusion. They assumed the humans walk on a ground plane and the image was captured by a camera looking down to the ground. Gao et al. [12] proposed that the bounding box with a set of binary variables each of which corresponded to a cell indicating whether the pixels in the cell belong to the object. The method was able to use segmentation reasoning to achieve improved detection results. OuYang et al. [8] developed a probabilistic framework to model the relationship between the configurations estimated by single and multi-human detectors, and to refine the single-human result with multi-human detection.

A method was first proposed to predict the occlusion information by constructing the occlusion likelihood map by Wang et al. [1]. They used the response of each block of HOG feature to the global detector, and then the map was segmented by mean-shift algorithm. After segmentation, the positive response referred to the unoccluded area while the negative response referred to the occluded one. When the segmented area was determined, the part detectors were applied on the unoccluded regions to achieve the final classification on the current scanning window. Arteta et al. [2] proposed a set of candidate regions, and then selected regions based on optimizing a global classification score, which subjected to the constraint that the selected regions are non-overlapping. The use of the model was particular attractive for biomedical images but the limitations of the proposed method appears when the instances become even denser than in the considered datasets. Felzenszwalb et al. [9] provided an elegant framework for object detection. The main contribution of the method was the use of deformable parts and the latent discriminative learning. DPM consisted of a global "root" filter and several part models. In detection stage, this method scanned the image using a sliding window, the score of the detection window is the score of the root filter plus the part filters. In training, they defined a generalization of SVM (LSVM) to learn the model.

3   The Proposed Algorithm

3.1   Weighted DPM

We start by reviewing a general framework for object detection with DPM proposed in [9]. The model is reviewed by a coarse root filter which covers the whole objects and several high resolution part filters which cover small part of the objects. The root filter location defines the detection window while the part filters are placed several levels down in the pyramid, so the HOG cells at that level have half the size of cells in the root level.

Using higher resolution feature for part filter is essential for the good performance of DPM [9]. The part filters capture higher resolution features and hence lead to more precise detection result than the root filter. Consider building a model for pedestrian, the root filter seizes the whole boundary of human and the part filter captures the details such as face, leg, arm and torso. However, not all the part features play the same function when estimating whether the sliding windows contain human. Face and leg part filters play a leading role for classification. Therefore, proper tuning the weights of different part filters can achieve better result.

Fig. 2. The image in the left shows the detection result for pedestrian, the red box is root filter and part filters are blue boxes. The picture in the middle is the root filter while the picture in the right is the high resolution part filters [9].

DPM model is defined by a root filter F_0 and a set of part filters $(P_1,...,P_n,b)$ where $P_i = (F_i, V_i, d_i, w_i)$. Here F_i is the i-th part filter. V_i is referred to a two-dimensional vector specifying the center for a box of possible positions for part i relative to the root position, and d_i is a four-dimensional vector specifying coefficients of a quadratic function defining a deformable cost for each possible placement of the part relative to the anchor position. w_i means that the different weight of the part filters. Fig .2. illustrates a human model.

The location of each filter is defined by $Z = (P_0,...,P_n)$, where $P_i = (x_i, y_i, l_i)$ specifies the level and position of the i-th filter. The score of a hypothesis is given by the scores of each filter at their respective locations minus a deformable cost that depends on the relative position with respect to the root, and then plus the biased.

$$score(P_0,...,P_n) = \sum_{i=0}^{n} w_i \times F_i^{'} \times \Theta(H,P_i) - \sum_{i=1}^{n} w_i \times d_i \times \Theta_d(dx_i,dy_i) + b \quad (1)$$

where

$$(dx_i,dy_i) = (x_i,y_i) - (2(x_0,y_0) + V_i) \quad (2)$$

$$\Theta_d(dx,dy) = (dx,dy,dx^2,dy^2) \quad (3)$$

In (1), $\Theta(H,P,w,h)$ denotes the vector obtained by concatenating the feature vectors in the $w \times h$ sub-window of H with top-left corner at P in row-major order.

Assume that $d = (0,0,1,1)$, the deformation cost for the i-th part is the square distance between its actual position and anchor position relative to the root. In general, the deformation cost is an arbitrary separable quadratic function of the displacements [9]. Then we can transform the equation in the following way:

$$score(P_0,...,P_n) = F_0^{'} \times \Theta(H,P_i) + \sum_{i=1}^{n} (w_i \times F_i \times \Theta(H,P_i) - w_i \times d_i \times \Theta_d(dx_i,dy_i)) + b \quad (4)$$

$$partscore(P_1,...,P_n) = \sum_{i=1}^{n} (w_i \times F_i \times \Theta(H,P_i) - w_i \times d_i \times \Theta_d(dx_i,dy_i)) \quad (5)$$

$$score(P_0,...,P_n) = F_0^{'} \times \Theta(H,P_i) + partscore(P_1,...,P_n) + b \quad (6)$$

$$w_i = a^{partscore(P_1,...,P_n)} \quad (7)$$

In (1)(4)(5), the variable w_i can be calculated through $partscore$, the weight of the part filter is determined by w_i. After adjusting the weights of part filters, we must normalize the weights to adjust to the threshold of the model.

3.2   Combine Root and Part Filters for Occlusion Handling

Due to the training of root filter and part filters are separated, we can fully use the information of linear root filter model. The whole framework is shown in Fig. 3.

If a part of pedestrian is occluded, the densely extracted blocks of features in occluded area respond to root classifier with negative products. In order to take advantage of the phenomenon, we separate the model into blocks, the classification score of each block infer whether the occlusion occurs and where it occurs [1]. The HOG vector of scanning windows is 2700 dimensions. The 2700 dimensional feature is constructed by HOG of 75 blocks. Then the 2700 dimensional feature can be transformed into the sum of 75 blocks. So the decision function of SVM is:

$$f(x) = \beta + \sum_{k=1}^{l} a_k <x, x_k> \tag{8}$$

x_k are the support vectors of the root filter. Considering the linearity of the scalar product, we will transform the decision function into:

$$f(x) = \beta + \sum_{k=1}^{l} a_k <x, x_k> = \beta + w^T x \tag{9}$$

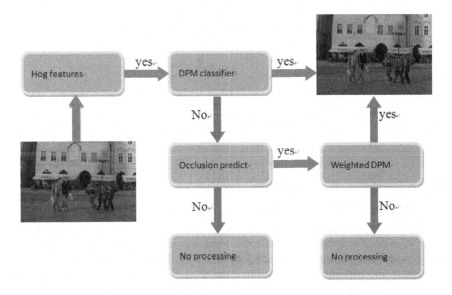

Fig. 3. Occlusion handling framework

In (9), the parameter w is defined as the support vectors and x means the feature captured by HOG. We can put the bias β into each block B_i, then contribution of the blocks are calculated by the summation of feature inner production subtract the corresponding bias. That is, to find a set of β_i such that $\beta = \sum \beta_i$, (9) will become:

$$f(x) = \beta + w^T x = \sum_{i=1}^{75} (\beta_i + w_i B_i) \tag{10}$$

The parameter β_i should be learnt from the training set of INRIA dataset by calculating the relative ratio of the bias β_i in each block to the total bias constant [1]. The set of positive example is: $p = 1,...,N^+$ while the set of negative example is: $q = 1,...,N^-$. Then the i-th block can be defined as $B_{p;i}^+$ and $B_{q;i}^-$, the following equation is the calculation all positive and negative classification score.

$$\sum_{p=1}^{N^+} f(x_p^+) = S^+ = N^+ \beta + \sum_{p=1}^{N^+} \sum_{i=1}^{75} w_i^T B_{p;i}^+ \tag{11}$$

$$\sum_{p=1}^{N^-} f(x_q^-) = S^- = N^- \beta + \sum_{p=1}^{N^-} \sum_{i=1}^{75} w_i^T B_{q;i}^- \tag{12}$$

denote $A = -\dfrac{S^-}{S^+}$, we can transform (11) (12) into:

$$AN^+ \beta + N^- \beta + \sum_{i=1}^{75} w_i^T (A\sum_{p=1}^{N^+} B_{p;i}^+ + \sum_{q=1}^{N^-} B_{q;i}^-) = 0 \tag{13}$$

denote $B = -\dfrac{1}{AN^+ + N^-}$, we have:

$$\beta_i = B \times w_i^T \times (A\sum_{p=1}^{N^+} B_{p;i}^+ + \sum_{q=1}^{N^-} B_{q;i}^-) \tag{14}$$

Through (14), we definitely distribute the bias β_i into each block. The negative block is denoted as B_i^-, similarly the positive block is denoted as B_i^+. If the geometric location of B_i^- are close to each other, while B_i^+ with the same nature, we tend to conclude the scanning window contains a human, who is partially occluded in the location. After the prediction, the mean-shift algorithm is applied to segment this binary window for each sliding window. Then the binary likelihood image can be segmented into different areas. Through the segmented area, we can classify if the part filter locate on the occluded area. Then manually adjust the weight of the part filter presented in the following:

$$partscore(P_1,...,P_n) = \sum_{i=1}^{n} (w_i \times F_i \times \Theta(H, P_i) - w_i \times d_i \times \Theta_d(dx_i, dy_i)) \tag{15}$$

$$score(P_0,...,P_n) = F_0' \times \Theta(H, P_i) + partscore(P_1,...,P_n) + b \tag{16}$$

$$w_i = \begin{cases} c_1 \times a^{partscore(P_1,...,P_n)} (area \in unoccluded) \\ c_2 \times a^{-partscore(P_1,...,P_n)} (area \in occluded) \end{cases} \tag{17}$$

$$c1 + c2 = 1 \tag{18}$$

In (15)(16)(17), the variable w_i can be calculated through the combination of *partscore* and occlusion prediction. In the sliding windows, the occluded area shares a lower weight, whereas the unoccluded one shares the high one. The weight of the part filters is determined by w_i. After adjust the weight of part filters, we must normalize the weights to adjust to the threshold of the model.

Our experiments on the INRIA dataset showed that the weighted DPM could effectively detect the partial occlusion example. Based on the localization of the occluded portion, the disturbance of the occluded object feature reduced due to changing the weights of part filters.

4   Experimental Results

In this section, two couple of experiments are taken to validate our assumptions. We firstly study the relationship of part filters, and then adjust to the weight of the part filters depending on its score. This method will compare with DPM in INRIA dataset. The second group experiment shows that the position of occlusion prediction in root template and weight adjustment are effective way to handle with partial occlusion. Due to INRIA dataset contains very few occluded pedestrian, we manually increase occlusion information in the dataset. After raising occlusion information, Some of the results are shown below in Fig. 4.

Fig. 4. Some samples in INRIA after gaining occlusion information

4.1   Weighted DPM Result

We use augmented HOG as the feature vector, and use LSVM to train the root filter and part filters for the human detection on INRIA dataset. The detection rate vs. False Positive Per Image (FPPI) criteria is used to evaluate the weighted DPM and the traditional DPM in INRIA dataset and occlusion INRIA dataset. Results are shown below in Fig. 5. and Fig. 6.

From Fig. 5. and Fig. 6. We can see weighted DPM method (the red line) is slightly better than traditional DPM method (the green line). Our approach achieves a detection rate of 93.4% at FPPI = 1 and 81.7% at FPPI = 0.1, while DPM method achieves a detection rate of 90.3% at FPPI = 1 and 80.4% at FPPI = 0.1. The detection rate is improved from 90.3% to 93.4% at FPPI = 1 from 80.4% to 81.7% at FPPI = 0.1. We can use the same method to analyze the occlusion INRIA dataset in Fig. 8. We find that the detection raises about 1.5%~3%. In Fig. 6., the difference between the red line and green line is small due to the weighted DPM model is not effective when dealing with occlusion problem.

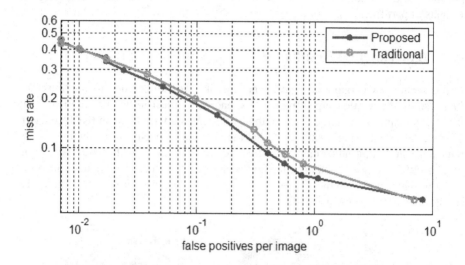

Fig. 5. Weighted DPM method and traditional DPM method in INRIA dataset

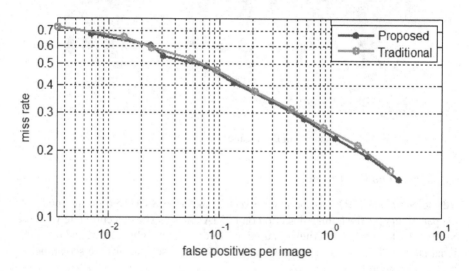

Fig. 6. Weighted DPM method and traditional DPM method in occlusion INRIA dataset

4.2   The Result of Weighted DPM with Occlusion Handling

In order to evaluate the proposed occlusion handling approach, we study the partial occlusion INRIA dataset. We gain the prediction of the position of the occlusion in sliding windows in weighted DPM. Then the detection rate vs. False Positive Per Image (FPPI) criteria is set to evaluate our approach and the traditional DPM in INRIA dataset and occlusion INRIA dataset.

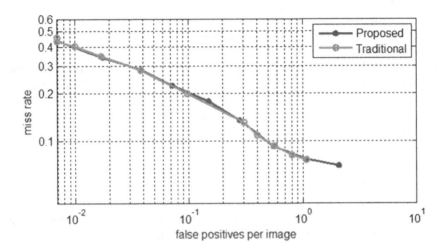

Fig. 7. Weighted DPM method with occlusion handling and traditional DPM method in INRIA dataset

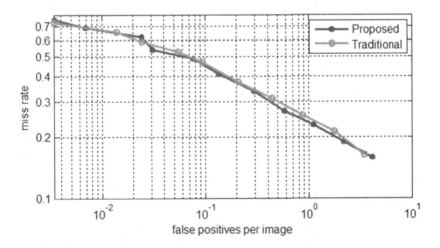

Fig. 8. Weighted DPM method with occlusion handling and traditional DPM method in occlusion INRIA dataset

From Fig. 7. we can see our method (the red line) is approximately the same as Fig. 5. in INRIA dataset. While in occlusion INRIA dataset From Fig. 8., the red line is slightly better than the green one. The detection rate raises from 74.1% to 76.5% at

FPPI = 1. The experiment result shows us that through the usage of occlusion prediction and weight adjustment, our approach is effective in handling occlusion problem.

5   Conclusion

Traditional DPM method does not explicitly handle occlusion and therefore it is sensitive to occlusion. To overcome this drawback, we, in this paper, have proposed a method of adjusting the score of part filters where the key part is strengthened while other parts are weakened. Moreover, through occlusion inference on sliding window classification results, we have proposed an approach to adjust the weight of part filters to reduce the influence of the occluded area. It has been demonstrated that our approach outperforms the state-of-the-art of DPM in INRIA dataset.

Acknowledgements. This work was supported in part by the National Basic Research Program of China 973 Program (Grant No. 2014CB340400), the National Natural Science Foundation of China (Grant Nos. 61172121, 61271412, and 61222109), and the Open Funding Project of State Key Laboratory of Virtual Reality Technology and Systems, Beihang University (Grant No. BUAA-VR-13KF)

References

1. Wang, X., Han, T.X., Yan, S.: An Hog-Lbp Human Detector With Partial Occlusion Handling. In: Proceedings of the International Conference on Computer Vision, pp. 32–39 (2009)
2. Arteta, C., Lempitsky, V., Noble, J., Zisserman, A.: Learning to Detect Partially Overlapping Instances. In: Proceedings of the IEEE Conference on Computer Vision And Pattern Recognition, pp. 3230–3237 (2013)
3. Wojek, C., Walk, S., Roth, S., Schindler, K., Schiele, B.: Monocular Visual Scene Understanding: Understanding Multi-Object Traffic Scenes. IEEE Transactions on Pattern Analysis and Machine Intelligence 35(4), 882–897 (2013)
4. Pepik, B., Stark, M., Gehler, P., Schiele, B.: Occlusion Patterns for Object Class Detection. In: Proceedings of the IEEE Conference on Computer Vision and Pattern Recognition, pp. 3286–3293 (2013)
5. Hsiao, E., Hebert, M.: Occlusion Reasoning for Object Detection Under Arbitrary Viewpoint. In: Proceedings of The IEEE Conference on Computer Vision and Pattern Recognition, pp. 3146–3153 (2012)
6. Rujikietgumjom, S., Collins, R.T.: Optimized Pedestrian Detection for Multiple and Occluded People. In: Proceedings of the IEEE Conference on Computer Vision and Pattern Recognition, pp. 3690–3697 (2013)
7. Maji, S., Shakhnarovich, G.: Part Discovery From Partial Correspondence. In: Proceedings of the IEEE Conference on Computer Vision and Pattern Recognition, pp. 931–938 (2013)
8. Ouyang And, W., Wang, X.: Single-Pedestrian Detection Aided By Multi-Pedestrian Detection. In: Proceedings of the IEEE Conference on Computer Vision and Pattern Recognition, pp. 3198–3205 (2013)

9. Felzenszwalb, P., Girshick, R., Mcallester, D., Ramanan, D.: Object Detection With Discriminatively Trained Part Based Models. IEEE Transactions on Pattern Analysis and Machine Intelligence 32(9), 1627–1645 (2010)

10. Dollár, P., Wojek, C., Schiele, B., Perona, P.: Pedestrian Detection: An Evaluation of the State of the Art. IEEE Transactions on Pattern Analysis and Machine Intelligence 34(4), 743–761 (2012)

11. Pang, Y., Yan, H., Yuan, Y., Wang, K.: Robust Cohog Feature Extraction in Human Centered Image/Video Management System. IEEE Transactions on Systems, Man and Cybernetics. Part B: Cybernetics 42(2), 458–468 (2012)

12. Gao, T., Packer, B., Koller, D.: A Segmentation-Aware Object Detection Model With Occlusion Handling. In: Proceedings of the IEEE Conference on Computer Vision and Pattern Recognition, pp. 3198–3205 (2013)

13. Felzenszwalb, P., Girshick, R., Mcallester, D.: Cascade Object Detection With Deformable Part Models. In: Proceedings of the IEEE Conference on Computer Vision and Pattern Recognition, pp. 2241–2248 (2010)

14. Pang, Y., Yuan, Y., Li, X., Pan, J.: Efficient Hog Human Detection. Signal Processing 91(4), 773–781 (2011)

15. Dalal And, N., Triggs, B.: Histograms of Oriented Gradients of Human Detection. In: Proceedings of The IEEE Conference on Computer Vision and Pattern Recognition, pp. 886–893 (2005)

16. Wu And, B., Nevatia, R.: Tracking of Multiple, Partially Occluded Humans Based on Static Body Part Detection. In: Proceedings of the IEEE Conference on Computer Vision and Pattern Recognition, pp. 951–958 (2006)

Contractive De-noising Auto-Encoder

Fu-qiang Chen, Yan Wu[*], Guo-dong Zhao, Jun-ming Zhang, Ming Zhu, and Jing Bai

College of Electronics and Information Engineering, Tongji University, Shanghai, China
yanwu@tongji.edu.cn

Abstract. De-noising auto-encoder (DAE) is an improved auto-encoder which is robust to the input by corrupting the original data first and then reconstructing the input by minimizing the error function. And contractive auto-encoder (CAE) is another kind of improved auto-encoder learning robust feature by introducing Frobenius norm of the Jacobean matrix of the learned feature with respect to the input. In this paper, we combine DAE and CAE, and propose contractive de-noising auto-encoder (CDAE), which is robust to both the original input and the learned feature. We stack CDAE to extract more abstract features and apply SVM for classification. The experiment on benchmark dataset MNIST shows that CDAE performed better than CAE and DAE.

Keywords: De-noising Auto-encoder, Contractive Auto-encoder, SVM.

1 Introduction

Neural network is composed of three layers, one input layer, one hidden layer and one output layer. For the input layer, we can input the original data to it. While for the hidden layer, we need compute a transformation of the input to get features from the input. For the output layer, we can get the values of the neurons as the label of the corresponding input.

In 1988, Bourlard and Kamp [1] first proposed auto-association, a kind of neural network which makes the output layer same as the input layer. And auto-association is the original auto-encoder. It can be learned by error back propagation (BP) [2]. In 2006, Hinton and Salakhutdinov [3] stacked auto-encoders to get deep auto-encoders, and they pre-trained the deep auto-encoder with restricted Boltzmann machine (RBM) [4].

In 2008, Vicent P. et al. [5] proposed de-noising auto-encoder (DAE), a kind of improved auto-encoder which first corrupts the original data and then transforms or encodes the corrupted input to the hidden layer and then decodes to the 'corrupted' output layer. In 2010, Vicent P. et al. stacked DAE (SDAE) to form deep neural network and learn useful representations [6], and they showed SDAE performed better than deep belief nets (DBN) in several cases. They found DAE could learn Gabor-like edge detectors from natural image patches and larger stroke detectors from digits [6].

Then in 2011, Rifai et al. [7] proposed another improved auto-encoder, contractive auto-encoder (CAE). They improved auto-encoder by introducing the Frobenius norm

---

[*] Corresponding author.

D.-S. Huang et al. (Eds.): ICIC 2014, LNCS 8588, pp. 776–781, 2014.
© Springer International Publishing Switzerland 2014

of the Jacobean matrix of all the learned features in the hidden layer with respect to the original input for the reconstruction loss function in AE.

In this paper, we combined DAE and CAE, and we proposed contractive de-noising auto-encoder (CDAE). CDAE is robust to the input and makes the learned feature robust by adding a penalty term to the reconstruction loss function. After extracting feature with CDAE, we apply SVM for classification. The experiment on benchmark dataset MNIST shows CDAE surpasses CAE and DAE, proving its effectiveness.

2   CDAE

In this section, we will present our proposed algorithm, CDAE.

2.1   AE

Let x represents a specific input or sample for AE, d_v the input layer size or the dimension for the input, d_h the size of the hidden layer or the number of neurons in the hidden layer. To encode x into the hidden layer, we can use the following activation function:

$$\mathbf{h} = sigmoid\left(\mathbf{Wx} + \mathbf{b}\right) \tag{1}$$

where $sigmoid\left(x\right) = \dfrac{1}{1 + e^{-x}}$.

Another commonly used activation function is $tanh\left(x\right) = \dfrac{e^{x} - e^{-x}}{e^{x} + e^{-x}}$. In formula (1), W is a matrix of $d_h \times d_v$ (usually $d_h < d_v$), while $\mathbf{b}$ is a $d_h \times 1$ bias vector for the hidden layer. Thence, $\mathbf{Wx+b}$ is a column vector, and the sigmoid function maps $\mathbf{Wx+b}$ element-wisely, then we can get a column vector of size $d_x \times 1$.

Then we decode the hidden layer to the output layer by the following formula:

$$\mathbf{x}_{rec} = sigmoid\left(W^{T}\mathbf{h} + \mathbf{c}\right)$$

where the subscript 'rec' means reconstruction considering that the output of an auto-encoder is the input itself. And $\mathbf{c}$ is a $d_x \times 1$ bias vector for the output layer.

Last, we compute and minimize the loss function with respect to W, $\mathbf{b}$ and $\mathbf{c}$ by:

$$\text{Minimize } L_{AE} = \frac{1}{N}\sum\nolimits_{x \in D}\left(\mathbf{x} - \mathbf{x}_{rec}\right)^{2} \tag{2}$$

where D is the set composed of all the input samples, while N is the cardinality of set D. Here the loss function is the squared error function, which is a widely used cost function in neural network. Another cost function (cross-entropy loss function) is:

$$L(\mathbf{x}, \mathbf{x}_{rec}) = L_{cor-en} = -\sum \mathbf{x}_i log\mathbf{x}_{rec_i} + (1 - \mathbf{x}_i)log(1 - \mathbf{x}_i).$$

For clarity, we show the architecture of the auto-encoder in Fig. 1.

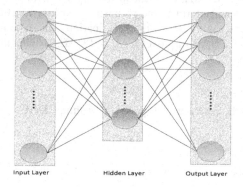

Fig. 1. The architecture of auto-encoder, in which the output is identical with the input

2.2   DAE

In DAE, we need to corrupt the original data firstly. There are two common ways to corrupt the original data, additive isotropic Gaussian noise and binary masking noise. Here, we denote the corrupted version of the original input sample $\mathbf{x}$ as $\mathbf{\tilde{x}}$. Then, for additive isotropic Gaussian noise,

$$\mathbf{\tilde{x}} = \mathbf{x} + N(0, \sigma^2 I)$$

where σ is a small constant maybe usually smaller than 0.5 [6]. And I is the identity matrix. For binary masking noise, a small fraction of the input sample $\mathbf{x}$ is set to be zero, and the fraction can take values on 10%, 25% etc.

For a specific input $\mathbf{x}$, first we need to get the corrupted version of $\mathbf{x}$, $\mathbf{\tilde{x}}$. Then we get

$$\mathbf{\tilde{h}} = sigmoid(\mathbf{W}\mathbf{\tilde{x}} + \mathbf{b}),$$

where W and b satisfies the same condition as in AE.

Subsequently, we can get the reconstructed version of input $\mathbf{x}$ by

$$\mathbf{\tilde{x}}_{rec} = sigmoid(W^T\mathbf{\tilde{h}} + \mathbf{c}).$$

Finally, we can get the object function of DAE as following:

$$\text{Minimize } L_{DAE} = \frac{1}{N}\sum_{x \in D}(\mathbf{x} - \mathbf{\tilde{x}}_{rec})^2 \qquad (3)$$

For clarity, we show the architecture of DAE in Fig. 2.

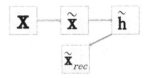

Fig. 2. The architecture of DAE

2.3   CAE

We can get CAE by adding one kind of penalty term to the original cost function (i.e. (2)) of AE. The added term is the Frobenius norm of the Jacobean matrix of the learned feature in the hidden layer with respect to the original input. The corresponding formula can be shown as following:

$$\text{Minimize } L_{CAE} = \frac{1}{N} \sum_{x \in D} \left((\mathbf{x} - \mathbf{x}_{rec})^2 + \lambda \| J_h(\mathbf{x}) \|_F^2 \right) \quad (4)$$

In formula (4), the λ is a parameter and it takes values of 0.1 in this paper [7]. For activation function $sigmoid$, the second term in formula (4) can be calculated by:

$$\| J_h(\mathbf{x}) \|_F^2 = \sum_{i=1}^{d_h} \left(\mathbf{h}_i (1 - \mathbf{h}_i) \right)^2 \sum_{j=1}^{dv} W_{ij}^2 ,$$

for activation function $tanh$, the second term in formula (4) can be calculated by:

$$\| J_h(\mathbf{x}) \|_F^2 = \sum_{i=1}^{d_h} \left((1 + \mathbf{h}_i)(1 - \mathbf{h}_i) \right)^2 \sum_{j=1}^{dv} W_{ij}^2 .$$

2.4   CDAE

From formulas (2-4), we can see that DAE modifies the traditional AE by corruption-reconstruction, which makes auto-encoder robust to the input, while CAE modifies the traditional AE by introducing a penalty term in the object function for every sample. CAE makes AE robust by changing the object function, which indeed makes some neurons in the hidden layer less active to learn more robust feature.

To combine the advantages of both CAE and DAE, we propose another improved AE. Here we call it contractive de-noising auto-encoder (CDAE). The object function of CDAE can be written as following:

$$L_{DAE} = \frac{1}{N} \sum_{x \in D} \left((\mathbf{x} - \mathbf{x}_{rec})^2 + \lambda \| J_h(\mathbf{x}) \|_F^2 \right) \quad (5)$$

We first corrupt the original data $\mathbf{x}$ to get $\tilde{\mathbf{x}}$. Then we can map $\tilde{\mathbf{x}}$ to the hidden layer and we can get $\tilde{\mathbf{h}}$.

Subsequently, we can get the reconstruction of the original sample $\mathbf{x}_{rec}$ by de-coding $\mathbf{\bar{h}}$. From the object function, we can see that CDAE is a directly improved version of CAE, by adding the corruption and de-noising. And similarly, CDAE is also a directly improved version of DAE by adding a penalty term to the object function.

3   Experiment

We apply support vector machine (SVM) [8] in our experiment. SVM is a commonly adopted classifier in various applications. It is based on kernel function, dual problem and convex quadratic programming. There are three common kernel functions in SVM: polynomial kernel function; Gaussian radial basis kernel function and *tanh* kernel function. The Gaussian radial basis kernel function is the most popular one. And we adopt Gaussian radial basis function in our experiment.

Our experiment data set is from MNIST [9] and to get the .mat form, the readers are referred to [10]. The experiment is implemented on MATLAB 2011b, win32, Intel(R) Core (TM) i3-2310, CPU @ 2.10GHz 2.00GB (RAM). Confined to the memory, we chose 18,000 of all the samples for our experiment, 1,800 for each of the number between 0 and 9.

For the 18,000 samples, we chose one half for training and another half for testing. And for each number between 0 and 9, we all chose half for training and half for testing. We applied one of the most common classifier, SVM [11] for classification.

In our experiment, the architecture of the deep auto-encoder is 784-200-50-200-784 [12]. And we chose the third layer to classify with SVM. The activation function we applied is *tanh*. And we corrupt the original input with masking noise, and we mask the original data as follows: we implement the masking by making the elements of the input with the index as 1:80:784. Besides, we initialize the weight matrix W with each element in the following range [13]:

$$\left(-\frac{\sqrt{6}}{\sqrt{d_v + d_h}}, +\frac{\sqrt{6}}{\sqrt{d_v + d_h}} \right)$$

And we initialize the bias vector $\mathbf{b}$ and $\mathbf{c}$ with $\mathbf{0}$.

For clarity, we show the architecture we use in our experiment (Fig. 3.).

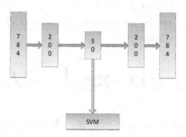

Fig. 3. The architecture of our model for 784-200-50-200-784

The experiment result is shown in the following table (Table 1). And we can see that our method performs best, beating AE, DAE and CAE.

Table 1. The experiment result of SVM classifying the feature extracted by AE, DAE, CAE and CDAE. The architecture is 784-200-50-200-784.

Method	Accuracy
AE	93.12%
DAE	93.28%
CAE	93.31%
CDAE	93.77%

4   Conclusion

In this paper, we proposed CDAE, a novel improved auto-encoder by combining CAE and DAE. CAE can make less neuron in the hidden layer active to learn more meaningful, abstract and robust features for classification, while DAE is robust to the input by first corrupting the original data and then minimizing the reconstruction loss function. The experiment shows that CDAE outperforms both CAE and DAE, which also shows the effectiveness of CDAE. Since CDAE performed better on MNIST, we'll try CDAE on other problems in future.

References

1. Bourlard, H., Kamp, Y.: Auto-association by multilayer perceptrons and singular value decomposition. Biological Cybernetics 59, 291 (1988)
2. Rumelhart, D.E., Hinton, G.E., Williams, R.J.: Learning representations by back-propagating errors. Nature 323, 533 (1986)
3. Hinton, G.E., Salakhutdinov, R.R.: Reducing the dimensionality of data with neural networks. Science 313, 504–507 (2006)
4. Hinton, G.E., Osindero, S., Teh, Y.W.: A fast learning algorithm for deep belief nets. Neural Computation 18, 1527 (2006)
5. Vincent, P., Larochelle, H., Bengio, Y., Manzagol, P.A.: Extracting and composing robust features with denoising autoencoders. In: ICML, pp. 1096–1103 (2008)
6. Vincent, P., Larochelle, H., Lajoie, I., Bengio, Y., Manzagol, P.A.: Stacked denoising autoencoders: Learning useful representations in a deep network with a local denoising criterion. The Journal of Machine Learning Research 11, 3371 (2010)
7. Rifai, S., Vincent, P., Muller, X., Glorot, X., Bengio, Y.: Contractive auto-encoders: Explicit invariance during feature Extraction. In: ICML, pp. 833–840 (2011)
8. Burges, C.J.: A tutorial on support vector machines for pattern recognition. Data Mining and Knowledge Discovery 2, 121 (1998)
9. http://yann.lecun.com/exdb/mnist/
10. http://www.cs.toronto.edu/~hinton/code/converter.m
11. Chang, C.C., Lin, C.J.: LIBSVM: a library for support vector machines. ACM Transactions on Intelligent Systems and Technology 2, 27 (2011)
12. Ngiam, J., Coates, A., Lahiri, A., Prochnow, B., Ng, A., Le, Q.V.: On optimization methods for deep learning. In: ICML, pp. 265–272 (2011)
13. Glorot, X., Bengio, Y.: Understanding the difficulty of training deep feedforward neural networks. In: AISTATS, pp. 249–256 (2010)

A Hybrid Algorithm to Improve the Accuracy of Support Vector Machines on Skewed Data-Sets

Jair Cervantes[1], De-Shuang Huang[2], Farid García-Lamont[1],
and Asdrúbal López Chau[1]

[1] Posgrado e Investigación UAEMEX (Autonomous University of Mexico State)
Av. Jardín Zumpango s/n, Fracc, El Tejocote, Texcoco, 56259, Mexico
[2] Department of Control Science & Engineering,
Tongji University Cao'an Road 4800, Shanghai, 201804 China

Abstract. Over the past few years, has been shown that generalization power of Support Vector Machines (SVM) falls dramatically on imbalanced data-sets. In this paper, we propose a new method to improve accuracy of SVM on imbalanced data-sets. To get this outcome, firstly, we used undersampling and SVM to obtain the initial SVs and a sketch of the hyperplane. These support vectors help to generate new artificial instances, which will take part as the initial population of a genetic algorithm. The genetic algorithm improves the population in artificial instances from one generation to another and eliminates instances that produce noise in the hyperplane. Finally, the generated and evolved data were included in the original data-set for minimizing the imbalance and improving the generalization ability of the SVM on skewed data-sets.

Keywords: Support Vector Machines, Hybrid, Imbalanced.

1 Introduction

Many real-world applications show imbalance in data-sets. In these problems the goal in classification problems is to find a function that best generalizes the minority class, usually it is the most significant one. Traditionally, classical classification methods do not perform well on imbalanced data-sets, because they were not designed to address such problems. Support Vector Machines (SVM) have shown excellent generalization power in classification problems. However, it has been shown that this generalization ability of SVM drops dramatically on skewed data-sets [7] [10]. The most widely used techniques to tackle this kind of problems are under-sampling, over-sampling and Synthetic Minority Over-sampling Technique (SMOTE) [2]. Under-sampling, gets the number of instances m in the minority class and selects randomly m instances in the majority class. Over-sampling technique eliminates the imbalance by replicating data instances in the minority class or generating artificial instances from the minority class. SMOTE Generates artificial instances over-sampling the minority class by taking each minority class instance and generating synthetic instances along the line segments, joining any or all of the k minority class nearest neighbors. It does not cause any information loss and could potentially find hidden minority regions.

D.-S. Huang et al. (Eds.): ICIC 2014, LNCS 8588, pp. 782–788, 2014.

However, it could introduce noise in the classifiers which could result in a loss of performance because the algorithm makes the assumption that the instance between a positive class instances and its nearest neighbors is also positive [8]. Several techniques inspired in SMOTE algorithm have been proposed [12][13][1][10][9]. Methods based on Evolve Algorithms (EA) have also proposed in the literature. [11] proposed an algorithm to expands the minority class boundary. The algorithm uses a Random Walk Over-Sampling approach (RWO-Sampling) to balancing different class samples by creating synthetic samples through randomly walking from the real data. In [14] is proposed a hybrid learning model to cope with the problem of imbalanced by evolving self-organizing maps. The authors use GA to evolve the subset of the minority examples into new stage that might discover novel knowledge from the limited and underrepresented minority class. In [3], the authors proposed a classification system in order to detect the most important rules, and the rules which perturb the performance of classifier. That system uses hierarchical fuzzy rules and a GA. Garcia et al. [4] implemented an algorithm which performs an optimized selection of examples from data sets. The learning algorithm based on the nested generalized exemplar method and GA to generate and select the best suitable data to enhance the classification performance over imbalanced domains.

In this paper, we present a new algorithm to generate artificial instances in order to improve the performance of SVM on imbalanced data-sets. In the proposed algorithm, GA is used to guide the search process of new artificial instances. The artificial instances are evolved using a GA, each generation the instances that best contribute to the performance of the data-set are improved by selection, cross and mutation operators, reducing the likelihood of add noise instances to the classifier.

2   SVM Classification via Genetic Algorithm

Formally, given a data set $\{(xi, yi)\}_{i=1}^{n}$ and separating hyperplane $f(x) = w_i^T x + b = 0$, the shortest distance from separating hyperplane to the closest positive example in the non separable case is

$$\gamma + = \min \gamma_i, \forall \gamma_i \in class + 1 \tag{1}$$

the shortest distance from separating hyperplane to the closest negative example is

$$\gamma - = \min \gamma_i, \forall \gamma_i \in class - 1 \tag{2}$$

where γ_i is given by

$$\frac{y_i \left(w_i^T K \langle x_i \cdot x_j \rangle + b_i \right)}{\|w\|} \tag{3}$$

The margin is

$$\gamma = \gamma_+ + \gamma_- \tag{4}$$

Methods based on the SMOTE algorithm introduce artificial instances in minority class in order to reduce the bias. However, SMOTE only introduces artificial instances between positive instances (one positive instance and its k nearest neighbors). Furthermore, the region with more information is between support vectors with different label. Introducing artificial instances in this region could help improve the performance of SVM. In order to avoid off noise by adding artificial instances and introduce new artificial data points with high discriminative features, we propose a novel algorithm based on a genetic algorithm. Figure 1 describes the general process of the proposed method.

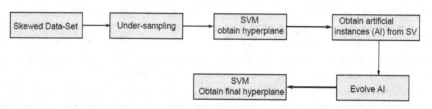

Fig. 1. Proposed method

2.1   Generating Artificial Instances

The proposed algorithm starts using under-sampling and obtaining an initial decision boundary by SVM learning. SVM is trained by X_{tr}^+ and X_{tr}^- to obtain support vectors. The hyperplane and the decision function obtained is skew due to the imbalanced data $\left(X_{tr}^+, X_{tr}^-\right)$. SV obtained in this first stage of SVM are used to generate the new population using a distance between positive SVs and negative SVs. For each SV in the minority class sv^+, the algorithm finds the k nearest neighbors in the

sv^- and calculate the distance between them for each dimension. The distance v_i is given by

$$v_i = x_{sv+}^i - x_{sv-}^j, i = 1,...,d \tag{5}$$

where x_{sv+}^i is the i-esime SV of X_{tr}^+, and x_{sv-}^j represent the j-esime sv^- nearestneighbors of x_{sv+}^i. Initial vector $v_i = 0, i = 1,...,d.$ and the algorithm pick one or more random entries out of an array. In our experiments we select only one. The artificial instance is obtained by

$$x_g = x_{sv+}^i + \varepsilon \cdot v_i \tag{6}$$

which is modified in just the i-esime dimension of x_{sv+}^i. Where the step size is ε; we select it between 0.1 and 0.001.

2.2   Genetic Algorithm

SV obtained are used to generate a population for a genetic algorithm. The artificial instances added to data-set could potentially find hidden regions in minority class because these instances were obtained with the most discriminative features of the entire data set. However, the principal disadvantage is that it could introduce noise for the classifier which provokes a loss of performance. On the other hand, classical methods cannot decide which artificial instances can improve the SVM performance in imbalanced data sets because the search space is often huge, complex or poorly understood. The GA objective is to find those evolved instances that improve the SVM's performance. Figure 2 illustrates how it is encoded and decoded each individual in the population. The proposed algorithm has two main parts, the first describes how the artificial data are generated and the second describes how are encoded, decoded and evaluated each individual to converge to a solution.

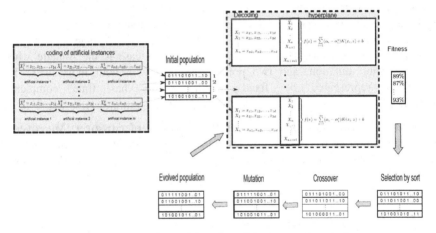

Fig. 2. Encoding and decoding of the proposed method

The initial population is made up by $X_{AI} = \left(x_{AI1}, x_{AI2}, ..., x_{AIp} \right)$ where p is the number of individual in the population and, $x_{AI1} = \left\{ x_{g1}, x_{g2}, ..., x_{gm} \right\}$ where $x_g \in R^d$ i.e. each individual in a population is a set of artificial instances in the search space. Figure 2 shows the proposed coded and decoded process. To locate the best artificial instances GA needs to encode $x_{AI1} = \left\{ x_{g1}, x_{g2}, ..., x_{gm} \right\}$ as binary strings, which are called chromosomes of GA, here all individual artificial instances x_{gi} are converted into binary numbers (we used gray code) and then the binary numbers are concatenated into one string of bits. Each individual in population is evaluated using a SVM learning adding the SVs found in the initial hyperplane *i.e.* the fitness function is obtained by training a SVM with $x_{svi}^- \cup x_{svj}^+ \cup x_{Alk}$.

Furthermore, each individual x_{Al1} is evaluated with the fitness function and manipulated using several genetic operators in order to evolve the population and optimize the solution of the problem. The stronger individuals among the population replace the weaker ones by competition. The survival of the fittest individuals among the population over consecutive generations. In order to prove the performance of the proposed algorithm we used G-mean measure as fitness function in our experiments. The final hyperplane is obtained until a stop criterion has been met. The proposed algorithm stops if there is no improvement in the fitness value for the best string during an interval of the three last generations or if the fitness value is 1.

3   Complexity of the Algorithm

The complexity of the proposed algorithm mainly depends on the fitness evaluation. Other factors that directly influence the complexity are data structures, population, genetic operators, and the implementation of the genetic operators. In the first stage of SVM, the complexity of SVM is O(n2), in our case, it is given for the under-sampled data-set m, then the algorithmic complexity here is O(m)2. In the stage of the GA we added 2 artificial instances for each SV. The simplest case is the roulette wheel selection, point mutation, and two point crossovers with both individuals. The populations represented by fixed length vectors has time complexity $O(gens \times q \times 3SV)$ where gens is the number of generations, $(q \times 3SV)$ is the complexity of point mutation and the time complexity of crossover. Moreover, a SVM is used in order to obtain the fitness function. The complexity of the proposed method does not depend on the entire input data, it just depends on the SVs obtained in the first stage. In the most cases, it is only a small part of the entire data set. The total complexity is given by

$$O(m)^2 + O(gens \times q \times 3m) + O\left(gens \times (3SV)^2\right)$$

Is clear that, the algorithmic complexity is the main disadvantage of the proposed method. However, to use some other learning algorithm to obtain fitness could drastically reduce the computational complexity.

4   Experimental Results

In this section, we show the experimental results obtained with the proposed algorithm. For the case of skewed data sets, is necessary to use a different performance measure to avoid wrong conclusions. In our experiments, we use the sensitivity $S_n^T = \left(T_p / T_p + F_N\right)$,   specificity $S_n^F = \left(T_N / T_N + F_P\right)$,   and

G-mean $= \sqrt{S_n^T + S_n^F}$ to evaluate the performance, where T_P , T_N , F_P and F_N are true positives, true negatives, false positives and false negatives respectively.

4.1 Selection Model and Data-Sets

In the experiments all data sets were normalized and 30 runs were executed in each experiment to obtain the results. In this paper, we use two point crossover and flip bit mutation. In the experiments, we used a crossover probability of $p_c = 0.9$, and a mutation probability of $p_m = 1/n$, where n is the string length for Gray coded. We use the radial basis function (RBF) as kernel, In order to select the optimal parameters of SVM, the hyper-parameter space is explored with some values of γ , and the regularization parameter C .

We use several benchmark data sets from the KEEL data-set repository (http:// sci2s.ugr.es/ keel/datasets.php). Table 1 shows the data sets used in the experiments. The approach was implemented in Matlab. We use 70% of data sets to training, 15% to obtain fitness from artificial data points and, 15% to testing in order to perform the classification of unseen examples. The results are reported in Table 1. In this Table we show results obtained with Under sampling, Over-sampling, Synthetic Minority Over-sampling Technique (SMOTE) and Proposed Method (PM).

Table 1. Detailed results table for the algorithm proposed

Data-set	Over-sampling S_n^T S_n^F G	Under-sampling S_n^T S_n^F G	SMOTE S_n^T S_n^F G	PM S_n^T S_n^F G
Liver_disorders	0.68 0.69 0.68	0.64 0.75 0.69	0.89 0.28 0.49	0.92 0.80 0.85
fourclass	0.78 0.78 0.78	0.81 0.8 0.80	0.91 0.71 0.80	1.00 1.00 1.00
pima	0.67 0.82 0.74	0.72 0.79 0.75	0.74 0.82 0.77	0.77 0.94 0.85
glass0	1.00 0.44 0.66	0.99 0.47 0.68	1.00 0.47 0.68	0.92 0.72 0.81
glass1	0.84 0.46 0.62	0.8 0.57 0.67	0.91 0.31 0.53	0.84 0.97 0.90
cleveland0vs4	0.95 0.70 0.81	0.20 0.99 0.44	0.60 0.99 0.77	1.00 0.92 0.95
segment	0.65 0.84 0.73	0.70 0.83 0.76	0.72 0.8 0.78	0.72 0.90 0.80
diabetes	0.74 .071 0.72	0.69 0.76 0.72	0.83 0.62 0.71	0.87 0.79 0.82
ecoli1	0.92 0.84 0.87	0.89 0.86 0.87	0.91 0.84 0.87	0.98 0.89 0.93
ecoli3	0.93 0.83 0.87	0.89 0.88 0.88	0.83 0.92 0.87	1.00 0.96 0.97
pagebloks13vs4	0.98 0.90 0.93	0.90 0.98 0.93	0.94 1.00 0.96	1.00 1.00 1.00
yeast4	0.75 0.87 0.80	0.71 0.88 0.79	0.54 0.97 0.72	1.00 0.89 0.94
yeast6	0.86 0.88 0.87	0.81 0.92 0.86	0.70 0.98 0.82	0.92 0.94 0.92
vehicle2	0.97 0.91 0.93	0.82 0.98 0.89	0.97 0.93 0.94	1.00 0.96 0.97
vehicle3	0.76 0.67 0.71	0.57 0.82 0.68	0.88 0.66 0.76	0.87 0.85 0.85
shuttle	0.81 0.88 0.84	0.78 0.89 0.83	0.84 0.87 0.85	0.87 0.92 0.89

In the results, we can observe that the proposed method obtains better classification results than the other techniques for almost all the imbalanced data-sets. SVM with GA has a positive synergy with the lateral searching of artificial instances, and leads to good global behavior, in the obtained results, when imbalance ratio is large, the performance achieved by the proposed method is better than the traditional methods.

5   Conclusions

In this paper, we proposed a novel method that enhances the performance of SVM on skewed data sets. The proposed method introduces artificial instances which are created modifying the SVs found in a first stage of SVM training. These artificial instances are evolved by using a genetic algorithm. According to the experiments, the proposed method produces noticeable results in comparison with actual implementations.

References

1. Akbani, R., Kwek, S.S., Japkowicz, N.: Applying Support Vector Machines to Imbalanced Datasets. In: Boulicaut, J.-F., Esposito, F., Giannotti, F., Pedreschi, D. (eds.) ECML 2004. LNCS (LNAI), vol. 3201, pp. 39–50. Springer, Heidelberg (2004)
2. Chawla, N.V., Bowyer, K.W., Kegelmeyer, W.P.: Smote: Synthetic minority over-sampling technique. J. of Artificial Intelligence Research 16, 321–357 (2002)
3. Fernández, A., de Jesus, M.J., Herrera, F.: Hierarchical fuzzy rule based classification systems with genetic rule selection for imbalanced data-sets. International Journal of Approximate Reasoning 50(3), 561–577 (2009)
4. García, S., Derrac, J., Triguero, I., Carmona, C.J., Herrera, F.: Evolutionary-based selection of generalized instances for imbalanced classification. Knowledge-Based Systems 25(1), 3–12 (2012)
5. Koknar-tezel, S., Latecki, L.J.: Improving SVM Classification on Imbalanced Data Sets in Distance Spaces. In: IEEE Int. Conference on Data Mining, pp. 259–267 (2009)
6. Wang, B.X., Japkowicz, N.: Boosting support vector machines for imbalanced data sets. In: An, A., Matwin, S., Raś, Z.W., Ślęzak, D. (eds.) ISMIS 2008. LNCS (LNAI), vol. 4994, pp. 38–47. Springer, Heidelberg (2008)
7. Wu, G., Chang, E.: KBA: Kernel boundary alignment considering imbalanced data distribution. IEEE Trans. Knowl. Data Eng. 17(6), 786–795 (2005)
8. Zeng, Z.-Q., Gao, J.: Improving SVM classification with imbalance data set. In: Leung, C.S., Lee, M., Chan, J.H. (eds.) ICONIP 2009, Part I. LNCS, vol. 5863, pp. 389–398. Springer, Heidelberg (2009)
9. Zhang, H., Li, M.: RWO-Sampling: A random walk over-sampling approach to imbalanced data classification. To appear in Information Fusion (2014)
10. Han, H., Wang, W.-Y., Mao, B.-H.: Borderline-SMOTE: A new over-sampling method in imbalanced data sets learning. In: Huang, D.-S., Zhang, X.-P., Huang, G.-B. (eds.) ICIC 2005. LNCS, vol. 3644, pp. 878–887. Springer, Heidelberg (2005)
11. Batista, G.E., Prati, R.C., Monard, M.C.: A study of the behavior of several methods for balancing machine learning training data. ACM SIGKDD Explor. Newsl. 6(1), 20–29 (2004)

Hyperspectral Data Dimensionality Reduction Based on Non-negative Sparse Semi-supervised Framework

Xuesong Wang, Yang Gao, and Yuhu Cheng

School of Information and Electrical Engineering
China University of Mining and Technology, Xuzhou, Jiangsu 221116, P.R. China
{wangxuesongcumt,chengyuhu}@163.com

Abstract. A non-negative sparse semi-supervised dimensionality reduction framework is proposed for hyperspectral data. The framework consists of two parts: 1) a discriminant item is designed to analyze the few labeled samples from the global viewpoint, which can assess the separability between each surface object; 2) a regularization term is used to build a non-negative sparse representation graph based on large scale unlabelled samples, which can adaptively find an adjacency graph for each sample and then find valuable samples from the original hyperspectral data. Based on the framework and the maximum margin criterion, a dimensionality reduction algorithm called non-negative sparse semi-supervised maximum margin criterion is proposed. Experimental results on the AVIRIS 92AV3C hyperspectral data show that the proposed algorithm can effectively utilize the unlabelled samples to obtain higher overall classification accuracy.

Keywords: Hyperspectral data, Semi-supervised dimensionality reduction, Non-negative sparse representation, Discriminant term, Regularization term.

1 Introduction

Hyperspectral sensing data are featured for many bands, huge data, data uncertainty and small sample classification, and the correlation between its bands is high and data redundancy is high, as directly reduces the accuracy and speed of hyperspectral sensing data classification. Therefore, prior to classification of hyperspectral images, dimensionality reduction (DR) is required for hyperspectral images so as to balance among efficiency, accuracy and adaptability.

In recent years, semi-supervised DR algorithms have been widely applied in face recognition, text classification and so on. For instance, Cai, et al. [1] proposed the semi-supervised discriminant analysis (SDA) and successfully applied it in the face recognition. Furthermore, it is a design concept to extend unsupervised and supervised DR algorithm by using the manifold regularization method, e.g., semi-supervised MMC (SSMMC) and semi-supervised LDA (SSLDA) proposed by Song, et al. [2] according to LDA and maximum margin criterion (MMC) [3]. Such semi-supervised DR algorithms can all be induced into one framework, i.e., using Graph

D.-S. Huang et al. (Eds.): ICIC 2014, LNCS 8588, pp. 789–796, 2014.

Laplacian method as the regularization term. Such graph-based semi-supervised dimensionality reduction algorithms have been successfully applied few labeled face recognition problems, but are still observed with some problems: (1) as Zhu [4] pointed out, such graph-focused semi-supervised algorithms have not yet been widely studied. Most of such algorithms as SDA build adjacency graphs using the nearest neighborhood method, but the nearest neighborhood criterion generally cannot acquire adequate discriminant information. (2) parameters in k neighborhood or sample-dependence neighborhood need to be artificially pre-defined and all the sampling points use the neighborhood number of a fixed sized or fixed neighborhood range. To resolve the foregoing problems, Qiao, et al. [5] build adjacency graphs using sparsity preserving projection instead of the conventional Graph Laplacian method and proposed sparsity preserving discriminant analysis (SPDA). However, this algorithm provides sparsity representation only for the regularization terms in the semi-supervised DR. Discriminant sparse neighborhood preserving embedding (DSNPE) proposed by Gui, et al. [6] proves that the classification effect of the sparse representation algorithm for labeled samples is far superior to LDA. These algorithms may also have a negative weight coefficient in the process of graphing. Wong proposed Non-negative sparseness preserving embedding (NSPE) [7], pointing out that the negative weight coefficient fails to rationally carry out the inter-sample information transfer. To resolve the foregoing problems, a semi-supervised dimensionality reduction framework is proposed and with reference to MMC, non-negative sparse semi-supervised maximum margin criterion (NS³MMC) is proposed and applied for hyperspectral data dimensionality reduction. The non-negative sparse semi-supervised dimensionality reduction framework includes two parts: discriminant term and regularization term.

2   Non-negative Sparse Semi-supervised Dimensionality Reduction Framework

In the semi-supervised problem, given is one high-dimensionality sample data of unknown distribution $\mathbf{O}_m = [\mathbf{X}, \mathbf{Y}]$. Assume the first l number of samples are labeled, while the remaining u number of samples are unlabeled, i.e., the high-dimensionality sample set $\mathbf{X} = (\mathbf{X}_L, \mathbf{X}_U)$, where $\mathbf{X}_L = (\mathbf{x}_1, \mathbf{x}_2, \cdots, \mathbf{x}_l)$, $\mathbf{X}_U = (\mathbf{x}_{l+1}, \mathbf{x}_{l+2}, \cdots, \mathbf{x}_{l+u})$, and $\mathbf{x}_i \in R^m$ is a m-dimensional vector. Mark the data $\mathbf{Y} = \mathbf{Y}_L = (y_1, y_2, \cdots, y_l)^T$, and relevant class label $y_i = \{1, \cdots, c, \cdots, C\}$. Make $n = l + u$, the semi-supervised learning DR aims to find one projection matrix $\mathbf{W}^* = (\mathbf{w}_1^*, \mathbf{w}_2^*, \cdots, \mathbf{w}_d^*) \in R^{m \times d} \ (d << m)$, by using labeled and unlabeled samples:

$$\mathbf{z} = \mathbf{W}^{*T} \mathbf{x} \in R^d \tag{1}$$

where, $\mathbf{z}$ is the low-dimensionality representation of high-dimensionality sample. To look for the optimal projection matrix $\mathbf{W}^*$, the whole objective function is:

$$J(\mathbf{W}^*) = J_D(\mathbf{W}_L^*) + \beta J_R(\mathbf{W}_U^*) \tag{2}$$

where, J_D and J_R are the discriminant term and the regularization term, β is the parameter controlling the weight correlation between the two terms.

2.1   Discriminant Term

The purpose of non-negative sparse representation is to use minimum elements (atoms) in over-complete dictionary $\mathbf{X}_L = [\mathbf{x}_1, \mathbf{x}_2, \cdots, \mathbf{x}_l] \in R^{m \times l}$ to represent hyperspectral data $\mathbf{x}_i$:

$$\min_{\mathbf{h}_i} \|\mathbf{h}_i\|_0 \tag{3}$$
$$\text{s.t. } \mathbf{x}_i = \mathbf{X}_L \mathbf{h}_i; \ \mathbf{h}_i \geq 0$$

where, $\|\mathbf{h}_i\|_0$ represents l_0 norm of $\mathbf{h}_i$, of which the value is the number of non-zero elements in $\mathbf{h}_i$, $\mathbf{h}_i = [h_{i1}, \cdots, h_{ii-1}, 0, h_{ii+1}, \cdots, h_{il}]^T \in R^l$, h_{ij} represents the contribution of sample $\mathbf{x}_j$ to reconstruction sample $\mathbf{x}_i$.

Eq. (3) is a NP-hard non-convex combinatorial optimization problem. The solution of l_0 minimization is approximate or equal to solution of l_1 minimization. l_1 minimization problem can be solved through LASSO [8] or Elastic Net [9].

$$\min_{\mathbf{h}_i} \|\mathbf{h}_i\|_1 \tag{4}$$
$$\text{s.t. } \mathbf{x}_i = \mathbf{X}_L^c \mathbf{h}_i; \ 1 = \mathbf{1}^T \cdot \mathbf{h}_i; \ \mathbf{h}_i \geq 0; \ y_i = c$$

where, $\mathbf{1} \in R^{l_c}$ is the all-1 vector and l_c is the number of samples labeled as class-c. Non-negative GLS (Generalized Least Squares) is used to solve Eq. (4) for obtaining the reconstruction error [10]:

$$\min_{\mathbf{h}_i} E(\mathbf{H}^c) = \min_{\mathbf{h}_i} \|\mathbf{X}_L^c - \mathbf{X}_L^c \mathbf{H}^c\|_2^2 + \gamma \|\mathbf{h}_i\|_1 \tag{5}$$
$$\text{s.t. } \mathbf{h}_i \geq 0; \ 1 = \mathbf{1}^T \mathbf{h}_i; \ y_i = c$$

where, γ is the non-negative coefficient. According to Eq. (5), compute the optimal non-negative sparseness reconstruction weight vector $\tilde{\mathbf{h}}_i \in R^{l_c}$ for each hyperspectral data point $\mathbf{x}_i$ in class-c, then class-c discriminant block non-negative sparseness reconstruction weight matrix is $\mathbf{H}^c = [\tilde{\mathbf{h}}_i]_{l_c \times l_c}$.

Discriminant non-negative sparseness reconstruction weight matrix $\mathbf{H}_D$ of hyperspectral data is represented as:

$$\mathbf{H}_{\mathrm{D}} = diag(\mathbf{H}^{1}, \cdots, \mathbf{H}^{c}, \cdots, \mathbf{H}^{C}) = [\tilde{\mathbf{h}}_{i}]_{\times l} \tag{6}$$

$\mathbf{H}_{\mathrm{D}}$ only considers construction between samples of one class as local and thus fails to ensure the excellent classification accuracy upon embedding into low-dimensionality subspace. To resolve this problem, non-local discriminant non-negative sparseness reconstruction weight matrix $\mathbf{H'}_{\mathrm{D}}$ is introduced to acquire the whole sample information in combination with whole and non-local discriminant information for the purpose of assessing the separability between different surface features. Discriminant term attempts to search one linear projection matrix: i.e., by maximizing non-local reconstruction matrix $\mathbf{S}_{\mathrm{N}}$, keep away the hyperspectral data of different categories with high similarity in output low-dimensionality subspace; by minimizing local reconstruction matrix $\mathbf{S}_{\mathrm{L}}$, keep close the hyperspectral data of one category with high similarity in output low-dimensionality subspace. It is defined as:

$$\mathbf{S}_{\mathrm{L}} = \sum_{i=1}^{l} (\mathbf{x}_{i} - \mathbf{X}_{\mathrm{L}}\tilde{\mathbf{h}}_{i})(\mathbf{x}_{i} - \mathbf{X}_{\mathrm{L}}\tilde{\mathbf{h}}_{i})^{\mathrm{T}} = \mathbf{X}_{\mathrm{L}}(\mathbf{I} - \mathbf{H}_{\mathrm{D}})^{\mathrm{T}}(\mathbf{I} - \mathbf{H}_{\mathrm{D}})\mathbf{X}_{\mathrm{L}}^{\mathrm{T}} \tag{7}$$

where, $\mathbf{I}$ is an unit matrix.

Making non-local discriminant non-negative sparseness reconstruction weight matrix $\mathbf{H}'_{\mathrm{D}} = \mathbf{I} - \mathbf{H}_{\mathrm{D}}$, non-local reconstruction matrix $\mathbf{S}_{\mathrm{N}}$ is:

$$\mathbf{S}_{\mathrm{N}} = \mathbf{X}_{\mathrm{L}}(\mathbf{I} - \mathbf{H}'_{\mathrm{D}})^{\mathrm{T}}(\mathbf{I} - \mathbf{H}'_{\mathrm{D}})\mathbf{X}_{\mathrm{L}}^{\mathrm{T}} \tag{8}$$

Similar to MMC, optimal project matrix $\mathbf{W}_{\mathrm{L}}^{*}$ can be obtained by solving:

$$J_{\mathrm{D}}(\mathbf{W}_{\mathrm{L}}^{*}) = \arg \max_{\mathbf{W}^{\mathrm{T}}\mathbf{W}=\mathbf{I}} tr(\mathbf{W}^{\mathrm{T}}(\mathbf{S}_{\mathrm{N}} - \alpha \mathbf{S}_{\mathrm{L}})\mathbf{W}) \tag{9}$$

2.2   Regularization Term

Since the conventional non-negative sparse representation has low reconstruction accuracy, and high content of unlabeled samples, regularization term is defined by using block non-negative sparse representation for the purpose of making more effective use of original samples. The block non-negative sparse representation method is mainly to divide over-complete dictionary $\mathbf{X}_{\mathrm{U}}$ into v blocks, i.e.:

$$\mathbf{X}_{\mathrm{U}} = \left[\underbrace{\mathbf{x}_{1}^{1}, \cdots \mathbf{x}_{b}^{1}, \cdots, \mathbf{x}_{B}^{1}}_{\mathbf{x}_{\mathrm{U}}^{1}}, \cdots, \underbrace{\mathbf{x}_{1}^{v}, \cdots \mathbf{x}_{b}^{v}, \cdots, \mathbf{x}_{B}^{v}}_{\mathbf{x}_{\mathrm{U}}^{v}}, \cdots, \underbrace{\mathbf{x}_{1}^{V}, \cdots \mathbf{x}_{b}^{V}, \cdots, \mathbf{x}_{B}^{V}}_{\mathbf{x}_{\mathrm{U}}^{V}} \right] \tag{10}$$

where, $u = VB$, $\mathbf{X}_{\mathrm{U}}^{v}$ is an over-complete block dictionary; if $\mathbf{x}_{b}^{v} = \mathbf{x}_{(v-1)B+b}$, hyperspectral data point $\mathbf{x}_{b}^{v}$ in No. v sub-block can be represented with over-complete block dictionary $\mathbf{X}_{\mathrm{U}}^{v}$ as:

$$\min_{\mathbf{h}_b^v} \quad \left\| \mathbf{h}_b^v \right\|_0$$

$$\text{s.t.} \quad \mathbf{x}_b^v = \mathbf{X}_U^v \mathbf{h}_b^v; \quad \mathbf{h}_b^v \geq 0 \tag{11}$$

where, $\mathbf{h}_b^v \in R^B$. Similar to Eqs. (3) and (4), Eq. (11) can be changed to:

$$\min_{\mathbf{h}_b^v} \quad \left\| \mathbf{h}_b^v \right\|_1$$

$$\text{s.t.} \quad \mathbf{x}_b^v = \mathbf{X}_U^v \mathbf{h}_b^v; \quad 1 = \mathbf{1}^T \cdot \mathbf{h}_b^v; \quad \mathbf{h}_b^v \geq 0 \tag{12}$$

Using non-negative GLS, reconstruction error is:

$$\min_{\mathbf{h}_b^v} E(\mathbf{H}^v) = \min_{\mathbf{h}_b^v} \left\| \mathbf{X}_U^v - \mathbf{X}_U^v \mathbf{H}^v \right\|_2^2 + \gamma \left\| \mathbf{h}_b^v \right\|_1$$

$$\text{s.t.} \quad \mathbf{h}_b^v \geq 0; \quad 1 = \mathbf{1}^T \mathbf{h}_i \tag{13}$$

According to Eq. (13), compute optimal non-negative sparseness reconstruction weight vector $\tilde{\mathbf{h}}_b^v \in R^B$ for each hyperspectral data point $\mathbf{x}_b^v$ in No. v sub-block, then block non-negative sparseness reconstruction weight matrix of No. v sub-block $\mathbf{H}^v = \left[\tilde{\mathbf{h}}_b^v \right]_{B \times B}$. The overall non-negative sparseness reconstruction weight matrix $\mathbf{H}_R$ of hyperspectral data is represented as:

$$H_R = diag\left(H^1, \cdots, H^v, \cdots, H^V \right) = \left[\tilde{h}_i \right]_{u \times u} \tag{14}$$

According to the overall non-negative sparseness reconstruction weight matrix $\mathbf{H}_R$, define the regularization term J_R, i.e., linear reconstruction can be carried out for each hyperspectral data by other hyperspectral data:

$$J_R = \sum_{i=1}^u \left\| \mathbf{z}_i - \sum_{j=1}^u \tilde{h}_i \mathbf{z}_j \right\|^2 \tag{15}$$

where, $\mathbf{z}_i = \mathbf{W}^T \mathbf{x}_i$, Eq. (15) can be changed to:

$$J_R(\mathbf{w}^*) = \arg\min_{\mathbf{w}} \left[-\frac{1}{u} \left(\sum_{i=1}^u \left\| \mathbf{w}^T \mathbf{x}_i - \mathbf{w}^T \mathbf{X}_U \tilde{\mathbf{h}}_i \right\|^2 \right) \right] \tag{16}$$

Since the solution of Eq. (16) is equal to the solution of objective function in NSPE, it can be obtained by simple algebraic operation:

$$J_R(\mathbf{W}_U^*) = \arg\min_{\mathbf{W}} (\mathbf{W}^T \mathbf{X}_U \mathbf{S}_R \mathbf{X}_U^T \mathbf{W}) \tag{17}$$

where, $\mathbf{S}_R = \dfrac{1}{u} (\mathbf{H}_R^T + \mathbf{H}_R - \mathbf{H}_R^T \mathbf{H}_R - \mathbf{I})$.

2.3   Non-negative Sparse Semi-supervised Dimensionality Reduction Algorithm

Based on the foregoing framework, one non-negative sparse semi-supervised maximum margin criterion (NS³MMC) is proposed. According to Eqs. (2), (9) and (17), the objective function of NS³MMC is defined as:

$$J(\mathbf{W}^*) = \arg\max_{\mathbf{W}^T\mathbf{W}=\mathbf{I}} tr(\mathbf{W}^T(\mathbf{S}_N - \alpha\mathbf{S}_L - \beta\mathbf{X}_U\mathbf{S}_R\mathbf{X}_U^T)\mathbf{W}) \tag{18}$$

Lagrange multiplier is introduced into Eq. (18) to obtain:

$$\Gamma(\mathbf{w}_j,\lambda_j) = \mathbf{w}_j^T(\mathbf{S}_N - \alpha\mathbf{S}_L - \beta\mathbf{X}_U\mathbf{S}_R\mathbf{X}_U^T)\mathbf{w}_j + \lambda_j(\|\mathbf{w}_j\|_2^2 - 1) \tag{19}$$

By derivation of $\mathbf{w}_j$ in Eq. (19), obtain:

$$(\mathbf{S}_N - \alpha\mathbf{S}_L - \beta\mathbf{X}_U\mathbf{S}_R\mathbf{X}_U^T)\mathbf{w}_j^* = \lambda_j^*\mathbf{w}_j^*, j=1,\cdots,d \tag{20}$$

where, λ_j^* and $\mathbf{w}_j^*$ are the eigenvalue and the corresponding eigenvector.

3   Experiments and Analysis

To assess the dimensionality reduction performance of NS³MMC, in the experiment, AVIRIS 92AV3C hyperspectral data is used to compare the algorithms, including SPDA, SSMMC, SSLDA, SDA, DSNPE, NSPE and PCA, and support vector machine (SVM) is used to classify hyperspectral data after dimensionality reduction. SVM uses Gauss kernel function, of which the width and penalty factor obtained by training with 5-fold cross-validation; in all algorithms, $\alpha = 0.1$, $\beta = 0.1$; in sparsity algorithm, $\gamma = 1$, $V = C$; in the regularization term of SSMMC, SSLDA and SDA, the neighborhood parameter is set as 5. Since SDA and SSLDA are proposed on the basis of LDA, the upper limit of dimensionality reduction is $C-1$. In simulation, each experiment is conducted twenty times, of which the mean value is taken.

In the experiment, the 92AV3C hyperspectral data is divided into three groups: 10% labeled samples, 20% unlabeled samples and 70% test samples. As for the unsupervised algorithm NSPE, all training samples are used to train the learner. As for the supervised algorithm DSNPE, only labeled samples are used to train the learner. For the other semi-supervised algorithms, all the training samples, including labeled and unlabeled samples, are used to train the learner.

Fig. 1 shows the relation between overall classification accuracy (OCA) and the low-dimensionality subspace dimension. From Fig. 1, it may be known: (1) in the case of different low-dimensionality subspace dimensions, NS³MMC can basically obtain the highest OCA. (2) As all types of surface features in 92AV3C hyperspectral data are unevenly distributed and a relatively low spectral difference is observed between certain surface features, in classification, most DR algorithms have a relatively low OCA and also show a high difference between each other in terms of OCA, but NS³MMC can acquire higher OCA, showing a more obviously and effective dimensionality reduction effect, higher applicability and data distribution more adaptive to complexity.

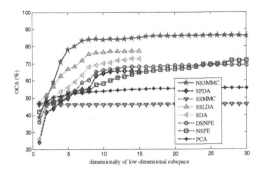

Fig. 1. The OCA versus the dimensionalty of low-dimensional subspace

4   Conclusions

For the purpose of making use of the available labeled samples as well as acquiring value discriminant information from the plentiful unlabeled samples, a non-negative sparse semi-supervised dimensionality reduction framework is proposed, which includes two parts: a discriminant term and a regularization term. By making adequate use of few labeled samples, the discriminant term preliminarily builds the separability model between different surface features. Using the block non-negative sparse representation with the natural discriminating capacity, the regularization term finds the samples with the discriminating capacity from the plentiful unlabeled samples and effectively improves the dimensionality reduction effect of the algorithm in combination with the separability model of the discriminant term. The common hyperspectral data experiment shows that the proposed algorithm can more properly acquire discriminate information and preserve the non-negative sparse structure.

Acknowledgements. This work is supported by Fundamental Research Funds for the Central Universities (2013RC10).

References

1. Cai, D., He, X.F., Han, J.W.: Semi-supervised discriminant analysis. In: Proceedings of the 11th IEEE International Conference on Computer Vision, pp. 222–228. IEEE, Brazil (2007)
2. Song, Y.Q., Nie, F.P., Zhang, C.S.: A Unified framework for semi-supervised dimensionality reduction. Pattern Recognition 41(9), 2789–2799 (2008)
3. Farcas, D., Marghes, C., Bouwmans, T.: Background subtraction via incremental maximum margin criterion: a discriminative subspace approach. Machine Vision and Applications 23(6), 1083–1101 (2012)
4. Zhu, X.: Semi-supervised learning literature survey. University of Wisconsin, Madison (2008)
5. Qiao, L.S., Chen, S.C., Tan, X.Y.: Sparsity Preserving discriminant analysis for single training image face recognition. Pattern Recognition Letters 31(5), 422–429 (2010)

6. Gui, J., Sun, Z.N., Jia, W.: Discriminant sparse neighborhood preserving embedding for face recognition. Pattern Recognition 45(8), 2884–2893 (2012)
7. Wong, W.K.: Discover latent discriminant information for dimensionality reduction: non-negative sparseness preserving embedding. Pattern Recognition 45(4), 1511–1523 (2012)
8. Lv, J.Q., Pawlak, M., Annakkage, U.D.: Prediction of the transient stability boundary using the lasso. IEEE Transactions on Power Systems 28(1), 281–288 (2013)
9. Zou, H., Hastie, T.: Regularization and variable selection via the elastic Net. Journal of the Royal Statistical Society Series B-Statistical Methodology 67(2), 301–320 (2005)
10. He, R., Hu, B.G., Zheng, W.S.: Two-stage sparse representation for robust recognition on large-scale database. In: Proceedings of the 24th AAAI Conference on Artificial Intelligence, pp. 475–480. American Association for Artificial Intelligence, Atlanta (2010)

Constrained Community Clustering

Ping He[*], Xiaohua Xu[*], Lei Zhang, Wei Zhang, Kanwen Li, and Heng Qian

Department of Computer Science, Yangzhou University, Yangzhou 225009, China
{angeletx,arterx}@gmail.com

Abstract. Constrained clustering uses pairwise constraints, i.e., pairs of data that belong to the same or different clusters, to indicate the user-desired contents. In this paper, we propose a new constrained clustering algorithm, which can utilize both must-link and cannot-link constraints. It first adaptively determines the influence range of each constrained data, and then performs clustering on the expanded range of data. The promising experiments on the real-world data sets demonstrate the effectiveness of our method.

Keywords: Constrained clustering, Must-link constraints, Cannot-link constraints.

1 Introduction

In the area of machine learning, constrained clustering refers to specifying pairs of data that belong to the same cluster (must-link constraints) and the pairs of data that belong to different clusters (cannot-link constraints) to facilitate the extraction of user-desired contents [1]. A common difficulty of existing constrained clustering algorithms is how to utilize the limited but informative pairwise constraints [2].

There are two major approaches to deal with constrained clustering. The first line adapts the unsupervised methods, including k-means [3], all-pairs shortest path 4], Gaussian mixtures models [5], and Spectral clustering [6-9], to satisfy the pairwise constraints in a smaller subset of solution space with additional requirements. However, some of them ignore the impact of the pairwise constraints on the unconstrained data, while the others either fail to handle the cannot-link constraints, or have problem in dealt with the multi-class clustering tasks. The second line derives an appropriate metric to adjust the must-linked data close and the cannot-linked data faraway [10]. The difficulty of the second lines lies in how to determine the scope of the metric.

In this paper, we propose a novel constrained clustering algorithm, which propagates the influence of the pairwise constraints to the unconstrained data in proportion to their similarities with the constrained data. To this end, we first determine the degree of influence that each unconstrained object receives from the constrained objects using the label propagation technique [11]. Then an intermediate structure is defined to include the factional data points that are affected by a constrained data, called "community", because it allows overlapping, similar to a community in the social network. With these communities, we can expand the influence of the pairwise constraints by directly enforcing them on the related communities.

---

[*] Corresponding author.

D.-S. Huang et al. (Eds.): ICIC 2014, LNCS 8588, pp. 797–802, 2014.

The remainder of this paper is organized as follows. We first introduce the label propagation technique in section 2, and then present our algorithm C^3 in section 3. They are followed by the performance evaluation in section 4. Section 5 concludes the whole paper.

2   Label Propagation

Label propagation is a well-known transductive learning algorithm [11]. Given a data set $S = (X_l, Y_l) \cup X_u$, where $X_l = \{x_1, \cdots, x_l\}$ is the labeled data subset, $Y_l = \{y_1, \cdots, y_l\}$ is the label subset of X_l, $X_u = \{x_{l+1}, \cdots, x_n\}$ is the unlabeled data subset, the aim of transductive learning is to predict Y_u, the label subset of X_u.

In correspondence to the partition of X_l and X_u, the transition probability matrix P is organized as follow,

$$P = \begin{bmatrix} P_{ll} & P_{lu} \\ P_{ul} & P_{uu} \end{bmatrix} \tag{1}$$

where P_{ll} and P_{uu} record the transition probabilities from X_l and X_u to themselves, while P_{lu} and P_{ul} record the transition probabilities between X_l and X_u.

The stationary distributions of the random walk is computed by

$$\hat{Y}^{t+1} = P\hat{Y}^t \tag{2}$$

$$\hat{Y} = \hat{Y}^\infty \tag{3}$$

Each column of $\hat{Y}_{n \times l}$ records the probabilities of all the particles absorbed by a different absorbing state when arriving stationary distribution.

To solve eq. (3), $\hat{Y}$ is partitioned into two parts,

$$\hat{Y} = \begin{bmatrix} \hat{Y}_l \\ \hat{Y}_u \end{bmatrix} \tag{4}$$

$\hat{Y}_l$ is called the class indicating matrix of X_l, while $\hat{Y}_u$ is called the class indicating matrix of X_u. Since Y_l is already known, we can determine $\hat{Y}_l$ in advance: $\hat{y}_{ij} = 1$ iff $y_i = j$, otherwise, $\hat{y}_{ij} = 0$.

The converged solution of $\hat{Y}_u$ can be proved as

$$\hat{Y}_u = (I - P_{uu})^{-1} P_{ul} \hat{Y}_l \tag{5}$$

In the end, each data is assigned with the class label that it most probably belongs to, i.e. $y_i = \arg\max_j \hat{y}_{ij}$.

3   Our Method

In constrained clustering, an unlabeled data set X and a pairwise constraint set $C = \{C_= \cup C_{\neq}\}$ are provided. $C_=$ is the must-link constraint subset, and $C_{\neq}$ is the cannot-link constraint subset. Both $\mathcal{X}$ and $\mathcal{C}$ are mapped onto a graph $\mathcal{G} = (\mathcal{V}, W)$, where $\mathcal{V} = \{v_1, \cdots, v_n\}$ is the vertex set, v_i corresponds to x_i, W is the similarity matrix of $\mathcal{V}$. The must-link constraint $(v_i, v_j, =) \in \mathcal{C}_=$ indicates that v_i and v_j belong to the same cluster, while the cannot-link constraint $(v_i, v_j, \neq) \in \mathcal{C}_{\neq}$ indicates that v_i and v_j belong to different clusters. The goal of constrained clustering is to partition $\mathcal{V}$ into p clusters and satisfy $\mathcal{C}$ as much as possible.

Let $\mathcal{V}_c$ be the vertex subset constrained by the pairwise constraints, and $\mathcal{V}_u$ be the unconstrained vertex set. We set $\mathcal{V}_c$ as the absorbing boundary of the vertex-level random walk, and the unconstrained vertices in $\mathcal{V}_u$ as the transitive states. Then label propagation is used to determine the influence of each constrained vertex on every unconstrained vertex. A $n \times |\mathcal{V}_c|$ matrix F is constructed to include the influence of each constrained vertex exerted on all the vertices including $\mathcal{V}_c$ and $\mathcal{V}_u$. Similar to the class indicating matrix in eq. (4), F is partitioned into two parts,

$$F = \begin{bmatrix} F_c \\ F_u \end{bmatrix} \tag{6}$$

where the subscript c indicates the constrained vertex subset $\mathcal{V}_c$, while the subscript u indicates the unconstrained vertex subset $\mathcal{V}_u$. The transition probability matrix P is reorganized accordingly,

$$P = \begin{bmatrix} P_{cc} & P_{cu} \\ P_{uc} & P_{uu} \end{bmatrix} \tag{7}$$

where P_{cc}, P_{uu}, P_{cu} and P_{uc} are the transition probability sub-matrices within or between $\mathcal{V}_c$ and $\mathcal{V}_u$.

According to the converged solution in eq. (5), the propagated influence of $\mathcal{V}_c$ on $\mathcal{V}_u$ is

$$F_u = (I - P_{uu})^{-1} P_{cu} F_c^0 \tag{8}$$

where F_c^0 is the initial state of F_c. In this paper, we set $F_c^0 = I$. F encodes the influence range of each constrained vertex, because the ith row of F indicates the degree of influence that v_i receives from different constrained vertices. We define the union of the fractional vertices affected by each constrained vertex (including itself) as a "community".

Now consider a higher-level random walk on the constructed communities. The pairwise community similarity matrix W_c is

$$W_c = F^T \tilde{W} F \tag{9}$$

whose element is the weighted sum of the pairwise vertex similarities, and the weight equals to the degree of influence that the pair of vertices receive from the constrained link. In order to ensure the similarities are within the range of [0,1], we normalize W_c in a symmetric manner.

$$\bar{W}_c = D_c^{-\frac{1}{2}} W_c D_c^{-\frac{1}{2}} \tag{10}$$

where $D_c = \text{diag}(W_c \mathbf{1}_{|\mathcal{T}|})$.

To enforce the pairwise constraints on the expanded influence range of the constrained vertices, we modify the normalized pairwise community similarity matrix $\tilde{W}_c$. $\forall \tilde{w}_c(i,j) \in \tilde{W}_c$,

$$\tilde{w}_c(i,j) = \begin{cases} \tilde{w}_c(i,j)^q & \text{if } \exists (v_i, v_j, =) \in \mathcal{C}_= \\ \tilde{w}_c(i,j)^{\frac{1}{q}} & \text{if } \exists (v_i, v_j, \neq) \in \mathcal{C}_{\neq} \\ \tilde{w}_c(i,j) & \text{otherwise} \end{cases} \qquad (11)$$

The parameter $q \in (0,1]$ in eq. (11) controls the power of the increase and the decrease of the original similarities.

By using the spectral clustering [12], we can partition the constrained communities into different clusters.

$$U = [u_1 \ u_2 \ \cdots \ u_p] \qquad (12)$$

Here $u_1, u_2, \ldots, u_p$ satisfy that $P_c u_i = \lambda_i u_i$, in which P_c is the transition probability matrix

$$P_c = \text{diag}(\tilde{W}_c \mathbf{1}_{|\mathcal{T}|})^{-1} \tilde{W}_c \qquad (13)$$

λ_i is the ith maximal eigenvalue of P_c.

Finally, we obtain the cluster indicating matrix for each vertex by multiplying F and U.

$$G = F \times U \qquad (14)$$

The cluster assignment of each vertex is obtained by projecting the row vectors of U onto a unit hypersphere [13],

$$\tilde{G} = \text{diag}(\text{diag}(GG^T))^{-\frac{1}{2}} G \qquad (15)$$

and then applying k-means algorithm on it.

4   Experiments

In this section, we evaluate the clustering performance of our algorithm ACC, short for Adaptive Constrained Clustering, in comparison with another three most related constrained clustering algorithms, including Spectral Learning (SL) [8], SS-Kernel-Kmeans (SSKK) [9], and Constrained Clustering with Spectral Regularization (CCSR) [2], on two real-world text document data sets, including 20Newsgroups[1] and TDT2[2].

All the algorithms are implemented in Matlab, and are provided with the same graphs. We construct 20NN (20-Nearest-Neighbors) sparse graphs on all the real-world data sets, where the pairwise similarities are computed using the cosine function. Both NMI and Rand Index are adopted to perform a comprehensive evaluation for the four constrained clustering algorithms.

---

[1] http://people.csail.mit.edu/jrennie/20Newsgroups/
20news-bydate-matlab.tgz
[2] http://www.zjucadcg.cn/dengcai/Data/TextData.html

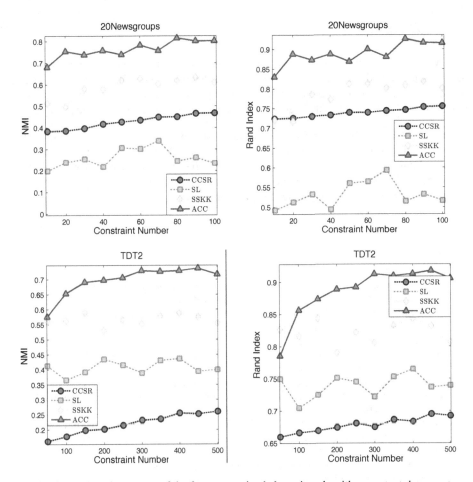

Fig. 1. The learning curves of the four constrained clustering algorithms on text documents

Fig. 1 illustrates the comparison result of the four constrained clustering algorithms on the two text document data sets. Our algorithm (ACC) exhibits the best clustering performance in respect of NMI and Rand Index on both the two document data sets. In Fig. 1, SSKK and SL fall behind of ACC because they fail to expand the constraint influence, but only exert the pairwise constraints on the constrained data. The inferiority of CCSR is due to the construction of pairwise similarity matrices. While our algorithm keeps the influence of each pairwise constraint within the influence range of constrained vertices, CCSR adapts the whole feature space to satisfy each pairwise constraint globally and uniformly. That disables CCSR from producing satisfactory clustering results on sparse graphs.

5   Conclusion

A new constrained clustering algorithm is proposed and demonstrated. Compared with the previous methods, our approach can adaptively determine the influence range of each constrained vertex, hence provides valuable insight to the underlying structures of clusters. Based on that, our algorithm can enforce the pairwise constraints on a wider range.

Acknowledgment. This work was supported in part by the Chinese National Natural Science Foundation under Grant nos. 61003180, 61379066 and 61103018, Natural Science Foundation of Education Department of Jiangsu Province under contracts 13KJB520026 and 09KJB20013, Natural Science Foundation of Jiangsu Province under contracts BK2010318 and BK2011442, and the New Century Talent Project of Yangzhou University.

References

1. Basu, S., Davidson, I., Wagstaff, K.L.: Constrained Clustering. In: Advances in Algorithms, Theory, and Applications. Chapman and Hall/CRC (2008)
2. Li, Z., Liu, J., Tang, X.: Constrained clustering via spectral regularization. In: Proceeding of Computer Vision and Pattern Recognition, pp. 421–428 (2009)
3. Wagsta, K., Cardie, C., Rogers, S., Schroedl, S.: Constrained k-means clustering with background knowledge. In: Proceedings of the Eighteenth International Conference on Machine Learning, pp. 577–584 (2001)
4. Klein, D., Kamvar, S., Manning, C.: From instance-level constraints to space-level constraints: Making the most of prior knowledge in data clustering. In: Proceedings of the Nineteenth International Conference on Machine Learning, pp. 307–314 (2002)
5. Shental, N., Bar-hillel, A., Hertz, T., Weinshall, D.: Computing gaussian mixture models with em using equivalence constraints. In: Advances in Neural Information Processing Systems 16, pp. 465–472 (2003)
6. Yu, S., Shi, J.: Segmentation given partial grouping constraints. IEEE Transactions on Pattern Analysis and Machine Intelligence 26(2), 173–180 (2004)
7. De Bie, T., Suykens, J., De Moor, B.: Learning from general label constraints. In: Fred, A., Caelli, T.M., Duin, R.P.W., Campilho, A.C., de Ridder, D. (eds.) SSPR&SPR 2004. LNCS, vol. 3138, pp. 671–679. Springer, Heidelberg (2004)
8. Kamvar, S., Klein, D., Manning, C.: Spectral learning. In: International Joint Conference on Artificial Intelligence, pp. 561–566 (2003)
9. Kulis, B., Basu, S., Dhillon, I., Mooney, R.: Semi-supervised graph clustering: A kernel approach. Machine Learning 74, 1–22 (2009)
10. Liu, W., Ma, S., Tao, D., Liu, J., Liu, P.: Semi-supervised sparse metric learning using alternating linearization optimization. In: Proceedings of the 16th ACM SIGKDD International Conference on Knowledge Discovery and Data Mining, pp. 1139–1148 (2010)
11. Zhu, X., Ghahramani, Z., Lafferty, J.: Semi-supervised learning using gaussian fields and harmonic functions. In: The 20th International Conference on Machine Learning (ICML), pp. 912–919 (2003)
12. Shi, J., Malik, J.: Normalized cuts and image segmentation. IEEE Transactions on Pattern Analysis and Machine Intelligence (PAMI) 6, 721–741 (2000)
13. Ng, A., Jordan, M., Weiss, Y.: On spectral clustering: Analysis and an algorithm. In: Advances in Neural Information Processing Systems 14, pp. 849–856 (2001)

Extended Embedding Function
for Model-Preserving Steganography

Saiful Islam and Phalguni Gupta

Department of Computer Science and Engineering,
Indian Institute of Technology Kanpur
Kanpur 208016, India
{sislam,pg}@cse.iitk.ac.in

Abstract. Existing model-preserving steganography techniques are based on embedding data by modifying least significant bits of a cover image. To keep invariant component of pixel intact only, Least Significant Bit Replacement (LSBR) function is used to modify non-invariant component of the pixel. LSBR is found to be weak against visual or statistical attacks and provide limited steganographic capacity. Least significant Bit Matching (LSBM) embedding provides better security in comparison LSBR but it is not suitable for most of the model-preserving steganographic techniques. This paper explores the possibility of securely embedding data in least two significant bits of the cover image. It is shown that the embedding in least two significant bits violates the assumption of the structural steganalysis tools and techniques available to detect presence of a message in a stego image. Therefore, structural and non-structural detectors fail to detect presence of data in a stego image. The proposed embedding functions result in improved security and steganographic capacity in comparison to LSBM.

Keywords: Model-preserving Steganography, Steganalysis, Support Vector Machine.

1 Introduction

Every steganographic technique requires an embedding function to hide the payload in the selected pixels of a cover image. Most of the existing steganographic techniques embed payload by modifying least significant bits (LSB) of the cover image. LSB steganography can broadly be classified into two types; LSB Replacement [1] and LSB Matching [2]. Security of steganographic technique is also related to the embedding functions used to embed payload. Such a modification function leads to characteristic distortion, artifacts, in statistical properties of the cover images. Targeted and blind steganalysis tools [3] classify between the cover and the stego image on the basis of the presence of these artifacts in the given image. One can estimate reliably the length of payload by these steganalysis tools [4].

Practical steganographic techniques can broadly be categorized into three classes; model-preserving [5][6], steganalysis-aware [7] and minimal-impact steganography [8][9]. LSBM is a preferred choice of embedding for techniques other than model-preserving steganographic approach. Fig. 1 shows embedding and extracting functions

D.-S. Huang et al. (Eds.): ICIC 2014, LNCS 8588, pp. 803–811, 2014.

of the model-preserving approach. Pixel bits of an image $\mathbf{X}$ are separated into two parts viz. invariant part, $\mathbf{X}inv$, and embedding part, Xemb, [10] where $\mathbf{X}inv$ part is invariant to embedding operation and Xemb are those bits of pixels that may be modified during embedding. Conditional probability Pr{Xemb|$\mathbf{X}inv$} representing the cover model is used to extract the message by the receiver. Therefore, the probability P r{Xemb|$\mathbf{X}inv$} depends on the invariant part $\mathbf{X}inv$.

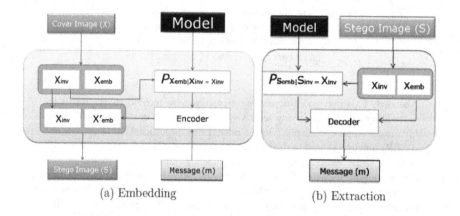

(a) Embedding   (b) Extraction

Fig. 1. Model-Preserving Steganography

LSBM is not suitable for most of the model-preserving steganographic techniques because the invariant part, $\mathbf{X}inv$, of a cover pixel may be affected by the embedding function. In other words, embedding impact of LSBM technique is not limited to only LSB as it may modify $\mathbf{X}inv$ part while LSBR keeps $\mathbf{X}inv$ part intact. Therefore, LSBR is only possible embedding function for model-preserving approach but it is easily detectable through histogram attack. To overcome statistical attacks against LSBR, entropy encoding technique is used to treat the message to preserve statistics of the cover image.

This paper proposes an embedding function of higher significant bits of the cover image. Embedding in steganography is restricted to insignificant parts of the cover image so that there is minimal visual and statistical distortion in a cover image. Embedding in extended bits preserves these characteristics of the cover image. It is shown that the embedding in extended bits violates the assumption of the structural steganalysis tools and techniques available to detect presence of a message in a steganogram. Therefore, structural and non-structural detectors fail to detect the presence of data in a stego image. It is observed that the proposed embedding functions result in the improved security and hiding capacity for model-preserving steganographic techniques in comparison to LSBR. Proposed embedding modification functions are based on the limitation of steganalysis techniques to detect messages in a steganogram. Effects of embedding in extended bits have been analyzed against visual and statistical attacks. The paper is organized as follows: In Section 2, the proposed embedding functions have been presented. The simulation results are analyzed in the next section. Conclusions are given in Section 4.

2   Proposed Embedding Function

In this section, an embedding function which provides better security has been proposed. It uses two bits of each pixel for embedding which helps to overcome structural detectors and blind steganalysis. For example, embedding in least two significant bits does not lead to pair of value (POV) [11] characteristics and modifies higher statistics in an unexpected way.

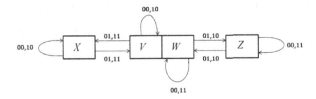

Fig. 2. State Diagram Sample Pair Analysis

Most of the statistical detectors are based on the concept of the Sample Pair (SP) [12] analysis. It uses second-order statistics to capture correlation among neighboring pixels. To capture the correlation among pixels pairs of neighboring pixels are considered for embedding. Let P be the set of all pairs of neighboring pixels. A pair $p \in$ P be (u, v) and this pair can accept four different forms, given below

$$(u, v) \in \{(2i, 2j), (2i, 2j + 1), (2i + 1, 2j), (2i + 1, 2j + 1)\} \tag{1}$$

Fig. 2 reproduces the state diagram of SP analysis. The set of all pairs of horizontally neighboring pixels P can be distributed into the following sets:

$$X = \{(u, v) \in P| \ (v \text{ is even and } u < v) \text{ or } (v \text{ is odd and } u > v) \}, \tag{2}$$

$$Y = \{(u, v) \in P \ | \ (v \text{ is even and } u > v) \text{ or } (v \text{ is odd and } u < v)\} \tag{3}$$

$$W = \{(u, v) \in P \ | \ u = 2k, \ v = 2k+1 \text{ or } u = 2k+1, \ v = 2k\} \tag{4}$$

$$V = Y - W \tag{5}$$

$$Z = \{(u, v) \in P \ | \ u = v\} \tag{6}$$

Labels, 00, 01, 10 and 11, in the state diagram indicate the modification pattern for a pixel pair (u, v). Modification pattern 00 implies both u and v remain unmodified, pattern 10 implies only u is modified. The remaining patterns are also defined in the similar way. It is assumed in SP analysis that the sets X and Y have same cardinality.

It is observed from BOWS2 database of 10000 images [13] that the cardinality of the sets X and Y is not equal for images without last bit plane. Structural detectors reliably detect LSB embedding but fail to detect LSB matching. This property makes the embedding in the last two bits more secure than LSB embedding. Therefore, it is expected that embedding in higher bits provides immunity against structural and

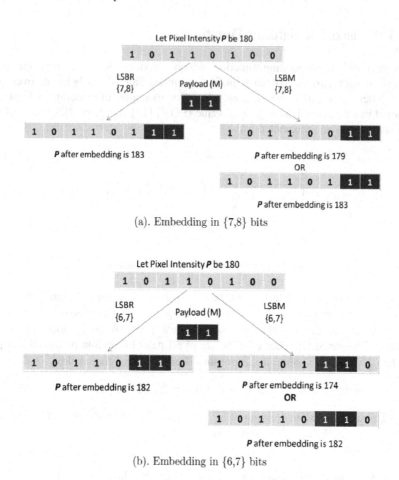

(a). Embedding in {7,8} bits

(b). Embedding in {6,7} bits

Fig. 3. An Example of Embedding Modification

non-structural detectors. It may lead to visual attacks which are countered by distributing secret message uniformly over the entire cover image. A stego key k is used to shuffle pixels of the cover image before embedding to simulate uniform distribution of message over the cover image. Embedding in higher bits can be seen as an extension of LSBR and LSBM embedding approach. In this section, two different embedding functions are proposed; embedding in {7, 8} and {6, 7} bits of pixel. Embedding in {7, 8} and {6, 7} may lead to change in pixel intensity by ± 3 and ± 6 respectively. Large change in pixel intensity is reduced by adjusting next higher and lower bits of the pixel. It is assumed that each pixel has depth of 8 bits. Payload is embedded in two continuous significant bits of the cover image through random walk generated by a stego key. Actual embedding is done through LSBR. Encrypted secret messages add to the security of the embedding process.

The proposed extended embedding functions have been illustrated with the help of an example shown in Fig. 3. Let p be a pixel with the intensity value 180 and two bits of payload M be 11. The pixel is modified at location {7, 8} and {6, 7} through LSB Replacement. Payload bits and modified bits are shown in dark blue color.

Modification of pixel's bits using LSBR in function {7, 8} and {6, 7} results in only one possible pixel intensity for each case. But embedding using LSB Matching has two possibilities in each case. It is because of randomly incrementing and decrementing the pixel intensity on mismatch between pixel's bit and a message bit.

3   Experimental Results

To analyze effects of embedding functions, the proposed function and its variant are implemented. These variants are compared with LSBR and LSBM. It can be noted that rLSBR and rLSBM are the modified form of LSBR and LSBM respectively. These functions are implemented to show that random selection of pixels over the entire cover image is immune to visual attacks. 12000 images from BOWS2 and NRCS [14] databases are used for validation purpose. Three different testing strategies are used for validation. Stego images are tested against first order statistics such as Histogram attacks, higher order statistics such as Sample Pair Analysis (SP) and Weighted Stego (WS) [15] and blind attack based on features extraction and machine learning such as SPAM.

Fig. 4 shows the results of histogram attack on stego images obtained through LSBR, LSBM and extended bits ({7, 8} and {6, 7}) for a relative payload of 0.20 bpp. As expected, all expect LSBR resist histogram attack. Probability of embedding for LSBM is close to 1 for initial 20% of pixels of cover image and negligible for the rest of the embedding modification functions. Probability of embedding less than 0.1 implies no-embedding and greater than 0.9 implies high possibility of embedding.

Table 1 compares the security of variants of the proposed embedding, LSBR and LSBM against visual attacks, Histogram attack, Sample Pair Analysis (SPA) and

Table 1. Security Comparison

Relative Payload	Algorithms	Visual Attack	Histogram Attack	Sample Pair Analysis	Weighted Stego
0.20	LSBR	Fail	Fail	Fail	Fail
	rLSBR	Pass	Fail	Fail	Fail
	LSBM	Fail	Pass	Fail	Fail
	rLSBM	Pass	Pass	Fail	Fail
	{6,7}	Pass	Pass	Pass	Pass
	{7,8}	Pass	Pass	Pass	Pass
0.60	LSBR	Fail	Fail	Fail	Fail
	rLSBR	Pass	Fail	Fail	Fail
	LSBM	Fail	Pass	Fail	Fail
	rLSBM	Pass	Pass	Fail	Fail
	{6,7}	Pass	Pass	Pass	Pass
	{7,8}	Pass	Pass	Pass	Pass
1.00	LSBR	Fail	Fail	ND[1]	Fail
	rLSBR	Pass	Pass	ND	Fail
	LSBM	Fail	Pass	ND	Fail
	rLSBM	Pass	Pass	ND	Fail
	{6,7}	Pass	Pass	Pass	Pass
	{7,8}	Pass	Pass	Pass	Pass

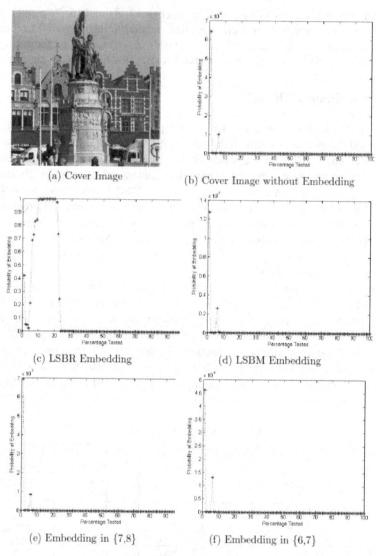

Fig. 4. Histogram Attack

Weighted Stego (WS). Histogram attack is based on POV analysis and it is considered to be successful if the false acceptance rate is less than 10%; otherwise, it fails to detect embedding. SPA analyzes pair of pixels through state diagram. WS is based on weights of neighboring pixels. It can be seen from Table 1 that the proposed function is robust against visual and statistical attacks for varying embedding capacity. If a embedding function clears the test, it implies that embedding is not detectable through the test. ND[1] for the relative payload implies that nothing can be concluded from analysis.

---

[1] Non deterministic results.

3.1   Steganalysis by Subtractive Pixel Adjacency Matrix (SPAM)

It is a machine learning based steganalysis technique which defines a feature set of 686 features from an image [16]. Analysis of these techniques are performed by taking features sets from their respective stego images and natural images. These features are used to train Support Vector Machine (SVM) to learn the difference in features caused by steganography [17]. Fig. 5 shows detection accuracy of the LSBR and LSBM variants of the proposed functions {6,7} and LSBM against SPAM. All functions resist SPAM steganalysis for small relative payload. But for large payload {6,7}, it is detected with good accuracy. Detection accuracy of LSBM and LSBR variant of {7,8} are close to LSBM is shown in Fig. 6 but LSBM variant of {7,8} is slightly better for all cases. LSBM variants of {6,7} and {7,8} perform better in comparison to their LSBR variants. Though, LSBR variants of the proposed function are inferior to their LSBM variants but they are quite better in comparison to original LSBR embedding function. Hence, these LSBR variants of the proposed function can replace LSBR embedding for the model-preserving steganographic techniques.

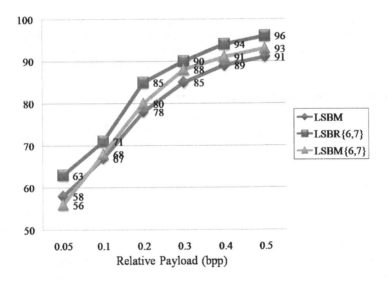

Fig. 5. Detection Accuracy of {6,7} variant of the proposed function

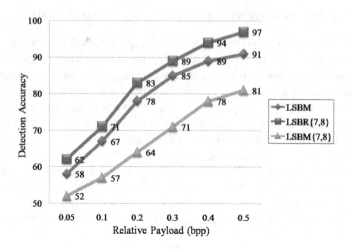

Fig. 6. Detection Accuracy of {7,8} variant of the proposed function

4   Conclusions

In this paper a steganographic function has been proposed on the basis of limitations of the current steganalysis techniques to detect embedding in images. The proposed function is more secure and has better embedding capacity for model-preserving steganography. It performs better or at par with LSBM for all the test cases. Though least two significant bits based steganography intuitively appears to be more vulnerable to steganalysis but in reality it is much more secure.

Acknowledgement. Authors like to acknowledge the support provided by the department of Information Technology, Government of India to carry out this research work.

References

1. Westfeld, A.: F5-A Steganographic Algorithm. In: Moskowitz, I.S. (ed.) IH 2001. LNCS, vol. 2137, pp. 289–302. Springer, Heidelberg (2001)
2. Mielikainen, J.: LSB matching revisited. IEEE Signal Processing Letters 13(5), 285–287 (2006)
3. Fridrich, J.J., Kodovský, J.: Rich Models for Steganalysis of Digital Images. IEEE Transactions on Information Forensics and Security 7(3), 868–882 (2012)
4. Pevný, T., Fridrich, J.J., Ker, A.D.: From Blind to Quantitative Steganalysis. IEEE Transactions on Information Forensics and Security 7(2), 445–454 (2012)
5. Provos, N., Honeyman, P.: Hide and seek: an introduction to steganography. IEEE Transactions on Security Privacy 1(3), 32–44 (2003)
6. Upham, D.: Jsteg (2013), ftp://ftp.funet.fi/pub/crypt/steganography/

7. Fridrich, J.J., Pevný, T., Kodovský, J.: Statistically Undetectable JPEG Steganography: Dead ends Challenges, and Opportunities. In: The 9th Workshop on Multimedia & Security (MM&Sec), Dallas, Texas, USA, pp. 3–14. ACM (2007)

8. Filler, T., Judas, J., Fridrich, J.: Minimizing Additive Distortion in Steganography using Syndrome-Trellis Codes. IEEE Transactions on Information Forensics and Security 6(3), 920–935 (2011)

9. Cox, I.J., Miller, M.L., Bloom, J.A., Fridrich, J., Kalker, T.: Digital Watermarking and Steganography, 2nd edn. Morgan Kaufmann (2008)

10. Lee, K., Westfeld, A., Lee, S.: Category Attack for LSB steganalysis of JPEG images. In: Shi, Y.Q., Jeon, B. (eds.) IWDW 2006. LNCS, vol. 4283, pp. 35–48. Springer, Heidelberg (2006)

11. Dumitrescu, S., Wu, X., Memon, N.: On steganalysis of random LSB embedding in continuous-tone images. In: International Conference on Image Processing, Rochester, New York, USA, vol. 3, pp. 641–644 (2002)

12. Ker, A.D., Böhme, R.: Revisiting weighted stego-image steganalysis. In: Electronic Imaging, Security, Forensics, Steganography, and Watermarking of Multimedia Contents X, SPIE, Orlando, Florida, USA, vol. 6819, pp. 5:1–5:17 (2008)

13. Pevný, T., Bas, P., Fridrich, J.: Steganalysis by subtractive pixel adjacency matrix. In: 11th ACM Workshop on Multimedia and Security, MM&Sec, New York, USA, pp. 75–84 (2009)

A Dynamic Slap Fingerprint Based Verification System

Puneet Gupta and Phalguni Gupta

Department of Computer Science and Engineering,
Indian Institute of Technology Kanpur
Kanpur 208016, India
{puneet,pg}@iitk.ac.in

Abstract. Slap image based verification system are highly secure and can be made more robust if a user's template represents all potential intra-class variations. In this paper, a template updation algorithm has been presented which helps to handle the problem of the temporal template aging and tries to capture pose variation. This uses the matched samples which are detected during slap image verification, to update the template. The proposed algorithm has been tested on challenging slap images. It shows that verification accuracy has been increased by more than 3.5% if the proposed template updation algorithm is incorporated in the verification system.

Keywords: Biometric template updation, Slap image verification, co- updation, Supertemplate Introduction.

1   Introduction

Fingerprint is one of the well known biometric traits because it has all necessary characteristics such as uniqueness, permanence and difficult to forge in an authentication system. Performance of a fingerprint based biometric system can be improved if multiple fingerprints are used [14]. For this, slap-image scanners which can acquire multiple fingerprints simultaneously are gaining importance [1]. From the slap image captured by the scanner, fingerprints can be extracted for verification.

A slap image based verification system has less inter-class variation because of the use of multiple fingerprints. But it contains large intra-class variation due to the partial capture of fingerprints or the capture of bad quality fingerprints which are unavoidable and as a result, it degrades the verification accuracy. In this paper, a template updation algorithm is used to handle the problem of intra-class variation in slap image. There exist various template improvement algorithms which make use of stored template of slap images to compensate variation in these images. This has improved the performance of the system.

The outline of this paper is as follows. Need of the template updation has been discussed in the next section. A template updation algorithm has been proposed in Section 3. Next section makes use of the template updation algorithm to design the slap fingerprint verification system. Its performance has been analyzed in Section 5. Conclusions are given in the last section.

D.-S. Huang et al. (Eds.): ICIC 2014, LNCS 8588, pp. 812–818, 2014.
© Springer International Publishing Switzerland 2014

2   Preliminaries

2.1   Requirement of Template Updation

At the enrolment stage, each biometric sample of a user is captured, processed and its feature template along with user identity is stored in the database. At the time of verification, the claimed identity of a user is tested by matching template of the acquired biometric sample with its enrolled template. Performance of the verification system is dependent on two types of errors, viz. false acceptance rate (FAR) and false rejection rate (FRR). Some of the critical issues which play an important role on the performance of the system are:

- There are few enrolment samples; as a result, they fail to capture all type of variations.
- Extracted features are changed against time (like cuts or scars on finger- prints).
- User may undergo some inappropriate interaction with the sensor (e.g., partial fingerprints).
- There can be some adverse environmental condition which can corrupt the acquired sample quality (like wet or dry fingerprints can occur due to humidity and temperature).

To handle these issues, template improvement algorithms can be used.

2.2   Template Improvement

Template improvement algorithms modify the stored templates by using matched samples such that large intraclass variability in fingerprint matching can be handled. To find matching samples during verification, matching score between a sample and a template is generated and if it is above a threshold then that sample is matched sample; otherwise it is a non-matched sample. Hence, the performance of a these algorithms is mainly depends on this threshold. If there is high value of threshold then few samples are used for template updation. This contradicts the aim of capturing variations in template updation. On the contrary, if low threshold is used, even dissimilar templates are used for template updation which can hamper the system performance. Also, one can either use multiple templates or can use a single template (referred as supertemplate) for a matching sample. Algorithms based on multiple templates consider and store templates with high similarities in the database. During verification, biometric sample is matched against all templates of a user which eliminate large intra- class variability during matching. To restrict the size of stored templates to a manageable amount, usually few templates which can capture large amount of variations are considered. The problem with these algorithms is the time and the space complexity. In contrast, algorithms based on supertemplate requires single template for updation. Thus these algorithms are less time and memory consuming. These algorithms can be divided into three techniques:

- **Periodic Template Update:** These are the supervised learning based template updation techniques in which template is updated by re-enrolling the user after a fixed time so as, to remove the temporal changes.

- **Template Selection:** In template selection, prototype templates are selected from given samples. In [12], two such algorithms are proposed. One algorithm uses a clustering technique while the other uses maximum similarity constraints to choose the stored templates.
- **Template Updation:** It uses the matched samples to upgrade the template. This type of techniques can be divided into two categories viz. self updating and co-updating algorithms [10]. In self-update algorithms [9], matched samples are classified by using only one biometric trait, even if multiple modalities are available. In contrast, co-update algorithms [3] use multiple modalities; thus, more accurate classification results are expected in this case. After classification, templates can be updated. In [6], minutiae points in the multiple suitable fingerprint samples are fused such that (i) corresponding minutiae in several samples are consolidating into a minutia; (ii) false minutiae are removed; and (iii) genuine minutiae are recovered. It is improved in [11] by using local fingerprint quality during template updation.

3   Proposed Template Updation Algorithm

This section presents an algorithm to update the template of the enrolled fingerprint in the database with the help of matched query sample. It assumes that each slap image consists of 4 distinct fingerprints and all single fingerprints present in the slap image are independent in nature. Corresponding to each user fingerprint, its template and skeleton [7] are stored in the database. By templates, we mean minutiae which are used during matching. Fingerprint skeletons are used to update the template by using fingerprint registration.

3.1   Database Creation during Enrollment

During enrollment, individual fingerprints from the slap image are extracted and labeled. Skeletons and minutiae of these fingerprints are extracted by [7] and [13] respectively. Extracted minutiae for each fingerprint forms a template. Corresponding to the user identity, individual fingerprints are stored in the database in the form of skeleton images and templates such that an ordering defined by index, middle, ring and little fingerprint is followed at the time of storing data.

3.2   Template Updation

Lets us assume that the query slap image Q corresponds to the subject B. Also, assume that ST_i and T_i (for i =1,2,3,4) are the stored skeletons and templates respectively of different fingers of B while F_i and X_i (for i =1,2,3,4) are the extracted skeletons and templates respectively from the query image Q. Since Q is matched with B, each T_i can be updated by using X_i such that minutiae present in T_i and X_i are present in the updated template. Also, there can be some minutiae that are present in both T_i and X_i but these are slightly deformed due to elasticity in the skin. Such minutiae should be added once and thus these should be accurately determined. For this, a fingerprint registration algorithm which can handle elastic deformation of skin is required. Also, it is observed that it is better to register skeleton of the fingerprint images [2], rather than minutiae for a better fingerprint registration. Further in some cases, even the genuine matched template can degrade the performance [8]. Thus the impact of template updation should be first determined and if it can improve the performance, then template can be updated. It has been observed that template updation can improve the performance of a verification system, if following two conditions are satisfied, which are given by:

- **Sufficient Uniform Structure:** It is observed that if ST_i and F_i have insufficient uniform ridge valley structure that is, have less foreground area or have bad quality due to dry or wet fingerprints, then template updation can degrade the performance. It occurs because some area of fingerprints have less features or discriminative power while bad quality fingerprints generate spurious flow of ridges and valleys. One can estimate foreground area and uniformity in ridge valley structure by using symmetric filters [4] which measures the uniformity in ridge-valley pattern, irrespective of linear or curved (near singular points) flow. Let the quality of ST_i and F_i be ST^q_i and F^q_i respectively. Thus, if ST^q_i and F^q_i are greater than th_1 (experimentally determined) then there is sufficient uniform structure.
- **Sufficient Overlapping Area:** In a template updation algorithm, template of biometric sample needs to be first registered with the matched template before updation. If there is sufficient overlapping area between the registered images then template updation can be accurately accomplished. In this, F_i with respect to ST_i, thin plate spline (TPS) model described in [2] is used. It uses non rigid transformations to handle the elastic nature of fingertip skin. By this, we obtain: (i) A which is the overlapping area between ST_i and F_i; (ii) T_F which is the transformed image of F_i such that T_F and ST_i are registered; and (iii) T_E which stores the transformation parameters used to register F_i with respect to ST_i. Thus, if A is greater than th_2 (experimentally determined) then conditions of sufficient overlapping area is satisfied.

If both the conditions are satisfied then T_i is updated with the help of X_i. For this, X_i is first transformed by using T_E and let it be give Y_i. All the minutiae present in Y_i and T_i are concatenated such that all the minutiae are unique and stored as T_i in the database. With this, minutiae present in both T_i and X_i are counted just once even if large deformations are introduced. Also, ST_i is updated by applying pixelwise binary *OR* operation on ST_i and T_F.

It may happen that during enrollment, captured biometric sample has poor quality and thus spurious template is stored in the database. It is better to replace such a stored template with the template of a good quality matched biometric sample. That is, if the quality of ST_i (ST^q_i) is less than the quality of F_i (F^q_i) then T_i is replaced by X_i. Further in this case, ST_i is replaced by F_i.

4   Slap Fingerprint Verification System

This section presents an efficient slap fingerprint verification system. It extracts individual fingerprints from the slap image and matches these with the templates stored in the database for slap image matching. Major steps of our slap fingerprint based verification system are:

4.1   Slap Fingerprint Segmentation

Individual fingerprints are extracted from the slap image I by using the algorithm proposed in [5]. Fingerprints are found to be extracted accurately even if there are some dull prints, large rotational angles or non-elliptical shape fingerprints in the slap image. Further, these fingerprints are labeled these as index, middle, ring and little finger of left/ right hand correctly. Let C_1, C_2, C_3 and C_4 be the fingerprints of index, middle, ring and little finger of I respectively.

4.2   Single Fingerprint Matching

Assume that the user has claimed the identity of Z and the templates corresponding to index, middle, ring and little fingerprints are represented by C'_1, C'_2, C'_3 and C'_4 respectively. Templates of each fingerprint query are matched with that of the corresponding finger of the claimed identity by using Bozorth's matcher [13].

4.3   Fusion of Matching Scores

Matching scores of the various fingers are fused to identify whether the slap image I is of Z or not. In order to do so, we consider two best matching scores and if both of these are above a threshold (experimentally determined) then I is of Z .

5   Experimental Results

The proposed algorithm has been evaluated on a database of size 1800 slap images of 150 users belonging to the rural area. Total 6 slap images per hand are collected per user. It is carried out in 2 separate sessions with a time gap of more than 2 months. Our experiments consider images from different hands as separate classes; thus 300 classes are used for our evaluation. Also, experimental results are given in terms of accuracy which is given by (100 - EER)% , where EER is the equal error rate.

The database is partitioned into six parts of equal size which are named as P_i, i = 1,2,..,6 such that each part has one slap image of each class. Without any loss of generality, P_1 is used as gallery while others are used for testing one by one, starting from P_2 to P_6. Experimental results are shown in Table 1. Accuracy of each fingerprint type is shown with and without template updation.

If template updation is not used then accuracies for fingerprints are almost stable. On the other hand, accuracy of each fingerprint increases when the proposed template updation algorithm is used. Reason for this is that template updation helps to minimize the problem of partial fingerprint capture and template aging. Further, probable reason of small deviation in case of P_4 is that P_1, P_2 and P_3 are captured in first session while P_4 is captured in second session and thus P_1, P_2 and P_3 are not capable enough to represent P_4 due to high temporal changes. Despite of this, one can see that accuracy parameters of individual fingerprints are increased if template updation is used. Similarly, it can be observed that by the use of template updation, accuracy of matching is improved.

Table 1. Comparative Results in terms of Accuracies (in %)

		Fingerprint Used	Database part used #				
			P_2	P_3	P_4	P_5	P_6
Template Updation	Without*	Index	84.37	84.13	83.62	83.70	83.05
		Middle	87.19	87.06	86.03	86.28	86.12
		Ring	86.82	86.57	85.76	85.09	85.39
		Little	80.26	80.23	78.71	78.52	78.57
		Combined**	89.79	89.48	88.26	88.73	88.51
	With+	Index	84.37	85.57	85.43	87.23	88.32
		Middle	87.19	88.42	88.16	89.58	90.71
		Ring	86.82	87.63	87.41	88.28	88.95
		Little	80.26	82.12	81.95	84.37	85.94
		Combined**	89.79	91.78	91.23	92.91	93.74

*: In this, template updation is not used.
+: This refers to the proposed template updation algorithm.
**: In this, all fingerprints are fused by using the proposed algorithm.
#: It is sequentially executed.

6   Conclusions

In this paper, a template updation algorithm has been introduced in the context of slap image verification. With this, accuracy of the verification system has been significantly increased because it reduces the temporal template aging and captures significant intra-class variations. The proposed algorithm has used only those samples, by which the template updation can increase the verification accuracy. Such samples have been effectively detected by using various like sufficient uniform structure and sufficient overlapping area. Further, it has been experimentally seen that the proposed algorithm performs better than well known algorithms.

Acknowledgement. Authors are thankful to the anonymous reviewers for their valuable suggestions to improve the quality of the paper. This work is partially supported by the Department of Information Technology (DIT), Government of India.

References

1. AADHAAR AUTOMATED: Uid enrolment proof-of-concept report, http://uidai.gov.in/images/FrontPageUpdates/uid_enrolment_poc_report.pdf
2. Bazen, A.M., Gerez, S.H.: Fingerprint matching by thin-plate spline modelling of elastic deformations. Pattern Recognition 36(8), 1859–1867 (2003)
3. Didaci, L., Marcialis, G.L., Roli, F.: Analysis of unsupervised template update in biometric recognition systems. Pattern Recognition Letters 37, 151–160 (2014)
4. Fronthaler, H., Kollreider, K., Bigun, J., Fierrez, J., Alonso-Fernandez, F., Ortega-Garcia, J., Gonzalez-Rodriguez, J.: Fingerprint image-quality estimation and its application to multialgorithm verification. IEEE Transactions on Information Forensics and Security 3(2), 331–338 (2008)

5. Gupta, P., Gupta, P.: Slap fingerprint segmentation. In: International Conference on Biometrics: Theory, Applications and Systems. IEEE (2012)
6. Jiang, X., Ser, W.: Online fingerprint template improvement. IEEE Transactions on Pattern Analysis and Machine Intelligence 24(8), 1121–1126 (2002)
7. Lam, L., Lee, S.W., Suen, C.Y.: Thinning methodologies-a comprehensive survey. IEEE Transactions on Pattern Analysis and Machine Intelligence 14(9), 869–885 (1992)
8. Rattani, A., Marcialis, G.L., Granger, E., Roli, F.: A dual-staged classification-selection approach for automated update of biometric templates. In: International Conference on Pattern Recognition. IEEE (2012)
9. Rattani, A., Marcialis, G.L., Roli, F.: Self adaptive systems: An experimental analysis of the performance over time. In: IEEE Workshop on Computational Intelligence in Biometrics and Identity Management. IEEE (2011)
10. Roli, F., Didaci, L., Marcialis, G.L.: Adaptive biometric systems that can improve with use. In: Advances in Biometrics. Springer (2008)
11. Ryu, C., Kim, H., Jain, A.K.: Template adaptation based fingerprint verification. In: International Conference on Pattern Recognition. IEEE (2006)
12. Uludag, U., Ross, A., Jain, A.: Biometric template selection and update: a case study in fingerprints. Pattern Recognition 37(7), 1533–1542 (2004)
13. Watson, C.I., Garris, M.D., Tabassi, E., Wilson, C.L., McCabe, R.M., Janet, S.: Users guide to nist fingerprint image software 2 (NFIS2). National Institute of Standards and Technology (2004)
14. Wilson, C., Hicklin, R., Korves, H., Ulery, B., Zoep, M., Bone, M., Grother, P., Micheals, R., Otto, S., Watson, C., et al.: Fingerprint vendor technology evaluation 2003: Summary of results and analysis report. US Dept. of Commerce, National Institute of Standards and Technology (2004)

Multimodal Personal Authentication Using Iris and Knuckleprint

Aditya Nigam and Phalguni Gupta

Department of Computer Science and Engineering,
Indian Institute of Technology Kanpur, Kanpur 208016, India
{naditya,pg}@cse.iitk.ac.in

Abstract. This paper proposes multi-modal authentication system by fusing iris and knuckleprint images. The iris and knuckleprint ROI's are preprocessed using the proposed LGBP method to obtain robust features. The corners features are extracted and matched using the proposed CIOF dissimilarity measure. The proposed approach has been tested on publicly available CASIA 4.0 Interval and Lamp iris along with PolyU knuckleprint databases. It is also tested on multimodal databases that are created by fusing iris and knuckleprint and has shown encouraging results.

Keywords: Multimodal, Biometrics, Knuckleprint, Iris, LBP.

1 Introduction

Biometrics can be used as an alternative to any token-based or knowledge based traditional authentication methods because it is easier to use and harder to circumvent. There are several popular traits such as face, iris, palmprint, fingerprint, knuckleprint, ear *etc.*, but none of them can be considered as the best because it depends on the type of application where it has to be applied. The performance of unimodal biometric system often got restricted due to varying environmental conditions, sensor precision and its reliability along-with several trait specific challenges like pose, expression, aging etc. Also the image quality [1,2] plays a significant role in the performance degradation. Hence fusing more than one biometric trait for superior performance can be a very useful idea.

The iris have abundance of micro-texture as crypts, furrows, ridges, corona, freckles and pigment spots which are randomly distributed. The major work in iris recognition is done in [3] using gabor responses and quantizing it to generate iris feature vector. On the other hand knuckleprint have line like rich patterns in vertical and horizontal directions. In [4], the region of interest using convex direction coding, local and global features are extracted and fused. The multimodal work is done only over chimeric databases. In [5], 2D discrete wavelets are used to extract features from iris and face for classification. In [6], face and iris traits are fused using SIFT feature vector. The classification is done using nearest neighborhood ratio matching. In [7], Near Infra Red NIR Iris and Face images are considered and PCA and gabor filter approach is used for face and iris recognition respectively.

D.-S. Huang et al. (Eds.): ICIC 2014, LNCS 8588, pp. 819–825, 2014.

In this paper iris and knuckleprint traits are considered for authentication using the proposed tracking based CIOF dissimilarity measure. The paper is organized as follows: Section 2 presents the proposed system. In Section 3 experimental results are discussed while the last section presents the concluding remarks.

2   Proposed System

Initially region of interests *(ROI)* is extracted from iris and knuckleprint samples. Segmentation of iris and knuckleprint is done using [8,4] respectively. Later ROI's are enhanced and transformed to obtain a robust representation that can tolerate small amount of illumination variations. To achieve robustness against affine transformations, constrained KL-tracking is used for matching.

[I] Enhancement: The abundant texture in a sample is enhanced to increase its richness. Any sample image is divided into blocks of size 8x8 and the mean of these blocks are considered as the coarse illumination of that block. This mean is expanded to the original size and is subtracted from the original image to obtain uniformly illuminated image. The contrast of uniformly illuminated image is enhanced using Contrast Limited Adaptive Histogram Equalization *(CLAHE)*.

[II] Transformation Using LGBP: The gradient of any edge pixel is positive if it lies on an edge created due to light to dark shade (*i.e. high to low gray value*) transition else it is having negative gradient value. Hence all the edge pixels can be divided into two classes of +*ve* and -*ve* gradient values. This sign augmented gradient information of each edge pixel can be more discriminative and robust. The proposed transformation uses this information to calculate a 8-bit code for each pixel using x and y-direction derivatives of its 8 neighboring pixels to compute *vcode* and *hcode* respectively. Let $P_{i,j}$ be the $(i,j)^{th}$ pixel of any biometric sample image P and *Neigh[l]*, $l = 1,2,...8$ are the gradients of 8 neighboring pixels centered at pixel $P_{i,j}$ obtained by applying *scharr* kernel, then the kth bit of the 8-bit code (termed as *lgbp_code*) is given by

$$lgbp\_code[k] = \begin{cases} 1 & \textbf{if } Neigh[k] > 0 \\ 0 & \textbf{otherwise} \end{cases} \qquad (1)$$

In *vcode* or *hcode* as shown in Fig. 1 every pixel is represented by its *lgbp_code*. Since *lgbp_code* of any pixel considers only the sign of the derivative it ensures the robustness of the proposed transformation under illumination variation.

[III] Feature Extraction: Corners have strong derivative in two orthogonal directions can provide enough information for tracking. The spectral decomposition of 2×2 autocorrelation matrix M is performed, for each pixel belonging to *vcode* or *hcode*. The matrix M can have two eigen values λ_1 and λ_2 such that $\lambda_1 \geq \lambda_2$. Finally all pixels having $\lambda_2 \geq T$ are considered as corner feature points [9].

[IV] Feature Matching: The image matching is performed using KL tracking algorithm [10] that uses three basic assumptions: *Brightness Consistency, Temporal Persistence,* and *Spatial Coherency.* Therefore one can estimate the motion of any pixel by assuming a local constant flow. The performance of KL tracking depends on

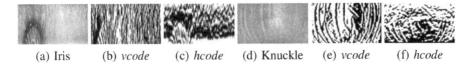

(a) Iris   (b) *vcode*   (c) *hcode*   (d) Knuckle   (e) *vcode*   (f) *hcode*

Fig. 1. Original and Transformed (*vcode, hcode*) for Iris and knuckleprint ROI's

how well these assumptions are satisfied. It is expected that the above mentioned assumptions are more likely to be satisfied for genuine matchings and degrades substantially for imposters. Therefore one can infer that the performance of KL tracking is good in genuine matching as compared with the imposter ones.

Potential corner matching pairs between two images are reported using KL tracking. The performance of KL tracking is estimated using the proposed dissimilarity measure *CIOF* (Corners having Inconsistent Optical Flow). The estimated optical flow direction is quantized into eight directions and the most consistent direction is selected as the one which has most number of successfully tracked corner features. Any corner matching pair having optical flow direction other than the most consistent direction is considered as false matching pairs and is discarded.

Matching Algorithm: Given two *vcode* I_A^v, I_B^v and two *hcode* I_A^h, I_B^h, Algorithm 1 has been presented which can be used to compare two sample images, *BSample$_a$* with *BSample$_b$* using *CIOF* dissimilarity measure. The Final score, *CIOF* (BSample$_a$, BSample$_b$) is obtained by using the sum rule fusion of horizontal and vertical matching scores. Any corner is considered as tracked successfully if it satisfies three constraints as defined below:

[i] Vicinity Constraints: Euclidean distance between any corner and its estimated tracked location should be less than or equal to an empirically selected threshold TH_d.

[ii] Patch-wise Dissimilarity: Tracking error defined as pixel-wise sum of absolute difference between a local patch centered at current corner and that of its estimated tracked location patch should be less than or equal to threshold TH_e.

[iii] Correlation Bound (only for knuckleprint): Phase only correlation [11] between a local patch centered at any feature and that of its estimated tracked location patch should be at-least equal to an empirically selected threshold TH_{cb}.

Out of all the successfully tracked corners (stc_{AB}^v, stc_{AB}^h) those that are having inconsistent optical flow are considered as false matches. The *CIOF* measure computes the total number of corners that are successfully tracked but have inconsistent optical flow. Average of *ciof$_{AB}$* and *ciof$_{BA}$* is computed to make the measure symmetric.

3   Experimental Results

3.1   Database

The proposed system is tested on two publicly available CASIA V4 Interval [12] and Lamp [12] iris databases along with the largest publicly available PolyU [13] knuckleprint database. Later two chimeric multimodal databases are also created for testing.

Algorithm 1 $CIOF(BSample_a, BSample_b)$

Require:
(a) The $vcode$ I_A^v, I_B^v and $hcode$ I_A^h, I_B^h of two sample images $BSample_a, BSample_b$ respectively.
(b) N_a^v, N_b^v, N_a^h and N_b^h are the number of corners in I_A^v, I_B^v, I_A^h, and I_B^h respectively.

Ensure: Return $CIOF(BSample_a, BSample_b)$.

1: Track all the corners of $vcode$ I_A^v in $vcode$ I_B^v and that of $hcode$ I_A^h in $hcode$ I_B^h.
2: **if** (Biometric Samples are of IRIS) **then**
3:    Obtain corner set successfully tracked in $vcode$ tracking (*i.e.* stc_{AB}^v) and $hcode$ tracking
     (*i.e.* stc_{AB}^h) having their tracked position and patch dissimilarity within TH_d and TH_e.
4: **else**
5:    Obtain the set of corners successfully tracked in $vcode$ tracking (*i.e.* stc_{AB}^v) and $hcode$
     tracking (*i.e.* stc_{AB}^h) that have their tracked position within TH_d and their local patch dis-
     similarity under TH_e also the patch-wise correlation is at-least equal to TH_{cb}.
6: **end if**
7: Calculate successfully tracked corners of I_B^v in I_A^v (*i.e.* stc_{BA}^v) and I_B^h in I_A^h (*i.e.* stc_{BA}^h).
8: Quantize optical flow direction for each successfully tracked corners into only eight direc-
   tions (*i.e.* at $\frac{\pi}{8}$ interval) and obtain 4 histograms $H_{AB}^v, H_{AB}^h, H_{BA}^v$ and H_{BA}^h using these four
   corner set $stc_{AB}^v, stc_{AB}^h, stc_{BA}^v$ and stc_{BA}^h respectively as computed above.
9: For each histogram, the bin (*i.e.* direction) having the maximum corners is considered as the
   consistent optical flow direction. The maximum value obtained from each histogram termed
   as corners having consistent optical flow represented as $cof_{AB}^v, cof_{AB}^h, cof_{BA}^v$ and cof_{BA}^h.
10: $ciof_{AB}^v = 1 - \frac{cof_{AB}^v}{N_a^v}; ciof_{BA}^v = 1 - \frac{cof_{BA}^v}{N_b^v}; ciof_{AB}^h = 1 - \frac{cof_{AB}^h}{N_a^h}; ciof_{BA}^h = 1 - \frac{cof_{BA}^h}{N_b^h};$
11: **return** $CIOF(BSample_a, BSample_b) = \frac{ciof_{AB}^v + ciof_{AB}^h + ciof_{BA}^v + ciof_{BA}^h}{4};$

UNI-MODAL: Casia V4 Interval database [12] contains 2,639 iris images collected from 249 subjects having 395 distinct irises with about 7 images per iris. Casia V4 Lamp [12] is a huge database consisting of 16,212 images collected from 411 subjects having 819 distinct irises with 20 images per iris. The PolyU Knuckleprint [13] database consisting of 7,920 *FKP* sample images obtained from 165 subjects in two sessions. In each session, 6 images from 4 fingers (distinct index and middle fingers of both hands) are collected.

MULTI-MODAL: Two multimodal databases are created by fusing the above mentioned iris and knuckleprint databases. In first **MM1** database, all of the Casia V4 Interval database images are considered along with the first 395 subject's knuckleprint samples from PolyU knuckleprint database. Hence a total of 2,639 iris images as well as knuckleprint sample images are considered from 395 distinct irises and knuckleprints. The second **MM2** database is also created in which all of the PolyU Knuckleprint database images are considered along with the first and last 6 images (*i.e.* a total of 12 samples) of initial 660 subject's iris samples from Casia V4 Lamp iris database. Hence a total of 7,920 iris as well as knuckleprint sample images are considered from 660 distinct irises as well as knuckleprints in multi-modal database **MM2**.

3.2 Performance Analysis

Testing Strategy (Unimodal): For testing Casia V4 Interval database, first session images are taken as training while remaining images are taken as testing. For testing

Table 1. Comparative Performance Analysis over Various Unimodel Databases

Systems	Interval			Lamp			PolyU		
	DI	CRR%	EER%	DI	CRR%	EER%	DI	CRR%	EER%
Daugman [3]	1.961	99.46	1.881	1.2420	98.90	5.59	-	-	-
Li Ma [18]	-	95.54	2.07	-	-	-	-	-	-
Masek [15]	1.99	99.58	1.09	-	-	-	-	-	-
K. Roy [14]	-	97.21	0.71	-	-	-	-	-	-
BOCV [16]	-	-	-	-	-	-	-	-	1.833
MTexcode [17]	-	-	-	-	-	-	-	-	1.816
Proposed	**2.0182**	**100**	**0.1093**	**1.5045**	**99.8722**	**1.3005**	**1.605**	**99.79**	**1.775**

Casia V4 Lamp database, first 10 images are considered as training and rest are taken as testing images. For testing PolyU Knuckleprint database, all first session 6 images are taken as training while second session images are taken as testing.

Testing Strategy (Multimodal): For testing chimerically fused iris Interval and knuckleprint PolyU database (**MM1**), same interval testing strategy is used. The first session iris images are fused with first 3 PolyU knuckleprint images and are considered as training data while remaining iris images of that subject are fused with the same number of second session PolyU knuckleprint images for testing. For testing chimerically fused iris Lamp and knuckleprint PolyU database (**MM2**), same PolyU Knuckleprint testing strategy is used. The first session knuckleprint images are fused with first 6 iris Lamp database images and are considered as training data while remaining knuckleprint images of that subject are fused with the last 6 iris Lamp database images for testing.

Parameterized Analysis: The parameter values for which iris system has found to be performing best are $TH_e=600$ with patch size of 5×5 and $TH_d=7$ for Lamp while $TH_e=600$ with patch size of 5×5 and $TH_d=10$ for Interval databases. These optimal values are obtained by testing over only first 100 subjects iris data. For knuckleprint, only left index first 50 subjects are included in validation dataset for parameterization. The parameter values for which system has found to be performing best are $TH_e=2000$ with patch size of 11×11 with $TH_d=22$ along with $TH_{cb}=0.4$ over PolyU Knuckleprint databases.

ROC Based Performance Analysis: The proposed system has been compared with the state of the art iris recognition systems [3,18,15,14]. For comparing with [3,15] we have coded their systems and with [18,14] we have used the results as stated in [14]. It is found that the *CRR* (**Rank 1** accuracy) of the proposed system is more than 99.77% for both iris databases. The comparison of the proposed system with other state of the art systems is shown in Table 1. Further, its *EER* is 0.108% for Interval and that for Lamp is 1.29% which is better than the reported systems. The proposed system has been compared with [16,17] knuckleprint systems and has shown better performance in terms of *EER* as shown in Table 1.

The proposed multimodal system is rigorously tested over chimerically self created multimodal databases *viz.* **MM1** and **MM2** by fusing iris samples with knuckleprint

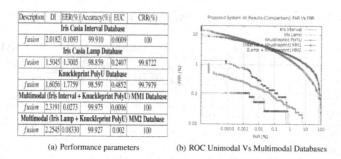

Description	DI	EER(%)	Accuracy(%)	EUC	CRR(%)
Iris Casia Interval Database					
fusion	2.0182	0.1093	99.910	0.0009	100
Iris Casia Lamp Database					
fusion	1.5045	1.3005	98.859	0.2407	99.8722
Knuckleprint PolyU Database					
fusion	1.6056	1.7759	98.597	0.4852	99.7979
Multimodal (Iris Interval + Knuckleprint PolyU) MM1 Database					
fusion	2.3191	0.0273	99.975	0.0006	100
Multimodal (Iris Lamp + Knuckleprint PolyU) MM2 Database					
fusion	2.2545	0.08330	99.927	0.002	100

(a) Performance parameters   (b) ROC Unimodal Vs Multimodal Databases

Fig. 2. Performance Analysis for various databases

samples. It is found that the *CRR* (**Rank 1** accuracy) of the proposed system is 100% over both databases. For both **MM1** and **MM2** databases, Receiver Operating Characteristics (*ROC*) curves are shown in Fig. 2(b). In Fig. 2(b), the *ROC* behavior of the proposed system over all five database *viz.* Casia V4 Interval, Casia V4 Lamp, PolyU Knuckle, **MM1** and **MM2** is plotted for comparative analysis. It is clearly evident that multimodal fusion enhances the system performance significantly. The comparison with some existing multimodal systems like [5,6,7] that are fusing, face and iris or knuckle and palm is not justified as their results are restricted due to their biometric modality selection hence it is not done.

One can clearly observe how well the proposed fusion based multimodal system has performed using Fig. 2(b) and Table 2(a). The verification accuracy is evaluated in terms of *EER* and is found to be 0.0273% over **MM1** and 0.0833% over **MM2**. These values signify that the proposed system performs only two and eight errors per ten thousand matchings over *MM1* and **MM2** respectively. The decidability index (d') is found to be 2.31 and 2.25 for **MM1** and **MM2** databases respectively. One can see that **MM2** is big database hence huge amount of matchings are performed (23,760 Genuine /15,657,840 Imposter) as compared with **MM1** (3,657 Genuine/1,272,636 Imposter). Also the iris images of Lamp database are severely affected due to lighting variation, but still after fusion with knuckleprint the system performance is tremendously improved as shown in Fig 2(b) and in Table 2(a). Apart from the performance, the system has shown tremendous scalability as the performance do not vary much with the size of the database. Negligible error under the *ROC* curve is obtained for both multimodal databases *viz.* **MM1** (*i.e. 6x10^{-4}*) and **MM2** (*i.e. 2x10^{-3}*) ensuring the robustness of the proposed multimodal system.

4   Conclusion

In this paper an accurate multimodal based authentication system has been proposed. The samples are processed and ROI's are extracted which are transformed using local gradient binary pattern (*LGBP*). The corner features are matched using the proposed dissimilarity measure (*CIOF*) that tracks corners using KL tracking. The proposed system has been tested on publicly available unimodal iris and knuckleprint databases as well as two chimerically created multimodal databases. It has been shown that the proposed system performed very well.

Acknowledgements. Authors like to acknowledge the support provided by the Department of Information Technology, Government of India to carry out this work.

References

1. Nigam, A., Gupta, P.: Quality assessment of knuckleprint biometric images. In: ICIP, pp. 4205–4209 (2013)
2. Nigam, A., Anvesh, T., Gupta, P.: Iris classification based on its quality. In: Huang, D.-S., Bevilacqua, V., Figueroa, J.C., Premaratne, P. (eds.) ICIC 2013. LNCS, vol. 7995, pp. 443–452. Springer, Heidelberg (2013)
3. Daugman, J.: High confidence visual recognition of persons by a test of statistical independence. IEEE Transactions on Pattern Analysis and Machine Intelligence
4. Zhang, L., Zhang, L., Zhang, D., Zhu, H.: Ensemble of local and global information for finger-knuckle-print recognition. Pattern Recognition 44(9), 1990–1998 (2011)
5. Son, B., Lee, Y.: Biometric authentication system using reduced joint feature vector of iris and face. In: Kanade, T., Jain, A., Ratha, N.K. (eds.) AVBPA 2005. LNCS, vol. 3546, pp. 513–522. Springer, Heidelberg (2005)
6. Rattani, A., Tistarelli, M.: Robust multi-modal and multi-unit feature level fusion of face and iris biometrics. In: Tistarelli, M., Nixon, M.S. (eds.) ICB 2009. LNCS, vol. 5558, pp. 960–969. Springer, Heidelberg (2009)
7. Zhang, Z., Wang, R., Pan, K., Li, S.Z., Zhang, P.: Fusion of near infrared face and iris biometrics. In: Lee, S.-W., Li, S.Z. (eds.) ICB 2007. LNCS, vol. 4642, pp. 172–180. Springer, Heidelberg (2007)
8. Bendale, A., Nigam, A., Prakash, S., Gupta, P.: Iris segmentation using improved hough transform. In: Huang, D.-S., Gupta, P., Zhang, X., Premaratne, P. (eds.) ICIC 2012. CCIS, vol. 304, pp. 408–415. Springer, Heidelberg (2012)
9. Shi, J.: Good features to track. In: Computer Vision and Pattern Recognition, pp. 593–600 (1994)
10. Lucas, B., Kanade, T.: An Iterative Image Registration Technique with an Application to Stereo Vision. In: International Joint Conference on Artificial Intelligence, pp. 674–679
11. Nigam, A., Gupta, P.: Finger knuckleprint based recognition system using feature tracking. In: Sun, Z., Lai, J., Chen, X., Tan, T. (eds.) CCBR 2011. LNCS, vol. 7098, pp. 125–132. Springer, Heidelberg (2011)
12. CASIA, The casia iris database, http://www.cbsr.ia.ac.cn
13. PolyU, The polyu knuckleprint database, http://www.comp.polyu.edu.hk/biometrics
14. Roy, K., Bhattacharya, P., Suen, C.Y.: Iris recognition using shape-guided approach and game theory. Pattern Analysis and Applications 14(4), 329–348 (2011)
15. Masek, L., Kovesi, P.: Matlab source code for a biometric identification system based on iris patterns. M.Tech Thesis
16. Guo, Z., Zhang, D., Zhang, L., Zuo, W.: Palmprint verification using binary orientation co-occurrence vector. Pattern Recognition Letters 30(13), 1219–1227 (2009)
17. Gao, G., Yang, J., Qian, J., Zhang, L.: Integration of multiple orientation and texture information for finger-knuckle-print verification. Neurocomputing 135, 180–191 (2014)
18. Ma, L., Tan, T., Wang, Y., Zhang, D.: Efficient iris recognition by characterizing key local variations. IEEE Transactions on Image Processing 13(6), 739–750 (2004)

Power-Aware Meta Scheduler with Non-linear Workload Prediction for Adaptive Virtual Machine Provisioning

Vasanth Ram Rajarathinam[1], Jeyarani Rajarathinam[2], and Himalaya Gupta[3]

[1] Advanced Micro Devices, Florida, USA
vasanthram.rajarathinam@amd.com
[2] Coimbatore Institute of Technology, Coimbatore, India
symphonyjr@gmail.com
[3] Aricent, Bangalore, India
himalaya.a@aricent.com

Abstract. Infrastructure cloud typically involves provisioning of dynamically scalable and virtualized resources to cloud users. It is a fact that the resource demand in the cloud is highly dynamic in nature. To meet the dynamic demand from the cloud consumers, over-provisioning of resources is the common solution. This ultimately increases power consumption when the demand is normal or drops below average. On the contrary, under-provisioning of resources may lead to Service Level Agreement (SLA) violations. To balance between power conservation and performance issues, it has been realized that forecasting the demand for computing resources is essential in cloud environment. Hence we proposed a prediction based adaptive resource provisioning methodology incorporating statistical predictor in our earlier work. In order to improve prediction accuracy, we have proposed a recurrent neural network called Non-linear Auto Regressive network with eXogenous input (NARX) based prediction in this paper. The proposed NARX predictor makes near-accurate run time estimate of resource demand as it is able to learn hidden patterns and trends in the historical data representing resource demand and hence it helps to realize better power conservation. This paper shows that the proposed predictor integrated with adaptive provisioning resulted in 27.25 % more power saving compared to its statistical counterpart.

Keywords: Cloud Computing, Power-aware Meta-Scheduler, Dynamic Workload Prediction, Artificial Neural Network, NARX Network.

1 Introduction

Infrastructure as a Service delivery model typically involves provisioning of dynamically scalable and often virtualized resources to cloud users. A study on cloud workload characteristics in a typical data center reveals that the workloads are highly dynamic in nature [9, 10]. Here the workload refers to Virtual Machine (VM) request. To manage the cloud resources efficiently, there is an important component called Meta-Scheduler and one of its key functionalities is Virtual Machine (VM) provisioning

D.-S. Huang et al. (Eds.): ICIC 2014, LNCS 8588, pp. 826–837, 2014.
© Springer International Publishing Switzerland 2014

[7, 8, 21]. The Meta-Scheduler has to deal with two important issues in resource management viz. over-provisioning of resources and under-provisioning of resources. First, over-provisioning of resources results increase in power consumption. Second, under-provisioning of resources results in delay in VM launching. When the powered-on servers are unable to meet the resource demands of incoming VM request, few cloud servers are initiated to power-on which generally involves significant amount of time. Failing towards timely resource allocation results in performance degradation of cloud data centers as responsiveness in VM deployment is a potential factor deciding market growth. The existing power conservation policy namely host on-off policy fails in making immediate VM allocation [1].

To balance power and performance issues, we proposed Power-Aware Meta-Scheduler (PAM) in our earlier work, which has employed prediction based adaptive resource provisioning approach [6 – 8, 11]. The PAM uses sleep state exploitation based resource provisioning policy and is efficient in power conservation without incurring significant latency. The PAM uses statistical predictor to make adaptive resource provisioning. As this approach necessitates near accurate demand prediction to improve power conservation and performance, in this paper we have enhanced PAM (EPAM) with neural network based prediction. It uses recurrent neural network called Non-linear Auto Regressive network with eXogenous input (NARX) based prediction [5, 20, 23]. A simulator for cloud known as CloudSim toolkit has been extended to incorporate the provisioning and allocation services of the Meta-Scheduler [22]. The proposed NARX based prediction service has been implemented in MatLAB and integrated with PAM [4].

The rest of the paper is organized as follows. Section 2 discusses related work in NARX based prediction. Section 3 describes problem formulation. Section 4 describes NARX Model. Section 5 Architecture of proposed Meta-Scheduler. Section 6 shows the design and training of proposed NARX network. Section 7 discuses training result analysis. Section 8 describes experimental set-up used. Section 9 describes the Integration of NARX predictor with PAM architecture. Section 10 concludes the paper suggesting the scope for future work.

2   Related Work

The authors Prodan and Nae [15 - 17] have proposed a prediction based method for dynamic provisioning and scaling of resources in distributed environment for the operation of multi-player online games. The load prediction service anticipates the future game world entity distribution from historical trace data using neural network based method. The author Shabri [18] compared time-series forecasting methods using neural networks and Box-Jenkins model, and showed that for the time series with seasonal patterns, both methods produce comparable results and also proved that neural networks represent a promising alternative method for forecasting. Sharda and Patil [19] also reported that simple neural network systems could forecast demand equivalent to Box-Jenkins system [2]. The authors Menezes-Junior and Barreto [12 - 14] have found the applicability of the NARX network to univariate time-series prediction. Univariate means that there is no exogenous variable in the NARX network and the prediction task is performed using the time series of a single variable. The proposed work focuses on the applicability of univariate time- series prediction for finding resource demand in cloud environment [3].

3   Problem Formulation

This section discuses about target cloud resource model and workload model.

3.1   Cloud Resource Model

Each server in the resource pool is a Multi-Chip Module (MCM) encompassing several Multi-Core Processors (MCP). The cores are homogeneous in processing speed. Different power saving states can be applied at core level, chip level as well as at the host level. The servers produced by different companies comply with ACPI specifications yet, they may have different hardware implementations internally [14]. For evaluating the proposed system, two generic power saving states called Shallow Sleep (SS) state and Deep Sleep (DS) state are considered. This benefits the algorithm to be robust as any desired sleep states of ACPI can be mapped to SS or DS state in realtime Meta-Scheduler deployment.

3.2   Workload Model

In the cloud environment, the VM request type is considered to be immediate and the process of switching on a new host for creating VM on it incur significant time. In the proposed system, the delay incurred is eliminated by keeping appropriate number of Processing Elements (PEs) in low-wakeup latency Shallow Sleep state and other PEs in Deep Sleep state to conserve power. This provisioning is adaptive as it is based on runtime prediction of resource demand. In the proposed system, the demand represents the number of (PEs) needed for the forthcoming batch of VM requests from the clients.

The data set is taken from Parallel Workload Archive [24]. The data set consists of 1,28,662 accounting records collected during the period Feb 2007 to June 2007 from Thunder system in LLNL cluster [24]. Thunder is a 1024- node Linux cluster. We generated a univariate time-series corresponding to LLNL cluster workload traces namely, LLNL containing resource demand for 5000 time instants. The generated time-series is a sequence of data points measured typically at successive time instants spaced at 30 minutes time intervals. Also exploratory data analysis has been done to uncover underlying structure of data and to find outliers and anomalies.

Run Sequence Plot. Time plot graphs are used to display time-value data pairs. Figure 1 summarizes univariate data set representing resource demand of LLNL cluster.

Pre-processing. The main purpose of pre-processing is to identify the hidden pattern in the original time-series. The generated LLNL time-series is first subjected to smoothing techniques such as quadratic-fit and moving-average with span-size of 500 and 100, respectively. It was performed using curve-fitting tool in MatLAB [20]. The hidden pattern in the original time-series was uncovered as a result of smoothing. The run sequence plot of resulting LLNL time-series has been depicted in Figure 2.

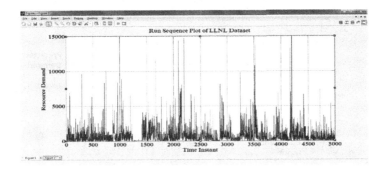

Fig. 1. Run Sequence plot of time-series data generated from LLNL workload log

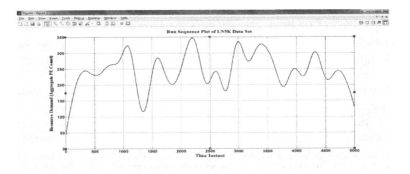

Fig. 2. Run sequence plot of LLNL workload after pre-processing

3.3   Formal Problem Definition

Given a time series representing the historical data about resource demand, find an optimal network structure for NARX network and train the network to minimize Mean Square Error. To evaluate the performance of the predictor, metric called Mean Square Error (MSE) is used. It is defined as the average of the squared differences between the forecasted and observed demands.

4   NARX Predictor Model

NARX is a dynamically driven recurrent network. It has global feedback from output neuron to the input layer. The input to the network is a sequential vector representing past resource demand, and the output is a single value representing future demand.

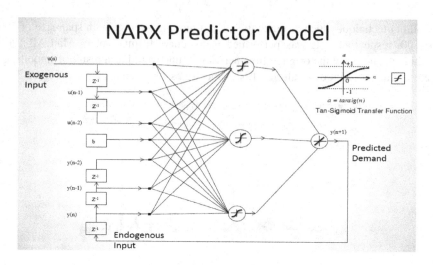

Fig. 3. NARX Predictor Model

The application of feedback connection in the networks makes them suitable devices for non-linear prediction [6]. The general architecture of NARX has been depicted in Figure 3, which incorporates a static multi-layer perceptron. The model has an exogenous input that is applied to a tapped-delay-line memory of q units. It has an endogenous output that is fed back as input via another tapped-delay-line memory of q units. The endogenous and exogenous inputs are used to feed the input layer of the multi-layer perceptron. The present input value of the model is denoted by $u(n)$, and the corresponding output value of the model is denoted by $y(n+1)$. With the absence of exogenous input, NARX reduces to Non-linear Autoregressive network (NAR). In the proposed work, the performance of NAR and NARX networks have been evaluated for demand prediction.

5   Architecture of Power-Aware Meta-Scheduler

Figure 4 shows the architecture of the proposed system, the position of the prediction component and its interaction with other components. The input represents resource requirement in batches and it is observed as non-linear time series. The predictor makes forecast of future demand, which is used by provisioning service for reserving the required resource in the resource pool. The resource availability information is also updated in Cloud Information Service (CIS). The resource allocation service takes a batch of jobs from the input queue and resource availability information from CIS and prepares a schedule for dispatch. The important component of the architecture is the feedback function that is used to control the provisioning of processors to maintain the overall resource usage.

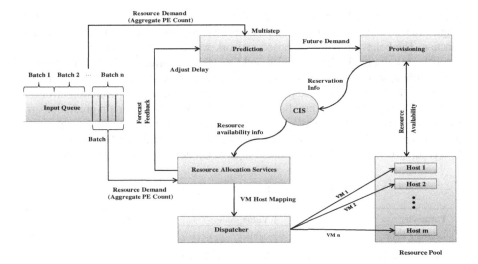

Fig. 4. Architecture of the proposed Meta-Scheduler

6   NARX-Based Demand Prediction

The actual-demand represents exogenous input and predicted-demand represents endogenous input. In this model, future resource demand is predicted with the help of endogenous and exogenous inputs. The LLNL workload trace was used to evaluate the effectiveness of the proposed NARX based approach in one-step-ahead as well as multi-step ahead prediction tasks. The same experiments were repeated with NAR network and the results were compared.

6.1   Network Parameter

As the neural networks are non-deterministic and the parameter namely embedding delay influences accuracy of prediction, experiments have been carried out many times for tuning the embedding delay until minimum mean square error is reached.

6.2   Network Type and Network Structure

The number of layers in the network was fixed to three, since the proposed recurrent network types hardly improve with more layers. The size of the input layer was set by tuning the parameter embedding-delay. The size of the output layer was just one. The optimal number of neurons in the hidden layer, for best prediction accuracy was found to be 10, empirically.

6.3   Transfer Function and Training Algorithm Used

The transfer function represents the composition of two functions, namely, the activation-function and the output-function. The activation-function represents the total stimuli a neuron receives and the output function determines the total stimulus a neuron generates. In the proposed system, all the neurons in the hidden layer use hyperbolic tangent functions and the neuron in the output layer use pure linear function as activation function. The behavior of the proposed network was tested with two different training algorithms, namely Gradient Descent and Levenberg-Marquardt (LM) back propagations. Although the later requires more memory, it is the fastest supervised training algorithm. The parameters used as shown in Table 1 gives lesser prediction-error than the Gradient Descent algorithm. For the LLNL data set, the minimum prediction error for NAR network with a delay of 14 is 0.085129 and that of NARX network with a delay of 10 is 1.77E-05. As a result of training, the architecture of NARX network for best prediction accuracy has been found to have ten neurons in the hidden layer.

Table 1. Parameters used in empirical analysis

Parameter	Value	Parameter	Value
Learning Rate	0.001	Minimum Gradient	1.0e-10
Training Algorithm	Levenberg-Marquardt	Epochs	10000
Maximum Validation Failures	6	Evaluation metric	MSE
Activation function for hidden Neurons	Tansig	Training goal (MSE)	1.0e-15

7   Performance of NARX in One-Step and Multi-step Prediction

The prediction module implemented in Matlab was tested with LLNL data set for one-step as well as multi-step ahead prediction. The results of NAR and NARX networks are depicted in Figure 5 and 6.   At 0th time step, the required number of Processing Elements (PEs) is 233.   The results show that NARX network based predictions are more accurate in finding the hidden pattern in the time-series compared to NAR network. The performance of the networks was tested for short-term multi-step ahead prediction. Both networks are able to perform multi-step ahead short term prediction for about 20 time instants. The NARX network gives more accurate results up to 14 time instants compared to NAR network because of the incorporation of exogenous input and are depicted in Figure 7 and Figure 8.

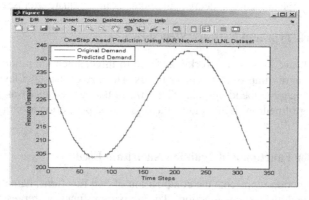

Fig. 5. One-step ahead prediction using NAR network in demand forecasting (LLNL)

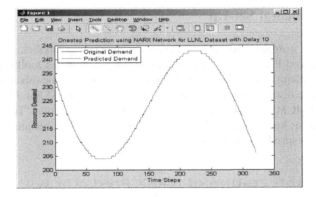

Fig. 6. One-step ahead prediction using NARX network in demand forecasting (LLNL)

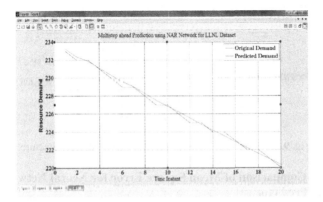

Fig. 7. Performance of NAR in making multi-step ahead short-term prediction (LLNL)

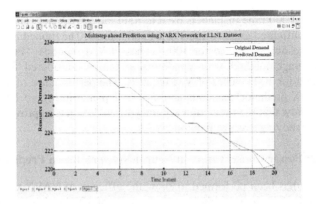

Fig. 8. Performance of NARX in making multi-step ahead short-term prediction (LLNL)

8   Experimental Set-Up

To evaluate and compare the proposed resource management policies with existing ones, the enhanced PAM architecture implemented in CloudSim toolkit is integrated with NARX implemented in MatLAB as shown in Figure 9. The submitted workloads require immediate resource allocation. The resource pool in EPAM has 15 hosts, where each host is an MCM with five chips. Each chip contains eight cores. The experiments are based on hosts built-up using Intel Xeon 7560 [25]. The cores are homogeneous and they run at a speed of 1000 MIPS.

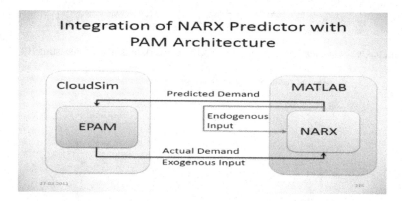

Fig. 9. Integration of NARX Predictor with PAM Architecture

Experiment 1: Comparison of Mean Square Error for Neural Network Predictor and Statistical Predictor

This experiment was conducted to show the prediction accuracy of NARX compared to Moving Average method. The LLNL data set representing the resource demand time series, with 1000 time instants is used to train the NARX net. The network with ten neurons is configured as shown in Table1 and trained. The trained network is used for forecasting resource demand for 100 batches of VM requests. For comparison, the moving average method with window size ten is considered for forecasting resource demand for 100 batches of VM requests. The performance of statistical predictor and NARX predictor was compared and is depicted in Figure 10. The experimental results show that the prediction accuracy of NARX gets improved by 46.16 % compared to the moving average method.

Experiment 2: Power Conservation in Neural Network Based Predictor

This experiment has been performed to show the efficacy of NARX predictor in the proposed EPAM. The input to the system consists of 100 batches of VM requests. The resource pool has 15 hosts, containing 600 PEs. Initially all PEs are in SS state. The EPAM gets the resource demand for next time interval from NARX predictor and provisions the needed PEs in SS state and the remaining idle PEs in DS state for power conservation. The window size of the moving-average method and the input delay of NARX predictor are set to six. So, for the first six batches, the power consumption is the same for both statistical predictor and neural-network-based predictor. The NARX based PAM requires 30% additional provisioning and the PAM with statistical predictor requires 160% additional provisioning to meet demand.

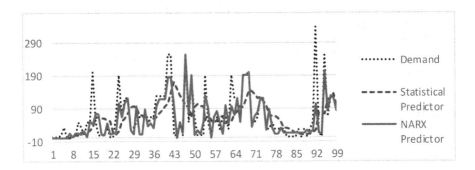

Fig. 10. Performance of NARX predictor compared to Statistical Predictor

The power consumption of PAM with statistical and NARX prediction is depicted in Figure 11. The experimental results show that the average power consumption of the PAM with statistical predictor and that of EPAM with neural network based predictor are 258.64 Watts and 187.94 Watts, respectively. The EPAM integrated with NARX predictor resulted in 27.25 % power saving compared to PAM with statistical predictor due to accuracy in resource demand prediction.

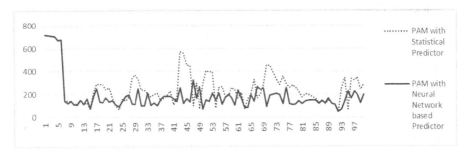

Fig. 11. Power Consumption of PAM with Statistical and Neural Network based Predictor

9   Conclusion

As power conservation is an important issue in large scale cloud environment, a novel neural-network-based prediction service has been proposed to support near accurate resource provisioning leading to the realization of green computing. As Artificial Neural Networks have tremendous ability to learn the hidden pattern and trends in a time-series data, its applicability in resource demand prediction in cloud data center is analyzed.

In this paper, we have given architecture of Enhanced Power-Aware Meta-Scheduler and shown the key role of neural network based Predictor in improving power conservation without performance degradation in terms of delay in VM launching. The feedback loop in the proposed architecture makes the system adaptive, as it reconfigures the NARX network according to change in the demand pattern in cloud data center. The multi-step ahead prediction plays a vital role in resource provisioning, as it is very much useful in handling the inherent issues such as over-provisioning and under-provisioning in cloud data center. Thus, neural network-based forecasting has given an enhanced Meta-Scheduler for realizing intelligent, adaptive and power-aware VM provisioning for cloud environment.

References

1. Beloglazov, A., Abawajy, J., Buyya, R.: Energy-Aware Resource Allocation Heuristics for Efficient Management of Data Centers for Cloud Computing. FGCS 28, 755–768 (2012)
2. Box, G.E.P., Jenkins, M.: Time Series Analysis. Pearson Education Publishers (2004)
3. Diaconescu, E.: The use of NARX Neural Networks to predict Chaotic Time Series. WSEAS Transactions on Computer Research 3 (2008)
4. Gilat, A.: MATLAB- An Introduction with Applications. John Wiley and Sons (2007)
5. Haykin, S.: Neural Networks-A Comprehensive Foundation. Pearson Prentice Hall (2006)
6. Jeyarani, R., Nagaveni, N., Vasanth Ram, R.: Design and Implementation of adaptive power-aware virtual machine provisioner (APA-VMP) using Swarm Intelligence. FGCS 28, 811–821 (2010)
7. Jeyarani, R., Nagaveni, N., Vasanth Ram, R.: Self- Adaptive Particle Swarm Optimization for Efficient Virtual Machine Provisioning in cloud. Intl. J. Intelligent Information Technologies 7, 25–44 (2011)
8. Jeyarani, R., Nagaveni, N., Sadasivam, S., Vasanth Ram, R.: Power Aware Meta-Scheduler for Adaptive VM Provisioning in IaaS Cloud. Intl. J. of Cloud Applications and Computing 1, 36–50 (2011)
9. Kant, K.: Data center evolution - A tutorial on state of the art issues and challenges. Computer Networks 53, 2939–2965 (2009)
10. Kaplan, J.M., Forest, W., Kindler, N.: Revolutionizing Data Centre Energy Efficiency. White Paper, McKinsay and Company (2008)
11. Khan, A., Yan, X., Tao, S., Anerousis, N.: Workload Characterization and Prediction in the Cloud: A Multiple Time Series Approach. In: IEEE/IFIP International Workshop on Cloud Management (2012)
12. Menezes Jr., J.M., Barreto, G.A.: A new look at Nonlinear Time Series Prediction with NARX Recurrent Neural Network. In: IX Brazillian Neural Networks Symposium (2006)
13. Menezes Jr., J.M., Barreto, G.A.: On the Prediction of Chaotic Time Series using the NARX Recurrent Neural Network: A New Approach. In: 9th Experimental Chaos Conference (2006)
14. Menezes Jr., J.M., Barreto, G.A.: Long-Term Load Prediction with NARX Network: An Empirical Evaluation. Neurocomputing 71, 3335–3343 (2008)
15. Nae, V., Iosup, A., Podlipnig, S., Prodan, R., Epema, D.H.J., Fahringer, T.: Efficient Management of Data Center Resources for Massively Multiplayer Online Games. In: ACM/IEEE Conference on Supercomputing, USA (2008)

16. Nae, V., Prodan, R., Fahringer, T.: Neural Network-Based Load Prediction for Highly Dynamic Distributed Online Games. In: Luque, E., Margalef, T., Benítez, D. (eds.) Euro-Par 2008. LNCS, vol. 5168, pp. 202–211. Springer, Heidelberg (2008)
17. Prodan, R., Nae, V.: Prediction-based real-time resource provisioning for massively multiplayer online games. Future Generation Computer Systems 25, 783–793 (2009)
18. Shabri, A.: Comparison of Time series Forecasting Methods Using Neural Networks and Box-Jenkins Model. Mathematika 17, 25–32 (2001)
19. Sharda, R., Patil, R.: Neural Network as Forecasting Experts: Empirical Test. In: Intl. Joint Conference on Neural Networks, vol. 1, pp. 491–494 (1990)
20. Sivanandam, S.N., Sumathi, S., Deepa, S.N.: Introduction to Neural Networks using MATLAB 6.0. Tata McGraw Hill Publishing Company (2006)
21. Smith, J.E., Nair, R.: Virtual Machines-Versatile platforms for Systems and Processes. Elsevier Margan Kaffmaun Publications, USA (2005)
22. Verma, A., Ahuja, P., Neogi, A.: pMapper: Power and Migration Cost aware application Placement in Virtualized Systems. In: Issarny, V., Schantz, R. (eds.) Middleware 2008. LNCS, vol. 5346, pp. 243–264. Springer, Heidelberg (2008)
23. Zurada, J.M.: Introduction to Artificial Neural Systems. Jaico Publishing House (1999)
24. Tsafrir, D., Feitelson, D.G.: Logs of Real Parallel Workloads from Production Systems, http://www.cs.huji.ac.il/labs/parallel/workload/logs.html
25. Intel Xeon Processor Specification, http://ark.intel.com/products/46499/Intel-Xeon-Processor-X7560-24M-Cache-2_26-GHz-6_40-GTs-Intel-QPI

A Fractional-Order Chaotic Circuit Based on Memristor and Its Generalized Projective Synchronization

Wenwen Shen, Zhigang Zeng, and Fang Zou

Huazhong University of Science and Technology, Wuhan, Hubei 430074, China
Wenwenshen86@gmail.com

Abstract. In this paper, we generalize the integer-order chua circuit model based on memristor into the fractional-order domain. The new fractional-order circuit can generate complex chaotic behavior. Based on the stability theory of fractional-order systems and active control, a controller for the synchronization of two commensurate fractional-order chaotic memristor based circuit is designed. This technique is applied to achieve generalized projective synchronization (GPS) between the fractional-order chaotic circuit. Numerical results demonstrate the effectiveness and feasibility of the proposed control technique.

Keywords: Memristor, Fractional-order, Generalized Projective Synchronization.

1 Introduction

The concept of memristor was introduced and named by Leon Chua in his seminal paper [1] in 1971. It represents a relationship between flux and charge. Memristor is a two-terminal element with variable resistance called memristance which depends on how much electric charge has been passed through it in a particular direction. The existence of the memristor as the fourth ideal electrical circuit element was predicted based on logical symmetry arguments, but it was not clearly experimentally demonstrated until April 30, 2008, that a team at Hewlett Packard Labs reported about discovery of a switching memristor in Nature [2]-[3]. In their work, Dr. Strukov et.al presented a physical device to illustrate their invention. It was a new nanometer-size two terminal device, which could "remember" its state after the voltage applied to it is turned off. The unique properties of memristors create great opportunities in future system design. Now, the research on chaotic circuits based on memristor is becoming a hot topic for researchers [4]-[7].

On the other hand, fractional calculus is a more than 300 years old topic. Although it has a long history, the development of models based on fractional order differential systems has recently gained popularity in the investigation of dynamical systems. Fractional derivatives provide an excellent instrument for the description of memory and hereditary properties of various materials and processes. The advantages of the real objects of the fractional order systems are that we have more degrees of freedom in the model and that a "memory" is included in the model. The chaotic dynamics of fractional order systems began to attract much attention in recent years [8]-[10]. It has been shown that fractional order systems, as generalizations of many well known

D.-S. Huang et al. (Eds.): ICIC 2014, LNCS 8588, pp. 838–844, 2014.
© Springer International Publishing Switzerland 2014

systems, can also behave chaotically, such as the fractional Duffing system, the fractional Chua system, the fractional Lorenz system, the fractional Liu system, the fractional Chen system, etc [11].These examples and many other similar samples perfectly clarify the importance of consideration and analysis of dynamical systems with fractional-order models. What's more, the long memory dependence of the fractional derivative can be a better fit for the memory character of memristor.

Meanwhile, recently studies about synchronization of chaotic fractional order systems have been investigated, such as Robust synchronization [12]-[14], projective synchronization [15]-[18], lag synchronization [19]-[21], Phase synchronization [22]-[23], complete synchronization [24]-[25]. In many literatures, synchronization among fractional order systems is only investigated through numerical simulations that are based on the stability criteria of linear fractional order systems[26]-[27], Linear Matrix Inequalities (LMI) methods[28], Laplace transform theory[8].

In this paper, a new fractional-order chaotic circuit based on memristor is presented, Which exhibits complex chaotic. Our aim is to synchronize two chaotic fractional-order circuit. To achieve this goal, we use active control methods, and the results between numerical emulation and circuit experimental simulation are in agreement with each other.

2   Preminaries

Fractional calculus is a generalization of integration and differentiation to non-integer-order fundamental operator $_aD_t^q$, where a and t are the bounds of the operation, and $q \in R$. The most usually used definitions of a fractional derivative of order α are, respectively, the Riemann–Liouville, Grünwald-Letnikov and Caputo formulations:

$$_a^{RL}D_t^q f(t) = \frac{1}{\Gamma(n-\alpha)} \frac{d^n}{dt^n} \int_a^t \frac{f(\tau)}{(t-\tau)^{\alpha-n+1}} d\tau, \ t > a \quad n-1 < \alpha < n \tag{1}$$

$$_a^{GL}D_t^q f(t) = \lim_{h \to 0} \frac{1}{h^\alpha} \sum_{k=0}^{[\frac{t-a}{h}]} (-1)^k \binom{\alpha}{k} f(t-kh), \ t > a \quad 0 < \alpha \tag{2}$$

$$_a^C D_t^q f(t) = \frac{1}{\Gamma(n-\alpha)} \int_a^t \frac{f^{(n)}(\tau)}{(t-\tau)^{\alpha-n+1}} d\tau, \ t > a \quad n-1 < \alpha < n \tag{3}$$

where $\Gamma()$ is Euler's gamma function, $[x]$ means the integer part of x, and h is the step time increment. In common, the well-known definitions, RL (1), GL (2), and Caputo (3) are equivalent. The relation for the explicit numerical approximation of the qth derivative at the points kh $(k = 1,2,...)$ has the following form[30]:

$$_{k-L_m/h}D_{t_k}^q f(t) \approx h^{-q} \sum_{i=0}^k c_i^{(q)} f(t_{k-i}) \tag{4}$$

where L_m is the "memory length," $t_k = kh$, h is the time step of calculation, and c_i^q $(i = 0,1,...)$ are binomial coefficients. For their calculation, we can use the following expression

$$c_0^q = 1, \quad c_i^q = (1 - \frac{1+q}{i})c_{i-1}^q \tag{5}$$

The general numerical solution of the fractional differential equation $D_t^q = f(y(t),t)$ can be expressed as

$$y_{t_k} = f(y(t_k),t_k)h^q - \sum_{i=0}^{k} c_i^{(q)} y(t_{k-i}) \tag{6}$$

3   Description of Memristor Based Hyper Chaotic Fractional-Order Circuit

Different from the other three fundamental circuit elements, the memristor is a nonlinear one and its value is not unique. Chua theoretically showed that the value of the memristor called memristance is defined as:

$$M(q) = \frac{d\varphi(q)}{dq} \quad \text{or} \quad W(\varphi) = \frac{dq(\varphi)}{d\varphi}$$

where q denotes electric charge through it and $\varphi(q)$ denotes the magnetic flux, $M(q)$ denotes memristance and $W(\varphi)$ denotes the menductance [1]. We can assume the memristance characterized by the "monotoneincreasing" and "piecewise-linear" [31]. One of possible circuit structure of the chaotic system is illustrated in Fig. 1., which is extended from Chua's hyperchaotic[7].

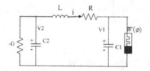

Fig. 1. Hyper chaotic circuit based on memristor

The new fractional-order chaotic system according to[7] is described as follows:

$$\begin{cases} D^q x_1 = a[(x_3 - W(x_4)x_1], \\ D^q x_2 = cx_2 - bx_3, \\ D^q x_3 = x_2 - x_1 - dx_3, \\ D^q x_4 = x_1, \end{cases} \tag{7}$$

and the piece-wise linear function $W(x_4)$ is given by

$$W(x_4) = \begin{cases} W_1, |x_4| < 1 \\ W_2, |x_4| > 1 \end{cases} \tag{8}$$

where W_1 represents the memristance when $x_4 < 1$ while W_2 represents the memristance when $x_4 > 1$. q is fractional-order which satisfies $0 < q < 1$. If we choose proper circuit parameters, for example we set $a = 5$, $b = 1$, $c = 0.7$, $d = 0.1$, $q = 0.95$. Thus, the simulation is done with the initial value $(0.08, 0.05, 0.07, 0.006)$ to system (7), the simulation results are shown in Fig. 2

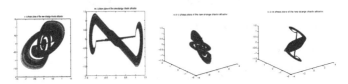

Fig. 2. Strange attractor of the fractional-order memristor-based circuit for parameters $a = 5$, $b = 1$, $c = 0.7$, $d = 0.1$, $W_1 = 0.1$, $W_2 = 3$,and orders $q = 0.95$

4   Generalized Projective Synchronization Use Active Control

The projective synchronization mens that the drive and response vectors synchronize up to a scaling factor θ, that is, the vectors become proportional. First, we define the GPS below.Consider the following chaotic system:

$$\begin{cases} \dot{x}_m = f(x_m), \\ \dot{x}_s = g(x_m, x_s), \end{cases} \tag{9}$$

where n-dimensional state vector $x_m, x_s \in R^n$. The low subscripts 'm' and 's' stand for the master and slave systems, respectively. $f : R^n \to R^n$ and $g : R^n \to R^n$ are vector fields in n-dimensional space. If there exists a constant $\theta (\theta \neq 0)$, such that

$$\lim_{t \to \infty} \| x_m - \theta x_s \| = 0 ,$$

then the GPS of the system (9) is achieved, and we call θ 'scaling factor'.

Now, in order to observe synchronization behavior in two coupled chaotic fractional order circuit based on memristor, we build the master and the slave fractional order circuit as

$$\begin{cases} D^q x_1^m = a[(x_3^m - W(x_4^m)x_1^m], \\ D^q x_2^m = cx_2^m - bx_3^m, \\ D^q x_3^m = x_2^m - x_1^m - dx_3^m, \\ D^q x_4^m = x_1^m, \end{cases} \tag{10}$$

$$\begin{cases} D^q x_1^s = a[(x_3^s - W(x_4^s)x_1^s] + u_1, \\ D^q x_2^s = c x_2^s - b x_3^s + u_2, \\ D^q x_3^s = x_2^s - x_1^s - d x_3^s + u_3, \\ D^q x_4^s = x_1^s + u_4. \end{cases} \tag{11}$$

Define the error vector as $e_1 = x_1^m - \theta x_1^s$, $e_2 = x_2^m - \theta x_2^s$, $e_3 = x_3^m - \theta x_3^s$, $e_4 = x_4^m - \theta x_4^s$. By subtracting Eqs. (10) from (11), we obtain

$$\begin{cases} D^q e_1 = a e_3 - a W(x_4^m)x_1^m + \theta a W(x_4^s)x_1^s - \theta u_1, \\ D^q e_2 = c e_2 - b e_3 - \theta u_2, \\ D^q e_3 = e_2 - e_1 - d e_3 - \theta u_3, \\ D^q e_4 = e_1 - \theta u_4. \end{cases} \tag{12}$$

Now, by letting $u_1 = a W(x_4^s)x_1^s - \dfrac{a}{\theta} W(x_4^m)x_1^m + k_1 e_1$, $u_2 = k_2 e_2$, $u_3 = k_3 e_3$, $u_4 = k_4 e_4$, then the error system (12) is reduced to $D^q e(t) = (C - \theta K)e(t)$, $e(t) = (e_1, e_2, e_3, e_4)$, where

$$C = \begin{pmatrix} 0 & 0 & a & 0 \\ 0 & c & -b & 0 \\ -1 & 1 & -d & 0 \\ 1 & 0 & 0 & 0 \end{pmatrix}, \qquad K = \begin{pmatrix} k_1 & 0 & 0 & 0 \\ 0 & k_2 & 0 & 0 \\ 0 & 0 & k_3 & 0 \\ 0 & 0 & 0 & k_4 \end{pmatrix}.$$

Lemma 1. [26] Let $A \in R^{n \times n}$, $0 < \alpha < 2$, the linear time-invariant system $D^\alpha x(t) = Ax(t)$ is asymptotically stable if $|arg(\lambda_i(A))| > \pi \alpha / 2$, where $\lambda_i(A)$, $i = 1, 2, ..., n$, denote the eigenvalues of A and $arg(\cdot)$ denotes the argument of a complex number.

Lemma 2. [28]-[29] Let $A \in R^{n \times n}$, and D a GLML region. A is D-stable if $\exists m$ matrices $X_k \in C^{n \times n}$ s.t.

$$\sum_{k=1}^{m} (\theta_k \otimes X_k + \theta_k^* \otimes X_k^* + \psi_k \otimes (AX_k) + \psi_k^* \otimes (AX_k)^*) < 0. \tag{13}$$

$$\sum_{k=1}^{m} (H_k \otimes X_k + J_k \otimes X_k^*) = 0_{nl \times nl}. \tag{14}$$

Theorem 1. From Lemma 2, the zero solution of error system (12) is asymptotically stable, and the response system in (11) can synchronize with the drive system (10) if we choose the proper parameters, the following conditions can be satisfied: there exists two Hermitian matrices $P_1 > 0$ and $P_2 > 0$, such that

$$\bar{r}P_1(C+\theta K)^T + r(C+\theta K)P_1 + rP_2(C+\theta K)^T + \bar{r}(C+\theta K)P_2 < 0, \tag{15}$$

where $r = e^{j(1-q)\frac{\pi}{2}}$, or there exist a Hermitian matrices $H > 0$, such that

$$(Re(rH))^T (C+\theta K)^T + (C+\theta K)(Re(rH)) < 0. \tag{16}$$

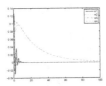

Fig. 3. The error state curves of the Generizad projective synchronization between systems (10) and (11) with $k_1 = 2$, $k_2 = 2$, $k_3 = 0$, $k_4 = 0.1$, $\theta = 0.5$

From Theorem 1, when $a = 5$, $b = 1$, $c = 0.7$, $d = 0.1$, $W_1 = 0.1$, $W_2 = 3$, and orders $q = 0.95$, $\theta = 0.5$, we can choose H as the unitary matrix, then by solving the LMI inequality (16), we get $k_1 = 1$, $k_2 = 1$, $k_3 = 0$, $k_4 = 0.1$, the state variables of system (12) are displayed by fig. 3.

Acknowledgments. The work is supported by the Natural Science Foundation of China under Grants 60974021 and 61125303, the 973 Program of China under Grant 2011CB710606, the Fund for Distinguished Young Scholars of Hubei Province under Grant 2010CDA081.

References

1. Chua, L.O.: Memristor-The missing circuit element. IEEE Trans. Circ. Theory 18, 507–519 (1971)
2. Strukov, D.B., Snider, G.S., Stewart, D.R., Williams, R.S.: The missing memristor found. Nature 453, 80–83 (2008)
3. Yang, J.J., Pickett, M.D., Li, X., Ohlberg, D.A.A., Stewart, D.R., Williams, R.S.: Memristive switching mechanism for metal/oxide/metal nanodevices. Nature Nano-technology 3, 429–433 (2008)
4. Hu, X., Chen, G., Duan, S., Feng, G.: A memristor-based chaotic system with boundary conditions. In: Memristor Networks, pp. 351–364. Springer International Publishing (2014)
5. Driscoll, T., Pershin, Y.V., Basov, D.N., Di, V.M.: Chaotic memristor. Applied Physics A 102(4), 885–889 (2011)
6. Muthuswamy, B.: Implementing memristor based chaotic circuits. International Journal of Bifurcation and Chaos 20(05), 1335–1350 (2010)
7. Wen, S., Zeng, Z., Huang, T.: Adaptive synchronization of memristor-based Chua's circuits. Physics Letters A 376(44), 2775–2780 (2012)
8. Miller, K.S., Ross, B.: An introduction to the fractional calculus and fractional differential equations (1993)
9. Sabatier, J., Agrawal, O.P., Machado, J.T.: Advances in fractional calculus. Springer, Dordrecht (2007)

10. Baleanu, D.: Fractional Calculus: Models and Numerical Methods. World Scientific (2012)
11. Petráš, I.: Fractional-order nonlinear systems: modeling, analysis and simulation. Springer (2011)
12. Zhang, R., Yang, S.: Robust synchronization of two different fractional-order chaotic systems with unknown parameters using adaptive sliding mode approach. Nonlinear Dynamics 71(1-2), 269–278 (2013)
13. Li, C., Su, K., Tong, Y., Li, H.: Robust synchronization for a class of fractional-order chaotic and hyperchaotic systems. Optik-International Journal for Light and Electron Optics 124(18), 3242–3245 (2013)
14. Asheghan, M.M., Hamidi Beheshti, M.T., Tavazoei, M.S.: Robust synchronization of perturbed Chen's fractional-order chaotic systems. Communications in Nonlinear Science and Numerical Simulation 16(2), 1044–1051 (2011)
15. Zhou, P., Zhu, W.: Function projective synchronization for fractional-order chaotic systems. Nonlinear Analysis: Real World Applications 12(2), 811–816 (2011)
16. Wang, S., Yu, Y.G.: Generalized projective synchronization of fractional order chaotic systems with different dimensions. Chinese Physics Letters 29(2), 020505 (2012)
17. Si, G., Sun, Z., Zhang, Y., Chen, W.: Projective synchronization of different fractional-order chaotic systems with non-identical orders. Nonlinear Analysis: Real World Applications 13(4), 1761–1771 (2012)
18. Wang, S., Yu, Y., Wen, G.: Hybrid projective synchronization of time-delayed fractional order chaotic systems. Nonlinear Analysis: Hybrid Systems 11, 129–138 (2014)
19. Chen, L., Chai, Y., Wu, R.: Lag projective synchronization in fractional-order chaotic (hyperchaotic) systems. Physics Letters A 375(21), 2099–2110 (2011)
20. Ruo-Xun, Z., Shi-Ping, Y.: Adaptive lag synchronization and parameter identification of fractional order chaotic systems. Chinese Physics B 20(9), 090512 (2011)
21. Sha, W., Yong-Guang, Y., Hu, W., Rahmani, A.: Function projective lag synchronization of fractional-order chaotic systems (2014)
22. Odibat, Z.: A note on phase synchronization in coupled chaotic fractional order systems. Nonlinear Analysis: Real World Applications 13(2), 779–789 (2012)
23. Taghvafard, H., Erjaee, G.H.: Phase and anti-phase synchronization of fractional order chaotic systems via active control. Communications in Nonlinear Science and Numerical Simulation 16(10), 4079–4088 (2011)
24. Li, H., Liao, X., Luo, M.: A novel non-equilibrium fractional-order chaotic system and its complete synchronization by circuit implementation. Nonlinear Dynamics 68(1-2), 137–149 (2012)
25. Razminia, A., Baleanu, D.: Complete synchronization of commensurate fractional order chaotic systems using sliding mode control. Mechatronics 23(7), 873–879 (2013)
26. Matignon, D.: Stability results for fractional differential equations with applications to control processing. Computational Engineering in Systems Applications 2, 963–968 (1996)
27. Chen, J., Zeng, Z., Jiang, P.: Global Mittag-Leffler stability and synchronization of memristor-based fractional-order neural networks. Neural Networks 51, 1–8 (2014)
28. Farges, C., Moze, M., Sabatier, J.: Pseudo-state feedback stabilization of commensurate fractional order systems. Automatica 46(10), 1730–1734 (2010)
29. Chilali, M.: Méthodes LMI pour l'Analyse et la Synthèse Multi-Critère (Doctoral dissertation) (1996)
30. Petráš, I.: Fractional-order memristor-based Chua's circuit. IEEE Transactions on Circuits and Systems II: Express Briefs 57(12), 975–979 (2010)
31. Itoh, M., Chua, L.O.: Memristor oscillators. International Journal of Bifurcation and Chaos 18(11), 3183–3206 (2008)

Author Index

Printed in the United States
By Bookmasters